STUDENT STUDY AND SOLUTIONS GUIDE

TO ACCOMPANY

MATHEMATICAL APPLICATIONS
FOR THE MANAGEMENT, LIFE, AND SOCIAL SCIENCES
SEVENTH EDITION

STUDENT STUDY AND SOLUTIONS GUIDE

TO ACCOMPANY

MATHEMATICAL APPLICATIONS
FOR THE MANAGEMENT, LIFE, AND SOCIAL SCIENCES
SEVENTH EDITION

HARSHBARGER/REYNOLDS

GORDON SHILLING
University of Texas at Arlington

HOUGHTON MIFFLIN COMPANY BOSTON NEW YORK

Sponsoring Editor: Lauren Schultz
Associate Editor: Marika Hoe
Manufacturing Manager: Florence Cadran
Senior Marketing Manager: Danielle Potvin

Printed in the U.S.A.

ISBN: 0-618-29370-1

5 6 7 8 9 –POO– 07 06 05

TABLE OF CONTENTS

SOLUTIONS

SUPPLEMENTARY EXERCISES

STUDENT STUDY AND SOLUTIONS GUIDE

TO ACCOMPANY

MATHEMATICAL APPLICATIONS

FOR THE MANAGEMENT, LIFE, AND SOCIAL SCIENCES

SEVENTH EDITION

Chapter 0: Algebraic Concepts

Exercise 0.1

1. $x \in \{x, y, z, a\}$

3. $12 \in \{1, 2, 3, 4, ...\}$

5. $6 \notin \{1, 2, 3, 4, 5\}$

7. $\{1, 2, 3, 4, 5, 6, 7\}$

9. $\{x: x$ is a natural number greater than 2 and less than 8$\}$

11. Yes. Every element of A is an element of B.

13. No. $c \in A$ but $c \notin B$.

15. $D \subseteq C$ since every element of D is an element of C.

17 $D \subseteq A$ since every element of D is an element of A.

19. $A \subseteq B$ and $B \subseteq A$. (Also $A = B$.)

21. $A \subseteq B$ and B $\subseteq$ A. Thus, $A = B$.

23. $D \neq E$ because $4 \in E$ and $4 \notin D$.

25. A and B are disjoint since they have no elements in common. B and D are disjoint since they have no elements in common. C and D are disjoint.

27. $A \cap B = \{4, 6\}$ since 4 and 6 are elements of each set.

29. $A \cap B = \varnothing$ since they have no common elements.

31. $A \cup B = \{1, 2, 3, 4, 5\}$

33. $A \cup B = \{1, 2, 3, 4\}$ or $A \cup B = B$.

For problems 35 - 45, we have
$U = \{1, 2, 3, \ldots, 9, 10\}$

35. $A' = \{4, 6, 9, 10\}$ since these are the only elements in U that are not elements of A.

37. $B' = \{1, 2, 5, 6, 7, 9\}$
$A \cap B' = \{1, 2, 5, 7\}$

39. $A \cup B = \{1, 2, 3, 4, 5, 7, 8, 10\}$
$(A \cup B)' = \{6, 9\}$

41. $A' = \{4, 6, 9, 10\}$
$B' = \{1, 2, 5, 6, 7, 9\}$
$A' \cup B' = \{1, 2, 4, 5, 6, 7, 9, 10\}$

43. $B' = \{1, 2, 5, 6, 7, 9\}$, $C' = \{1, 3, 5, 7, 9\}$
$A \cap B' = \{1, 2, 3, 5, 7, 8\} \cap \{1, 2, 5, 6, 7, 9\}$
$= \{1, 2, 5, 7\}$
$(A \cap B') \cup C' = \{1, 2, 3, 5, 7, 9\}$

45. $B' = \{1, 2, 5, 6, 7, 9\}$,
$A \cap B' = \{1, 2, 3, 5, 7, 8\} \cap \{1, 2, 5, 6, 7, 9\}$
$= \{1, 2, 5, 7\}$
$(A \cap B')' \cap C$
$= \{3, 4, 6, 8, 9, 10\} \cap \{2, 4, 6, 8, 10\}$
$= \{4, 6, 8, 10\}$

47. $A - B = \{1, 3, 7, 9\} - \{3, 5, 8, 9\} = \{1, 7\}$

49. $A - B = \{2, 1, 5\} - \{1, 2, 3, 4, 5, 6\} = \varnothing$

51. **a.** $L = \{94, 95, 96, 97, 98, 99, 00\}$
$H = \{92, 93, 94, 95, 96, 97, 98, 99, 00\}$
$C = \{90, 91, 95, 96, 97, 98, 99\}$
b. $L \subseteq H$
c. C' is the years when the percentage change (from low to high) was less than or equal to 25%.
d. $H' = \{90, 91\}$ $C' = \{92, 93, 94, 00\}$
$H' \cup C' = \{90, 91, 92, 93, 94, 00\}$ $H' \cup C'$ are the years when the high was less than or equal to 3300 or the percentage gain was less than or equal to 25%.
e. $L' = \{90, 91, 92, 93\}$ $L' \cap C = \{90, 91\}$ $L' \cap C$ are the years when the low was less than or equal to 3300 and the percentage gain was more than 25%.

53. **a.** From the table, there are 100 white Republicans and 30 non-white Republicans who favor national health care, for a total of 130.
b. From the table, there are 350 + 40 Republicans, and 250 + 200 Democrats who favor national health care, for a total of 840.
c. From the table, there are 350 white Republicans, and 150 white Democrats and 20 non-whites who oppose national health care, for a total of 520.

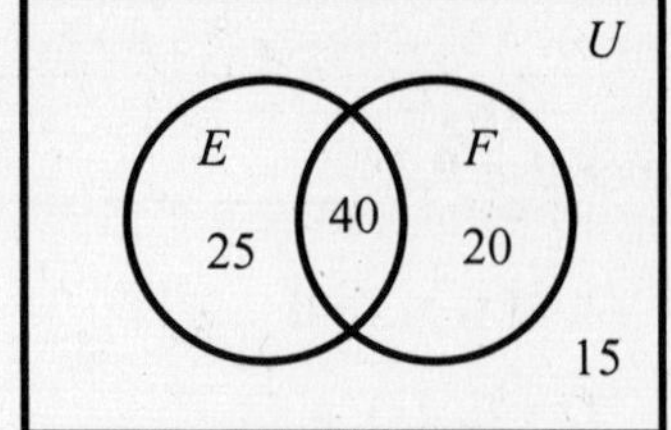

55. a. The key to solving this problem is to work from "the inside out". There are 40 aides in $E \cap F$. This leaves $65 - 40 = 25$ aides who speak English but do not speak French. Also we have $60 - 40 = 20$ aides who speak French but do not speak English. Thus there are $40 + 25 + 20 = 85$ aides who speak English or French. This means there are 15 aides who do not speak English or French.

b. From the Venn diagram $E \cap F$ has 40 aides.

c. From the Venn diagram $E \cup F$ has 85 aides.

d. From the Venn diagram $E \cap F'$ has 25 aides.

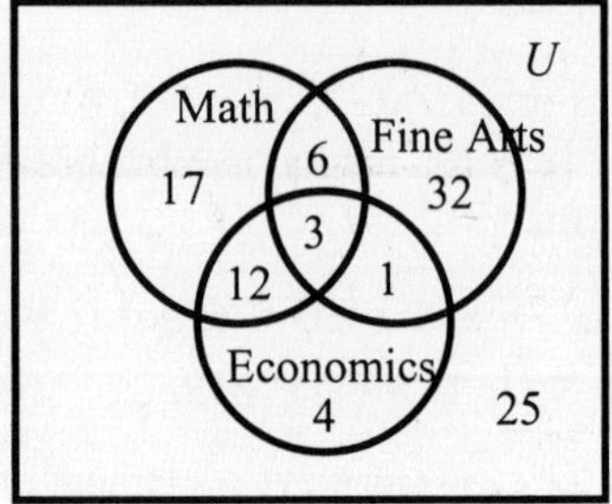

57. Since 12 students take M and E but not FA, and 15 take M and E, 3 take all three classes. Since 9 students take M and FA and we have already counted 3, there are 6 taking M and FA which are not taking E. Since 4 students take E and FA and we have already counted 3, there is only 1 taking E and FA but not taking M also. Since 20 students take E and we already have 16 enrolled in E, this leaves 4 taking only E. Since 42 students take FA and we already have 10 enrolled in FA, this leaves 32 taking only FA. Since 38 students take M and we already have 21 enrolled in M, this leaves 17 taking only M.

a. In the union of the 3 courses we have $17 + 12 + 3 + 6 + 32 + 1 + 4 = 75$ students enrolled. Thus, there are $100 - 75 = 25$ students who are not enrolled in any of these courses.

b. In $M \cup E$ we have $17 + 12 + 3 + 6 + 1 + 4 = 43$ enrolled.

c. We have $17 + 32 + 4 = 53$ students enrolled in exactly one of the courses.

59. (a) and (b)

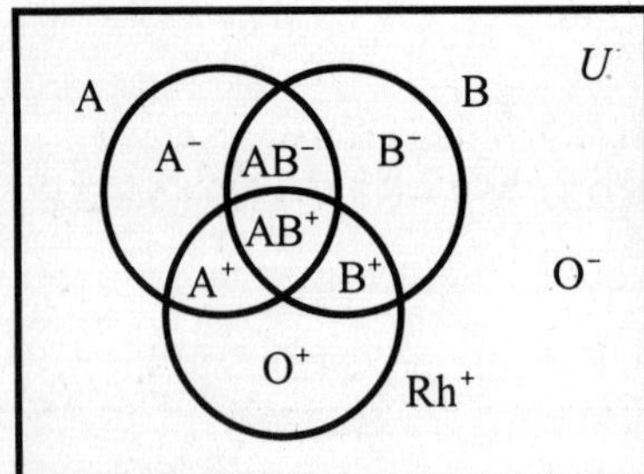

Exercise 0.2

1. a. Note that $-\frac{\pi}{10} = \pi \cdot \left(-\frac{1}{10}\right)$, where π is irrational and $-\frac{1}{10}$ is rational. The product of a rational number and an irrational number is an irrational number.

b. –9 is rational and an integer.

c. $\frac{9}{3} = \frac{3}{1} = 3$. This is a natural number, an integer, and a rational number.

d. Division by zero is meaningless.

3. a. Commutative

b. Distributive

c. Multiplicative identity

5. $-14 < -3$

7. $0.333 < \frac{1}{3}\left(\frac{1}{3} = 0.3333\cdots\right)$

9. $8 > 2$

11. $-3^2 + 10 \cdot 2 = -3^2 + 20 = -9 + 20 = 11$

13. $\frac{4+2^2}{2} = \frac{4+4}{2} = \frac{8}{2} = 4$

15. $\frac{16-(-4)}{8-(-2)} = \frac{16+4}{8+2} = \frac{20}{10} = 2$

17. $\frac{|5-2|-|-7|}{|5-2|} = \frac{|3|-|-7|}{|3|} = \frac{3-7}{3} = -\frac{4}{3}$

19. $\frac{(-3)^2 - 2\cdot 3 + 6}{4 - 2^2 + 3} = \frac{9-6+6}{4-4+3} = \frac{9}{3} = 3$

21. $\frac{-4^2+5-2\cdot 3}{5-4^2}=\frac{-16+5-6}{5-16}=\frac{-17}{-11}=\frac{17}{11}$

23. The entire line

25. (1, 3]

27. (2, 10)

29. $x \le 5$

31. $x > 4$

33. $(-\infty, 4) \cap (-3, \infty) = (-3, 4)$

35. $x > 4$ and $x \ge 0 = (4, \infty)$

37. $[0, \infty) \cup [-1, 5] = [-1, \infty)$

39. $(-\infty, 0) \cup (7, \infty)$

41. –0.000038585

43. 9122.387471

45. $\frac{2500}{[(1.1^6)-1]}=\frac{2500}{0.77156}=3240.184509$

47. a. \$300.00 + \$788.91 = \$1088.91

b. 0.25[1088.91 – 0.05(1088.91)] = \$258.62

c. Retirement: 0.05(1088.91) = \$54.45

State tax = Retirement = \$54.45

Local tax = 0.01(1088.91) = \$10.89

Federal tax = 0.25(1088.91 – 54.45) = \$258.62

Soc. Sec. tax = 0.0765(1088.91) = \$83.30

Total Withholding = \$461.71

Take-home pay = 1088.91 – 461.71
= \$627.20

49. a. 1996:

$C = 1.2393(16) + 4.2594 = \$24.09$

$C = 0.0232(16)^2 + 0.7984(16) + 5.8271$
$= \$24.54$

1997:

$C = 1.2393(17) + 4.2594 = \$25.33$

$C = 0.0232(17)^2 + 0.7984(17) + 5.8271$
$= \$26.10$

The second formula is more accurate.

b. $C = 0.0232(25)^2 + 0.7984(25) + 5.8271$
$= \$40.29$

Exercise 0.3

1. $(-4)^3 = (-4)(-4)(-4) = -64$

3. $-2^4 = -1\cdot 2\cdot 2\cdot 2\cdot 2 = -16$

5. $3^{-2} = \frac{1}{3^2} = \frac{1}{9}$

7. $-\left(\frac{3}{2}\right)^2 = (-1)\left(\frac{3}{2}\right)\left(\frac{3}{2}\right) = -\frac{9}{4}$

9. $6^5 \cdot 6^3 = 6^{5+3} = 6^8$

11. $\frac{10^8}{10^9} = 10^{8-9} = 10^{-1} = \frac{1}{10}$

13. $(3^3)^3 = 3^{3\cdot 3} = 3^9$

15. $\left(\frac{2}{3}\right)^{-2} = \left(\frac{3}{2}\right)^2 = \frac{9}{4}$

17. $(x^2)^{-3} = x^{2(-3)} = x^{-6} = \frac{1}{x^6}$

19. $xy^{-2}z^0 = x\cdot\frac{1}{y^2}\cdot 1 = \frac{x}{y^2}$

21. $x^3 \cdot x^4 = x^{3+4} = x^7$

23. $x^{-5}\cdot x^3 = x^{-5+3} = x^{-2} = \frac{1}{x^2}$

25. $\frac{x^8}{x^4} = x^{8-4} = x^4$

27. $\frac{y^5}{y^{-7}} = y^{5-(-7)} = y^{12}$

29. $(x^4)^3 = x^{3\cdot 4} = x^{12}$

31. $(xy)^2 = x^2y^2$

33. $\left(\frac{2}{x}\right)^4 = \frac{2^4}{x^4} = \frac{16}{x^4}$

35. $(2x^{-2}y)^{-4} = 2^{-4}x^8y^{-4} = \frac{x^8}{16y^4}$

37. $(-8a^{-3}b^2)(2a^5b^{-4}) = -16a^{-3+5}b^{2-4}$

$= -16a^2b^{-2} = -\dfrac{16a^2}{b^2}$

39. $2x^{-2} \div x^{-1}y^2 = \dfrac{2}{x^2} \div \dfrac{y^2}{x} = \dfrac{2}{x^2} \cdot \dfrac{x}{y^2} = \dfrac{2}{xy^2}$

41. $\left(\dfrac{x^3}{y^{-2}}\right)^{-3} = \dfrac{x^{-9}}{y^6} = \dfrac{1}{x^9} \cdot \dfrac{1}{y^6} = \dfrac{1}{x^9y^6}$

43. $\left(\dfrac{a^{-2}b^{-1}c^{-4}}{a^4b^{-3}c^0}\right)^{-3} = \left(\dfrac{b^2}{a^6c^4}\right)^{-3} = \left(\dfrac{a^6c^4}{b^2}\right)^3$

$= \dfrac{a^{18}c^{12}}{b^6}$

45. **a.** $\dfrac{2x^{-2}}{(2x)^2} = 2 \cdot \dfrac{1}{x^2} \cdot \dfrac{1}{(2x)^2} = 2 \cdot \dfrac{1}{x^2} \cdot \dfrac{1}{4x^2} = \dfrac{1}{2x^4}$

b. $\dfrac{(2x)^{-2}}{(2x)^2} = \dfrac{1}{(2x)^2} \cdot \dfrac{1}{(2x)^2} = \dfrac{1}{4x^2} \cdot \dfrac{1}{4x^2} = \dfrac{1}{16x^4}$

c. $\dfrac{2x^{-2}}{2x^2} = 2 \cdot \dfrac{1}{x^2} \cdot \dfrac{1}{2x^2} = \dfrac{1}{x^4}$

d. $\dfrac{2x^{-2}}{(2x)^{-2}} = 2 \cdot \dfrac{1}{x^2} \cdot (2x)^2 = 2 \cdot \dfrac{1}{x^2} \cdot 4x^2 = 8$

47. $\dfrac{1}{x} = x^{-1}$

49. $(2x)^3 = 2^3x^3 = 8x^3$

51. $\dfrac{1}{(4x^2)} = \dfrac{1}{4} \cdot \dfrac{1}{x^2} = \dfrac{1}{4}x^{-2}$

53. $\left(\dfrac{-x}{2}\right)^3 = \dfrac{-x^3}{2^3} = -\dfrac{1}{8}x^3$

55. 1.2 $\boxed{y^x}$ 4 $\boxed{=}$ 2.0736

57. 1.5 $\boxed{y^x}$ −5 $\boxed{=}$ 0.1316872428

59. $P = 1200$, $i = 0.12$, $n = 5$

$S = P(1+i)^n$

$= 1200(1+0.12)^5$

$= 1200(1.12)^5$

$= \$2114.81$

$I = S - P = 2114.81 - 1200 = \$914.81$

61. $P = 5000$, $i = 0.115$, $n = 6$

$S = P(1+i)^n$

$= 5000(1+0.115)^6$

$= 5000(1.115)^6$

$= \$9607.70$

$I = S - P = 9607.70 - 5000 = \$4607.70$

63. $S = 15{,}000$, $n = 6$, $i = 0.115$

$P = S(1+i)^{-n}$

$= 15{,}000(1+0.115)^{-6}$

$= 15{,}000(1.115)^{-6}$

$= \$7806.24$

65. $T = 246.47(1.07134)^t$

	Year	1970	1980	1990	1999
a.	t-value	20	30	40	49
b.	Tax	\$977.94	\$1947.98	\$3880.22	\$7214.44

c. $T = 246.47(1.07134)^{55} = \$10{,}908.54$

67. $D = 1.63(1.087)^t$

	Number of years	40	45	91	101
a.	Debt in billions	\$48.85	\$69.59	\$3229.06	\$7436.55

b. The end of WWar II was a factor.

69. $H = 28.8(1.1045)^t$

a. $t = 1970 - 1960 = 10$

b. $H = 28.8(1.1045)^{10} = \$77.81$ billion

c. $H = 28.8(1.1045)^{33} = \$765.4$ billion

d. $H = 28.8(1.1045)^{45} = \$2522.6$ billion

Exercise 0.4

1. Since $\left(\dfrac{16}{3}\right)^2 = \dfrac{256}{9}$ we have $\sqrt{\dfrac{256}{9}} = \dfrac{16}{3}$.

3. $\sqrt[5]{-32^3} = \sqrt[5]{-1 \cdot 32^3} = -\sqrt[5]{32^3} = -(32)^{3/5}$

$= -\left(\sqrt[5]{32}\right)^3 = -(2)^3 = -8$

5. $16^{3/4} = \left(\sqrt[4]{16}\right)^3 = 2^3 = 8$

7. $(-16)^{-3/2} = \left(\sqrt{-16}\right)^{-3}$

The square root of a negative number is not real.

9. $\left(\frac{8}{27}\right)^{-2/3} = \left(\frac{27}{8}\right)^{2/3} = \left(\sqrt[3]{\frac{27}{8}}\right)^2 = \left(\frac{3}{2}\right)^2 = \frac{9}{4}$

11. a. $8^{2/3} = \left(\sqrt[3]{8}\right)^2 = 2^2 = 4$

b. $(-8)^{-2/3} = \frac{1}{(-8)^{2/3}} = \frac{1}{\left(\sqrt[3]{-8}\right)^2} = \frac{1}{(-2)^2} = \frac{1}{4}$

13. $\sqrt[9]{(6.12)^4} = (6.12)^{4/9} \approx 2.2370$

15. $\sqrt{m^3} = m^{3/2}$

17. $\sqrt[4]{m^2n^5} = \left(m^2n^5\right)^{1/4} = m^{2/4}n^{5/4} = m^{1/2}n^{5/4}$

19. $x^{7/4} = \sqrt[4]{x^7}$

21. $-\left(\frac{1}{4}\right)x^{-5/4} = -\frac{1}{4}\cdot\frac{1}{x^{5/4}} = \frac{-1}{4\sqrt[4]{x^5}}$

23. $y^{1/4}\cdot y^{1/2} = y^{(1/4)+(1/2)} = y^{3/4}$

25. $z^{3/4}\cdot z^4 = z^{(3/4)+(16/4)} = z^{19/4}$

27. $y^{-3/2}\cdot y^{-1} = y^{(-3/2)-(2/2)} = y^{-5/2} = \frac{1}{y^{5/2}}$

29. $\frac{x^{1/3}}{x^{-2/3}} = x^{(1/3)-(-2/3)} = x^{3/3} = x$

31. $\frac{y^{-5/2}}{y^{-2/5}} = y^{(-5/2)-(-2/5)}$

$= y^{(-25/10)+(4/10)}$

$= y^{-21/10}$

$= \frac{1}{y^{21/10}}$

33. $(x^{2/3})^{3/4} = x^{(2/3)(3/4)} = x^{2/4} = x^{1/2}$

35. $(x^{-1/2})^2 = x^{-1} = \frac{1}{x}$

37. $\sqrt{64x^4} = 8x^2$

39. $\sqrt{128x^4y^5} = \sqrt{64x^4y^4\cdot 2y}$

$= \sqrt{64}\cdot\sqrt{x^4}\cdot\sqrt{y^4}\cdot\sqrt{2y}$

$= 8x^2y^2\sqrt{2y}$

41. $\sqrt[3]{40x^8y^5} = \sqrt[3]{8x^6y^3\cdot 5x^2y^2}$

$= \sqrt[3]{8}\cdot\sqrt[3]{x^6}\cdot\sqrt[3]{y^3}\cdot\sqrt[3]{5x^2y^2}$

$= 2x^2y\sqrt[3]{5x^2y^2}$

43. $\sqrt{12x^3y}\cdot\sqrt{3x^2y} = \sqrt{36x^5y^2}$

$= \sqrt{36}\cdot\sqrt{x^5}\cdot\sqrt{y^2}$

$= 6x^2y\sqrt{x}$

45. $\sqrt{63x^5y^3}\cdot\sqrt{28x^2y} = \sqrt{9x^4y^2\cdot 7xy}\cdot\sqrt{4x^2\cdot 7y}$

$= 3x^2y\sqrt{7xy}\cdot 2x\sqrt{7y}$

$= 42x^3y^2\sqrt{x}$

47. $\frac{\sqrt{12x^3y^{12}}}{\sqrt{27xy^2}} = \sqrt{\frac{4x^2y^{10}}{9}} = \frac{2xy^5}{3}$

49. $\frac{\sqrt[4]{32a^9b^5}}{\sqrt[4]{162a^{17}}} = \sqrt[4]{\frac{16b^4}{81a^8}\cdot\frac{b}{1}} = \frac{2b}{3a^2}\sqrt[4]{b}$

51. $(A^9)^x = A^{9x}$

$A^{9x} = A^1$

$9x = 1$

$x = \frac{1}{9}$

53. $\left(\sqrt[7]{R}\right)^x = R^{x/7}$

$R^{x/7} = R^1$

$\frac{x}{7} = 1$

$x = 7$

55. $\sqrt{\frac{2}{3}}\cdot\frac{\sqrt{3}}{\sqrt{3}} = \frac{\sqrt{2}\cdot\sqrt{3}}{\sqrt{3}\cdot\sqrt{3}} = \frac{\sqrt{6}}{3}$

57. $\frac{\sqrt{m^2x}}{\sqrt{mx^2}} = \frac{\sqrt{m}}{\sqrt{x}} = \frac{\sqrt{m}\cdot\sqrt{x}}{\sqrt{x}\cdot\sqrt{x}} = \frac{\sqrt{mx}}{x}$

59. $\frac{\sqrt[3]{m^2x}}{\sqrt[3]{mx^5}} = \frac{\sqrt[3]{m}}{\sqrt[3]{x^4}} = \frac{\sqrt[3]{m}}{\sqrt[3]{x^3}\cdot\sqrt[3]{x}}\cdot\frac{\sqrt[3]{x^2}}{\sqrt[3]{x^2}}$

$= \frac{\sqrt[3]{mx^2}}{x\sqrt[3]{x^3}} = \frac{\sqrt[3]{mx^2}}{x^2}$

61. $\frac{-2}{3\sqrt[3]{x^2}} = \frac{-2}{3}\cdot\frac{1}{x^{2/3}} = -\frac{2}{3}x^{-2/3}$

63. $3x\sqrt{x} = 3x\cdot x^{1/2} = 3x^{3/2}$

65. $\frac{3}{2}x^{1/2} = \frac{3}{2}\sqrt{x}$

67. $\frac{1}{2}x^{-1/2} = \frac{1}{2}\cdot\frac{1}{x^{1/2}} = \frac{1}{2\sqrt{x}}$

69. a. $R = 8.5 = \frac{17}{2}$ $I = 10^{17/2} = \sqrt{10^{17}}$

b. $I = \sqrt{10^{17}} = 10^{8.5} = 316,227,766$

c. $\frac{I_{06}}{I_{89}} = \frac{10^{8.25}}{10^{7.10}} = 10^{1.15} \approx 14.125$

71. $k = 25, t = 10, q_0 = 98$

$q = q_0(2^{-t/k})$

$= 98(2^{-10/25})$

$= 98(2^{-2/5})$

≈ 74 kg

73. $P = P_0(2.5)^{ht}$

$= 30,000(2.5)^{0.03(10)}$

$= 30,000(2.5)^{0.3}$

$\approx 39,491$

75. a. $N = 500(0.02)^{(0.7)^t}$; at $t = 0$ we have $(0.7)^0 = 1$. Thus, $N = 500(0.02)^1 = 10$.

b. $N = 500(0.02)^{(0.7)^5} = 500(0.02)^{0.16807} = 259$

Exercise 0.5

1. $10 - 3x - x^2$

a. The largest exponent is 2. The degree of the polynomial is 2.

b. The coefficient of x^2 is –1.

c. The constant term is 10.

d. It is a polynomial of one variable x.

3. $7x^2y - 14xy^3z$

a. The sum of the exponents in each term is 3 and 5, respectively. The degree of the polynomial is 5.

b. The coefficient of xy^3 is –14.

c. The constant term is zero.

d. It is a polynomial of three variables; x, y, and z.

5. $2x^5 - 3x^2 - 5$

a. a_nx^n means $2 = a_5$.

b. $a_3 = 0$ (Term is $0x^3$)

c. $-3 = a_2$

d. $a_0 = -5$, the constant term.

7. $4x - x^2$

When $x = -2$,

$4x - x^2 = 4(-2) - (-2)^2 = -8 - 4 = -12$

9. $\frac{2x - y}{x^2 - 2y}$

When $x = -5$ and $y = -3$,

$\frac{2(-5) - (-3)}{(-5)^2 - 2(-3)} = \frac{-10 + 3}{25 + 6} = -\frac{7}{31}$.

11. $(16pq - 7p^2) + (5pq + 5p^2) = 21pq - 2p^2$

13. $(4m^2 - 3n^2 + 5) - (3m^2 + 4n^2 + 8)$

$= 4m^2 - 3n^2 + 5 - 3m^2 - 4n^2 - 8$

$= m^2 - 7n^2 - 3$

15. $-[8 - 4(q + 5) + q] = -[8 - 4q - 20 + q]$

$= -[-12 - 3q]$

$= 12 + 3q$

17. $x^2 - [x - (x^2 - 1) + 1 - (1 - x^2)] + x$

$= x^2 - [x - x^2 + 1 + 1 - 1 + x^2] + x$

$= x^2 - [x + 1] + x = x^2 - x - 1 + x$

$= x^2 - 1$

19. $(5x^3)(7x^2) = 35x^{3+2} = 35x^5$

21. $(39r^3s^2) \div (13r^2s) = 3r^{3-2}s^{2-1} = 3rs$

23. $ax^2(2x^2 + ax + ab) = 2ax^4 + a^2x^3 + a^2bx^2$

25. $(3y + 4)(2y - 3) = 6y^2 - 9y + 8y - 12$

$= 6y^2 - y - 12$

27. $(1-2x^2)(2-x^2)=2-x^2-4x^2+2x^4$
$=2x^4-5x^2+2$

29. $(4x+3)^2=16x^2+2(4x)(3)+9=16x^2+24x+9$

31. $\left(x^2-\dfrac{1}{2}\right)^2=x^4+2(x^2)\left(-\dfrac{1}{2}\right)+\left(-\dfrac{1}{2}\right)^2$
$=x^4-x^2+\dfrac{1}{4}$

33. $(2x+1)(2x-1)=(2x)^2-1^2=4x^2-1$

35. $(0.1-4x)(0.1+4x)=(0.1)^2-(4x)^2$
$=0.01-16x^2$

37.
$$\begin{array}{r}
x^2+2x+4\\
\underline{\qquad x-2}\\
-2x^2-4x-8\\
\underline{x^3+2x^2+4x\qquad}\\
x^3\qquad\qquad -8
\end{array}$$

39.
$$\begin{array}{r}
x^5-2x^3+5\\
\underline{x^3+5x}\\
5x^6-10x^4\qquad +25x\\
\underline{x^8-2x^6\qquad +5x^3\qquad\qquad}\\
x^8+3x^6-10x^4+5x^3+25x
\end{array}$$

41. $\dfrac{18m^2n+6m^3n+12m^4n^2}{6m^2n}$
$=\dfrac{18m^2n}{6m^2n}+\dfrac{6m^3n}{6m^2n}+\dfrac{12m^4n^2}{6m^2n}$
$=3+m+2m^2n$

43. $\dfrac{24x^8y^4+15x^5y-6x^7y}{9x^5y^2}$
$=\dfrac{24x^8y^4}{9x^5y^2}+\dfrac{15x^5y}{9x^5y^2}-\dfrac{6x^7y}{9x^5y^2}$
$=\dfrac{8x^3y^2}{3}+\dfrac{5}{3y}-\dfrac{2x^2}{3y}$

45. $(x+1)^3=x^3+3(x^2)(1)+3(x)(1)^2+1^3$
$=x^3+3x^2+3x+1$

47. $(2x-3)^3=(2x)^3-3(2x)^2(3)+3(2x)(3)^2-3^3$
$=8x^3-36x^2+54x-27$

49. $(0.1x-2)(x+0.05)=0.1x^2+0.005x-2x-0.10$
$=0.1x^2-1.995x-0.10$

51.
$$\begin{array}{r}
x^2-2x+5\\
x+2\overline{)x^3\qquad\qquad +x-1}\\
\underline{x^3+2x^2\qquad\qquad}\\
-2x^2+\ x-1\\
\underline{-2x^2-4x\quad}\\
5x-\ 1\\
\underline{5x+10}\\
-11
\end{array}$$

Quotient: $x^2-2x+5-\dfrac{11}{x+2}$

53.
$$\begin{array}{r}
x^2+3x-1\qquad\\
x^2+1\overline{)x^4+3x^3\qquad -x+1}\\
\underline{x^4\qquad +x^2\qquad\qquad}\\
3x^3-x^2-x+1\\
\underline{3x^3\qquad +3x\quad}\\
-x^2-4x+1\\
\underline{-x^2\qquad -1}\\
-4x+2
\end{array}$$

Quotient: $x^2+3x-1+\dfrac{-4x+2}{x^2+1}$

55. $x^{1/2}(x^{1/2}+2x^{3/2})=x^{2/2}+2x^{4/2}=x+2x^2$

57. $(x^{1/2}+1)(x^{1/2}-2)=x-2x^{1/2}+x^{1/2}-2$
$=x-x^{1/2}-2$

59. $(\sqrt{x}+3)(\sqrt{x}-3)=(\sqrt{x})^2-(3)^2=x-9$

61. $(2x+1)^{1/2}[(2x+1)^{3/2}-(2x+1)^{-1/2}]$
$=(2x+1)^2-(2x+1)^0$
$=4x^2+4x+1-1$
$=4x^2+4x$

63. **a.** $(3x-2)^2-3x-2(3x-2)+5$
$=9x^2-12x+4-3x-6x+4+5$
$=9x^2-21x+13$

b. $(3x-2)^2-(3x-2)(3x-2)+5$
$=(3x-2)^2-(3x-2)^2+5$
$=5$

65. $R=55x$

67. **a.** $4000-x$
b. $0.10x$
c. $0.08(4000-x)$
d. $0.10x+0.08(4000-x)$ or $320+0.02x$

69. $V=x(15-2x)(10-2x)$

71. **a.** Lengths decrease by one.
Thus (A3) = A2 − 1.

b. Width = 50 − Length

c. (B2) = 50 − A2

d. (C2) = A2 · B2

e. Your spreadsheet will give Length = 33, Width = 17.

73. $P = 6.02t + 3.53$

a.

Year	Internet Households
1995	3.53
1996	9.55
1997	15.57
1998	21.59
1999	27.61
2000	33.63
2001	39.65
2002	45.67

Year	Internet Households
2003	51.69
2004	57.71
2005	63.73
2006	69.75
2007	75.77
2008	81.79
2009	87.81
2010	93.83

b. $6.02t + 3.53 > 75$

$6.02t > 71.47$

$t > 11.87$

In 2007, P is greater than 75%

c. The formula is no longer valid when $P > 100\%$.

Exercise 0.6

1. $9ab - 12a^2b + 18b^2 = 3b(3a - 4a^2 + 6b)$

3. $4x^2 + 8xy^2 + 2xy^3 = 2x(2x + 4y^2 + y^3)$

5. $(7x^3 - 14x^2) + (2x - 4) = 7x^2(x-2) + 2(x-2)$
$= (x-2)(7x^2 + 2)$

7. $6x - 6m + xy - my = (6x - 6m) + (xy - my)$
$= 6(x-m) + y(x-m)$
$= (x-m)(6+y)$

9. $x^2 + 8x + 12 = (x+6)(x+2)$

11. $x^2 - x - 6 = (x-3)(x+2)$

13. $7x^2 - 10x - 8$
$7x^2 \cdot 8 = 56x^2$
The factors $-14x$ and $+4x$ give a sum of $-10x$.
$7x^2 - 10x - 8 = 7x^2 - 14x + 4x - 8$
$= 7x(x-2) + 4(x-2)$
$= (x-2)(7x+4)$

15. $x^2 - 10x + 25 = x^2 - 2 \cdot 5x + 5^2 = (x-5)^2$

17. $49a^2 - 144b^2 = (7a)^2 - (12b)^2$
$= (7a + 12b)(7a - 12b)$

19. **a.** $9x^2 + 21x - 8$
$9x^2(-8) = -72x^2$
The factors $24x$ and $-3x$ give a sum of $21x$.
$9x^2 + 21x - 8 = 9x^2 + 24x - 3x - 8$
$= 3x(3x+8) - 1(3x+8)$
$= (3x+8)(3x-1)$

b. $9x^2 + 22x + 8$
$9x^2 \cdot 8 = 72x^2$
The factors $18x$ and $4x$ give a sum of $22x$.
$9x^2 + 22x + 8 = 9x^2 + 18x + 4x + 8$
$= 9x(x+2) + 4(x+2)$
$= (x+2)(9x+4)$

21. $4x^2 - x = x(4x-1)$

23. $x^3 + 4x^2 - 5x - 20 = x^2(x+4) - 5(x+4)$
$= (x+4)(x^2 - 5)$

25. $2x^2 - 8x - 42 = 2(x^2 - 4x - 21) = 2(x-7)(x+3)$

27. $2x^3 - 8x^2 + 8x = 2x(x^2 - 4x + 4)$
$= 2x(x^2 - 2 \cdot 2x + 2^2)$
$= 2x(x-2)^2$

29. $2x^2+x-6$
$2x^2\cdot(-6)=-12x^2$
The factors $4x$ and $-3x$ give a sum of x.
$$\begin{aligned}2x^2+x-6&=2x^2+4x-3x-6\\&=2x(x+2)-3(x+2)\\&=(2x-3)(x+2)\end{aligned}$$

31. $3x^2+3x-36=3(x^2+x-12)=3(x+4)(x-3)$

33. $2x^3-8x=2x(x^2-4)=2x(x+2)(x-2)$

35. $10x^2+19x+6$
$10x^2\cdot 6=60x^2$
The factors $4x$ and $15x$ give a sum of $19x$.
$$\begin{aligned}10x^2+19x+6&=10x^2+4x+15x+6\\&=2x(5x+2)+3(5x+2)\\&=(5x+2)(2x+3)\end{aligned}$$

37. $9-47x+10x^2$
$9\cdot 10x^2=90x^2$
The factors $-45x$ and $-2x$ give a sum of $-47x$.
$$\begin{aligned}9-47x+10x^2&=9-45x-2x+10x^2\\&=9(1-5x)-2x(1-5x)\\&=(1-5x)(9-2x)\\&\text{or } (5x-1)(2x-9)\end{aligned}$$

39. $$\begin{aligned}y^4-16x^4&=(y^2)^2-(4x^2)^2\\&=(y^2-4x^2)(y^2+4x^2)\\&=(y-2x)(y+2x)(y^2+4x^2)\end{aligned}$$

41. $$\begin{aligned}x^4-8x^2+16&=(x^2)^2-2\cdot 4x^2+4^2=(x^2-4)^2\\&=[(x-2)(x+2)]^2\\&=(x-2)^2(x+2)^2\end{aligned}$$

43. $$\begin{aligned}4x^4-5x^2+1&=(4x^2-1)(x^2-1)\\&=(2x+1)(2x-1)(x+1)(x-1)\end{aligned}$$

45. $x^3+3x^2+3x+1=(x+1)^3$

47. $$\begin{aligned}x^3-12x^2+48x-64&=x^3-3(4x^2)+3(16x)-4^3\\&=x^3-3x^2(4)+3x(4)^2-4^3\\&=(x-4)^3\end{aligned}$$

49. $x^3-64=x^3-4^3=(x-4)(x^2+4x+16)$

51. $27+8x^3=3^3+(2x)^3=(3+2x)(9-6x+4x^2)$

53. $x^{3/2}+x^{1/2}=x^{1/2}(x^{2/2}+1)=x^{1/2}(x+1)$
$?=(x+1)$

55. $x^{-3}+x^{-2}=x^{-3}(1+x^1)=x^{-3}(1+x)$
$?=1+x$

57. $$\begin{aligned}&(-x^3+x)(3-x^2)^{-1/2}+2x(3-x^2)^{1/2}\\&=(3-x^2)^{-1/2}[(-x^3+x)+2x(3-x^2)^{2/2}]\\&=(3-x^2)^{-1/2}[-x^3+x+6x-2x^3]\\&=(3-x^2)^{-1/2}[7x-3x^3]\end{aligned}$$
$?=7x-3x^3$

59. $P+Prt=P(1+rt)$

61. $S=cm-m^2=m(c-m)$

63. **a.** In the form px we have $p(10{,}000-100p)$.
$x=10{,}000-100p$
b. If $p=38$, then
$x=10{,}000-100\cdot 38=6200$.

Exercise 0.7

1. $\dfrac{18x^3y^3}{9x^3z}=\dfrac{2x^3y^3}{x^3z}=\dfrac{2y^3}{z}$

3. $\dfrac{x-3y}{3x-9y}=\dfrac{1(x-3y)}{3(x-3y)}=\dfrac{1}{3}$

5. $\dfrac{x^2-2x+1}{x^2-4x+3}=\dfrac{(x-1)(x-1)}{(x-3)(x-1)}=\dfrac{x-1}{x-3}$

7. $\dfrac{6x^3y^3-15x^2y}{3x^2y^2+9x^2y}=\dfrac{3x^2y(2xy^2-5)}{3x^2y(y+3)}=\dfrac{2xy^2-5}{y+3}$

9. $\dfrac{6x^3}{8y^3}\cdot\dfrac{16x}{9y^2}\cdot\dfrac{15y^4}{x^3}=\dfrac{6}{y^3}\cdot\dfrac{2x}{9y^2}\cdot\dfrac{15y^4}{1}=\dfrac{2}{1}\cdot\dfrac{2x}{3y^2}\cdot\dfrac{15y}{1}=\dfrac{2}{1}\cdot\dfrac{2x}{y}\cdot\dfrac{5}{1}=\dfrac{20x}{y}$

11. $\dfrac{8x-16}{x-3}\cdot\dfrac{4x-12}{3x-6}=\dfrac{8(x-2)}{x-3}\cdot\dfrac{4(x-3)}{3(x-2)}=\dfrac{8\cdot 4}{3}=\dfrac{32}{3}$

15. $\frac{x^2-x-2}{2x^2-8}\cdot\frac{18-2x^2}{x^2-5x+4}\cdot\frac{x^2-2x-8}{x^2-6x+9}=\frac{(x-2)(x+1)}{2(x^2-4)}\cdot\frac{-2(x^2-9)}{(x-4)(x-1)}\cdot\frac{(x-4)(x+2)}{(x-3)(x-3)}$

$=\frac{(x-2)(x+1)}{2(x-2)(x+2)}\cdot\frac{-2(x-3)(x+3)}{(x-1)}\cdot\frac{(x+2)}{(x-3)(x-3)}=-\frac{(x+1)(x+3)}{(x-1)(x-3)}$

17. $\frac{15ac^2}{7bd}\div\frac{4a}{14b^2d}=\frac{15ac^2}{7bd}\cdot\frac{14b^2d}{4a}=\frac{15c^2}{1}\cdot\frac{2b}{4}=\frac{15bc^2}{2}$

19. $\frac{y^2-2y+1}{7y^2-7y}\div\frac{y^2-4y+3}{35y^2}=\frac{y^2-2y+1}{7y(y-1)}\cdot\frac{35y^2}{y^2-4y+3}=\frac{(y-1)(y-1)}{7y(y-1)}\cdot\frac{35y^2}{(y-3)(y-1)}=\frac{5y}{y-3}$

21. $\frac{x^2-x-6}{1}\div\frac{9-x^2}{x^2-3x}=\frac{x^2-x-6}{1}\cdot\frac{x^2-3x}{-1(x^2-9)}=\frac{(x-3)(x+2)}{1}\cdot\frac{x(x-3)}{-1(x-3)(x+3)}=\frac{-x(x-3)(x+2)}{x+3}$

23. $\frac{2x}{x^2-x-2}-\frac{x+2}{x^2-x-2}=\frac{2x-x-2}{(x-2)(x+1)}=\frac{x-2}{(x-2)(x+1)}=\frac{1}{x+1}$

25. $\frac{a}{a-2}-\frac{a-2}{a}=\frac{a}{a-2}\cdot\frac{a}{a}-\frac{a-2}{a}\cdot\frac{a-2}{a-2}=\frac{a^2-(a^2-4a+4)}{a(a-2)}=\frac{4a-4}{a(a-2)}=\frac{4(a-1)}{a(a-2)}$

27. $\frac{x}{x+1}-x+1=\frac{x}{x+1}-\frac{x}{1}\cdot\frac{x+1}{x+1}+\frac{1}{1}\cdot\frac{x+1}{x+1}=\frac{x-x^2-x+x+1}{x+1}=\frac{-x^2+x+1}{x+1}$

29. $\frac{4a}{3x+6}+\frac{5a^2}{4x+8}=\frac{4a}{3(x+2)}+\frac{5a^2}{4(x+2)}=\frac{4a}{3(x+2)}\cdot\frac{4}{4}+\frac{5a^2}{4(x+2)}\cdot\frac{3}{3}=\frac{16a+15a^2}{12(x+2)}$

31. $\frac{3x-1}{2x-4}+\frac{4x}{3x-6}-\frac{x-4}{5x-10}=\frac{3x-1}{2(x-2)}+\frac{4x}{3(x-2)}-\frac{x-4}{5(x-2)}$

$=\frac{3x-1}{2(x-2)}\cdot\frac{3\cdot5}{3\cdot5}+\frac{4x}{3(x-2)}\cdot\frac{2\cdot5}{2\cdot5}-\frac{x-4}{5(x-2)}\cdot\frac{3\cdot2}{3\cdot2}=\frac{(45x-15)+40x-6x+24}{30(x-2)}=\frac{79x+9}{30(x+2)}$

33. $\frac{1}{x^2-4y^2}-\frac{1}{x^2-4xy+4y^2}=\frac{1}{(x-2y)(x+2y)}-\frac{1}{(x-2y)(x-2y)}$

$=\frac{1}{(x-2y)(x+2y)}\cdot\frac{x-2y}{x-2y}-\frac{1}{(x-2y)(x-2y)}\cdot\frac{x+2y}{x+2y}=\frac{(x-2y)-(x+2y)}{(x-2y)^2(x+2y)}=\frac{-4y}{(x-2y)^2(x+2y)}$

35. $\frac{x}{x^2-4}+\frac{4}{x^2-x-2}-\frac{x-2}{x^2+3x+2}=\frac{x}{(x+2)(x-2)}+\frac{4}{(x-2)(x+1)}-\frac{x-2}{(x+2)(x+1)}$

$=\frac{x}{(x+2)(x-2)}\cdot\frac{x+1}{x+1}+\frac{4}{(x-2)(x+1)}\cdot\frac{x+2}{x+2}-\frac{x-2}{(x+2)(x+1)}\cdot\frac{x-2}{x-2}$

$=\frac{(x^2+x)+(4x+8)-(x^2-4x+4)}{(x+2)(x+1)(x-2)}=\frac{9x+4}{(x+2)(x+1)(x-2)}$

37. $\frac{-x^3+x}{\sqrt{3-x^2}}+\frac{2x\sqrt{3-x^2}}{1}=\frac{-x^3+x}{\sqrt{3-x^2}}+\frac{2x\sqrt{3-x^2}}{1}\cdot\frac{\sqrt{3-x^2}}{\sqrt{3-x^2}}=\frac{-x^3+x+2x(3-x^2)}{\sqrt{3-x^2}}$

$=\frac{-x^3+x+6x-2x^3}{\sqrt{3-x^2}}=\frac{7x-3x^3}{\sqrt{3-x^2}}$

39. $\dfrac{\frac{3}{1}-\frac{2}{3}}{\frac{14}{1}}\cdot\dfrac{3}{3}=\dfrac{9-2}{14(3)}=\dfrac{7}{14(3)}=\dfrac{1}{6}$

41. $\dfrac{x+y}{\frac{1}{x}+\frac{1}{y}}=\dfrac{(x+y)}{\frac{1}{x}+\frac{1}{y}}\cdot\dfrac{xy}{xy}=\dfrac{xy(x+y)}{y+x}=xy$

43. $\dfrac{2-\frac{1}{x}}{2x-\frac{3x}{x+1}}=\dfrac{\frac{2}{1}-\frac{1}{x}}{\frac{2x}{1}-\frac{3x}{x+1}}\cdot\dfrac{x(x+1)}{x(x+1)}=\dfrac{2x(x+1)-1(x+1)}{2x^2(x+1)-3x(x)}=\dfrac{2x^2+x-1}{2x^3-x^2}=\dfrac{(2x-1)(x+1)}{x^2(2x-1)}=\dfrac{x+1}{x^2}$

45. $\dfrac{\sqrt{a}-\frac{b}{\sqrt{a}}}{a-b}=\dfrac{\frac{\sqrt{a}}{1}-\frac{b}{\sqrt{a}}}{\frac{a-b}{1}}\cdot\dfrac{\sqrt{a}}{\sqrt{a}}=\dfrac{a-b}{\sqrt{a}(a-b)}=\dfrac{1}{\sqrt{a}}$ or $\dfrac{\sqrt{a}}{a}$

47. $\dfrac{\sqrt{x^2+9}-\frac{13}{\sqrt{x^2+9}}}{x^2-x-6}=\dfrac{\frac{\sqrt{x^2+9}}{1}-\frac{13}{\sqrt{x^2+9}}}{(x-3)(x+2)}\cdot\dfrac{\sqrt{x^2+9}}{\sqrt{x^2+9}}=\dfrac{x^2+9-13}{(x-3)(x+2)\sqrt{x^2+9}}=\dfrac{(x-2)(x+2)}{(x-3)(x+2)\sqrt{x^2+9}}=\dfrac{x-2}{(x-3)\sqrt{x^2+9}}$

49. a. $(2^{-2}-3^{-1})^{-1}=\left(\dfrac{1}{2^2}-\dfrac{1}{3}\right)^{-1}=\left(-\dfrac{1}{12}\right)^{-1}=-12$

b. $(2^{-1}+3^{-1})^2=\left(\dfrac{1}{2}+\dfrac{1}{3}\right)^2=\left(\dfrac{5}{6}\right)^2=\dfrac{25}{36}$

Hint: Work inside ( ) first when adding or subtracting is involved.

51. $\dfrac{2a^{-1}-b^{-1}}{(ab)^{-1}}=\dfrac{\frac{2}{a}-\frac{1}{b}}{\frac{1}{ab}}\cdot\dfrac{ab}{ab}=\dfrac{2b-a}{1}$ or $2b-a$

53. $\dfrac{xy^{-2}+x^{-2}y}{x+y}=\dfrac{\frac{x}{y^2}+\frac{y}{x^2}}{(x+y)}\cdot\dfrac{x^2y^2}{x^2y^2}=\dfrac{x^3+y^3}{x^2y^2(x+y)}=\dfrac{(x^2-xy+y^2)(x+y)}{x^2y^2(x+y)}=\dfrac{x^2-xy+y^2}{x^2y^2}$

55. $\dfrac{1-\sqrt{x}}{1+\sqrt{x}}=\dfrac{1-\sqrt{x}}{1+\sqrt{x}}\cdot\dfrac{1-\sqrt{x}}{1-\sqrt{x}}=\dfrac{1-2\sqrt{x}+x}{1-x}$

57. $\dfrac{\sqrt{x+h}-\sqrt{x}}{h}=\dfrac{\sqrt{x+h}-\sqrt{x}}{h}\cdot\dfrac{\sqrt{x+h}+\sqrt{x}}{\sqrt{x+h}+\sqrt{x}}=\dfrac{(x+h)-(x)}{h\left(\sqrt{x+h}+\sqrt{x}\right)}=\dfrac{h}{h\left(\sqrt{x+h}+\sqrt{x}\right)}=\dfrac{1}{\sqrt{x+h}+\sqrt{x}}$

59. $\dfrac{1}{a}+\dfrac{1}{b}+\dfrac{1}{c}=\dfrac{1}{a}\cdot\dfrac{bc}{bc}+\dfrac{1}{b}\cdot\dfrac{ac}{ac}+\dfrac{1}{c}\cdot\dfrac{ab}{ab}=\dfrac{bc+ac+ab}{abc}$

61. a. Avg. cost $=\dfrac{4000}{x}+\dfrac{55}{1}+\dfrac{0.1x}{1}=\dfrac{4000+55x+0.1x^2}{x}$

b. Total cost = (Avg. cost)(number of units) $=4000+55x+0.1x^2$

63. $SV=1+\dfrac{3}{t+3}-\dfrac{18}{(t+3)^2}=\dfrac{(t+3)^2+3(t+3)-18}{(t+3)^2}=\dfrac{t^2+6t+9+3t+9-18}{(t+3)^2}=\dfrac{t^2+9t}{(t+3)^2}$

Review Exercises

1. $B=\{1, 2, 3, 4, 5, 6, 7, 8\}$. Since every element of A is also an element of B, A is a subset of B.

2. No. $3\notin\{x:x>3\}$

3. A and B are not disjoint since each set contains the element 1.

4. $A=\{1, 2, 3, 9\}$ $B'=\{2, 4, 9\}$
$A\cup B'=\{1, 2, 3, 4, 9\}$

5. $\{4, 5, 6, 7, 8, 10\}\cap\{1, 3, 5, 6, 7, 8, 10\}$
$=\{5, 6, 7, 8, 10\}$

6. $A = \{1, 2, 3, 9\}$ $\quad B = \{1, 3, 5, 6, 7, 8, 10\}$
$A' = \{4, 5, 6, 7, 8, 10\}$
$A' \cap B = \{5, 6, 7, 8, 10\}$
$(A' \cap B)' = \{1, 2, 3, 4, 9\}$

7. $\{4, 5, 6, 7, 8, 10\} \cup \{2, 4, 9\}$
$= \{2, 4, 5, 6, 7, 8, 9, 10\}$
$(A' \cup B')' = \{2, 4, 5, 6, 7, 8, 9, 10\}' = \{1, 3\}$
$A \cap B = \{1, 2, 3, 9\} \cap \{1, 3, 5, 6, 7, 8, 10\}$
$= \{1, 3\}$ Yes.

8. **a.** $6 + \frac{1}{3} = \frac{1}{3} + 6$ illustrates the commutative property of addition.
b. $2(3 \cdot 4) = (2 \cdot 3)4$ illustrates the associative property of multiplication.
c. $\frac{1}{3}(6+9) = 2+3$ illustrates the distributive property.

9. **a.** irrational
b. rational, integer
c. meaningless

10. **a.** $\pi > 3.14$
b. $-100 < 0.1$
c. $-3 > -12$

11. $|5-11| = |-6| = -(-6) = 6$

12. $44 \div 2 \cdot 11 - 10^2 = 22 \cdot 11 - 100 = 242 - 100 = 142$

13. $(-3)^2 - (-1)^3 = 9 - (-1) = 10$

14. $\frac{(3)(2)(15)-(5)(8)}{(4)(10)} = \frac{90-40}{40} = \frac{50}{40} = \frac{5}{4}$

15.
$$2 - [3 - (2 - |-3|)] + 11 = 2 - [3 - (2-3)] + 11$$
$$= 2 - [3 - (-1)] + 11$$
$$= 2 - [3+1] + 11$$
$$= 2 - 4 + 11$$
$$= 9$$

16. $-4^2 - (-4)^2 + 3 = -16 - 16 + 3 = -32 + 3 = -29$

17. $\frac{4+3^2}{4} = \frac{4+9}{4} = \frac{13}{4}$

18. $\frac{(-2.91)^5}{\sqrt{3.29^5}} \approx \frac{-208.6724}{19.6331} \approx -10.6286$

19. **a.** [0, 5], closed

b. [−3, 7), half open

c. (−4, 0) open

20. **a.** $(-1, 16)$
$-1 < x < 16$
b. $[-12, 8]$
$-12 \le x \le 8$
c. $x < -1$

21. **a.** $\left(\frac{3}{8}\right)^0 = 1$
b. $2^3 \cdot 2^{-5} = 2^{-2} = \frac{1}{2^2} = \frac{1}{4}$
c. $\frac{4^9}{4^3} = 4^6 = 4096$
d. $\left(\frac{1}{7}\right)^3 \left(\frac{1}{7}\right)^{-4} = \left(\frac{1}{7}\right)^{-1} = 7$

22. **a.** $x^5 \cdot x^{-7} = x^{5+(-7)} = x^{-2} = \frac{1}{x^2}$
b. $\frac{x^8}{x^{-2}} = x^{8-(-2)} = x^{10}$
c. $(x^3)^3 = x^{3 \cdot 3} = x^9$
d. $(y^4)^{-2} = y^{(4)(-2)} = y^{-8} = \frac{1}{y^8}$
e. $(-y^{-3})^{-2} = y^{(-3)(-2)} = y^6$

There are other correct methods of working problems 23–28.

23. $\frac{-(2xy^2)^{-2}}{(3x^{-2}y^{-3})^2} = \frac{(-1)(2)^{-2}x^{-2}y^{-4}}{3^2x^{-4}y^{-6}} = \frac{(-1)x^4y^6}{2^2 \cdot 3^2 x^2 y^4} = -\frac{x^2y^2}{36}$

24. $\left(\frac{2}{3}x^2y^{-4}\right)^{-2} = \left(\frac{2}{3}\right)^{-2}\left(x^2\right)^{-2}\left(y^{-4}\right)^{-2} = \left(\frac{3}{2}\right)^2\left(x^{-4}\right)\left(y^8\right) = \left(\frac{9}{4}\right)\left(\frac{1}{x^4}\right)\left(y^8\right) = \frac{9y^8}{4x^4}$

25. $\left(\frac{x^{-2}}{2y^{-1}}\right)^2 = \left(\frac{y}{2x^2}\right)^2 = \frac{y^2}{4x^4}$

26. $\frac{(-x^4y^{-2}z^2)^0}{-(x^4y^{-2}z^2)^{-2}} = \frac{1}{-(x^4)^{-2}(y^{-2})^{-2}(z^2)^{-2}} = \frac{1}{-x^{-8}y^4z^{-4}} = \frac{-x^8z^4}{y^4}$

27. $\left(\frac{x^{-3}y^4z^{-2}}{3x^{-2}y^{-3}z^{-3}}\right)^{-1} = \left(\frac{y^{4-(-3)}z^{-2-(-3)}}{3x^{-2-(-3)}}\right)^{-1} = \left(\frac{y^7z}{3x}\right)^{-1} = \frac{3x}{y^7z}$

28. $\left(\frac{x}{2y}\right)\left(\frac{y}{x^2}\right)^{-2} = \left(\frac{x}{2y}\right)\left(\frac{x^2}{y}\right)^2 = \left(\frac{x}{2y}\right)\left(\frac{(x^2)^2}{y^2}\right) = \left(\frac{x}{2y}\right)\left(\frac{x^4}{y^2}\right) = \frac{x^5}{2y^3}$

29. a. $-\sqrt[3]{-64} = -\sqrt[3]{(-4)^3} = -(-4) = 4$

b. $\sqrt{\frac{4}{49}} = \sqrt{\frac{2^2}{7^2}} = \frac{2}{7}$

c. $\sqrt[7]{1.9487171} = 1.1$

30. a. $\sqrt{x} = x^{1/2}$

b. $\sqrt[3]{x^2} = x^{2/3}$

c. $1/\sqrt[4]{x} = \frac{1}{x^{1/4}} = x^{-1/4}$

31. a. $x^{2/3} = \sqrt[3]{x^2}$

b. $x^{-1/2} = \frac{1}{\sqrt{x}} = \frac{\sqrt{x}}{x}$

c. $-x^{3/2} = -x\sqrt{x}$

32. a. $\frac{5xy}{\sqrt{2x}} \cdot \frac{\sqrt{2x}}{\sqrt{2x}} = \frac{5xy\sqrt{2x}}{2x} = \frac{5y\sqrt{2x}}{2}$

b. $\frac{y}{x\sqrt[3]{xy^2}} \cdot \frac{\sqrt[3]{x^2y}}{\sqrt[3]{x^2y}} = \frac{y\sqrt[3]{x^2y}}{x\sqrt[3]{x^3y^3}} = \frac{y\sqrt[3]{x^2y}}{x(xy)}$

$= \frac{y\sqrt[3]{x^2y}}{x^2y} = \frac{\sqrt[3]{x^2y}}{x^2}$

33. $x^{1/2} \cdot x^{1/3} = x^{(3/6)+(2/6)} = x^{5/6}$

34. $\frac{y^{-3/4}}{y^{-7/4}} = y^{-3/4-(-7/4)} = y^{4/4} = y$

35. $x^4 \cdot x^{1/4} = x^{(16/4)+(1/4)} = x^{17/4}$

36. $\frac{1}{x^{-4/3} \cdot x^{-7/3}} = \frac{1}{x^{-11/3}} = x^{11/3}$

37. $(x^{4/5})^{1/2} = x^{(4/5)(1/2)} = x^{2/5}$

38. $(x^{1/2}y^2)^4 = (x^{1/2})^4(y^2)^4 = x^2y^8$

39. $\sqrt{12x^3y^5} = \sqrt{4x^2y^4 \cdot 3xy} = 2xy^2\sqrt{3xy}$

40. $\sqrt{1250x^6y^9} = \sqrt{625x^6y^8 \cdot 2y} = 25x^3y^4\sqrt{2y}$

41. $\sqrt[3]{24x^4y^4} \cdot \sqrt[3]{45x^4y^{10}}$

$= \sqrt[3]{8x^3y^3 \cdot 3xy} \cdot \sqrt[3]{9x^3y^9 \cdot 5xy}$

$= 2xy\sqrt[3]{3xy} \cdot xy^3\sqrt[3]{9 \cdot 5xy}$

$= 2x^2y^4\sqrt[3]{27 \cdot 5x^2y^2}$

$= 6x^2y^4\sqrt[3]{5x^2y^2}$

42. $\sqrt{16a^2b^3} \cdot \sqrt{8a^3b^5} = \sqrt{128a^5b^8}$

$= \sqrt{64a^4b^8 \cdot 2a}$

$= 8a^2b^4\sqrt{2a}$

43. $\frac{\sqrt{52x^3y^6}}{\sqrt{13xy^4}} = \sqrt{4x^2y^2} = 2xy$

44. $\frac{\sqrt{32x^4y^3}}{\sqrt{6xy^{10}}} = \sqrt{\frac{16x^3}{3y^7}} = \frac{4x\sqrt{x}}{y^3\sqrt{3y}} \cdot \frac{\sqrt{3y}}{\sqrt{3y}} = \frac{4x\sqrt{3xy}}{3y^4}$

45. $(3x+5)-(4x+7) = 3x+5-4x-7 = -x-2$

46. $x(1-x)+x[x-(2+x)] = x-x^2+x(-2)$

$= -x^2 - x$

47. $(3x^3-4xy-3)+(5xy+x^3+4y-1)$

$= 4x^3+xy+4y-4$

48. $(4xy^3)(6x^4y^2) = 24x^{1+4}y^{3+2} = 24x^5y^5$

49. $(3x-4)(x-1) = 3x^2-3x-4x+4 = 3x^2-7x+4$

50. $(3x-1)(x+2) = 3x^2+6x-x-2 = 3x^2+5x-2$

51. $(4x+1)(x-2) = 4x^2 - 8x + x - 2 = 4x^2 - 7x - 2$

52. $(3x-7)(2x+1) = 6x^2 + 3x - 14x - 7$
$= 6x^2 - 11x - 7$

53. $(2x-3)^2 = (2x)^2 - 2(2x)(3) + 3^2 = 4x^2 - 12x + 9$

54. $(4x+3)(4x-3) = 16x^2 - 9$
Difference of two squares

55.

$$\begin{array}{r} x^2 + x - 3 \\ 2x^2 + 1 \\ \hline x^2 + x - 3 \\ 2x^4 + 2x^3 - 6x^2 \qquad\quad \\ \hline 2x^4 + 2x^3 - 5x^2 + x - 3 \end{array}$$

56. $(2x-1)^3 = 8x^3 - 12x^2 + 6x - 1$ Binomial cubed

57.

$$\begin{array}{r} x^2 + xy + y^2 \\ x - y \\ \hline -x^2y - xy^2 - y^3 \\ x^3 + x^2y + xy^2 \qquad \\ \hline x^3 \qquad\qquad\quad - y^3 \end{array}$$

Difference of two cubes

58. $\dfrac{4x^2y - 3x^3y^3 - 6x^4y^2}{2x^2y^2} = \dfrac{2}{y} - \dfrac{3xy}{2} - 3x^2$

59.

$$\begin{array}{r} 3x^2 + 2x - 3 \qquad\qquad \\ x^2 + 1 \overline{\smash{\big)}\, 3x^4 + 2x^3 \qquad\quad - x + 4} \\ \underline{3x^4 \qquad + 3x^2 \qquad\qquad} \\ 2x^3 - 3x^2 - x + 4 \\ \underline{2x^3 \qquad + 2x \qquad} \\ -3x^2 - 3x + 4 \\ \underline{-3x^2 \qquad - 3} \\ -3x + 7 \end{array}$$

Quotient is $3x^2 + 2x - 3 + \dfrac{7-3x}{x^2+1}$.

60.

$$\begin{array}{r} x^3 - x^2 + 2x + 7 \\ x - 3 \overline{\smash{\big)}\, x^4 - 4x^3 + 5x^2 + \ x} \\ \underline{x^4 - 3x^3 \qquad\qquad\quad} \\ -x^3 + 5x^2 \qquad \\ \underline{-x^3 + 3x^2 \qquad} \\ 2x^2 + \ x \\ \underline{2x^2 - 6x} \\ 7x \qquad \\ \underline{7x - 21} \\ 21 \end{array}$$

Quotient is $x^3 - x^2 + 2x + 7 + \dfrac{21}{x-3}$.

61. $x^{4/3}(x^{2/3} - x^{-1/3}) = x^{6/3} - x^{3/3} = x^2 - x$

62. $\left(\sqrt{x} + \sqrt{a-x}\right)\left(\sqrt{x} - \sqrt{a-x}\right) = \left(\sqrt{x}\right)^2 - \left(\sqrt{a-x}\right)^2$
$= x - (a - x)$
$= x - a + x$
$= 2x - a$

63. $2x^4 - x^3 = x^3(2x-1)$

64. $4(x^2+1)^2 - 2(x^2+1)^3 = 2(x^2+1)^2[2 - (x^2+1)]$
$= 2(x^2+1)^2(2 - x^2 - 1)$
$= 2(x^2+1)^2(1 - x^2)$
$= 2(x^2+1)^2(1+x)(1-x)$

65. $4x^2 - 4x + 1 = (2x)^2 - 2(2x) + 1^2 = (2x-1)^2$

66. $16 - 9x^2 = (4+3x)(4-3x)$

67. $2x^4 - 8x^2 = 2x^2(x^2 - 4) = 2x^2(x+2)(x-2)$

68. $x^2 - 4x - 21 = (x-7)(x+3)$

69. $3x^2 - x - 2 = (3x+2)(x-1)$

70. $12x^2 - 23x - 24$
Two expressions whose product is $12x^2(-24) = -288x^2$ and whose sum is $-23x$ are $-32x$ and $9x$. So,
$12x^2 - 23x - 24 = 12x^2 + 9x - 32x - 24$
$= 3x(4x+3) - 8(4x+3)$
$= (4x+3)(3x-8)$.

71. $16x^4-72x^2+81=(4x^2)^2-2(4x^2\cdot 9)+9^2$
$=(4x^2-9)^2$
$=[(2x+3)(2x-3)]^2$
$=(2x+3)^2(2x-3)^2$

72. $x^{-2/3}+x^{-4/3}=x^{-4/3}(?)$
$x^{-2/3}+x^{-4/3}=x^{-4/3}(x^{2/3}+1)$
$?=x^{2/3}+1$

73. a. $\dfrac{2x}{2x+4}=\dfrac{2x}{2(x+2)}=\dfrac{x}{x+2}$

b. $\dfrac{4x^2y^3-6x^3y^4}{2x^2y^2-3xy^3}=\dfrac{2x^2y^3(2-3xy)}{xy^2(2x-3y)}$
$=\dfrac{2xy(2-3xy)}{2x-3y}$

74. $\dfrac{x^2-4x}{x^2+4}\cdot\dfrac{x^4-16}{x^4-16x^2}=\dfrac{x(x-4)}{x^2+4}\cdot\dfrac{(x^2-4)(x^2+4)}{x^2(x^2-16)}$
$=\dfrac{(x-4)(x+2)(x-2)}{x(x-4)(x+4)}$
$=\dfrac{(x+2)(x-2)}{x(x+4)}$

75. $\dfrac{x^2+6x+9}{x^2-7x+12}\cdot\dfrac{x^2-3x-4}{x^2+4x+3}$
$=\dfrac{(x+3)(x+3)}{(x-4)(x-3)}\cdot\dfrac{(x-4)(x+1)}{(x+3)(x+1)}$
$=\dfrac{x+3}{x-3}$

76. $\dfrac{x^4-2x^3}{3x^2-x-2}\div\dfrac{x(x^2-4)}{9x^2-4}$
$=\dfrac{x^3(x-2)}{(3x+2)(x-1)}\cdot\dfrac{(3x+2)(3x-2)}{x(x+2)(x-2)}$
$=\dfrac{x^2(3x-2)}{(x-1)(x+2)}$

77. $1+\dfrac{3}{2x}-\dfrac{1}{6x^2}=\dfrac{1}{1}\cdot\dfrac{6x^2}{6x^2}+\dfrac{3}{2x}\cdot\dfrac{3x}{3x}-\dfrac{1}{6x^2}$
$=\dfrac{6x^2+9x-1}{6x^2}$

78. $\dfrac{1}{x-2}-\dfrac{x-2}{4}=\dfrac{1\cdot 4-(x-2)(x-2)}{4(x-2)}=\dfrac{4x-x^2}{4(x-2)}$

79. $\dfrac{x+2}{x(x-1)}-\dfrac{x^2+4}{(x-1)(x-1)}+\dfrac{1}{1}$
$=\dfrac{(x+2)(x-1)-(x^2+4)x+x(x-1)(x-1)}{x(x-1)(x-1)}$
$=\dfrac{x^2+x-2-x^3-4x+x^3-2x^2+x}{x(x-1)^2}$
$=\dfrac{-(x^2+2x+2)}{x(x-1)^2}$

80. $\dfrac{x-1}{x^2-x-2}-\dfrac{x}{x^2-2x-3}+\dfrac{1}{x-2}=\dfrac{x-1}{(x-2)(x+1)}-\dfrac{x}{(x-3)(x+1)}+\dfrac{1}{x-2}$
$=\dfrac{(x-1)(x-3)}{(x-2)(x+1)(x-3)}-\dfrac{x(x-2)}{(x-2)(x+1)(x-3)}+\dfrac{(x+1)(x-3)}{(x-2)(x+1)(x-3)}=\dfrac{x^2-4x+3-x^2+2x+x^2-2x-3}{(x-2)(x+1)(x-3)}$
$=\dfrac{x^2-4x}{(x-2)(x+1)(x-3)}=\dfrac{x(x-4)}{(x-2)(x+1)(x-3)}$

81. $\dfrac{\frac{x-1}{1}-\frac{x-1}{x}}{\frac{1}{x-1}+1}\cdot\dfrac{x(x-1)}{x(x-1)}=\dfrac{x(x-1)^2-(x-1)^2}{x+x(x-1)}=\dfrac{(x-1)^2(x-1)}{x^2}=\dfrac{(x-1)^3}{x^2}$

82. $\dfrac{x^{-2}-x^{-1}}{x^{-2}+x^{-1}}=\dfrac{\frac{1}{x^2}-\frac{1}{x}}{\frac{1}{x^2}+\frac{1}{x}}\cdot\dfrac{x^2}{x^2}=\dfrac{1-x}{1+x}$

83. $\dfrac{3x-3}{\sqrt{x}-1}\cdot\dfrac{\sqrt{x}+1}{\sqrt{x}+1}=\dfrac{3(x-1)(\sqrt{x}+1)}{x-1}=3(\sqrt{x}+1)$

84. $\dfrac{\sqrt{x}-\sqrt{x-4}}{2}\cdot\dfrac{\sqrt{x}+\sqrt{x-4}}{\sqrt{x}+\sqrt{x-4}}=\dfrac{x-(x-4)}{2(\sqrt{x}+\sqrt{x-4})}=\dfrac{x-x+4}{2(\sqrt{x}+\sqrt{x-4})}=\dfrac{4}{2(\sqrt{x}+\sqrt{x-4})}=\dfrac{2}{\sqrt{x}+\sqrt{x-4}}$

85. a. R: Recognized
C: Involved
E: Exercised

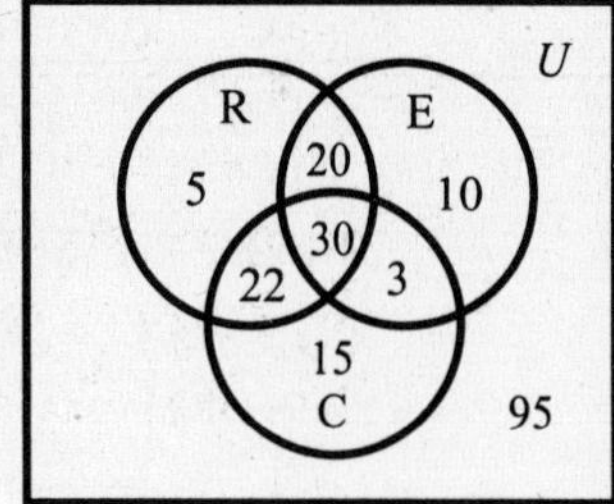

Numbered statement indicates solution for that question.
1. 30
2. 50 – 30 = 20
3. 52 – 30 = 22
4. 30 + 22 + 20 + 5 = 77
5. 37 – 22 = 15
6. 77 + 15 + 3 = 95

b. $200-(95+5+22+30+20+3+15)=10$ So, 10 exercised only.

c. $63+70-(3+30)=100$ So, 100 exercised or were involved in the community.

86. $S=100\left[\dfrac{(1.0075)^n-1}{0.0075}\right]$

a. $S(36)=100\left[\dfrac{(1.0075)^{36}-1}{0.0075}\right]$

$\approx 100\left[\dfrac{0.30865}{0.0075}\right]$

$\approx \$4115.27$

b. $S(240)=100\left[\dfrac{(1.0075)^{240}-1}{0.0075}\right]$

$\approx 100\left[\dfrac{5.00915}{0.0075}\right]$

$\approx \$66,788.69$

87. a. $R=10,000\left[\dfrac{0.0065}{1-(1.0065)^{-n}}\right]$

$=10,000\left[\dfrac{0.0065}{1-\frac{1}{1.0065^n}}\right]\cdot\dfrac{1.0065^n}{1.0065^n}$

$=10,000\left[\dfrac{0.0065(1.0065)^n}{1.0065^n-1}\right]$

$=\dfrac{65(1.0065)^n}{1.0065^n-1}$

b. $R=10,000\left[\dfrac{0.0065}{1-(1.0065)^{-48}}\right]\approx 243.19$

$R=\dfrac{65(1.0065)^{48}}{1.0065^{48}-1}\approx 243.19$

88. $S=kA^{1/3}$

a. $S=k\sqrt[3]{A}$

b. Let S_1 be the number of species on 20,000 acres. Then $S_1=k\sqrt[3]{20,000}$. Let S_2 be the number of species on 45,000 acres. Then

$S_2=k\sqrt[3]{45000}$

$=\sqrt[3]{2.25\cdot 20,000}$

$=\sqrt[3]{2.25}\cdot k\sqrt[3]{20,000}$

$=\sqrt[3]{2.25}\cdot S_1$

$S_2\approx 1.31S_1$

89. a. $C=\dfrac{540,000}{100-p}-\dfrac{5400}{1}$

$=\dfrac{540,000-5400(100-p)}{100-p}$

$=\dfrac{5400p}{100-p}$

b. If $p=0$, $C=\dfrac{5400(0)}{100-0}=\dfrac{0}{100}=0$.
The cost of removing no pollution is zero.

c. $C=\dfrac{5400(98)}{100-98}=\$264,600$

d. The formula is not defined when $p = 100$. We are dividing by zero. The cost increases as p approaches 100. It is cost prohibitive to remove all of the pollution.

Chapter Test

1. **a.** $A = \{6, 8\}$ $B' = \{3, 4, 6\}$
$A \cup B' = \{3, 4, 6, 8\}$
b. $\{3, 4\}$, $\{3, 6\}$, and $\{4, 6\}$ are disjoint from B.
c. $\{6\}$, and $\{8\}$ are non-empty subsets of A.

2. $(4-2^3)^2 - 3^4 \cdot 0^{15} + 12 \div 3 + 1 = (-4)^2 - 0 + 4 + 1$
$= 16 + 4 + 1 = 21$

3. **a.** $x^4 \cdot x^4 = x^8$
b. $x^0 = 1$, if $x \neq 0$
c. $\sqrt{x} = x^{1/2}$
d. $(x^{-5})^2 = x^{-10}$ or $\dfrac{1}{x^{10}}$
e. $a^{27} \div a^{-3} = a^{27-(-3)} = a^{30}$
f. $x^{1/2} \cdot x^{1/3} = x^{5/6}$
g. $\dfrac{1}{\sqrt[3]{x^2}} = \dfrac{1}{x^{2/3}}$
h. $\dfrac{1}{x^3} = x^{-3}$

4. **a.** $x^{1/4} = \sqrt[4]{x}$
b. $x^{-3/4} = \sqrt[4]{x^{-3}}$ or $\left(\sqrt[4]{x}\right)^{-3}$ or $\dfrac{1}{\sqrt[4]{x^3}}$

5. **a.** $x^{-5} = \dfrac{1}{x^5}$
b. $\left(\dfrac{x^{-8}y^2}{x^{-1}}\right)^{-3} = \dfrac{x^{24}y^{-6}}{x^3} = \dfrac{x^{21}}{y^6}$

6. **a.** $\dfrac{x}{\sqrt{5x}} \cdot \dfrac{\sqrt{5x}}{\sqrt{5x}} = \dfrac{x\sqrt{5x}}{5x} = \dfrac{\sqrt{5x}}{5}$
b. $\sqrt{24a^2b} \cdot \sqrt{a^3b^4} = 2a\sqrt{6b} \cdot ab^2\sqrt{a}$
$= 2a^2b^2\sqrt{6ab}$
c. $\dfrac{1-\sqrt{x}}{1+\sqrt{x}} \cdot \dfrac{1-\sqrt{x}}{1-\sqrt{x}} = \dfrac{1-2\sqrt{x}+x}{1-x}$

7. $2x^3 - 7x^5 - 5x - 8$
a. Degree is 5.
b. Constant is –8.
c. Coefficient of x is –5.

8. In interval notation, $(-2,\infty) \cap (-\infty, 3] = (-2, 3]$

−2 −1 0 1 2 3 4 5 6

9. **a.** $8x^3 - 2x^2 = 2x^2(4x-1)$
b. $x^2 - 10x - 24 = (x-12)(x+2)$
c. $6x^2 - 13x + 6 = (2x-3)(3x-2)$
d. $2x^3 - 32x^5 = 2x^3(1-16x^2)$
$= 2x^3(1-4x)(1+4x)$

10. A quadratic polynomial has degree two. (c) is the quadratic.
$4 - x - x^2 = 4 - (-3) - (-3)^2$
$= 4 + 3 - 9$
$= -2$, when $x = -3$

11.
$$\begin{array}{r|l} & 2x+1 \\ x^2-1 & 2x^3 + x^2 \qquad -7 \\ & \underline{2x^3 \qquad -2x} \\ & x^2 + 2x - 7 \\ & \underline{x^2 \qquad\quad -1} \\ & 2x - 6 \end{array}$$

Quotient: $2x + 1 + \dfrac{2x-6}{x^2-1}$

12. **a.** $4y - 5(9-3y) = 4y - 45 + 15y = 19y - 45$
b. $-3t^2(2t^4 - 3t^7) = -6t^6 + 9t^9$
c.
$$\begin{array}{r} x^2 - 5x + 2 \\ \underline{4x - 1} \\ -x^2 + 5x - 2 \\ \underline{4x^3 - 20x^2 + 8x } \\ 4x^3 - 21x^2 + 13x - 2 \end{array}$$
d. $(6x-1)(2-3x) = 12x - 18x^2 - 2 + 3x$
$= -18x^2 + 15x - 2$
e. $(2m-7)^2 = 4m^2 - 28m + 49$
f. $\dfrac{x^6}{x^2-9} \cdot \dfrac{x-3}{3x^2} = \dfrac{x^4}{(x+3)(x-3)} \cdot \dfrac{(x-3)}{3}$
$= \dfrac{x^4}{3(x+3)}$
g. $\dfrac{x^4}{9} \div \dfrac{9x^3}{x^6} = \dfrac{x^4}{9} \cdot \dfrac{x^6}{9x^3} = \dfrac{x^7}{81}$
h. $\dfrac{4}{x-8} - \dfrac{x-2}{x-8} = \dfrac{4-x+2}{x-8} = \dfrac{6-x}{x-8}$

i. $\frac{x-1}{x^2-2x-3}-\frac{3}{x^2-3x}$

$=\frac{x-1}{(x-3)(x+1)}-\frac{3}{x(x-3)}$

$=\frac{x(x-1)-3(x+1)}{x(x-3)(x+1)}$

$=\frac{x^2-x-3x-3}{x(x-3)(x+1)}$

$=\frac{x^2-4x-3}{x(x-3)(x+1)}$

13. $\frac{\frac{1}{x}-\frac{1}{y}}{\frac{1}{x}+y}\cdot\frac{xy}{xy}=\frac{y-x}{y+xy^2}$ or $\frac{y-x}{y(1+xy)}$

14. Construct a Venn diagram:

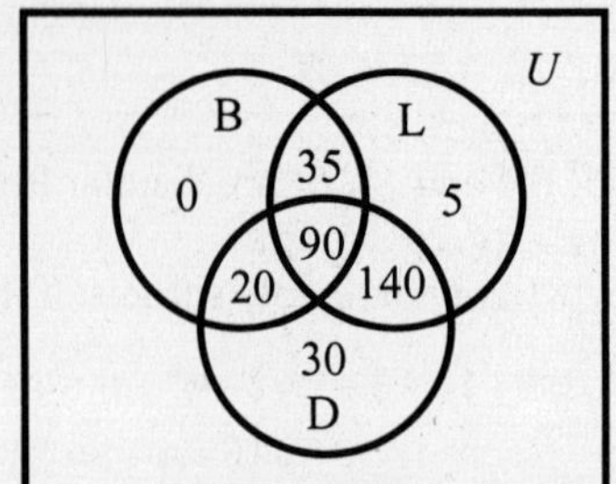

a. 0 students ate only breakfast.

b. $320 - 145 = 175$
175 students skipped breakfast.

15. $S=1000\left(1+\frac{0.08}{4}\right)^{4x}=1000\left(1+\frac{0.08}{4}\right)^{4(20)}$

$=1000(1+0.02)^{80}=1000(1.02)^{80}\approx 4875.44$

In 20 years, the future value will be about \$4875.44.

Chapter 1: Linear Equations and Functions

Exercise 1.1

1. $4x-7=8x+2$

$4x-7+7-8x=8x+2+7-8x$

$-4x=9$

$x=-\frac{9}{4}$

3. $x+8=8(x+1)$

$x+8=8x+8$

$x-8x=8-8$

$-7x=0$

$x=0$

5. $-\frac{3x}{4}=24$

$-3x=4(24)=96$

$x=-32$

7. $\frac{5x}{6}=\frac{8}{3}$

$6\left(\frac{5}{6}x\right)=6\left(\frac{8}{3}\right)$

$5x=16$

$x=\frac{16}{5}$

9. $2(x-7)=5(x+3)-x$

$2x-14=5x+15-x$

$2x-5x+x=15+14$

$-2x=29$

$x=-\frac{29}{2}$

11. $\frac{5x}{2}-4=\frac{2x-7}{6}$

$6\left(\frac{5x}{2}-4\right)=6\left(\frac{2x-7}{6}\right)$

$15x-24=2x-7$

$15x-2x=24-7$

$13x=17$

$x=\frac{17}{13}$

13. $x+\frac{1}{3}=2\left(x-\frac{2}{3}\right)-6x$

$x+\frac{1}{3}=2x-\frac{4}{3}-6x$

$3x+1=6x-4-18x$

$3x+18x-6x=-4-1$

$15x=-5$

$x=\frac{-5}{15}=-\frac{1}{3}$

15. $(5x)\frac{33-x}{5x}=5x(2)$

$33-x=10x$

$-x-10x=-33$

$-11x=-33$

$x=3$

Check: $\frac{33-3}{5(3)}\stackrel{?}{=}2$

$\frac{30}{15}\stackrel{?}{=}2$

$2=2$

$x=3$ is the solution.

17. $\frac{2x}{2x+5}=\frac{2}{3}-\frac{5}{2(2x+5)}$

Multiply each term by $6(2x+5)$.

$12x=(8x+20)-15$

$12x-8x=20-15$

$4x=5$ or $x=\frac{5}{4}$

Check: $\frac{2(\frac{5}{4})}{2(\frac{5}{4})+5}\stackrel{?}{=}\frac{2}{3}-\frac{5}{4(\frac{5}{4})+10}$

$\frac{10}{10+20}\stackrel{?}{=}\frac{2}{3}-\frac{5}{15}$

$\frac{10}{30}=\frac{1}{3}$ and $\frac{2}{3}-\frac{5}{15}=\frac{1}{3}$

$x=\frac{5}{4}$ is the solution.

19. $\frac{2x}{x-1}+\frac{1}{3}=\frac{5}{6}+\frac{2}{x-1}$

$\frac{2x-2}{x-1}+\frac{1}{3}=\frac{5}{6}$

$\frac{2(x-1)}{x-1}+\frac{1}{3}=\frac{5}{6}$

$2+\frac{1}{3}\neq\frac{5}{6}$

There is no solution.

21. $3.259x-8.638=-3.8(8.625x+4.917)$

$3.259x-8.638=-32.775x-18.6846$

$3.259x+32.775x=8.638-18.6846$

$36.034x=-10.0466$

$x=\frac{-10.0466}{36.034}\approx-0.279$

23. $0.000316x+9.18=2.1(3.1-0.0029x)-4.68$

$0.000316x+9.18=6.51-0.00609x-4.68$

$0.000316x+0.00609x=6.51-4.68-9.18$

$0.006406x=-7.35$

$x=\frac{-7.35}{0.006406}\approx-1147.362$

25. $3x-4y=15$

$-4y=-3x+15$

$y=\frac{-3x}{-4}+\frac{15}{-4}$

$y=\frac{3}{4}x-\frac{15}{4}$

27. $2\left(9x+\frac{3}{2}y\right)=2(11)$

$18x+3y=22$

$3y=-18x+22$

$y=-6x+\frac{22}{3}$

29. $I=Prt$

$\frac{I}{rt}=\frac{Prt}{rt}$

$P=\frac{I}{rt}$

31. $S=P+Prt$

$Prt=S-P$

$t=\frac{S-P}{Pr}$

33. $y=648,000-1800x$

$387,000=648,000-1800x$

$1800x=648,000-387,000=261,000$

$x=\frac{261,000}{1800}=145$ months

35. $A=P+Prt,\ A=6000,\ r=0.1,\ t=5$

$6000=P+P(0.1)(5)$

$6000=1.5P$

$P=\frac{6000}{1.5}=\$4000$

37. $\frac{I}{175.393}+0.663=r$

$\frac{I}{175.393}+0.663=19.8$

$\frac{I}{175.393}=19.8-0.663=19.137$

$I=19.137(175.393)$

$I=\$3356.50$

39. $R=C$ for breakeven point

$20x=2x+7920$

$18x=7920$

$x=440$ packs or 220,000 CD's

41. $170,500=5.76x$

$x=\frac{170,500}{5.76}=\$29,600$

43. $S=241.33+29t$

a. 2000 is represented by $t=12$

b. $1000=241.33+29t$

$29t=758.67$

$t=26.16$

c. From (b), $1988+26.16=2014.16$.

45. $53x-2p=126$

$53x-2(50)=126$

$x=\frac{126+100}{53}=4.3$

50% will use Internet during $4.3+1995\approx2000$.

47. $\frac{93+69+89+97+FE+FE}{6}=90$

$2FE+348=540$

$2FE=192$

$FE=96$

A 96 is the lowest grade that can be earned on the final.

49. x = total served
x = active + dropouts

$$x = 6000 + \frac{1}{3}x$$
$$3x = 18,000 + x$$
$$2x = 18,000$$
$x = 9000$ youths served

51. **a.**
$$7n - 12T = 52$$
$$-12T = -7n + 52$$
$$T = \frac{-7n+52}{-12}$$
$$T = \frac{7n-52}{12}$$

b. 28 chirps in 15 seconds means 112 chirps in 1 minute (or 60 seconds).
$$T = \frac{7(112)-52}{12} = 61$$
Temperature is 61°.

53. x = amount in safe fund
$120,000 - x$ = amount in risky fund
Yield: $0.09x + 0.13(120,000 - x) = 12,000$
$$0.09x + 15,600 - 0.13x = 12,000$$
$$-0.04x = -3600$$
$$x = 90,000$$
x = \$90,000 in 9% fund
120,000 – 90000 = \$30,000 in 13% fund.

55. Reduced salary: 2000 – 0.10(2000) = \$1800
Increased salary: 1800 + 0.20(1800) = \$2160
160 = R% of 2000
$$R = \frac{160}{2000} = \frac{8}{100}$$
\$160 is an 8% increase.

57. $C + MU = SP$ (MU = Profit)
$$x_1 + 0.2x_1 = 480$$
$$1.2x_1 = 480$$
$$x_1 = 400$$
$$x_2 - 0.2x_2 = 480$$
$$0.8x_2 = 480$$
$$x_2 = 600$$
Profit is \$80 on x_1. Loss is \$120 on x_2.
The collector lost \$40 on the transaction.

59. Cost + Markup = Selling price
$214.90 + 0.3x = x$
$$214.9 = x - 0.3x = 0.7x$$
$$x = \frac{214.9}{0.7} = 307$$
The selling price is \$307.

61. Wholesaler
$154.98 + 0.1W = W$
$$0.9W = 154.98$$
$$W = \$172.20$$
Retailer
$172.20 + 0.3R = R$
$$0.7R = 172.20$$
$$R = \$246.00$$

Exercise 1.2

1. **a.** For each value of x there is only one y.
b. $D = \{-7, -1, 0, 3, 4.2, 9, 11, 14, 18, 22\}$
$R = \{0, 1, 5, 9, 11, 22, 35, 60\}$

3. $f(0) = 1,\ f(11) = 35$

5. This is a function, since for each x there is only one y. $D = \{1, 2, 3, 8, 9\}$, $R = \{-4, 5, 16\}$

7. The vertical line test shows that graph (a) is a function of x, and that graph (b) is not a function of x.

9. If $y = 3x^3$, then y is a function of x.

11. If $y^2 = 3x$, then y is not a function. If, for example, $x = 3$, there are two possible values for y.

13. $R(x) = 8x - 10$
a. $R(0) = 8(0) - 10 = -10$
b. $R(2) = 8(2) - 10 = 6$
c. $R(-3) = 8(-3) - 10 = -34$
d. $R(1.6) = 8(1.6) - 10 = 2.8$

15. $C(x) = 4x^2 - 3$

a. $C(0) = 4(0)^2 - 3 = -3$

b. $C(-1) = 4(-1)^2 - 3 = 1$

c. $C(-2) = 4(-2)^2 - 3 = 13$

d. $C\left(-\frac{3}{2}\right) = 4\left(-\frac{3}{2}\right)^2 - 3 = 6$

17. $f(x) = x^3 - \frac{4}{x}$

a. $f\left(-\frac{1}{2}\right) = \left(-\frac{1}{2}\right)^3 - \frac{4}{-\frac{1}{2}} = -\frac{1}{8} + 8 = \frac{63}{8}$

b. $f(2) = 2^3 - \frac{4}{2} = 8 - 2 = 6$

c. $f(-2) = (-2)^3 - \frac{4}{-2} = -8 + 2 = -6$

19. $f(x) = 1 + x + x^2$

a. $f(2+1) = f(3) = 1 + 3 + 3^2 = 13$
$f(2) + f(1) = 7 + 3 = 10$
$f(2) + f(1) \neq f(2+1)$

b. $f(x+h) = 1 + (x+h) + (x+h)^2$

c. No. This is equivalent to a.

d. $f(x) + h = 1 + x + x^2 + h$
No. $f(x+h) \neq f(x) + h$

e.
$$f(x+h) = 1 + (x+h) + (x+h)^2$$
$$= 1 + x + h + x^2 + 2xh + h^2$$
$$f(x) = 1 + x + x^2$$
$$f(x+h) - f(x) = h + 2xh + h^2$$
$$= h(1 + 2x + h)$$
$$\frac{f(x+h) - f(x)}{h} = 1 + 2x + h$$

21. $f(x) = x - 2x^2$

a. $f(x+h) = (x+h) - 2(x+h)^2$
$= -2x^2 - 4xh - 2h^2 + x + h$

b. $f(x+h) - f(x)$
$= (x+h) - 2(x+h)^2 - (x - 2x^2)$
$= x + h - 2x^2 - 4xh - 2h^2 - x + 2x^2$
$= h - 4xh - 2h^2$
$$\frac{f(x+h) - f(x)}{h} = \frac{h - 4xh - 2h^2}{h}$$
$$= 1 - 4x - 2h$$

23. Since (9, 10) and (5, 6) are points on the graph,

a. $f(9) = 10$

b. $f(5) = 6$

25. a. The ordered pair (a, b) satisfies the equation. Thus $b = a^2 - 4a$.

b. The coordinates of $Q = (1, -3)$. Since the point is on the curve, the coordinates satisfy the equation.

c. The coordinates of $R = (3, -3)$. They satisfy the equation.

d. The x values are 0 and 4. These values are also solutions of $x^2 - 4x = 0$.

27. $f(x) = 3x,\ g(x) = x^3$

a. $(f+g)(x) = 3x + x^3$

b. $(f-g)(x) = 3x - x^3$

c. $(f \cdot g)(x) = 3x \cdot x^3 = 3x^4$

d. $\left(\frac{f}{g}\right)(x) = \frac{3x}{x^3} = \frac{3}{x^2}$

29. $f(x) = \sqrt{2x}$, $g(x) = x^2$

a. $(f+g)(x) = \sqrt{2x} + x^2$

b. $(f-g)(x) = \sqrt{2x} - x^2$

c. $(f \cdot g)(x) = \sqrt{2x} \cdot x^2 = x^2\sqrt{2x}$

d. $\left(\frac{f}{g}\right)(x) = \frac{\sqrt{2x}}{x^2}$

31. $f(x) = (x-1)^3$, $g(x) = 1 - 2x$

a. $(f \circ g)(x) = f(1-2x) = (1-2x-1)^3 = -8x^3$

b. $(g \circ f)(x) = g((x-1)^3) = 1 - 2(x-1)^3$

c. $f(f(x)) = f((x-1)^3) = [(x-1)^3 - 1]^3$

d. $(f \cdot f)(x) = (x-1)^3 \cdot (x-1)^3 = (x-1)^6$
$[(f \cdot f)(x) \neq f(f(x))]$

33. $f(x) = 2\sqrt{x}$, $g(x) = x^4 + 5$

a. $(f \circ g)(x) = f(x^4 + 5) = 2\sqrt{x^4 + 5}$

b. $(g \circ f)(x) = g(2\sqrt{x}) = (2\sqrt{x})^4 + 5$
$= 16x^2 + 5$

c. $f(f(x)) = f(2\sqrt{x}) = 2\sqrt{2\sqrt{x}}$

d. $(f \cdot f)(x) = 2\sqrt{x} \cdot 2\sqrt{x} = 4x$
$[(f \cdot f)(x) \neq f(f(x))]$

35. $y = x^2 + 4$
There is no division by zero or square roots. Domain is all the reals, i.e., $\{x : x \in \text{Reals}\}$. Since $x^2 \geq 0, x^2 + 4 \geq 4$, the range is reals ≥ 4 or $\{y : y \geq 4\}$.

37. $y = \sqrt{x+4}$
There is no division by zero. To get a real number *y*, we must have $x + 4 \geq 0$ or $x \geq -4$. Domain: $x \geq -4$. The square root is always nonnegative.
Thus, the range is $\{y : y \in \text{reals}, y \geq 0\}$.

39. *D*: $\{x : x \geq 1,\ x \neq 2\}$

41. *D*: $\{x : -7 \leq x \leq 7\}$

43. **a.** $f(20) = 103{,}000$ means it will take 20 years to pay off a debt of \$103,000 (at \$800 per month and 7.5% compounded monthly.)
b. $f(5+5) = f(10) = 69{,}000$;
$f(5) + f(5) = 80{,}000$; No.
c. It will take 15 years to pay off the debt, i.e., $89{,}000 = f(15)$.

45. **a.** $f(1950) = 16.5$ means that in 1950 there were 16.5 workers supporting each person receiving Social Security benefits.
b. $f(1990) = 3.4$
c. The points based on known data must be the same and those based on projections might be the same.
d. Domain: $1950 \leq t \leq 2050$
Range: $1.9 \leq n \leq 16.5$

47. **a.** $f(95) = 1{,}000{,}000 \quad g(95) = 600{,}000$
b. $f(100) = 1{,}200{,}000$. This was the number of prisoners in 2000.
c. $g(90) = 500{,}000$. This was the number of parolees in 1990.
d. $(f - g)(100) = 1{,}200{,}000 - 700{,}000$
$= 500{,}000$. There were this many more in prison than were out on parole.
e. $(f - g)(93) = 900{,}000 - 600{,}000 = 300{,}000$
$(f - g)(98) = 1{,}100{,}000 - 600{,}000 = 500{,}000$
$(f - g)(98)$ is greater. Possible reason is increased prison capacity but parolee level is constant.

49. **a.** Since the wind speed cannot be negative, $s \geq 0$.
b. $f(10) = 45.694 + 1.75(10) - 29.26\sqrt{10} = -29.33$
At a temperature of –5°F and a wind speed of 10 mph, the temperature feels like -29.33°F.
c. $f(0) = 45.694$, but $f(0)$ should equal the air temperature, –5°F.

51. $C = \frac{5}{9}F - \frac{160}{9}$
a. *C* is a function of *F*.
b. Mathematically, the domain is all reals.
c. Domain: $\{F : 32 \leq F \leq 212\}$
Range: $\{C : 0 \leq C \leq 100\}$
d. $C(40) = \frac{5}{9}(40) - \frac{160}{9}$
$= \frac{200 - 160}{9} = \frac{40}{9} = 4.44°C$

53. $C(p) = \frac{7300p}{100 - p}$
a. Domain: $\{p : 0 \leq p < 100\}$
b. $C(45) = \frac{7300(45)}{100-45} = \frac{328{,}500}{55} = \$5972.73$
c. $C(90) = \frac{7300(90)}{100-90} = \frac{657{,}000}{10} = \$65{,}700$
d. $C(99) = \frac{7300(99)}{100-99} = \frac{722{,}700}{1} = \$722{,}700$
e. $C(99.6) = \frac{7300(99.6)}{100-99.6} = \frac{727{,}080}{0.4} = \$1{,}817{,}700$
In each case above, to remove *p*% of the particulate pollution would cost *C*(*p*).

55. **a.** *A* is a function of *x*.
b. $A(2) = 2(50 - 2) = 96$ sq ft
$A(30) = 30(50 - 30) = 600$ sq ft
c. For the problem to have meaning we have $0 < x < 50$.

57. **a.** $P(q(t)) = P(1000 + 10t)$
$= 180(1000 + 10t) - \frac{(1000 + 10t)^2}{100} - 200$
$= 169{,}800 + 1600t - t^2$
b. $q(15) = 1000 + 10(15) = 1150$
$P(q(15)) = \$193{,}575$

59. $R = f(C) \quad C = g(A)$

a. $(f \circ g)(x) = f(C) = R$

b. $(g \circ f)(x)$ is not defined.

c. A is the independent variable and R is the dependent variable. Revenue depends on money spent for advertising.

61. length = x
width = y
$L = 2x + 2y$

$1600 = xy$ or $y = \dfrac{1600}{x}$

$$L = 2x + 2\left(\frac{1600}{x}\right) = 2x + \frac{3200}{x}$$

63. Revenue = (no. of people)(price per person)
Example: $R = 30 \times 10$

$R = 31 \times 9.80$

$R = 32 \times 9.60$

Solution: $R = (30 + x)(10 - 0.20x)$

Exercise 1.3

1. $3x + 4y = 12$

x-intercept: $y = 0$ then $x = 4$.
y-intercept: $x = 0$ then $y = 3$.

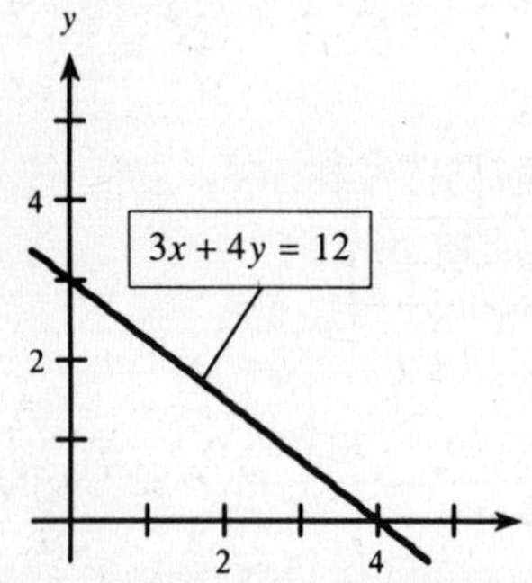

3. $2x - 3y = 12$

x-intercept: $y = 0$ then $x = 6$.
y-intercept: $x = 0$ then $y = -4$

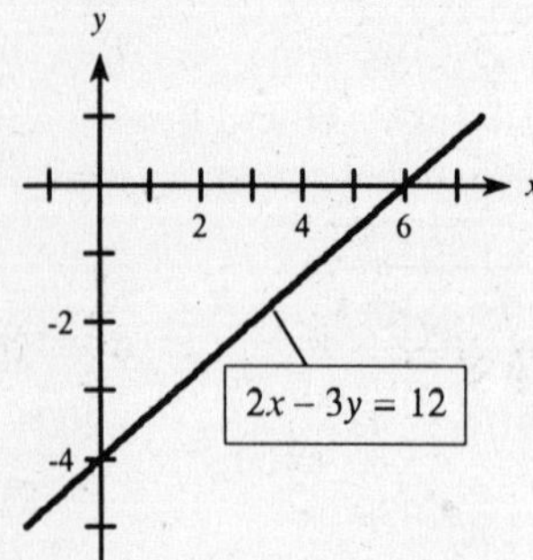

5. $3x + 2y = 0$

x-intercept: $y = 0$ then $x = 0$
Likewise, y-intercept is $y = 0$.

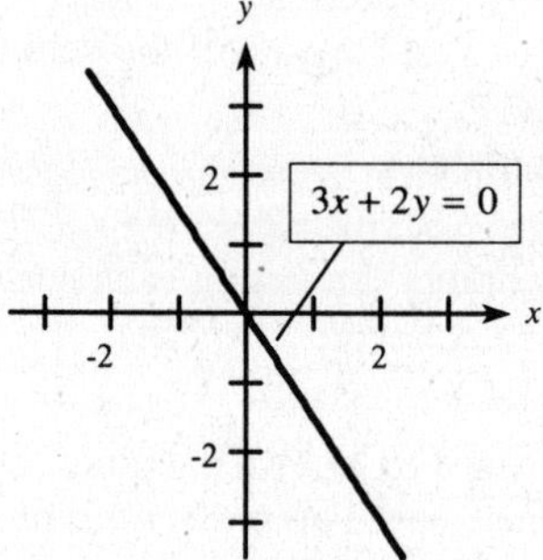

7. (2, 1) and (3, –4)

$$m = \frac{y_2 - y_1}{x_2 - x_1} = \frac{1-(-4)}{2-3} = -5$$

9. (3, 2) and (–1, 2)

$$m = \frac{y_2 - y_1}{x_2 - x_1} = \frac{2-2}{3-(-1)} = \frac{0}{4} = 0$$

11. $(11,-5)$ and $(-9,-4)$

$$m = \frac{y_2 - y_1}{x_2 - x_1} = \frac{-4-(-5)}{-9-11} = -\frac{1}{20}$$

13. A horizontal line has a slope of 0 .

15. **a.** Slope is negative.
b. Slope is undefined.

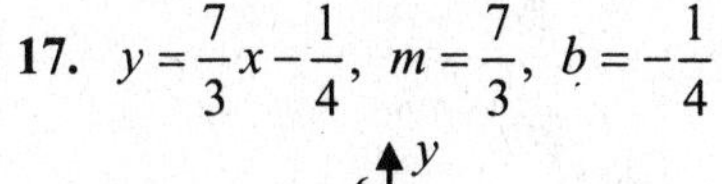

17. $y = \frac{7}{3}x - \frac{1}{4}$, $m = \frac{7}{3}$, $b = -\frac{1}{4}$

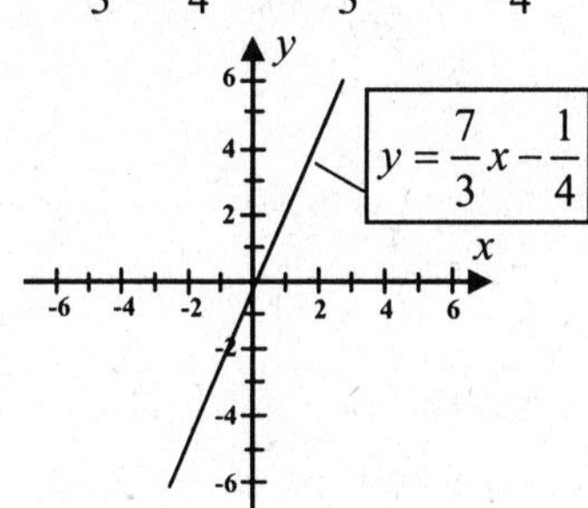

19. $y = 3$ or $y = 0x + 3$, $m = 0$, $b = 3$

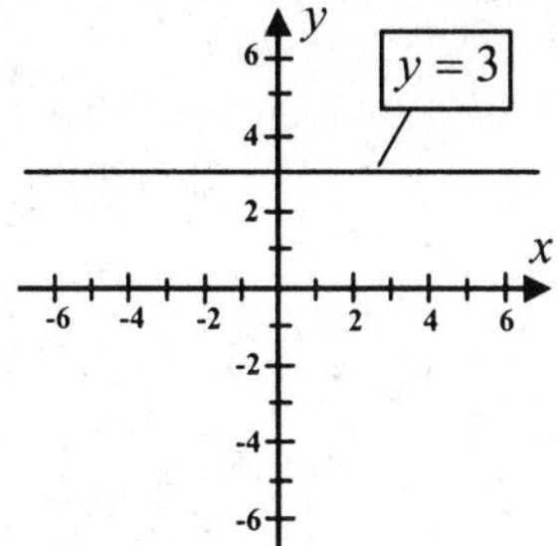

21. $x = -8$
Slope is undefined.
There is no y-intercept.

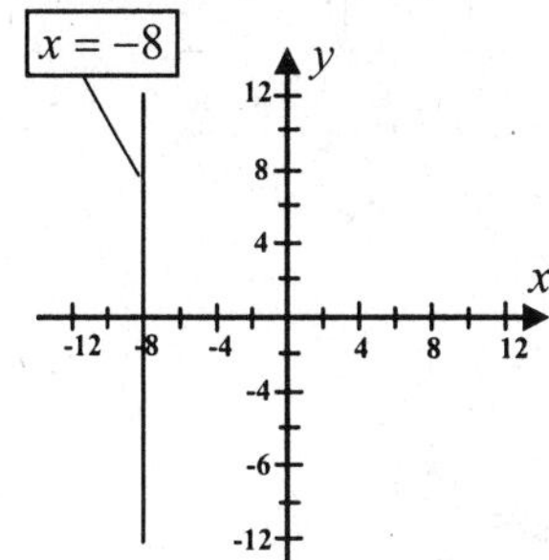

23. $2x + 3y = 6$ or $y = -\frac{2}{3}x + 2, m = -\frac{2}{3}, b = 2$.

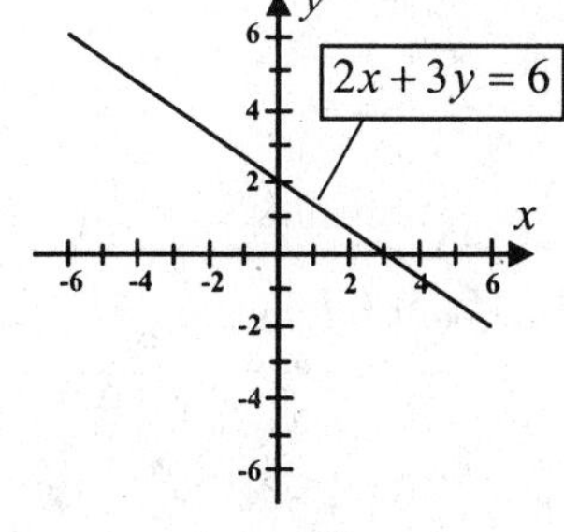

25. $m = \frac{1}{2}, b = 3$

$y = \frac{1}{2}x + 3$

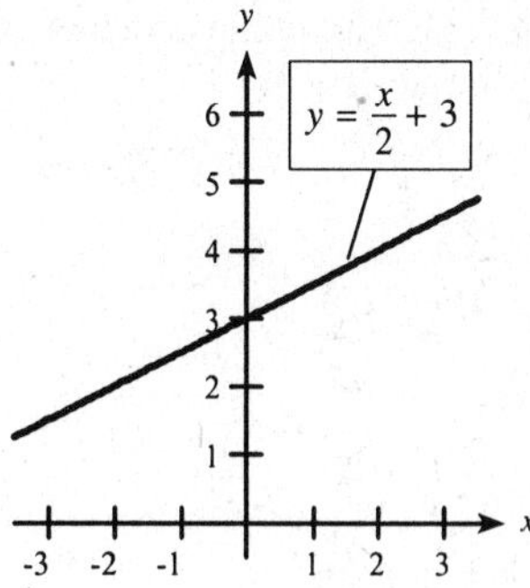

27. $m = -2,\ b = \frac{1}{2}$

$y = -2x + \frac{1}{2}$

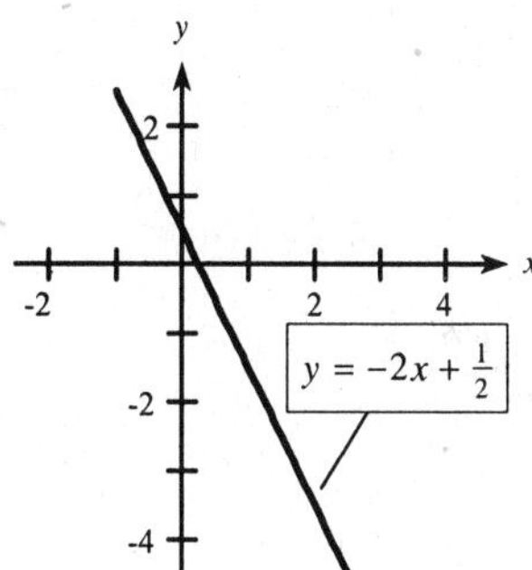

29. $P(2, 0), m = \frac{1}{2}$

$y - y_1 = m(x - x_1)$

$y = \frac{1}{2}x - 1$

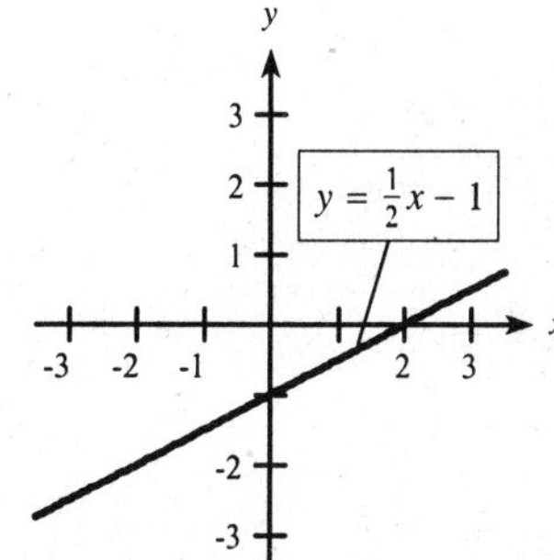

31. $P(-1, 3)$, $m = -2$

$y-3=-2(x-(-1))$

$y=-2x+1$

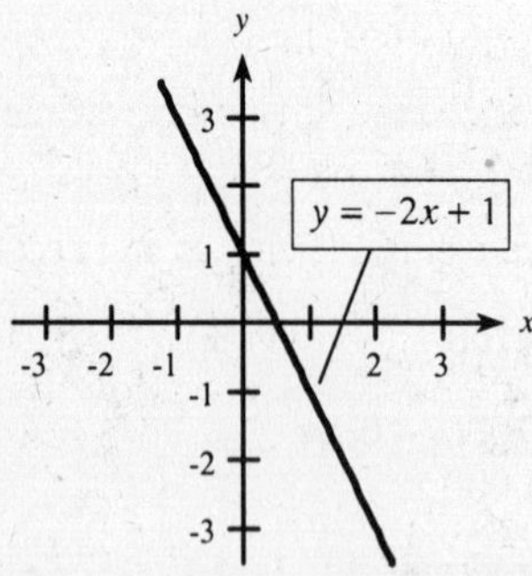

33. $P(-1, 1)$, m is undefined

$x=-1$

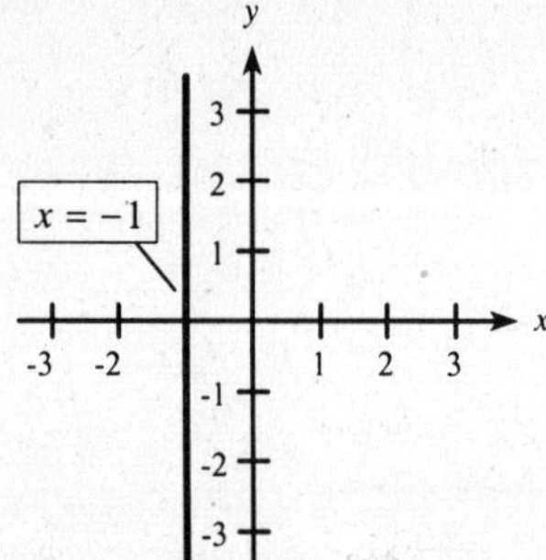

35. $P_1=(3,2), P_2=(-1,-6)$

$m=\dfrac{-6-2}{-1-3}=2$

$y-2=2(x-3)$

$y=2x-4$

37. $P_1=(7,3), P_2=(-6,2)$

$m=\dfrac{2-3}{-6-7}=\dfrac{-1}{-13}=\dfrac{1}{13}$

$y-3=\dfrac{1}{13}(x-7)$

$y=\dfrac{1}{13}x-\dfrac{7}{13}+3$

$y=\dfrac{1}{13}x+\dfrac{32}{13}$ or $-x+13y=32$

39. $3x+2y=6$ $\quad$ $2x-3y=6$

$y=-\dfrac{3}{2}x+3$ $\quad$ $y=\dfrac{2}{3}x-2$

Lines are perpendicular since $\left(-\dfrac{3}{2}\right)\left(\dfrac{2}{3}\right)=-1$.

41. $6x-4y=12$ $\quad$ $3x-2y=6$

$y=\dfrac{6}{4}x-\dfrac{12}{4}$ $\quad$ $y=\dfrac{3}{2}x-3$

or $y=\dfrac{3}{2}x-3$

Lines are the same.

43. If $3x + 5y = 11$, then $y=-\dfrac{3}{5}x+\dfrac{11}{5}$. So, $m=-\dfrac{3}{5}$. A line parallel will have the same slope. Thus, $m=-\dfrac{3}{5}$ and $P = (-2, -7)$ gives $y-(-7)=-\dfrac{3}{5}(x-(-2))$ which simplifies to $y=-\dfrac{3}{5}x-\dfrac{41}{5}$.

45. If $5x - 6y = 4$, then $y=\dfrac{5}{6}x-\dfrac{4}{6}$. Slope of the perpendicular line is $-\dfrac{6}{5}$. Thus $m=-\dfrac{6}{5}$ and $P = (3, 1)$ gives $y-1=-\dfrac{6}{5}(x-3)$ which simplifies to $y=-\dfrac{6}{5}x+\dfrac{23}{5}$.

47. **a.**

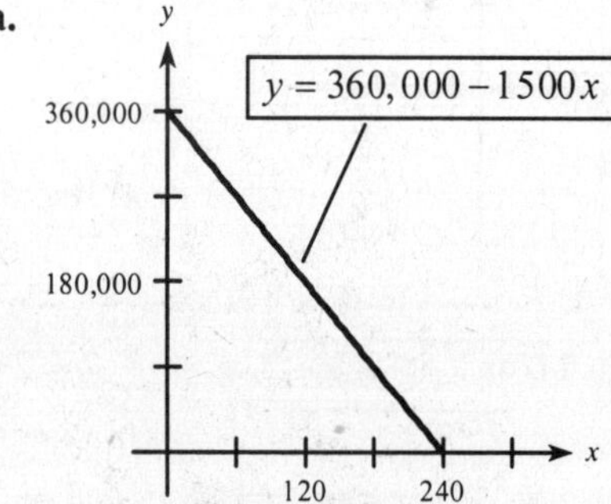

b. $0 = 360{,}000 - 1500x$

$x=\dfrac{360000}{1500}=240$ months

In 240 months, the building will be completely depreciated.

c. (60, 270,000) means that after 60 months the value of the building will be $270,000.

49. $y = 0.1369x - 5.091255$

a. $m = 0.1369 \qquad b = -5.091255$

b. $x = 0$ means that there was a negative amount of transactions. Restrictions are $x > 0$ and $y \geq 0$.

c.-d. With an increase of 1 (thousand) terminals, the amount of transactions increases by \$0.1369 (billion).

51. $M(x) = -0.762x + 85.284$

a. $m = -0.762 \quad b = 85.284$

b. In 1950, 85% of the unmarried women became married.

c. The annual rate of change is -0.762%. For each passing year the percent of unmarried women who get married decreases by 0.762%.

53. $F = 0.518M + 2.775$

a. $m = 0.518$

b. For each \$1000 increase in male earnings, the female's earning increases only by \$518.

c. $F(30) = 0.518(30) + 2.775$

$= 15.540 + 2.775$

$= 18.315$ thousands $= \$18,315$

55. $y = 0.0838x + 4.95$ Both units are in dollars.

57. a. $B - 14.397 = 0.6324(W - 22.375)$

$B = 0.6324W + 0.24705$

b. $B(35) = 0.6324(35) + 0.24705$

$= \$22.38105$ thousand $= \$22,381$

59. a. $m = \dfrac{68.5 - 18.1}{1990 - 1890} = 0.504$

$M - 18.1 = 0.504(YR - 1890)$

$M = 0.504(YR) - 934.46$

b. For each year change, the number of men employed increases 504,000.

61. (x, p) is the reference. (0, 85000) is one point.

$m = \dfrac{-1700}{1} = -1700$

$p - 85,000 = -1700(x - 0)$ or
$p = -1700x + 85,000$

63. (t, R) is the ordered pair.

$P_1 = \left(\dfrac{7}{2}, 11\right)$, $P_2 = (6, 19)$

$m = \dfrac{19 - 11}{6 - \frac{7}{2}} = \dfrac{8}{\frac{5}{2}} = \dfrac{16}{5} = 3.2$

$R - 19 = 3.2(t - 6)$ or $R = 3.2t - 19.2 + 19$ or
$R = 3.2t - 0.2$

65. $P_1 = (200, 25)$ $P_2 = (250, 49)$

$m = \dfrac{49 - 25}{250 - 200} = \dfrac{24}{50} = 0.48$

$y - 25 = 0.48(x - 200)$ or $y = 0.48x - 71$

Exercise 1.4

1.

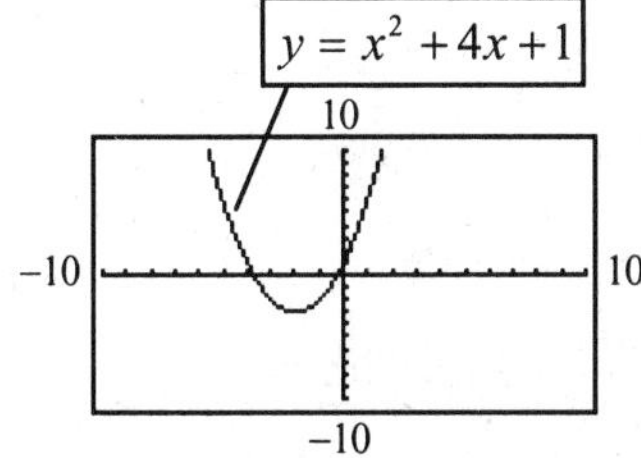

3.

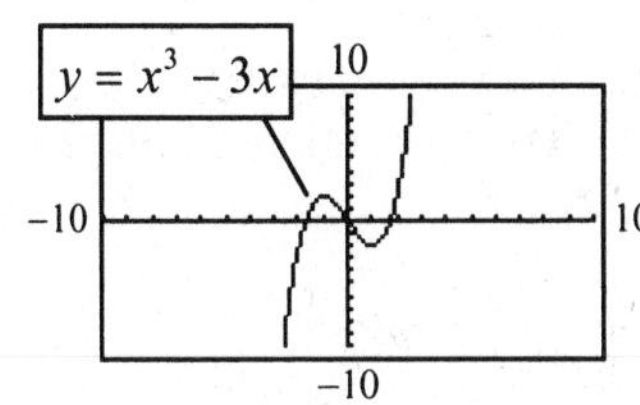

5.

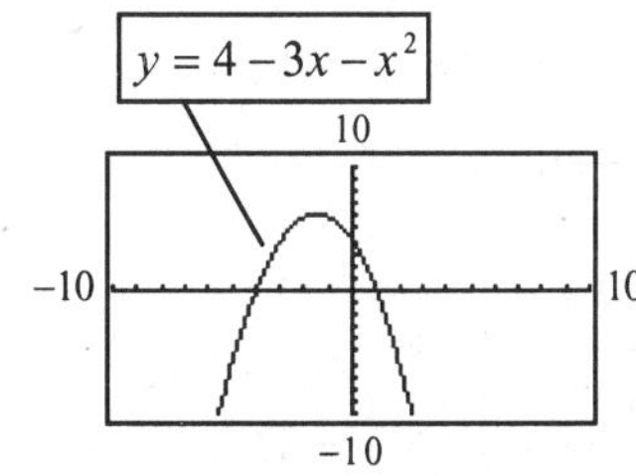

7.

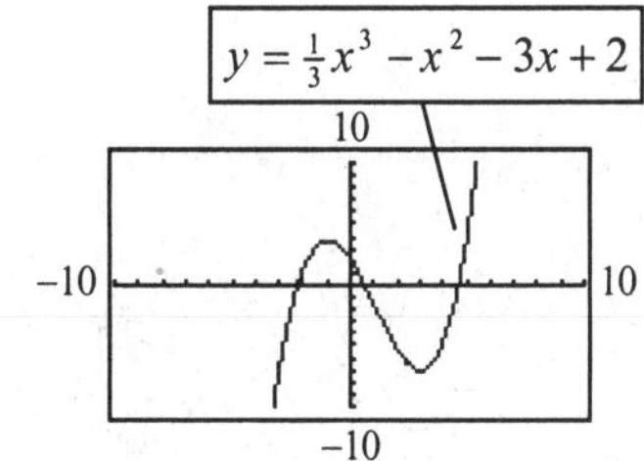

9.

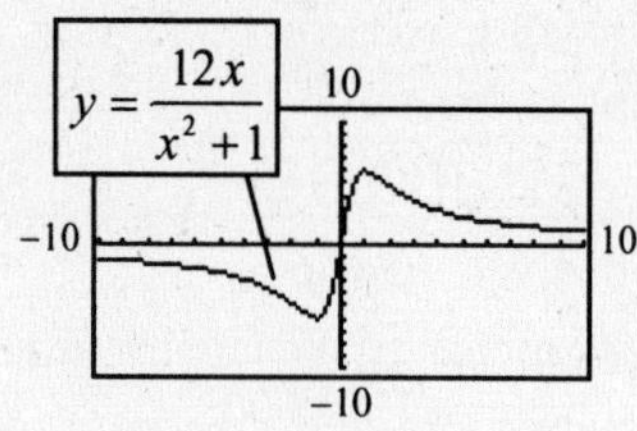

11.

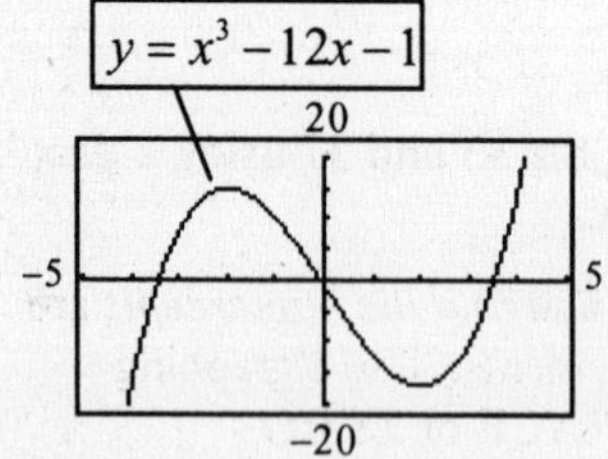

13. $y = 0.01x^3 + 0.3x^2 - 72x + 150$

a.

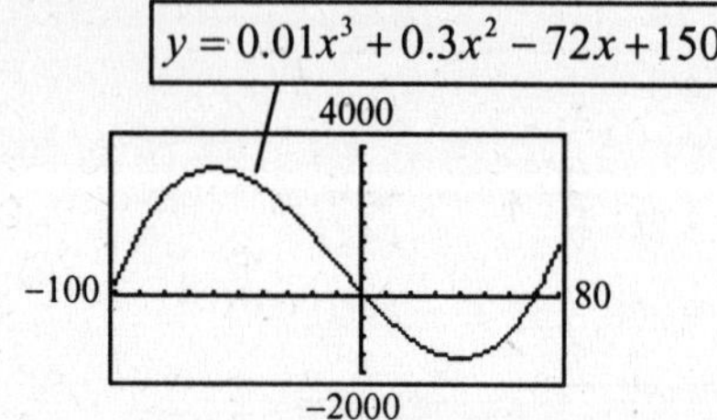

b.

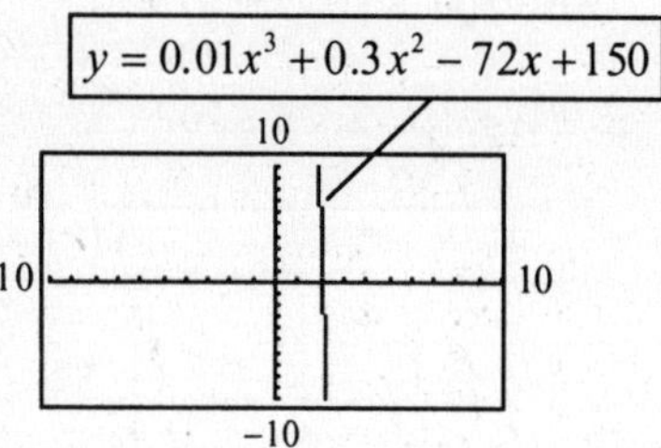

15. a.

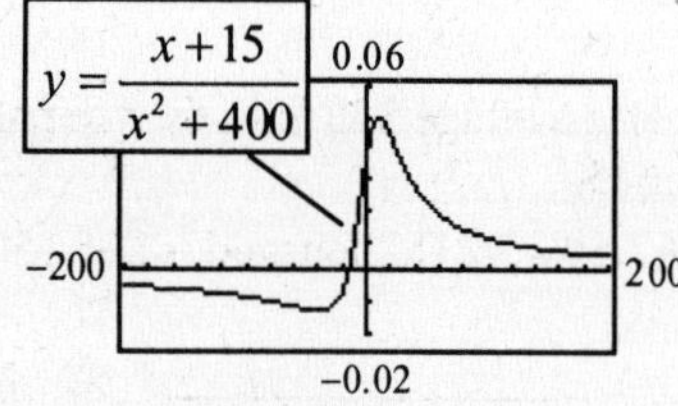

b. Standard Window

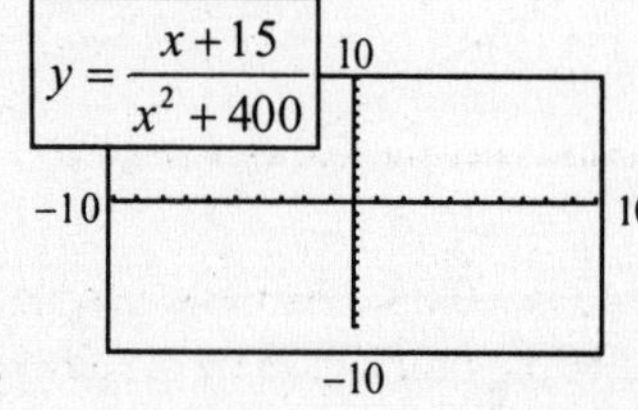

17. a. y-intercept $= -0.03$

x-intercept: $0.001x = 0.03$

$x = 30$

b.

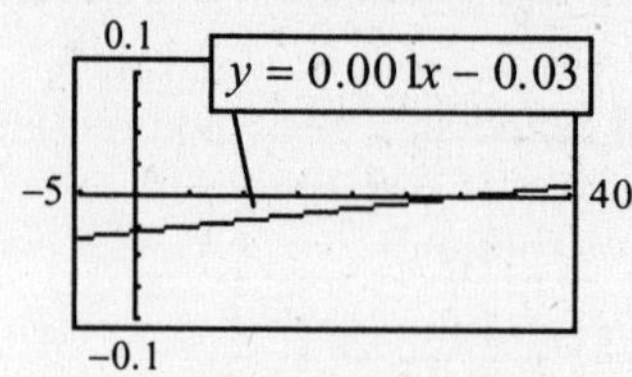

19 – 21. Complete graphs can be seen with different windows. A hint is to look at the equation and try to determine the max and/or min of *y*. Also, find the *x*-intercepts.

19. $y = -0.15(x - 10.2)^2 + 10$

There is no min.

Max value of $y = 10$.

x-intercepts: $0 = -0.15(x - 10.2)^2 + 10$

$$(x - 10.2)^2 = \frac{10}{0.15} = 66.66$$

$$x - 10.2 = \pm\sqrt{66.66} \approx \pm 8$$

$x = 10.2 \pm 8$ or $x = 2.2$ or 18.2

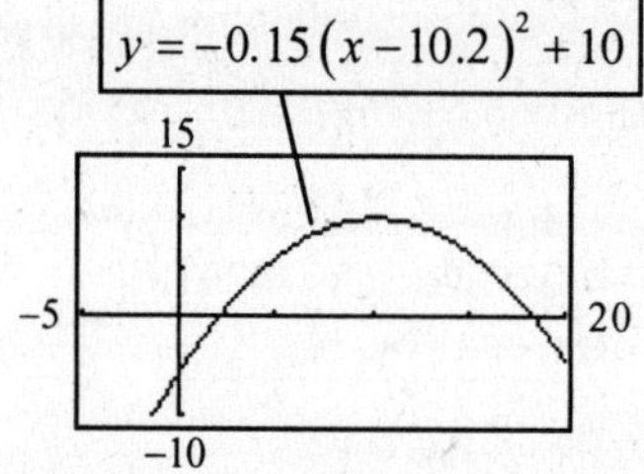

21. If $x = 0$, $y = -42$.

A suggested window is shown below.

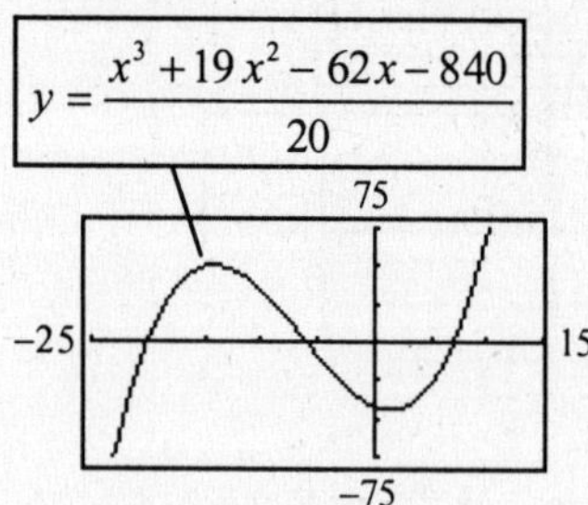

23. $4x - y = 8$

$y = 4x - 8$

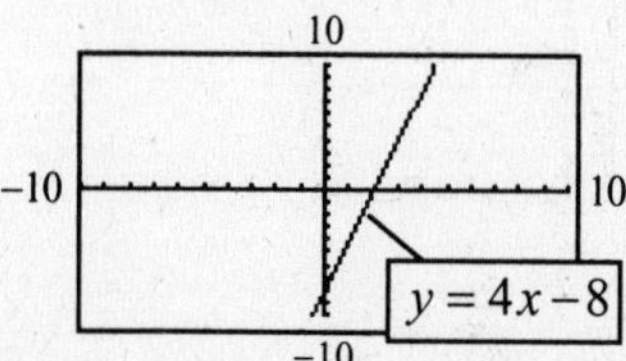

25. $4x^2 + 2y = 5$

$$2y = -4x^2 + 5$$

$$y = -2x^2 + \frac{5}{2}$$

10 $y = -2x^2 + \frac{5}{2}$

−10 10

−10

27. $x^2 + 2y = 6$

$$2y = -x^2 + 6$$

$$y = -\frac{1}{2}x^2 + 3$$

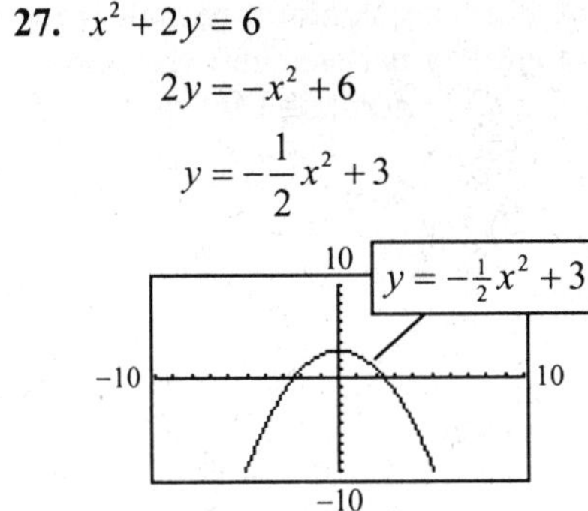

29. $f(x) = x^3 - 3x^2 + 2$

$$f(-2) = (-2)^3 - 3(-2)^2 + 2 = -8 - 12 + 2 = -18$$

$$f\left(\frac{3}{4}\right) = \left(\frac{3}{4}\right)^3 - 3\left(\frac{3}{4}\right)^2 + 2 = 0.734375$$

Use your graphing calculator, and evaluate the function at these two points. If either of your answers differ, can you explain the difference?

31. As x gets large, y approaches 12.
When $x = 0$, $y = -12$, x intercepts at ± 1.

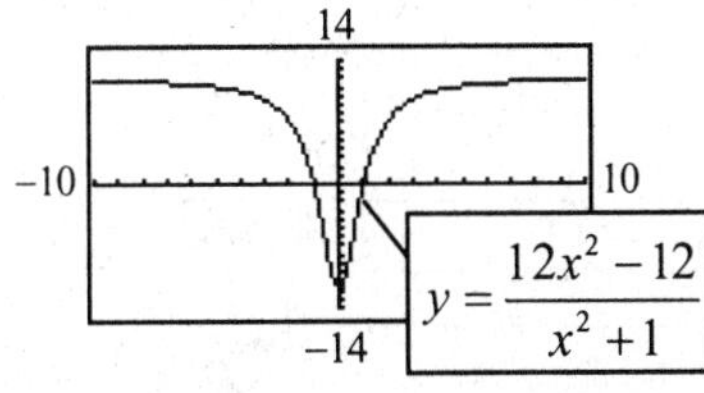

33. $y = \frac{x^2 - x - 6}{x^2 + 5x + 6} = \frac{(x-3)(x+2)}{(x+3)(x+2)}$

What happens to y as x approaches −3? −2?

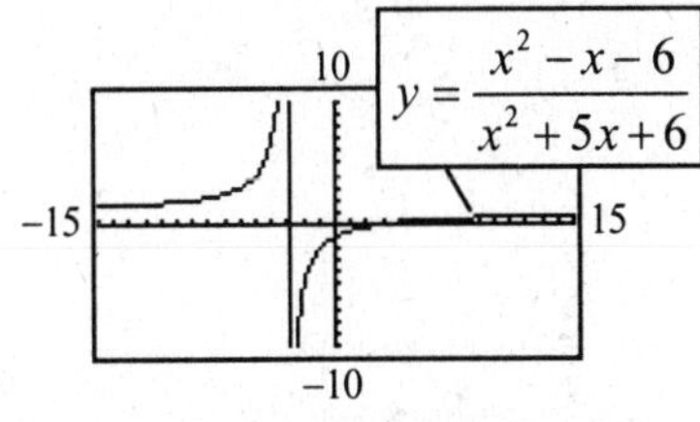

35. $6x - 21 = 0$

$$6x = 21$$

$$x = \frac{21}{6} = \frac{7}{2}$$

37. $x^2 - 3x - 10 = 0$

$$(x-5)(x+2) = 0$$

$$x = -2 \text{ or } 5$$

Compare answers for 35 and 37 using a graphing calculator.

39. *Find the zeros* and *find the x-intercepts* are equivalent statements. Use a graphing calculator's **TRACE** or **ZERO**.

a.-b. Graphing calculator approximation is $x = -1.1098$ and 8.1098.

41. a.

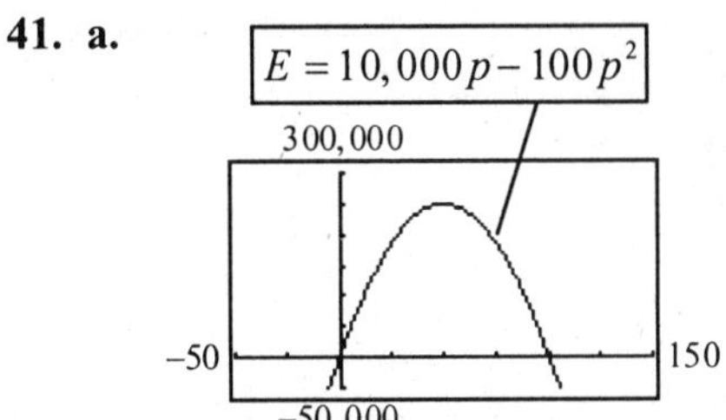

b.  $E \geq 0$ *when* $0 \leq p \leq 100$.

43. a.

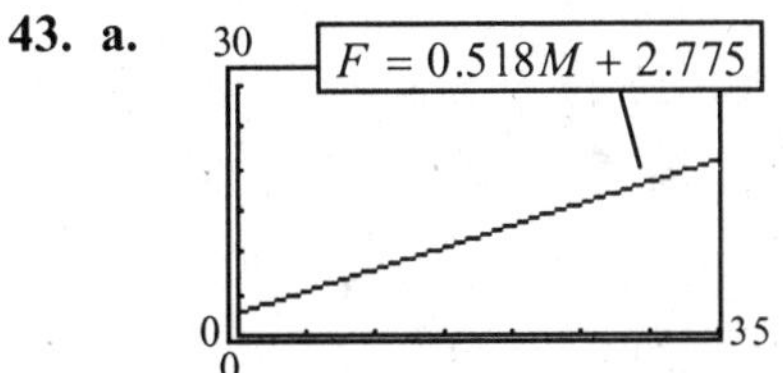

b. If males average $50,000, then females earn $28,675.

c. $F(62.5) = 35.15$ thousand = $35,150

45. a.

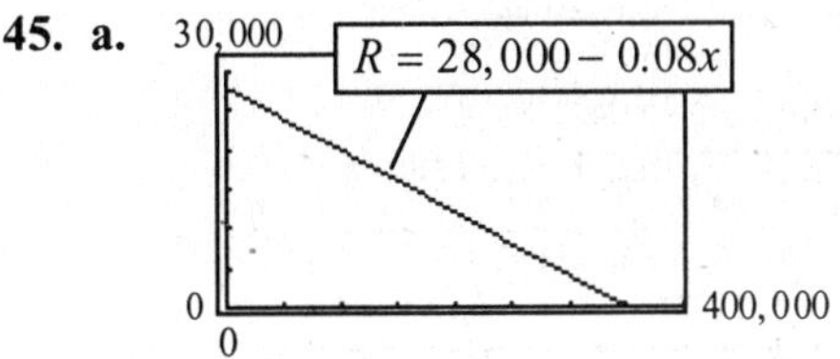

b. The rate is −0.08. As more people become aware of the product, there are fewer to learn about it.

47. a.

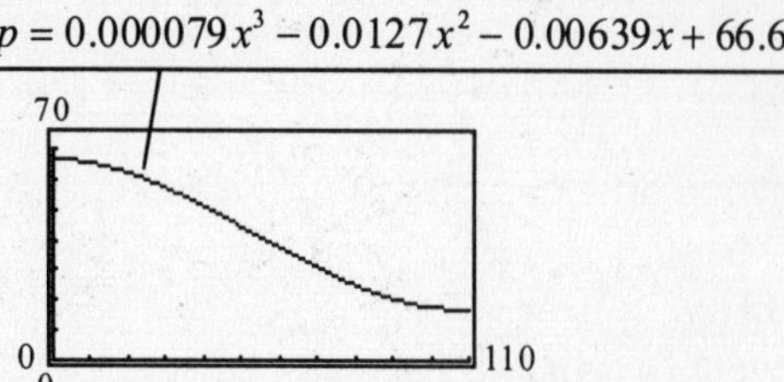

b.

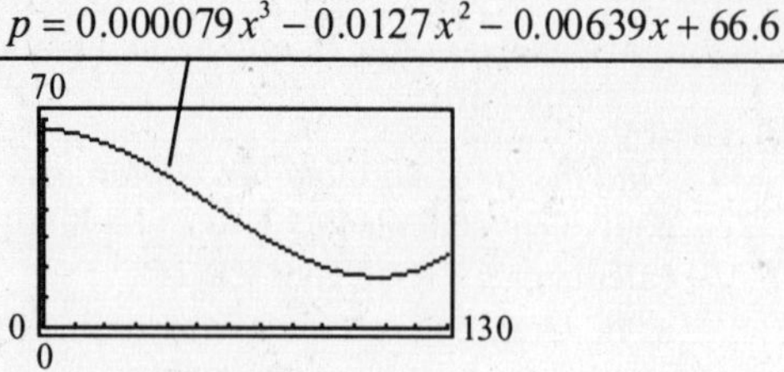

c. The percentage decreases before 2000 and increases after 2000.

49. a.

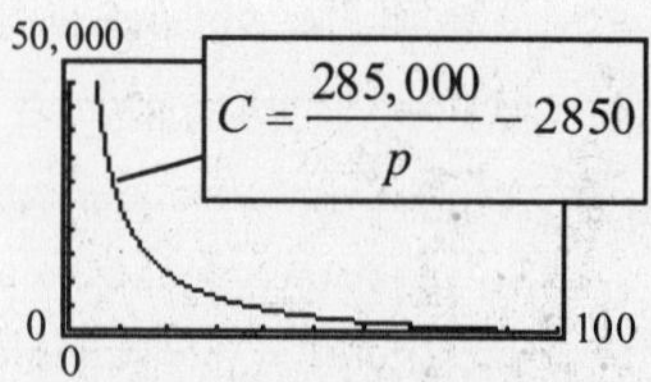

b. Near $p = 0$, cost grows without bound.

c. The coordinates of the point mean that the cost of obtaining stream water with 1% of the current pollution levels would cost $282,150.

d. The p-intercept means that the cost of stream water with 100% of the current pollution levels would cost $0.

51. a.

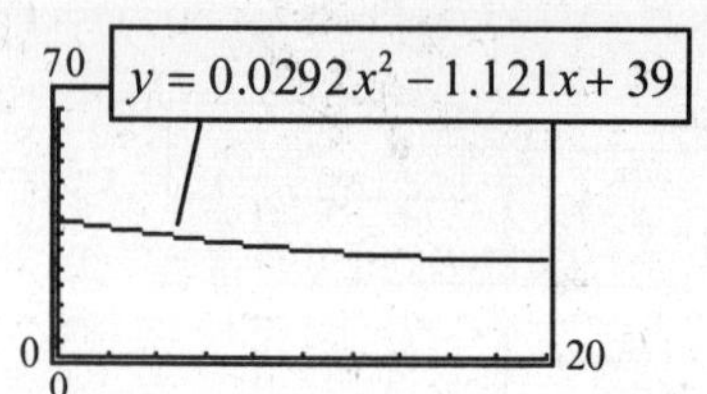

b. Graph is decreasing. The percent of Canadian smokers is declining.

Exercise 1.5

1. Solution: $(2,2)$

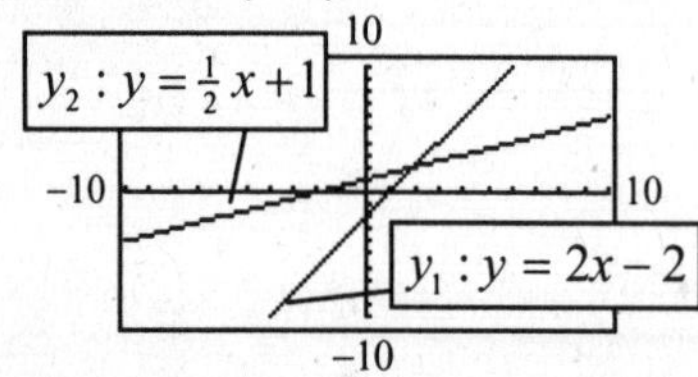

3. Solution: No solution since the graphs do not intersect.

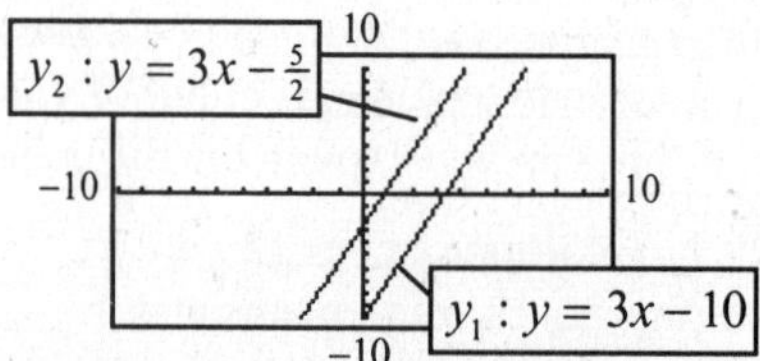

5. $3x - 2y = 6$

$4y = 8$ Solve for y. $y = \frac{8}{4} = 2$

Substitute for this variable in first equation and solve for the other variable.

$3x - 2(2) = 6$

$3x = 6 + 4 = 10$

$x = \frac{10}{3}$

The solution of the system is $x = \frac{10}{3}$ and $y = 2$, or $\left(\frac{10}{3}, 2\right)$.

7. $2x - y = 2$

$3x + 4y = 6$ Solve for y. $y = 2x - 2$

Substitute for this variable in 2nd equation and solve for the other variable.

$3x + 4(2x - 2) = 6$

$3x + 8x - 8 = 6$

$11x = 14$

$x = 14/11$

Solve for y: $y = 2\left(\frac{14}{11}\right) - 2 = \frac{6}{11}$ The solution of the system is $x = \frac{14}{11}$ and $y = \frac{6}{11}$, or $\left(\frac{14}{11}, \frac{6}{11}\right)$.

9. $3x+4y=1$ Multiply 1st equation by 3. $9x+12y=3$

$2x-3y=12$ Multiply 2nd equation by 4. $8x-12y=48$

Add the two equations. $17x \quad =51$

Solve for the variable. $x=3$

Substitute for this variable in either original equation and solve for the other variable.

$$3(3)+4y=1$$
$$4y=-8$$
$$y=-2$$

The solution of the system is $x=3$ and $y=-2$, or $(3,-2)$.

11. $-4x+3y=-5$ Multiply first equation by 3. $-12x+9y=-15$

$3x-2y=4$ Multiply second equation by 4. $12x-8y=16$

Add the two equations. $y=1$

Substitute for this variable in either original equation and solve for the other variable.

$$-4x+3(1)=-5$$
$$-4x=-8$$
$$x=2$$

The solution of the system is $x = 2$ and $y = 1$.

13. $0.2x-0.3y=4$ $\qquad 0.20x-0.3y=4$

$2.3x-\quad y=1.2$ Multiply 2nd equation by 0.3. $0.69x-0.3y=0.36$

Subtract the two equations. $-0.49x \quad =3.64$

Solve for the variable. $x=-\frac{52}{7}$

Substitute, solve for y. $y=-\frac{128}{7}$

The solution of the system is $x=-\dfrac{52}{7}$ and $y=-\dfrac{128}{7}$, or $\left(-\dfrac{52}{7},-\dfrac{128}{7}\right)$.

15. $\frac{5}{2}x-\frac{7}{2}y=-1$ Multiply first equation by 6. $15x-21y=-6$

$8x+3y=11$ Multiply second equation by 7. $56x+21y=77$

Add the two equations. $71x \quad =71$

Substitute for this variable in either original equation and solve for the other variable.

$$x=1$$
$$8(1)+3y=11$$
$$3y=3$$
$$y=1$$

The solution of the system is $x = 1$ and $y = 1$, or $(1, 1)$.

17. $4x+6y=4$ $\qquad 4x+6y=4$

$2x+3y=2$ Multiply second equation by -2. $-4x-6y=-4$

Add the two equations: $0=0$

There are infinitely many solutions. The system is dependent. Solve for one of the variables in terms of the remaining variable: $y=\dfrac{2}{3}-\dfrac{2}{3}x$. Then a general solution is $\left(c,\dfrac{2}{3}-\dfrac{2}{3}c\right)$, where any value of c will give a particular solution.

19. $\frac{x+y}{4} = 2$ Before we can solve this system $x + y = 8$

$\frac{y-1}{x} = 6$ we must arrange each equation $6x - y = -1$

in $ax + by = c$ form. $7x \quad = 7$

Add and solve for x. $x = 1$

Substitute in an original equation $\frac{y-1}{1} = 6$

and solve for remaining variable. $y = 7$

The solution is $x = 1$ and $y = 7$, or (1, 7).

21–23 Use the standard window and graph each equation. Use the TRACE or INTERSECT feature to find the solution.

21. $\begin{cases} y = 8 - \dfrac{3x}{2} \\ y = \dfrac{3x}{4} - 1 \end{cases}$

$y_2 = -\frac{3}{2}x + 8$

10

−10 10

−10

$y_1 = \frac{3}{4}x - 1$

Solution: $(4, 2)$

23. $\begin{cases} y_1 : 5x + 3y = -2 \\ y_2 : 3x + 7y = 4 \end{cases}$

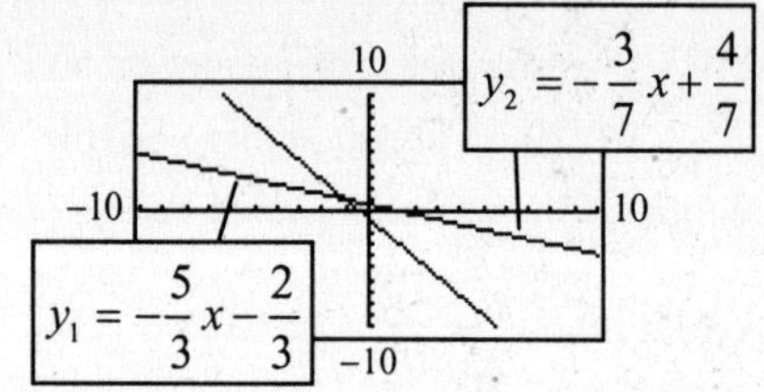

Solution: (–1, 1)

25. Eq. 1 $x + 2y + z = 2$ Steps 1, 2, and 3 of the systematic

Eq. 2 $-y + 3z = 8$ procedure are completed.

Eq. 3 $2z = 10$ Step 4: $z = 5$

From Eq. 2 $-y + 3(5) = 8$ or $y = 7$

From Eq. 1 $x + 2(7) + 5 = 2$ or $x = -17$

The solution is $x = -17, y = 7, z = 5$.

27. Eq. 1 $x - y - 8z = 0$ Steps 1 and 2 of the systematic

Eq. 2 $y + 4z = 8$ procedure are completed.

Eq. 3 $3y + 14z = 22$

Step 3: $(-3)\times$ Eq 2 added to Eq. 3 gives $2z = -2$ or $z = -1$.

From Eq. 2 $y + 4(-1) = 8$ or $y = 12$

From Eq. 1 $x - 12 - 8(-1) = 0$ or $x = 4$

The solution is $x = 4, y = 12, z = -1$ or (4, 12, –1).

29. Eq. 1 $x+4y-2z=9$ Step 1 is completed.

Eq. 2 $x+5y+2z=-2$

Eq. 3 $x+4y-28z=22$

Step 2:

$x+4y-2z=9$ Eq. 1

Eq. 4 $y+4z=-11$ $(-1)\times$ Eq. 1 added to Eq. 2

Eq. 5 $-26z=13$ $(-1)\times$ Eq. 1 added to Eq. 3

Step 3 is also completed.

Step 4: $z=-\frac{1}{2}$ from Eq. 5.

From Eq. 4 $y+4\left(-\frac{1}{2}\right)=-11$ or $y=-9$

From Eq. 1 $x+4(-9)-2\left(-\frac{1}{2}\right)=9$ or $x=44$

The solution is $x=44$, $y=-9$, $z=-\frac{1}{2}$ or $\left(44,-9,-\frac{1}{2}\right)$.

31. $y=-5.68x+676.173$ Navy personnel, $y=-11.997x+847.529$ Air Force personnel

a. $-5.68x+676.173=-11.997x+847.529$

$6.317x=171.356$

$x=27.13$ years

$1960+27.13\approx 1988$ When Navy personnel equals Air Force personnel.

b. $y=-5.68(27.13)+676.173=522.075$ thousand

Equal personnel occur with 522,075 in each branch.

33. a. $x+y=1800$ Total number of tickets

b. $20x=$ revenue from \$20 tickets

c. $30y=$ revenue from \$30 tickets

d. $20x+30y=42{,}000$ Total Revenue

e. Multiply equation from part (a) by -20.

$-20x-20y=-36000$

$20x+30y=42000$

$10y=6000$

$y=600$

Substitution into equation from part (a) gives $x=1200$.
Sell 1200 of the \$20 tickets and 600 of the \$30 tickets.

35. x = amount of safe investment.
y = amount of risky investment.
$x+y=145{,}600$ Total amount invested
$0.1x+0.18y=20{,}000$ Income from investments
The solution is the solution of the above system of equations.

$x+\ y=145{,}600$

$x+1.8y=200{,}000$ (10) × second equation

$0.8y=54{,}400$ Subtract equations

$y=68{,}000$ Solve for y or amount of risky investment.

Substituting $y=68{,}000$ into one of the original equations we have $x+68{,}000=145{,}600$ or $x=\$77{,}600$.
Solution: Put \$77,600 in a safe investment and \$68,000 in a risky investment.

37. x = amount invested at 10%.
y = amount invested at 12%.
$x + y = 235{,}000$ Total amount invested
$0.10x + 0.12y = 25{,}500$ Investment income
Solve the system of equation:

$$\begin{array}{ll} x + \quad y = 235{,}000 & \\ \underline{x + 1.2y = 255{,}000} & (10)\times \text{ second equation} \\ 0.2y = 20{,}000 & \text{Subtract equations} \\ y = 100{,}000 & \text{Solve for } y \end{array}$$

Substituting into the first equation we have $x + 100{,}000 = 235{,}000$ or $x = 135{,}000$.
Thus, \$135,000 is invested at 10% and \$100,000 is invested at 12%.

39. A = ounces of substance A.
B = ounces of substance B.

Required ratio $\frac{A}{B} = \frac{3}{5}$ gives $5A - 3B = 0$.

Required nutrition is 5%A + 12%B = 100%. This gives $5A + 12B = 100$.
The % notation can be trouble. Be careful! Now we can solve the system.

$$\begin{array}{ll} 5A - \ 3B = 0 & \\ \underline{5A + 12B = 100} & \\ 15B = 100 & \text{Subtract first equation from second.} \\ B = \frac{100}{15} = \frac{20}{3} & \end{array}$$

Substituting into the original equation gives $5A - 3\left(\frac{20}{3}\right) = 0$ or $A = 4$.

The solution is 4 ounces of substance A and $6\frac{2}{3}$ ounces of substance B.

41. x = population of species A.
y = population of species B.
$2x + y = 10{,}600$ units of first nutrient
$3x + 4y = 19{,}650$ units of second nutrient

$$\begin{array}{ll} 8x + 4y = 42{,}400 & (4)\times \text{ first equation} \\ \underline{3x + 4y = 19{,}650} & \\ 5x \quad = 22{,}750 & \text{Subtract} \\ x \quad = 4550 & \text{Solve for } x \end{array}$$

Substituting $x = 4550$ into an original equation we have $2(4550) + y = 10{,}600$.
So, $y = 1500$. Solution is 4550 of species A and 1500 of species B.

43. x = amount of 20% concentration.
y = amount of 5% concentration.
$x + y = 10$ amount of solution
$0.20x + 0.05y = 0.155(10)$ concentration of medicine
Solving this system of equations:

$$\begin{array}{ll} x + \quad y = 10 & \\ \underline{x + 0.25y = 7.75} & (5)\times \text{ second equation} \\ 0.75y = 2.25 & \text{Subtract equations} \\ y = 3 & \text{Solve for } y \end{array}$$

Substituting into the first equation we have $x + 3 = 10$ or $x = 7$.
The solution is 3 cc of 5% concentration and 7 cc of 20% concentration.

45. x = number of \$20 tickets.
y = number of \$30 tickets.
$x + y = 16{,}000$ total number of tickets
$20x + 30y = 380{,}000$ total revenue
To solve the system of equations multiply Eq. 1 by 30.

$$\begin{aligned} 30x + 30y &= 480{,}000 \\ 20x + 30y &= 380{,}000 \\ \hline 10x \quad &= 100{,}000 \text{ or } x = 10{,}000 \end{aligned}$$

Substituting into the first equation gives $y = 6000$.
Sell 10,000 tickets for \$20 each and 6000 tickets for \$30 each.

47. x = amount of 20% solution to be added.
$0.20x$ = concentration of nutrient in 20% solution.
$0.02(100) = 2$ is the concentration of nutrient in 2% solution.

$$0.20x + 2 = 0.10(x + 100)$$
$$0.20x + 2 = 0.10x + 10$$
$$0.1x = 8 \text{ or } x = 80\text{cc of 20\% solution is needed.}$$

49. x = ounces of substance A,
y = ounces of substance B, and
z = ounces of substance C.
$5x + 15y + 12z = 100$ Nutrition requirements
$x = z$ Digestive restrictions
$y = \frac{1}{5}z$ Digestive restrictions
Since both x and y are in terms of z, we can substitute in the first equation and solve for z.
So, $5z + 3z + 12z = 100$ or $20z = 100$. So, $z = 5$. Now, since $x = z$, we have $x = 5$.
Since $y = \dfrac{1}{5}z$, we have $y = 1$. The solution is 5 ounces of substance A, 1 ounce of substance B, and 5 ounces of substance C.

51. A = number of A type clients, B = number of B type clients, C = number of C type clients.
$A + B + C = 500$ Total clients
$200A + 500B + 300C = 150{,}000$ Counseling costs
$300A + 200B + 100C = 100{,}000$ Food and shelter
To find the solution we must solve the system of equations.

Eq. 1	$A + B + C = 500$	
Eq. 2	$2A + 5B + 3C = 1500$	Original equation divided by 100
Eq. 3	$3A + 2B + C = 1000$	Original equation divided by 100

	$A + B + C = 500$	Eq. 1
Eq. 4	$3B + C = 500$	$(-2)\times$ Eq. 1 added to Eq. 2
Eq. 5	$-B - 2C = -500$	$(-3)\times$ Eq. 1 added to Eq. 3

$A + B + C = 500$	Eq. 1
$3B + C = 500$	Eq. 4
$-\frac{5}{3}C = \frac{-1000}{3}$	$\frac{1}{3}\times$ Eq. 4 added to Eq. 5
$C = \frac{1000}{3} \cdot \frac{3}{5} = 200$	

Substituting $C = 200$ into Eq. 4 gives $3B + 200 = 500$ or $3B = 300$. So, $B = 100$.
Substituting $C = 200$ and $B = 100$ into Eq. 1 gives $A + 100 + 200 = 500$. So, $A = 200$.
Thus, the solution is 200 type A clients, 100 type B clients, and 200 type C clients.

Exercise 1.6

1. **a.** $P(x) = R(x) - C(x)$
$= 34x - (17x + 3400)$
$= 17x - 3400$

b. $P(300) = 17(300) - 3400 = \$1700$

3. **a.** $P(x) = R(x) - C(x)$
$= 80x - (43x + 1850)$
$= 37x - 1850$

b. $P(30) = 37(30) - 1850 = -\$740$
The total costs are more than the revenue.

c. $P(x) = 0$ or $37x - 1850 = 0$
So, $x = \frac{1850}{37} = 50$ units is the break-even point.

5. $C(x) = 5x + 250$

a. $m = 5$, C-intercept: 250

b. $\overline{MC} = 5$ means that each additional unit produced costs \$5.

c. Slope = marginal cost.
C-intercept =fixed costs.

d. \$5, \$5 ($\overline{MC} = 5$ at every point)

7. $R = 27x$

a. $m = 27$

b. 27; each additional unit sold yields \$27 in revenue.

c. In each case, one more unit yields \$27.

9. $R(x) = 27x$, $C(x) = 5x + 250$

a. $P(x) = 27x - (5x + 250) = 22x - 250$

b. $m = 22$

c. Marginal profit is 22.

d. Each additional unit sold gives a profit of \$22. To maximize profit sell all that you can produce. Note that this is not always true.

11. (x, P) is the correct form.
$P_1 = (200, 3100)$
$P_2 = (250, 6000)$
$m = \frac{6000 - 3100}{250 - 200} = 58$
$P - 3100 = 58(x - 200)$ or $P = 58x - 8500$
The marginal profit is 58.

13. **a.** $TC = 35H + 6600$

b. $TR = 60H$

c. $P = R - C$
$= 60H - (35H + 6600)$
$= 25H - 6600$

d. $C(200) = 35(200) + 6600$
$= \$13,600$ cost of 200 helmets
$R(200) = 60(200)$
$= \$12,000$ revenue from 200 helmets
$P(200) = R(200) - C(200)$
$= \$12,000 - 13,600$
$= -\$1600$ loss from 200 helmets

e. $C(300) = 35(300) + 6600$
$= \$17,100$ cost of 300 helmets
$R(300) = 60(300)$
$= \$18,000$ revenue from 300 helmets
$P(300) = R(300) - C(300)$
$= 18,000 - 17,100$
$= \$900$ profit from 300 helmets

f. The marginal profit is 25. Each additional helmet sold gives a profit of \$25.

15. **a.** The revenue function is the graph that passes through the origin.

b. At a production of zero the fixed costs are \$2000.

c. From the graph, the break-even point is 400 units and \$3000 in revenue or costs.

d. Marginal cost $= \frac{3000 - 2000}{400 - 0} = 2.5$
Marginal revenue $= \frac{3000 - 0}{400 - 0} = 7.5$

17. $R(x) = C(x) = 85x = 35x + 1650$ or $50x = 1650$ or $x = 33$.
Thus, 33 necklaces must be sold to break even.

19. **a.** $R(x) = 12x$, $C(x) = 8x + 1600$

b. $R(x) = C(x)$ if $12x = 8x + 1600$ or $4x = 1600$ or $x = 400$.
It takes 400 units to break even.

21. **a.** $P(x) = R(x) - C(x)$
$= 12x - (8x + 1600)$
$= 4x - 1600$

b. By setting $P(x) = 0$ we get $x = 400$.
Same as 19(b).

23. a. $TC = 4.50x + 1045$

b. $TR = 10x$

c. $P = R - C$

$= 10x - (4.50x + 1045)$

$= 5.50x - 1045$

d. Breakeven also means $P = 0$.

$5.50x - 1045 = 0$

$5.50x = 1045$

$x = 190$ units to break even

25. a. $R(x) = 54.90x$

b. $P_1 = (2000, 50000)$

$P_2 = (800, 32120)$

$$m = \frac{32,120 - 50,000}{800 - 2000} = \frac{-17,880}{-1200} = 14.90$$

$y - 50,000 = 14.90(x - 2000)$ or

$y = 14.90x + 20,200 = C(x)$

c. From $54.90x = 14.90x + 20,200$ we have $x = 505$ units to break even.

27. If price increases, then the demand for the product decreases.

29. a. If $p = \$100$, then $q = 600$ (approximately).

b. If $p = \$100$, then $q = 300$.

c. There is a shortage since more is demanded.

31. Demand: $2p + 5q = 200$

$2(60) + 5q = 200$

$5q = 80$

$q = 16$

Supply: $p - 2q = 10$

$60 - 2q = 10$

$2q = 50$

$q = 25$

There will be a surplus of 9 units at a price of \$60.00.

33. Remember that (q, p) is the correct form.

$P_1 = (240, 900)$

$P_2 = (315, 850)$

$$m = \frac{850 - 900}{315 - 240} = -\frac{50}{75} = -\frac{2}{3}$$

Note: $m < 0$ for demand equations.

$p - 900 = -\frac{2}{3}(q - 240)$ or

$p = -\frac{2}{3}q + 1060$

35. (q, p) is the correct form.

$P_1 = (10000, 1.50)$

$P_2 = (5000, 1.00)$

$$m = \frac{1 - 1.50}{5000 - 10000} = \frac{-0.50}{-5000} = 0.0001$$

Note: $m > 0$ for supply equations.

$p - 1 = 0.0001(q - 5000)$ or $p = 0.0001q + 0.5$

37. a. The decreasing function is the demand curve. The increasing function is the supply curve.

b. Reading the graph, we have equilibrium at $q = 30$ and $p = 25$.

39. a. Reading the graph, at $p = 20$ we have 20 units supplied.

b. Reading the graph, at $p = 20$ we have 40 units demanded.

c. At $p = 20$ there is a shortage of 20 units.

41. By observing the graph in the figure, we see that a price below the equilibrium price results in a shortage.

43. $-\frac{1}{2}q + 28 = \frac{1}{3}q + \frac{34}{3}$ Required condition.

$-3q + 168 = 2q + 68$ Multiply both sides by 6 to simplify.

$-5q = -100$

$q = 20$

Substituting into one of the original equations gives $p = -\frac{1}{2}(20) + 28 = 18$.

Thus, the equilibrium point is $(q, p) = (20, 18)$.

45. $-4q+220=15q+30$ Required condition.

$190=19q$

$q=10$ Solve for q.

Substituting q = 10 into one of the original equations gives p = 180.
Thus, the equilibrium point is $(q, p) = (10, 180)$.

47. Demand: (80, 350) and (120, 300) are two points. $m=\dfrac{350-300}{80-120}=-\dfrac{5}{4}$

$p-p_1=m(q-q_1)$ or $p-300=-\frac{5}{4}(q-120)$ or $p=-\frac{5}{4}q+450$

Supply: (60, 280) and (140, 370) are two points. $m=\dfrac{280-370}{60-140}=\dfrac{9}{8}$

$p-p_1=m(q-q_1)$ or $p-280=\dfrac{9}{8}(q-60)$ or $p=\dfrac{9}{8}q+212.5$

Now, set these two equations for p equal to each other and solve for q.

$\frac{9}{8}q+212.5=-\frac{5}{4}q+450$ Required for equilibrium.

$9q+1700=-10q+3600$ Multiply both sides by 8 to simplify.

$19q=1900$

$q=100$

Substituting q = 100 into one of the original equations gives p = 325.
Thus, the equilibrium point is $(q, p) = (100, 325)$.

49. **a.** Reading the graph, we have that the tax is $15.
b. From the graph, the original equilibrium was (100, 100).
c. From the graph, the new equilibrium is (50, 110).
d. The supplier suffers because the increased price reduces the demand.

51. New supply price: $p = 15q + 30 + 38 = 15q + 68$

$15q+68=-4q+220$ Required condition

$19q=152$

$q=8$

Substituting q = 8 into one of the original equations gives p = 188.
Thus, the new equilibrium point is $(q, p) = (8, 188)$.

53. New supply price: $p=\frac{q}{20}+10+5=\frac{q}{20}+15$

$\frac{q}{20}+15=-\frac{q}{20}+65$ Required condition

$q+300=-q+1300$

$2q=1000$

$q=500$ Thus, $p=\frac{500}{20}+15=40$.

The new equilibrium point is (500, 40).

55. Demand: $p=\dfrac{-q+2100}{60}$ Supply: $p=\dfrac{q+540}{120}$

New supply: $p=\dfrac{q+540}{120}+\dfrac{1}{2}=\dfrac{q+540}{120}+\dfrac{60}{120}=\dfrac{q+600}{120}$

$\frac{q+600}{120}=\frac{-q+2100}{60}$ Required condition

$q+600=-2q+4200$ Multiply both sides by 120

$3q=3600$

$q=1200$ Thus, $p=\frac{1200+600}{120}=15$.

The new equilibrium quantity is 1200. The new equilibrium price is $15.

Review Exercises

For this set of exercises we will not give reasons for any steps or list any formulas.

1. $x+7=14$

$x=14-7$

$x=7$

2. $3x-8=23$

$3x=31$

$x=\frac{31}{3}$

3. $2x-8=3x+5$

$-x=13$

$x=-13$

4. $\frac{6x+3}{6}=\frac{5(x-2)}{9}$

$18\left(\frac{6x+3}{6}\right)=18\left(\frac{5(x-2)}{9}\right)$

$3(6x+3)=10(x-2)$

$18x+9=10x-20$

$8x=-29$

$x=-\frac{29}{8}$

5. $2x+\frac{1}{2}=\frac{x}{2}+\frac{1}{3}$

$12x+3=3x+2$

$9x=-1$

$x=-\frac{1}{9}$

6. $0.6x+4=x-0.02$

$4.02=0.4x$

$10.05=x$

7. $\frac{6}{3x-5}=\frac{6}{2x+3}$

$6(2x+3)=6(3x-5)$

$2x+3=3x-5$

$3+5=3x-2x$

$x=8$

8. $\frac{2x+5}{x+7}=\frac{1}{3}+\frac{x-11}{2(x+7)}$

$6(2x+5)=2(x+7)+3(x-11)$

$12x+30=2x+14+3x-33$

$12x-2x-3x=14-33-30$

$7x=-49$

$x=-7$

There is no solution since we have division by zero when $x=-7$.

9. $3y-6=-2x-10$

$3y=-2x-4$

$y=\frac{-2x-4}{3}$

$y=-\frac{2}{3}x-\frac{4}{3}$

10. Yes.

11. $y^2=9x$, is not a function of x. If $x=1$, then $y=\pm 3$.

12. Yes.

13. $y=\sqrt{9-x}$

Domain: $9-x\geq 0$ or $9\geq x$ or $x\leq 9$.

Range: Positive square root means $y\geq 0$.

14. $f(x)=x^2+4x+5$

a. $f(-3)=(-3)^2+4(-3)+5=9-12+5=2$

b. $f(4)=(4)^2+4(4)+5=16+16+5=37$

c. $f\left(\frac{1}{2}\right)=\left(\frac{1}{2}\right)^2+4\left(\frac{1}{2}\right)+5=\frac{1}{4}+2+5=\frac{29}{4}$

15. $g(x)=x^2+\frac{1}{x}$

a. $g(-1)=(-1)^2+\frac{1}{-1}=1-1=0$

b. $g\left(\frac{1}{2}\right)=\left(\frac{1}{2}\right)^2+\frac{1}{\frac{1}{2}}=\frac{1}{4}+2=2\frac{1}{4}$

c. $g(0.1)=(0.1)^2+\frac{1}{0.1}=0.01+10=10.01$

16. $f(x) = 9x - x^2$

$f(x+h) = 9(x+h) - (x+h)^2$

$= 9x + 9h - x^2 - 2xh - h^2$

$f(x) = 9x - x^2$

$f(x+h) - f(x) = 9h - 2xh - h^2$

$= h(9 - 2x - h)$

$\dfrac{f(x+h) - f(x)}{h} = 9 - 2x - h$

17. y is a function of x. (Use vertical line test.)

18. No, the graph fails vertical line test.

19. $f(2) = 4$

20. $x = 0, x = 4$

21. a. $D = \{-2, -1, 0, 1, 3, 4\}$, $R = \{-3, 2, 4, 7, 8\}$

b. $f(4) = 7$

c. $f(x) = 2$ if $x = -1, 3$

d.

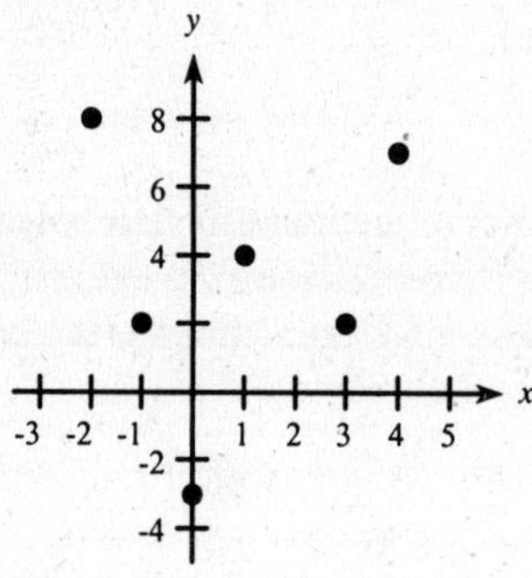

e. No. For $y = 2$, there are two values of x.

22. $f(x) = 3x + 5,\ g(x) = x^2$

a. $(f + g)x = (3x + 5) + x^2 = x^2 + 3x + 5$

b. $\left(\dfrac{f}{g}\right)x = \dfrac{3x + 5}{x^2}$ or $\dfrac{3x}{x^2} + \dfrac{5}{x^2} = \dfrac{3}{x} + \dfrac{5}{x^2}$

c. $f(g(x)) = f(x^2) = 3x^2 + 5$

d. $(f \circ f)x = f(3x + 5)$

$= 3(3x + 5) + 5$

$= 9x + 20$

23. $5x + 2y = 10$

x-intercept: If $y = 0$, $x = 2$

y-intercept: If $x = 0$, $y = 5$

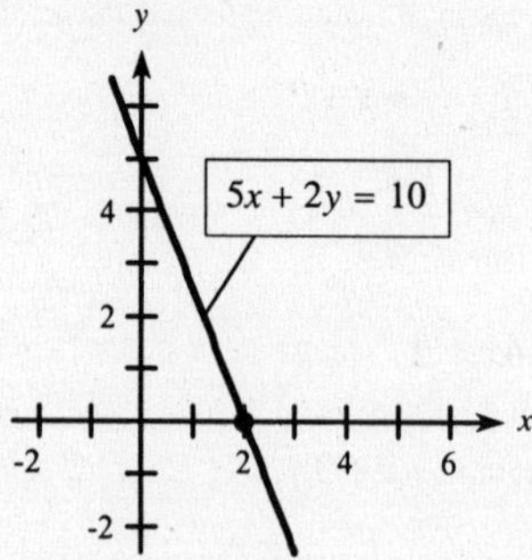

24. $6x + 5y = 9$

x-intercept: If $y = 0$, $x = \dfrac{9}{6} = \dfrac{3}{2}$

y-intercept: If $x = 0$ or $y = \dfrac{9}{5}$

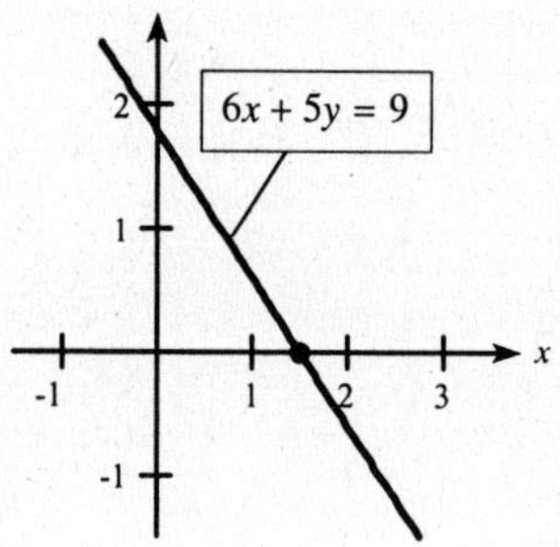

25. $x = -2$

x-intercept: $x = -2$

There is no y-intercept.

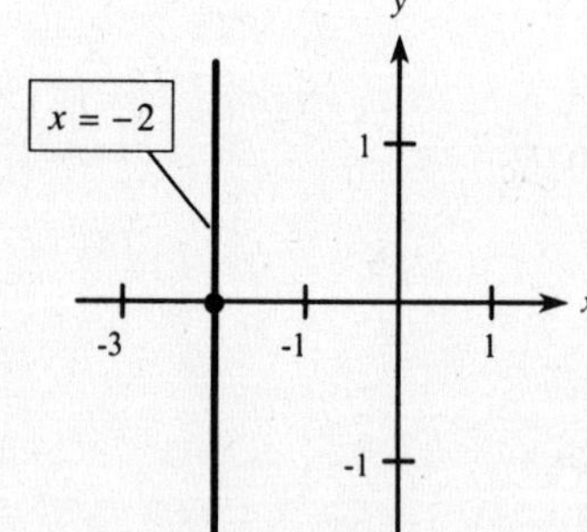

26. $P_1(2, -1);\ P_2(-1, -4)$

$m = \dfrac{-4 - (-1)}{-1 - 2} = \dfrac{-3}{-3} = 1$

27. $(-3.8, -7.16)$ and $(-3.8, 1.16)$

$m = \dfrac{-7.16 - 1.16}{-3.8 - (-3.8)} = \dfrac{-8.32}{0}$

Slope is undefined.

28. $2x + 5y = 10$

$y = -\frac{2}{5}x + 2,\ m = -\frac{2}{5},\ b = 2$

29. $x = -\frac{3}{4}y + \frac{3}{2}$ or $y = -\frac{4}{3}x + 2$

$m = -\frac{4}{3},\ b = 2$

30. $m = 4, b = 2, y = 4x + 2$

31. $m = -\frac{1}{2},\ b = 3,\ y = -\frac{1}{2}x + 3$

32. $P = (-2, 1),\ m = \frac{2}{5}$

$y - 1 = \frac{2}{5}(x + 2)$ or $y = \frac{2}{5}x + \frac{9}{5}$

33. $(-2, 7)$ and $(6, -4)$

$m = \frac{-4-7}{6-(-2)} = \frac{-11}{8}$

$y - 7 = \frac{-11}{8}(x - (-2))$ or

$y = \frac{-11}{8}x + \frac{17}{4}$

34. $P_1(-1, 8);\ P_2(-1, -1)$
The line is vertical since the x-coordinates are the same. Equation: $x = -1$

35. Parallel to $y = 4x - 6$ means $m = 4$.
$y - 6 = 4(x - 1)$ or $y = 4x + 2$

36. $P(-1, 2)$; $\perp$ to $3x + 4y = 12$

or

$y = -\frac{3}{4}x + 3$

$m = \frac{4}{3}$

$y - 2 = \frac{4}{3}(x + 1)$ or

$y = \frac{4}{3}x + \frac{10}{3}$

37. $x^2 + y - 2x - 3 = 0$

$y = -x^2 + 2x + 3$

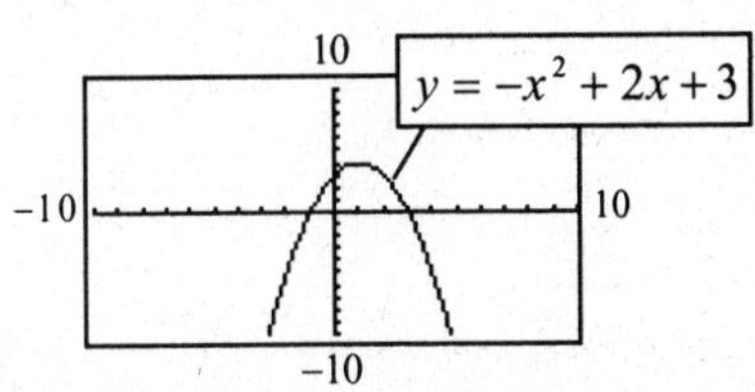

38.

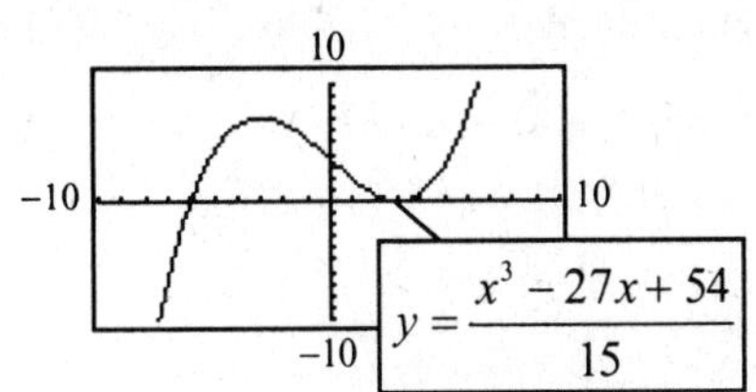

39. a.

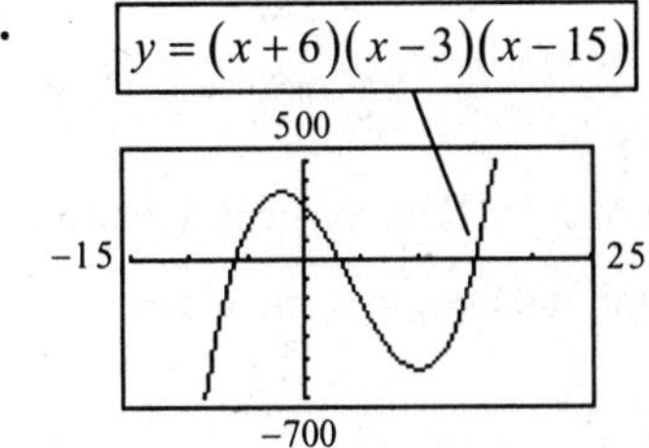

b.

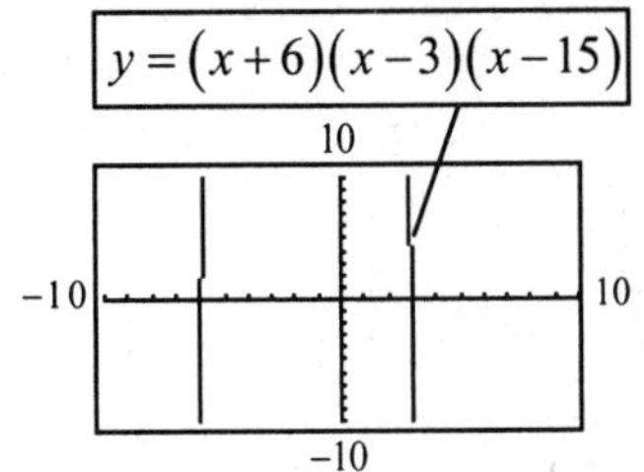

c. The graph in (a) shows the complete graph. The graph in (b) shows a piece that rises toward the high point and a piece between the high and low points.

40. $y = x^2 - x - 42$ is a parabola opening upward.

a.

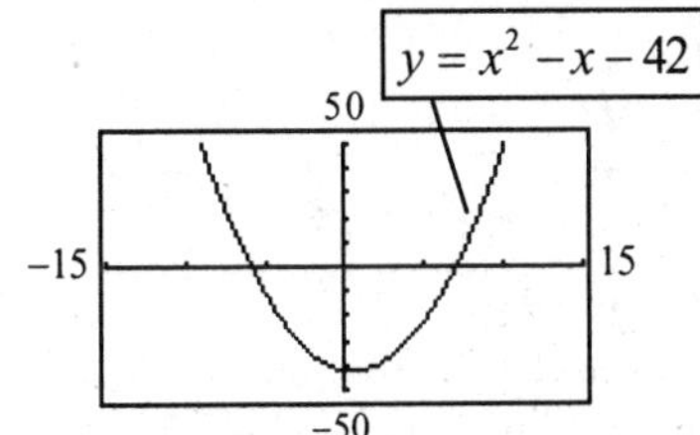

b.

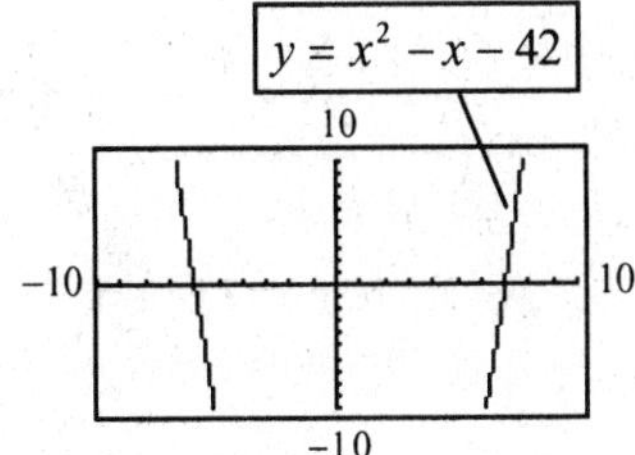

c. (a) shows the complete graph. The y-min is too large in absolute value for (b) to get a complete graph.

41. $y = \frac{\sqrt{x+3}}{x}$

$x \neq 0;\ x + 3 \geq 0$ or $x \geq -3;$

Domain: $x \neq 0,\ x \geq -3$

42. Trace approximates $x = -7.2749$, $x = 0.2749$

43.
$$\begin{aligned} 4x - 2y &= 6 \\ 3x + 3y &= 9 \end{aligned}$$
Then,
$$\begin{aligned} 12x - 6y &= 18 \\ 6x + 6y &= 18 \\ \hline 18x \qquad &= 36 \\ x &= 2 \\ 4(2) - 2y &= 6 \\ -2y &= -2 \\ y &= 1 \end{aligned}$$
Solution: (2, 1)

44.
$$\begin{aligned} 2x + \ y &= 19 \\ x - 2y &= 12 \end{aligned}$$
Then,
$$\begin{aligned} 4x + 2y &= 38 \\ x - 2y &= 12 \\ \hline 5x \qquad &= 50 \\ x &= 10 \\ 2(10) + y &= 19 \\ y &= -1 \end{aligned}$$
Solution: (10, −1)

45.
$$\begin{aligned} 3x + 2y &= 5 \\ 2x - 3y &= 12 \end{aligned}$$
Then,
$$\begin{aligned} 9x + 6y &= 15 \\ 4x - 6y &= 24 \\ \hline 13x \qquad &= 39 \\ x &= 3 \\ 3(3) + 2y &= 5 \\ 2y &= -4 \\ y &= -2 \end{aligned}$$
Solution: (3, −2)

46.
$$\begin{aligned} 6x + 3y &= 1 \\ y &= -2x + 1 \\ 6x + 3(-2x + 1) &= 1 \\ 6x - 6x + 3 &= 1 \\ 3 &= 1 \end{aligned}$$
No solution.

47.
$$\begin{aligned} 4x - 3y &= 253 \\ 13x + 2y &= -12 \end{aligned} \qquad \begin{aligned} 8x - 6y &= 506 \\ 39x + 6y &= -36 \\ \hline 47x \qquad &= 470 \\ x \qquad &= 10 \end{aligned} \qquad \begin{aligned} 4(10) - 3y &= 253 \\ -3y &= 213 \\ y &= -71 \end{aligned}$$
Solution: (10, −71)

48.
$$\begin{aligned} x + 2y + 3z &= 5 \\ y + 11z &= 21 \\ 5y + \ 9z &= 13 \end{aligned}$$
Steps 1 and 2: Nothing to be done.

Step 3:
$$\begin{aligned} x + 2y + \ 3z &= 5 \\ y + 11z &= 21 \\ -46z &= -92 \end{aligned}$$
Step 4: $z = 2$
$$\begin{aligned} y + 11(2) &= 21 \\ y &= -1 \end{aligned} \qquad \begin{aligned} x + 2(-1) + 3(2) &= 5 \\ x &= 1 \end{aligned}$$
Solution is $x = 1$, $y = -1$, $z = 2$.

49.
$$\begin{aligned} x + \ y - \ z &= 12 \\ 2y - 3z &= -7 \\ 3x + 3y - 7z &= 0 \\ x + \ y - \ z &= 12 \\ 2y - 3z &= -7 \\ -4z &= -36 \end{aligned}$$
Thus $z = 9$

$2y - 27 = -7$

$2y = 20$ or $y = 10$

$x + 10 - 9 = 12$

$x = 11$

Solution: (11, 10, 9)

50. $A = 3.303x - 18.59$

a. $x = 7$ corresponds to 1997

b. 2002 gives $x = 2002 - 1990 = 12$

c. $27.652 = 3.303x - 18.59$

$$3.303x = 46.242$$

$$x = 14$$

27.652 is reached in 14+1990=2004

51. Student has total points of 91 + 82 + 88 + 50 + 42 + 42 = 395.
Total of possible points is 300 + 150 + 200 = 650.
To earn an A students need at least 0.9(650) = 585 points.
Student must earn 585 – 395 = 190 points on the final. This is the same as 95%.

52. Diesel: $C = 0.24x + 38{,}000$

Gas: $C = 0.30x + 35{,}600$

$$0.24x + 38{,}000 = 0.30x + 35{,}600$$

$$0.06x = 2400$$

$$x = 40{,}000$$

Costs are equal at 40,000 miles. A truck is used more than 40,000 miles in 5 years. Buy the diesel.

53. a. Yes

b. No

c. $f(300) = 4$

54. a. $f(80) = 565.44$

b. The monthly payment on a $70,000 loan is $494.75.

55. $P(x) = 180x - \frac{x^2}{100} - 200 \quad x = q(t) = 1000 + 10t$

a. $(P \circ q)(t) = P(1000 + 10t) = 180(1000 + 10t) - \frac{(1000 + 10t)^2}{100} - 200$

b. $x = q(15) = 1000 + 10(15) = 1150$ units produced

$$P(1150) = 180(1150) - \frac{(1150)^2}{100} - 200 = \$193{,}575$$

56. $W(L) = kL^3,\ L(t) = 50 - \frac{(t-20)^2}{10},\ 0 \le t \le 20$

$$(W \circ L)(t) = W\left(50 - \frac{(t-20)^2}{10}\right) = 0.02\left(50 - \frac{(t-20)^2}{10}\right)^3$$

57. a.

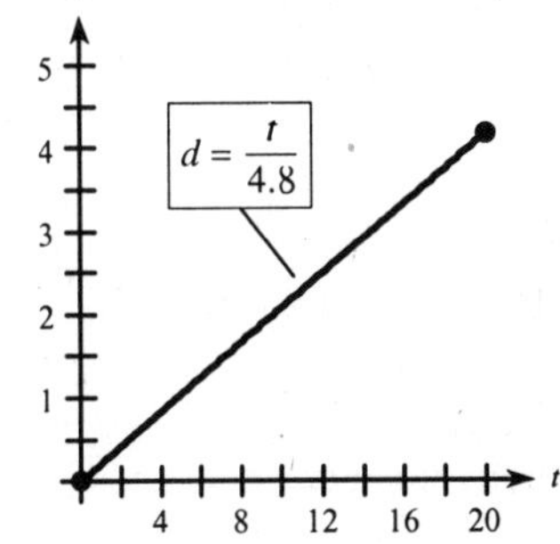

b. (9.6, 2) means that the thunderstorm is two miles away if the flash and thunder are 9.6 seconds apart.

58.

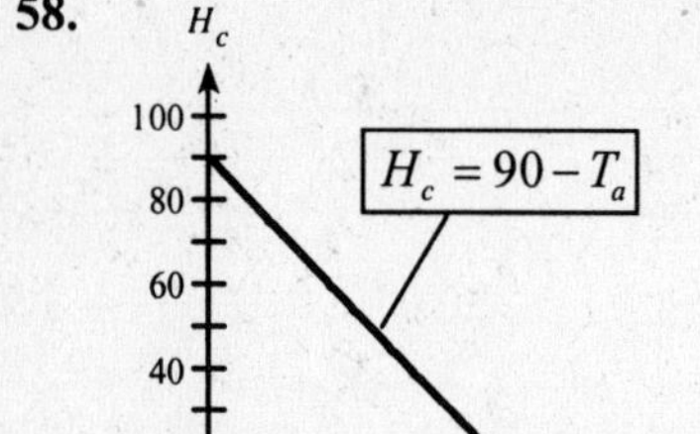

59. a. (*x*, *P*) is the required form.

$P_1 = (200, 3100)$, $P_2 = (250, 6000)$

$$m = \frac{6000-3100}{250-200} = \frac{2900}{50} = 58$$

$P - 3100 = 58(x - 200)$ or

$P(x) = 58x - 8500$

b. For each additional unit sold the profit increases by \$58.

60. $T = 1.834x + 20.37$

a. *T* is linear since it is in the form $y = mx + b$.

b. $m = 1.834$ and $b = 20.37$

c. In 1980 the tax burden per capita was \$2037.

d. The tax burden increases \$183.40 each year.

61. Use (*C*, *F*) to get equation.

$P_1(0, 32)$; $P_2(100, 212)$

$$m = \frac{212-32}{100-0} = \frac{180}{100} = \frac{9}{5}$$

$F - 32 = \frac{9}{5}(C - 0)$ or $F = \frac{9}{5}C + 32$.

Also $C = \frac{5}{9}(F - 32)$.

62. a.

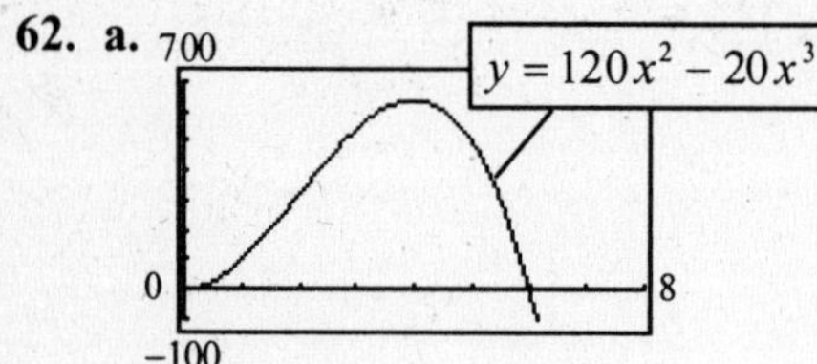

b. Algebraically, $y \ge 0$ if

$120x^2 - 20x^3 = 20x^2(x-6) \ge 0$.

Answer: $0 \le x \le 6$

63. a. $v^2 = 1960(h + 10)$

$$h + 10 = \frac{v^2}{1960}$$

$$h = \frac{v^2}{1960} - 10$$

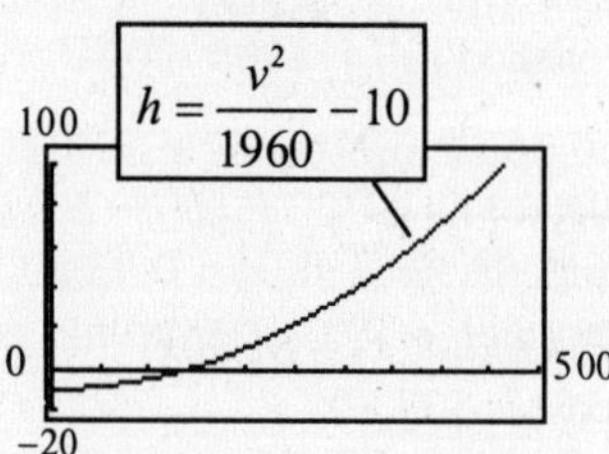

b. $h(210) = \frac{210^2}{1960} - 10 = 12.5$ cm

64. x = amount of safer investment and y = amount of other investment.

$$x + y = 150000$$

$$0.095x + 0.11y = 15000$$

Solving the system:

$$0.11x + 0.11y = 16500$$

$$0.095x + 0.11y = 15000$$

$$0.015x = 1500$$

$$x = 100000$$

Then $y = 50000$. Thus, invest \$100,000 at 9.5% and \$50,000 at 11%.

65. x = liters of 20% solution

y = liters of 70% solution

$$x + y = 4$$

$$0.2x + 0.7y = 1.4$$

$$x + y = 4$$

$$x + 3.5y = 7$$

$$2.5y = 3 \quad y = 1.2$$

$$x + 1.2 = 4$$

$$x = 2.8$$

Answer: 2.8 liters of 20%, 1.2 of 70%.

66. $S: p = 4q + 5$, $D: p = -2q + 81$

a. $S: 53 = 4q + 5$ $D: 53 = -2q + 81$

$4q = 48$ $2q = 28$

$q = 12$ $q = 14$

b. Demand is greater.
There is a shortfall.

c. Price is likely to increase.

67. a.-c.

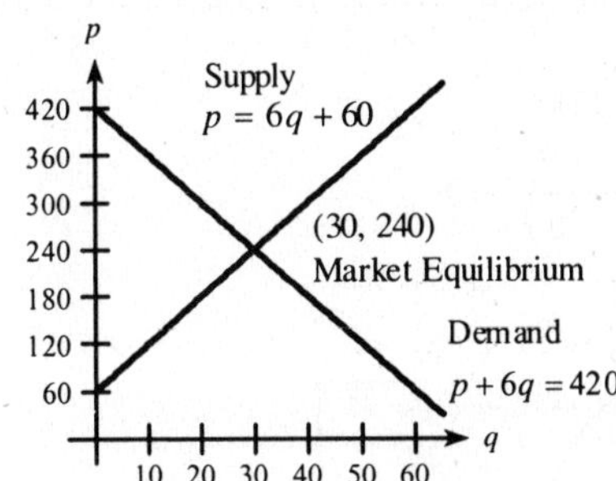

68. $C(x) = 38.80x + 4500$, $R(x) = 61.30x$

a. Marginal cost is \$38.80.

b. Marginal revenue is \$61.30.

c. Marginal profit is \$61.30 – 38.80 = \$22.50.

d. $61.30x = 38.80x + 4500$

$22.50x = 4500$

$x = 200$ units to break even.

69. $FC = \$1500$, $VC = \$22$ per unit, $R = \$52$ per unit

a. $C(x) = 22x + 1500$

b. $R(x) = 52x$

c. $P = R - C = 30x - 1500$

d. $\overline{MC} = 22$

e. $\overline{MR} = 52$

f. $\overline{MP} = 30$

g. Break even means $30x - 1500 = 0$ or $x = 50$.

70. Supply: $m = \frac{100-200}{200-400} = \frac{1}{2}$ Demand: $m = \frac{200-0}{200-600} = -\frac{1}{2}$

$p - 100 = \frac{1}{2}(q - 200)$ $\quad$ $p - 0 = -\frac{1}{2}(q - 600)$

$p = \frac{1}{2}q$ $\quad$ $p = -\frac{1}{2}q + 300$

So, $\frac{1}{2}q = -\frac{1}{2}q + 300$ or $q = 300$. The equilibrium price is $p = \frac{1}{2}(300) = \$150$.

71. New supply equation: $p = \frac{q}{10} + 8 + 2 = \frac{q}{10} + 10$

Demand:

$$p = \frac{-q + 1500}{10} = -\frac{q}{10} + 150$$

$$\frac{q}{10} + 10 = -\frac{q}{10} + 150$$

$$\frac{2q}{10} = 140 \text{ or } q = 700$$

$$p = \frac{700}{10} + 10 = 80$$

Solution: (700, 80)

Chapter Test

1. $4x - 3 = \frac{x}{2} + 6$

$8x - 6 = x + 12$

$7x = 18$

$x = \frac{18}{7}$

2. $\frac{3}{x} + 4 = \frac{4x}{x+1}$

$3(x+1) + 4x(x+1) = 4x(x)$

$3x + 3 + 4x^2 + 4x = 4x^2$

$7x = -3$

$x = -\frac{3}{7}$

3. $\frac{3x-1}{4x-9} = \frac{5}{7}$

$7(3x-1) = 5(4x-9)$

$21x - 7 = 20x - 45$

$x = -38$

4. $f(x) = 7 + 5x - 2x^2$

$f(x+h) = 7 + 5(x+h) - 2(x+h)^2$

$= 7 + 5x + 5h - 2x^2 - 4xh - 2h^2$

$f(x) = 7 + 5x - 2x^2$

$f(x+h) - f(x) = 5h - 4xh - 2h^2$

$\frac{f(x+h) - f(x)}{h} = 5 - 4x - 2h$

5. $5x - 6y = 30$
x-intercept: 6
y-intercept: –5

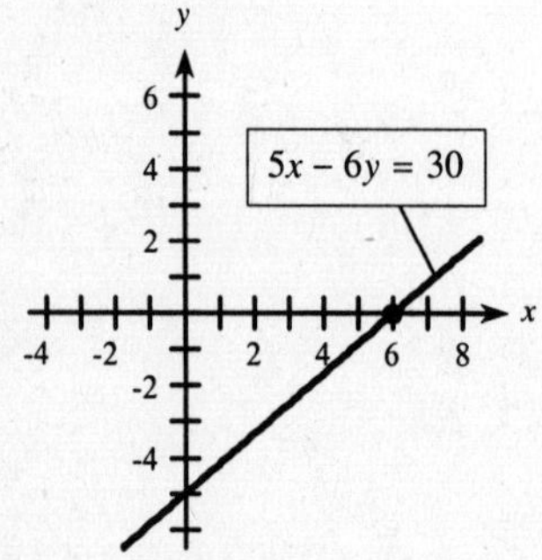

6. $7x + 5y = 21$
x-intercept: 3
y-intercept: $\frac{21}{5}$

7. $f(x) = \sqrt{4x+16}$

a. $4x + 16 \geq 0$
$4x \geq -16$
Domain: $x \geq -4$; Range: $y \geq 0$
For range, note square root is positive.

b. $f(3) = \sqrt{12+16} = 2\sqrt{7}$

c. $f(5) = \sqrt{20+16} = 6$

8. (–1, 2) and (3, –4)

$m = \frac{-4-2}{3-(-1)} = \frac{-6}{4} = \frac{-3}{2}$

$y - 2 = \frac{-3}{2}(x - (-1))$

$y = \frac{-3}{2}x + \frac{1}{2}$

9. $5x + 4y = 15$

$y = -\frac{5}{4}x + \frac{15}{4}$

$m = -\frac{5}{4},\ b = \frac{15}{4}$

10. Point (–3, –1)

a. Undefined slope means vertical line. $x = -3$

b. $\perp$ to $y = \frac{1}{4}x + 2$ means $m = -4$.
Thus, $y + 1 = -4(x + 3)$ or $y = -4x - 13$.

11. a. is not a function since for some x-values there are two y's.

b. is a function since for each x there is only one y.

c. is not a function for same reason as (a).

12. $3x+2y=-2$

$4x+5y=2$

$12x+\ 8y=-8$

$\underline{12x+15y=6}$

$-7y=-14$

$y=2$

$3x+2(2)=-2$

$3x=-6$

$x=-2$

Solution: $(-2, 2)$

13. $f(x)=5x^2-3x,\ g(x)=x+1$

a. $(fg)(x)=(5x^2-3x)(x+1)$

b. $g(g(x))=g(x+1)=(x+1)+1=x+2$

c. $(f\circ g)(x)=f(x+1)$

$=5(x+1)^2-3(x+1)$

$=5x^2+10x+5-3x-3$

$=5x^2+7x+2$

14. $R(x)=38x$, $C(x)=30x+1200$

a. $\overline{MC}=\$30$

b. $P(x)=38x-(30x+1200)$

$=8x-1200$

c. Break-even means $P(x)=0$.
$8x=1200$ or $x=150$ units

d. $\overline{MP}=\$8$. Each additional unit sold increases the profit by \$8.

15. a. $R(x)=50x$

b. $C(100)=10(100)+18000$

$=\$19,000$

It costs \$19,000 to make 100 units.

c. $50x=10x+18000$

$40x=18000$

$x=450$ units

16. $S: p=5q+1500,\ D: p=-3q+3100$
$5q+1500=-3q+3100$

$8q=1600$ or $q=200$

$p(200)=5(200)+1500=\$2500$

17. $y=360000-1500x$

a. $b=360,000$
The original value is \$360,000.

b. $m=-1500$.
The building is depreciating \$1500 each month.

18. x = number of reservations
$0.90x=360$

$x=400$

Accept 400 reservations.

19. x = amount invested at 9%
y = amount invested at 6%
$x+y=20000$ Amount
$0.09x+0.06y=1560$ Interest
$0.09x+0.09y=1800$

$\underline{0.09x+0.06y=1560}$

$0.03y=240$

$y=\$8000$

Invest \$8000 at 6% and \$12000 at 9%.

Chapter 2: Special Functions

Exercise 2.1

All problems must be in the form of $ax^2 + bx + c = 0$ before solutions can be found.

The Quadratic formula, $x = \frac{-b \pm \sqrt{b^2 - 4ac}}{2a}$, is used when factoring is difficult or not possible.

1. $2x^2 + 3 = x^2 - 2x + 4$

$x^2 + 2x - 1 = 0$

3. $(y+1)(y+2) = 4$

$y^2 + 3y + 2 = 4$

$y^2 + 3y - 2 = 0$

5. $9 - 4x^2 = 0$

$(3+2x)(3-2x) = 0$

$3 + 2x = 0$ or $3 - 2x = 0$

Solution: $x = -\frac{3}{2}, \frac{3}{2}$

7. $x = x^2$

$x^2 - x = 0$

$x(x-1) = 0$

Solution: $x = 0, 1$

Never divide by a variable. A root is lost if you divide.

9. $x^2 + 5x = 21 + x$

$x^2 + 4x - 21 = 0$

$(x+7)(x-3) = 0$

$x + 7 = 0$ or $x - 3 = 0$

Solution: $x = -7, 3$

11. $4t^2 - 4t + 1 = 0$

$(2t-1)(2t-1) = 0$

$2t - 1 = 0$

Solution: $t = \frac{1}{2}$

13. $\frac{w^2}{8} - \frac{w}{2} - 4 = 0$

$w^2 - 4w - 32 = 0$

$(w-8)(w+4) = 0$

$w - 8 = 0$ or $w + 4 = 0$

Solution: $w = 8, -4$

15. $\frac{x^2}{4} + \frac{3x}{2} + 2 = 0$

$x^2 + 6x + 8 = 0$

$(x+4)(x+2) = 0$

$x + 4 = 0$ or $x + 2 = 0$

Solution: $x = -4, -2$

17. $(x-1)(x+5) = 7$

$x^2 + 4x - 5 = 7$

$x^2 + 4x - 12 = 0$

$(x+6)(x-2) = 0$

Solution: $x = -6, 2$

19. $(4x-1)(x-3) = 15$

$4x^2 - 13x + 3 = 15$

$4x^2 - 13x - 12 = 0$

$(4x+3)(x-4) = 0$

Solution: $x = -\frac{3}{4}, 4$

21. $x + \frac{8}{x} = 9$

$x^2 + 8 = 9x$

$x^2 - 9x + 8 = 0$

$(x-8)(x-1) = 0$

Solution: $x = 1, 8$

23. $\frac{x}{x-1} = 2x + \frac{1}{x-1}$

$x = (2x^2 - 2x) + 1$

$2x^2 - 3x + 1 = 0$

$(2x-1)(x-1) = 0$

Solution: $x = \frac{1}{2}$

1 is not a root since division by zero is not defined.

25. a. $x^2 - 4x - 4 = 0$

$a = 1, b = -4, c = -4$

$$x = \frac{-(-4) \pm \sqrt{(-4)^2 - 4(1)(-4)}}{2(1)}$$

$$= \frac{4 \pm \sqrt{32}}{2} = \frac{4 \pm 4\sqrt{2}}{2} = 2 \pm 2\sqrt{2}$$

b. Since $\sqrt{2} \approx 1.414$, the solutions are approximately 4.83, –0.83.

27. $2w^2 + w + 1 = 0$

$a = 2, b = 1, c = 1$

$$w = \frac{-1 \pm \sqrt{1-8}}{4} = \frac{-1 \pm \sqrt{-7}}{4}$$

There are no real solutions.

29. a. $16z^2 + 16z - 21 = 0$

$a = 16, b = 16, c = -21$

$$z = \frac{-16 \pm \sqrt{256 + 1344}}{32}$$

$$= \frac{-16 \pm 40}{32} = \frac{3}{4} \text{ or } -\frac{7}{4}$$

b. 0.75, –1.75

31. a. $5x^2 = 2x + 6$ or $5x^2 - 2x - 6 = 0$

$a = 5, b = -2, c = -6$

$$x = \frac{2 \pm \sqrt{4 + 120}}{10} = \frac{1 \pm \sqrt{31}}{5}$$

b. $\frac{1 \pm \sqrt{31}}{5} \approx \frac{1 \pm 5.57}{5} \approx 1.31, -0.91$

33. $y^2 = 7$

$y = \pm\sqrt{7}$

35. $y^2 + 9 = 0$

$y^2 + 3^2 = 0$

The sum of two squares cannot be factored. There are no real solutions.

37. $(x+4)^2 = 25$

$x + 4 = \pm 5$

$x = -4 \pm 5$

Solution: $x = 1, -9$

39. $(x+8)^2 + 3(x+8) + 2 = 0$

$[(x+8)+2][(x+8)+1] = 0$

$(x+8)+2 = 0$ or $(x+8)+1 = 0$

Solution: $x = -10, -9$

41. $21x + 70 - 7x^2 = 0$

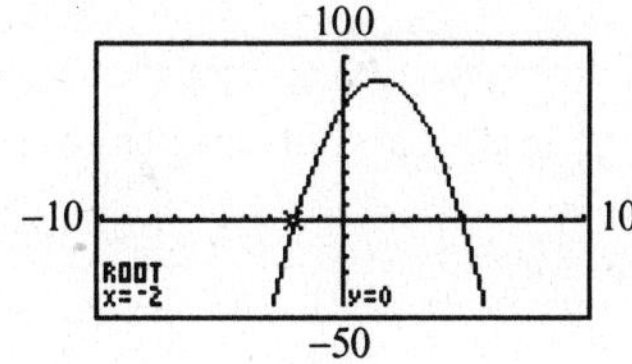

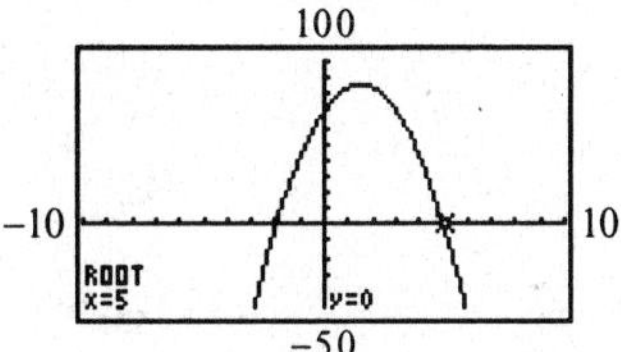

Divide by –7 and rearrange.

$x^2 - 3x - 10 = 0$

$(x-5)(x+2) = 0$

Solution: $x = -2, 5$

43. $300 - 2x - 0.01x^2 = 0$

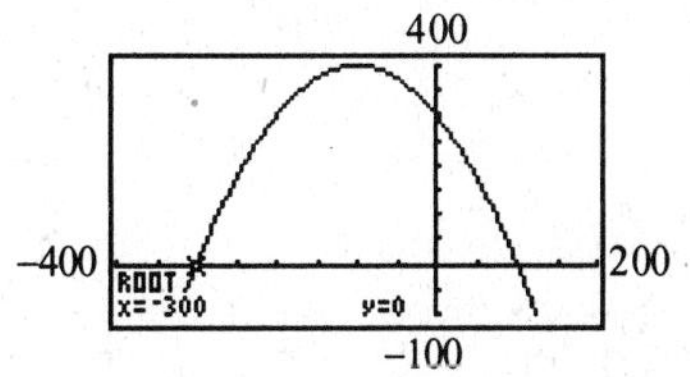

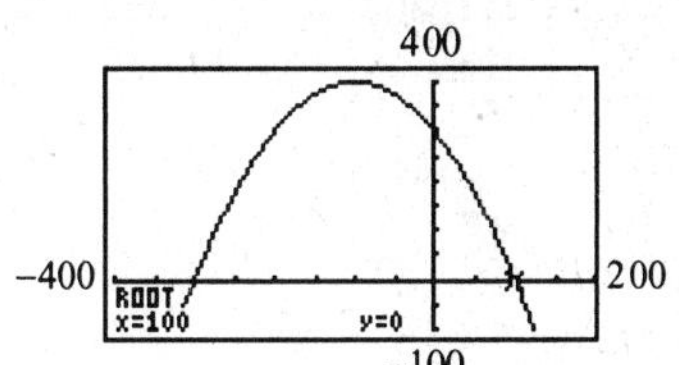

$a = -0.01, b = -2, c = 300$

$$x = \frac{2 \pm \sqrt{4 + 12}}{-0.02} = \frac{2 \pm 4}{-0.02}$$

$= -300$ or 100

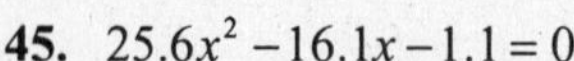

45. $25.6x^2 - 16.1x - 1.1 = 0$

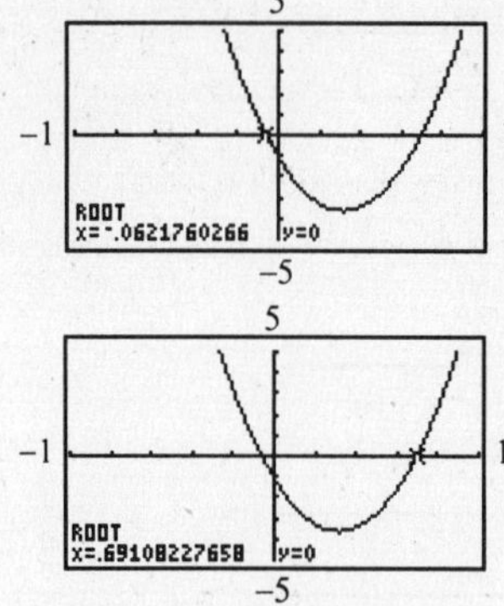

$a = 25.6,\ b = -16.1,\ c = -1.1$

$$x = \frac{16.1 \pm \sqrt{259.21 + 112.64}}{51.2}$$

$$= \frac{16.1 \pm \sqrt{371.85}}{51.2}$$

$$\approx 0.69 \text{ or } -0.06$$

47. $P = -x^2 + 90x - 200$

$1200 = -x^2 + 90x - 200$

$0 = x^2 - 90x + 1400$

$0 = (x - 20)(x - 70)$

A profit of \$1200 is earned at $x = 20$ units or $x = 70$ units of production.

49. a. $P = -18x^2 + 6400x - 400$

$61{,}800 = -18x^2 + 6400x - 400$

$18x^2 - 6400x + 62{,}200 = 0$

Factoring appears difficult, so let us apply the quadratic formula.

$$x = \frac{6400 \pm \sqrt{6400^2 - 4(18)(62{,}200)}}{36}$$

$$= \frac{6400 \pm \sqrt{36{,}481{,}600}}{36}$$

$$= \frac{6400 \pm 6040}{36} = 10 \text{ or } 345.56$$

So, a profit of \$61,800 is earned for 10 units or for 345.56 units.

b. Yes. Maximum profit occurs at vertex as seen using the graphing calculator.

51. $5 = 100 + 96t - 16t^2$

$100 = 100 + 96t - 16t^2$

$0 = 96t - 16t^2 = 16t(6 - t)$

The ball is 100 feet high 6 seconds later.

53. $p = 25 - 0.01s^2$

a. $0 = 25 - 0.01s^2$

$= (5 + 0.1s)(5 - 0.1s)$

$p = 0$ if $5 - 0.1s = 0$ or $s = 50$.

b. $s \geq 0$. $p = 0$ means there is no particulate pollution.

55. $v = k\left(R^2 - r^2\right)$

$v = 2\left(0.01 - r^2\right)$

a. $0.02 = 2\left(0.01 - r^2\right)$

$0.01 = 0.01 - r^2$

$r^2 = 0$ or $r = 0$

b. $0.015 = 2\left(0.01 - r^2\right)$

$0.0075 = 0.01 - r^2$

$r^2 = 0.0025$ or $r = 0.05$

c. $0 = 2\left(0.01 - r^2\right)$

$r^2 = 0.01$ or $r = 0.1$

57. $t = 0.01\left(0.115x^2 + 1.164x + 123.48\right)$

$5.5(100) = 0.115x^2 + 1.164x + 123.48$

$0 = 0.115x^2 + 1.164x - 426.52$

Use the graphing calculator and a varying range. The BMW is going approximately 56 mph.

59. $y = 0.003x^2 - 0.42x + 19.8$

a. $y = 0.003(40)^2 - 0.42(40) + 19.8$

$= 4.8 - 16.80 + 19.8 = 7.8$

b. $7.8 = 0.003x^2 - 0.42x + 19.8$

$0 = 0.003x^2 - 0.42x + 12$

Now $12 = 40(0.3)$

$0 = (x - 40)(0.003x - 0.3)$

Population is 7.8% foreign born when $0.003x - 0.3 = 0$ or $x = 100$. In 2000 the population is 7.8% foreign born.

61. $p = f(t) = 0.121t^2 - 4.180t + 79.620$

$44.4 = 0.121t^2 - 4.180t + 79.620$

$0 = 0.121t^2 - 4.180t + 35.220$

Use the graphing calculator and vary the range. This will give $x = 19.97$ or $x = 14.58$. In 2000 the percentage will reach 44.4, and it also reached that percentage in 1995.

63. $C =$ cost
profit = selling price – cost = 144 – C
percentage of profit = $\dfrac{144-C}{C}\cdot 100$
$$C = 100\left(\frac{144-C}{C}\right)$$
$$C^2 + 100C - 14{,}400 = 0$$
$$(C-80)(C+180) = 0$$
The store paid \$80.

65. $y = 11.786x^2 - 142.214x + 493$
$$812 = 11.786x^2 - 142.214x + 493$$
$$0 = 11.786x^2 - 142.214x - 319$$
Vary the range on the graphing calculator to see where the graph crosses the x-axis. This is approximately $x = 14$. In 2004 there will be 812 million users.

Exercise 2.2

1. $y = \frac{1}{2}x^2 + x$ $a > 0$, so vertex is a minimum.
$$x = \frac{-b}{2a} = \frac{-1}{2(1/2)} = -1$$
$$y = \frac{1}{2}(-1)^2 + (-1) = -\frac{1}{2}$$
Vertex is at $\left(-1, -\frac{1}{2}\right)$.

3. $y = 8 + 2x - x^2$ $a < 0$, so vertex is a maximum.
$$x = \frac{-b}{2a} = \frac{-2}{2(-1)} = 1$$
$$y = 8 + 2(1) - (1)^2 = 9$$
Vertex is at (1, 9).

5. $f(x) = 6x - x^2$
The maximum point is at the vertex.
a. $a = -1, b = 6$
$$x = \frac{-b}{2a} = \frac{-6}{-2} = 3.$$
$f(x)$ has a maximum at $x = 3$.
b. The maximum value is the y-coordinate of the vertex.
$f(3) = 18 - 3^2 = 9$ is the maximum value of the function.

7. $f(x) = x^2 + 2x - 3$
The minimum point is at the vertex.
a. $a = 1, b = 2$
$$x = \frac{-b}{2a} = \frac{-2}{2} = -1.$$
$f(x)$ has a minimum at $x = -1$.
b. The minimum value is the y-coordinate of the vertex.
$f(-1) = 1 - 2 - 3 = -4$ is the minimum value of the function.

9. $y = x^2 - 4$
Vertex is a minimum point since $a > 0$.
V: $x = \frac{-b}{2a} = \frac{0}{2} = 0$
$$y = 0^2 - 4 = -4$$
Zeros: $x^2 - 4 = 0$
$$(x+2)(x-2) = 0$$
$$x = -2, 2$$

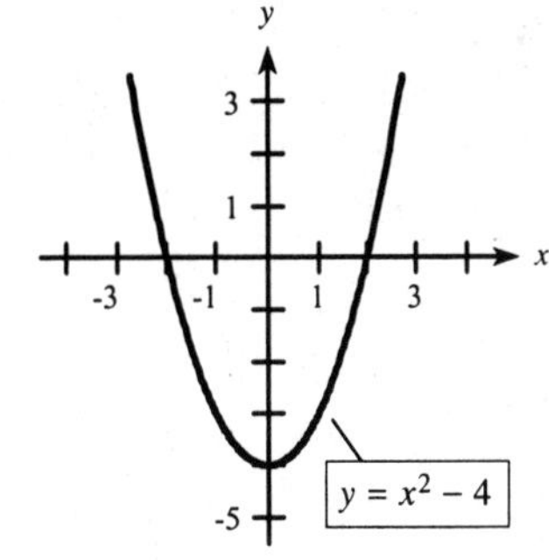

11. $y = -\frac{1}{4}x^2 + x$

Vertex is a maximum point since $a < 0$.

V: $x = \frac{-b}{2a} = \frac{-1}{2(-1/4)} = 2$

$y = -\frac{1}{4}(2)^2 + 2 = 1$

Zeros: $-\frac{1}{4}x^2 + x = 0$

$x\left(-\frac{1}{4}x + 1\right) = 0$

$x = 0, 4$

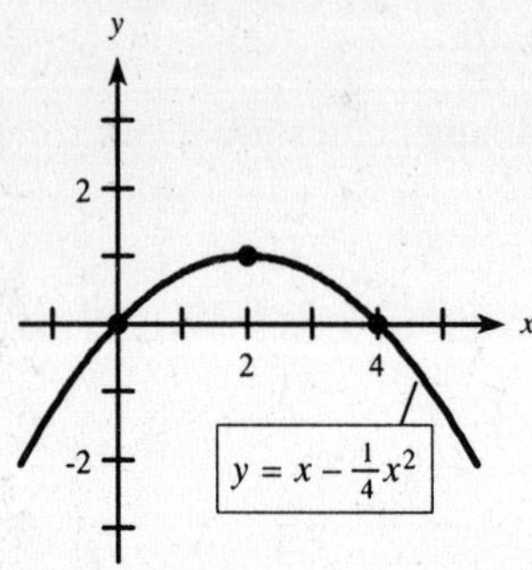

13. $y = x^2 + 4x + 4$

Vertex is a minimum point since $a > 0$.

V: $x = \frac{-b}{2a} = \frac{-4}{2(1)} = -2$

$y = (-2)^2 + 4(-2) + 4 = 0$

Zeros: $x^2 + 4x + 4 = (x+2)(x+2) = 0$

$x = -2$

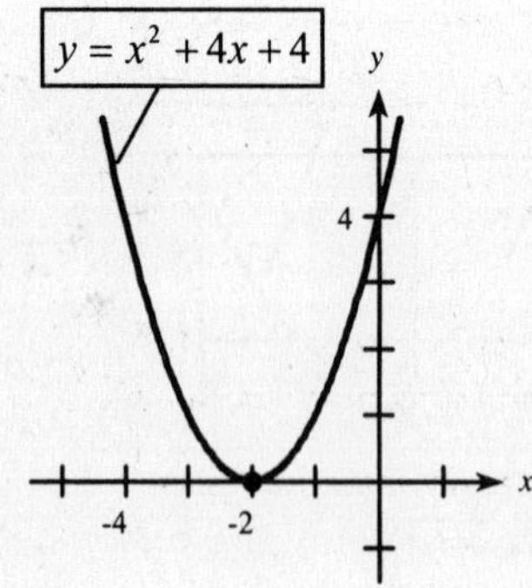

15. $y = \frac{1}{2}x^2 + x - 3$

Vertex is a minimum point since $a > 0$.

V: $x = \frac{-b}{2a} = \frac{-1}{2(1/2)} = -1$

$y = \frac{1}{2}(-1)^2 + (-1) - 3 = -\frac{7}{2}$

Zeros: $\frac{1}{2}x^2 + x - 3 = 0 \rightarrow x^2 + 2x - 6 = 0$

$x = \frac{-2 \pm \sqrt{4+24}}{2} = \frac{-2 \pm 2\sqrt{7}}{2} = -1 \pm \sqrt{7}$

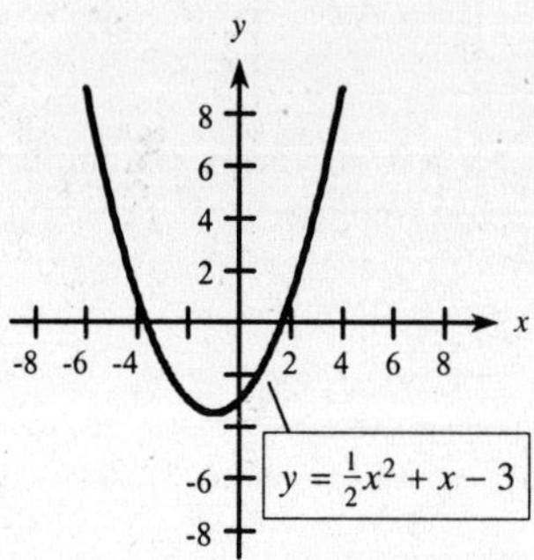

17. $y = (x-3)^2 + 1$

a. Graph is shifted 3 units to the right and 1 unit up.

b.

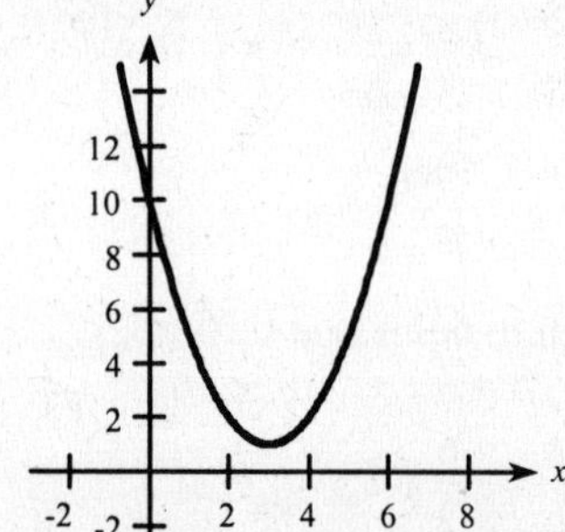

19. $y = (x-10)^2 + 12$

a. Graph is shifted 10 units to the right and 12 units up.

b.

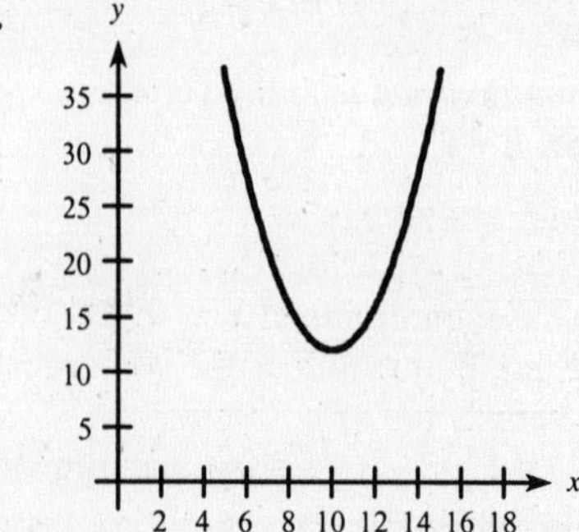

21. $y=\frac{1}{2}x^2-x-\frac{15}{2}$

V: $x=\frac{-b}{2a}=\frac{-(-1)}{2(1/2)}=1$

$y=\frac{1}{2}(1)^2-1-\frac{15}{2}=-8$

Zeros: $x^2-2x-15=(x-5)(x+3)=0$

$x=5,-3$

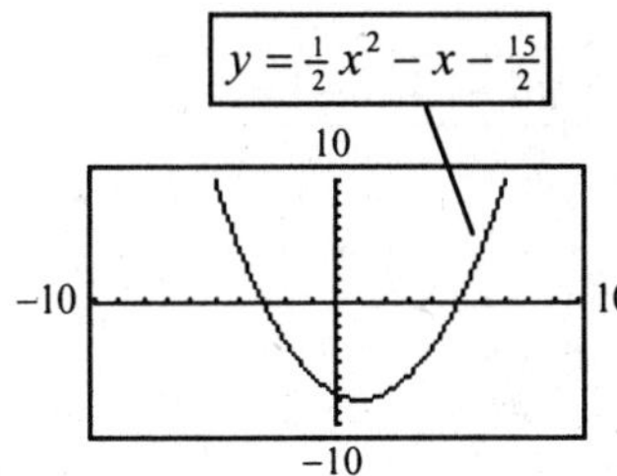

23. $y=-x^2-5x$

V: $x=\frac{-b}{2a}=\frac{-(-5)}{2(-1)}=-\frac{5}{2}$

$y=-\left(-\frac{5}{2}\right)^2-5\left(-\frac{5}{2}\right)=\frac{25}{4}$

Zeros: $-x^2-5x=-x(x+5)=0$

$x=0,-5$

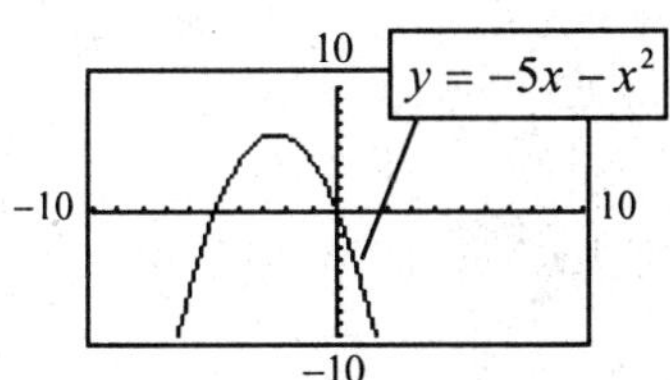

25. $y=\frac{1}{4}x^2+3x+12$

V: $x=\frac{-b}{2a}=\frac{-3}{2\left(\frac{1}{4}\right)}=-6$

$y=\frac{1}{4}(-6)^2+3(-6)+12=3$

Zeros: $x^2+12x+48=0$

$b^2-4ac=144-192<0$

There are no zeros.

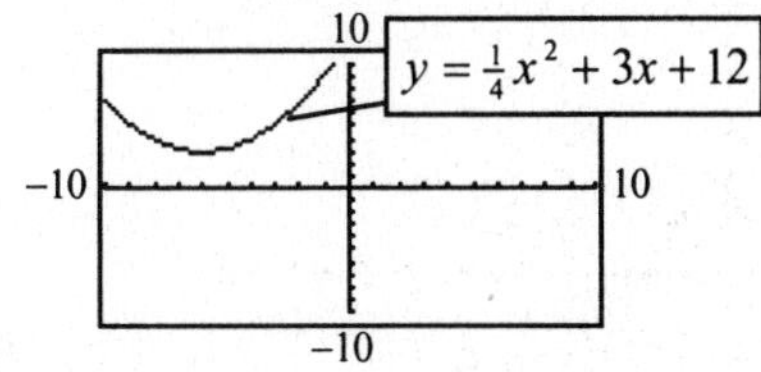

27. $y=63+0.2x-0.01x^2$

V: $x=\frac{-0.2}{-0.02}=10$

$y=63+2-1=64$

Zeros: $x^2-20x-6300=(x-90)(x+70)=0$

$x=90,-70$

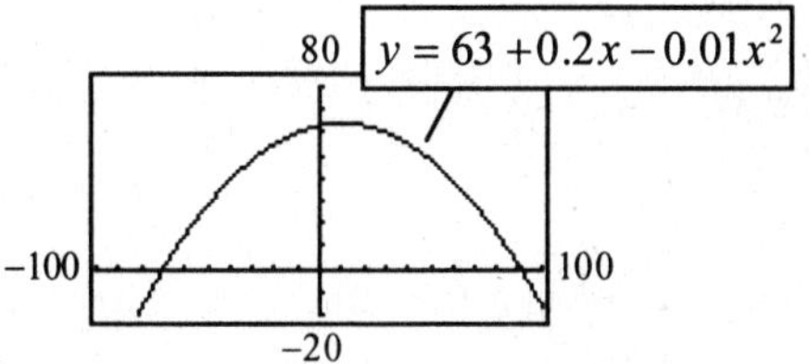

29. $y=0.0001x^2-0.01$

V: $x=\frac{-0}{2(0.0001)}=0$

$y=0-0.01=-0.01$

Zeros: $0.0001x^2-0.01=0.01(0.01x^2-1)=0$

$0.01(0.1x+1)(0.1x-1)=0$

$x=-10,10$

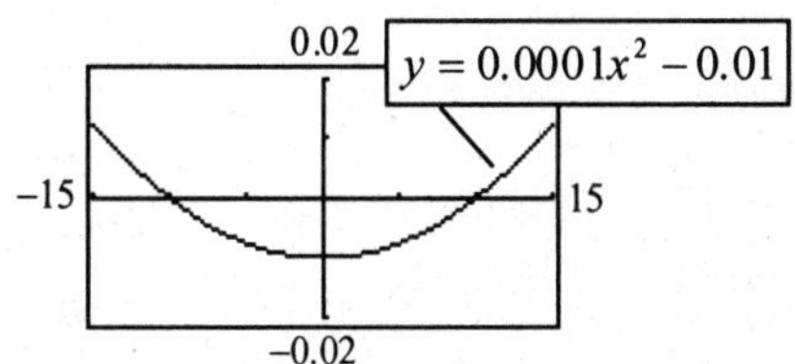

31. $f(x)=8x^2-16x-16$

a. $x=\frac{-b}{2a}=\frac{16}{16}=1$ and $f(1)=-24$

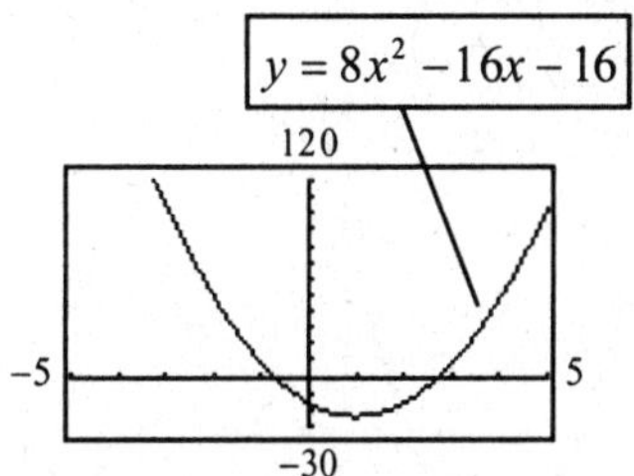

b. Graphical approximation gives $x=-0.73,\ 2.73$

33. $f(x)=3x^2-8x+4$

a. The TRACE gives $x=2$ as a solution.

b. $(x-2)$ is a factor.

c. $3x^2-8x+4=(x-2)(3x-2)$

d. $(x-2)(3x-2)=0$

$x-2=0$ or $3x-2=0$

Solution is $x=2,\ 2/3$.

35. $P = -0.1x^2 + 16x - 100$
The vertex coordinates are the answers to the questions.

a. $a = -0.1, b = 16$

$$x = \frac{-b}{2a} = \frac{-16}{-0.2} = 80$$

Profit is maximized at a production level of 80 units.

b. $P(80) = -0.1(80)^2 + 16(80) - 100 = \$540$ is the maximum profit.

37. $Y = 800x - x^2$
Opens down so maximum Y is at vertex.

$$\text{V: } x = \frac{-800}{-2} = 400$$

Maximum yield occurs at $x = 400$ trees.

39. $S = 1000x - x^2$
Maximum sensitivity occurs at vertex.

$$\text{V: } x = \frac{-1000}{-2} = 500$$

The dosage for maximum sensitivity is 500.

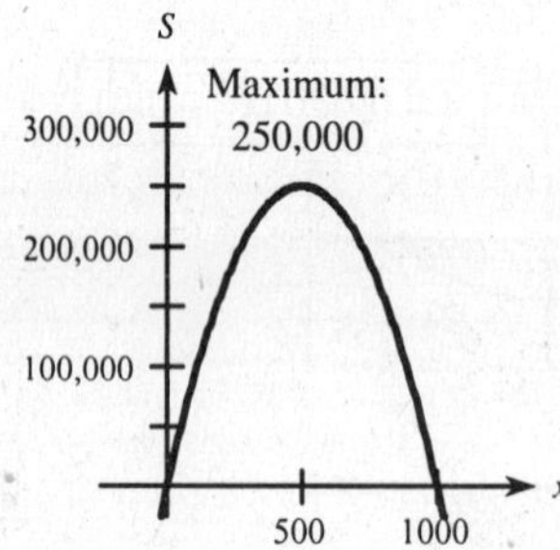

41. $R = 270x - 90x^2$
Maximum rate occurs at vertex.

$$\text{V: } x = \frac{-270}{2(-90)} = \frac{3}{2} \text{ (lumens)}$$

is the intensity for maximum rate.

43. a. $y = -0.0013x^2 + x + 10$

$$\text{V: } x = \frac{-1}{-0.0026} = 384.62;$$

$$y = -0.0013(384.62)^2 + 384.62 + 10 = 202.31$$

b. $y = -\frac{1}{81}x^2 + \frac{4}{3}x + 10$

$$\text{V: } x = \frac{-\frac{4}{3}}{\frac{-2}{81}} = 54;$$

$$y = -\frac{1}{81}(54)^2 + \frac{4}{3}(54) + 10 = 46$$

Projectile **a.** goes $202.31 - 46 = 156.31$ feet higher.

45. a.

No. of Apts	Rent	Total Revenue
50	\$600	\$30,000
49	\$620	\$30,380
48	\$640	\$30,720

b. Revenue increases \$720

c. $R = (50 - x)(600 + 20x)$

d. $R = -20x^2 + 400x + 30{,}000$

R is maximized at $x = \frac{-400}{2(-20)} = 10$.

Rent would be $\$600 + \$200 = \$800$.

47. a. quadratic

b. $a > 0$ since graph opens upward.

49. a.

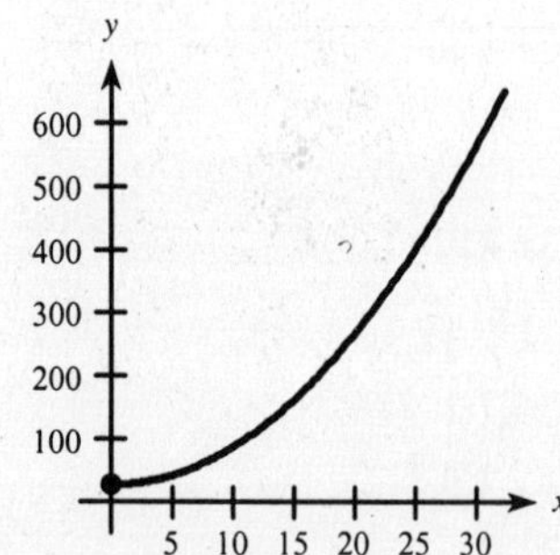

b. The shape appears to be quadratic.

c. The graph is of the form $y = ax^2 + 26.7$.

Using $a \le 1$ gives $y = 0.78x^2 + 26.7$.

d. $y = 0.78(40)^2 + 26.7 = \$1275$ million

51.

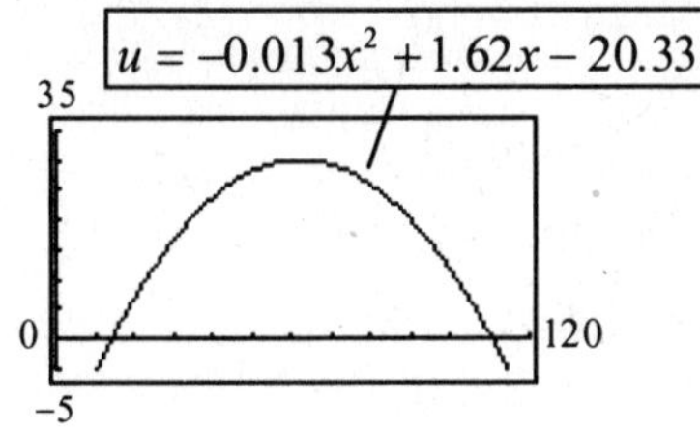

53. **a.** 1914 and 2010
b. 2011 (% is negative).

Exercise 2.3

1. **a.** Supply: $p = \frac{1}{4}q^2 + 10$ (see below)

b. Demand: $p = 86 - 6q - 3q^2$ (see below)

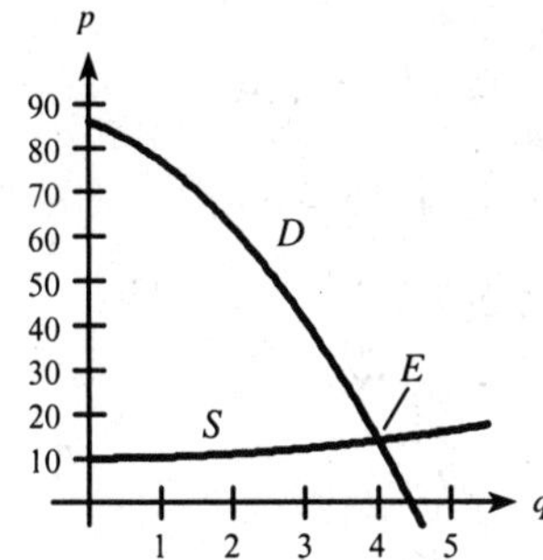

c. See E on graph.

d. $\frac{1}{4}q^2 + 10 = 86 - 6q - 3q^2$

$$q^2 + 40 = 344 - 24q - 12q^2$$
$$0 = 13q^2 + 24q - 304$$
$$0 = (q-4)(13q+76)$$

$q = 4$ must be positive.

$p = \frac{1}{4}(4)^2 + 10 = 14$

E: (4, 14)

3. **a.** Supply: $p = 0.2q^2 + 0.4q + 1.8$ (see below)

b. Demand: $p = 9 - 0.2q - 0.1q^2$ (see below)

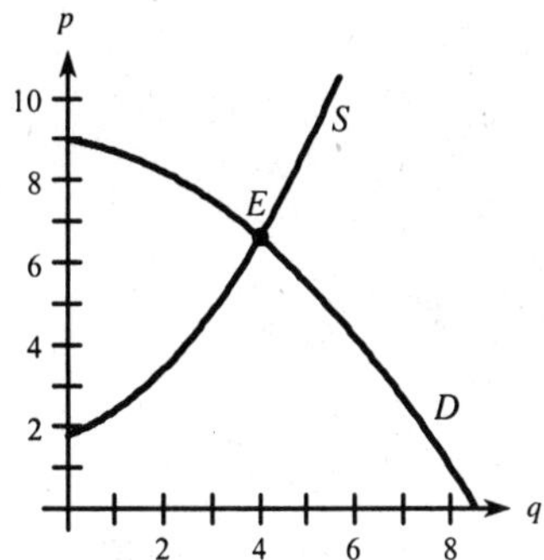

c. See E on graph.

d. $0.2q^2 + 0.4q + 1.8 = 9 - 0.2q - 0.1q^2$

$$0.3q^2 + 0.6q - 7.2 = 0$$
$$q^2 + 2q - 24 = 0$$
$$(q+6)(q-4) = 0$$

$q = 4$ positive root only

$p = 0.2(4)^2 + 0.4(4) + 1.8 = 6.60$

E: (4, 6.60)

5. $p = q^2 + 8q + 16$

$p = -3q^2 + 6q + 436$

$$q^2 + 8q + 16 = -3q^2 + 6q + 436$$
$$4q^2 + 2q - 420 = 0$$
$$2q^2 + q - 210 = 0$$
$$(2q+21)(q-10) = 0$$
$$q = 10$$

$p = 10^2 + 8(10) + 16 = 196$

E: (10, 196)

7. $p^2 + 4q = 1600$

$300 - p^2 + 2q = 0$

$(300 + 2q) + 4q = 1600$

$6q = 1300$

$q = 216\frac{2}{3}$

$p^2 + 4\left(\frac{1300}{6}\right) = 1600 \text{ or } p^2 = 733.33 \text{ or } p = 27.08$

E: $\left(216\frac{2}{3}, 27.08\right)$

9. $q = p - 10$

$q = \frac{1200}{p}$

$\frac{1200}{p} = p - 10$

$p^2 - 10p - 1200 = 0$

$(p - 40)(p + 30) = 0$

$p = 40$

$q = 40 - 10 = 30$

E: (30, 40)

11. $2p - q - 10 = 0$

$(p + 10)(q + 30) = 7200$

So, $(p + 10)(2p - 10 + 30) = 7200$

$p^2 + 20p + 100 = 3600$

$p^2 + 20p - 3500 = 0$

$(p + 70)(p - 50) = 0$

$p = 50$

$q = 2(50) - 10 = 90$

E: $(q, p) = (90, 50)$

13. $p = \frac{1}{2}q + 5 + 22 = \frac{1}{2}q + 27$

So, $\left(\frac{1}{2}q + 27 + 10\right)(q + 30) = 7200$

$(q + 74)(q + 30) = 14,400$

$q^2 + 104q - 12,180 = 0$

$(q + 174)(q - 70) = 0$

$p = \frac{1}{2}(70) + 27 = 62$

E: (70, 62)

15. $C(x) = x^2 + 40x + 2000$

$R(x) = 130x$

$x^2 + 40x + 2000 = 130x$

$x^2 - 90x + 2000 = 0$

$(x - 40)(x - 50) = 0$

$x = 40$ or $x = 50$

Break-even values are at $x = 40$ and 50 units.

17. $C(x) = 15,000 + 35x + 0.1x^2$

$R(x) = 385x - 0.9x^2$

$15,000 + 35x + 0.1x^2 = 385x - 0.9x^2$

$x^2 - 350x + 15,000 = 0$

$(x - 300)(x - 50) = 0$

$x = 300$ or $x = 50$

Break-even values are at $x = 50$ and 300 units.

19. $C(x) = 150 + x + 0.09x^2$

$R(x) = 12.5x - 0.01x^2$

$x < 75$

$150 + x + 0.09x^2 = 12.5x - 0.01x^2$

$0.1x^2 - 11.5x + 150 = 0$

$x^2 - 115x + 1500 = 0$

$(x - 100)(x - 15) = 0$

$x = 100$ or $x = 15$

Break-even value is at $x = 15$ units since we were given that $x < 75$.

21. $R(x) = 385x - 0.9x^2$

$a = -0.9, b = 385$

Maximum revenue is at the vertex.

V: $x = \frac{-385}{-1.8} = 213.89$ or 214 total units

$R(214) = 385(214) - 0.9(214)^2 = \$41,173.60$

23. $R(x) = x(175 - 0.50x) = 175x - 0.5x^2$

$a = -0.50, b = 175$

Revenue is a maximum at $x = \frac{-175}{-1} = 175$.

Price that will maximize revenue is $p = 175 - 87.50 = \$87.50$.

25. $P(x) = -x^2 + 110x - 1000$

Maximum profit is at the vertex or when

$x = \frac{-110}{-2} = 55$.

$P(55) = \$2025$.

27. $R(x) = 385x - 0.9x^2$

$C(x) = 15,000 + 35x + 0.1x^2$

a. $$P(x) = 385x - 0.9x^2 - (15,000 + 35x + 0.1x^2)$$
$$= -x^2 + 350x - 15,000$$

At the vertex we have $x = \dfrac{-350}{-2} = 175.$

So, $P(175) = \$15,625$.

b. No. More units are required to maximize revenue.

c. The break-even values and zeros of $P(x)$ are the same.

29. a. $C(x) = 28,000 + \left(\dfrac{2}{5}x + 222\right)x = \dfrac{2}{5}x^2 + 222x + 28,000$

$R(x) = \left(1250 - \dfrac{3}{5}x\right)x = 1250x - \dfrac{3}{5}x^2$

(The key is "per unit x.")

$$R(x) = C(x)$$
$$1250x - \frac{3}{5}x^2 = \frac{2}{5}x^2 + 222x + 28,000$$

$x^2 - 1028x + 28,000 = 0$

$(x - 1000)(x - 28) = 0$

Break-even values are at $x = 28$ and $x = 1000$.

b. Maximum revenue occurs at $x = \dfrac{-1250}{-\frac{6}{5}} = 1042$ (rounded).

$R(1042) = \$651,041.60$ is the maximum revenue.

c. $P(x) = 1250x - \dfrac{3}{5}x^2 - \left(\dfrac{2}{5}x^2 + 222x + 28,000\right) = -x^2 + 1028x - 28,000$

Maximum profit is at $x = \dfrac{-1028}{-2} = 514.$ $P(514) = \$236,196$ is the maximum profit.

d. Price that will maximize profit is $p = 1250 - \dfrac{3}{5}(514) = \$941.60$.

31. $R(t) = 0.253t^2 - 4.03t + 76.84$

a. Minimum occurs at vertex where $t = \dfrac{-(-4.03)}{2(0.253)} = 7.96 \approx 8$. The minimum revenue is in 1995.

$R(8) = \$60.79$ million.

b. The data shows that the minimum revenue was in 1994.

c.

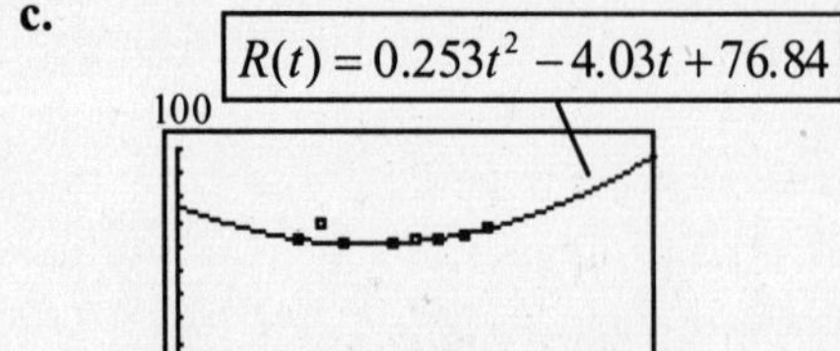

d. Except for 1993, the model and the data are a good fit.

33. a. $P = R - C = (-0.031t^2 + 0.776t + 0.179) - (-0.012t^2 + 0.492t + 0.725) = -0.019t^2 + 0.284t - 0.546$

b. V: $t = \dfrac{-0.284}{-0.038} = 7.47 \approx 7$ years or 1994

c.

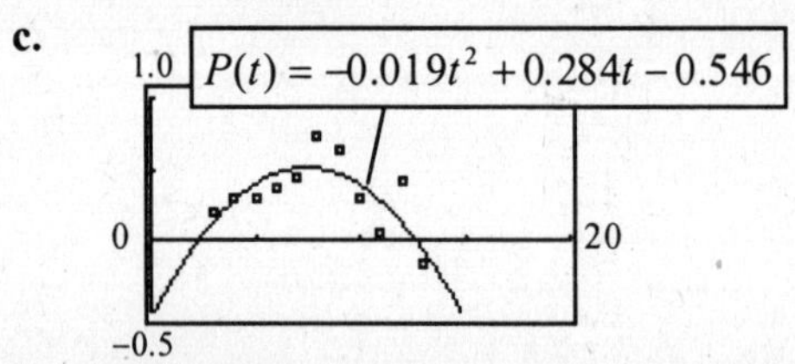

d. The model projects decreasing profits. Except for 1999, the data supports the model.

e. Revenue must be increased and/or costs must be decreased.

Exercise 2.4

Problems 1–27 Helpful methods:

- **a.** Set $x = 0$.
- **b.** Set $y = 0$.
- **c.** As the denominator gets close to zero does y increase without bound?
- **d.** As x gets large what happens to y?

1. l

3. k

5. b

7. f

9. g

11. j

13. i

15. cubic

17. quartic

19. $y = x^3 - x = x(x+1)(x-1)$: j

21. $y = 16x^2 - x^4 = x^2(4+x)(4-x)$: g

23. $y = x^2 + 7x = x(x+7)$: a

25. $y = x^4 - 3x^2 - 4 = (x^2 - 4)(x^2 + 1)$: f

27. $y = \dfrac{x-3}{x+1}$: d

29. $F(x) = \dfrac{x^2 - 1}{x}$

- **a.** $F\left(-\dfrac{1}{3}\right) = \dfrac{\frac{1}{9} - 1}{-\frac{1}{3}} = \dfrac{8}{3}$
- **b.** $F(10) = \dfrac{100 - 1}{10} = \dfrac{99}{10}$
- **c.** $F(0.001) = \dfrac{0.000001 - 1}{0.001} = \dfrac{-0.999999}{0.001} = -999.999$
- **d.** $F(0)$ is not defined–division by zero.

31. $f(x) = x^{3/2}$

- **a.** $f(16) = (\sqrt{16})^3 = 64$
- **b.** $f(1) = (\sqrt{1})^3 = 1$
- **c.** $f(100) = (\sqrt{100})^3 = 1000$
- **d.** $f(0.09) = (\sqrt{0.09})^3 = 0.027$

33. $k(x) = \begin{cases} 2 & \text{if } x < 0 \\ x + 4 & \text{if } 0 \le x < 1 \\ 1 - x & \text{if } x \ge 1 \end{cases}$

- **a.** $k(-5) = 2$ since $x < 0$.
- **b.** $k(0) = 0 + 4 = 4$ since $x = 0$.
- **c.** $k(1) = 1 - 1 = 0$ since $x = 1$.
- **d.** $k(-0.001) = 2$ since $x < 0$.

35. $y = 1.6x^2 - 0.1x^4$

a.

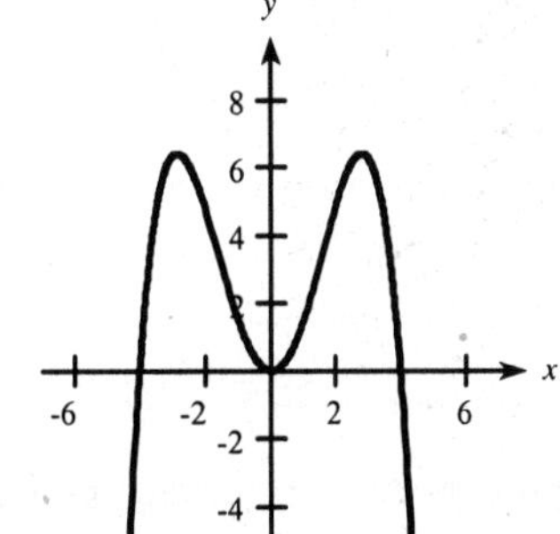

- **b.** polynomial
- **c.** no asymptotes
- **d.** turning points at $x = 0$ and approximately $x = -2.8$ and $x = 2.8$

37. $y = \dfrac{2x+4}{x+1}$

a.

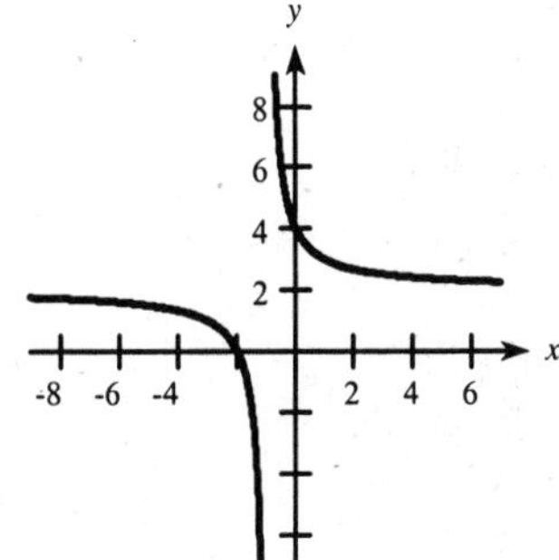

- **b.** rational
- **c.** vertical: $x = -1$
  horizontal: $y = 2$
- **d.** no turning points

39. $f(x) = \begin{cases} -x & \text{if } x < 0 \\ 5x & \text{if } x \geq 0 \end{cases}$

a.

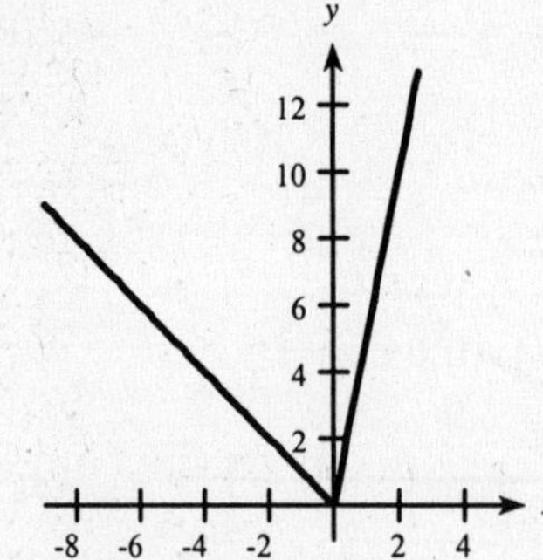

b. piecewise
c. no asymptotes
d. turning point at $x = 0$.

41. $V = V(x) = x^2(108 - 4x)$

a. $V(10) = 100(68) = 6800$ cubic inches
$V(20) = 400(28) = 11{,}200$ cubic inches

b. $108 - 4x > 0$

$-4x > -108$

$0 < x < 27$

43. $f(x) = 105.095x^{1.5307}$

a. This is a power function graph turning upward.

b.

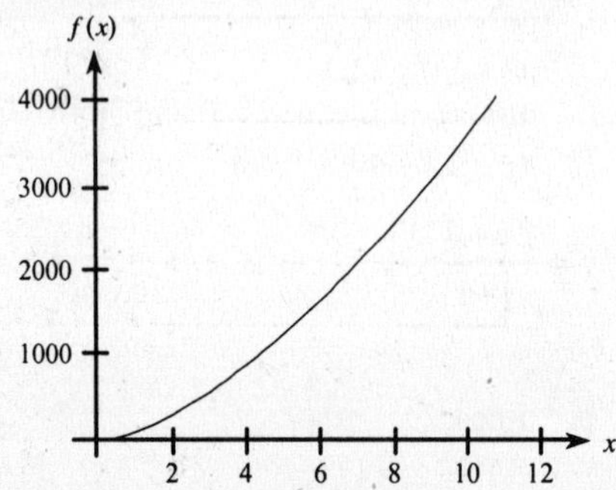

a. From the graph, assets reached \$4000 billion in 10.7 years, or late 2000.

45. $C(p) = \dfrac{7300p}{100 - p}$

a. $0 \leq p < 100$

b. $C(45) = \dfrac{7300 \cdot 45}{100 - 45} = \$5972.73$

c. $C(90) = \dfrac{7300 \cdot 90}{100 - 90} = \$65{,}700$

d. $C(99) = \dfrac{7300 \cdot 99}{100 - 99} = \$722{,}700$

e. $C(99.6) = \dfrac{7300(99.6)}{100 - 99.6} = \$1{,}817{,}700$

f. To remove p% of the pollution would cost $C(p)$. Note how cost increases as p (the percent of pollution removed) increases.

47. $A = A(x) = x(50 - x)$

a. $A(2) = 2 \cdot 48 = 96$ square feet
$A(30) = 30 \cdot 20 = 600$ square feet

b. $0 < x < 50$ in order to have a rectangle.

49. a. $P(t) = \begin{cases} 1.965t - 5.65 & 5 \leq t \leq 20 \\ 0.095t^2 - 2.925t + 54.15 & t > 20 \end{cases}$

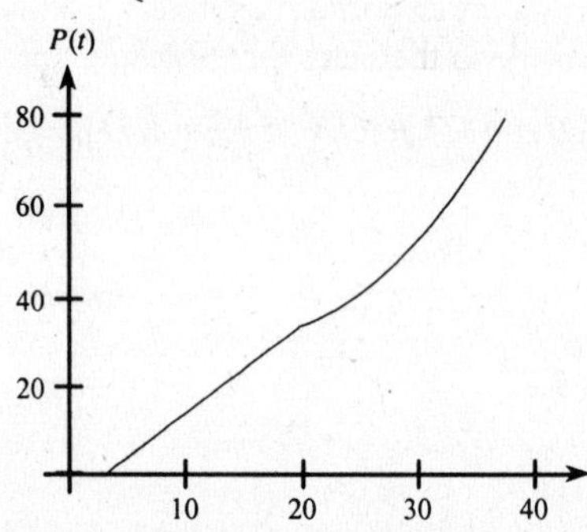

b. From the graph, $P(20) = \$33.65$ billion .

c. From the graph, $P(38) = \$80.18$ billion (approx.)

51. a. $P(w) = \begin{cases} 0.37 & \text{if } 0 < x \leq 1 \\ 0.60 & \text{if } 1 < x \leq 2 \\ 0.83 & \text{if } 2 < x \leq 3 \\ 1.06 & \text{if } 3 < x \leq 4 \end{cases}$

b. The postage for a 1.2 ounce letter is \$0.60.

c. $D = \{x : 0 < x \leq 4\}$,
$R = \{0.37, 0.60, 0.83, 1.06\}$

d. 2 ounces cost is \$0.60, 2.01 ounces cost is \$0.83

53. $p = \dfrac{200}{2 + 0.1x}$

a.

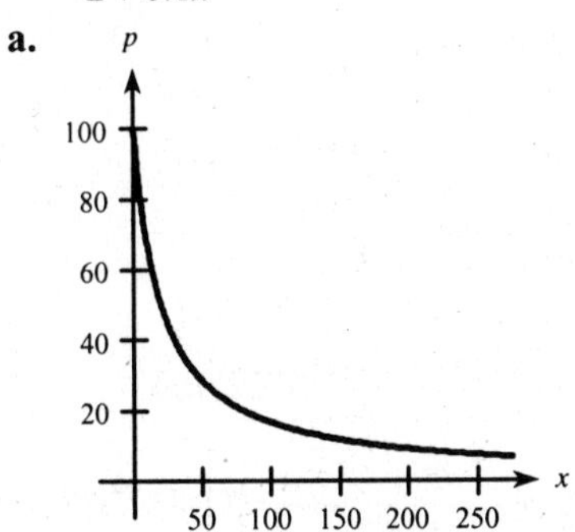

b. No.

55. $C(x) = 30(x-1) + \dfrac{3000}{x+10}$

a.

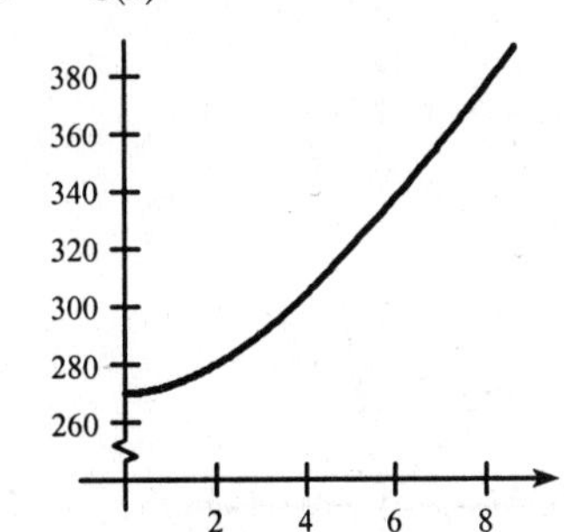

b. A turning point indicates a minimum or maximum cost.

c. This is the fixed cost of production.

Exercise 2.5

1. linear The points are in a straight line.

3. quadratic The points appear to fit a parabola.

5. quartic The graph crosses the x-axis four times. Also there are three bends.

7. quadratic There is one bend. A parabola is the best fit.

9. $y = 2x - 3$ is the best fit.

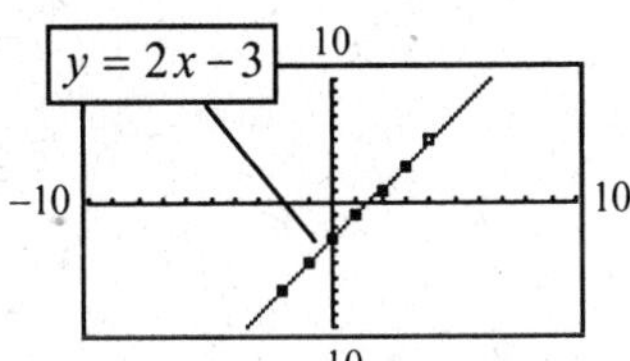

11. $y = 2x^2 - 1.5x - 4$ is the best fit.

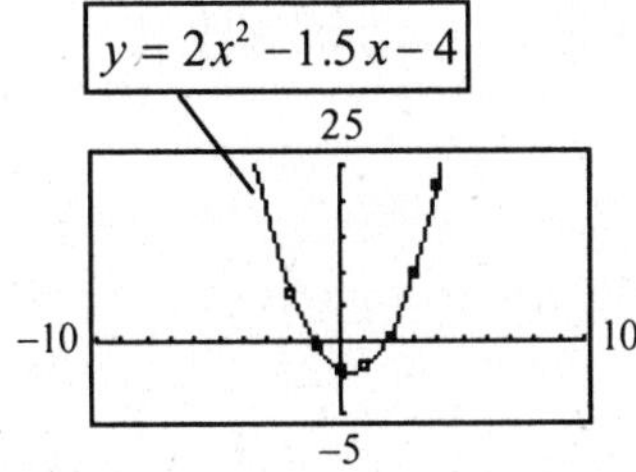

13. $y = x^3 - x^2 - 3x - 4$ is the best fit.

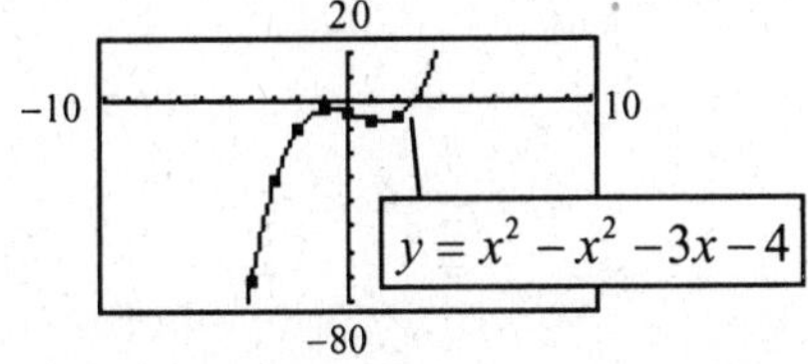

15. $y = 2x^{1/2}$ is the best fit.

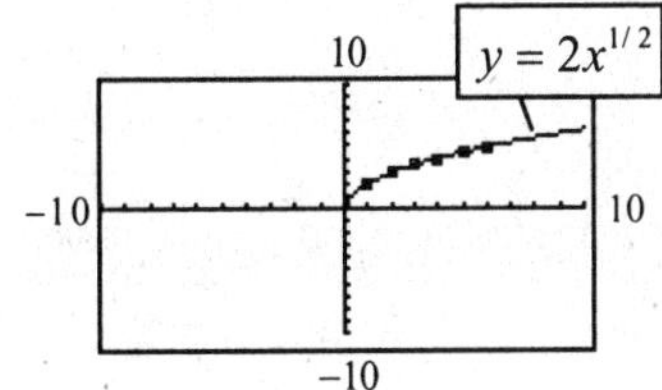

17. a.

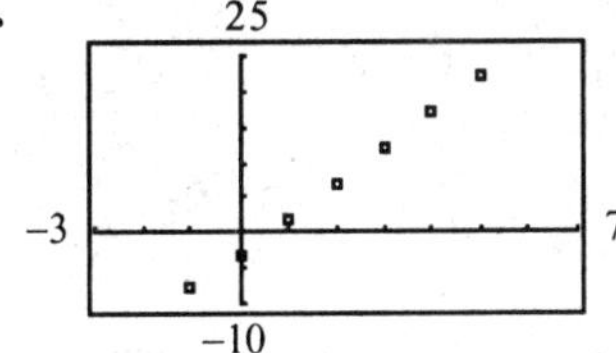

b. linear

c. $y = 5x - 3$

19. a.

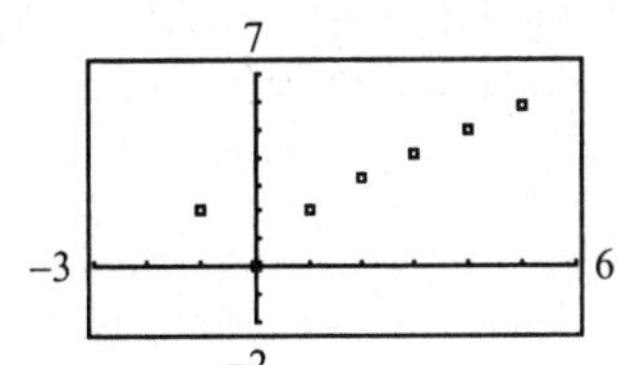

b. quadratic

c. $y = 0.0959x^2 + 0.4656x + 1.4758$

21. a.

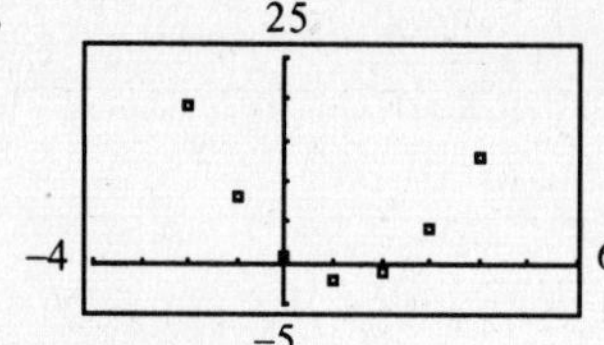

b. quadratic

c. $y = 2x^2 - 5x + 1$

23. a.

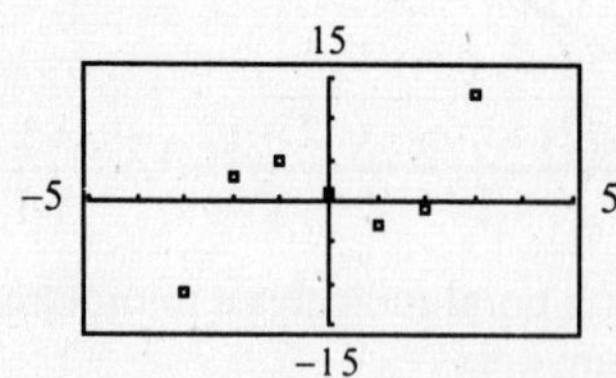

b. cubic

c. $y = x^3 - 5x + 1$

25. a.

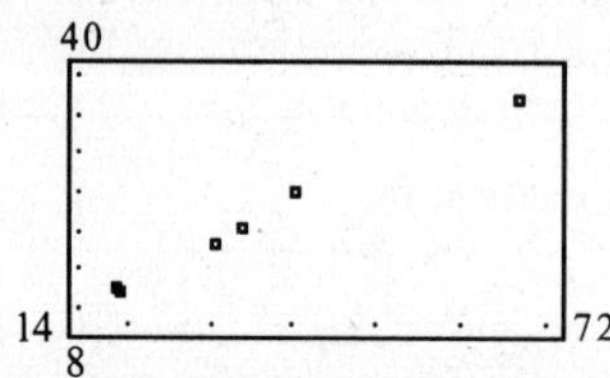

b. $y = 0.518x + 2.775$

c. $m = 0.518$; for each $1000 of annual earnings for males, females earned $518 per year.

27. a. $y = 0.5807x^2 - 8.867x + 49.514$

b. $x = 7.64 \approx 8$; the year for $x = 8$ (1988) has the minimum for the data.

29. a.

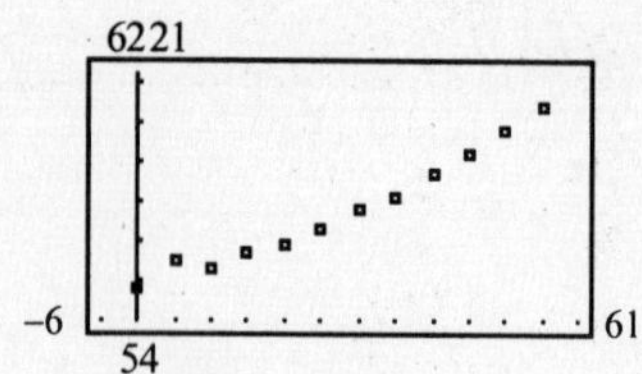

b. $y = 80.309x + 624.5$

c. $y = 0.875x^2 + 32.161x + 1025.74$

d.

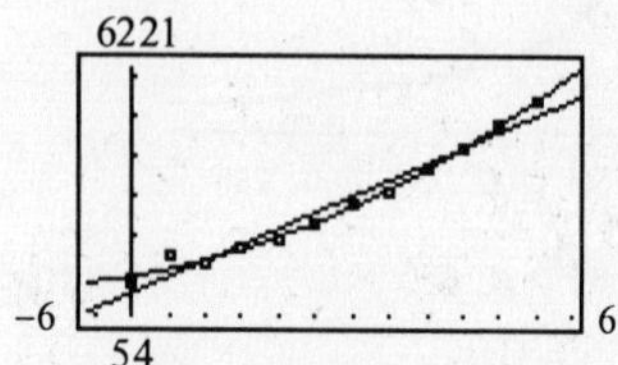

The quadratic model is the better fit.

31. a. $y_1 = 5.362x^2 - 5.839x - 262.5$,

$y_2 = -0.7266x^3 + 17.35x^2 - 56.34x - 226.5$

b. The cubic model is a slightly better fit.
2001: cubic: $y \approx 286.01$
quadratic: $y \approx 322.07$
2002: cubic: $y \approx 340.26$
quadratic: $y \approx 439.56$

c. cubic: $x \approx 7.25$, quadratic: $x \approx 7.56$
Both predict a zero deficit in 1998.

33. a. $y = 154.13x^{-0.4919}$

b. 26.8%

c. 22.5%

d. It is unreliable. Sept. 11, 2001 can change how people feel about government.

35. a.

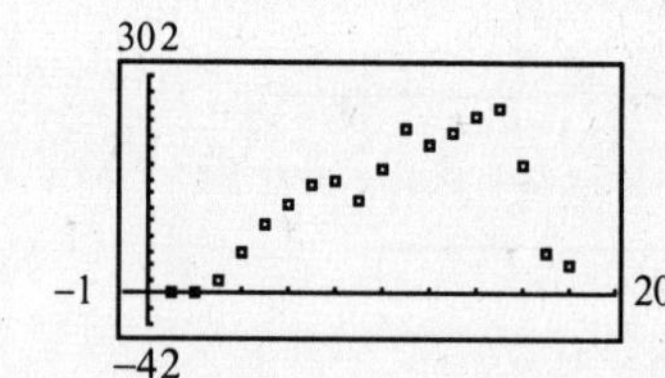

b. $y = -0.278x^3 + 5.543x^2 - 7.898x + 9.101$

c. $x \approx 12.5$ This would occur during 1993.

Review Exercises

1. $3x^2 + 10x = 5x$

$3x^2 + 5x = 0$

$x(3x + 5) = 0$

$x = 0 \text{ or } x = -\frac{5}{3}$

2. $4x - 3x^2 = 0$

$x(4 - 3x) = 0$

$x = 0 \text{ or } x = \frac{4}{3}$

3. $x^2 + 5x + 6 = 0$

$(x + 3)(x + 2) = 0$

$x = -3 \text{ or } x = -2$

4. $11 - 10x - 2x^2 = 0$

$a = -2,\ b = -10,\ c = 11$

$x = \frac{10 \pm \sqrt{100 + 88}}{-4} = \frac{-5 \pm \sqrt{47}}{2}$

5. $(x - 1)(x + 3) = -8$

$x^2 + 2x - 3 = -8$

$x^2 + 2x + 5 = 0$

$b^2 - 4ac < 0$

No real solution

6. $4x^2 = 3$

$x^2 = \frac{3}{4}$

$x = \pm\sqrt{\frac{3}{4}} = \pm\frac{\sqrt{3}}{2}$

7. $20x^2 + 3x = 20 - 15x^2$

$35x^2 + 3x - 20 = 0$

$(7x - 5)(5x + 4) = 0$

$x = \frac{5}{7} \text{ or } x = -\frac{4}{5}$

8. $8x^2 + 8x = 1 - 8x^2$

$16x^2 + 8x - 1 = 0$

$a = 16,\ b = 8,\ c = -1$

$x = \frac{-8 \pm \sqrt{64 + 64}}{32} = \frac{-1 \pm \sqrt{2}}{4}$

9. $0.02x^2 - 2.07x + 7 = 0$

$a = 0.02,\ b = -2.07,\ c = 7$

$x = \frac{2.07 \pm \sqrt{4.2849 - 0.56}}{0.04} = \frac{2.07 \pm 1.93}{0.04}$

$= 100 \text{ or } 3.5$

10. $46.3x - 117 - 0.5x^2 = 0$

$a = -0.5,\ b = 46.3,\ c = -117$

$x = \frac{-46.3 \pm \sqrt{2143.69 + (-234)}}{-1} = \frac{-46.3 \pm 43.7}{-1}$

$= 90 \text{ or } 2.6$

11. $4z^2 + 25 = 0$

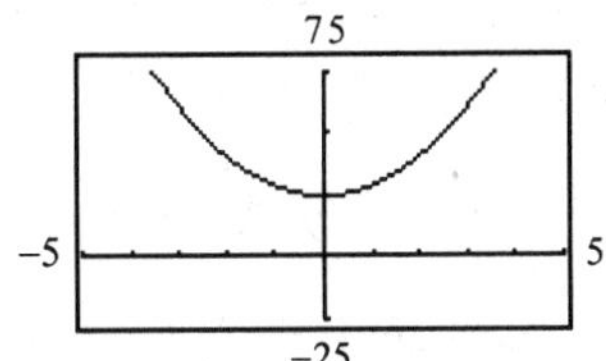

$4z^2 + 5^2 = 0$

The sum of 2 squares cannot be factored. There are no real solutions.

12. $f(z) = z^2 + 6z - 27$

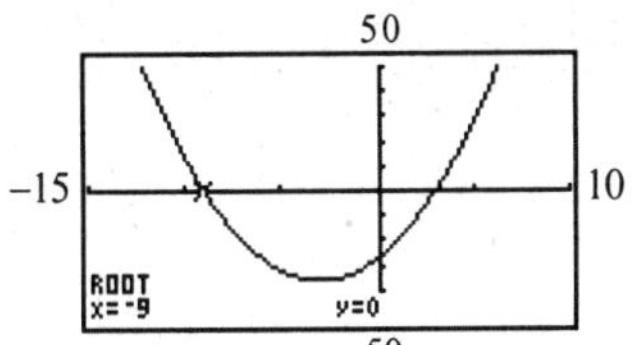

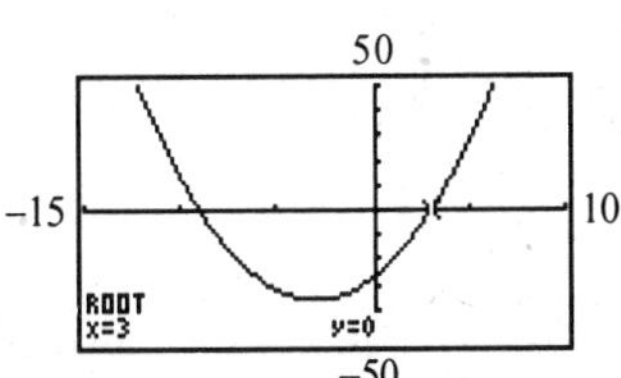

From the graph, the zeros are −9 and 3.

Algebraic solution:

$z(z + 6) = 27$

$z^2 + 6z - 27 = 0$

$(z + 9)(z - 3) = 0$

$z = -9 \text{ or } z = 3$

13. $3x^2 - 18x - 48 = 0$

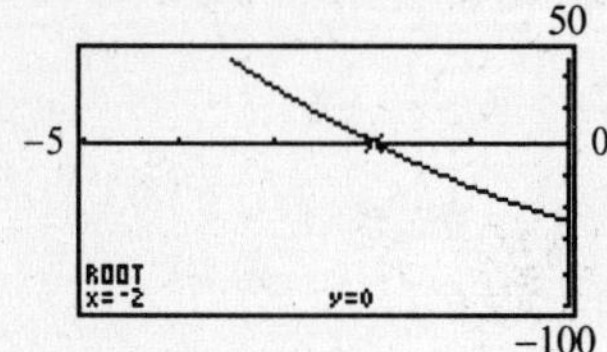

$3(x^2 - 6x - 16) = 0$

$3(x - 8)(x + 2) = 0$

$x = -2, x = 8$

14. $f(x) = 3x^2 - 6x - 9$

$3x^2 - 6x - 9 = 0$

$3(x^2 - 2x - 3) = 0$

$3(x - 3)(x + 1) = 0$

$x = 3,\ x = -1$

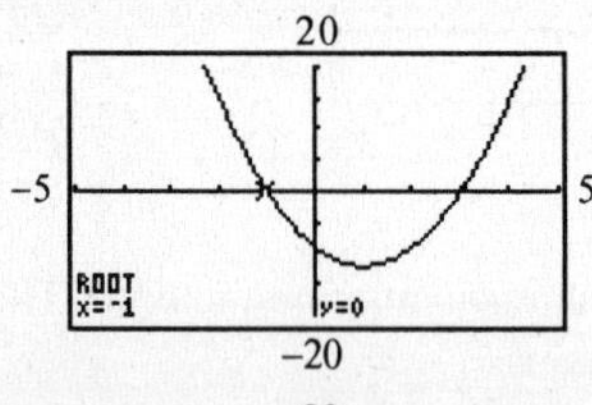

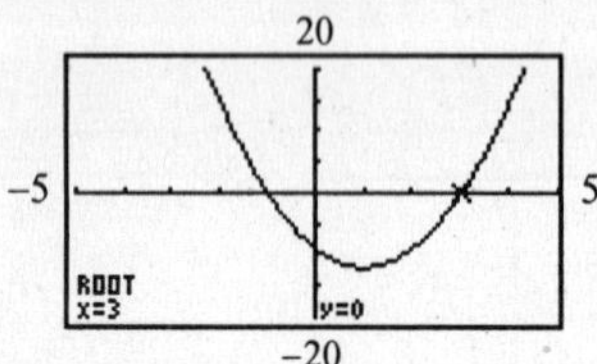

15. $x^2 + ax + b = 0$
To apply the quadratic formula we have "a" = 1, "b" = a, and "c" = b.

$$x = \frac{-a \pm \sqrt{a^2 - 4b}}{2}$$

16. $xr^2 - 4ar - x^2c = 0$
To solve for r, use the quadratic formula with "a" = x, "b" = $-4a$, and "c" = $-x^2c$.

$$r = \frac{4a \pm \sqrt{16a^2 + 4x(x^2c)}}{2x} = \frac{4a \pm \sqrt{16a^2 + 4x^3c}}{2x}$$

$$= \frac{4a \pm 2\sqrt{4a^2 + x^3c}}{2x} = \frac{2a \pm \sqrt{4a^2 + x^3c}}{x}$$

17. $-0.002x^2 - 14.1x + 23.1 = 0$

$$x = \frac{14.1 \pm \sqrt{198.81 + 0.1848}}{-0.004} = \frac{14.1 \pm 14.107}{-0.004}$$

$= -7051.64,\ 1.64,$ or 1.75 (using 14.107)

18. $1.03x^2 + 2.02x - 1.015 = 0$
$a = 1.03,\ b = 2.02,\ c = -1.015$

$$x = \frac{-2.02 \pm \sqrt{4.0804 + 4.1818}}{2.06} = \frac{-2.02 \pm 2.87}{2.06}$$

$= -2.38$ or 0.41

19. $y = \frac{1}{2}x^2 + 2x$
$a > 0$, thus vertex is a minimum.

V: $x = \dfrac{-2}{2\left(\frac{1}{2}\right)} = -2$

$y = \frac{1}{2}(-2)^2 + 2(-2) = -2$

Zeros: $\frac{1}{2}x^2 + 2x = 0$

$x\left(\frac{1}{2}x + 2\right) = 0$

$x = 0, -4$

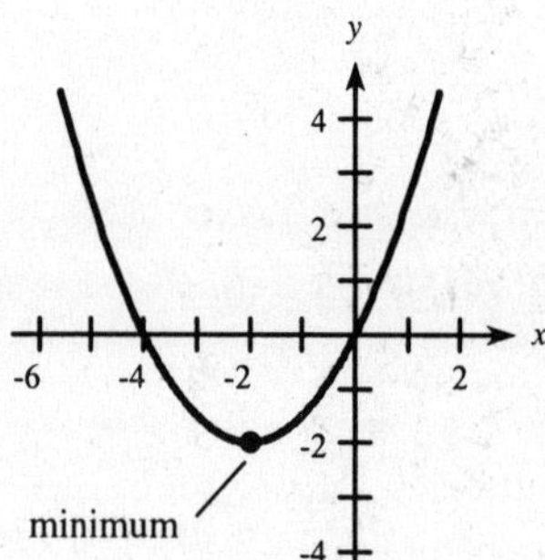

20. $y = 4 - \frac{1}{4}x^2$
V: x-coordinate = 0
y-coordinate = 4
(0, 4) is a maximum point
Zeros are $x = \pm 4$.

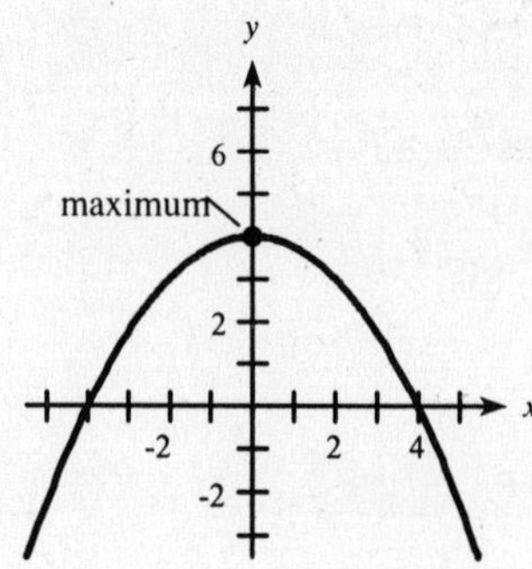

21. $y = 6 + x - x^2$

$a < 0$, thus vertex is a maximum.

V: $x = \frac{-1}{2(-1)} = \frac{1}{2}$

$y = 6 + \frac{1}{2} - \left(\frac{1}{2}\right)^2 = \frac{25}{4}$

Zeros: $6 + x - x^2 = 0$

$(3 - x)(2 + x) = 0$

$x = -2, 3$

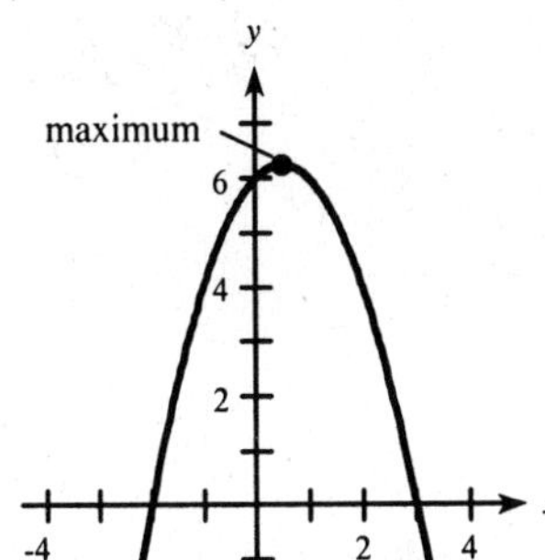

22. $y = x^2 - 4x + 5$

V: x-coordinate $= \frac{4}{2} = 2$

y-coordinate $= 2^2 - 4(2) + 5 = 1$

(2, 1) is a minimum point.

Zeros: Since the minimum point is above the x-axis, there are no zeros.

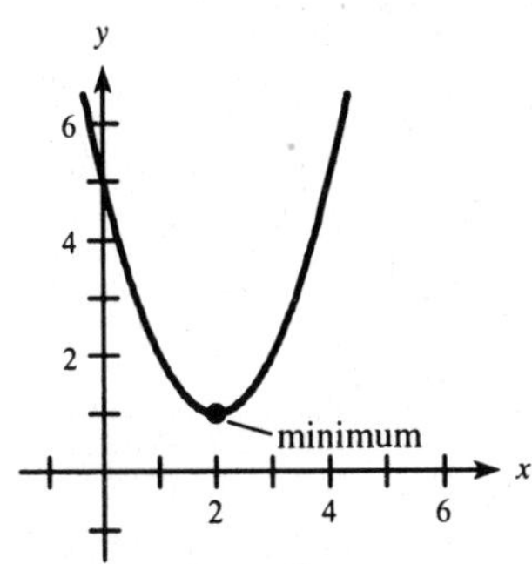

23. $y = x^2 + 6x + 9$

$a > 0$, thus vertex is a minimum.

V: $x = \frac{-6}{2(1)} = -3$

$y = (-3)^2 + 6(-3) + 9 = 0$

Zeros: $x^2 + 6x + 9 = 0$

$(x + 3)(x + 3) = 0$

$x = -3$

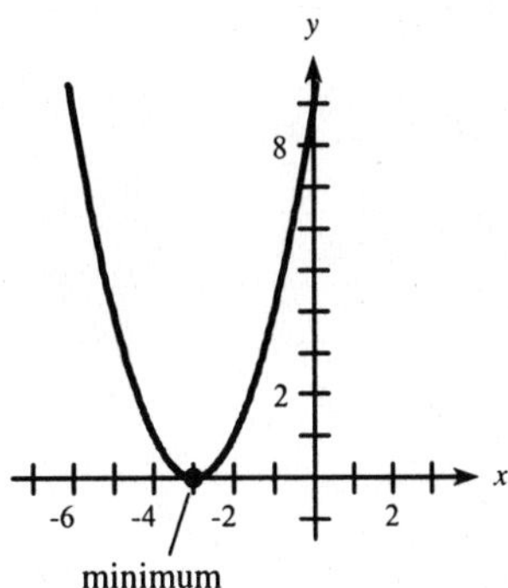

24. $y = 12x - 9 - 4x^2$

V: x-coordinate $= -\frac{12}{-8} = \frac{3}{2}$

y-coordinate $= 12\left(\frac{3}{2}\right) - 9 - 4\left(\frac{3}{2}\right)^2 = 0$

$\left(\frac{3}{2}, 0\right)$ is a maximum point.

Zeros: From the vertex we have that $x = \frac{3}{2}$ is the only zero.

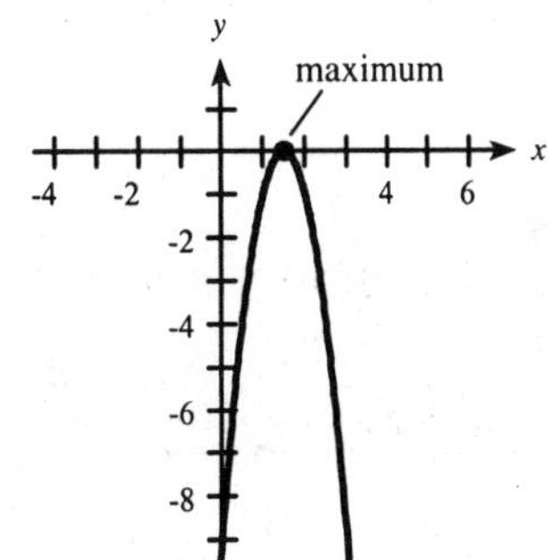

25. $y = \frac{1}{3}x^2 - 3$

V: (0, –3)

Zeros: $\frac{1}{3}x^2 - 3 = 0$

$x^2 = 9$

$x = \pm 3$

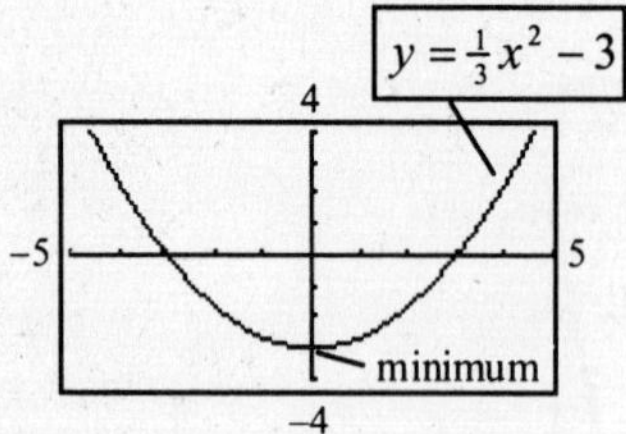

26. $y = \frac{1}{2}x^2 + 2$

Vertex: (0, 2) ← minimum

No zeros.

The graph using x-min = –4 y-min = 0
x-max = 4 y-max = 6

is shown below.

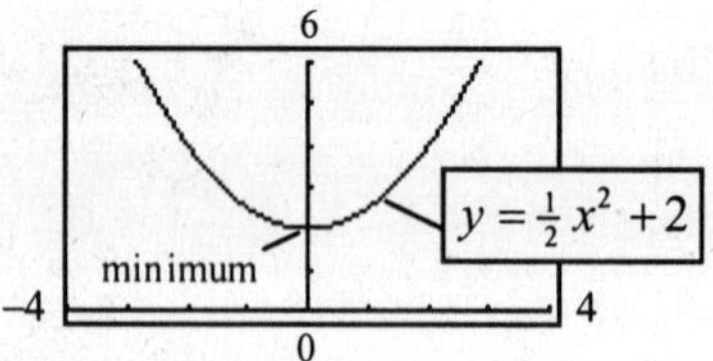

27. $y = x^2 + 2x + 5$

V: (–1, 4)

There are no real zeros.

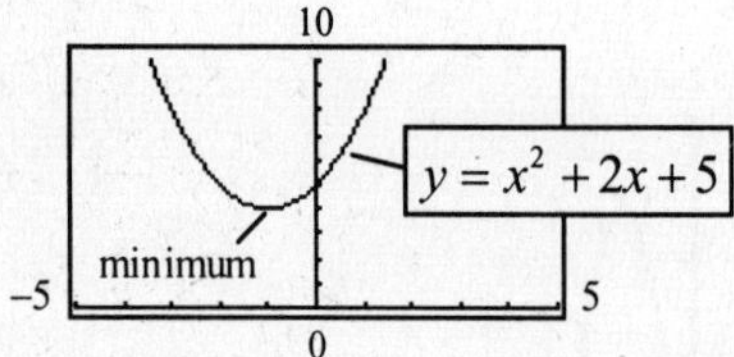

28. $y = -10 + 7x - x^2$

Vertex: $\left(\frac{7}{2}, \frac{9}{4}\right)$ ← maximum

Zeros: $x^2 - 7x + 10 = 0$

$(x - 5)(x - 2) = 0$

$x = 5$ or $x = 2$

Graph using x-min = 0 y-min = –5
x-max = 8 y-max = 5

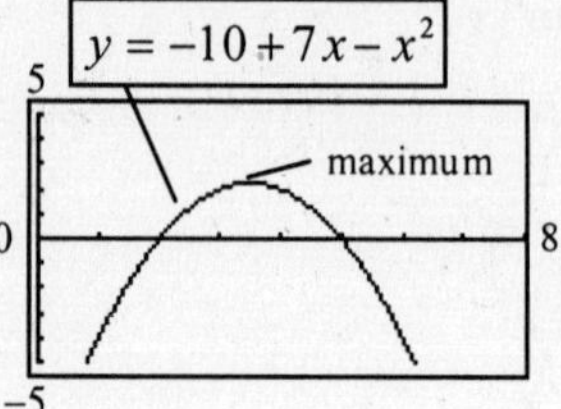

29. $y = 20x - 0.1x^2$

Zeros: $x(20 - 0.1x) = 0$

$x = 0, 200$

(This is an alternative method of getting the vertex.)

The x-coordinate of the vertex is halfway between the zeros.

V: (100, 1000)

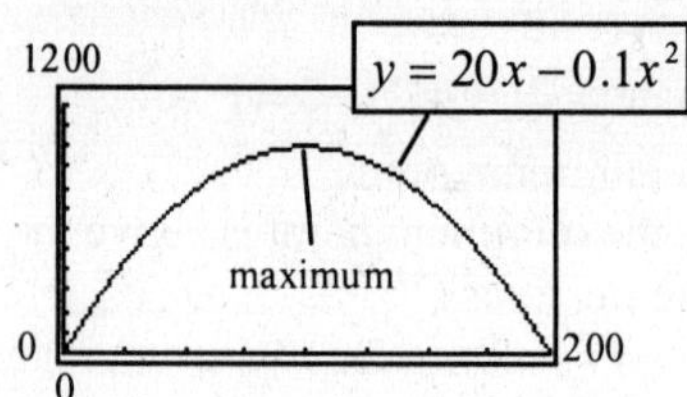

30. $y = 50 - 1.5x + 0.01x^2$

Vertex: (75, –6.25) ← minimum

Zeros: $0.01x^2 - 1.5x + 50 = 0$

$0.01(x^2 - 150x + 5000) = 0$

$0.01(x - 50)(x - 100) = 0$

$x = 50$ or $x = 100$

Graph using x-min = 0 y-min = –10
x-max = 125 y-max = 10

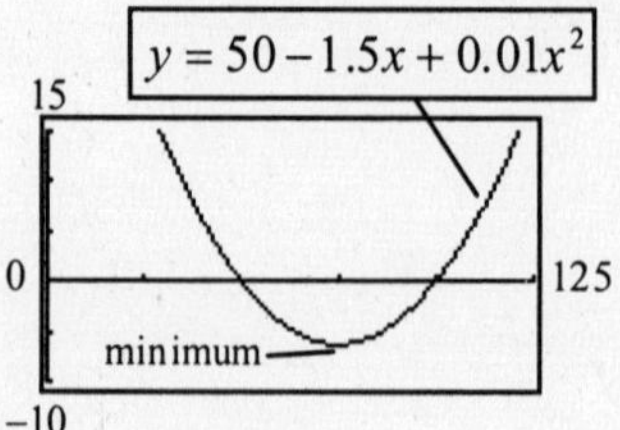

31. a. The vertex is halfway between the zeros. So, the vertex is $\left(1, -4\frac{1}{2}\right)$.

b. The zeros are where the graph crosses the x-axis. $x = -2, 4$.

c. The graph matches B.

32. From the graph,

a. Vertex is (0, 49)

b. Zeros are $x = \pm 7$.

c. Matches with D.

33. a. The vertex is halfway between the zeros. So, the vertex is (7, 24.5).

b. Zeros are $x = 0, 14$.

c. The graph matches A.

34. From the graph,

a. Vertex is (–1, 9).

b. Zeros are $x = -4$ and $x = 2$.

c. Matches with C.

35. a. $f(x) = x^2$

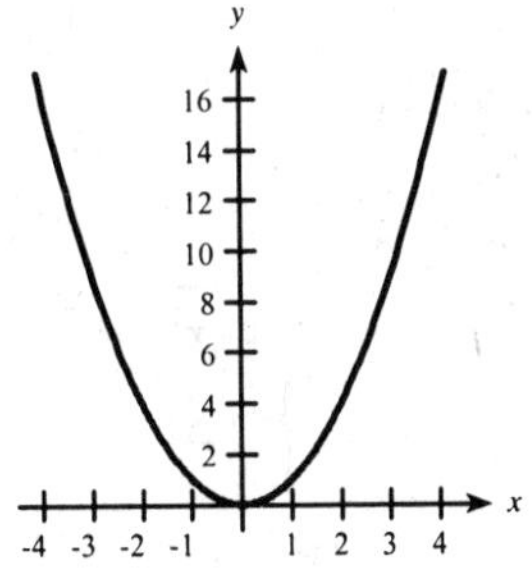

b. $f(x) = \frac{1}{x}$

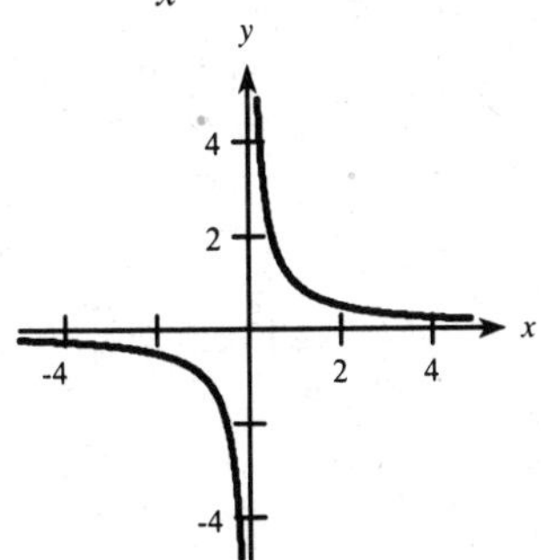

c. $f(x) = x^{1/4}$

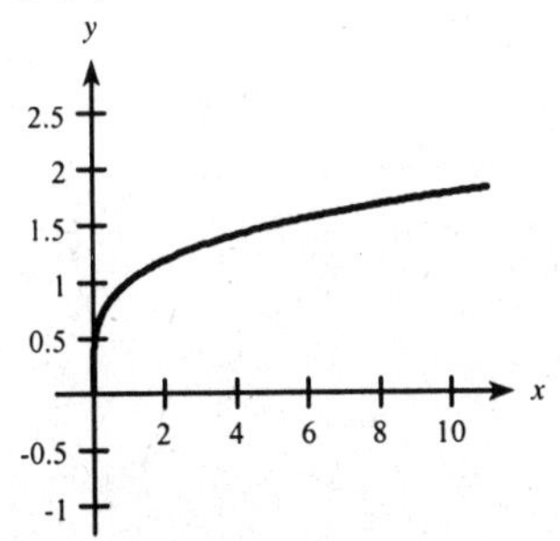

36. $f(x) = \begin{cases} -x^2 & \text{if } x \le 0 \\ \frac{1}{x} & \text{if } x > 0 \end{cases}$

a. $f(0) = -(0^2) = 0$

b. $f(0.0001) = \frac{1}{0.0001} = 10{,}000$

c. $f(-5) = -(-5)^2 = -25$

d. $f(10) = \frac{1}{10} = 0.1$

37. $f(x) = \begin{cases} x & \text{if } x \le 1 \\ 3x - 2 & \text{if } x > 1 \end{cases}$

a. $f(-2) = -2$

b. $f(0) = 0$

c. $f(1) = 1$

d. $f(2) = 3 \cdot 2 - 2 = 4$

38. $f(x) = \begin{cases} x & \text{if } x \le 1 \\ 3x - 2 & \text{if } x > 1 \end{cases}$

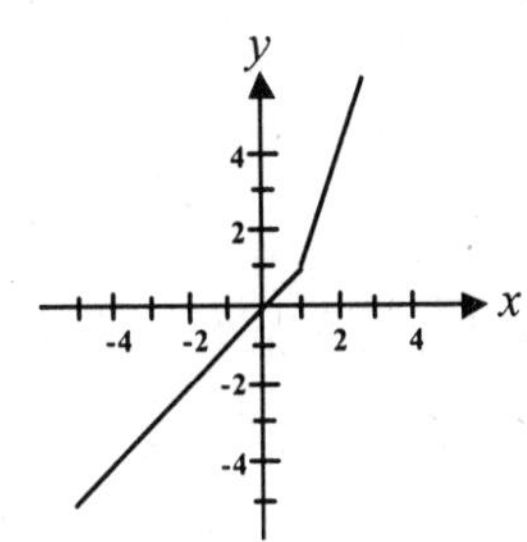

39. a. $f(x) = (x-2)^2$

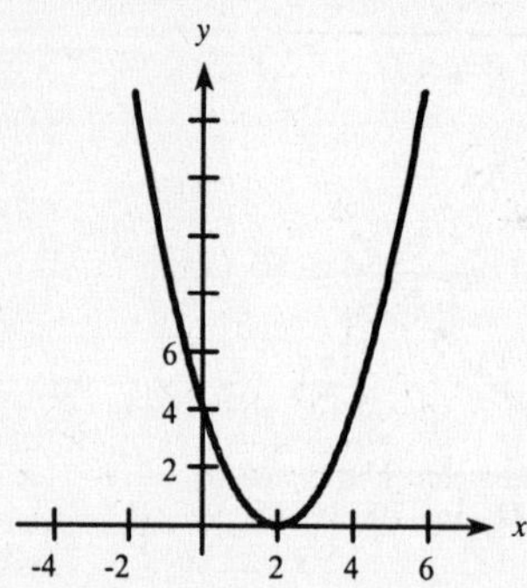

b. $f(x) = (x+1)^3$

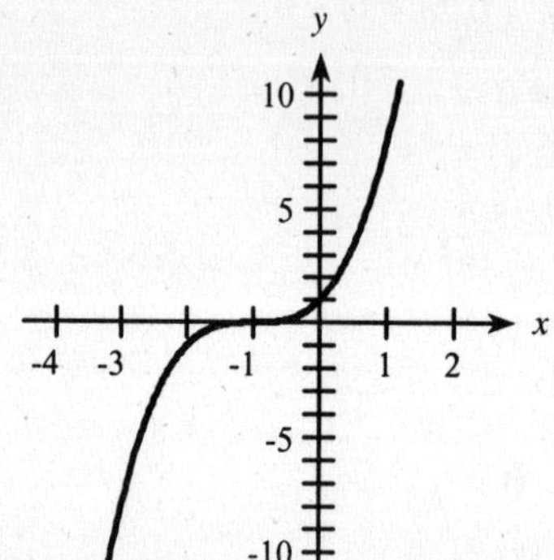

40. $y = x^3 + 3x^2 - 9x$

Using x-min = –10, x-max = 10, y-min = –10, y-max = 35, the turning points are at $x = -3$ and 1.

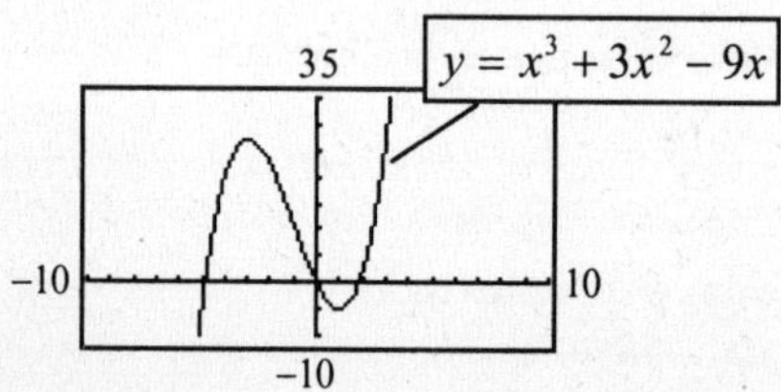

41. $y = x^3 - 9x$

Using x-min = –4.7, x-max = 4.7, y-min = –15, y-max = 15, the turning points are at $x = \pm 1.732$.

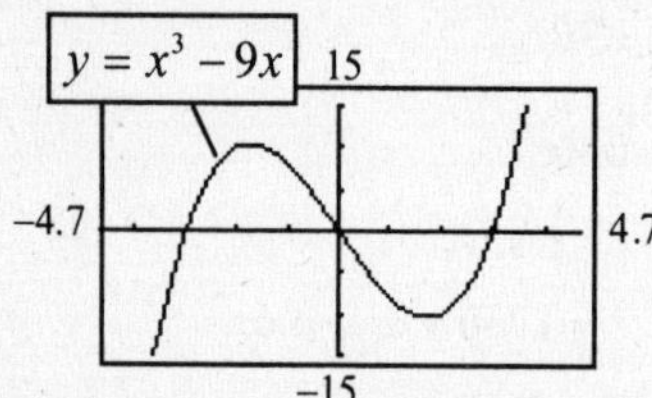

Note: Your turning points in 40.–41. may vary depending on your scale.

42. $y = \dfrac{1}{x-2}$

There is a vertical asymptote $x = 2$.
There is a horizontal asymptote $y = 0$.

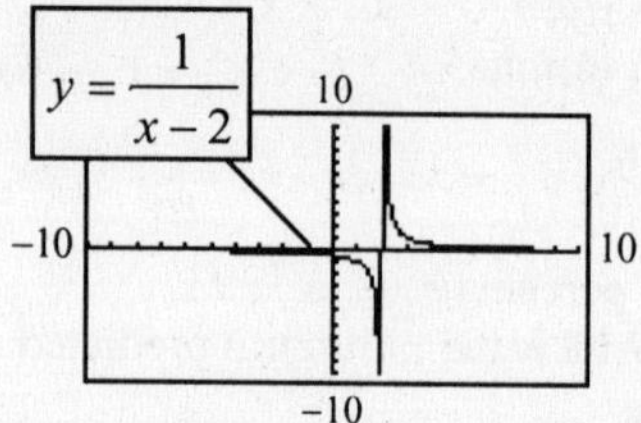

43. $y = \dfrac{2x-1}{x+3} = \dfrac{2 - \frac{1}{x}}{1 + \frac{3}{x}}$

Vertical asymptote is $x = -3$.
Horizontal asymptote is $y = 2$.

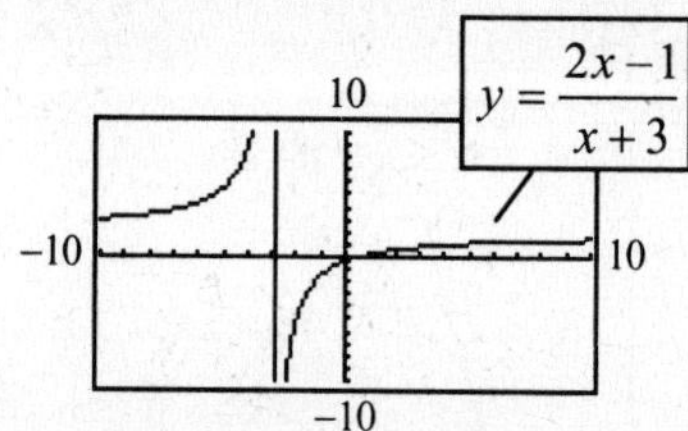

44. a. Left to reader.

b. $y = -2.179x + 159.857$ is a good fit to the data.

c. $y = -0.082x^2 - 0.214x + 153.310$ is a slightly better fit.

45. a. Left to reader.

b. $y = 2.141x + 34.391$ is a good fit to the data.

c. $y = -0.045x^2 + 3.607x + 26.610$ is a slightly better fit.

46. $S = 96 + 32t - 16t^2$

a. $16(6 + 2t - t^2) = 0$

$$t = \frac{-2 \pm \sqrt{4+24}}{-2}$$

$t \approx -1.65$ or $t \approx 3.65$

b. $t \ge 0$ Use $t = 3.65$

c. After 3.65 seconds

47. $P(x) = -0.10x^2 + 82x - 1600$

$(-0.10x + 80)(x - 20) = 0$

Break-even at $x = 20, 800$

48. $f(t) = 0.0241904t^2 - 4.47459t + 216.074$

a. $0.0241904t^2 - 4.47459t + (216.074 - 15) = 0$
Using the calculator and the quadratic formula.
$t = 76.9$ or 108
Percentage will be 15% in 1977 or 2008.

b. Percentage is a minimum at $t = \dfrac{-b}{2a} = 92.5$
Minimum percentage is in 1993.
$f(92.5) = 9.15$ is the minimum predicted percentage.

49. $A = -\dfrac{3}{4}x^2 + 300x$

a. V: $x = \dfrac{-300}{-\dfrac{3}{2}} = 200$ ft

b. $A = -\dfrac{3}{4}(200)^2 + 300(200) = 30{,}000$ sq ft

50. $p = 2q^2 + 4q + 6$

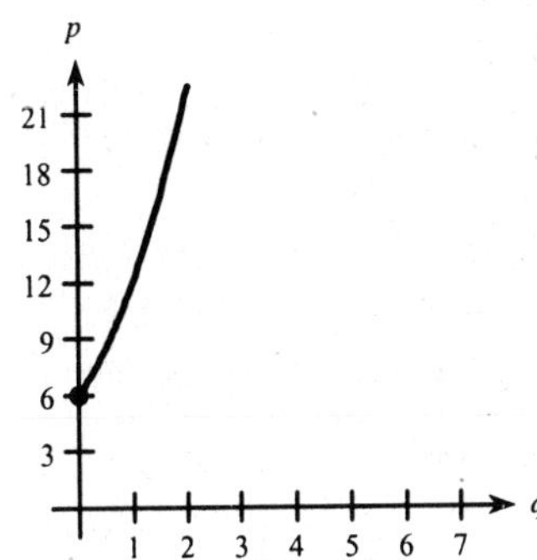

51.

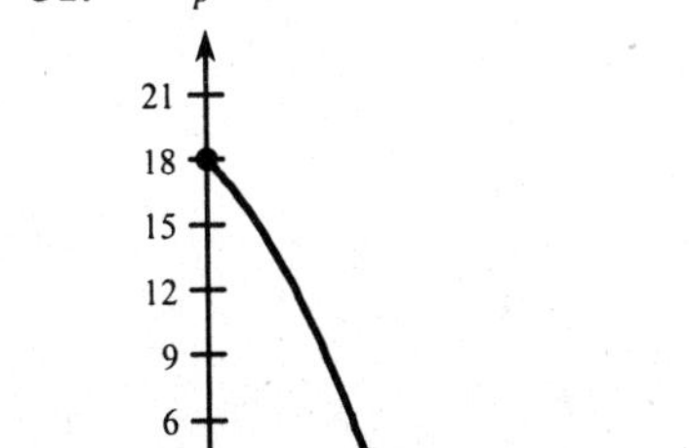

52. a.

$p = 85 - 0.2q - 0.1q^2$

$p = 0.1q^2 + 1$

b.
$$0.1q^2 + 1 = 85 - 0.2q - 0.1q^2$$
$$0.2q^2 + 0.2q - 84 = 0$$
$$0.2(q^2 + q - 420) = 0$$
$$0.2(q - 20)(q + 21) = 0$$
$q = 20$ (only positive value)
$$p = 0.1(20)^2 + 1 = 41$$

53. $p = q^2 + 300$
$p = -q + 410$
$$q^2 + 300 = -q + 410$$
$$q^2 + q - 110 = 0$$
$$(q + 11)(q - 10) = 0$$
$$q = 10$$
$$p = -10 + 410 = 400$$
So, E: (10, 400).

54. D: $p^2 + 5q = 200 \rightarrow p^2 = 200 - 5q$
S: $40 - p^2 + 3q = 0$
Substitute $200 - 5q$ for p^2 in the second equation and solve for q.
$$40 - (200 - 5q) + 3q = 0$$
$$-160 = -8q$$
$$q = 20$$
$$p^2 = 200 - 5(20)$$
$p^2 = 100$ or $p = 10$

55. $R(x) = 100x - 0.4x^2$
$C(x) = 1760 + 8x + 0.6x^2$
$$100x - 0.4x^2 = 1760 + 8x + 0.6x^2$$
$$x^2 - 92x + 1760 = 0$$
$$x = \frac{92 \pm \sqrt{1424}}{2} = 46 \pm 2\sqrt{89} \approx 64.87, 27.13$$
$(\sqrt{1424} = \sqrt{16 \cdot 89})$

56. $C(x) = 900 + 25x$

$R(x) = 100x - x^2$

$900 + 25x = 100x - x^2$

$x^2 - 75x + 900 = 0$

$(x-60)(x-15) = 0$

$x = 60$ or $x = 15$

$R(60) = 2400$; $R(15) = 1275$

(60, 2400) and (15, 1275)

57. $R(x) = 100x - x^2$

V: $x = \dfrac{-100}{-2} = 50$

$R(50) = 100(50) - 50^2$

$= \$2500$ max revenue

$P(x) = (100x - x^2) - (900 + 25x)$

$= -x^2 + 75x - 900$

V: $x = \dfrac{-75}{-2} = 37.5$

$P(37.5) = \$506.25$ max profit

58. $P(x) = 1.3x - 0.01x^2 - 30$

x-coordinate of the vertex $= \dfrac{1.3}{0.02} = 65$

$P(65) = 1.3(65) - 0.01(65)^2 - 30 = 12.25 \leftarrow$ max

Break-even points:

$0 = 1.3x - 0.01x^2 - 30$

$0 = -0.01(x^2 - 130x + 3000)$

$0 = -0.01(x-30)(x-100)$

$x = 30$ or $x = 100$

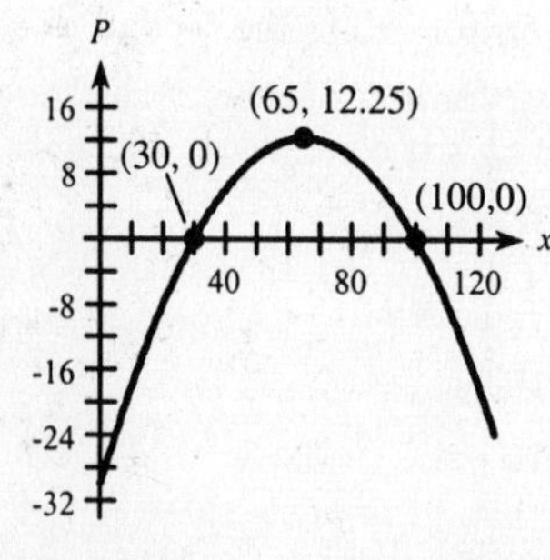

59. $P(x) = (50x - 0.2x^2) - (360 + 10x + 0.2x^2)$

$= -0.4x^2 + 40x - 360$

V: $x = \dfrac{-40}{-0.8} = 50$ units for maximum profit.

$P(50) = -0.4(50)^2 + 40(50) - 360$

$= \$640$ maximum profit.

60. a. $C(x) = 15{,}000 + (140 + 0.04x)x$

$= 15{,}000 + 140x + 0.04x^2$

$R(x) = (300 - 0.06x)x$

$= 300x - 0.06x^2$

b. $15{,}000 + 140x + 0.04x^2 = 300x - 0.06x^2$

$0.10x^2 - 160x + 15{,}000 = 0$

$0.1(x^2 - 1600x + 150{,}000) = 0$

$0.1(x-100)(x-1500) = 0$

$x = 100$ or $x = 1500$

c. Maximum revenue:

x-coordinate: $-\dfrac{300}{-0.12} = 2500$

d. $P(x) = R(x) - C(x)$

$= -0.10x^2 + 160x - 15{,}000$

x-coordinate of max $= -\dfrac{160}{-0.20} = 800$

e. $P(2500) = \$240{,}000$ loss

$P(800) = \$49{,}000$ profit

61. $H(t) = 0.099t^{1.969}$

a. Power function

b. $H(25) = 0.099(25)^{1.969} = 55.998845$

or 5,599,885 cases.

$H(t)$ is in hundreds of thousands.

c. $H(15) = 0.099(15)^{1.969} = 20.481364$

There will be 2,048,136 cases in 1995.

62. $y = 120x^2 - 20x^3$, for $x \ge 0$.

a.

700

$y = 120x^2 - 20x^3$

0

8

−100

b. $y = 20x^2(6 - x)$.

Domain: $0 \le x \le 6$

63. $C(p) = \dfrac{4800p}{100 - p}$

a. Rational function

b. Domain: $0 \le p < 100$

c. $C(0) = 0$ means there is no cost if no pollution is removed.

d. $C(99) = \dfrac{4800(99)}{100-99} = \$475{,}200$

64. $C(x) = \begin{cases} 1.557x & 0 \le x \le 100 \\ 155.7 + 1.04(x-100) & 100 < x \le 1000 \\ 1091.7 + 0.689(x-1000) & x > 1000 \end{cases}$

a. $C(12) = 1.557(12) = \$18.68$

b. $C(825) = 155.70 + 1.04(825 - 100) =$ $909.70

65. a. The best fit is the power function. $y = 17.3969x^{0.5094}$.

b.

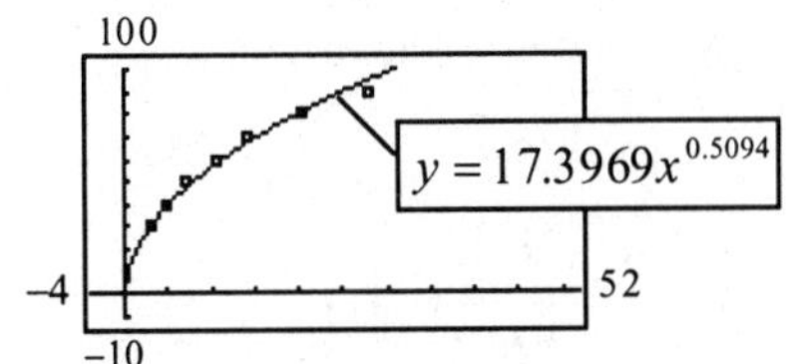

c. $f(5) = 17.3969(5)^{0.5094} = 39.5$ mph

d. Use the TRACE KEY. It will take 18. 3 seconds.

66. a. The best fit is the quadratic function $g = 1.8155x^2 - 11.1607x + 29.1845$.

b. For 2000 – 2010 $x = 11$.
$C(11) = 126$ million cu. yards.

67. a.

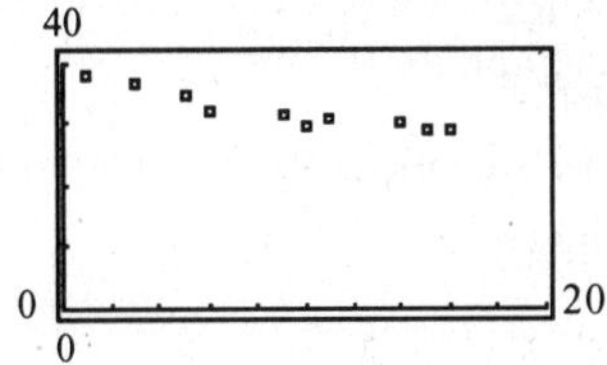

b. $y = -0.001x^3 + 0.065x^2 - 1.447x + 39.847$

c. The model fits the data very well.

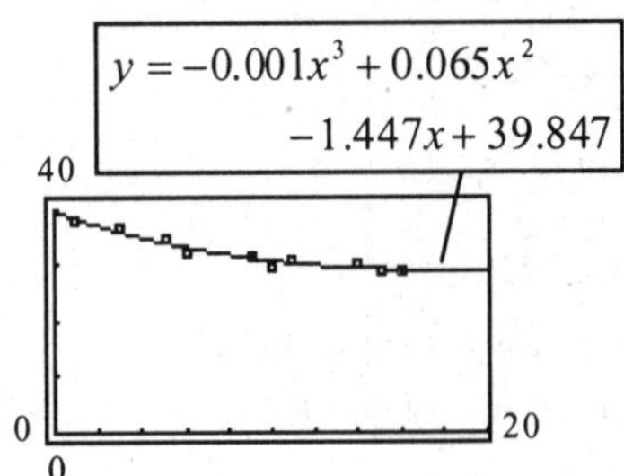

d. No. $y(20) \approx 28.9$. The actual may be a little less than the model value.

Chapter Test

1. a. $f(x) = x^4$

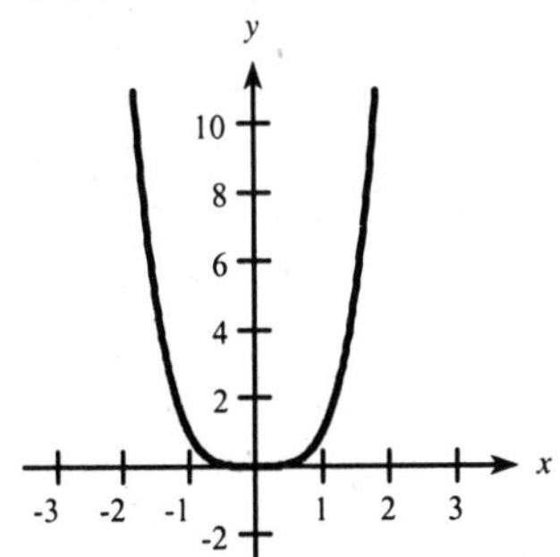

b. $g(x) = |x|$

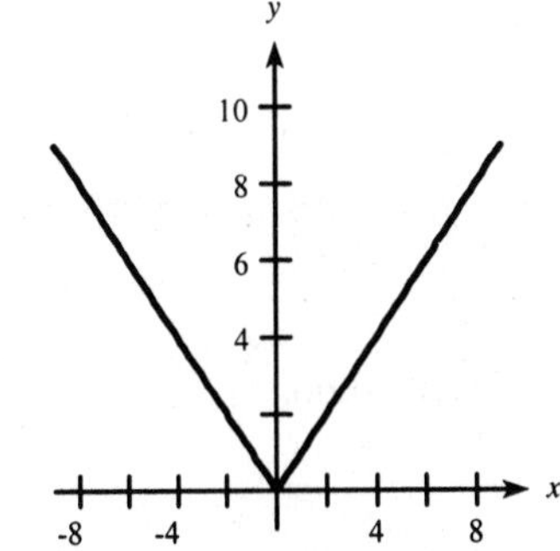

c. $h(x) = -1$

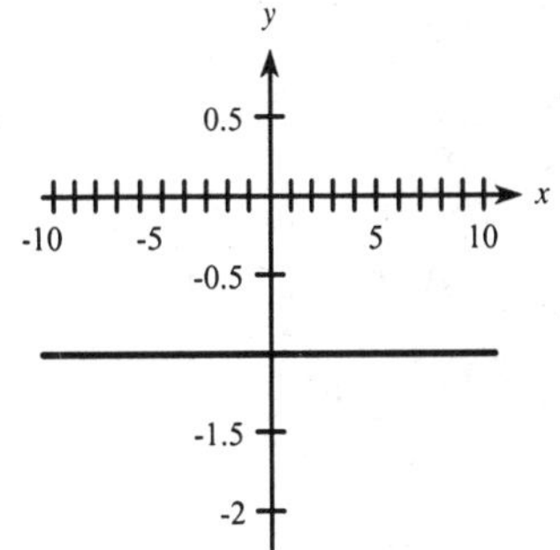

d. $k(x) = \sqrt{x}$

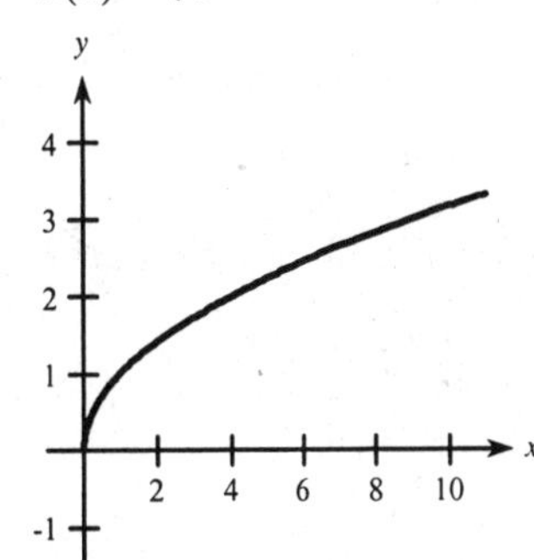

2. figure b is the graph for $b > 1$.
figure a is the graph for $0 < b < 1$.

3. $f(x) = ax^2 + bx + c$ and $a < 0$ is a parabola opening downward.

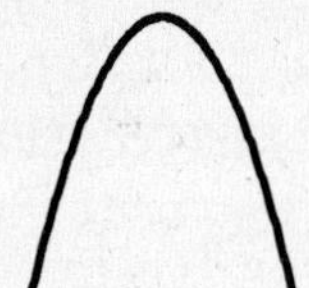

4. **a.** $f(x) = (x+1)^2 - 1$

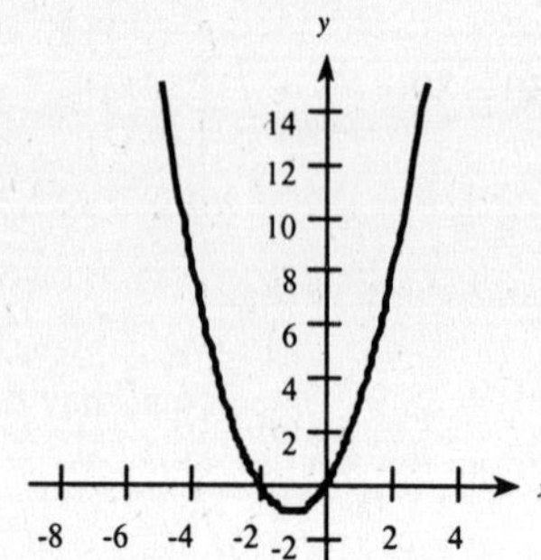

b. $f(x) = (x-2)^3 + 1$

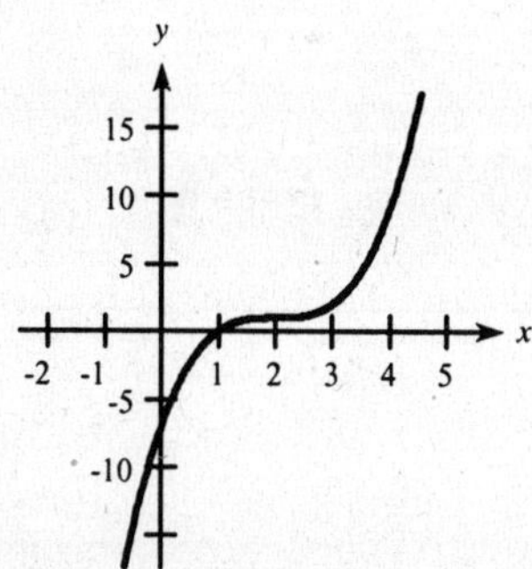

5. $f(x) = x^3 - 4x^2 = x^2(x-4)$.
a. and **b.** are the cubic choices. $f(x) < 0$ if $0 < x < 4$. Answer: b

6. $f(x) = \begin{cases} 8x + \dfrac{1}{x} & \text{if } x < 0 \\ 4 & \text{if } 0 \le x \le 2 \\ 6 - x & \text{if } x > 2 \end{cases}$

a. $f(16) = 6 - 16 = -10$

b. $f(-2) = 8(-2) + \dfrac{1}{-2} = -16\dfrac{1}{2}$

c. $f(13) = 6 - 13 = -7$

7. $g(x) = \begin{cases} x^2 & \text{if } x \le 1 \\ 4 - x & \text{if } x > 1 \end{cases}$

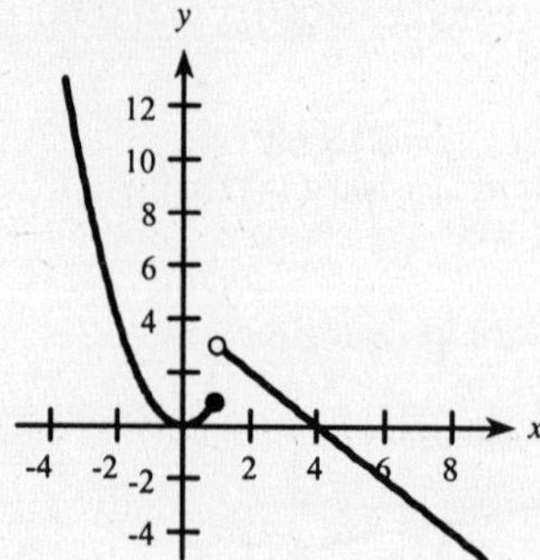

8. $f(x) = 21 - 4x - x^2 = (7+x)(3-x)$

Vertex: $x = \dfrac{-b}{2a} = \dfrac{-(-4)}{2(-1)} = -2$

Point: $(-2, 25)$
Zeros: $f(x) = 0$ at $x = -7$ or 3.

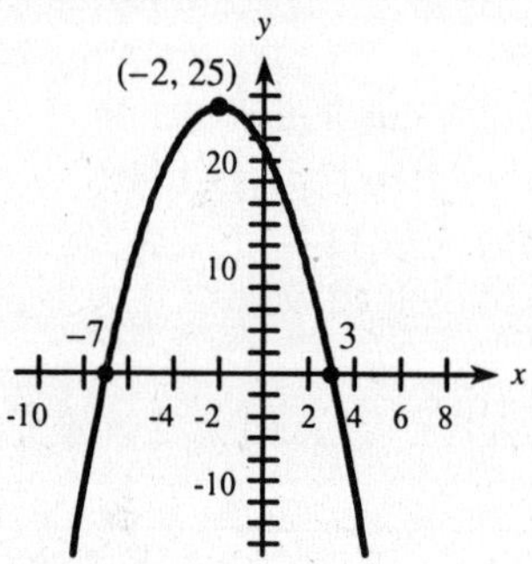

9.
$$3x^2 + 2 = 7x$$
$$3x^2 - 7x + 2 = 0$$
$$(3x-1)(x-2) = 0$$
$$3x - 1 = 0 \text{ or } x - 2 = 0$$
$$x = \frac{1}{3}, 2$$

10. $2x^2 + 6x - 9 = 0$

$$x = \frac{-6 \pm \sqrt{36+72}}{4} = \frac{-6 \pm 6\sqrt{3}}{4} = \frac{-3 \pm 3\sqrt{3}}{2}$$

11. $\left(\dfrac{1}{x} + 2x = \dfrac{1}{3} + \dfrac{x+1}{x}\right)3x$

$$3 + 6x^2 = x + 3x + 3$$
$$6x^2 - 4x = 0$$
$$2x(3x-2) = 0$$

$x = \dfrac{2}{3}$ is the only solution.

12. $g(x) = \dfrac{3(x-4)}{x+2}$

Vertical asymptote at $x = -2$.
$g(4) = 0$
Answer: c

13. a. quartic
b. cubic

14. a. $f(x) = -0.3577x + 19.9227$

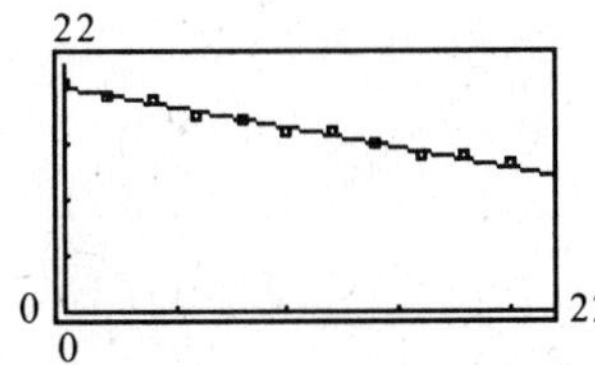

b. $f(40) = 5.6$

c. $f(x) = 0$ if $x = \dfrac{19.9227}{0.3577} \approx 55.7$

15. S: $p = \dfrac{1}{6}q + 30$

D: $p = \dfrac{30,000}{q} - 20$

$$\left(\frac{1}{6}q + 30 = \frac{30,000}{q} - 20\right)6q$$
$$q^2 + 180q = 180,000 - 120q$$
$$q^2 + 300q - 180,000 = 0$$
$$(q+600)(q-300) = 0$$
$$E_q : q = 300$$
$$E_p : p = 50 + 30 = 80$$

16. $R(x) = 285x - 0.9x^2$

$C(x) = 15,000 + 35x + 0.1x^2$

a. $P(x) = 285x - 0.9x^2 - (15,000 + 35x + 0.1x^2)$
$= -x^2 + 250x - 15,000$
$= (100 - x)(x - 150)$

b. Maximum profit is at vertex.
$x = \dfrac{-250}{2(-1)} = 125$
Maximum profit = $P(125)$ = \$625

c. Break-even means $P(x) = 0$.
From **a.**, $x = 100, 150$.

17. $$WC = f(s) = \begin{cases} 0 & \text{if } 0 \le s \le 4 \\ 91.4 - 7.46738(5.81 + 3.7\sqrt{s} - 0.25s) & \text{if } 4 < s \le 45 \\ -55 & \text{if } s > 45 \end{cases}$$

a. Use middle rule for $s = 15$. $f(15) = -31$
If the air temperature is 0°F and the wind speed is 15 mph, the air temperature feels like –31°F. In winter, the TV weather report usually gives the wind chill temperature.

b. $f(48) = -55$ (last rule, $s > 45$)

18. a.

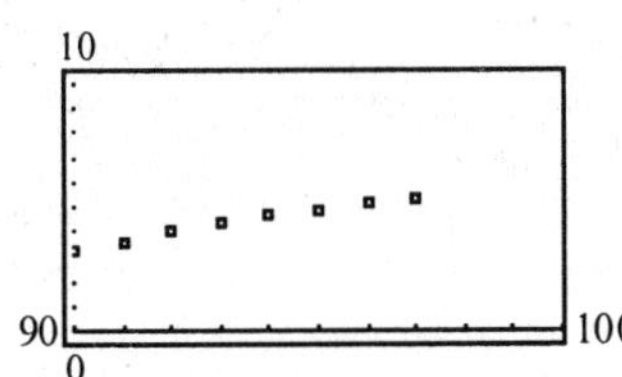

b. $y = 0.276x - 21.409$

c. No; slope is positive.

d. The predicted debt continues to increase.

19. a. quadratic

a. $f(x) = -0.01607x^2 + 3.31845x - 165.27798$

b. $f(98)$ = \$5.59 trillion
$f(99)$ = \$5.75 trillion

c. The maximum occurs at $x = \dfrac{-b}{2a} = 103.2$
After 2003, the debt begins to decline.

Chapter 3: Matrices

Exercise 3.1

1. Matrix B has 3 rows.

3. $-F = \begin{bmatrix} -1 & -2 & -3 \\ 1 & 0 & -1 \\ -2 & 3 & 4 \end{bmatrix}$

5. A, C, D, F, and Z are square.

7. A, F, and Z are 3×3. C and D are 2×2.

9. $a_{23} = 1$

11. $A^T = \begin{bmatrix} 1 & 3 & 4 \\ 0 & 2 & 0 \\ 2 & 1 & 3 \end{bmatrix}$

13. No, M and N are not the same order.

15. $M+(-M) = \begin{bmatrix} 0 & 0 & 0 \\ 0 & 0 & 0 \\ 0 & 0 & 0 \end{bmatrix}$

17. $C + D = \begin{bmatrix} 5+4 & 3+2 \\ 1+3 & 2+5 \end{bmatrix} = \begin{bmatrix} 9 & 5 \\ 4 & 7 \end{bmatrix}$

19. $A - F = \begin{bmatrix} 1-1 & 0-2 & 2-3 \\ 3+1 & 2-0 & 1-1 \\ 4-2 & 0+3 & 3+4 \end{bmatrix} = \begin{bmatrix} 0 & -2 & -1 \\ 4 & 2 & 0 \\ 2 & 3 & 7 \end{bmatrix}$

21. $A + A^T = \begin{bmatrix} 1 & 0 & 2 \\ 3 & 2 & 1 \\ 4 & 0 & 3 \end{bmatrix} + \begin{bmatrix} 1 & 3 & 4 \\ 0 & 2 & 0 \\ 2 & 1 & 3 \end{bmatrix} = \begin{bmatrix} 2 & 3 & 6 \\ 3 & 4 & 1 \\ 6 & 1 & 6 \end{bmatrix}$

23. Because the orders are different, $D - E$ is undefined.

25. $A + E^T = \begin{bmatrix} 1 & 0 & 2 \\ 3 & 2 & 1 \\ 4 & 0 & 3 \end{bmatrix} + \begin{bmatrix} 2 & 1 & 5 \\ 1 & 0 & 1 \\ 1 & 4 & 0 \end{bmatrix} = \begin{bmatrix} 3 & 1 & 7 \\ 4 & 2 & 2 \\ 5 & 4 & 3 \end{bmatrix}$

27. $\begin{bmatrix} x & 1 & 0 \\ 0 & y & z \\ w & 2 & 1 \end{bmatrix} = \begin{bmatrix} 3 & 1 & 0 \\ 0 & 2 & 3 \\ 4 & 2 & 1 \end{bmatrix}$

By setting corresponding elements equal we have $x = 3$, $y = 2$, $z = 3$, $w = 4$.

29. $\begin{bmatrix} x & 3 & 2x-1 \\ y & 4 & 4y \end{bmatrix} = \begin{bmatrix} 2x-4 & z & 7 \\ 1 & w+1 & 3y+1 \end{bmatrix}$

$2x - 4 = x$ or $x = 4$, $z = 3$, $y = 1$,
$w + 1 = 4$ or $w = 3$.

31.

$$\begin{array}{lll} x+2x=6 & y+3y=8 & z-4z=9 \\ 3x=6 & 4y=8 & -3z=9 \\ x=2 & y=2 & z=-3 \end{array}$$

33. a. $A = \begin{bmatrix} 63 & 78 & 14 & 10 & 70 \\ 9 & 14 & 22 & 8 & 44 \end{bmatrix}$

$B = \begin{bmatrix} 251 & 175 & 64 & 8 & 11 \\ 17 & 6 & 15 & 1 & 0 \end{bmatrix}$

b. $A + B = \begin{bmatrix} 314 & 253 & 78 & 18 & 81 \\ 26 & 20 & 37 & 9 & 44 \end{bmatrix}$

c. $B - A = \begin{bmatrix} 188 & 97 & 50 & -2 & -59 \\ 8 & -8 & -7 & -7 & -44 \end{bmatrix}$

The negative entries tell us how many more endangered or threatened species are in the US.

35. a. Capital Expenditures + Gross Operating Costs= $\begin{bmatrix} 11041.7 & 8978.4 & 6461.0 \\ 8739.8 & 9159.6 & 6877.3 \\ 9798.1 & 9086.7 & 6448.4 \\ 9696.6 & 8926.7 & 6109.5 \end{bmatrix}$

b. R3, C1 was the most expensive

37. a. $A+B=\begin{bmatrix}825 & 580 & 1560\\ 810 & 650 & 350\end{bmatrix}$ **b.** $B-A=\begin{bmatrix}-75 & 20 & -140\\ 10 & -50 & 50\end{bmatrix}$

39.

$$A=\begin{matrix} & \text{M} & \text{F}\\ & \begin{bmatrix}54.4 & 55.6\\ 62.1 & 66.6\\ 67.4 & 74.1\\ 70.7 & 78.1\\ 72.9 & 79.4\\ 74.6 & 79.9\end{bmatrix}\end{matrix} \quad B=\begin{matrix} & \text{M} & \text{F}\\ & \begin{bmatrix}45.5 & 45.2\\ 51.5 & 54.9\\ 61.1 & 67.4\\ 63.8 & 72.5\\ 64.5 & 73.6\\ 67.8 & 74.7\end{bmatrix}\end{matrix} \quad A-B=\begin{matrix} & \text{M} & \text{F}\\ & \begin{bmatrix}8.9 & 10.4\\ 10.6 & 11.7\\ 6.3 & 6.7\\ 6.9 & 5.6\\ 8.4 & 5.8\\ 6.8 & 5.2\end{bmatrix}\end{matrix}$$

41. a.

Group	wt	length
I	140	5.5
II	151	5.7
III	141	5.5

b.

Group	wt	length
I	250	12.5
II	215	11.8
III	190	9.8

43. a. $\begin{bmatrix}0 & 1 & 1 & 0\\ 1 & 0 & 1 & 1\\ 0 & 0 & 0 & 1\\ 0 & 1 & 1 & 0\end{bmatrix}$

b. $\begin{bmatrix}0 & 1 & 1 & 1\\ 1 & 0 & 1 & 1\\ 0 & 1 & 0 & 1\\ 1 & 1 & 1 & 0\end{bmatrix}$

c. $A+B=\begin{bmatrix}0 & 2 & 2 & 1\\ 2 & 0 & 2 & 2\\ 0 & 1 & 0 & 2\\ 1 & 2 & 2 & 0\end{bmatrix}$

Sum in Row 1 = 5, Row 2 = 6, Row 3 = 3, Row 4 = 5. Person 2 is the most active.

45. a. $A+B=\begin{bmatrix}50+30 & 30+45\\ 36+22 & 44+62\end{bmatrix}=\begin{bmatrix}80 & 75\\ 58 & 106\end{bmatrix}$

b. $A+B+C=\begin{bmatrix}80+96 & 75+52\\ 58+81 & 106+37\end{bmatrix}=\begin{bmatrix}176 & 127\\ 139 & 143\end{bmatrix}$

c. $A-D=\begin{bmatrix}50–40 & 30–26\\ 36–29 & 44–42\end{bmatrix}=\begin{bmatrix}10 & 4\\ 7 & 2\end{bmatrix}$

d. $B–D=\begin{bmatrix}30–40 & 45–26\\ 22–29 & 62–42\end{bmatrix}=\begin{bmatrix}-10 & 19\\ -7 & 20\end{bmatrix}$

Shortage; take from inventory.

47. a. Row 7: 3, 4, 5, 6
b. Row 1: 1

49. For each row: (C1 + C2 + C3 + C4) ÷ 4 = average efficiency

W1: .9625	W2: .9375	W3: .9125
W4: .8875	W5: .85	W6: .875
W7: .90	W8: .925	W9: .95

W5 is least efficient. He works best at center 5.

Exercise 3.2

1. $\begin{bmatrix}1 & 2 & 3\end{bmatrix}\begin{bmatrix}4\\5\\6\end{bmatrix} = [1\cdot4+2\cdot5+3\cdot6] = [32]$ (not 32)

3. $\begin{bmatrix}1 & 2\end{bmatrix}\begin{bmatrix}3 & 5\\4 & 6\end{bmatrix} = \begin{bmatrix}1\cdot3+2\cdot4 & 1\cdot5+2\cdot6\end{bmatrix}$
$= \begin{bmatrix}11 & 17\end{bmatrix}$

5. $3A = \begin{bmatrix}3(1) & 3(0) & 3(2)\\3(3) & 3(2) & 3(1)\\3(4) & 3(0) & 3(3)\end{bmatrix} = \begin{bmatrix}3 & 0 & 6\\9 & 6 & 3\\12 & 0 & 9\end{bmatrix}$

7. $4C+2D = \begin{bmatrix}4(5) & 4(3)\\4(1) & 4(2)\end{bmatrix} + \begin{bmatrix}2(4) & 2(2)\\2(3) & 2(5)\end{bmatrix}$
$= \begin{bmatrix}20 & 12\\4 & 8\end{bmatrix} + \begin{bmatrix}8 & 4\\6 & 10\end{bmatrix}$
$= \begin{bmatrix}28 & 16\\10 & 18\end{bmatrix}$

9. $2A-3B$
Matrices are not the same size and cannot be combined.

11. $CD = \begin{bmatrix}5 & 3\\1 & 2\end{bmatrix}\cdot\begin{bmatrix}4 & 2\\3 & 5\end{bmatrix}$
$= \begin{bmatrix}20+9 & 10+15\\4+6 & 2+10\end{bmatrix}$
$= \begin{bmatrix}29 & 25\\10 & 12\end{bmatrix}$

13. $DC = \begin{bmatrix}4 & 2\\3 & 5\end{bmatrix}\cdot\begin{bmatrix}5 & 3\\1 & 2\end{bmatrix}$
$= \begin{bmatrix}20+2 & 12+4\\15+5 & 9+10\end{bmatrix}$
$= \begin{bmatrix}22 & 16\\20 & 19\end{bmatrix}$
Note $CD \neq DC$

15. $DF = \begin{bmatrix}4 & 2\\3 & 5\end{bmatrix}\cdot\begin{bmatrix}1 & 0 & -1 & 3\\2 & -1 & 3 & -4\end{bmatrix} = \begin{bmatrix}4+4 & 0-2 & -4+6 & 12-8\\3+10 & 0-5 & -3+15 & 9-20\end{bmatrix} = \begin{bmatrix}8 & -2 & 2 & 4\\13 & -5 & 12 & -11\end{bmatrix}$

17. $AB = \begin{bmatrix}1 & 0 & 2\\3 & 2 & 1\\4 & 0 & 3\end{bmatrix}\begin{bmatrix}1 & 1 & 3 & 0\\4 & 2 & 1 & 1\\3 & 2 & 0 & 1\end{bmatrix} = \begin{bmatrix}1+0+6 & 1+0+4 & 3+0+0 & 0+0+2\\3+8+3 & 3+4+2 & 9+2+0 & 0+2+1\\4+0+9 & 4+0+6 & 12+0+0 & 0+0+3\end{bmatrix} = \begin{bmatrix}7 & 5 & 3 & 2\\14 & 9 & 11 & 3\\13 & 10 & 12 & 3\end{bmatrix}$

19. BA is undefined.

21. $EB = \begin{bmatrix}1 & 0 & 4\\5 & 1 & 0\end{bmatrix}\begin{bmatrix}1 & 1 & 3 & 0\\4 & 2 & 1 & 1\\3 & 2 & 0 & 1\end{bmatrix} = \begin{bmatrix}1+0+12 & 1+0+8 & 3+0+0 & 0+0+4\\5+4+0 & 5+2+0 & 15+1+0 & 0+1+0\end{bmatrix} = \begin{bmatrix}13 & 9 & 3 & 4\\9 & 7 & 16 & 1\end{bmatrix}$

23. $EA^T = \begin{bmatrix}1 & 0 & 4\\5 & 1 & 0\end{bmatrix}\begin{bmatrix}1 & 3 & 4\\0 & 2 & 0\\2 & 1 & 3\end{bmatrix} = \begin{bmatrix}1+0+8 & 3+0+4 & 4+0+12\\5+0+0 & 15+2+0 & 20+0+0\end{bmatrix} = \begin{bmatrix}9 & 7 & 16\\5 & 17 & 20\end{bmatrix}$

25. $A^2 = \begin{bmatrix}1 & 0 & 2\\3 & 2 & 1\\4 & 0 & 3\end{bmatrix}\begin{bmatrix}1 & 0 & 2\\3 & 2 & 1\\4 & 0 & 3\end{bmatrix} = \begin{bmatrix}1+0+8 & 0+0+0 & 2+0+6\\3+6+4 & 0+4+0 & 6+2+3\\4+0+12 & 0+0+0 & 8+0+9\end{bmatrix} = \begin{bmatrix}9 & 0 & 8\\13 & 4 & 11\\16 & 0 & 17\end{bmatrix}$

27. $C^3 = \begin{bmatrix}5 & 3\\1 & 2\end{bmatrix}\cdot\left(\begin{bmatrix}5 & 3\\1 & 2\end{bmatrix}\cdot\begin{bmatrix}5 & 3\\1 & 2\end{bmatrix}\right) = \begin{bmatrix}5 & 3\\1 & 2\end{bmatrix}\cdot\begin{bmatrix}25+3 & 15+6\\5+2 & 3+4\end{bmatrix}$
$= \begin{bmatrix}5 & 3\\1 & 2\end{bmatrix}\cdot\begin{bmatrix}28 & 21\\7 & 7\end{bmatrix} = \begin{bmatrix}140+21 & 105+21\\28+14 & 21+14\end{bmatrix} = \begin{bmatrix}161 & 126\\42 & 35\end{bmatrix}$

29. $AA^T = \begin{bmatrix} 1 & 0 & 2 \\ 3 & 2 & 1 \\ 4 & 0 & 3 \end{bmatrix} \begin{bmatrix} 1 & 3 & 4 \\ 0 & 2 & 0 \\ 2 & 1 & 3 \end{bmatrix} = \begin{bmatrix} 5 & 5 & 10 \\ 5 & 14 & 15 \\ 10 & 15 & 25 \end{bmatrix}$

$(AA^T)^T = \begin{bmatrix} 5 & 5 & 10 \\ 5 & 14 & 15 \\ 10 & 15 & 25 \end{bmatrix}$ $\quad A^T A = \begin{bmatrix} 1 & 3 & 4 \\ 0 & 2 & 0 \\ 2 & 1 & 3 \end{bmatrix} \begin{bmatrix} 1 & 0 & 2 \\ 3 & 2 & 1 \\ 4 & 0 & 3 \end{bmatrix} = \begin{bmatrix} 26 & 6 & 17 \\ 6 & 4 & 2 \\ 17 & 2 & 14 \end{bmatrix}$

No, they are not equal.

31. No. See problems 11 and 13.

33. $AB = \begin{bmatrix} 2 & 5 & 4 \\ 1 & 4 & 3 \\ 1 & -3 & -2 \end{bmatrix} \begin{bmatrix} -1 & 2 & 1 \\ -5 & 8 & 2 \\ -7 & 11 & -3 \end{bmatrix} = \begin{bmatrix} -2-25-28 & 4+40+44 & 2+10-12 \\ -1-20-21 & 2+32+33 & 1+8-9 \\ -1+15+14 & 2-24-22 & 1-6+6 \end{bmatrix} = \begin{bmatrix} -55 & 88 & 0 \\ -42 & 67 & 0 \\ 28 & -44 & 1 \end{bmatrix}$

35. $CD = \begin{bmatrix} 3 & 0 & 4 \\ 1 & 7 & -1 \\ 3 & 0 & 4 \end{bmatrix} \begin{bmatrix} 4 & 4 & -8 \\ -1 & -1 & 2 \\ -3 & -3 & 6 \end{bmatrix} = \begin{bmatrix} 12+0-12 & 12+0-12 & -24+0+24 \\ 4-7+3 & 4-7+3 & -8+14-6 \\ 12+0-12 & 12+0-12 & -24+0+24 \end{bmatrix} = \begin{bmatrix} 0 & 0 & 0 \\ 0 & 0 & 0 \\ 0 & 0 & 0 \end{bmatrix}$

37. $F^2 = \begin{bmatrix} 0 & 1 & 2 \\ 0 & 0 & -4 \\ 0 & 0 & 0 \end{bmatrix} \begin{bmatrix} 0 & 1 & 2 \\ 0 & 0 & -4 \\ 0 & 0 & 0 \end{bmatrix} = \begin{bmatrix} 0+0+0 & 0+0+0 & 0-4+0 \\ 0+0+0 & 0+0+0 & 0+0+0 \\ 0+0+0 & 0+0+0 & 0+0+0 \end{bmatrix} = \begin{bmatrix} 0 & 0 & -4 \\ 0 & 0 & 0 \\ 0 & 0 & 0 \end{bmatrix}$

$F^3 = \begin{bmatrix} 0 & 0 & -4 \\ 0 & 0 & 0 \\ 0 & 0 & 0 \end{bmatrix} \begin{bmatrix} 0 & 1 & 2 \\ 0 & 0 & -4 \\ 0 & 0 & 0 \end{bmatrix} = \begin{bmatrix} 0+0+0 & 0+0+0 & 0+0+0 \\ 0+0+0 & 0+0+0 & 0+0+0 \\ 0+0+0 & 0+0+0 & 0+0+0 \end{bmatrix} = \begin{bmatrix} 0 & 0 & 0 \\ 0 & 0 & 0 \\ 0 & 0 & 0 \end{bmatrix}$

39. $AI = A$ See Example 6.

41. $Z(CI) = ZC = Z \qquad (ZC)I = ZI = Z$
Product of zero matrix and A is the zero matrix.

43. Problems 35 and 37 show that a product can be "zero" and neither of the factors is zero.

45. We will use $\dfrac{1}{ad-bc}AB$ and $\dfrac{1}{ad-bc}BA.$

$AB = \begin{bmatrix} a & b \\ c & d \end{bmatrix} \begin{bmatrix} d & -b \\ -c & a \end{bmatrix} = \begin{bmatrix} ad-bc & -ab+ba \\ cd-dc & -bc+ad \end{bmatrix} = \begin{bmatrix} ad-bc & 0 \\ 0 & ad-bc \end{bmatrix}$

$BA = \begin{bmatrix} d & -b \\ -c & a \end{bmatrix} \begin{bmatrix} a & b \\ c & d \end{bmatrix} = \begin{bmatrix} ad-bc & bd-bd \\ -ac+ac & -bc+ad \end{bmatrix} = \begin{bmatrix} ad-bc & 0 \\ 0 & ad-bc \end{bmatrix}$

$\dfrac{1}{ad-bc}AB = \dfrac{1}{ad-bc}BA = \begin{bmatrix} 1 & 0 \\ 0 & 1 \end{bmatrix}.$

47. $\begin{bmatrix} 1 & 2 & 1 \\ 3 & 4 & -2 \\ 2 & 0 & -1 \end{bmatrix} \begin{bmatrix} 2 \\ -1 \\ 2 \end{bmatrix} = \begin{bmatrix} 2-2+2 \\ 6-4-4 \\ 4+0-2 \end{bmatrix} = \begin{bmatrix} 2 \\ -2 \\ 2 \end{bmatrix}$

These values are the solution.

49. $\begin{bmatrix} 1 & 1 & 2 \\ 4 & 0 & 1 \\ 2 & 1 & 1 \end{bmatrix} \begin{bmatrix} 1 \\ 2 \\ 1 \end{bmatrix} = \begin{bmatrix} 1+2+2 \\ 4+0+1 \\ 2+2+1 \end{bmatrix} = \begin{bmatrix} 5 \\ 5 \\ 5 \end{bmatrix}$

These values are the solution.

51. a. $AB = \begin{bmatrix} 1 & 2 & 3 \\ 2 & -2 & -1 \\ 3 & 0 & 2 \end{bmatrix} \begin{bmatrix} 1 & 2 & 1 \\ 2 & 4 & 5 \\ 2 & -2 & 3 \end{bmatrix} = \begin{bmatrix} 11 & 4 & 20 \\ -4 & -2 & -11 \\ 7 & 2 & 9 \end{bmatrix}$

$BA = \begin{bmatrix} 1 & 2 & 1 \\ 2 & 4 & 5 \\ 2 & -2 & 3 \end{bmatrix} \begin{bmatrix} 1 & 2 & 3 \\ 2 & -2 & -1 \\ 3 & 0 & 2 \end{bmatrix} = \begin{bmatrix} 8 & -2 & 3 \\ 25 & -4 & 12 \\ 7 & 8 & 14 \end{bmatrix}$

b. $AB \neq BA$ Matrix multiplication is not commutative.

53. The graphing calculator should show 1's on the main diagonal and 0's everywhere else. (1 may be 0.9999 and 0 may be 0.00001.)

55. a. $A = \begin{bmatrix} 376 & 342 \\ 603 & 493 \\ 731 & 520 \\ 777 & 565 \\ 738 & 505 \\ 537 & 378 \end{bmatrix}$ **b.** $(1.2)A = \begin{bmatrix} 451.2 & 410.4 \\ 723.6 & 591.6 \\ 877.2 & 624 \\ 932.4 & 678 \\ 885.6 & 606 \\ 644.4 & 453.6 \end{bmatrix}$

57. a. $1.05A$ means multiply each element of A by 1.05.

$1.05A = \begin{bmatrix} 23.10 & 42.00 & 105.00 & 5.25 \\ 21.00 & 42.00 & 21.00 & 0.00 \\ 29.40 & 73.50 & 47.25 & 0.00 \\ 15.75 & 73.50 & 21.00 & 10.50 \\ 21.00 & 0.00 & 105.00 & 5.25 \end{bmatrix}.$

b. In this case, we find $1.1A$.

$1.1A = \begin{bmatrix} 24.20 & 44.00 & 110.00 & 5.50 \\ 22.00 & 44.00 & 22.00 & 0.00 \\ 30.80 & 77.00 & 49.50 & 0.00 \\ 16.50 & 77.00 & 22.00 & 11.00 \\ 22.00 & 0.00 & 110.00 & 5.50 \end{bmatrix}.$

59. $\begin{bmatrix} 0.88 & 0 \\ 0 & 0.85 \end{bmatrix} \begin{bmatrix} 25,000 & 28,000 \\ 36,000 & 42,000 \end{bmatrix} = \begin{bmatrix} 22,000 & 24,640 \\ 30,600 & 35,700 \end{bmatrix}$

The product is the price the dealer pays for the cars

61. a. $A = \begin{bmatrix} 2703 & 2378 \\ 2383 & 2723 \\ 4376 & 5114 \\ 1443 & 1581 \\ 3152 & 2490 \\ 1289 & 1245 \\ 916 & 824 \end{bmatrix}$ $B^T = \begin{bmatrix} 0.961 & 0.481 & 0.794 & 0.99 & 0.773 & 0.658 & 0.895 \\ 1.31 & 0.478 & 0.868 & 1.18 & 1.07 & 0.642 & 0.963 \end{bmatrix}$

b. $\frac{1}{12}AB^T$ is a 7×7 matrix.

c. $\frac{1}{12}[(2703)(0.961) + (2378)(1.31)] \approx \$476$ million gives the total value of production for Alabama in 1995 and 2000. The other states are likewise represented on the diagonal entries.

63. TC is the only product that is defined.

$\begin{bmatrix} 50 & 30 & 5 \\ 40 & 60 & 30 \\ 45 & 40 & 60 \end{bmatrix} \begin{bmatrix} 20 \\ 9 \\ 6 \end{bmatrix} = \begin{bmatrix} 1300 \\ 1520 \\ 1620 \end{bmatrix} \begin{matrix} \text{Teens} \\ \text{Single} \\ \text{Married} \end{matrix}$

65. T h e blank d i e blank i s blank c a s t

20 8 5 27 4 9 5 27 9 19 27 3 1 19 20

$$\begin{bmatrix} 5 & 9 \\ 6 & 11 \end{bmatrix}\begin{bmatrix} 20 & 5 & 4 & 5 & 9 & 27 & 1 & 20 \\ 8 & 27 & 9 & 27 & 19 & 3 & 19 & 27 \end{bmatrix} = \begin{bmatrix} 172 & 268 & 101 & 268 & 216 & 162 & 176 & 343 \\ 208 & 327 & 123 & 327 & 263 & 195 & 215 & 417 \end{bmatrix}$$

So, 172, 208, 268, 327, 101, 123, 268, 327, 216, 263, 162, 195, 176, 215, 343, 417 is the message sent.

67. We need one-half the sum of the two matrices. This represents Male and Female per 100,000.

Answer: $$\begin{bmatrix} 2137 & 84 & 41.5 & 128.5 & 158.5 & 317 & 738 \\ 1285.5 & 64 & 30.5 & 115 & 136 & 229 & 590 \\ 969.5 & 46.5 & 24 & 98 & 139 & 224 & 476.5 \\ 852.5 & 45 & 23 & 97.5 & 142.5 & 236.5 & 463 \\ 809 & 44.5 & 22.5 & 99 & 141.5 & 240.5 & 455.5 \end{bmatrix}$$

69. a. A 5% decrease means multiply each entry by 0.95 to get the new budget.

Answer: $$\begin{bmatrix} 0.665 & 8.075 & 9.690 & 1.045 & 5.320 & 3.420 \\ 0.475 & 0.190 & 5.795 & 1.235 & 0.190 & 0.950 \\ 2.090 & 0.380 & 8.360 & 1.140 & 1.140 & 4.560 \\ 239.210 & 60.230 & 77.520 & 33.440 & 51.585 & 136.990 \\ 28.500 & 0.950 & 0.950 & 0.950 & 0.950 & 0.950 \\ 749.455 & 0 & 0 & 0 & 0 & 0 \end{bmatrix}$$

b. An 8% increase means multiply each entry by 1.08 to get the new budget.

Answer: $$\begin{bmatrix} 0.756 & 9.180 & 11.016 & 1.188 & 6.048 & 3.888 \\ 0.540 & 0.216 & 6.588 & 1.404 & 0.216 & 1.080 \\ 2.376 & 0.432 & 9.504 & 1.296 & 1.296 & 5.184 \\ 271.944 & 68.472 & 88.128 & 38.016 & 58.644 & 155.736 \\ 32.400 & 1.080 & 1.080 & 1.080 & 1.080 & 1.080 \\ 852.012 & 0 & 0 & 0 & 0 & 0 \end{bmatrix}$$

71. To get new efficiencies, multiply each entry by 0.9 to reflect the 10% decrease.

$$\begin{bmatrix} .900 & .855 & .855 & .855 & .855 & .765 & .765 & .765 & .765 \\ .765 & .900 & .855 & .855 & .855 & .855 & .765 & .765 & .765 \\ .765 & .765 & .900 & .855 & .855 & .855 & .855 & .765 & .765 \\ .765 & .765 & .765 & .900 & .855 & .855 & .855 & .855 & .765 \\ .765 & .765 & .765 & .765 & .900 & .855 & .855 & .855 & .855 \\ .855 & .765 & .765 & .765 & .765 & .900 & .855 & .855 & .855 \\ .855 & .855 & .765 & .765 & .765 & .765 & .900 & .855 & .855 \\ .855 & .855 & .855 & .765 & .765 & .765 & .765 & .900 & .855 \\ .855 & .855 & .855 & .855 & .765 & .765 & .765 & .765 & .900 \end{bmatrix}$$

Exercise 3.3

1. $\left[\begin{array}{ccc|c} 1 & -2 & -1 & -7 \\ 3 & 1 & 2 & 0 \\ 4 & 2 & 2 & 1 \end{array}\right] \; \underset{\rightarrow}{-3R_1 + R_2 \rightarrow R_2}$

$\left[\begin{array}{ccc|c} 1 & -2 & -1 & -7 \\ 0 & 7 & 5 & 21 \\ 4 & 2 & 2 & 1 \end{array}\right]$

3. $\left[\begin{array}{ccc|c} 3 & 2 & 4 & 0 \\ 2 & -1 & 2 & 0 \\ 1 & -2 & -4 & 0 \end{array}\right]$

5. $\left[\begin{array}{ccc|c} 1 & -3 & 4 & 2 \\ 2 & 0 & 2 & 1 \\ 1 & 2 & 1 & 1 \end{array}\right]$

7. $x = 2,\ y = \frac{1}{2},\ z = -5$

9. $x = 18, y = 10, z = 0$

11. $\left[\begin{array}{ccc|c} 1 & 1 & 2 & -1 \\ 0 & 3 & 1 & 7 \\ 0 & -2 & 4 & 0 \end{array}\right] \; \overset{\rightarrow}{R_3 + R_2 \rightarrow R_2} \; \left[\begin{array}{ccc|c} 1 & 1 & 2 & -1 \\ 0 & 1 & 5 & 7 \\ 0 & -2 & 4 & 0 \end{array}\right] \; \begin{array}{c} -R_2 + R_1 \rightarrow R_1 \\ \rightarrow \\ 2R_2 + R_3 \rightarrow R_3 \end{array}$

$\left[\begin{array}{ccc|c} 1 & 0 & -3 & -8 \\ 0 & 1 & 5 & 7 \\ 0 & 0 & 14 & 14 \end{array}\right] \; \begin{array}{c} \rightarrow \\ \frac{1}{14}R_3 \rightarrow R_3 \end{array} \; \left[\begin{array}{ccc|c} 1 & 0 & -3 & -8 \\ 0 & 1 & 5 & 7 \\ 0 & 0 & 1 & 1 \end{array}\right] \; \begin{array}{c} 3R_3 + R_1 \rightarrow R_1 \\ -5R_3 + R_2 \rightarrow R_2 \end{array} \; \left[\begin{array}{ccc|c} 1 & 0 & 0 & -5 \\ 0 & 1 & 0 & 2 \\ 0 & 0 & 1 & 1 \end{array}\right]$

Solution: $x = -5,\ y = 2,\ z = 1$

13. $\left[\begin{array}{ccc|c} 1 & 2 & 5 & -4 \\ 2 & -2 & 4 & -2 \\ 0 & 1 & -3 & 7 \end{array}\right] \; \overset{\rightarrow}{-2R_1 + R_2 \rightarrow R_2} \; \left[\begin{array}{ccc|c} 1 & 2 & 5 & -4 \\ 0 & -6 & -6 & 6 \\ 0 & 1 & -3 & 7 \end{array}\right] \; \overset{R_2 \leftrightarrow R_3}{\rightarrow} \; \left[\begin{array}{ccc|c} 1 & 2 & 5 & -4 \\ 0 & 1 & -3 & 7 \\ 0 & -6 & -6 & 6 \end{array}\right] \; \begin{array}{c} -2R_2 + R_1 \rightarrow R_1 \\ \rightarrow \\ 6R_2 + R_3 \rightarrow R_3 \end{array}$

$\left[\begin{array}{ccc|c} 1 & 0 & 11 & -18 \\ 0 & 1 & -3 & 7 \\ 0 & 0 & -24 & 48 \end{array}\right] \; \begin{array}{c} \rightarrow \\ -\frac{1}{24}R_3 \rightarrow R_3 \end{array} \; \left[\begin{array}{ccc|c} 1 & 0 & 11 & -18 \\ 0 & 1 & -3 & 7 \\ 0 & 0 & 1 & -2 \end{array}\right] \; \begin{array}{c} -11R_3 + R_1 \rightarrow R_1 \\ 3R_3 + R_2 \rightarrow R_2 \\ \rightarrow \end{array} \; \left[\begin{array}{ccc|c} 1 & 0 & 0 & 4 \\ 0 & 1 & 0 & 1 \\ 0 & 0 & 1 & -2 \end{array}\right]$

Solution: $x = 4,\ y = 1,\ z = -2$

15. $\left[\begin{array}{cc|c} 7 & -2 & -1 \\ 3 & 6 & 11 \end{array}\right] \; \begin{array}{c} -2R_2 + R_1 \rightarrow R_1 \\ \rightarrow \end{array} \; \left[\begin{array}{cc|c} 1 & -14 & -23 \\ 3 & 6 & 11 \end{array}\right] \; \begin{array}{c} \rightarrow \\ -3R_1 + R_2 \rightarrow R_2 \end{array}$

$\left[\begin{array}{cc|c} 1 & -14 & -23 \\ 0 & 48 & 80 \end{array}\right] \; \begin{array}{c} \rightarrow \\ \frac{1}{48}R_2 \rightarrow R_2 \end{array} \; \left[\begin{array}{cc|c} 1 & -14 & -23 \\ 0 & 1 & 5/3 \end{array}\right] \; \begin{array}{c} 14R_2 + R_1 \rightarrow R_1 \\ \rightarrow \end{array} \; \left[\begin{array}{cc|c} 1 & 0 & 1/3 \\ 0 & 1 & 5/3 \end{array}\right]$

Solution: $x = \frac{1}{3},\ y = \frac{5}{3}$

17. $\left[\begin{array}{ccc|c} 1 & 1 & -1 & 0 \\ 1 & 2 & 3 & -5 \\ 2 & -1 & -13 & 17 \end{array}\right] \; \begin{array}{c} \rightarrow \\ -R_1 + R_2 \rightarrow R_2 \\ -2R_1 + R_3 \rightarrow R_3 \end{array} \; \left[\begin{array}{ccc|c} 1 & 1 & -1 & 0 \\ 0 & 1 & 4 & -5 \\ 0 & -3 & -11 & 17 \end{array}\right] \; \begin{array}{c} -R_2 + R_1 \rightarrow R_1 \\ \rightarrow \\ 3R_2 + R_3 \rightarrow R_3 \end{array}$

$\left[\begin{array}{ccc|c} 1 & 0 & -5 & 5 \\ 0 & 1 & 4 & -5 \\ 0 & 0 & 1 & 2 \end{array}\right] \; \begin{array}{c} 5R_3 + R_1 \rightarrow R_1 \\ -4R_3 + R_2 \rightarrow R_2 \\ \rightarrow \end{array} \; \left[\begin{array}{ccc|c} 1 & 0 & 0 & 15 \\ 0 & 1 & 0 & -13 \\ 0 & 0 & 1 & 2 \end{array}\right]$

Solution: $x = 15,\ y = -13,\ z = 2$

19. $$\left[\begin{array}{ccc|c} 2 & -6 & -12 & 6 \\ 3 & -10 & -20 & 5 \\ 2 & 0 & -17 & -4 \end{array}\right] \xrightarrow{\frac{1}{2}R_1 \to R_1} \left[\begin{array}{ccc|c} 1 & -3 & -6 & 3 \\ 3 & -10 & -20 & 5 \\ 2 & 0 & -17 & -4 \end{array}\right] \begin{array}{c} \to \\ -3R_1 + R_2 \to R_2 \\ -2R_1 + R_3 \to R_3 \end{array}$$

$$\left[\begin{array}{ccc|c} 1 & -3 & -6 & 3 \\ 0 & -1 & -2 & -4 \\ 0 & 6 & -5 & -10 \end{array}\right] \begin{array}{c} \to \\ -R_2 \to R_2 \\ \end{array} \left[\begin{array}{ccc|c} 1 & -3 & -6 & 3 \\ 0 & 1 & 2 & 4 \\ 0 & 6 & -5 & -10 \end{array}\right] \begin{array}{c} 3R_2 + R_1 \to R_1 \\ \to \\ -6R_2 + R_3 \to R_3 \end{array}$$

$$\left[\begin{array}{ccc|c} 1 & 0 & 0 & 15 \\ 0 & 1 & 2 & 4 \\ 0 & 0 & -17 & -34 \end{array}\right] \begin{array}{c} \\ \to \\ -\frac{1}{17}R_3 \to R_3 \end{array} \left[\begin{array}{ccc|c} 1 & 0 & 0 & 15 \\ 0 & 1 & 2 & 4 \\ 0 & 0 & 1 & 2 \end{array}\right] \begin{array}{c} \to \\ -2R_3 + R_2 \to R_2 \\ \end{array} \left[\begin{array}{ccc|c} 1 & 0 & 0 & 15 \\ 0 & 1 & 0 & 0 \\ 0 & 0 & 1 & 2 \end{array}\right]$$

Solution: $x = 15,\ y = 0,\ z = 2$

21. $$\left[\begin{array}{ccc|c} 1 & -3 & 3 & 7 \\ 1 & 2 & -1 & -2 \\ 3 & 2 & 4 & 5 \end{array}\right] \begin{array}{c} \to \\ -R_1 + R_2 \to R_2 \\ -3R_1 + R_3 \to R_3 \end{array} \left[\begin{array}{ccc|c} 1 & -3 & 3 & 7 \\ 0 & 5 & -4 & -9 \\ 0 & 11 & -5 & -16 \end{array}\right] \begin{array}{c} \\ \to \\ -2R_2 + R_3 \to R_3 \end{array}$$

$$\left[\begin{array}{ccc|c} 1 & -3 & 3 & 7 \\ 0 & 5 & -4 & -9 \\ 0 & 1 & 3 & 2 \end{array}\right] \xrightarrow{R_2 \leftrightarrow R_3} \left[\begin{array}{ccc|c} 1 & -3 & 3 & 7 \\ 0 & 1 & 3 & 2 \\ 0 & 5 & -4 & -9 \end{array}\right] \begin{array}{c} 3R_2 + R_1 \to R_1 \\ \to \\ -5R_2 + R_3 \to R_3 \end{array}$$

$$\left[\begin{array}{ccc|c} 1 & 0 & 12 & 13 \\ 0 & 1 & 3 & 2 \\ 0 & 0 & -19 & -19 \end{array}\right] \begin{array}{c} \\ \to \\ -\frac{1}{19}R_3 \to R_3 \end{array} \left[\begin{array}{ccc|c} 1 & 0 & 12 & 13 \\ 0 & 1 & 3 & 2 \\ 0 & 0 & 1 & 1 \end{array}\right] \begin{array}{c} -12R_3 + R_1 \to R_1 \\ -3R_3 + R_2 \to R_2 \\ \to \end{array} \left[\begin{array}{ccc|c} 1 & 0 & 0 & 1 \\ 0 & 1 & 0 & -1 \\ 0 & 0 & 1 & 1 \end{array}\right]$$

Solution: $x = 1,\ y = -1,\ z = 1$

23.

$$\left[\begin{array}{rrrr|r} 1 & 3 & 2 & 2 & 3 \\ 1 & 1 & 3 & 0 & 4 \\ 2 & 0 & 2 & -3 & 4 \\ 1 & -3 & 0 & 0 & 1 \end{array}\right] \begin{array}{c} \to \\ -R_1 + R_2 \to R_2 \\ -2R_1 + R_3 \to R_3 \\ -R_1 + R_4 \to R_4 \end{array} \left[\begin{array}{rrrr|r} 1 & 3 & 2 & 2 & 3 \\ 0 & -2 & 1 & -2 & 1 \\ 0 & -6 & -2 & -7 & -2 \\ 0 & -6 & -2 & -2 & -2 \end{array}\right] \begin{array}{c} \\ \text{*See note} \\ -3R_4 + R_3 \to R_3 \\ \to \end{array}$$

$$\left[\begin{array}{rrrr|r} 1 & 3 & 2 & 2 & 3 \\ 0 & -2 & 1 & -2 & 1 \\ 0 & 12 & 4 & -1 & 4 \\ 0 & -6 & -2 & -2 & -2 \end{array}\right] \begin{array}{c} \\ \to \\ -R_3 \to R_3 \\ \text{then } R_3 \leftrightarrow R_4 \end{array} \left[\begin{array}{rrrr|r} 1 & 3 & 2 & 2 & 3 \\ 0 & -2 & 1 & -2 & 1 \\ 0 & -6 & -2 & -2 & -2 \\ 0 & -12 & -4 & 1 & -4 \end{array}\right] \begin{array}{c} -2R_4 + R_1 \to R_1 \\ 2R_4 + R_2 \to R_2 \\ 2R_4 + R_3 \to R_3 \\ \to \end{array}$$

$$\left[\begin{array}{rrrr|r} 1 & 27 & 10 & 0 & 11 \\ 0 & -26 & -7 & 0 & -7 \\ 0 & -30 & -10 & 0 & -10 \\ 0 & -12 & -4 & 1 & -4 \end{array}\right] \begin{array}{c} \to \\ -3R_2 \to R_2 \\ \\ -2R_3 \to R_3 \end{array} \left[\begin{array}{rrrr|r} 1 & 27 & 10 & 0 & 11 \\ 0 & 78 & 21 & 0 & 21 \\ 0 & 60 & 20 & 0 & 20 \\ 0 & -12 & -4 & 1 & -4 \end{array}\right] \begin{array}{c} \to \\ -R_3 + R_2 \to R_2 \\ \\ \\ \end{array}$$

$$\left[\begin{array}{rrrr|r} 1 & 27 & 10 & 0 & 11 \\ 0 & 18 & 1 & 0 & 1 \\ 0 & 60 & 20 & 0 & 20 \\ 0 & -12 & -4 & 1 & -4 \end{array}\right] \begin{array}{c} -10R_2 + R_1 \to R_1 \\ \to \\ -20R_2 + R_3 \to R_3 \\ 4R_2 + R_4 \to R_4 \end{array} \left[\begin{array}{rrrr|r} 1 & -153 & 0 & 0 & 1 \\ 0 & 18 & 1 & 0 & 1 \\ 0 & -300 & 0 & 0 & 0 \\ 0 & 60 & 0 & 1 & 0 \end{array}\right] \begin{array}{c} \\ \to \\ -\frac{1}{300}R_3 \to R_3 \\ \text{then } R_2 \leftrightarrow R_3 \end{array}$$

$$\left[\begin{array}{rrrr|r} 1 & -153 & 0 & 0 & 1 \\ 0 & 1 & 0 & 0 & 0 \\ 0 & 18 & 1 & 0 & 1 \\ 0 & 60 & 0 & 1 & 0 \end{array}\right] \begin{array}{c} 153R_2 + R_1 \to R_1 \\ \to \\ -18R_2 + R_3 \to R_3 \\ -60R_2 + R_4 \to R_4 \end{array} \left[\begin{array}{rrrr|r} 1 & 0 & 0 & 0 & 1 \\ 0 & 1 & 0 & 0 & 0 \\ 0 & 0 & 1 & 0 & 1 \\ 0 & 0 & 0 & 1 & 0 \end{array}\right]$$

Solution: $x_1 = 1, x_2 = 0, x_3 = 1, x_4 = 0$

*Note: The author's step by step method always works. However, it is often necessary to introduce fractions. This problem has been worked "backwards" to illustrate an alternate method and avoid fractions.

25.

$$\left[\begin{array}{rrrr|r} 1 & -2 & 3 & 1 & -2 \\ 1 & -3 & 1 & -1 & -7 \\ 1 & -1 & 0 & 0 & -2 \\ 1 & 0 & 1 & 1 & 2 \end{array}\right] \begin{array}{c} \to \\ -R_1 + R_2 \to R_2 \\ -R_1 + R_3 \to R_3 \\ -R_1 + R_4 \to R_4 \end{array} \left[\begin{array}{rrrr|r} 1 & -2 & 3 & 1 & -2 \\ 0 & -1 & -2 & -2 & -5 \\ 0 & 1 & -3 & -1 & 0 \\ 0 & 2 & -2 & 0 & 4 \end{array}\right] \begin{array}{c} \to \\ -1R_2 \to R_2 \\ \\ \\ \end{array}$$

$$\left[\begin{array}{rrrr|r} 1 & -2 & 3 & 1 & -2 \\ 0 & 1 & 2 & 2 & 5 \\ 0 & 1 & -3 & -1 & 0 \\ 0 & 2 & -2 & 0 & 4 \end{array}\right] \begin{array}{c} 2R_2 + R_1 \to R_1 \\ \to \\ -R_2 + R_3 \to R_3 \\ -2R_2 + R_4 \to R_4 \end{array} \left[\begin{array}{rrrr|r} 1 & 0 & 7 & 5 & 8 \\ 0 & 1 & 2 & 2 & 5 \\ 0 & 0 & -5 & -3 & -5 \\ 0 & 0 & -6 & -4 & -6 \end{array}\right] \begin{array}{c} \\ \to \\ -\frac{1}{5}R_3 \to R_3 \\ \\ \end{array}$$

$$\left[\begin{array}{rrrr|r} 1 & 0 & 7 & 5 & 8 \\ 0 & 1 & 2 & 2 & 5 \\ 0 & 0 & 1 & 3/5 & 1 \\ 0 & 0 & -6 & -4 & -6 \end{array}\right] \begin{array}{c} -7R_3 + R_1 \to R_1 \\ -2R_3 + R_2 \to R_2 \\ \to \\ 6R_3 + R_4 \to R_4 \end{array} \left[\begin{array}{rrrr|r} 1 & 0 & 0 & 4/5 & 1 \\ 0 & 1 & 0 & 4/5 & 3 \\ 0 & 0 & 1 & 3/5 & 1 \\ 0 & 0 & 0 & 2/5 & 0 \end{array}\right] \begin{array}{c} \text{Now take } \frac{5}{2}R_4 \\ \text{and get 0's} \\ \text{in column 4;} \\ \text{Result is} \end{array} \left[\begin{array}{rrrr|r} 1 & 0 & 0 & 0 & 1 \\ 0 & 1 & 0 & 0 & 3 \\ 0 & 0 & 1 & 0 & 1 \\ 0 & 0 & 0 & 1 & 0 \end{array}\right]$$

Solution: $x = 1,\ y = 3,\ z = 1,\ w = 0$

27. There is no solution since the last row of the reduced matrix says $0x + 0y + 0z = 1$, which is not possible.

29. From first row: $x-\frac{2}{3}z=\frac{11}{3}$

From second row: $y+\frac{1}{3}z=-\frac{1}{3}$

General solution: $x=\frac{11}{3}+\frac{2}{3}z=\frac{11+2z}{3}$, $y=-\frac{1}{3}-\frac{1}{3}z=\frac{-1-z}{3}$ for any real number z.

31.

$$\left[\begin{array}{ccc|c}1&1&1&0\\2&-1&-1&0\\-1&2&2&0\end{array}\right]\begin{array}{c}\to\\-2R_1+R_2\to R_2\\R_1+R_3\to R_3\end{array}\left[\begin{array}{ccc|c}1&1&1&0\\0&-3&-3&0\\0&3&3&0\end{array}\right]\begin{array}{c}\\ \to\\R_2+R_3\to R_3\end{array}$$

$$\left[\begin{array}{ccc|c}1&1&1&0\\0&-3&-3&0\\0&0&0&0\end{array}\right]\begin{array}{c}\to\\-\frac{1}{3}R_2\to R_2\\ \\ \end{array}\left[\begin{array}{ccc|c}1&1&1&0\\0&1&1&0\\0&0&0&0\end{array}\right]\begin{array}{c}-R_2+R_1\to R_1\\ \to\\ \\ \end{array}\left[\begin{array}{ccc|c}1&0&0&0\\0&1&1&0\\0&0&0&0\end{array}\right]$$

General solution: $x=0,\ y=-z$

33.

$$\left[\begin{array}{ccc|c}3&2&1&0\\1&1&2&1\\2&1&-1&-1\end{array}\right]\begin{array}{c}R_1\leftrightarrow R_2\\ \to\\ \\ \end{array}\left[\begin{array}{ccc|c}1&1&2&1\\3&2&1&0\\2&1&-1&-1\end{array}\right]\begin{array}{c}\to\\-3R_1+R_2\to R_2\\-2R_1+R_3\to R_3\end{array}$$

$$\left[\begin{array}{ccc|c}1&1&2&1\\0&-1&-5&-3\\0&-1&-5&-3\end{array}\right]\begin{array}{c}\to\\ \\-R_2+R_3\to R_3\end{array}\left[\begin{array}{ccc|c}1&1&2&1\\0&-1&-5&-3\\0&0&0&0\end{array}\right]\begin{array}{c}R_2+R_1\to R_1\\ \to\\ \\ \end{array}$$

$$\left[\begin{array}{ccc|c}1&0&-3&-2\\0&-1&-5&-3\\0&0&0&0\end{array}\right]\begin{array}{c}\to\\-R_2\to R_2\\ \\ \end{array}\left[\begin{array}{ccc|c}1&0&-3&-2\\0&1&5&3\\0&0&0&0\end{array}\right]$$

General solution: $x=3z-2,\ y=3-5z$

35.

$$\left[\begin{array}{ccc|c}2&2&1&2\\1&-2&2&1\\-1&2&-2&-1\end{array}\right]\begin{array}{c}R_1\leftrightarrow R_2\\ \to\\ \\ \end{array}\left[\begin{array}{ccc|c}1&-2&2&1\\2&2&1&2\\-1&2&-2&-1\end{array}\right]\begin{array}{c}\to\\-2R_1+R_2\to R_2\\R_1+R_3\to R_3\end{array}$$

$$\left[\begin{array}{ccc|c}1&-2&2&1\\0&6&-3&0\\0&0&0&0\end{array}\right]\begin{array}{c}\to\\\frac{1}{6}R_2\to R_2\\ \\ \end{array}\left[\begin{array}{ccc|c}1&-2&2&1\\0&1&-1/2&0\\0&0&0&0\end{array}\right]\begin{array}{c}2R_2+R_1\to R_1\\ \to\\ \\ \end{array}\left[\begin{array}{ccc|c}1&0&1&1\\0&1&-1/2&0\\0&0&0&0\end{array}\right]$$

General solution: $x=1-z,\ y=\frac{1}{2}z$

37.

$$\left[\begin{array}{ccc|c}2&-5&1&-9\\1&4&-6&2\\3&-4&-2&-10\end{array}\right]\begin{array}{c}R_1\leftrightarrow R_2\\ \to\\ \\ \end{array}\left[\begin{array}{ccc|c}1&4&-6&2\\2&-5&1&-9\\3&-4&-2&-10\end{array}\right]\begin{array}{c}\to\\-2R_1+R_2\to R_2\\-3R_1+R_3\to R_3\end{array}\left[\begin{array}{ccc|c}1&4&-6&2\\0&-13&13&-13\\0&-16&16&-16\end{array}\right]\begin{array}{c}\to\\-\frac{1}{13}R_2\to R_2\\ \\ \end{array}$$

$$\left[\begin{array}{ccc|c}1&4&-6&2\\0&1&-1&1\\0&-16&16&-16\end{array}\right]\begin{array}{c}-4R_2+R_1\to R_1\\ \to\\16R_2+R_3\to R_3\end{array}\left[\begin{array}{ccc|c}1&0&-2&-2\\0&1&-1&1\\0&0&0&0\end{array}\right]$$

General solution: $x=2z-2,\ y=z+1$

39. $\left[\begin{array}{ccc|c} 1 & 1 & 1 & 3 \\ 1 & -1 & 1 & 4 \end{array}\right] \begin{array}{c} \to \\ -R_1+R_2 \to R_2 \end{array} \left[\begin{array}{ccc|c} 1 & 1 & 1 & 3 \\ 0 & -2 & 0 & 1 \end{array}\right] \begin{array}{c} \to \\ -\frac{1}{2}R_2 \to R_2 \end{array}$

$\left[\begin{array}{ccc|c} 1 & 1 & 1 & 3 \\ 0 & 1 & 0 & -1/2 \end{array}\right] \begin{array}{c} -R_2+R_1 \to R_1 \\ \to \end{array} \left[\begin{array}{ccc|c} 1 & 0 & 1 & 7/2 \\ 0 & 1 & 0 & -1/2 \end{array}\right]$

General Solution: $x = \dfrac{7}{2} - z$, $y = -\dfrac{1}{2}$, $z = z$

41. $\left[\begin{array}{ccc|c} 3 & -2 & 5 & 14 \\ 2 & -3 & 4 & 8 \end{array}\right] \begin{array}{c} -R_2+R_1 \to R_1 \\ \to \end{array} \left[\begin{array}{ccc|c} 1 & 1 & 1 & 6 \\ 2 & -3 & 4 & 8 \end{array}\right] \begin{array}{c} \to \\ -2R_1+R_2 \to R_2 \end{array}$

$\left[\begin{array}{ccc|c} 1 & 1 & 1 & 6 \\ 0 & -5 & 2 & -4 \end{array}\right] \begin{array}{c} \to \\ -\frac{1}{5}R_2 \to R_2 \end{array} \left[\begin{array}{ccc|c} 1 & 1 & 1 & 6 \\ 0 & 1 & -\frac{2}{5} & \frac{4}{5} \end{array}\right] \begin{array}{c} -R_2+R_1 \to R_1 \\ \to \end{array} \left[\begin{array}{ccc|c} 1 & 0 & \frac{7}{5} & \frac{26}{5} \\ 0 & 1 & -\frac{2}{5} & \frac{4}{5} \end{array}\right]$

General Solution: $x = \dfrac{26}{5} - \dfrac{7}{5}z$, $y = \dfrac{4}{5} + \dfrac{2}{5}z$

43. $\left[\begin{array}{ccc|c} -0.6 & 0.1 & 0.3 & 0 \\ 0.4 & -0.7 & 0.2 & 0 \\ 0.2 & 0.6 & -0.5 & 0 \end{array}\right] \begin{array}{c} 10R_1 \to R_1 \\ 10R_2 \to R_2 \\ 10R_3 \to R_3 \end{array} \left[\begin{array}{ccc|c} -6 & 1 & 3 & 0 \\ 4 & -7 & 2 & 0 \\ 2 & 6 & -5 & 0 \end{array}\right] \begin{array}{c} -\frac{1}{6}R_1 \to R_1 \\ -2R_3+R_2 \to R_2 \\ \to \end{array}$

$\left[\begin{array}{ccc|c} 1 & -\frac{1}{6} & -\frac{1}{2} & 0 \\ 0 & -19 & 12 & 0 \\ 2 & 6 & -5 & 0 \end{array}\right] \begin{array}{c} \to \\ -\frac{1}{19}R_2 \to R_2 \\ -2R_1+R_3 \to R_3 \end{array} \left[\begin{array}{ccc|c} 1 & -\frac{1}{6} & -\frac{1}{2} & 0 \\ 0 & 1 & -\frac{12}{19} & 0 \\ 0 & \frac{19}{3} & -4 & 0 \end{array}\right] \begin{array}{c} \\ \to \\ -\frac{19}{3}R_2+R_3 \to R_3 \end{array}$

$\left[\begin{array}{ccc|c} 1 & -\frac{1}{6} & -\frac{1}{2} & 0 \\ 0 & 1 & -\frac{12}{19} & 0 \\ 0 & 0 & 0 & 0 \end{array}\right] \begin{array}{c} \frac{1}{6}R_2+R_1 \to R_1 \\ \\ \\ \end{array} \left[\begin{array}{ccc|c} 1 & 0 & -\frac{23}{38} & 0 \\ 0 & 1 & -\frac{12}{19} & 0 \\ 0 & 0 & 0 & 0 \end{array}\right]$

General solution: $x_1 = \dfrac{23}{38}x_3$, $x_2 = \dfrac{12}{19}x_3$

45. $\left[\begin{array}{cccc|c} 3 & 2 & 1 & -1 & 3 \\ 1 & -1 & -2 & 2 & 2 \\ 2 & 3 & -1 & 1 & 1 \\ -1 & 1 & 2 & -2 & -2 \end{array}\right] \begin{array}{c} R_1 \leftrightarrow R_2 \\ \to \\ \\ \\ \end{array} \left[\begin{array}{cccc|c} 1 & -1 & -2 & 2 & 2 \\ 3 & 2 & 1 & -1 & 3 \\ 2 & 3 & -1 & 1 & 1 \\ -1 & 1 & 2 & -2 & -2 \end{array}\right] \begin{array}{c} \to \\ -3R_1+R_2 \to R_2 \\ -2R_1+R_3 \to R_3 \\ R_1+R_4 \to R_4 \end{array}$

$\left[\begin{array}{cccc|c} 1 & -1 & -2 & 2 & 2 \\ 0 & 5 & 7 & -7 & -3 \\ 0 & 5 & 3 & -3 & -3 \\ 0 & 0 & 0 & 0 & 0 \end{array}\right] \begin{array}{c} \\ \to \\ -R_2+R_3 \to R_3 \\ \\ \end{array} \left[\begin{array}{cccc|c} 1 & -1 & -2 & 2 & 2 \\ 0 & 5 & 7 & -7 & -3 \\ 0 & 0 & -4 & 4 & 0 \\ 0 & 0 & 0 & 0 & 0 \end{array}\right] \begin{array}{c} \to \\ \frac{1}{5}R_2 \to R_2 \\ -\frac{1}{4}R_3 \to R_3 \\ \\ \end{array}$

$\left[\begin{array}{cccc|c} 1 & -1 & -2 & 2 & 2 \\ 0 & 1 & \frac{7}{5} & -\frac{7}{5} & -\frac{3}{5} \\ 0 & 0 & 1 & -1 & 0 \\ 0 & 0 & 0 & 0 & 0 \end{array}\right] \begin{array}{c} R_2+R_1 \to R_1 \\ -\frac{7}{5}R_3+R_2 \to R_2 \\ \to \\ \\ \end{array} \left[\begin{array}{cccc|c} 1 & 0 & -\frac{3}{5} & \frac{3}{5} & \frac{7}{5} \\ 0 & 1 & 0 & 0 & -\frac{3}{5} \\ 0 & 0 & 1 & -1 & 0 \\ 0 & 0 & 0 & 0 & 0 \end{array}\right] \begin{array}{c} \frac{3}{5}R_3+R_1 \to R_1 \\ \to \\ \\ \\ \end{array} \left[\begin{array}{cccc|c} 1 & 0 & 0 & 0 & \frac{7}{5} \\ 0 & 1 & 0 & 0 & -\frac{3}{5} \\ 0 & 0 & 1 & -1 & 0 \\ 0 & 0 & 0 & 0 & 0 \end{array}\right]$

General solution: $x = \frac{7}{5}$, $y = -\frac{3}{5}$, $z = w$

47. $\left[\begin{array}{cccc|c} 1 & 2 & -1 & 1 & 3 \\ 1 & 3 & 4 & 1 & -2 \\ 2 & 5 & 2 & 2 & 1 \\ 2 & 3 & -6 & 2 & 3 \end{array}\right] \begin{array}{c} \to \\ -R_1+R_2\to R_2 \\ -2R_1+R_3\to R_3 \\ -2R_1+R_4\to R_4 \end{array} \left[\begin{array}{cccc|c} 1 & 2 & -1 & 1 & 3 \\ 0 & 1 & 5 & 0 & -5 \\ 0 & 1 & 4 & 0 & -5 \\ 0 & -1 & -4 & 0 & -3 \end{array}\right] \begin{array}{c} -2R_2+R_1\to R_1 \\ \to \\ -R_2+R_3\to R_3 \\ R_2+R_4\to R_4 \end{array} \left[\begin{array}{cccc|c} 1 & 0 & -11 & 1 & 13 \\ 0 & 1 & 5 & 0 & -5 \\ 0 & 0 & -1 & 0 & 0 \\ 0 & 0 & 1 & 0 & -8 \end{array}\right]$

The last two rows state that $x_3 = 0$ and $x_3 = -8$. Thus, there is no solution.

49. $\left[\begin{array}{cccc|c} 1 & 3 & 4 & -1 & 1 \\ 1 & -1 & -2 & 1 & 2 \\ 1 & 2 & 3 & 1 & 0 \\ 2 & 2 & 2 & 0 & 3 \end{array}\right] \begin{array}{c} \to \\ -R_1+R_2\to R_2 \\ -R_1+R_3\to R_3 \\ -2R_1+R_4\to R_4 \end{array} \left[\begin{array}{cccc|c} 1 & 3 & 4 & -1 & 1 \\ 0 & -4 & -6 & 2 & 1 \\ 0 & -1 & -1 & 2 & -1 \\ 0 & -4 & -6 & 2 & 1 \end{array}\right] \; -R_2+R_4\to R_4$

$\left[\begin{array}{cccc|c} 1 & 3 & 4 & -1 & 1 \\ 0 & -4 & -6 & 2 & 1 \\ 0 & -1 & -1 & 2 & -1 \\ 0 & 0 & 0 & 0 & 0 \end{array}\right] \begin{array}{c} \to \\ -5R_3+R_2\to R_2 \\ \\ \\ \end{array} \left[\begin{array}{cccc|c} 1 & 3 & 4 & -1 & 1 \\ 0 & 1 & -1 & -8 & 6 \\ 0 & -1 & -1 & 2 & -1 \\ 0 & 0 & 0 & 0 & 0 \end{array}\right] \begin{array}{c} -3R_2+R_1\to R_1 \\ \to \\ R_2+R_3\to R_3 \\ \\ \end{array}$

$\left[\begin{array}{cccc|c} 1 & 0 & 7 & 23 & -17 \\ 0 & 1 & -1 & -8 & 6 \\ 0 & 0 & -2 & -6 & 5 \\ 0 & 0 & 0 & 0 & 0 \end{array}\right] \begin{array}{c} \\ \to \\ -\frac{1}{2}R_3\to R_3 \\ \\ \end{array} \left[\begin{array}{cccc|c} 1 & 0 & 7 & 23 & -17 \\ 0 & 1 & -1 & -8 & 6 \\ 0 & 0 & 1 & 3 & -\frac{5}{2} \\ 0 & 0 & 0 & 0 & 0 \end{array}\right] \begin{array}{c} -7R_3+R_1\to R_1 \\ R_3+R_2\to R_2 \\ \to \\ \\ \end{array} \left[\begin{array}{cccc|c} 1 & 0 & 0 & 2 & \frac{1}{2} \\ 0 & 1 & 0 & -5 & \frac{7}{2} \\ 0 & 0 & 1 & 3 & -\frac{5}{2} \\ 0 & 0 & 0 & 0 & 0 \end{array}\right]$

General solution: $x_1 = \frac{1}{2} - 2x_4,\ x_2 = \frac{7}{2} + 5x_4,\ x_3 = -\frac{5}{2} - 3x_4$

51. $\left[\begin{array}{cc|c} a_1 & b_1 & c_1 \\ a_2 & b_2 & c_2 \end{array}\right] \begin{array}{c} \frac{1}{a_1}R_1\to R_1 \\ \to \end{array} \left[\begin{array}{cc|c} 1 & \frac{b_1}{a_1} & \frac{c_1}{a_1} \\ a_2 & b_2 & c_2 \end{array}\right] \begin{array}{c} \to \\ -a_2R_1+R_2\to R_2 \end{array} \left[\begin{array}{cc|c} 1 & \frac{b_1}{a_1} & \frac{c_1}{a_1} \\ 0 & \frac{b_2a_1-b_1a_2}{a_1} & \frac{c_2a_1-c_1a_2}{a_1} \end{array}\right]$

$\begin{array}{c} -\frac{b_1}{a_1}\cdot\frac{a_1}{b_2a_1-b_1a_2}R_2+R_1\to R_1 \\ \to \end{array} \left[\begin{array}{cc|c} 1 & 0 & \frac{b_2c_1-c_2b_1}{a_1b_2-a_2b_1} \\ 0 & \frac{b_2a_1-b_1a_2}{a_1} & \frac{c_2a_1-c_1a_2}{a_1} \end{array}\right]$ Thus, $x = \frac{b_2c_1-c_2b_1}{a_1b_2-a_2b_1}$.

Fractions and messy computations cannot always be avoided.

53. Let $x = 15\%$ investment amount. Let $y = 16\%$ investment amount.

$x + y = 23{,}500$ total investment

$0.15x + 0.16y = 3{,}625$ investment income

Use the results for x in #51. $x = \dfrac{23{,}500(0.16) - 1(3625)}{1(0.16) - 1(0.15)} = \dfrac{135}{0.01} = 13{,}500$

Then $13{,}500 + y = 23{,}500$ gives $y = 10{,}000$. Invest \$13,500 at 15% and \$10,000 at 16%.

Income: $0.15(13{,}500) = \$2{,}025$ and $0.16(10{,}000) = \$1{,}600$

55. Let x = cups of Beef. Let y = cups of Sirloin Burger.
Fat: $4x + 9y = 80$, Cholesterol: $25x + 20y = 210$

Use the result for x in #51. $x = \dfrac{20(80) - 210(9)}{4(20) - 25(9)} = \dfrac{-290}{-145} = 2$

Then $4(2) + 9y = 80$ gives $y = 8$. Use 2 cups of Beef and 8 cups of Sirloin Burger.

57. Let x = ounces of AF, y = ounces of FP, z = ounces of NMG
We start directly with the augmented matrix.

$$\begin{matrix}\text{Calories}\\ \text{Fat}\\ \text{Carbohydrates}\end{matrix}\left[\begin{array}{ccc|c}50 & 108 & 127 & 443\\ 0 & 0.1 & 5.5 & 5.7\\ 22 & 25.7 & 18 & 113.4\end{array}\right]\begin{matrix}\frac{1}{50}R_1\to R_1\\ 10R_2\to R_2\\ \to\end{matrix}\left[\begin{array}{ccc|c}1 & 2.16 & 2.54 & 8.86\\ 0 & 1 & 55 & 57\\ 22 & 25.7 & 18 & 113.4\end{array}\right]\begin{matrix}\\ \to\\ -22R_1+R_3\to R_3\end{matrix}$$

$$\left[\begin{array}{ccc|c}1 & 2.16 & 2.54 & 8.86\\ 0 & 1 & 55 & 57\\ 0 & -21.82 & -37.88 & -81.52\end{array}\right]\begin{matrix}-2.16R_2+R_1\to R_1\\ \to\\ 21.82R_2+R_3\to R_3\end{matrix}\left[\begin{array}{ccc|c}1 & 0 & -116.26 & -114.26\\ 0 & 1 & 55 & 57\\ 0 & 0 & 1162.22 & 1162.22\end{array}\right]\begin{matrix}\\ \to\\ \frac{1}{1162.22}R_3\to R_3\end{matrix}$$

$$\left[\begin{array}{ccc|c}1 & 0 & -116.26 & -114.26\\ 0 & 1 & 55 & 57\\ 0 & 0 & 1 & 1\end{array}\right]\begin{matrix}116.26R_3+R_1\to R_1\\ -55R_3+R_2\to R_2\\ \to\end{matrix}\left[\begin{array}{ccc|c}1 & 0 & 0 & 2\\ 0 & 1 & 0 & 2\\ 0 & 0 & 1 & 1\end{array}\right]$$

Use 2 ounces of AF, 2 ounces of FP, and 1 ounce of NMG.

59. x = # porfolio I $\quad 2x+4y+2z=12$
y= #portfolio II $\quad x+2y+2z=6$
z = #portfolio III $\quad 3y+3z=6$

$$\left[\begin{array}{ccc|c}2 & 4 & 2 & 12\\ 1 & 2 & 2 & 6\\ 0 & 3 & 3 & 6\end{array}\right]\begin{matrix}R_1\to R_2\\ \to\\ \end{matrix}$$

$$\left[\begin{array}{ccc|c}1 & 2 & 2 & 6\\ 2 & 4 & 2 & 12\\ 0 & 3 & 3 & 6\end{array}\right]\begin{matrix}\to\\ -2R_1+R_2\to R_2\\ \end{matrix}\left[\begin{array}{ccc|c}1 & 2 & 2 & 6\\ 0 & 0 & -2 & 0\\ 0 & 3 & 3 & 6\end{array}\right]\begin{matrix}\to\\ -\frac{1}{2}R_2\to R_2\\ \frac{1}{3}R_3\to R_3\end{matrix}\left[\begin{array}{ccc|c}1 & 2 & 2 & 6\\ 0 & 0 & 1 & 0\\ 0 & 1 & 1 & 2\end{array}\right]\begin{matrix}-2R_2+R_1\to R_1\\ \to\\ -R_2+R_3\to R_3\end{matrix}$$

$$\left[\begin{array}{ccc|c}1 & 2 & 0 & 6\\ 0 & 0 & 1 & 0\\ 0 & 1 & 0 & 2\end{array}\right]\begin{matrix}R_2\leftrightarrow R_3\\ \to\end{matrix}\left[\begin{array}{ccc|c}1 & 2 & 0 & 6\\ 0 & 1 & 0 & 2\\ 0 & 0 & 1 & 0\end{array}\right]\begin{matrix}-2R_2+R_1\to R_1\\ \to\\ \end{matrix}\left[\begin{array}{ccc|c}1 & 0 & 0 & 2\\ 0 & 1 & 0 & 2\\ 0 & 0 & 1 & 0\end{array}\right]$$

2 units each of portfolio I and portfolio II.

61. $0.1S+3.4M+2.2B=12.1$ iron needed
$8.5S+22M+10B=97$ protein needed
$1S+20M+12B=70$ carbohydrates needed

$$\left[\begin{array}{ccc|c}0.1 & 3.4 & 2.2 & 12.1\\ 8.5 & 22 & 10 & 97\\ 1 & 20 & 12 & 70\end{array}\right]\begin{matrix}10R_1\to R_1\\ \to\\ \end{matrix}$$

$$\left[\begin{array}{ccc|c}1 & 34 & 22 & 121\\ 8.5 & 22 & 10 & 97\\ 1 & 20 & 12 & 70\end{array}\right]\begin{matrix}\to\\ -8.5R_1+R_2\to R_2\\ -R_1+R_3\to R_3\end{matrix}\left[\begin{array}{ccc|c}1 & 34 & 22 & 121\\ 0 & -267 & -177 & -931.5\\ 0 & -14 & -10 & -51\end{array}\right]\begin{matrix}\to\\ -19R_3+R_2\to R_2\\ \end{matrix}$$

$$\left[\begin{array}{ccc|c}1 & 34 & 22 & 121\\ 0 & -1 & 13 & 37.5\\ 0 & -14 & -10 & -51\end{array}\right]\begin{matrix}34R_2+R_1\to R_1\\ \to\\ -14R_2+R_3\to R_3\end{matrix}\left[\begin{array}{ccc|c}1 & 0 & 464 & 1396\\ 0 & -1 & 13 & 37.5\\ 0 & 0 & -192 & -576\end{array}\right]\begin{matrix}\\ \to\\ -\frac{1}{192}R_3\to R_3\end{matrix}$$

$$\left[\begin{array}{ccc|c}1 & 0 & 464 & 1396\\ 0 & -1 & 13 & 37.5\\ 0 & 0 & 1 & 3\end{array}\right]\begin{matrix}-464R_3+R_1\to R_1\\ -13R_3+R_2\to R_2\\ \to\end{matrix}\left[\begin{array}{ccc|c}1 & 0 & 0 & 4\\ 0 & -1 & 0 & -1.5\\ 0 & 0 & 1 & 3\end{array}\right]\begin{matrix}-R_2\to R_2\\ \to\end{matrix}\left[\begin{array}{ccc|c}1 & 0 & 0 & 4\\ 0 & 1 & 0 & 1.5\\ 0 & 0 & 1 & 3\end{array}\right]$$

4 glasses of milk, 1.5 quarter-pound servings of meat, and 3 2-slice servings of bread. Note that this is $3/8$ pounds of meat and 6 slices of bread.

63. x = Type I bags; y = Type II bags; z = Type III bags; w = Type IV bags

$$\left[\begin{array}{cccc|c} 5 & 5 & 10 & 5 & 10{,}000 \\ 10 & 5 & 30 & 10 & 20{,}000 \\ 5 & 15 & 10 & 25 & 20{,}000 \end{array}\right] \begin{array}{c} \frac{1}{5}R_1 \to R_1 \\ \frac{1}{5}R_2 \to R_2 \\ \frac{1}{5}R_3 \to R_3 \end{array} \left[\begin{array}{cccc|c} 1 & 1 & 2 & 1 & 2000 \\ 2 & 1 & 6 & 2 & 4000 \\ 1 & 3 & 2 & 5 & 4000 \end{array}\right] \begin{array}{c} \to \\ -2R_1 + R_2 \to R_2 \\ -R_1 + R_3 \to R_3 \end{array}$$

$$\left[\begin{array}{cccc|c} 1 & 1 & 2 & 1 & 2000 \\ 0 & -1 & 2 & 0 & 0 \\ 0 & 2 & 0 & 4 & 2000 \end{array}\right] \begin{array}{c} R_2 + R_1 \to R_1 \\ \to \\ 2R_2 + R_3 \to R_3 \end{array} \left[\begin{array}{cccc|c} 1 & 0 & 4 & 1 & 2000 \\ 0 & -1 & 2 & 0 & 0 \\ 0 & 0 & 4 & 4 & 2000 \end{array}\right] \begin{array}{c} \to \\ -R_2 \to R_2 \\ \frac{1}{4}R_3 \to R_3 \end{array}$$

$$\left[\begin{array}{cccc|c} 1 & 0 & 4 & 1 & 2000 \\ 0 & 1 & -2 & 0 & 0 \\ 0 & 0 & 1 & 1 & 500 \end{array}\right] \begin{array}{c} -4R_3 + R_1 \to R_1 \\ 2R_3 + R_2 \to R_2 \\ \to \end{array} \left[\begin{array}{cccc|c} 1 & 0 & 0 & -3 & 0 \\ 0 & 1 & 0 & 2 & 1000 \\ 0 & 0 & 1 & 1 & 500 \end{array}\right]$$

Solution: $x = 3w$, $y = 1000 - 2w$, $z = 500 - w$, where w is any non-negative amount ≤ 500.

65. $$\left[\begin{array}{cccc|c} 100 & 40 & 30 & 20 & 102{,}300 \\ 10 & 3.6 & 6 & 2.4 & 15{,}345 \end{array}\right] \begin{array}{c} 0.01R_1 \to R_1 \\ \to \end{array} \left[\begin{array}{cccc|c} 1 & 0.40 & 0.30 & 0.20 & 1023 \\ 10 & 3.6 & 6 & 2.4 & 15{,}345 \end{array}\right] \begin{array}{c} \to \\ -10R_1 + R_2 \to R_2 \end{array}$$

$$\left[\begin{array}{cccc|c} 1 & 0.40 & 0.30 & 0.20 & 1023 \\ 0 & -0.40 & 3 & 0.40 & 5115 \end{array}\right] \begin{array}{c} R_2 + R_1 \to R_1 \\ \to \end{array} \left[\begin{array}{cccc|c} 1 & 0 & 3.3 & 0.6 & 6138 \\ 0 & -0.4 & 3 & 0.4 & 5115 \end{array}\right] \begin{array}{c} \to \\ -\frac{5}{2}R_2 \to R_2 \end{array}$$

$$\left[\begin{array}{cccc|c} 1 & 0 & 3.3 & 0.6 & 6138 \\ 0 & 1 & -7.5 & -1 & -12{,}787.5 \end{array}\right]$$

Oil = $6138 - 3.30C - 0.60R$ and Bank = $7.5C + R - 12{,}787.50$, where C = Computer and R = Retail. We leave it to the reader to explain and solve $-3.30C - 0.60R + 6138 \ge 0$

$$7.5C + R - 12{,}787.50 \ge 0$$

67.

Nutrient	Species x_1	x_2	x_3	
A	1	2	2	5100 units
B	1	0	3	6900 units
C	2	2	5	12,000 units

$$\left[\begin{array}{ccc|c} 1 & 2 & 2 & 5100 \\ 1 & 0 & 3 & 6900 \\ 2 & 2 & 5 & 12{,}000 \end{array}\right] \begin{array}{c} \to \\ -R_1 + R_2 \to R_2 \\ -2R_1 + R_3 \to R_3 \end{array} \left[\begin{array}{ccc|c} 1 & 2 & 2 & 5100 \\ 0 & -2 & 1 & 1800 \\ 0 & -2 & 1 & 1800 \end{array}\right] \begin{array}{c} R_2 + R_1 \to R_1 \\ \to \\ -R_2 + R_3 \to R_3 \end{array} \left[\begin{array}{ccc|c} 1 & 0 & 3 & 6900 \\ 0 & -2 & 1 & 1800 \\ 0 & 0 & 0 & 0 \end{array}\right]$$

Species $x_1 = 6900 - 3x_3$ and species $x_2 = -900 + \frac{1}{2}x_3$, with any amount of x_3 between 1800 and 2300.

69. x = units of Portfolio I $\quad 2x + 4y + 2z = 16$ common stock blocks
y = units of Portfolio II $\quad x + 2y + z = 8$ municipal bond blocks
z = units of Portfolio III $\quad 3y + 3z = 6$ preferred stock blocks

$$\left[\begin{array}{ccc|c} 2 & 4 & 2 & 16 \\ 1 & 2 & 1 & 8 \\ 0 & 3 & 3 & 6 \end{array}\right] \begin{array}{c} R_1 \leftrightarrow R_2 \\ \to \\ \end{array}$$

$$\left[\begin{array}{ccc|c} 1 & 2 & 1 & 8 \\ 2 & 4 & 2 & 16 \\ 0 & 3 & 3 & 6 \end{array}\right] \begin{array}{c} \to \\ -2R_1 + R_2 \to R_2 \\ \frac{1}{3}R_3 \to R_3 \end{array} \left[\begin{array}{ccc|c} 1 & 2 & 1 & 8 \\ 0 & 0 & 0 & 0 \\ 0 & 1 & 1 & 2 \end{array}\right] \begin{array}{c} -2R_3 + R_1 \to R_1 \\ \to \\ \end{array} \left[\begin{array}{ccc|c} 1 & 0 & -1 & 4 \\ 0 & 0 & 0 & 0 \\ 0 & 1 & 1 & 2 \end{array}\right]$$

Thus, $x = z + 4$ and $y = -z + 2$. Note that z = 0, 1, 2 only.

Possible Offerings	I	II	III
	4	2	0
	5	1	1
	6	0	2

Exercise 3.4

1. The product is the 3×3 identity matrix.

3. $\begin{bmatrix}1&2&1\\0&0&3\\1&0&1\end{bmatrix}\begin{bmatrix}0&-1/3&1\\1/2&0&-1/2\\0&1/3&0\end{bmatrix}=\begin{bmatrix}1&0&0\\0&1&0\\0&0&1\end{bmatrix}$

   Yes, $B = A^{-1}$.

5. $A^{-1}=\begin{bmatrix}1/3&0&0\\0&1/3&0\\0&0&1/3\end{bmatrix}$

   The inverse of a diagonal matrix is a diagonal matrix. Each entry on the diagonal is the multiplicative inverse of the original corresponding entry.

7. $A^{-1}=\dfrac{1}{(4)(2)-7(1)}\begin{bmatrix}2&-7\\-1&4\end{bmatrix}=\begin{bmatrix}2&-7\\-1&4\end{bmatrix}$

9. $ad-bc=2(2)-(-4)(-1)=0$, so no inverse exists.

11. $A^{-1}=\dfrac{1}{2(5)-2(4)}\begin{bmatrix}5&-2\\-4&2\end{bmatrix}=\begin{bmatrix}5/2&-1\\-2&1\end{bmatrix}$

13. $A^{-1}=\dfrac{1}{4(1)-7(2)}\begin{bmatrix}1&-7\\-2&4\end{bmatrix}=\begin{bmatrix}-1/10&7/10\\1/5&-2/5\end{bmatrix}$

15. $\left[\begin{array}{ccc|ccc}0&1&0&1&0&0\\1&1&0&0&1&0\\0&1&1&0&0&1\end{array}\right]\begin{array}{c}R_1\leftrightarrow R_2\\ \to\\ \end{array}\left[\begin{array}{ccc|ccc}1&1&0&0&1&0\\0&1&0&1&0&0\\0&1&1&0&0&1\end{array}\right]\begin{array}{c}-R_2+R_1\to R_1\\ \to\\ -R_2+R_3\to R_3\end{array}\left[\begin{array}{ccc|ccc}1&0&0&-1&1&0\\0&1&0&1&0&0\\0&0&1&-1&0&1\end{array}\right]$

    Inverse $=\begin{bmatrix}-1&1&0\\1&0&0\\-1&0&1\end{bmatrix}$

17. $\left[\begin{array}{ccc|ccc}3&1&2&1&0&0\\1&2&3&0&1&0\\1&1&1&0&0&1\end{array}\right]\begin{array}{c}R_1\leftrightarrow R_3\\ \to\\ \end{array}\left[\begin{array}{ccc|ccc}1&1&1&0&0&1\\1&2&3&0&1&0\\3&1&2&1&0&0\end{array}\right]\begin{array}{c}\to\\ -R_1+R_2\to R_2\\ -3R_1+R_3\to R_3\end{array}$

    $\left[\begin{array}{ccc|ccc}1&1&1&0&0&1\\0&1&2&0&1&-1\\0&-2&-1&1&0&-3\end{array}\right]\begin{array}{c}-R_2+R_1\to R_1\\ \to\\ 2R_2+R_3\to R_3\end{array}\left[\begin{array}{ccc|ccc}1&0&-1&0&-1&2\\0&1&2&0&1&-1\\0&0&3&1&2&-5\end{array}\right]\begin{array}{c}\to\\ \\ \frac{1}{3}R_3\to R_3\end{array}$

    $\left[\begin{array}{ccc|ccc}1&0&-1&0&-1&2\\0&1&2&0&1&-1\\0&0&1&1/3&2/3&-5/3\end{array}\right]\begin{array}{c}R_3+R_1\to R_1\\ -2R_3+R_2\to R_2\\ \to\end{array}\left[\begin{array}{ccc|ccc}1&0&0&1/3&-1/3&1/3\\0&1&0&-2/3&-1/3&7/3\\0&0&1&1/3&2/3&-5/3\end{array}\right]$

    Inverse $=\begin{bmatrix}1/3&-1/3&1/3\\-2/3&-1/3&7/3\\1/3&2/3&-5/3\end{bmatrix}$

19. $\left[\begin{array}{ccc|ccc} 1 & 3 & 5 & 1 & 0 & 0 \\ -1 & -1 & 2 & 0 & 1 & 0 \\ 1 & 5 & 12 & 0 & 0 & 1 \end{array}\right] \xrightarrow[\substack{R_1+R_2\to R_2 \\ -R_1+R_3\to R_3}]{} \left[\begin{array}{ccc|ccc} 1 & 3 & 5 & 1 & 0 & 0 \\ 0 & 2 & 7 & 1 & 1 & 0 \\ 0 & 2 & 7 & -1 & 0 & 1 \end{array}\right] \xrightarrow[-R_2+R_3\to R_3]{} \left[\begin{array}{ccc|ccc} 1 & 3 & 5 & 1 & 0 & 0 \\ 0 & 2 & 7 & 1 & 1 & 0 \\ 0 & 0 & 0 & -2 & -1 & 1 \end{array}\right]$

There is no inverse since there is a row of 0's in the original matrix.

21. $\left[\begin{array}{ccc|ccc} 1 & 2 & 4 & 1 & 0 & 0 \\ 1 & -1 & -3 & 0 & 1 & 0 \\ 2 & 1 & 1 & 0 & 0 & 1 \end{array}\right] \xrightarrow[\substack{-R_1+R_2\to R_2 \\ -2R_1+R_3\to R_3}]{} \left[\begin{array}{ccc|ccc} 1 & 2 & 4 & 1 & 0 & 0 \\ 0 & -3 & -7 & -1 & 1 & 0 \\ 0 & -3 & -7 & -2 & 0 & 1 \end{array}\right] \xrightarrow[-R_2+R_3\to R_3]{} \left[\begin{array}{ccc|ccc} 1 & 2 & 4 & 1 & 0 & 0 \\ 0 & -3 & -7 & -1 & 1 & 0 \\ 0 & 0 & 0 & -1 & -1 & 1 \end{array}\right]$

There is no inverse since there is a row of 0's in the original matrix.

23. $C^{-1} = \begin{bmatrix} 2 & 2 & 0 & 2 & 2 \\ 1 & 0 & 2 & 2 & 1 \\ 0 & 1 & 0 & 2 & 1 \\ 2 & 0 & 2 & 2 & 1 \\ 1 & 0 & 0 & 0 & 2 \end{bmatrix}$ Use a graphing calculator to find the inverse.

25. $AX = \begin{bmatrix} 3 \\ 2 \end{bmatrix} \quad X = A^{-1}\begin{bmatrix} 3 \\ 2 \end{bmatrix} = \begin{bmatrix} 3 & 2 \\ 1 & 1 \end{bmatrix}\begin{bmatrix} 3 \\ 2 \end{bmatrix} = \begin{bmatrix} 13 \\ 5 \end{bmatrix}$

27. $AX = \begin{bmatrix} 3 \\ -1 \\ 2 \end{bmatrix} \quad X = A^{-1}\begin{bmatrix} 3 \\ -1 \\ 2 \end{bmatrix} = \begin{bmatrix} 3 & 2 & 1 \\ 1 & 1 & 2 \\ 1 & 2 & 1 \end{bmatrix} \cdot \begin{bmatrix} 3 \\ -1 \\ 2 \end{bmatrix} = \begin{bmatrix} 9 \\ 6 \\ 3 \end{bmatrix}$

29. $A \cdot \begin{bmatrix} x \\ y \\ z \end{bmatrix} = \begin{bmatrix} 1 \\ 2 \\ 3 \end{bmatrix} \quad \begin{bmatrix} x \\ y \\ z \end{bmatrix} = A^{-1}\begin{bmatrix} 1 \\ 2 \\ 3 \end{bmatrix} = \begin{bmatrix} -1 & 1 & 0 \\ 1 & 0 & 0 \\ -1 & 0 & 1 \end{bmatrix} \cdot \begin{bmatrix} 1 \\ 2 \\ 3 \end{bmatrix} = \begin{bmatrix} 1 \\ 1 \\ 2 \end{bmatrix}$ So, $\begin{array}{l} x = 1 \\ y = 1 \\ z = 2 \end{array}$.

31. $A = \begin{bmatrix} 1 & 2 \\ 3 & 4 \end{bmatrix}$, $A^{-1} = \frac{1}{-2}\begin{bmatrix} 4 & -2 \\ -3 & 1 \end{bmatrix} = \begin{bmatrix} -2 & 1 \\ 3/2 & -1/2 \end{bmatrix}$; $\begin{bmatrix} x \\ y \end{bmatrix} = A^{-1}\begin{bmatrix} 4 \\ 10 \end{bmatrix} = \begin{bmatrix} -2 & 1 \\ 3/2 & -1/2 \end{bmatrix}\begin{bmatrix} 4 \\ 10 \end{bmatrix} = \begin{bmatrix} 2 \\ 1 \end{bmatrix}$

So, $x = 2, y = 1$.

33. $A = \begin{bmatrix} 2 & 1 \\ 3 & 1 \end{bmatrix}$, $A^{-1} = \frac{1}{-1}\begin{bmatrix} 1 & -1 \\ -3 & 2 \end{bmatrix} = \begin{bmatrix} -1 & 1 \\ 3 & -2 \end{bmatrix}$; $\begin{bmatrix} x \\ y \end{bmatrix} = A^{-1}\begin{bmatrix} 4 \\ 5 \end{bmatrix} = \begin{bmatrix} -1 & 1 \\ 3 & -2 \end{bmatrix}\begin{bmatrix} 4 \\ 5 \end{bmatrix} = \begin{bmatrix} 1 \\ 2 \end{bmatrix}$

So, $x = 1, y = 2$.

35. $\left[\begin{array}{ccc|ccc} 6 & 3 & -8 & 1 & 0 & 0 \\ 1 & 0 & -3 & 0 & 1 & 0 \\ 20 & -12 & 2 & 0 & 0 & 1 \end{array}\right] \xrightarrow{R_1 \leftrightarrow R_2} \left[\begin{array}{ccc|ccc} 1 & 0 & -3 & 0 & 1 & 0 \\ 6 & 3 & -8 & 1 & 0 & 0 \\ 20 & -12 & 2 & 0 & 0 & 1 \end{array}\right] \xrightarrow[\substack{-6R_1+R_2\to R_2 \\ -20R_1+R_3\to R_3}]{} \left[\begin{array}{ccc|ccc} 1 & 0 & -3 & 0 & 1 & 0 \\ 0 & 3 & 10 & 1 & -6 & 0 \\ 0 & -12 & 62 & 0 & -20 & 1 \end{array}\right]$

These numbers tell us that the inverse will have messy fractions. The graphing calculator yields

$A^{-1} \approx \begin{bmatrix} 0.1176 & -0.2941 & 0.0294 \\ 0.2026 & -0.5621 & -0.0327 \\ 0.0392 & -0.4314 & 0.0098 \end{bmatrix} \quad \begin{bmatrix} x \\ y \\ z \end{bmatrix} = A^{-1}\begin{bmatrix} 1 \\ -2 \\ 10 \end{bmatrix} = \begin{bmatrix} 1 \\ 1 \\ 1 \end{bmatrix}$

So, $x = 1, y = 1, z = 1$.

37. $\left[\begin{array}{ccc|ccc} 1 & 1 & 1 & 1 & 0 & 0 \\ 2 & 1 & 1 & 0 & 1 & 0 \\ 2 & 2 & 1 & 0 & 0 & 1 \end{array}\right] \begin{array}{c} \to \\ -2R_1+R_2 \to R_2 \\ -2R_1+R_3 \to R_3 \end{array} \left[\begin{array}{ccc|ccc} 1 & 1 & 1 & 1 & 0 & 0 \\ 0 & -1 & -1 & -2 & 1 & 0 \\ 0 & 0 & -1 & -2 & 0 & 1 \end{array}\right] \begin{array}{c} \to \\ -R_2 \to R_2 \\ -R_3 \to R_3 \end{array}$

$\left[\begin{array}{ccc|ccc} 1 & 1 & 1 & 1 & 0 & 0 \\ 0 & 1 & 1 & 2 & -1 & 0 \\ 0 & 0 & 1 & 2 & 0 & -1 \end{array}\right] \begin{array}{c} -R_2+R_1 \to R_1 \\ \to \\ \end{array} \left[\begin{array}{ccc|ccc} 1 & 0 & 0 & -1 & 1 & 0 \\ 0 & 1 & 1 & 2 & -1 & 0 \\ 0 & 0 & 1 & 2 & 0 & -1 \end{array}\right] \begin{array}{c} \to \\ -R_3+R_2 \to R_2 \\ \end{array}$

$\left[\begin{array}{ccc|ccc} 1 & 0 & 0 & -1 & 1 & 0 \\ 0 & 1 & 0 & 0 & -1 & 1 \\ 0 & 0 & 1 & 2 & 0 & -1 \end{array}\right] \qquad \begin{bmatrix} x \\ y \\ z \end{bmatrix} = \begin{bmatrix} -1 & 1 & 0 \\ 0 & -1 & 1 \\ 2 & 0 & -1 \end{bmatrix} \begin{bmatrix} 3 \\ 4 \\ 5 \end{bmatrix} = \begin{bmatrix} 1 \\ 1 \\ 1 \end{bmatrix}$

So, $x = 1,\ y = 1,\ z = 1$.

39. $\left[\begin{array}{ccc|ccc} 1 & 1 & 2 & 1 & 0 & 0 \\ 2 & 1 & 1 & 0 & 1 & 0 \\ 2 & 2 & 1 & 0 & 0 & 1 \end{array}\right] \begin{array}{c} \to \\ -2R_1+R_2 \to R_2 \\ -2R_1+R_3 \to R_3 \end{array} \left[\begin{array}{ccc|ccc} 1 & 1 & 2 & 1 & 0 & 0 \\ 0 & -1 & -3 & -2 & 1 & 0 \\ 0 & 0 & -3 & -2 & 0 & 1 \end{array}\right] \begin{array}{c} R_2+R_1 \to R_1 \\ \to \\ -\frac{1}{3}R_3 \to R_3 \end{array}$

$\left[\begin{array}{ccc|ccc} 1 & 0 & -1 & -1 & 1 & 0 \\ 0 & -1 & -3 & -2 & 1 & 0 \\ 0 & 0 & 1 & 2/3 & 0 & -1/3 \end{array}\right] \begin{array}{c} R_3+R_1 \to R_1 \\ \to \\ 3R_3+R_2 \to R_2 \end{array} \left[\begin{array}{ccc|ccc} 1 & 0 & 0 & -1/3 & 1 & -1/3 \\ 0 & -1 & 0 & 0 & 1 & -1 \\ 0 & 0 & 1 & 2/3 & 0 & -1/3 \end{array}\right] \begin{array}{c} -R_2 \to R_2 \\ \to \end{array}$

$\left[\begin{array}{ccc|ccc} 1 & 0 & 0 & -1/3 & 1 & -1/3 \\ 0 & 1 & 0 & 0 & -1 & 1 \\ 0 & 0 & 1 & 2/3 & 0 & -1/3 \end{array}\right] \qquad \begin{bmatrix} x \\ y \\ z \end{bmatrix} = \begin{bmatrix} -1/3 & 1 & -1/3 \\ 0 & -1 & 1 \\ 2/3 & 0 & -1/3 \end{bmatrix} \begin{bmatrix} 8 \\ 7 \\ 10 \end{bmatrix} = \begin{bmatrix} 1 \\ 3 \\ 2 \end{bmatrix}$

So, $x = 1,\ y = 3,\ z = 2$..

41. The graphing calculator yields

$$A^{-1} = \begin{bmatrix} 3 & 2 & 2 & 4 & 3 \\ 1 & 0 & 2 & 2 & 1 \\ 0.5 & 1 & 1 & 3 & 1.5 \\ 2 & 0 & 2 & 2 & 1 \\ 1 & 0 & 0 & 0 & 2 \end{bmatrix}; \quad X = A^{-1} \begin{bmatrix} 0.7 \\ -1.6 \\ 1.275 \\ 1.15 \\ -0.15 \end{bmatrix} = \begin{bmatrix} 5.6 \\ 5.4 \\ 3.25 \\ 6.1 \\ 0.4 \end{bmatrix}$$

43. $\begin{vmatrix} 1 & 2 \\ 3 & 4 \end{vmatrix} = 1(4) - 3(2) = -2$

45. $\begin{vmatrix} 3 & -1 \\ 2 & 4 \end{vmatrix} = 3(4) - 2(-1) = 14$

47. Using technology, $\begin{vmatrix} 3 & 2 & 1 \\ -1 & 0 & 2 \\ 0 & 1 & 1 \end{vmatrix} = -5$

49. Using technology, $\begin{vmatrix} 0 & 1 & 2 \\ 3 & 1 & 1 \\ 4 & -1 & 3 \end{vmatrix} = -19$

51. Since $\det \begin{bmatrix} 2 & 3 \\ -1 & 4 \end{bmatrix} = 11 \neq 0$, $\begin{bmatrix} 2 & 3 \\ -1 & 4 \end{bmatrix}$ has an inverse.

53. Since $\det \begin{bmatrix} 1 & 3 & -2 \\ 2 & -1 & 5 \\ 3 & 2 & 3 \end{bmatrix} = 0$, $\begin{bmatrix} 1 & 3 & -2 \\ 2 & -1 & 5 \\ 3 & 2 & 3 \end{bmatrix}$ does not have an inverse.

55. $A^{-1} = \begin{bmatrix} 11 & -9 \\ -6 & 5 \end{bmatrix}$, $\begin{bmatrix} 11 & -9 \\ -6 & 5 \end{bmatrix} \cdot \begin{bmatrix} 49 & 133 & 270 & 313 \\ 59 & 161 & 327 & 381 \end{bmatrix} = \begin{bmatrix} 8 & 14 & 27 & 14 \\ 1 & 7 & 15 & 27 \end{bmatrix}$

8 1 14 7 27 15 14 27

H A N G O N

57. $A^{-1} = \begin{bmatrix} 1 & 1 & -2 \\ 2 & 1 & -3 \\ -3 & -2 & 6 \end{bmatrix}$, $\begin{bmatrix} 1 & 1 & -2 \\ 2 & 1 & -3 \\ -3 & -2 & 6 \end{bmatrix} \cdot \begin{bmatrix} 47 & 28 & 63 & 56 & 17 \\ 22 & 87 & 66 & 44 & 14 \\ 34 & 46 & 55 & 43 & 15 \end{bmatrix} = \begin{bmatrix} 1 & 23 & 19 & 14 & 1 \\ 14 & 5 & 27 & 27 & 3 \\ 19 & 18 & 9 & 2 & 11 \end{bmatrix}$

1 14 19 23 5 18 19 27 9 14 27 2 1 3 11

A N S W E R S I N B A C K

59. $\begin{bmatrix} 1900 \\ 1700 \end{bmatrix} = \begin{bmatrix} 2/3 & 1/4 \\ 1/3 & 3/4 \end{bmatrix} \begin{bmatrix} x_0 \\ y_0 \end{bmatrix}$ For ease in calculations, use a graphing calculator.

$$\begin{bmatrix} x_0 \\ y_0 \end{bmatrix} = \begin{bmatrix} 2/3 & 1/4 \\ 1/3 & 3/4 \end{bmatrix}^{-1} \begin{bmatrix} 1900 \\ 1700 \end{bmatrix} = \begin{bmatrix} 9/5 & -3/5 \\ -4/5 & 8/5 \end{bmatrix} \begin{bmatrix} 1900 \\ 1700 \end{bmatrix} = \begin{bmatrix} 2400 \\ 1200 \end{bmatrix}$$

61. $6A + 2B = 50.6$ or $3A + B = 25.3$ Patient I

$6A + 2B = 92.0$ or $3A + B = 46.0$ Patient II

$\frac{A}{B} = \frac{5}{8}$ or $8A - 5B = 0$

The inverse of $\begin{bmatrix} 3 & 1 \\ 8 & -5 \end{bmatrix}$ is $-\frac{1}{23}\begin{bmatrix} -5 & -1 \\ -8 & 3 \end{bmatrix} = \begin{bmatrix} 5/23 & 1/23 \\ 8/23 & -3/23 \end{bmatrix}$

Patient 1: $\begin{bmatrix} A \\ B \end{bmatrix} = \begin{bmatrix} 5/23 & 1/23 \\ 8/23 & -3/23 \end{bmatrix} \begin{bmatrix} 25.3 \\ 0.0 \end{bmatrix} = \begin{bmatrix} 5.5 \\ 8.8 \end{bmatrix}$ 5.5 mg of A, 8.8 mg of B for Patient I.

Patient II: $\begin{bmatrix} A \\ B \end{bmatrix} = \begin{bmatrix} 5/23 & 1/23 \\ 8/23 & -3/23 \end{bmatrix} \begin{bmatrix} 46 \\ 0 \end{bmatrix} = \begin{bmatrix} 10 \\ 16 \end{bmatrix}$ 10 mg of A, 16 mg of B for Patient II.

63. x = amount invested at 10%.

y = amount invested at 18%.

$x + y = 145{,}600$ Total investment

$0.10x + 0.18y = 20{,}000$ Total income

$$\begin{bmatrix} x \\ y \end{bmatrix} = \begin{bmatrix} 1 & 1 \\ 0.10 & 0.18 \end{bmatrix}^{-1} \begin{bmatrix} 145{,}600 \\ 20{,}000 \end{bmatrix} = \begin{bmatrix} 77{,}600 \\ 68{,}000 \end{bmatrix}$$

\$77,600 at 10% and \$68,000 at 18%

Note - Use the graphing utility to find the inverse, then multiply.

65. Use the graphing utility to find the inverse.

$$\begin{bmatrix} \text{Deluxe} \\ \text{Premium} \\ \text{Ultimate} \end{bmatrix} = \begin{bmatrix} 1.6 & 2 & 2.4 \\ 2 & 3 & 4 \\ 0.5 & 0.5 & 1 \end{bmatrix}^{-1} \cdot \begin{bmatrix} 96 \\ 156 \\ 37 \end{bmatrix} = \begin{bmatrix} 2.5 & -2 & 2 \\ 0 & 1 & -4 \\ -1.25 & 0.5 & 2 \end{bmatrix} \cdot \begin{bmatrix} 96 \\ 156 \\ 37 \end{bmatrix} = \begin{bmatrix} 2 \\ 8 \\ 32 \end{bmatrix}$$

Produce 2 deluxe, 8 premium, and 32 ultimate models.

67. $x = 8\%$ investment, $y = 10\%$ investment, $z = 6\%$ investment

$x + y + z = 1{,}000{,}000$ Total investment

$0.08x + 0.10y + 0.06z = 86{,}000$ Investment income

$y = x + z$ Third condition

Use the graphing utility to find the inverse.

$$\begin{bmatrix} x \\ y \\ z \end{bmatrix} = \begin{bmatrix} 1 & 1 & 1 \\ 0.08 & 0.10 & 0.06 \\ 1 & -1 & 1 \end{bmatrix}^{-1} \cdot \begin{bmatrix} 1{,}000{,}000 \\ 86{,}000 \\ 0 \end{bmatrix} = \begin{bmatrix} -4 & 50 & 1 \\ 0.5 & 0 & -0.5 \\ 4.5 & -50 & -0.5 \end{bmatrix} \cdot \begin{bmatrix} 1{,}000{,}000 \\ 86{,}000 \\ 0 \end{bmatrix} = \begin{bmatrix} 300{,}000 \\ 500{,}000 \\ 200{,}000 \end{bmatrix}$$

69. a. $$\begin{bmatrix} a & b & c \\ d & e & f \\ g & h & i \end{bmatrix} \begin{bmatrix} n_t \\ n_{t+1} \\ n_{t+2} \end{bmatrix} = \begin{bmatrix} n_{t+1} \\ n_{t+2} \\ n_t + n_{t+1} + n_{t+2} \end{bmatrix}$$

$an_t + bn_{t+1} + cn_{t+2} = n_{t+1}$, etc.

implies $a = 0,\ b = 1,\ c = 0,\ d = 0,\ e = 0,\ f = 1,\ g = h = i = 1$

$$M = \begin{bmatrix} 0 & 1 & 0 \\ 0 & 0 & 1 \\ 1 & 1 & 1 \end{bmatrix}$$

b. $$M^{-1} = \begin{bmatrix} -1 & -1 & 1 \\ 1 & 0 & 0 \\ 0 & 1 & 0 \end{bmatrix} \quad \begin{bmatrix} n_t \\ n_{t+1} \\ n_{t+2} \end{bmatrix} = M^{-1} \begin{bmatrix} 191 \\ 346 \\ 645 \end{bmatrix} = \begin{bmatrix} 108 \\ 191 \\ 346 \end{bmatrix}$$

There were 108 visitors on the day before there were 191 visitors.

71. a. See exercise 69 where M was found to be $\begin{bmatrix} 0 & 1 & 0 \\ 0 & 0 & 1 \\ 1 & 1 & 1 \end{bmatrix}$.

b. $$M^{-1} = \begin{bmatrix} -1 & -1 & 1 \\ 1 & 0 & 0 \\ 0 & 1 & 0 \end{bmatrix}$$

At the end of 3 successive 1-hour periods, the value for N was $N = \begin{bmatrix} 200 \\ 370 \\ 600 \end{bmatrix}$.

$$M^{-1}N = \begin{bmatrix} -1 & -1 & 1 \\ 1 & 0 & 0 \\ 0 & 1 & 0 \end{bmatrix} \begin{bmatrix} 200 \\ 370 \\ 600 \end{bmatrix} = \begin{bmatrix} 30 \\ 200 \\ 370 \end{bmatrix}$$

So, the population was 30 one hour before it was 200.

Exercise 3.5

1. a. Row 3, column 2 = 0.15 100(0.15) = 15
b. Row 4, column 1 = 0.10 40(0.10) = 4

3. 1000(0.008) = 8.

5. 1000(0.040) = 40.

7. Most dependent would be the largest entry on the main diagonal. Raw materials is the most self dependent. Likewise, Fuels is least dependent.

9. The largest entries in Row 2 give the industries most affected by a rise in raw material cost. These are Raw materials, Manufacturing, and Service industries.

11. $10M = 50$ $M = 5$ units of manufacturing
$8A - 4(5) = 60$ $A = 10$ units of agriculture
$U - 2(10) - 1(5) = 80$ $U = 105$ units of utilities

13. $D = \begin{bmatrix} 96 \\ 8 \end{bmatrix}$. $(I - A)X = D$ or $\begin{bmatrix} 0.5 & -0.1 \\ -0.1 & 0.7 \end{bmatrix}\begin{bmatrix} P \\ M \end{bmatrix} = \begin{bmatrix} 96 \\ 8 \end{bmatrix}$

$$\begin{bmatrix} P \\ M \end{bmatrix} = \begin{bmatrix} 0.5 & -0.1 \\ -0.1 & 0.7 \end{bmatrix}^{-1} \cdot \begin{bmatrix} 96 \\ 8 \end{bmatrix} = \frac{1}{0.34} \cdot \begin{bmatrix} 0.7 & 0.1 \\ 0.1 & 0.5 \end{bmatrix}\begin{bmatrix} 96 \\ 8 \end{bmatrix} = \frac{1}{0.34}\begin{bmatrix} 68 \\ 13.6 \end{bmatrix} = \begin{bmatrix} 200 \\ 40 \end{bmatrix}$$

$P = 200$ units of farm products and $M = 40$ units of machinery.

15. $D = \begin{bmatrix} 0 \\ 610 \end{bmatrix}$. $(I - A)X = D$ or $\begin{bmatrix} 0.7 & -0.1 \\ -0.2 & 0.9 \end{bmatrix}\begin{bmatrix} AP \\ OP \end{bmatrix} = \begin{bmatrix} 0 \\ 610 \end{bmatrix}$

$$\begin{bmatrix} AP \\ OP \end{bmatrix} = \begin{bmatrix} 0.7 & -0.1 \\ -0.2 & 0.9 \end{bmatrix}^{-1} \cdot \begin{bmatrix} 0 \\ 610 \end{bmatrix} = \frac{1}{0.61}\begin{bmatrix} 0.9 & 0.1 \\ 0.2 & 0.7 \end{bmatrix}\begin{bmatrix} 0 \\ 610 \end{bmatrix} = \begin{bmatrix} 100 \\ 700 \end{bmatrix} \begin{matrix} \text{Ag. Products} \\ \text{Oil Products} \end{matrix}$$

17. $D = \begin{bmatrix} 80 \\ 180 \end{bmatrix}$. $(I - A)X = D$ or $\begin{bmatrix} 0.7 & -0.15 \\ -0.3 & 0.6 \end{bmatrix}\begin{bmatrix} U \\ M \end{bmatrix} = \begin{bmatrix} 80 \\ 180 \end{bmatrix}$

$$\begin{bmatrix} U \\ M \end{bmatrix} = \begin{bmatrix} 0.7 & -0.15 \\ -0.3 & 0.6 \end{bmatrix}^{-1} \cdot \begin{bmatrix} 80 \\ 180 \end{bmatrix} = \frac{1}{0.375}\begin{bmatrix} 0.6 & 0.15 \\ 0.3 & 0.7 \end{bmatrix}\begin{bmatrix} 80 \\ 180 \end{bmatrix} = \begin{bmatrix} 200 \\ 400 \end{bmatrix} \begin{matrix} \text{Utility} \\ \text{Manufacturing} \end{matrix}$$

19. $D = \begin{bmatrix} 36 \\ 278 \end{bmatrix}$. $(I - A)X = D$ or $\begin{bmatrix} 0.8 & -0.4 \\ -0.3 & 0.7 \end{bmatrix}\begin{bmatrix} \text{MINE} \\ \text{MFG} \end{bmatrix} = \begin{bmatrix} 36 \\ 278 \end{bmatrix}$

$$\begin{bmatrix} \text{MINE} \\ \text{MFG} \end{bmatrix} = \begin{bmatrix} 0.8 & -0.4 \\ -0.3 & 0.7 \end{bmatrix}^{-1} \cdot \begin{bmatrix} 36 \\ 278 \end{bmatrix} = \frac{1}{0.44}\begin{bmatrix} 0.7 & 0.4 \\ 0.3 & 0.8 \end{bmatrix}\begin{bmatrix} 36 \\ 278 \end{bmatrix} = \begin{bmatrix} 310 \\ 530 \end{bmatrix} \begin{matrix} \text{Mining} \\ \text{Manufacturing} \end{matrix}$$

21. $D = \begin{bmatrix} 648 \\ 16 \end{bmatrix}$ $(I - A)X = D$ or $\begin{bmatrix} 0.7 & -0.6 \\ -0.2 & 0.8 \end{bmatrix}\begin{bmatrix} \text{EC} \\ \text{COMP} \end{bmatrix} = \begin{bmatrix} 648 \\ 16 \end{bmatrix}$

$$\begin{bmatrix} \text{EC} \\ \text{COMP} \end{bmatrix} = \begin{bmatrix} 0.7 & -0.6 \\ -0.2 & 0.8 \end{bmatrix}^{-1} \cdot \begin{bmatrix} 648 \\ 16 \end{bmatrix} = \frac{1}{0.44}\begin{bmatrix} 0.8 & 0.6 \\ 0.2 & 0.7 \end{bmatrix}\begin{bmatrix} 648 \\ 16 \end{bmatrix} = \begin{bmatrix} 1200 \\ 320 \end{bmatrix} \begin{matrix} \text{Electronics components} \\ \text{Computers} \end{matrix}$$

23. $D = \begin{bmatrix} 20 \\ 1090 \end{bmatrix}$. $(I - A)X = D$ or $\begin{bmatrix} 0.7 & -0.04 \\ -0.35 & 0.9 \end{bmatrix}\begin{bmatrix} \text{Fish} \\ \text{Oil} \end{bmatrix} = \begin{bmatrix} 20 \\ 1090 \end{bmatrix}$

$$\begin{bmatrix} \text{Fish} \\ \text{Oil} \end{bmatrix} = \begin{bmatrix} 0.7 & -0.04 \\ -0.35 & 0.9 \end{bmatrix}^{-1} \cdot \begin{bmatrix} 20 \\ 1090 \end{bmatrix} = \frac{1}{0.616}\begin{bmatrix} 0.9 & 0.04 \\ 0.35 & 0.7 \end{bmatrix}\begin{bmatrix} 20 \\ 1090 \end{bmatrix} = \begin{bmatrix} 100 \\ 1250 \end{bmatrix} \begin{matrix} \text{Fishing} \\ \text{Oil} \end{matrix}$$

25. $\begin{bmatrix} x \\ y \end{bmatrix} = \begin{bmatrix} 0 & 0.05 \\ 0.1 & 0 \end{bmatrix}\begin{bmatrix} x \\ y \end{bmatrix} + \begin{bmatrix} 20,400 \\ 9900 \end{bmatrix}$ or $\begin{bmatrix} 1 & -0.05 \\ -0.1 & 1 \end{bmatrix}\begin{bmatrix} x \\ y \end{bmatrix} = \begin{bmatrix} 20,400 \\ 9900 \end{bmatrix}$

$$\begin{bmatrix} x \\ y \end{bmatrix} = \frac{1}{1 - 0.005}\begin{bmatrix} 1 & 0.05 \\ 0.1 & 1 \end{bmatrix}\begin{bmatrix} 20,400 \\ 9900 \end{bmatrix} = \frac{1}{0.995}\begin{bmatrix} 20,895 \\ 11,940 \end{bmatrix} = \begin{bmatrix} 21,000 \\ 12,000 \end{bmatrix}$$

Costs for development are \$21,000. Costs for promotional are \$12,000.

27. $\begin{bmatrix} \text{E} \\ \text{C} \end{bmatrix} = \begin{bmatrix} 11,750 \\ 10,000 \end{bmatrix} + \begin{bmatrix} 0 & 0.25 \\ 0.2 & 0 \end{bmatrix}\begin{bmatrix} \text{E} \\ \text{C} \end{bmatrix}$

$$\begin{bmatrix} 1 & -0.25 \\ -0.2 & 1 \end{bmatrix}\begin{bmatrix} \text{E} \\ \text{C} \end{bmatrix} = \begin{bmatrix} 11,750 \\ 10,000 \end{bmatrix} \text{ or } \begin{bmatrix} \text{E} \\ \text{C} \end{bmatrix} = \frac{1}{1 - 0.05}\begin{bmatrix} 1 & 0.25 \\ 0.2 & 1 \end{bmatrix}\begin{bmatrix} 11,750 \\ 10,000 \end{bmatrix} = \begin{bmatrix} 15,000 \\ 13,000 \end{bmatrix}$$

Engineering costs are \$15,000. Computer costs are \$13,000.

29. $D=\begin{bmatrix}110\\50\\50\end{bmatrix}$. $(I-A)X=D$ or $X=(I-A)^{-1}\cdot D$

$$X=\begin{bmatrix}0.5&-0.1&-0.1\\-0.3&0.5&-0.2\\-0.1&-0.3&0.6\end{bmatrix}^{-1}\cdot\begin{bmatrix}110\\50\\50\end{bmatrix}=\begin{bmatrix}2.79&1.05&0.81\\2.33&3.37&1.51\\1.63&1.86&2.56\end{bmatrix}\cdot\begin{bmatrix}110\\50\\50\end{bmatrix}=\begin{bmatrix}400\\500\\400\end{bmatrix}$$

400 units of fishing output, 500 units of agricultural goods and 400 units of mining goods are needed.

31. $D=\begin{bmatrix}100\\272\\200\end{bmatrix}$. $(I-A)X=D$ or $\left[\begin{array}{ccc|c}0.4&-0.2&-0.2&100\\-0.1&0.6&-0.5&272\\-0.1&-0.2&0.8&200\end{array}\right]$ is the required augmented matrix.

By reducing this to $\left[\begin{array}{ccc|c}1&0&0&1240\\0&1&0&1260\\0&0&1&720\end{array}\right]$ we have 1240 electronics, 1260 steel, and 720 autos.

33. $D=\begin{bmatrix}24\\62\\32\end{bmatrix}$. $(I-A)X=D$ or $X=(I-A)^{-1}\cdot D$

$$X=\begin{bmatrix}0.6&-0.1&-0.1\\-0.2&0.5&-0.2\\-0.2&-0.1&0.7\end{bmatrix}^{-1}\cdot\begin{bmatrix}24\\62\\32\end{bmatrix}=\begin{bmatrix}1.964&0.476&0.417\\1.071&2.381&0.833\\0.714&0.476&1.667\end{bmatrix}\cdot\begin{bmatrix}24\\62\\32\end{bmatrix}=\begin{bmatrix}90\\200\\100\end{bmatrix}=\begin{bmatrix}S\\M\\A\end{bmatrix}$$

35. $0.5P+0.1M+0.2H=P$ or $-0.5P+0.1M+0.2H=0$
$0.1P+0.3M+0.0H=M$ $0.1P-0.7M+0.0H=0$
$0.4P+0.6M+0.8H=H$ $0.4P+0.6M-0.2H=0$

$$\left[\begin{array}{ccc|c}1&-7&0&0\\-5&1&2&0\\4&6&-2&0\end{array}\right]\begin{array}{c}\rightarrow\\5R_1+R_2\rightarrow R_2\\-4R_1+R_3\rightarrow R_3\end{array}\left[\begin{array}{ccc|c}1&-7&0&0\\0&-34&2&0\\0&34&-2&0\end{array}\right]\begin{array}{c}\rightarrow\\R_2+R_3\rightarrow R_3\end{array}$$

$$\left[\begin{array}{ccc|c}1&-7&0&0\\0&-34&2&0\\0&0&0&0\end{array}\right]\begin{array}{c}\rightarrow\\-\tfrac{1}{34}R_2\rightarrow R_2\end{array}\left[\begin{array}{ccc|c}1&-7&0&0\\0&1&-1/17&0\\0&0&0&0\end{array}\right]\begin{array}{c}7R_2+R_1\rightarrow R_1\\\rightarrow\end{array}\left[\begin{array}{ccc|c}1&0&-7/17&0\\0&1&-1/17&0\\0&0&0&0\end{array}\right]$$

So, Farm Products = $\dfrac{7}{17}$ Households and Farm Machinery = $\dfrac{1}{17}$ Households.

37. $0.4G+0.2I+0.2H=G$ or $-0.6G+0.2I+0.2H=0$
$0.2G+0.3I+0.3H=I$ $0.2G-0.7I+0.3H=0$
$0.4G+0.5I+0.5H=H$ $0.4G+0.5I-0.5H=0$

$$\left[\begin{array}{ccc|c}2&-7&3&0\\-6&2&2&0\\4&5&-5&0\end{array}\right]\begin{array}{c}\\3R_1+R_2\rightarrow R_2\\-2R_1+R_3\rightarrow R_3\end{array}\left[\begin{array}{ccc|c}2&-7&3&0\\0&-19&11&0\\0&19&-11&0\end{array}\right]\begin{array}{c}\rightarrow\\\\R_2+R_3\rightarrow R_3\end{array}\left[\begin{array}{ccc|c}2&-7&3&0\\0&-19&11&0\\0&0&0&0\end{array}\right]\begin{array}{c}\\-\tfrac{1}{19}R_2\rightarrow R_2\\\rightarrow\end{array}$$

$$\left[\begin{array}{ccc|c}2&-7&3&0\\0&1&-11/19&0\\0&0&0&0\end{array}\right]\begin{array}{c}7R_2+R_1\rightarrow R_1\\\rightarrow\end{array}\left[\begin{array}{ccc|c}2&0&-20/19&0\\0&1&-11/19&0\\0&0&0&0\end{array}\right]\begin{array}{c}\tfrac{1}{2}R_1\rightarrow R_1\\\rightarrow\end{array}\left[\begin{array}{ccc|c}1&0&-10/19&0\\0&1&-11/19&0\\0&0&0&0\end{array}\right]$$

Government = $\frac{10}{19}$ Households. Industry = $\frac{11}{19}$ Households.

39. $0.5M + 0.4U + 0.3H = M$ or $-0.5M + 0.4U + 0.3H = 0$
$0.4M + 0.5U + 0.3H = U$ $\quad 0.4M - 0.5U + 0.3H = 0$
$0.1M + 0.1U + 0.4H = H$ $\quad 0.1M + 0.1U - 0.6H = 0$

$$\left[\begin{array}{ccc|c} 1 & 1 & -6 & 0 \\ 4 & -5 & 3 & 0 \\ -5 & 4 & 3 & 0 \end{array}\right] \begin{array}{l} \\ -4R_1 + R_2 \to R_2 \\ 5R_1 + R_3 \to R_3 \end{array} \left[\begin{array}{ccc|c} 1 & 1 & -6 & 0 \\ 0 & -9 & 27 & 0 \\ 0 & 9 & -27 & 0 \end{array}\right] \begin{array}{l} \to \\ \\ R_2 + R_3 \to R_3 \end{array} \left[\begin{array}{ccc|c} 1 & 1 & -6 & 0 \\ 0 & -9 & 27 & 0 \\ 0 & 0 & 0 & 0 \end{array}\right] \begin{array}{l} \to \\ -\frac{1}{9}R_2 \to R_2 \\ \end{array}$$

$$\left[\begin{array}{ccc|c} 1 & 1 & -6 & 0 \\ 0 & 1 & -3 & 0 \\ 0 & 0 & 0 & 0 \end{array}\right] \begin{array}{l} -R_2 + R_1 \to R_1 \\ \to \\ \end{array} \left[\begin{array}{ccc|c} 1 & 0 & -3 & 0 \\ 0 & 1 & -3 & 0 \\ 0 & 0 & 0 & 0 \end{array}\right]$$

Manufacturing = 3 Households and Utilities = 3 Households.

41.
$$\left[\begin{array}{cccccc|c} 1 & 0 & 0 & 0 & 0 & 0 & 24 \\ -4 & 1 & 0 & 0 & 0 & 0 & 0 \\ -1 & 0 & 1 & 0 & 0 & 0 & 0 \\ 0 & -1 & -1 & 1 & 0 & 0 & 0 \\ 0 & -4 & -4 & 0 & 1 & 0 & 12 \\ -20 & -24 & -24 & 0 & 0 & 1 & 96 \end{array}\right] \begin{array}{l} \to \\ 4R_1 + R_2 \to R_2 \\ R_1 + R_3 \to R_3 \\ \\ \\ 20R_1 + R_6 \to R_6 \end{array} \left[\begin{array}{cccccc|c} 1 & 0 & 0 & 0 & 0 & 0 & 24 \\ 0 & 1 & 0 & 0 & 0 & 0 & 96 \\ 0 & 0 & 1 & 0 & 0 & 0 & 24 \\ 0 & -1 & -1 & 1 & 0 & 0 & 0 \\ 0 & -4 & -4 & 0 & 1 & 0 & 12 \\ 0 & -24 & -24 & 0 & 0 & 1 & 576 \end{array}\right] \begin{array}{l} \\ \to \\ \\ R_2 + R_4 \to R_4 \\ 4R_2 + R_5 \to R_5 \\ 24R_2 + R_6 \to R_6 \end{array}$$

$$\left[\begin{array}{cccccc|c} 1 & 0 & 0 & 0 & 0 & 0 & 24 \\ 0 & 1 & 0 & 0 & 0 & 0 & 96 \\ 0 & 0 & 1 & 0 & 0 & 0 & 24 \\ 0 & 0 & -1 & 1 & 0 & 0 & 96 \\ 0 & 0 & -4 & 0 & 1 & 0 & 396 \\ 0 & 0 & -24 & 0 & 0 & 1 & 2880 \end{array}\right] \begin{array}{l} \\ \to \\ \\ R_3 + R_4 \to R_4 \\ 4R_3 + R_5 \to R_5 \\ 24R_3 + R_6 \to R_6 \end{array} \left[\begin{array}{cccccc|c} 1 & 0 & 0 & 0 & 0 & 0 & 24 \\ 0 & 1 & 0 & 0 & 0 & 0 & 96 \\ 0 & 0 & 1 & 0 & 0 & 0 & 24 \\ 0 & 0 & 0 & 1 & 0 & 0 & 120 \\ 0 & 0 & 0 & 0 & 1 & 0 & 492 \\ 0 & 0 & 0 & 0 & 0 & 1 & 3456 \end{array}\right]$$

Therefore, 120 sheets, 492 braces, and 3456 bolts are required to fill the order.

43. $D = \begin{bmatrix} 10 \\ 0 \\ 0 \\ 6 \\ 0 \\ 6 \\ 100 \end{bmatrix}$ $X = \begin{bmatrix} x_1 \\ x_2 \\ x_3 \\ x_4 \\ x_5 \\ x_6 \\ x_7 \end{bmatrix} = \begin{bmatrix} \text{total sawhorses} \\ \text{total tops} \\ \text{total leg pairs} \\ \text{total } 2 \times 4\text{s} \\ \text{total braces} \\ \text{total clamps} \\ \text{total nails} \end{bmatrix}$ Solve $(I - A)X = D$

$$\left[\begin{array}{ccccccc|c} 1 & 0 & 0 & 0 & 0 & 0 & 0 & 10 \\ -1 & 1 & 0 & 0 & 0 & 0 & 0 & 0 \\ -2 & 0 & 1 & 0 & 0 & 0 & 0 & 0 \\ 0 & -1 & -2 & 1 & 0 & 0 & 0 & 6 \\ 0 & 0 & -1 & 0 & 1 & 0 & 0 & 0 \\ 0 & 0 & -1 & 0 & 0 & 1 & 0 & 6 \\ -4 & 0 & -8 & 0 & 0 & 0 & 1 & 100 \end{array}\right] \begin{array}{l} \\ R_1 + R_2 \to R_2 \\ 2R_1 + R_3 \to R_3 \\ \\ \to \\ \\ 4R_1 + R_7 \to R_7 \end{array} \left[\begin{array}{ccccccc|c} 1 & 0 & 0 & 0 & 0 & 0 & 0 & 10 \\ 0 & 1 & 0 & 0 & 0 & 0 & 0 & 10 \\ 0 & 0 & 1 & 0 & 0 & 0 & 0 & 20 \\ 0 & -1 & -2 & 1 & 0 & 0 & 0 & 6 \\ 0 & 0 & -1 & 0 & 1 & 0 & 0 & 0 \\ 0 & 0 & -1 & 0 & 0 & 1 & 0 & 6 \\ 0 & 0 & -8 & 0 & 0 & 0 & 1 & 140 \end{array}\right] \begin{array}{l} \\ \to \\ \\ R_2 + R_4 \to R_4 \\ \\ \\ \\ \end{array}$$

$$\left[\begin{array}{ccccccc|c} 1 & 0 & 0 & 0 & 0 & 0 & 0 & 10 \\ 0 & 1 & 0 & 0 & 0 & 0 & 0 & 10 \\ 0 & 0 & 1 & 0 & 0 & 0 & 0 & 20 \\ 0 & 0 & -2 & 1 & 0 & 0 & 0 & 16 \\ 0 & 0 & -1 & 0 & 1 & 0 & 0 & 0 \\ 0 & 0 & -1 & 0 & 0 & 0 & 10 & 6 \\ 0 & 0 & -8 & 0 & 0 & 0 & 1 & 140 \end{array}\right] \begin{array}{l} \\ \to \\ \\ 2R_3 + R_4 \to R_4 \\ R_3 + R_5 \to R_5 \\ R_3 + R_6 \to R_6 \\ 8R_3 + R_7 \to R_7 \end{array} \left[\begin{array}{ccccccc|c} 1 & 0 & 0 & 0 & 0 & 0 & 0 & 10 \\ 0 & 1 & 0 & 0 & 0 & 0 & 0 & 10 \\ 0 & 0 & 1 & 0 & 0 & 0 & 0 & 20 \\ 0 & 0 & 0 & 1 & 0 & 0 & 0 & 56 \\ 0 & 0 & 0 & 0 & 1 & 0 & 0 & 20 \\ 0 & 0 & 0 & 0 & 0 & 1 & 0 & 26 \\ 0 & 0 & 0 & 0 & 0 & 0 & 1 & 300 \end{array}\right]$$

To fill the order, 56 2×4s , 20 braces, 26 clamps, and 300 nails.

Review Exercises

1. $a_{12} = 4$

2. $b_{23} = 0$

3. A and B

4. None

5. D, F, G, I

6. The negative of B is $\begin{bmatrix} -2 & 5 & 11 & -8 \\ -4 & 0 & 0 & -4 \\ 2 & 2 & -1 & -9 \end{bmatrix}$.

7. Zero matrix

8. Two matrices can be added if they have the same order.

9. $A+B$

$$= \begin{bmatrix} 4 & 4 & 2 & -5 \\ 6 & 3 & -1 & 0 \\ 0 & 0 & -3 & 5 \end{bmatrix} + \begin{bmatrix} 2 & -5 & -11 & 8 \\ 4 & 0 & 0 & 4 \\ -2 & -2 & 1 & 9 \end{bmatrix}$$

$$= \begin{bmatrix} 6 & -1 & -9 & 3 \\ 10 & 3 & -1 & 4 \\ -2 & -2 & -2 & 14 \end{bmatrix}$$

10. $C-E = \begin{bmatrix} 4 & -2 \\ 5 & 0 \\ 6 & 0 \\ 1 & 3 \end{bmatrix} - \begin{bmatrix} 1 & 1 \\ 1 & 1 \\ 4 & 6 \\ 0 & 5 \end{bmatrix} = \begin{bmatrix} 3 & -3 \\ 4 & -1 \\ 2 & -6 \\ 1 & -2 \end{bmatrix}$

11. $D^T - I = \begin{bmatrix} 3 & 1 \\ 5 & 2 \end{bmatrix} - \begin{bmatrix} 1 & 0 \\ 0 & 1 \end{bmatrix} = \begin{bmatrix} 2 & 1 \\ 5 & 1 \end{bmatrix}$

12. $3C = 3\begin{bmatrix} 4 & -2 \\ 5 & 0 \\ 6 & 0 \\ 1 & 3 \end{bmatrix} = \begin{bmatrix} 12 & -6 \\ 15 & 0 \\ 18 & 0 \\ 3 & 9 \end{bmatrix}$

13. $4I = \begin{bmatrix} 4 & 0 \\ 0 & 4 \end{bmatrix}$

14. $-2F = -2\begin{bmatrix} -1 & 6 \\ 4 & 11 \end{bmatrix} = \begin{bmatrix} 2 & -12 \\ -8 & -22 \end{bmatrix}$

15. $4D - 3I = \begin{bmatrix} 12 & 20 \\ 4 & 8 \end{bmatrix} - \begin{bmatrix} 3 & 0 \\ 0 & 3 \end{bmatrix} = \begin{bmatrix} 9 & 20 \\ 4 & 5 \end{bmatrix}$

16. $F + 2D = \begin{bmatrix} -1 & 6 \\ 4 & 11 \end{bmatrix} + \begin{bmatrix} 6 & 10 \\ 2 & 4 \end{bmatrix} = \begin{bmatrix} 5 & 16 \\ 6 & 15 \end{bmatrix}$

17. $3A - 5B$

$$= \begin{bmatrix} 12 & 12 & 6 & -15 \\ 18 & 9 & -3 & 0 \\ 0 & 0 & -9 & 15 \end{bmatrix} - \begin{bmatrix} 10 & -25 & -55 & 40 \\ 20 & 0 & 0 & 20 \\ -10 & -10 & 5 & 45 \end{bmatrix}$$

$$= \begin{bmatrix} 2 & 37 & 61 & -55 \\ -2 & 9 & -3 & -20 \\ 10 & 10 & -14 & -30 \end{bmatrix}$$

18. $AC = \begin{bmatrix} 4 & 4 & 2 & -5 \\ 6 & 3 & -1 & 0 \\ 0 & 0 & -3 & 5 \end{bmatrix} \begin{bmatrix} 4 & -2 \\ 5 & 0 \\ 6 & 0 \\ 1 & 3 \end{bmatrix} = \begin{bmatrix} 43 & -23 \\ 33 & -12 \\ -13 & 15 \end{bmatrix}$

19. $CD = \begin{bmatrix} 4 & -2 \\ 5 & 0 \\ 6 & 0 \\ 1 & 3 \end{bmatrix} \begin{bmatrix} 3 & 5 \\ 1 & 2 \end{bmatrix} = \begin{bmatrix} 10 & 16 \\ 15 & 25 \\ 18 & 30 \\ 6 & 11 \end{bmatrix}$

20. $DF = \begin{bmatrix} 3 & 5 \\ 1 & 2 \end{bmatrix} \begin{bmatrix} -1 & 6 \\ 4 & 11 \end{bmatrix} = \begin{bmatrix} 17 & 73 \\ 7 & 28 \end{bmatrix}$

21. $FD = \begin{bmatrix} -1 & 6 \\ 4 & 11 \end{bmatrix} \begin{bmatrix} 3 & 5 \\ 1 & 2 \end{bmatrix} = \begin{bmatrix} 3 & 7 \\ 23 & 42 \end{bmatrix}$

22. $FI = F$

23. $IF = F$

24. $DG^T = \begin{bmatrix} 3 & 5 \\ 1 & 2 \end{bmatrix} \begin{bmatrix} 2 & -1 \\ -5 & 3 \end{bmatrix} = \begin{bmatrix} -19 & 12 \\ -8 & 5 \end{bmatrix}$

25. $DG = \begin{bmatrix} 3 & 5 \\ 1 & 2 \end{bmatrix} \begin{bmatrix} 2 & -5 \\ -1 & 3 \end{bmatrix} = \begin{bmatrix} 1 & 0 \\ 0 & 1 \end{bmatrix} = I$

So, $(DG)F = F$.

26. $\left[\begin{array}{ccc|c} 1 & 1 & 2 & 5 \\ 4 & 0 & 1 & 5 \\ 2 & 1 & 1 & 5 \end{array}\right] \begin{array}{c} \rightarrow \\ -4R_1 + R_2 \rightarrow R_2 \\ -2R_1 + R_3 \rightarrow R_3 \end{array} \left[\begin{array}{ccc|c} 1 & 1 & 2 & 5 \\ 0 & -4 & -7 & -15 \\ 0 & -1 & -3 & -5 \end{array}\right] \begin{array}{c} R_2 \leftrightarrow R_3 \\ \rightarrow \\ \ \end{array} \left[\begin{array}{ccc|c} 1 & 1 & 2 & 5 \\ 0 & -1 & -3 & -5 \\ 0 & -4 & -7 & -15 \end{array}\right] \begin{array}{c} \rightarrow \\ -R_2 \rightarrow R_2 \\ \ \end{array}$

$\left[\begin{array}{ccc|c} 1 & 1 & 2 & 5 \\ 0 & 1 & 3 & 5 \\ 0 & -4 & -7 & -15 \end{array}\right] \begin{array}{c} -R_2 + R_1 \rightarrow R_1 \\ \rightarrow \\ 4R_2 + R_3 \rightarrow R_3 \end{array} \left[\begin{array}{ccc|c} 1 & 0 & -1 & 0 \\ 0 & 1 & 3 & 5 \\ 0 & 0 & 5 & 5 \end{array}\right] \begin{array}{c} \rightarrow \\ \ \\ \frac{1}{5}R_3 \rightarrow R_3 \end{array} \left[\begin{array}{ccc|c} 1 & 0 & -1 & 0 \\ 0 & 1 & 3 & 5 \\ 0 & 0 & 1 & 1 \end{array}\right] \begin{array}{c} R_3 + R_1 \rightarrow R_1 \\ -3R_3 + R_2 \rightarrow R_2 \\ \rightarrow \end{array}$

$\left[\begin{array}{ccc|c} 1 & 0 & 0 & 1 \\ 0 & 1 & 0 & 2 \\ 0 & 0 & 1 & 1 \end{array}\right]$ The solution is (1, 2, 1).

27. $\left[\begin{array}{cc|c} 1 & -2 & 4 \\ -3 & 10 & 24 \end{array}\right] \begin{array}{c} \rightarrow \\ 3R_1 + R_2 \rightarrow R_2 \end{array} \left[\begin{array}{cc|c} 1 & -2 & 4 \\ 0 & 4 & 36 \end{array}\right] \begin{array}{c} \rightarrow \\ \frac{1}{4}R_2 \rightarrow R_2 \end{array} \left[\begin{array}{cc|c} 1 & -2 & 4 \\ 0 & 1 & 9 \end{array}\right] \begin{array}{c} 2R_2 + R_1 \rightarrow R_1 \\ \rightarrow \end{array} \left[\begin{array}{cc|c} 1 & 0 & 22 \\ 0 & 1 & 9 \end{array}\right]$

Solution: $x = 22, y = 9$

28. $\left[\begin{array}{ccc|c} 1 & 1 & 1 & 4 \\ 3 & 4 & -1 & -1 \\ 2 & -1 & 3 & 3 \end{array}\right] \begin{array}{c} \rightarrow \\ -3R_1 + R_2 \rightarrow R_2 \\ -2R_1 + R_3 \rightarrow R_3 \end{array} \left[\begin{array}{ccc|c} 1 & 1 & 1 & 4 \\ 0 & 1 & -4 & -13 \\ 0 & -3 & 1 & -5 \end{array}\right] \begin{array}{c} -R_2 + R_1 \rightarrow R_1 \\ \rightarrow \\ 3R_2 + R_3 \rightarrow R_3 \end{array}$

$\left[\begin{array}{ccc|c} 1 & 0 & 5 & 17 \\ 0 & 1 & -4 & -13 \\ 0 & 0 & -11 & -44 \end{array}\right] \begin{array}{c} \rightarrow \\ \ \\ -\frac{1}{11}R_3 \rightarrow R_3 \end{array} \left[\begin{array}{ccc|c} 1 & 0 & 5 & 17 \\ 0 & 1 & -4 & -13 \\ 0 & 0 & 1 & 4 \end{array}\right] \begin{array}{c} -5R_3 + R_1 \rightarrow R_1 \\ 4R_3 + R_2 \rightarrow R_2 \\ \rightarrow \end{array} \left[\begin{array}{ccc|c} 1 & 0 & 0 & -3 \\ 0 & 1 & 0 & 3 \\ 0 & 0 & 1 & 4 \end{array}\right]$

Solution: $x = -3, \ y = 3, \ z = 4$

29. $\left[\begin{array}{ccc|c} -1 & 1 & 1 & 3 \\ 3 & 0 & -1 & 1 \\ 2 & -3 & -4 & -2 \end{array}\right] \begin{array}{c} \rightarrow \\ 3R_1 + R_2 \rightarrow R_2 \\ 2R_1 + R_3 \rightarrow R_3 \end{array} \left[\begin{array}{ccc|c} -1 & 1 & 1 & 3 \\ 0 & 3 & 2 & 10 \\ 0 & -1 & -2 & 4 \end{array}\right] \begin{array}{c} R_3 + R_1 \rightarrow R_1 \\ 3R_3 + R_2 \rightarrow R_2 \\ \rightarrow \end{array} \left[\begin{array}{ccc|c} -1 & 0 & -1 & 7 \\ 0 & 0 & -4 & 22 \\ 0 & -1 & -2 & 4 \end{array}\right] \begin{array}{c} -R_1 \rightarrow R_1 \\ \rightarrow \\ R_2 \leftrightarrow R_3 \end{array}$

$\left[\begin{array}{ccc|c} 1 & 0 & 1 & -7 \\ 0 & -1 & -2 & 4 \\ 0 & 0 & -4 & 22 \end{array}\right] \begin{array}{c} \rightarrow \\ -R_2 \rightarrow R_2 \\ -\frac{1}{4}R_3 \rightarrow R_3 \end{array} \left[\begin{array}{ccc|c} 1 & 0 & 1 & -7 \\ 0 & 1 & 2 & -4 \\ 0 & 0 & 1 & -11/2 \end{array}\right] \begin{array}{c} -R_3 + R_1 \rightarrow R_1 \\ -2R_3 + R_2 \rightarrow R_2 \\ \rightarrow \end{array} \left[\begin{array}{ccc|c} 1 & 0 & 0 & -3/2 \\ 0 & 1 & 0 & 7 \\ 0 & 0 & 1 & -11/2 \end{array}\right]$

Solution: $x = -\frac{3}{2}, y = 7, z = -\frac{11}{2}$

30. $\left[\begin{array}{ccc|c} 1 & 1 & -2 & 5 \\ 3 & 2 & 5 & 10 \\ -2 & -3 & 15 & 2 \end{array}\right] \begin{array}{c} \rightarrow \\ -3R_1 + R_2 \rightarrow R_2 \\ 2R_1 + R_3 \rightarrow R_3 \end{array} \left[\begin{array}{ccc|c} 1 & 1 & -2 & 5 \\ 0 & -1 & 11 & -5 \\ 0 & -1 & 11 & 12 \end{array}\right] \begin{array}{c} \rightarrow \\ \ \\ -R_2 + R_3 \rightarrow R_3 \end{array} \left[\begin{array}{ccc|c} 1 & 1 & -2 & 5 \\ 0 & -1 & 11 & -5 \\ 0 & 0 & 0 & 17 \end{array}\right]$

No solution.

31. $\left[\begin{array}{ccc|c} 1 & -1 & 0 & 3 \\ 1 & 1 & 4 & 1 \\ 2 & -3 & -2 & 7 \end{array}\right] \begin{array}{c} \rightarrow \\ -R_1 + R_2 \rightarrow R_2 \\ -2R_1 + R_3 \rightarrow R_3 \end{array} \left[\begin{array}{ccc|c} 1 & -1 & 0 & 3 \\ 0 & 2 & 4 & -2 \\ 0 & -1 & -2 & 1 \end{array}\right] \begin{array}{c} \rightarrow \\ \frac{1}{2}R_2 \rightarrow R_2 \\ \ \end{array}$

$\left[\begin{array}{ccc|c} 1 & -1 & 0 & 3 \\ 0 & 1 & 2 & -1 \\ 0 & -1 & -2 & 1 \end{array}\right] \begin{array}{c} R_2 + R_1 \rightarrow R_1 \\ \rightarrow \\ R_2 + R_3 \rightarrow R_3 \end{array} \left[\begin{array}{ccc|c} 1 & 0 & 2 & 2 \\ 0 & 1 & 2 & -1 \\ 0 & 0 & 0 & 0 \end{array}\right]$

Solution: $x = 2 - 2z, \ y = -1 - 2z$

32. $\left[\begin{array}{cccc|c} 1 & 1 & 1 & 1 & 3 \\ 1 & -2 & 1 & -4 & -5 \\ 1 & 0 & -1 & 1 & 0 \\ 0 & 1 & 1 & 1 & 2 \end{array}\right] \begin{array}{c} \rightarrow \\ -R_1 + R_2 \rightarrow R_2 \\ -R_1 + R_3 \rightarrow R_3 \\ \end{array} \left[\begin{array}{cccc|c} 1 & 1 & 1 & 1 & 3 \\ 0 & -3 & 0 & -5 & -8 \\ 0 & -1 & -2 & 0 & -3 \\ 0 & 1 & 1 & 1 & 2 \end{array}\right] \begin{array}{c} R_2 \leftrightarrow R_4 \\ \rightarrow \end{array}$

$\left[\begin{array}{cccc|c} 1 & 1 & 1 & 1 & 3 \\ 0 & 1 & 1 & 1 & 2 \\ 0 & -1 & -2 & 0 & -3 \\ 0 & -3 & 0 & -5 & -8 \end{array}\right] \begin{array}{c} -R_2 + R_1 \rightarrow R_1 \\ \rightarrow \\ R_2 + R_3 \rightarrow R_3 \\ 3R_2 + R_4 \rightarrow R_4 \end{array} \left[\begin{array}{cccc|c} 1 & 0 & 0 & 0 & 1 \\ 0 & 1 & 1 & 1 & 2 \\ 0 & 0 & -1 & 1 & -1 \\ 0 & 0 & 3 & -2 & -2 \end{array}\right] \begin{array}{c} \rightarrow \\ -R_3 \rightarrow R_3 \end{array}$

$\left[\begin{array}{cccc|c} 1 & 0 & 0 & 0 & 1 \\ 0 & 1 & 1 & 1 & 2 \\ 0 & 0 & 1 & -1 & 1 \\ 0 & 0 & 3 & -2 & -2 \end{array}\right] \begin{array}{c} \rightarrow \\ -R_3 + R_2 \rightarrow R_2 \\ \\ -3R_3 + R_4 \rightarrow R_4 \end{array} \left[\begin{array}{cccc|c} 1 & 0 & 0 & 0 & 1 \\ 0 & 1 & 0 & 2 & 1 \\ 0 & 0 & 1 & -1 & 1 \\ 0 & 0 & 0 & 1 & -5 \end{array}\right] \begin{array}{c} \rightarrow \\ -2R_4 + R_2 \rightarrow R_2 \\ R_4 + R_3 \rightarrow R_3 \end{array} \left[\begin{array}{cccc|c} 1 & 0 & 0 & 0 & 1 \\ 0 & 1 & 0 & 0 & 11 \\ 0 & 0 & 1 & 0 & -4 \\ 0 & 0 & 0 & 1 & -5 \end{array}\right]$

Solution: $x_1 = 1,\ x_2 = 11,\ x_3 = -4,\ x_4 = -5$

33. $DG = \begin{bmatrix} 3 & 5 \\ 1 & 2 \end{bmatrix} \begin{bmatrix} 2 & -5 \\ -1 & 3 \end{bmatrix} = \begin{bmatrix} 1 & 0 \\ 0 & 1 \end{bmatrix}$ Yes, D and G are inverse matrices.

34. $\begin{bmatrix} 7 & -1 \\ -10 & 2 \end{bmatrix}^{-1} = \dfrac{1}{14-10} \begin{bmatrix} 2 & 1 \\ 10 & 7 \end{bmatrix} = \begin{bmatrix} 1/2 & 1/4 \\ 5/2 & 7/4 \end{bmatrix}$

35. $\left[\begin{array}{ccc|ccc} 1 & 0 & 2 & 1 & 0 & 0 \\ 3 & 4 & -1 & 0 & 1 & 0 \\ 1 & 1 & 0 & 0 & 0 & 1 \end{array}\right] \begin{array}{c} \rightarrow \\ -3R_1 + R_2 \rightarrow R_2 \\ -R_1 + R_3 \rightarrow R_3 \end{array} \left[\begin{array}{ccc|ccc} 1 & 0 & 2 & 1 & 0 & 0 \\ 0 & 4 & -7 & -3 & 1 & 0 \\ 0 & 1 & -2 & -1 & 0 & 1 \end{array}\right] \begin{array}{c} R_2 \leftrightarrow R_3 \\ \rightarrow \end{array}$

$\left[\begin{array}{ccc|ccc} 1 & 0 & 2 & 1 & 0 & 0 \\ 0 & 1 & -2 & -1 & 0 & 1 \\ 0 & 4 & -7 & -3 & 1 & 0 \end{array}\right] \begin{array}{c} \rightarrow \\ \\ -4R_2 + R_3 \rightarrow R_3 \end{array} \left[\begin{array}{ccc|ccc} 1 & 0 & 2 & 1 & 0 & 0 \\ 0 & 1 & -2 & -1 & 0 & 1 \\ 0 & 0 & 1 & 1 & 1 & -4 \end{array}\right] \begin{array}{c} -2R_3 + R_1 \rightarrow R_1 \\ 2R_3 + R_2 \rightarrow R_2 \\ \rightarrow \end{array}$

$\left[\begin{array}{ccc|ccc} 1 & 0 & 0 & -1 & -2 & 8 \\ 0 & 1 & 0 & 1 & 2 & -7 \\ 0 & 0 & 1 & 1 & 1 & -4 \end{array}\right]$ Answer: $\begin{bmatrix} -1 & -2 & 8 \\ 1 & 2 & -7 \\ 1 & 1 & -4 \end{bmatrix}$

36. $\left[\begin{array}{ccc|ccc} 3 & 3 & 2 & 1 & 0 & 0 \\ -1 & 4 & 2 & 0 & 1 & 0 \\ 2 & 5 & 3 & 0 & 0 & 1 \end{array}\right] \begin{array}{c} 2R_2 + R_1 \rightarrow R_1 \\ \rightarrow \\ 2R_2 + R_3 \rightarrow R_3 \end{array} \left[\begin{array}{ccc|ccc} 1 & 11 & 6 & 1 & 2 & 0 \\ -1 & 4 & 2 & 0 & 1 & 0 \\ 0 & 13 & 7 & 0 & 2 & 1 \end{array}\right] \begin{array}{c} \rightarrow \\ R_1 + R_2 \rightarrow R_2 \end{array}$

$\left[\begin{array}{ccc|ccc} 1 & 11 & 6 & 1 & 2 & 0 \\ 0 & 15 & 8 & 1 & 3 & 0 \\ 0 & 13 & 7 & 0 & 2 & 1 \end{array}\right] \begin{array}{c} \rightarrow \\ -R_3 + R_2 \rightarrow R_2 \end{array} \left[\begin{array}{ccc|ccc} 1 & 11 & 6 & 1 & 2 & 0 \\ 0 & 2 & 1 & 1 & 1 & -1 \\ 0 & 13 & 7 & 0 & 2 & 1 \end{array}\right] \begin{array}{c} -6R_2 + R_1 \rightarrow R_1 \\ \rightarrow \\ -7R_2 + R_3 \rightarrow R_3 \end{array}$

$\left[\begin{array}{ccc|ccc} 1 & -1 & 0 & -5 & -4 & 6 \\ 0 & 2 & 1 & 1 & 1 & -1 \\ 0 & -1 & 0 & -7 & -5 & 8 \end{array}\right] \begin{array}{c} R_2 \leftrightarrow R_3 \\ \rightarrow \\ (-1)\text{ new } R_2 \end{array} \left[\begin{array}{ccc|ccc} 1 & -1 & 0 & -5 & -4 & 6 \\ 0 & 1 & 0 & 7 & 5 & -8 \\ 0 & 2 & 1 & 1 & 1 & -1 \end{array}\right] \begin{array}{c} R_2 + R_1 \rightarrow R_1 \\ \rightarrow \\ -2R_2 + R_3 \rightarrow R_3 \end{array}$

$\left[\begin{array}{ccc|ccc} 1 & 0 & 0 & 2 & 1 & -2 \\ 0 & 1 & 0 & 7 & 5 & -8 \\ 0 & 0 & 1 & -13 & -9 & 15 \end{array}\right]$ Answer: $\begin{bmatrix} 2 & 1 & -2 \\ 7 & 5 & -8 \\ -13 & -9 & 15 \end{bmatrix}$

37. Use the inverse of problem 35.

$$\begin{bmatrix} x \\ y \\ z \end{bmatrix} = \begin{bmatrix} 1 & 0 & 2 \\ 3 & 4 & -1 \\ 1 & 1 & 0 \end{bmatrix}^{-1} \begin{bmatrix} 5 \\ 2 \\ -3 \end{bmatrix} = \begin{bmatrix} -1 & -2 & 8 \\ 1 & 2 & -7 \\ 1 & 1 & -4 \end{bmatrix} \begin{bmatrix} 5 \\ 2 \\ -3 \end{bmatrix} = \begin{bmatrix} -33 \\ 30 \\ 19 \end{bmatrix}$$

38. Use the inverse of problem 36.

$$\begin{bmatrix} x \\ y \\ z \end{bmatrix} = \begin{bmatrix} 3 & 3 & 2 \\ -1 & 4 & 2 \\ 2 & 5 & 3 \end{bmatrix}^{-1} \begin{bmatrix} 1 \\ -10 \\ -6 \end{bmatrix} = \begin{bmatrix} 2 & 1 & -2 \\ 7 & 5 & -8 \\ -13 & -9 & 15 \end{bmatrix} \begin{bmatrix} 1 \\ -10 \\ -6 \end{bmatrix} = \begin{bmatrix} 4 \\ 5 \\ -13 \end{bmatrix}$$

39.

$$\left[\begin{array}{ccc|ccc} 1 & 3 & 1 & 1 & 0 & 0 \\ 1 & 4 & 3 & 0 & 1 & 0 \\ 2 & -1 & -11 & 0 & 0 & 1 \end{array}\right] \begin{array}{c} \to \\ -R_1 + R_2 \to R_2 \\ -2R_1 + R_3 \to R_3 \end{array} \left[\begin{array}{ccc|ccc} 1 & 3 & 1 & 1 & 0 & 0 \\ 0 & 1 & 2 & -1 & 1 & 0 \\ 0 & -7 & -13 & -2 & 0 & 1 \end{array}\right] \begin{array}{c} -3R_2 + R_1 \to R_1 \\ \to \\ 7R_2 + R_3 \to R_3 \end{array}$$

$$\left[\begin{array}{ccc|ccc} 1 & 0 & -5 & 4 & -3 & 0 \\ 0 & 1 & 2 & -1 & 1 & 0 \\ 0 & 0 & 1 & -9 & 7 & 1 \end{array}\right] \begin{array}{c} 5R_3 + R_1 \to R_1 \\ -2R_3 + R_2 \to R_2 \\ \to \end{array} \left[\begin{array}{ccc|ccc} 1 & 0 & 0 & -41 & 32 & 5 \\ 0 & 1 & 0 & 17 & -13 & -2 \\ 0 & 0 & 1 & -9 & 7 & 1 \end{array}\right]$$

$$\begin{bmatrix} x \\ y \\ z \end{bmatrix} = \begin{bmatrix} -41 & 32 & 5 \\ 17 & -13 & -2 \\ -9 & 7 & 1 \end{bmatrix} \begin{bmatrix} 0 \\ 2 \\ -12 \end{bmatrix} = \begin{bmatrix} 4 \\ -2 \\ 2 \end{bmatrix}$$

40. The determinant of the matrix is 0. The matrix does not have an inverse.

41. Total production = $N + M = \begin{bmatrix} 250 & 140 \\ 480 & 700 \end{bmatrix}$

42. $M + P - S = \begin{bmatrix} 1030 & 800 \\ 700 & 1200 \end{bmatrix}$

43. June: $\begin{bmatrix} 100 & 0 \\ 0 & 120 \end{bmatrix} \cdot M = \begin{bmatrix} 15{,}000 & 8000 \\ 33{,}600 & 36{,}000 \end{bmatrix}$ (columns A, B; totals 48,600 and 44,000)

July: $\begin{bmatrix} 100 & 0 \\ 0 & 120 \end{bmatrix} \cdot N = \begin{bmatrix} 10{,}000 & 6000 \\ 24{,}000 & 48{,}000 \end{bmatrix}$ (columns A, B; totals 34,000 and 54,000)

a. Production was higher at Plant A in June.
b. Production was higher at Plant B in July.

44.

$$\begin{array}{c} \\ R \\ H \end{array}\begin{array}{c} \begin{array}{ccc} S & M & L \end{array} \\ \begin{bmatrix} 25 & 40 & 45 \\ 10 & 10 & 10 \end{bmatrix} \end{array} \begin{array}{c} \begin{array}{cc} M & W \end{array} \\ \begin{bmatrix} 1 & 14 \\ 12 & 10 \\ 8 & 3 \end{bmatrix} \end{array}\begin{array}{c} \\ S \\ M \\ L \end{array} = \begin{array}{c} \begin{array}{cc} M & W \end{array} \\ \begin{bmatrix} 865 & 885 \\ 210 & 270 \end{bmatrix} \end{array}\begin{array}{c} \\ R \\ H \end{array}$$

BA gives costs of robes and hoods for men and women.

45.

$$\begin{bmatrix} 25 & 40 & 45 \\ 10 & 10 & 10 \end{bmatrix} \begin{bmatrix} 1 & 14 \\ 12 & 10 \\ 8 & 3 \end{bmatrix} = \begin{bmatrix} 865 & 885 \\ 210 & 270 \end{bmatrix}$$

$$\begin{bmatrix} 865 & 885 \\ 210 & 270 \end{bmatrix} \begin{bmatrix} 1 \\ 1 \end{bmatrix} = \begin{bmatrix} 1750 \\ 480 \end{bmatrix} \begin{array}{c} R \\ H \end{array}$$

The cost of new robes is \$1750.
The cost of new hoods is \$480.

46. a. $\begin{bmatrix} 30 & 20 & 10 \\ 20 & 10 & 20 \end{bmatrix}\begin{bmatrix} 300 & 280 \\ 150 & 100 \\ 150 & 200 \end{bmatrix} = \begin{bmatrix} 9000+3000+1500 & 8400+2000+2000 \\ 6000+1500+3000 & 5600+1000+4000 \end{bmatrix} = \begin{bmatrix} 13,500 & 12,400 \\ 10,500 & 10,600 \end{bmatrix}$

b. Column 1 is Ace's price & column 2 is Kink's price. Dept. A buys from Kink and Dept. B buys from Ace.

47. a. $W = \begin{bmatrix} 0.20 & 0.30 & 0.50 \end{bmatrix}$

b. $R = \begin{bmatrix} 0.013469 \\ 0.013543 \\ 0.006504 \end{bmatrix}$

c. $WR = 0.0100087$

d. The historical return is the same as the estimated expected monthly return, i.e., 1%.

48. x = fast food shares; y = software shares; z = pharmaceutical shares

$$50x + 20y + 80z = 50,000$$
$$0.115(50x) + 0.15(20y) + 0.10(80z) = 0.12(50,000)$$
$$x = 2z$$

Rearranged and simplified:

$$x - 2z = 0$$
$$5x + 2y + 8z = 5000$$
$$5.75x + 3y + 8z = 6000$$

$$\left[\begin{array}{ccc|c} 1 & 0 & -2 & 0 \\ 5 & 2 & 8 & 5000 \\ 5.75 & 3 & 8 & 6000 \end{array}\right] \xrightarrow[-5.75R_1 + R_3 \to R_3]{-5R_1 + R_2 \to R_2} \left[\begin{array}{ccc|c} 1 & 0 & -2 & 0 \\ 0 & 2 & 18 & 5000 \\ 0 & 3 & 19.5 & 6000 \end{array}\right] \xrightarrow{\frac{1}{2}R_2 \to R_2} \left[\begin{array}{ccc|c} 1 & 0 & -2 & 0 \\ 0 & 1 & 9 & 2500 \\ 0 & 3 & 19.5 & 6000 \end{array}\right]$$

$$\xrightarrow{-3R_2 + R_3 \to R_3} \left[\begin{array}{ccc|c} 1 & 0 & -2 & 0 \\ 0 & 1 & 9 & 2500 \\ 0 & 0 & -7.5 & -1500 \end{array}\right] \xrightarrow{-\frac{2}{15}R_3 \to R_3} \left[\begin{array}{ccc|c} 1 & 0 & -2 & 0 \\ 0 & 1 & 9 & 2500 \\ 0 & 0 & 1 & 200 \end{array}\right] \xrightarrow[-9R_3 + R_2 \to R_2]{2R_3 + R_1 \to R_1} \left[\begin{array}{ccc|c} 1 & 0 & 0 & 400 \\ 0 & 1 & 0 & 700 \\ 0 & 0 & 1 & 200 \end{array}\right]$$

Buy 400 shares of fast food company, 700 shares of software company, and 200 shares of pharmaceutical company.

49.

$A + B + 2C = 2000$ Units of I

$3A + 4B + 10C = 8000$ Units of II

$A + 2B + 6C = 4000$ Units of III

$$\left[\begin{array}{ccc|c} 1 & 1 & 2 & 2000 \\ 3 & 4 & 10 & 8000 \\ 1 & 2 & 6 & 4000 \end{array}\right] \xrightarrow[-R_1 + R_3 \to R_3]{-3R_1 + R_2 \to R_2} \left[\begin{array}{ccc|c} 1 & 1 & 2 & 2000 \\ 0 & 1 & 4 & 2000 \\ 0 & 1 & 4 & 2000 \end{array}\right] \xrightarrow[-R_2 + R_3 \to R_3]{-R_2 + R_1 \to R_1} \left[\begin{array}{ccc|c} 1 & 0 & -2 & 0 \\ 0 & 1 & 4 & 2000 \\ 0 & 0 & 0 & 0 \end{array}\right]$$

Solution: $A = 2C,\ B = 2000 - 4C$

50.

$$\left[\begin{array}{ccc|c} 100 & 100 & 100 & 1100 \\ 150 & 20 & 350 & 1930 \\ 20 & 65 & 35 & 460 \end{array}\right] \begin{array}{c} \frac{1}{100}R_1 \to R_1 \\ \frac{1}{10}R_2 \to R_2 \\ \frac{1}{5}R_3 \to R_3 \end{array} \left[\begin{array}{ccc|c} 1 & 1 & 1 & 11 \\ 15 & 2 & 35 & 193 \\ 4 & 13 & 7 & 92 \end{array}\right] \xrightarrow[-4R_1 + R_3 \to R_3]{-15R_1 + R_2 \to R_2} \left[\begin{array}{ccc|c} 1 & 1 & 1 & 11 \\ 0 & -13 & 20 & 28 \\ 0 & 9 & 3 & 48 \end{array}\right] \xrightarrow{\frac{1}{3}R_3 \to R_3}$$

$$\left[\begin{array}{ccc|c} 1 & 1 & 1 & 11 \\ 0 & -13 & 20 & 28 \\ 0 & 3 & 1 & 16 \end{array}\right] \xrightarrow{4R_3 + R_2 \to R_2} \left[\begin{array}{ccc|c} 1 & 1 & 1 & 11 \\ 0 & -1 & 24 & 92 \\ 0 & 3 & 1 & 16 \end{array}\right] \xrightarrow[3R_2 + R_3 \to R_3]{R_2 + R_1 \to R_1} \left[\begin{array}{ccc|c} 1 & 0 & 25 & 103 \\ 0 & -1 & 24 & 92 \\ 0 & 0 & 73 & 292 \end{array}\right] \xrightarrow[\frac{1}{73}R_3 \to R_3]{-R_2 \to R_2}$$

$$\left[\begin{array}{ccc|c} 1 & 0 & 25 & 103 \\ 0 & 1 & -24 & -92 \\ 0 & 0 & 1 & 4 \end{array}\right] \xrightarrow[24R_3 + R_2 \to R_2]{-25R_3 + R_1 \to R_1} \left[\begin{array}{ccc|c} 1 & 0 & 0 & 3 \\ 0 & 1 & 0 & 4 \\ 0 & 0 & 1 & 4 \end{array}\right]$$

Use 3 passenger, 4 transport, and 4 jumbo.

51. $D=\begin{bmatrix}4720\\40\end{bmatrix}$ $(I-A)X=D$ or $\begin{bmatrix}0.8 & -0.1\\-0.1 & 0.8\end{bmatrix}\begin{bmatrix}S\\A\end{bmatrix}=\begin{bmatrix}4720\\40\end{bmatrix}$

$$\begin{bmatrix}S\\A\end{bmatrix}=\begin{bmatrix}0.8 & -0.1\\-0.1 & 0.8\end{bmatrix}^{-1}\begin{bmatrix}4720\\40\end{bmatrix}=\frac{1}{0.63}\begin{bmatrix}0.8 & 0.1\\0.1 & 0.8\end{bmatrix}\begin{bmatrix}4720\\40\end{bmatrix}=\begin{bmatrix}6000\\800\end{bmatrix}$$

52. $D=\begin{bmatrix}850\\275\end{bmatrix}$ $(I-A)X=D$ or $\begin{bmatrix}0.9 & -0.1\\-0.2 & 0.95\end{bmatrix}\begin{bmatrix}S\\C\end{bmatrix}=\begin{bmatrix}850\\275\end{bmatrix}$

$$\begin{bmatrix}S\\C\end{bmatrix}=\begin{bmatrix}0.9 & -0.1\\-0.2 & 0.95\end{bmatrix}^{-1}\begin{bmatrix}850\\275\end{bmatrix}=\frac{1}{0.835}\begin{bmatrix}0.95 & 0.1\\0.2 & 0.9\end{bmatrix}\begin{bmatrix}850\\275\end{bmatrix}=\begin{bmatrix}1000\\500\end{bmatrix}$$

53. $(I-A)X=D$ or $\begin{bmatrix}0.6 & -0.4 & -0.2\\-0.2 & 0.6 & -0.2\\-0.1 & -0.2 & 0.6\end{bmatrix}\begin{bmatrix}\text{Min}\\\text{Mfg}\\\text{Fuel}\end{bmatrix}=\begin{bmatrix}10\\40\\140\end{bmatrix}$

$$\left[\begin{array}{ccc|c}1 & 2 & -6 & -1400\\-2 & 6 & -2 & 400\\6 & -4 & -2 & 100\end{array}\right]\begin{array}{c}\to\\2R_1+R_2\to R_2\\-6R_1+R_3\to R_3\end{array}\left[\begin{array}{ccc|c}1 & 2 & -6 & -1400\\0 & 10 & -14 & -2400\\0 & -16 & 34 & 8500\end{array}\right]\begin{array}{c}\to\\\frac{1}{2}R_2\to R_2\\\to\end{array}\left[\begin{array}{ccc|c}1 & 2 & -6 & -1400\\0 & 5 & -7 & -1200\\0 & -16 & 34 & 8500\end{array}\right]$$

$$\begin{array}{c}\to\\ \\3R_2+R_3\to R_3\end{array}\left[\begin{array}{ccc|c}1 & 2 & -6 & -1400\\0 & 5 & -7 & -1200\\0 & -1 & 13 & 4900\end{array}\right]\begin{array}{c}2R_3+R_1\to R_1\\5R_3+R_2\to R_2\\\to\end{array}\left[\begin{array}{ccc|c}1 & 0 & 20 & 8400\\0 & 0 & 58 & 23{,}300\\0 & -1 & 13 & 4900\end{array}\right]\begin{array}{c}\to\\\frac{1}{58}R_2\to R_2\\ \end{array}$$

$$\left[\begin{array}{ccc|c}1 & 0 & 20 & 8400\\0 & 0 & 1 & 401.7\\0 & -1 & 13 & 4900\end{array}\right]\begin{array}{c}-20R_2+R_1\to R_1\\\to\\-13R_2+R_3\to R_3\end{array}\left[\begin{array}{ccc|c}1 & 0 & 0 & 366\\0 & 0 & 1 & 401.7\\0 & -1 & 0 & -322\end{array}\right]\begin{array}{c}R_2\leftrightarrow R_3\\\to\\(-1)\text{ new }R_2\end{array}\left[\begin{array}{ccc|c}1 & 0 & 0 & 366\\0 & 1 & 0 & 322\\0 & 0 & 1 & 402\end{array}\right]$$

366 units of mined goods, 322 units of manufactured goods, and 402 units of fuel.

54. A closed Leontief model must be solved by the Gauss-Jordan elimination method. A graphing calculator is strongly suggested.

$G=\frac{64}{93}H,\ A=\frac{59}{93}H,\ M=\frac{40}{93}H$

Chapter Test

1. $A^T + B = \begin{bmatrix} 1 & 3 & 4 \\ -2 & 2 & 1 \end{bmatrix} + \begin{bmatrix} 2 & -2 & 1 \\ 3 & 1 & 5 \end{bmatrix} = \begin{bmatrix} 3 & 1 & 5 \\ 1 & 3 & 6 \end{bmatrix}$

2. $B - C = \begin{bmatrix} 2 & -2 & 1 \\ 3 & 1 & 5 \end{bmatrix} - \begin{bmatrix} 3 & -4 & -1 \\ 2 & 2 & -1 \end{bmatrix} = \begin{bmatrix} -1 & 2 & 2 \\ 1 & -1 & 6 \end{bmatrix}$

3. $CD = \begin{bmatrix} 3 & -4 & -1 \\ 2 & 2 & -1 \end{bmatrix} \begin{bmatrix} 1 & 2 & 4 \\ 3 & 5 & 41 \\ 3 & 2 & 3 \end{bmatrix} = \begin{bmatrix} -12 & -16 & -155 \\ 5 & 12 & 87 \end{bmatrix}$

4. $DA = \begin{bmatrix} 1 & 2 & 4 \\ 3 & 5 & 41 \\ 3 & 2 & 3 \end{bmatrix} \begin{bmatrix} 1 & -2 \\ 3 & 2 \\ 4 & 1 \end{bmatrix} = \begin{bmatrix} 23 & 6 \\ 182 & 45 \\ 21 & 1 \end{bmatrix}$

5. $BA = \begin{bmatrix} 2 & -2 & 1 \\ 3 & 1 & 5 \end{bmatrix} \begin{bmatrix} 1 & -2 \\ 3 & 2 \\ 4 & 1 \end{bmatrix} = \begin{bmatrix} 0 & -7 \\ 26 & 1 \end{bmatrix}$

6. $ABD = \begin{bmatrix} 1 & -2 \\ 3 & 2 \\ 4 & 1 \end{bmatrix} \begin{bmatrix} 2 & -2 & 1 \\ 3 & 1 & 5 \end{bmatrix} \begin{bmatrix} 1 & 2 & 4 \\ 3 & 5 & 41 \\ 3 & 2 & 3 \end{bmatrix} = \begin{bmatrix} -4 & -4 & -9 \\ 12 & -4 & 13 \\ 11 & -7 & 9 \end{bmatrix} \begin{bmatrix} 1 & 2 & 4 \\ 3 & 5 & 41 \\ 3 & 2 & 3 \end{bmatrix} = \begin{bmatrix} -43 & -46 & -207 \\ 39 & 30 & -77 \\ 17 & 5 & -216 \end{bmatrix}$

7. $\begin{bmatrix} 1 & 3 \\ 2 & 4 \end{bmatrix}^{-1} = \frac{1}{-2} \begin{bmatrix} 4 & -3 \\ -2 & 1 \end{bmatrix} = \begin{bmatrix} -2 & 3/2 \\ 1 & -1/2 \end{bmatrix}$

8. $\left[\begin{array}{ccc|ccc} 1 & 2 & 4 & 1 & 0 & 0 \\ 1 & 2 & 2 & 0 & 1 & 0 \\ 1 & 1 & 4 & 0 & 0 & 1 \end{array}\right] \begin{array}{l} \rightarrow \\ -R_1 + R_2 \rightarrow R_2 \\ -R_1 + R_3 \rightarrow R_3 \end{array}$ $\left[\begin{array}{ccc|ccc} 1 & 2 & 4 & 1 & 0 & 0 \\ 0 & 0 & -2 & -1 & 1 & 0 \\ 0 & -1 & 0 & -1 & 0 & 1 \end{array}\right] \begin{array}{l} \rightarrow \\ -R_3 \rightarrow R_3 \\ R_2 \leftrightarrow \text{new } R_3 \end{array}$

$\left[\begin{array}{ccc|ccc} 1 & 2 & 4 & 1 & 0 & 0 \\ 0 & 1 & 0 & 1 & 0 & -1 \\ 0 & 0 & -2 & -1 & 1 & 0 \end{array}\right] \begin{array}{l} -2R_2 + R_1 \rightarrow R_1 \\ \rightarrow \\ -\frac{1}{2}R_3 \rightarrow R_3 \end{array}$ $\left[\begin{array}{ccc|ccc} 1 & 0 & 4 & -1 & 0 & 2 \\ 0 & 1 & 0 & 1 & 0 & -1 \\ 0 & 0 & 1 & 1/2 & -1/2 & 0 \end{array}\right] \begin{array}{l} -4R_3 + R_1 \rightarrow R_1 \\ \rightarrow \\ \end{array}$

$\left[\begin{array}{ccc|ccc} 1 & 0 & 0 & -3 & 2 & 2 \\ 0 & 1 & 0 & 1 & 0 & -1 \\ 0 & 0 & 1 & 1/2 & -1/2 & 0 \end{array}\right]$ $\quad \text{Inverse} = \begin{bmatrix} -3 & 2 & 2 \\ 1 & 0 & -1 \\ 1/2 & -1/2 & 0 \end{bmatrix}$

9.
$$AX = B$$
$$A^{-1}AX = A^{-1}B$$
$$X = A^{-1}B$$
$$X = \begin{bmatrix} 1 & 2 & 0 \\ 3 & 1 & 2 \\ 4 & 1 & 1 \end{bmatrix} \begin{bmatrix} 3 \\ 1 \\ 2 \end{bmatrix} = \begin{bmatrix} 5 \\ 14 \\ 15 \end{bmatrix}$$

10. Begin with the augmented matrix.

$$\left[\begin{array}{ccc|c} 1 & -1 & 2 & 4 \\ 1 & 4 & 1 & 4 \\ 2 & 2 & 4 & 10 \end{array}\right] \xrightarrow{\frac{1}{2}R_3 \to R_3} \left[\begin{array}{ccc|c} 1 & -1 & 2 & 4 \\ 1 & 4 & 1 & 4 \\ 1 & 1 & 2 & 5 \end{array}\right] \xrightarrow[-R_1+R_3 \to R_3]{-R_1+R_2 \to R_2} \left[\begin{array}{ccc|c} 1 & -1 & 2 & 4 \\ 0 & 5 & -1 & 0 \\ 0 & 2 & 0 & 1 \end{array}\right] \xrightarrow[R_2 \leftrightarrow \text{new } R_3]{\frac{1}{2}R_3 \to R_3}$$

$$\left[\begin{array}{ccc|c} 1 & -1 & 2 & 4 \\ 0 & 1 & 0 & 1/2 \\ 0 & 5 & -1 & 0 \end{array}\right] \xrightarrow[-5R_2+R_3 \to R_3]{R_2+R_1 \to R_1} \left[\begin{array}{ccc|c} 1 & 0 & 2 & 9/2 \\ 0 & 1 & 0 & 1/2 \\ 0 & 0 & -1 & -5/2 \end{array}\right] \xrightarrow[-R_3 \to R_3]{2R_3+R_1 \to R_1} \left[\begin{array}{ccc|c} 1 & 0 & 0 & -1/2 \\ 0 & 1 & 0 & 1/2 \\ 0 & 0 & 1 & 5/2 \end{array}\right]$$

Solution: $x = -\frac{1}{2},\ y = \frac{1}{2},\ z = \frac{5}{2}$

11. Begin with the augmented matrix.

$$\left[\begin{array}{ccc|c} 1 & -1 & 2 & 4 \\ 1 & 4 & 1 & 4 \\ 2 & 3 & 3 & 8 \end{array}\right] \xrightarrow[-2R_1+R_3 \to R_3]{-R_1+R_2 \to R_2} \left[\begin{array}{ccc|c} 1 & -1 & 2 & 4 \\ 0 & 5 & -1 & 0 \\ 0 & 5 & -1 & 0 \end{array}\right] \xrightarrow[-5R_2+R_3 \to R_3]{\frac{1}{5}R_2 \to R_2} \left[\begin{array}{ccc|c} 1 & -1 & 2 & 4 \\ 0 & 1 & -1/5 & 0 \\ 0 & 0 & 0 & 0 \end{array}\right] \xrightarrow{R_2+R_1 \to R_1}$$

$$\left[\begin{array}{ccc|c} 1 & 0 & 9/5 & 4 \\ 0 & 1 & -1/5 & 0 \\ 0 & 0 & 0 & 0 \end{array}\right]$$ Solution: $x = 4 - \frac{9}{5}z,\ y = \frac{1}{5}z,\ z = z$

12. Begin with augmented matrix.

$$\left[\begin{array}{ccc|c} 1 & -1 & 3 & 4 \\ 1 & 5 & 2 & 3 \\ 2 & 4 & 5 & 8 \end{array}\right] \xrightarrow[-2R_1+R_3 \to R_3]{-R_1+R_2 \to R_2} \left[\begin{array}{ccc|c} 1 & -1 & 3 & 4 \\ 0 & 6 & -1 & -1 \\ 0 & 6 & -1 & 0 \end{array}\right] \xrightarrow{-R_3+R_2 \to R_2} \left[\begin{array}{ccc|c} 1 & -1 & 3 & 4 \\ 0 & 0 & 0 & -1 \\ 0 & 6 & -1 & 0 \end{array}\right]$$

There is no solution. In R_2 we have $0 = -1$.

13. Your calculator will give, in decimal notation, A^{-1}.

$$A^{-1} = \begin{bmatrix} 1 & 2 & -1 & 0 \\ -4/3 & -2/3 & 1 & -2/3 \\ 2/3 & -2/3 & 0 & 1/3 \\ 1 & 1 & -1 & 1 \end{bmatrix}; \quad \begin{bmatrix} x \\ y \\ z \\ w \end{bmatrix} = A^{-1} \cdot \begin{bmatrix} 4 \\ 4 \\ 10 \\ 0 \end{bmatrix} = \begin{bmatrix} 2 \\ 2 \\ 0 \\ -2 \end{bmatrix}$$

14. Your calculator will say that the coefficient matrix has no inverse. We will use the Gauss-Jordan method, give the matrices, and ask you to use the same steps with your calculator.

$$\left[\begin{array}{cccc|c} 1 & -1 & 2 & -1 & 4 \\ 1 & 4 & 1 & 1 & 4 \\ 2 & 2 & 4 & 2 & 10 \\ 0 & -1 & 1 & 2 & 2 \end{array}\right] \xrightarrow[-2R_1+R_3 \to R_3]{-R_1+R_2 \to R_2} \left[\begin{array}{cccc|c} 1 & -1 & 2 & -1 & 4 \\ 0 & 5 & -1 & 2 & 0 \\ 0 & 4 & 0 & 4 & 2 \\ 0 & -1 & 1 & 2 & 2 \end{array}\right] \begin{array}{c} -R_4+R_1 \to R_1 \\ 5R_4+R_2 \to R_2 \\ 4R_4+R_3 \to R_3 \\ \to \end{array} \left[\begin{array}{cccc|c} 1 & 0 & 1 & -3 & 2 \\ 0 & 0 & 4 & 12 & 10 \\ 0 & 0 & 4 & 12 & 10 \\ 0 & -1 & 1 & 2 & 2 \end{array}\right]$$

$$\xrightarrow[-R_4 \to R_4]{-R_2+R_3 \to R_3} \left[\begin{array}{cccc|c} 1 & 0 & 1 & -3 & 2 \\ 0 & 0 & 4 & 12 & 10 \\ 0 & 0 & 0 & 0 & 0 \\ 0 & 1 & -1 & -2 & -2 \end{array}\right] \begin{array}{l} \text{1. Interchange } R_3 \text{ and } R_4. \\ \text{2. Interchange } R_2 \text{ and } R_3. \\ \text{3. } \frac{1}{4}R_3 \to R_3 \end{array} \left[\begin{array}{cccc|c} 1 & 0 & 1 & -3 & 2 \\ 0 & 1 & -1 & -2 & -2 \\ 0 & 0 & 1 & 3 & 5/2 \\ 0 & 0 & 0 & 0 & 0 \end{array}\right] \begin{array}{c} -R_3+R_1 \to R_1 \\ R_3+R_2 \to R_2 \\ \to \end{array}$$

$$\left[\begin{array}{cccc|c} 1 & 0 & 0 & -6 & -1/2 \\ 0 & 1 & 0 & 1 & 1/2 \\ 0 & 0 & 1 & 3 & 5/2 \\ 0 & 0 & 0 & 0 & 0 \end{array}\right]$$ Solution: $x = 6w - \frac{1}{2},\ y = -w + \frac{1}{2},\ z = -3w + \frac{5}{2}$

15. a. $AB = \begin{bmatrix} 0.08 & 0.22 & 0.12 \\ 0.10 & 0.08 & 0.19 \\ 0.05 & 0.07 & 0.09 \\ 0.10 & 0.26 & 0.15 \\ 0.12 & 0.04 & 0.24 \end{bmatrix}$

b. Plant type 1:
0.08, 0.22, 0.12 are consumed by carnivores 1, 2, 3, respectively.

c. Plant 5 by 1; Plant 4 by 2; Plant 5 by 3.

16. a. $\begin{bmatrix} 1000 & 4000 & 2000 & 1000 \end{bmatrix}$

b. $\begin{bmatrix} 1000 & 4000 & 2000 & 1000 \end{bmatrix} \begin{bmatrix} 10 & 5 & 0 & 0 \\ 5 & 0 & 20 & 10 \\ 5 & 20 & 0 & 10 \\ 5 & 10 & 10 & 10 \end{bmatrix} = \begin{bmatrix} 45,000 & 55,000 & 90,000 & 70,000 \end{bmatrix}$

c. $\begin{bmatrix} 5 \\ 3 \\ 4 \\ 4 \end{bmatrix}$

d. $\begin{bmatrix} 45,000 & 55,000 & 90,000 & 70,000 \end{bmatrix} \begin{bmatrix} 5 \\ 3 \\ 4 \\ 4 \end{bmatrix} = \begin{bmatrix} 1,030,000 \end{bmatrix}$

e. $\begin{bmatrix} 10 & 5 & 0 & 0 \\ 5 & 0 & 20 & 10 \\ 5 & 20 & 0 & 10 \\ 5 & 10 & 10 & 10 \end{bmatrix} \begin{bmatrix} 5 \\ 3 \\ 4 \\ 4 \end{bmatrix} = \begin{bmatrix} 65 \\ 145 \\ 125 \\ 135 \end{bmatrix}$

17. Using the calculator and solving with the inverse is more efficient. This method shows the inefficiency of the Gauss-Jordan method.

x = Growth shares $\quad 30x = 100y + 50z$
y = Blue-chip shares $\quad 4.6x + 11y + 5z = 0.13(120,000)$
z = Utility shares $\quad 30x + 100y + 50z = 120,000$

$$\left[\begin{array}{ccc|c} 30 & -100 & -50 & 0 \\ 4.6 & 11 & 5 & 15,600 \\ 30 & 100 & 50 & 120,000 \end{array}\right] \xrightarrow{-R_1 + R_3 \to R_3} \left[\begin{array}{ccc|c} 30 & -100 & -50 & 0 \\ 4.6 & 11 & 5 & 15,600 \\ 0 & 200 & 100 & 120,000 \end{array}\right] \xrightarrow{\frac{1}{30}R_1 \to R_1}$$

$$\left[\begin{array}{ccc|c} 1 & -10/3 & -5/3 & 0 \\ 4.6 & 11 & 5 & 15,600 \\ 0 & 200 & 100 & 120,000 \end{array}\right] \xrightarrow{-4.6R_1 + R_2 \to R_2} \left[\begin{array}{ccc|c} 1 & -10/3 & -5/3 & 0 \\ 0 & 79/3 & 38/3 & 15,600 \\ 0 & 200 & 100 & 120,000 \end{array}\right] \xrightarrow{R_2 \leftrightarrow R_3}$$

$$\left[\begin{array}{ccc|c} 1 & -10/3 & -5/3 & 0 \\ 0 & 200 & 100 & 120,000 \\ 0 & 79/3 & 38/3 & 15,600 \end{array}\right] \xrightarrow{\frac{1}{200}R_2 \to R_2} \left[\begin{array}{ccc|c} 1 & -10/3 & -5/3 & 0 \\ 0 & 1 & 1/2 & 600 \\ 0 & 79/3 & 38/3 & 15,600 \end{array}\right] \begin{array}{l} \frac{10}{3}R_2 + R_1 \to R_1 \\ \to \\ -\frac{79}{3}R_2 + R_3 \to R_3 \end{array}$$

$$\left[\begin{array}{ccc|c} 1 & 0 & 0 & 2000 \\ 0 & 1 & 1/2 & 600 \\ 0 & 0 & -1/2 & -200 \end{array}\right] \xrightarrow{-2R_3 \to R_3} \left[\begin{array}{ccc|c} 1 & 0 & 0 & 2000 \\ 0 & 1 & 1/2 & 600 \\ 0 & 0 & 1 & 400 \end{array}\right] \xrightarrow{-\frac{1}{2}R_3 + R_2 \to R_2} \left[\begin{array}{ccc|c} 1 & 0 & 0 & 2000 \\ 0 & 1 & 0 & 400 \\ 0 & 0 & 1 & 400 \end{array}\right]$$

Solution: $x = 2000,\ y = 400,\ z = 400$

18. The technological equation is $(I-A)X = D$.

$$I - A = \begin{bmatrix} 0.6 & -0.2 \\ -0.1 & 0.7 \end{bmatrix};\ (I-A)^{-1} = \frac{1}{0.42-0.02}\begin{bmatrix} 0.7 & 0.2 \\ 0.1 & 0.6 \end{bmatrix}$$

$$\begin{bmatrix} \text{Ag} \\ \text{M} \end{bmatrix} = \frac{1}{0.40}\begin{bmatrix} 0.7 & 0.2 \\ 0.1 & 0.6 \end{bmatrix}\begin{bmatrix} 140 \\ 140 \end{bmatrix} = \frac{1}{0.4}\begin{bmatrix} 126 \\ 98 \end{bmatrix} = \begin{bmatrix} 315 \\ 245 \end{bmatrix}$$

19. The technological equation is $(I-A)X = 0$.

The augmented matrix to be solved is

$$\left[\begin{array}{ccc|c} 0.6 & -0.3 & -0.4 & 0 \\ -0.2 & 0.6 & -0.2 & 0 \\ -0.4 & -0.3 & 0.6 & 0 \end{array}\right] \begin{array}{c} 10R_1 \to R_1 \\ 5R_2 \to R_2 \\ 10R_3 \to R_3 \end{array} \left[\begin{array}{ccc|c} 6 & -3 & -4 & 0 \\ -1 & 3 & -1 & 0 \\ -4 & -3 & 6 & 0 \end{array}\right] \begin{array}{c} 6R_2 + R_1 \to R_1 \\ \to \\ -4R_2 + R_3 \to R_3 \end{array}$$

$$\left[\begin{array}{ccc|c} 0 & 15 & -10 & 0 \\ -1 & 3 & -1 & 0 \\ 0 & -15 & 10 & 0 \end{array}\right] \begin{array}{c} \to \\ -R_2 \to R_2 \\ R_1 + R_3 \to R_3 \end{array} \left[\begin{array}{ccc|c} 0 & 15 & -10 & 0 \\ 1 & -3 & 1 & 0 \\ 0 & 0 & 0 & 0 \end{array}\right] \begin{array}{c} \frac{1}{15}R_1 \text{ and} \\ \text{then interchange} \to \\ R_1 \text{ and } R_2 \end{array}$$

$$\left[\begin{array}{ccc|c} 1 & -3 & 1 & 0 \\ 0 & 1 & -2/3 & 0 \\ 0 & 0 & 0 & 0 \end{array}\right] \begin{array}{c} 3R_2 + R_1 \to R_1 \\ \to \\ \ \end{array} \left[\begin{array}{ccc|c} 1 & 0 & -1 & 0 \\ 0 & 1 & -2/3 & 0 \\ 0 & 0 & 0 & 0 \end{array}\right]$$

Profit = Households Non-profit = $\frac{2}{3}$Households

20.

	Ag	M	F	S
Ag	0.2	0.1	0.1	0.1
Mach	0.3	0.2	0.2	0.2
Fuel	0.2	0.2	0.3	0.3
Steel	0.1	0.4	0.2	0.2

21. The technological equation is $(I-A)X = D$.

$$(I-A) = \begin{bmatrix} 0.8 & -0.1 & -0.1 & -0.1 \\ -0.3 & 0.8 & -0.2 & -0.2 \\ -0.2 & -0.2 & 0.7 & -0.3 \\ -0.1 & -0.4 & -0.2 & 0.8 \end{bmatrix} \qquad X = (I-A)^{-1}\begin{bmatrix} 1700 \\ 1900 \\ 900 \\ 300 \end{bmatrix}$$

$$\text{Then } X = \begin{bmatrix} 1.74 & 0.68 & 0.62 & 0.62 \\ 1.30 & 2.30 & 1.18 & 1.18 \\ 1.39 & 1.55 & 2.50 & 1.50 \\ 1.22 & 1.62 & 1.29 & 2.29 \end{bmatrix}\begin{bmatrix} 1700 \\ 1900 \\ 900 \\ 300 \end{bmatrix} = \begin{bmatrix} 5000 \\ 8000 \\ 8000 \\ 7000 \end{bmatrix}$$

Use the graphing utility to find $(I-A)^{-1}$.

22. The Gauss-Jordan method must be used since the equation is $(I - A)X = 0$. We begin with the augmented matrix and will use fractions. The student is encouraged to solve with the graphing calculator and work with decimals.

$$\left[\begin{array}{cccc|c} 3/4 & -1/4 & -3/10 & -1/10 & 0 \\ -3/10 & 3/4 & -2/10 & -4/10 & 0 \\ -3/20 & -2/10 & 9/10 & -3/10 & 0 \\ -3/10 & -3/10 & -4/10 & 8/10 & 0 \end{array}\right] \begin{array}{c} \rightarrow \\ -4R_1 + R_2 \rightarrow R_2 \\ -3R_1 + R_3 \rightarrow R_3 \\ 8R_1 + R_4 \rightarrow R_4 \end{array} \left[\begin{array}{cccc|c} 3/4 & -1/4 & -3/10 & -1/10 & 0 \\ -33/10 & 7/4 & 1 & 0 & 0 \\ -48/20 & 11/20 & 18/10 & 0 & 0 \\ 57/10 & -23/10 & -28/10 & 0 & 0 \end{array}\right]$$

$$\begin{array}{c} \frac{3}{10}R_2 + R_1 \rightarrow R_1 \\ \rightarrow \\ -\frac{18}{10}R_2 + R_3 \rightarrow R_3 \\ \frac{28}{10}R_2 + R_4 \rightarrow R_4 \end{array} \left[\begin{array}{cccc|c} -24/100 & 11/40 & 0 & -1/10 & 0 \\ -33/10 & 7/4 & 1 & 0 & 0 \\ 354/100 & -104/40 & 0 & 0 & 0 \\ -354/100 & 104/40 & 0 & 0 & 0 \end{array}\right] \begin{array}{c} \\ \rightarrow \\ \\ R_3 + R_4 \rightarrow R_4 \end{array} \left[\begin{array}{cccc|c} -24/100 & 11/40 & 0 & -1/10 & 0 \\ -33/10 & 7/4 & 1 & 0 & 0 \\ 354/100 & -104/40 & 0 & 0 & 0 \\ 0 & 0 & 0 & 0 & 0 \end{array}\right]$$

$$\begin{array}{c} \rightarrow \\ \\ -\frac{40}{104}R_3 \rightarrow R_3 \\ \\ \end{array} \left[\begin{array}{cccc|c} -24/100 & 11/40 & 0 & -1/10 & 0 \\ -33/10 & 7/4 & 1 & 0 & 0 \\ -177/130 & 1 & 0 & 0 & 0 \\ 0 & 0 & 0 & 0 & 0 \end{array}\right] \begin{array}{c} -\frac{11}{40}R_3 + R_1 \rightarrow R_1 \\ -\frac{7}{4}R_3 + R_2 \rightarrow R_2 \\ \rightarrow \\ \\ \end{array} \left[\begin{array}{cccc|c} 699/5200 & 0 & 0 & -1/10 & 0 \\ -477/520 & 0 & 1 & 0 & 0 \\ -177/130 & 1 & 0 & 0 & 0 \\ 0 & 0 & 0 & 0 & 0 \end{array}\right]$$

Our solution yields: $\text{Hhold} = \dfrac{699}{520}\text{Ag}$, $\text{Fuel} = \dfrac{477}{520}\text{Ag}$, $\text{Steel} = \dfrac{177}{130}\text{Ag}$

Now we must solve in terms of Hholds to yield the desired form.

$$\text{Ag} = \frac{520}{699}\text{Hhold}$$

$$\text{Steel} = \frac{177}{130}\left(\frac{520}{699}\text{Hhold}\right) = \frac{236}{233}\text{Hhold}$$

$$\text{Fuel} = \frac{477}{520}\left(\frac{520}{699}\text{Hhold}\right) = \frac{159}{233}\text{Hhold.}$$

This was a difficult problem. The authors' step by step method always works. Sometimes it is not the shortest method.

Chapter 4: Inequalities and Linear Programming

Exercise 4.1

1. $2x+1 < x-3$

$2x-x<-3-1$

$x<-4$ or $(-\infty,-4)$

3. $3(x-1)<2x-1$

$3x-3<2x-1$

$3x-2x<-1+3$

$x<2$ or $(-\infty, 2)$

5. $1-2x>9$

$-2x>9-1=8$

$x<-4$ or $(-\infty,-4)$

7. $\frac{3(x-1)}{2} \le x-2$

$3(x-1)\le 2x-4$

$3x-3\le 2x-4$

$x\le -1$ or $(-\infty,-1]$

9. $2.9(2x-4)\ge \frac{3.6x-8.5}{5}$

$14.5(2x-4)\ge 3.6x-8.5$

$29x-58\ge 3.6x-8.5$

$25.4x\ge 49.5$

$x\ge \frac{495}{254}$ or $\left[\frac{495}{254},\infty\right)$

or $x\ge 1.949$

11. $3(x-1)<2x$

$3x-3<2x$

$x<3$

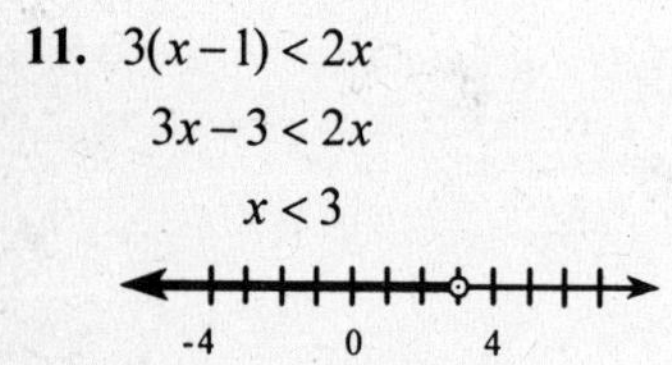

13. $2(x-1)-3>4x+1$

$2x-2-3>4x+1$

$2x-4x>2+3+1$

$-2x>6$ or $x<-3$

-6 -4 -2 0 x

15. $\frac{x}{3}>x-1$

$x>3x-3$

$x-3x>-3$

$-2x>-3$

$x<\frac{3}{2}$

$\frac{3}{2}$

-2 0 2

17. $-\frac{3x}{2}>9$

$x<9\left(-\frac{2}{3}\right)$

$x<-6$

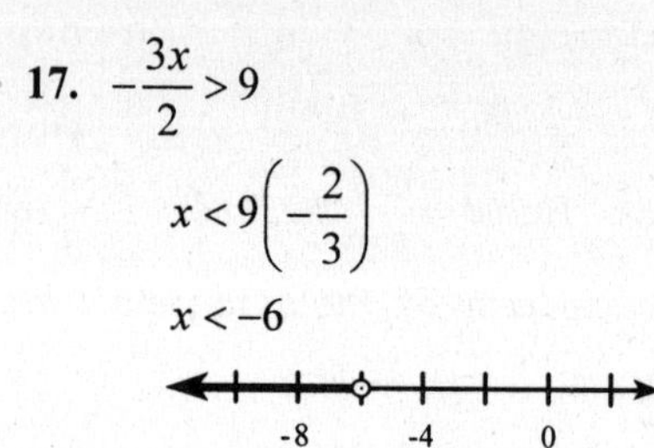

19. $\frac{3}{4}x-\frac{1}{6}<x-\frac{2(x-1)}{3}$

$9x-2<12x-8(x-1)$

$9x-2<12x-8x+8$

$5x<10$

$x<2$

0 2 4

21. $\frac{3}{4}x-\frac{1}{3}<1-\frac{2}{3}\left(x-\frac{1}{2}\right)$

$\frac{3}{4}x-\frac{1}{3}<1-\frac{2}{3}x+\frac{1}{3}$

$9x-4<12-8x+4$

$17x<20$

$x<\frac{20}{17}$

0 $\frac{20}{17}$ 2

23. $\left(-\frac{1}{2},3\right]$ $-\frac{1}{2}<x\le 3$

25. $(1, 4)$ $1<x<4$

27. From graph: $4<x$ or $x>4$

29. From graph: $-50\le x<-22$

31. (1, 3] half-open

33. (2,10) open

35. [–3, 2] closed

37. (–4, 3) open

39. [4, 6] closed

41. $40x-(20x+1600)>0$

$$20x-1600>0$$
$$20x>1600$$
$$x>80$$

0 20 40 60 80

43. $695+5.75x\le 900$

$$5.75x\le 205$$
$$x\le 35$$

(nearest whole number)

45. a.

$$0\le I\le 27{,}050$$
$$27{,}051\le I\le 65{,}550$$
$$65{,}551\le I\le 136{,}750$$
$$136{,}751\le I\le 297{,}350$$
$$297{,}351\le I$$

b. $0\le T\le 0.15(27{,}050)=4057.50$

$4057.50\le T<0.275(65{,}551-27{,}050)+4057.50$ or $4057.50\le T<14{,}645$

$14{,}645\le T<0.305(136{,}751-65{,}550)+14{,}645$ or $14{,}645\le T<36{,}361$

$36{,}361\le T<0.355(297{,}351-136{,}750)+36{,}361$ or $36{,}361\le T<93{,}374$

47. $A=90.2+41.3h$

a. $110\le 90.2+41.3h$

$$h\ge\frac{19.8}{41.3}\approx 0.479$$

$0.479\le h\le 1$

If $h\ge 0.479$ then apparent temperature will be at least 110°F.

b. $100>90.2+41.3h$

$$h<\frac{9.8}{41.3}\approx 0.237$$

If humidity is less than 0.237 then apparent temperature is less than 100°F.

49.

$$F\ge 98.6$$
$$\frac{9}{5}C+32\ge 98.6$$
$$9C+160\ge 493$$
$$9C\ge 333$$
$$C\ge 37$$

51. a. $x_4\le 1100, x_4\le 1000, x_4\le 900, x_4\ge 0$

b. $0\le x_4\le 900$ satisfies these inequalities.

c.

$x_1=x_4+100$	$x_2=x_4+200$	$x_3=x_4+300$
$x_1\le 900+100$	$x_2\le 900+200$	$x_3\le 900+300$
$x_1\le 1000$	$x_2\le 1100$	$x_3\le 1200$
$x_1\in[100,1000]$	$x_2\in[200,1100]$	$x_3\in[300, 1200]$

Exercise 4.2

1. $y \le 2x - 1$

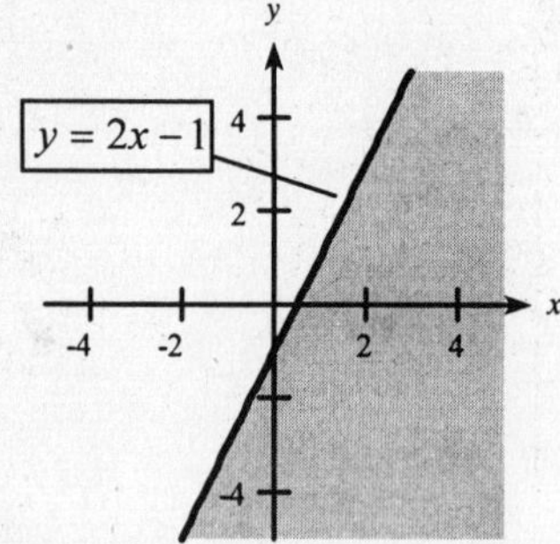

3. $\frac{x}{2} + \frac{y}{4} < 1$

$2x + y < 4$

$y < -2x + 4$

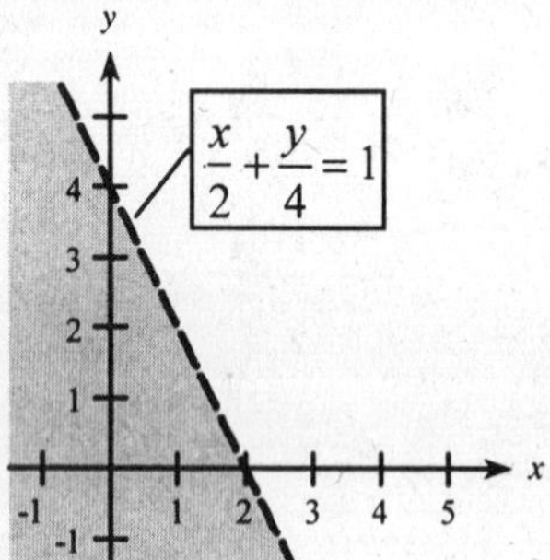

5. $2(x - y) < y + 3$

$2x - 2y < y + 3$

$-3y < -2x + 3$

$y > \frac{2}{3}x - 1$

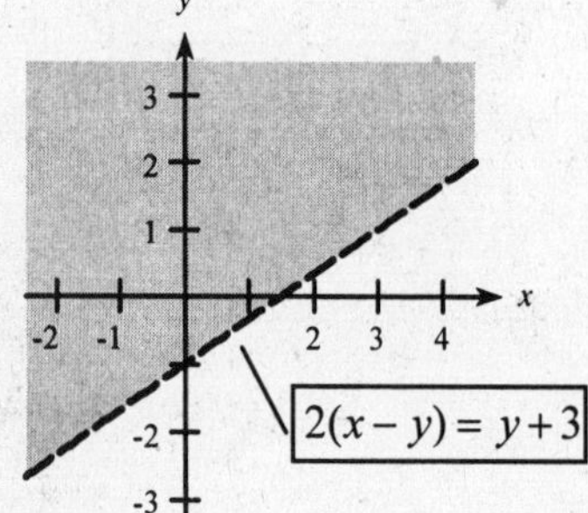

7. $0.4x \ge 0.8$

$x \ge 2$

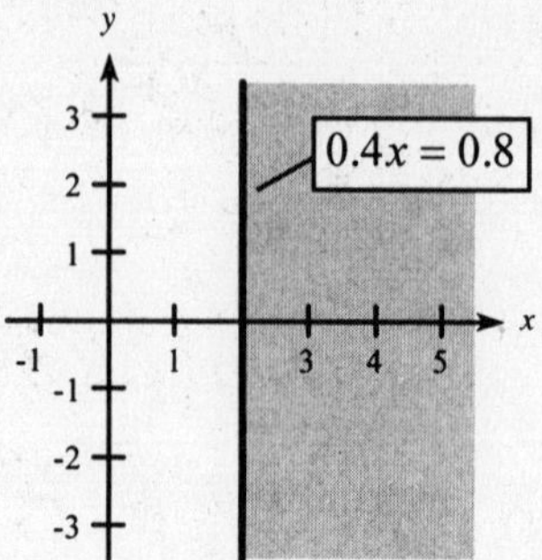

9. **a.**

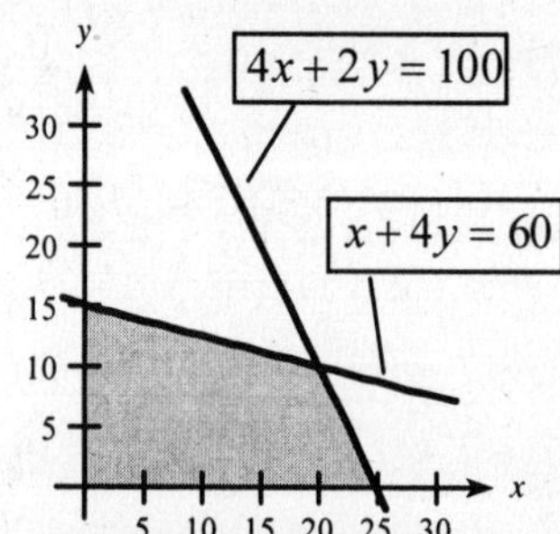

b. From the graph we read intercept corners (0, 15), (0, 0), (25, 0).

$4x + 2y = 100 \qquad 8x + 4y = 200$

$x + 4y = 60 \qquad x + 4y = 60$

$7x = 140$

$x = 20$

$20 + 4y = 60$ or $y = 10$

Other corner is (20, 10).

11. a.

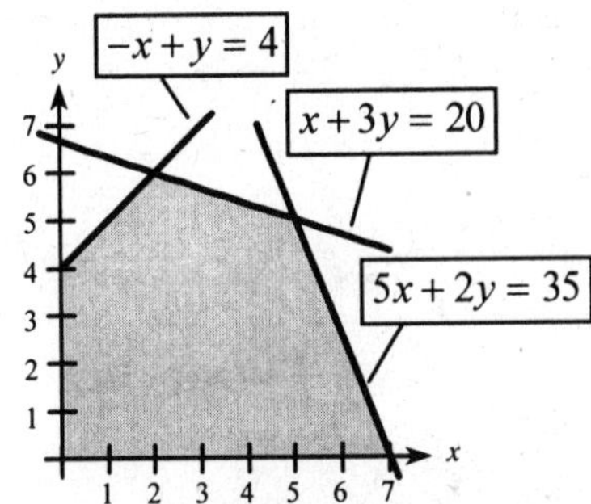

b. From the graph we read intercept corners (0, 4), (0, 0), (7, 0).

$$\begin{aligned} -x+y&=4 \\ x+3y&=20 \\ \hline 4y&=24 \\ y&=6 \end{aligned} \qquad \begin{aligned} -x+6&=4 \\ x&=2 \end{aligned}$$

Corner is (2, 6).

$$\begin{aligned} x+3y&=20 \\ 5x+2y&=35 \end{aligned} \qquad \begin{aligned} 5x+15y&=100 \\ 5x+2y&=35 \\ \hline 13y&=65 \\ y&=5 \end{aligned}$$

$x + 3(5) = 20$ or $x = 5$ Corner is (5, 5)

13. a.

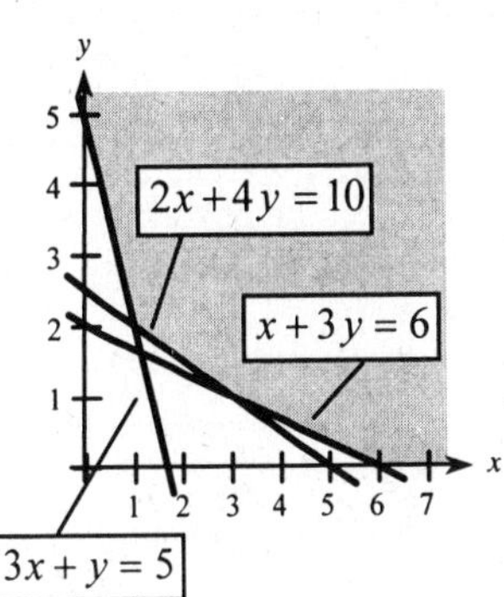

b. From the graph we read intercept corners (0, 5), (6, 0).

$$\begin{aligned} 3x+y&=5 \\ 2x+4y&=10 \end{aligned} \qquad \begin{aligned} 12x+4y&=20 \\ 2x+4y&=10 \\ \hline 10x&=10 \\ x&=1 \end{aligned}$$

$3(1) + y = 5$ or $y = 2$
Corner is (1, 2).

$$\begin{aligned} 2x+4y&=10 \\ x+3y&=6 \end{aligned} \qquad \begin{aligned} 2x+4y&=10 \\ 2x+6y&=12 \\ \hline 2y&=2 \\ y&=1 \end{aligned}$$

$x + 3(1) = 6$ or $x = 3$
Corner is (3, 1)

15. $\begin{cases} y<2x \\ y>x-1 \end{cases}$

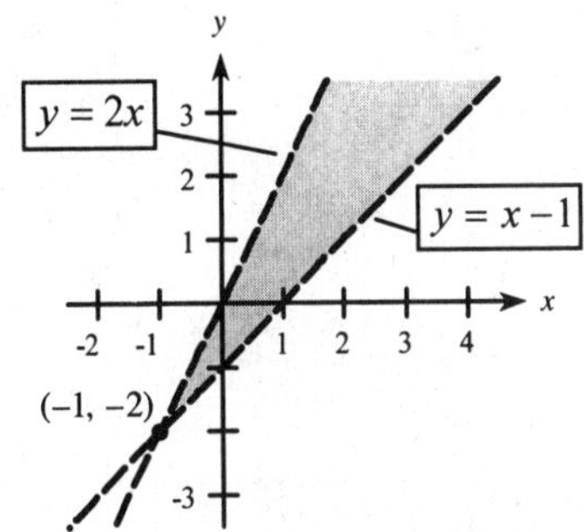

17. $\begin{cases} y<-2x+3 \\ y\le \frac{1}{2}x+\frac{1}{2} \end{cases}$

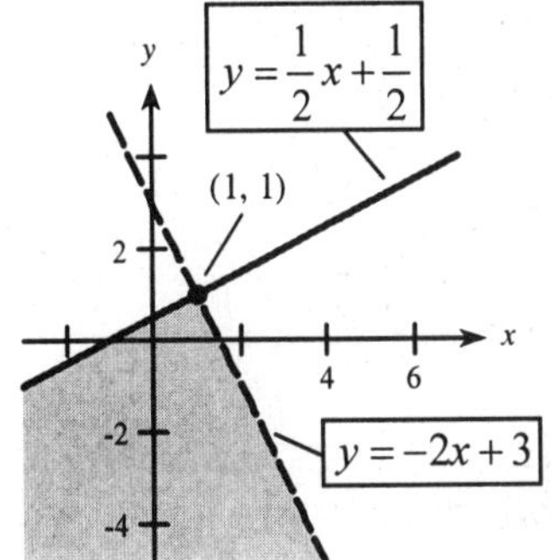

19.

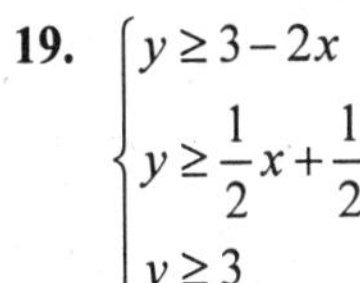

$\begin{cases} y\ge 3-2x \\ y\ge \frac{1}{2}x+\frac{1}{2} \\ y\ge 3 \end{cases}$

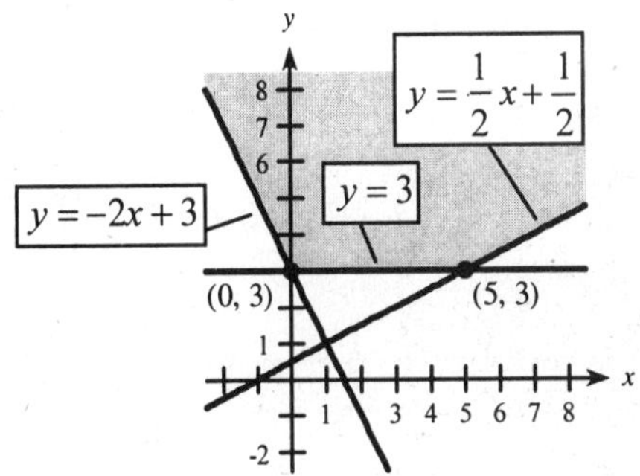

21. $\begin{cases} y \le -\frac{1}{5}x + 40 \\ y \le -\frac{2}{3}x + \frac{134}{3} \\ x \ge 0, y \ge 0 \end{cases}$

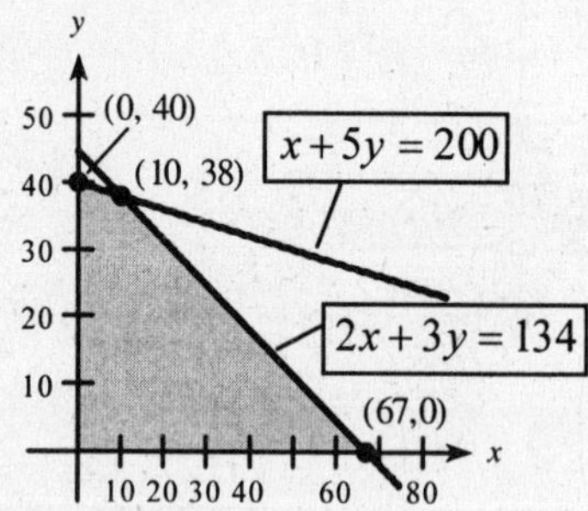

23. $\begin{cases} y \le -\frac{1}{2}x + 24 \\ y \le -x + 30 \\ y \le -2x + 50 \\ x \ge 0,\ y \ge 0 \end{cases}$

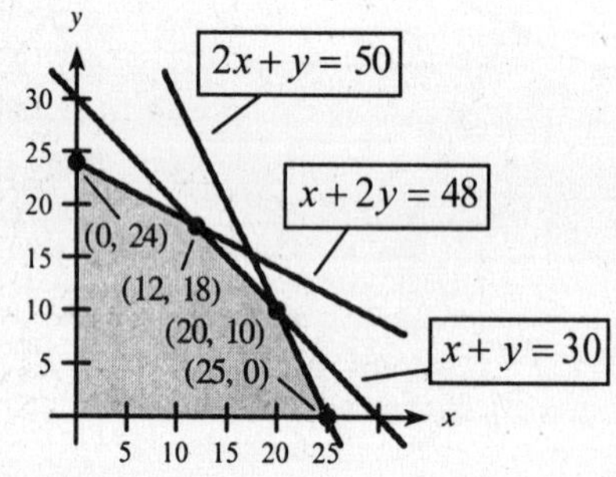

25. $\begin{cases} y \ge -\frac{1}{2}x + \frac{19}{2} \\ y \ge -\frac{3}{2}x + \frac{29}{2} \\ x \ge 0, y \ge 0 \end{cases}$

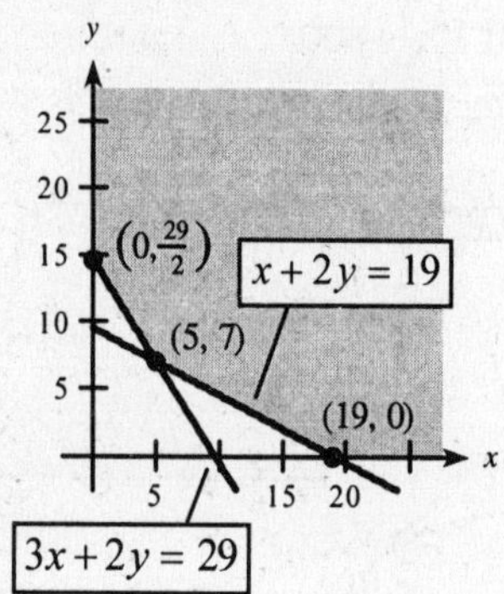

27. $\begin{cases} y \ge -\frac{1}{3}x + 1 \\ y \ge -\frac{2}{3}x + \frac{5}{3} \\ y \ge -2x + 3 \\ x \ge 0, y \ge 0 \end{cases}$

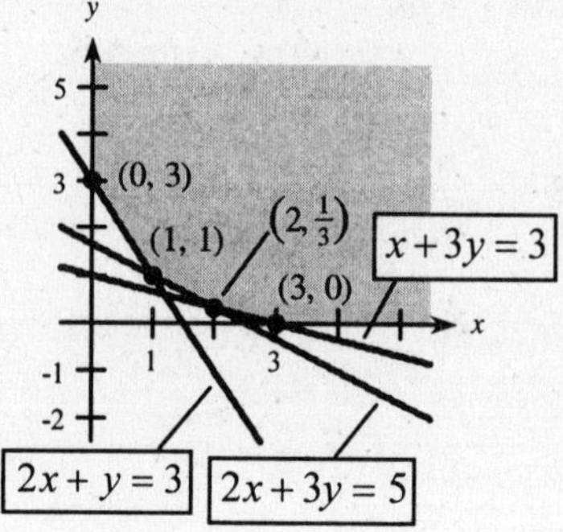

29. $\begin{cases} y \ge -\frac{1}{2}x + 10 \\ y \le \frac{3}{2}x + 2 \\ x \ge 12 \\ x \ge 0,\ y \ge 0 \end{cases}$

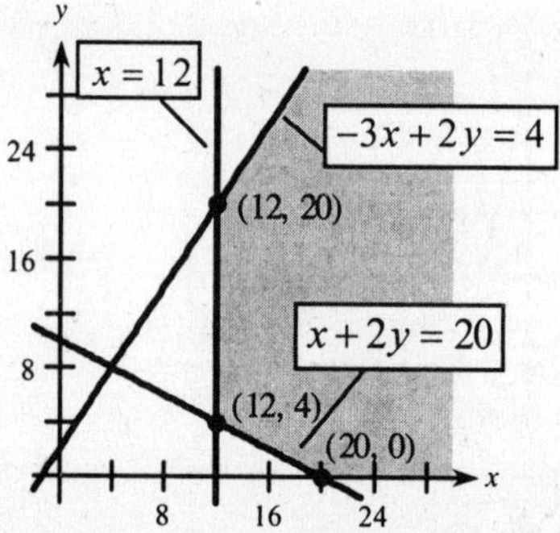

31. Let x = number of deluxe models.
Let y = number of economy models.

a. $x, y \ge 0 \quad 3x + 2y \le 24 \quad 0.5x + y \le 8$

b. Points of Intersection:

$$\begin{array}{l} 3x + 2y = 24 \\ \underline{2(0.5x + y = 8)} \\ 2x \quad = 8 \\ x \quad = 4 \end{array}$$

$0.5(4) + y = 8$ or $y = 6$

Feasible corners are $(0,0)$, $(8,0)$, $(4,6)$, and $(0,8)$.

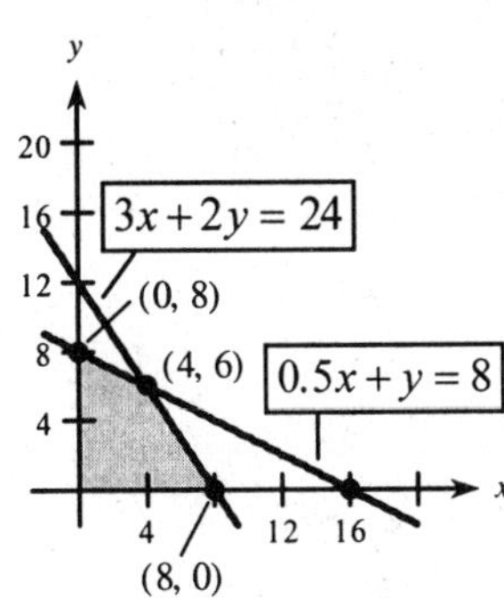

33. Let x = number of cord type models and y = number of cordless models.

a.

$x + y \le 300$	Packing department
$2x+4y \le 800$	Manufacturing
$x \ge 0$	
$y \ge 0$	

b.

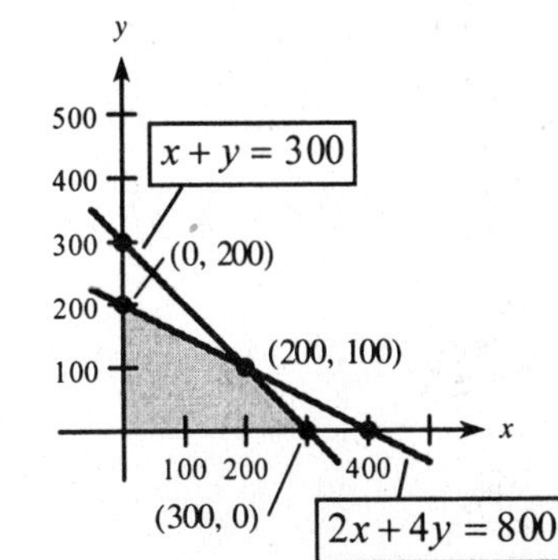

35. Let x = minutes of business/investment program commercials
Let y = minutes of sporting events commercials

a.

$7x + 2y \ge 30$	women reached
$4x + 12y \ge 28$	men reached
$x \ge 0, \quad y \ge 0$	

b.

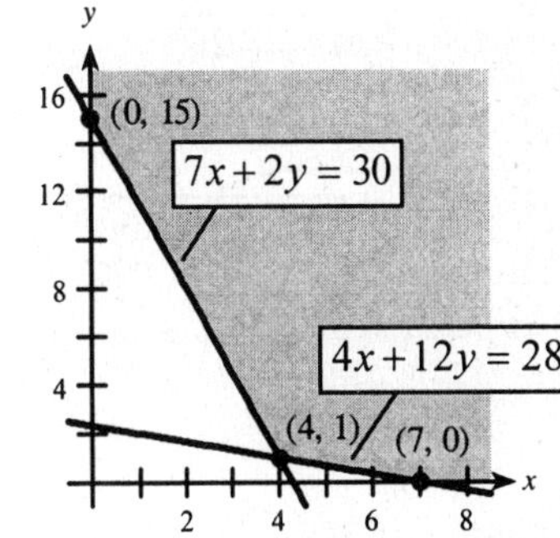

37. x = radio minutes, y = television minutes.

a.

$x + y \ge 80$	Advertising time
$0.006x+0.09y \ge 2.16$	People reached
$x, y \ge 0$	

(Selecting a good scale is the most difficult part of the problem.)

b.

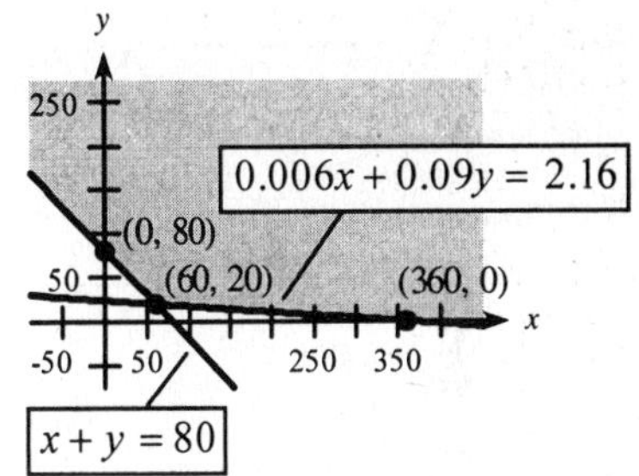

39. x = pounds of regular meat
y = pounds of all beef

a.

$0.18x + 0.75y \le 1020$	(beef)
$0.20x+0.20y \ge 500$	(spices)
$0.30x \le 600$	(pork)
$x, y \ge 0$	

b.

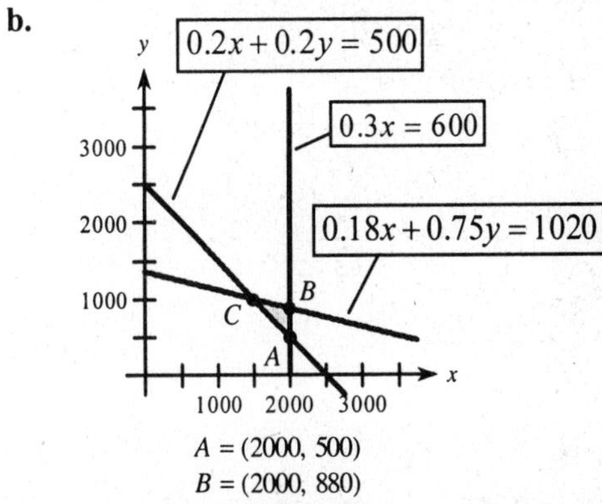

$A = (2000, 500)$
$B = (2000, 880)$
$C = (1500, 1000)$

Exercise 4.3

1.

Corner	$f=6x+3y$
$(0,0)$	$6\cdot0+3\cdot0=0$
$(0,8)$	$6\cdot0+3\cdot8=24$
$(12,0)$	$6\cdot12+3\cdot0=72$
$(6,5)$	$6\cdot6+3\cdot5=51$

Maximum = 72 at $(12,0)$

3.

Corner	$C=2x+3y$
$(0,0)$	0
$(7,0)$	14
$(6,2)$	18
$(4,4)$	20
$(0,3)$	9

Maximum = 20 at $(4,4)$

Minimum = 0 at $(0,0)$

5.

Corner	$C=5x+2y$
$(6,1)$	32
$(12,4)$	68
$(6,4)$	38
$(2,2)$	14

Maximum = 68 at $(12,4)$.

Minimum = 14 at $(2,2)$.

7.

Corner	$f=3x+4y$
$(0,6)$	24
$(1,3)$	15
$(2,1)$	10
$(4,0)$	12

Maximum: None

Minimum = 10 at $(2,1)$.

9. $x+5y=100$ If $x=0$, then $y=20$.

$2x+y=40$ If $y=0$, then $x=20$.

$8x+5y=170$

$$\begin{array}{rl} 10x+5y &= 200 \\ 8x+5y &= 170 \\ \hline 2x &= 30 \\ x &= 15 \\ y &= 10 \end{array} \qquad \begin{array}{rl} x+5y &= 100 \\ 8x+5y &= 170 \\ \hline 7x &= 70 \\ x &= 10 \\ y &= 18 \end{array}$$

Corners: (15, 10) and (10, 18)

Corner	$f=3x+2y$
$(0,0)$	0
$(0,20)$	40
$(20,0)$	60
$(15,10)$	65
$(10,18)$	66

Maximum = 66 at $(10,18)$

11. $3x+y=60$

$4x+10y=280$

$x+y=40$

If $x=0$, then $y=60$

If $y=0$, then $x=70$

$$\begin{array}{rl} 10x+10y &= 400 \\ 4x+10y &= 280 \\ \hline 6x &= 120 \\ x &= 20 \\ y &= 20 \end{array} \qquad \begin{array}{rl} x+y &= 40 \\ 3x+y &= 60 \\ \hline 2x &= 20 \\ x &= 10 \\ y &= 30 \end{array}$$

Corners: (20, 20) and (10, 30)

Corner	$g=3x+2y$
$(0,60)$	120
$(70,0)$	210
$(20,20)$	100
$(10,30)$	90

Minimum = 90 at $(10,30)$.

13. Refer to Problem 21 in Section 4.2 to obtain the corners

Corner	$f=4x+9y$
(0, 40)	360
(10, 38)	382
(67, 0)	268

Maximum = 382 at (10, 38).

15. Refer to Problem 23 in Section 4.2, to obtain the corners (0, 24) and (25, 0).

$$\begin{aligned} 2x+y&=50 \\ x+y&=30 \\ \hline x\quad&=20 \end{aligned}$$

(20, 10) is a point of intersection.

$$\begin{aligned} x+2y&=48 \\ x+y&=30 \\ \hline y&=18 \end{aligned}$$

(12,18) is a feasible corner.

Corner	$f = 3x + 2y$
(0, 24)	48
(25, 0)	75
(20, 10)	80
(12, 18)	72

Maximum = 80 at (20, 10).

17. Refer to Problem 25 in Section 4.2 to obtain the corners $\left(0, \frac{29}{2}\right)$ and (19, 0).

$$\begin{aligned} x+2y&=19 \\ 3x+2y&=29 \\ \hline 2x&=10 \\ x&=5 \\ y&=7 \end{aligned}$$

Other corner is (5, 7)

Corner	$g = 9x + 10y$
$\left(0, \frac{29}{2}\right)$	145
$(19,0)$	171
$(5,7)$	115

Minimum = 115 at $(5,7)$.

19. Refer to Problem 27 in Section 4.2 to obtain the corners (3, 0) and (0, 3).

$$\begin{aligned} 2x+y&=3 \\ 2x+3y&=5 \\ \hline 2y&=2 \\ y&=1 \end{aligned} \qquad \begin{aligned} x+3y&=3 \\ 2x+3y&=5 \\ \hline x\quad&=2 \end{aligned}$$

Corner: (1, 1) Corner: $(2, \frac{1}{3})$

Corner	$g = 12x + 48y$
$(3,0)$	36
$(0,3)$	144
$(1,1)$	60
$(2,\frac{1}{3})$	40

Minimum = 36 at (3, 0).

21. Note: Problem 21 is worked by an alternate method. No graph is involved. The method is

(1) Find all corners.
(2) Is it a feasible corner? (Does it satisfy all constraints?)
(3) Evaluate the objective function at the feasible corners.
(4) Write the maximum or minimum as requested.

$$x+2y=4 \qquad \begin{aligned} 2x+4y&=8 \end{aligned}$$
$$2x+y=4 \qquad \begin{aligned} 2x+y&=4 \\ \hline 3y&=4 \end{aligned}$$
$$y=\frac{4}{3},\ x=\frac{4}{3}$$

$$\begin{aligned} x+2y&=4 \\ x\quad&=0 \\ \hline &(0, 2) \end{aligned} \qquad \begin{aligned} 2x+y&=4 \\ x\quad&=0 \\ \hline &(0, 4) \end{aligned}$$

$$\begin{aligned} x+2y&=4 \\ y&=0 \\ \hline &(4, 0) \end{aligned} \qquad \begin{aligned} 2x+y&=4 \\ y&=0 \\ \hline &(2, 0) \end{aligned}$$

Corner	Feasible	$f = x + 3y$
$(\frac{4}{3}, \frac{4}{3})$	yes	$\frac{16}{3}$
$(0,2)$	yes	6
$(4,0)$	no	
$(0,4)$	no	
$(2,0)$	yes	2
$(0,0)$	yes	0

Maximum = 6 at (0, 2).

23. The graph has enough accuracy to read the feasible corners from the graph.

Corner	$f = 3x + 4y$
(0, 4)	16
(2, 4)	22
(4, 2)	20
(5, 0)	15
(0, 0)	0

Maximum = 22 at (2, 4).

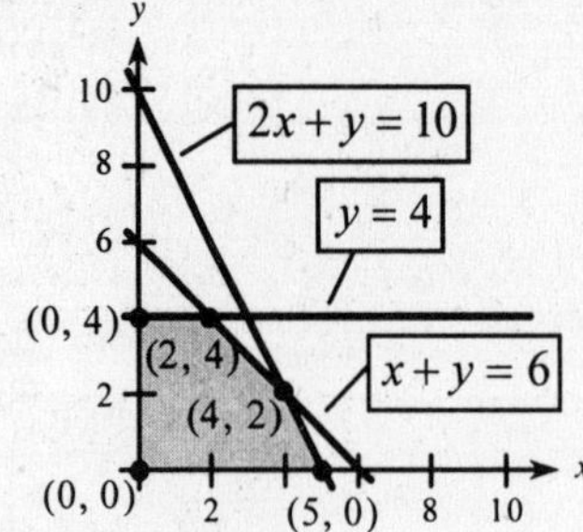

25. The feasible corners can be read directly from the graph.

Corner	$f = 2x + 6y$
(0, 5)	30
(3, 4)	30
(5, 2)	22
(6, 0)	12
(0, 0)	0

Maximum = 30 at any point on line from (0, 5) to (3, 4)

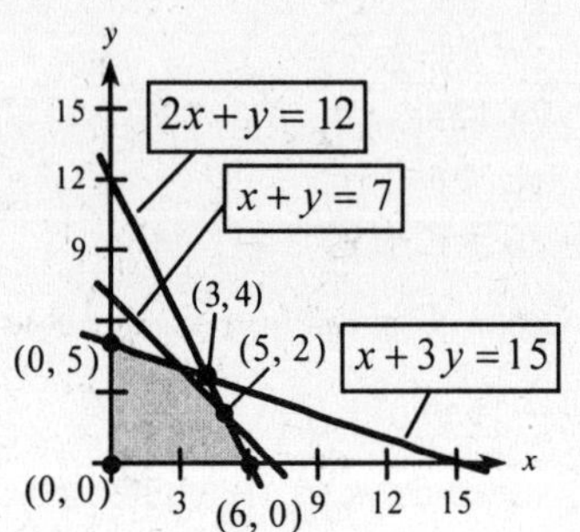

27. The feasible corners can be read directly from the graph.

Corner	$g = 7x + 6y$
(9, 0)	63
(2, 3)	32
(0, 8)	48

Minimum = 32 at (3, 2).

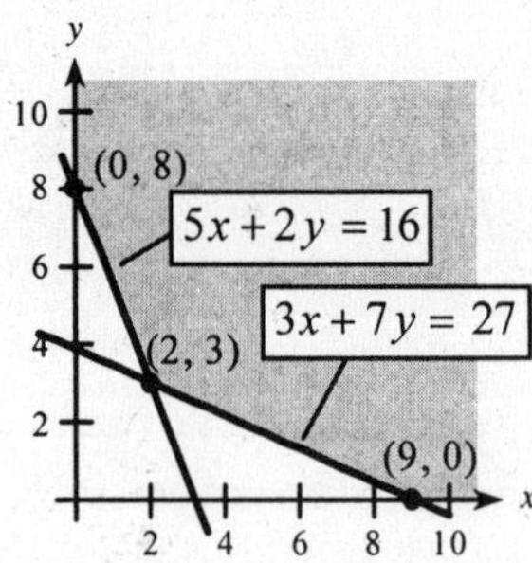

29. $3x + 2y = 12$ $3x + 2y = 12$

$\underline{4x + y = 11}$ $\underline{8x + 2y = 22}$

$5x = 10$

$x = 2$

$y = 3$

(2, 3) is a feasible corner.

Corner	$g = 3x + y$
(4, 0)	12
(0, 11)	11
(2, 3)	9

Minimum = 9 at (2, 3).

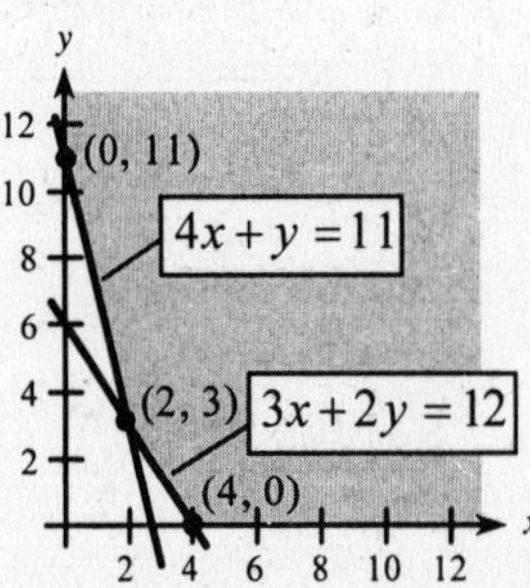

31.

$$\begin{array}{r} y = 30 \\ 3x + 2y = 75 \\ \hline 3x = 15 \\ x = 5 \end{array} \qquad \begin{array}{r} y = 30 \\ -3x + 5y = 30 \\ \hline -3x = -120 \\ x = 40 \end{array}$$

(5, 30) and (40, 30) are feasible corners.

$$\begin{array}{r} 3x + 2y = 75 \\ -3x + 5y = 30 \\ \hline 7y = 105 \\ y = 15 \end{array}$$

(15, 15) is a feasible corner.

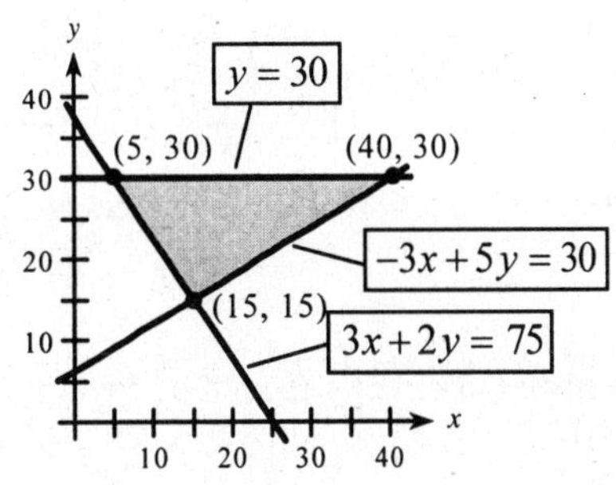

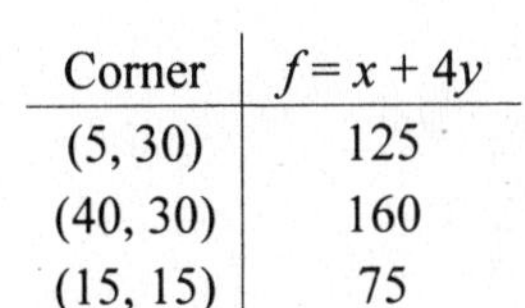

Corner	$f = x + 4y$
(5, 30)	125
(40, 30)	160
(15, 15)	75

Minimum = 75 at (15, 15).

33. The feasible corners can be read directly from the graph.

Corner	$f = x + 2y$
(0, 4)	8
(2, 4)	10
(4, 0)	4

Maximum = 10 at (2, 4).

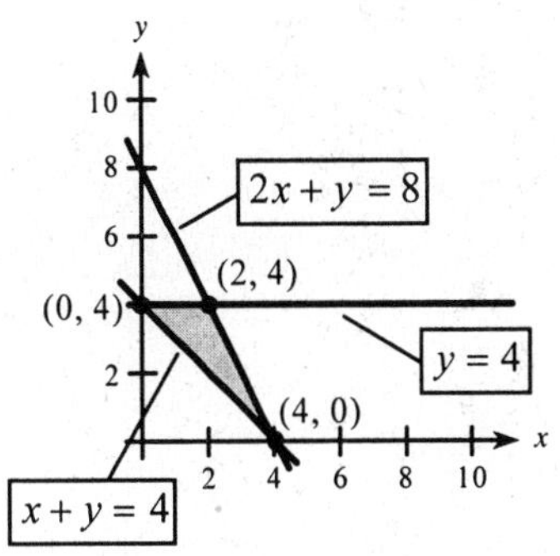

35.

$$\begin{array}{r} x + y = 100 \\ -x + y = 20 \\ \hline 2y = 120 \\ y = 60 \\ x = 40 \end{array} \qquad \begin{array}{r} x + y = 100 \\ -2x + 3y = 30 \\ \\ 2x + 2y = 200 \\ -2x + 3y = 30 \\ \hline 5y = 230 \\ y = 46 \\ x = 54 \end{array}$$

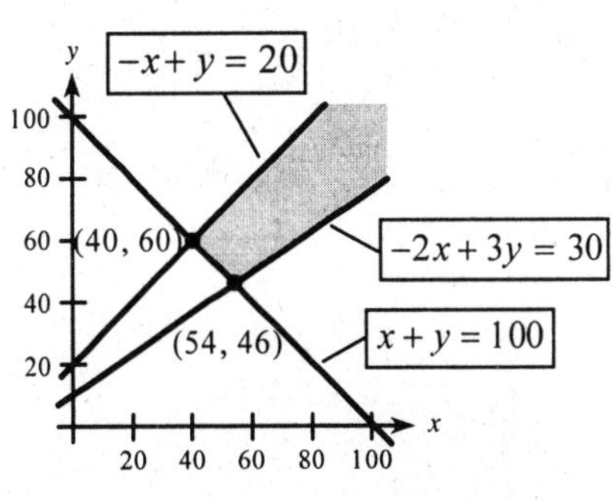

(40, 60) and (54, 46) are the feasible corners.

Corner	$g = 40x + 25y$	
(40,60)	3100	← minimum
(54,46)	3310	

37. From the graph of Problem 31, Section 4.2, the corners along the axes are (0, 0), (0, 8), and (8, 0). The other corner (see graph) is (4, 6).

Corner	$P = 15x + 12y$	
(0, 8)	96	
(8, 0)	120	
(4, 6)	132	← Maximum profit of $132 at (4, 6).
(0, 0)	0	

39. From the graph of Problem 33, Section 4.2, the corners are (0, 200), (0, 0), (300, 0), and (200, 100).

Corner	$S = 30x + 40y$	
(0, 200)	8000	
(300, 0)	9000	
(200, 100)	10,000	← Maximum sales of $10,000 at 200 cord type and 100 cordless type.
(0, 0)	0	

41. x = number of bass, y = number of trout

Maximize $f = x + y$

$2x + 5y \le 800$ Food A

$4x + 2y \le 800$ Food B

$x, y \ge 0$

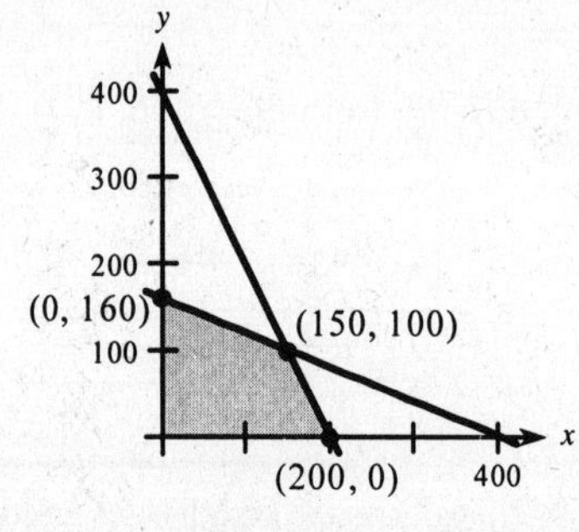

$2x + 5y = 800$ $\qquad$ $4x + 10y = 1600$

$4x + 2y = 800$ $\qquad$ $4x + 2y = 800$

$$8y = 800$$

$$y = 100$$

$$x = 150$$

Corner	$f = x + y$
(0, 160)	160
(150, 100)	250
(200, 0)	200

Maximum number of fish is 250 with 150 bass and 100 trout.

43. x = satellite branches, y = full service
Objective function: $R = 10{,}000x + 18{,}000y$

$100{,}000x + 140{,}000y \le 2{,}980{,}000$	Construction costs
$3x + 6y \le 120$	Employees
$x + y \le 25$	Number of branches
$x, y \ge 0$	

$10x + 14y = 298 \qquad 15x + 21y = 447$
$3x + 6y = 120 \qquad 15x + 30y = 600$
$9y = 153$
$y = 17$
$x = 6$

$10x + 14y = 298 \qquad 10x + 14y = 298$
$x + y = 25 \qquad 10x + 10y = 250$
$4y = 48$
$y = 12$
$x = 13$

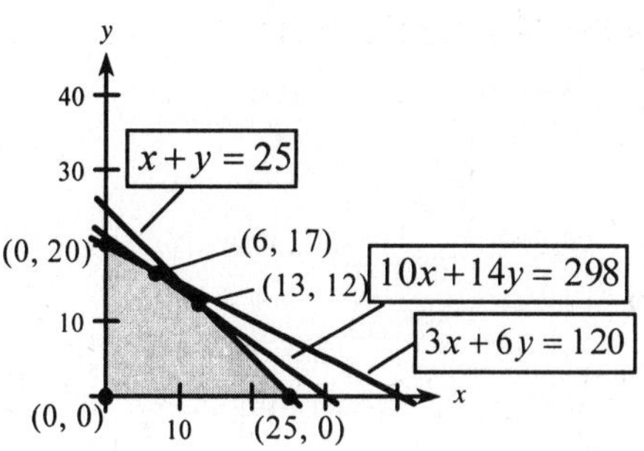

Corner	$R = 1000(10x + 18y)$	
(13, 12)	346,000	
(6, 17)	366,000	← Maximum revenue of \$366,000 with 6 satellite and 17 full service branches
(25, 0)	250,000	
(0, 20)	360,000	

45. From the graph of Problem 37, Section 4.2, we have corners (0, 80), (60, 20), and (360, 0).

Corner	$C = 100x + 500y$	
(0, 80)	40,000	
(360, 0)	36,000	
(60, 20)	16,000	← Minimum cost =\$16,000 with 60 minutes of radio time and 20 minutes of TV time.

47. x = days for Factory 1 to operate, y = days for Factory 2 to operate
Objective function: $F = 10{,}000x + 20{,}000y$
Constraints: $x, y \geq 0$

$80x + 20y \geq 1600$ #1

$10x + 10y \geq 500$ #2

$20x + 70y \geq 2000$ #3

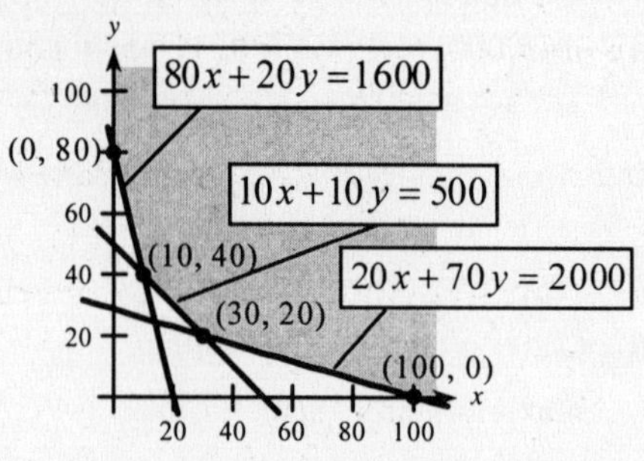

From the graph (10, 40) and (30, 20) are the two feasible corners not on the coordinate axes.

Corner	$F = 10{,}000x + 20{,}000y$
(0, 80)	1,600,000
(10, 40)	900,000
(30, 20)	700,000
(100, 0)	1,000,000

Minimum cost is $700,000 when operating Factory 1 for 30 days and Factory 2 for 20.

49. x = days at location I, y = days at location II
Objective function: $F = 500x + 800y$
Constraints: $x, y \geq 0$

$10x + 20y \geq 2000$ Deluxe

$20x + 50y \geq 4200$ Better

$13x + 6y \geq 1200$ Standard

$x + 2y = 200$ $\quad$ $13x + 6y = 1200$

$2x + 5y = 420$ $\quad$ $x + 2y = 200$

$A(160, 20)$ $\quad$ $B(60, 70)$

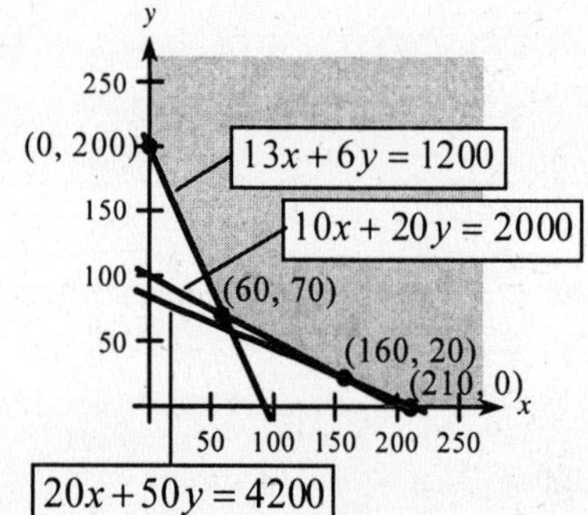

Corner	$F = 500x + 800y$
(160, 20)	96,000
(60, 70)	86,000
(0, 200)	160,000
(210, 0)	105,000

Minimum costs of $86,000 occurs with 60 days at location I and 70 days at location II.

51. See the graph for Problem 39, in Section 4.2, From the graph, feasible corners are (2000, 500), (2000, 880), and (1500, 1000).

Corner	$P = 0.4x + 0.6y$
(1500, 1000)	1200
(2000, 500)	1100
(2000, 880)	1328

Maximum profit = $1328 with 2000 lbs of regular and 880 lbs of all beef.

53. Let x = number from P to B; let y = number from P to Y.
Then $35 - x$ = number from E to B; $40 - y$ = number from E to Y.
Objective function: $C = 18x + 20(35 - x) + 22y + 25(40 - y)$

$$= 1700 - 2x - 3y$$

Constraints:

$x, y \ge 0$	P inventory
$x + y \le 60$	or
$(35 - x) + (40 - y) \le 30$	E inventory
$x + y \ge 45$	

$y \le 40$ maximum order size

$x \le 35$ maximum order size

Drawing the graph and obtaining the feasible corners.

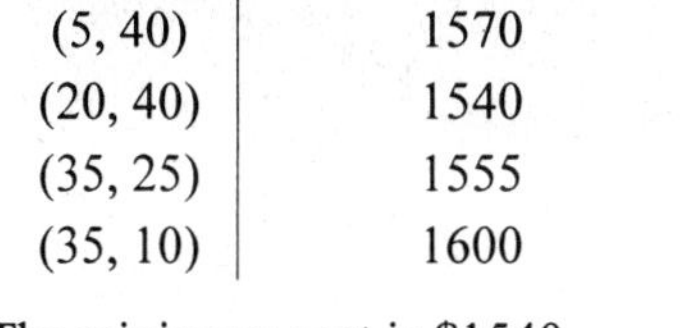

Corner	$C = 1700 - 2x - 3y$
(5, 40)	1570
(20, 40)	1540
(35, 25)	1555
(35, 10)	1600

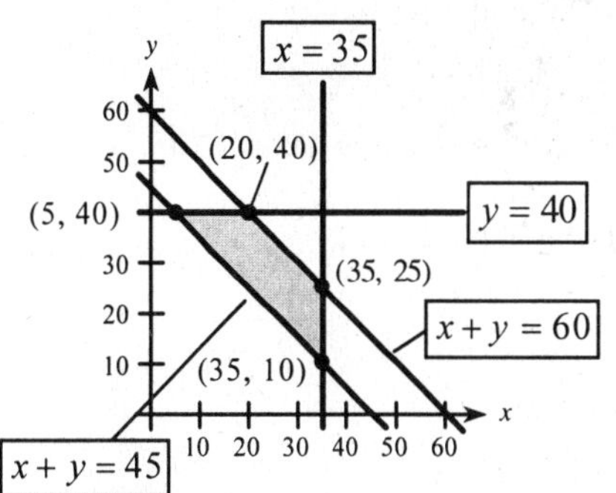

The minimum cost is $1540.

Ship From/To	B	Y
P	20	40
E	15	0

Exercise 4.4

1. $3x + 5y + s_1 = 15 \quad 3x + 6y + s_2 = 20$

3.

$$\begin{array}{ccccc} x & y & s_1 & s_2 & f \end{array}$$

$$\left(\begin{array}{ccccc|c} 2 & 5 & 1 & 0 & 0 & 30 \\ 1 & 5 & 0 & 1 & 0 & 25 \\ \hline -2 & -4 & 0 & 0 & 1 & 0 \end{array}\right)$$

5.

$$\begin{array}{ccccc} x & y & s_1 & s_2 & f \end{array}$$

$$\left(\begin{array}{ccccc|c} 1 & 5 & 1 & 0 & 0 & 200 \\ 2 & 3 & 0 & 1 & 0 & 134 \\ \hline -4 & -9 & 0 & 0 & 1 & 0 \end{array}\right)$$

7.

$$\begin{array}{ccccccc} x & y & z & s_1 & s_2 & s_3 & f \end{array}$$

$$\left(\begin{array}{ccccccc|c} 2 & 7 & 9 & 1 & 0 & 0 & 0 & 100 \\ 6 & 5 & 1 & 0 & 1 & 0 & 0 & 145 \\ 1 & 2 & 7 & 0 & 0 & 1 & 0 & 90 \\ \hline -2 & -5 & -2 & 0 & 0 & 0 & 1 & 0 \end{array}\right)$$

9. $\left(\begin{array}{rrrrr|r} 2 & 4 & 1 & 0 & 0 & 24 \\ 1 & \mathbf{1} & 0 & 1 & 0 & 5 \\ \hline -4 & -11 & 0 & 0 & 1 & 0 \end{array}\right)$

– 11 is the most negative.

$R_1 : \frac{24}{4} = 6 \quad R_2 : \frac{5}{1} = 5$

Thus, pivot on the 1.

11. $\left(\begin{array}{rrrrrr|r} 10 & \mathbf{27} & 1 & 0 & 0 & 0 & 200 \\ 4 & 51 & 0 & 1 & 0 & 0 & 400 \\ 15 & 27 & 0 & 0 & 1 & 0 & 350 \\ \hline -6 & -7 & 0 & 0 & 0 & 1 & 0 \end{array}\right)$

– 7 is the most negative.

$R_1 : \frac{200}{27} \approx 7.41 \qquad R_2 : \frac{400}{51} \approx 7.84 \qquad R_3 : \frac{350}{27} \approx 12.96$ Thus, pivot on the 27 in R_1.

13. $\left(\begin{array}{rrrrr|r} 2 & 0 & 1 & -\frac{3}{4} & 0 & 12 \\ \mathbf{3} & 1 & 0 & \frac{1}{3} & 0 & 15 \\ \hline -4 & 0 & 0 & 3 & 1 & 15 \end{array}\right)$

Not complete due to – 4.

$R_1 : \frac{12}{2} = 6 \qquad R_2 : \frac{15}{3} = 5$ Pivot on 3.

15. Solution is complete because all indicators are 0 or greater.

17. x-column and y-column indicators are negative and equal.

$R_1 : \frac{12}{4} = 3 \quad \frac{12}{4} = 3$

$R_2 : \frac{4}{2} = 2 \quad \frac{4}{4} = 1$

Pivot about $(R, C) = (2, 1) = 2$ or $(R, C) = (2, 2) = 4$. Pivot only one of them.

19. – 5 is the most negative indicator.

$R_1 : \frac{5}{-1} < 0, \ R_2 : \frac{12}{0}$ is not defined, $R_3 : \frac{6}{(-3)} < 0$

There is no solution because there are no nonnegative quotients.

21. f is maximized at 20 when $x = 11$ and $y = 9$.

23. f is maximized at 525 when $y = 14$, $z = 11$, and $x = 0$.

25. f is maximized at 100 when $x = 50$, and $y = 10$.
Since multiple solutions are possible, pivot on R_3, C_5 entry.

27. $$\left(\begin{array}{ccccc|c} 14 & \mathbf{7} & 1 & 0 & 0 & 35 \\ 5 & 5 & 0 & 1 & 0 & 50 \\ \hline -3 & -10 & 0 & 0 & 1 & 0 \end{array}\right)$$

– 10 is the most negative. Pivot on the 7.

$$\left[\begin{array}{ccccc|c} 2 & 1 & \frac{1}{7} & 0 & 0 & 5 \\ 5 & 5 & 0 & 1 & 0 & 50 \\ \hline -3 & -10 & 0 & 0 & 1 & 0 \end{array}\right] \begin{array}{c} \to \\ -5R_1+R_2\to R_2 \\ 10R_1+R_3\to R_3 \end{array} \left[\begin{array}{ccccc|c} 2 & 1 & \frac{1}{7} & 0 & 0 & 5 \\ -5 & 0 & -\frac{5}{7} & 1 & 0 & 25 \\ \hline 17 & 0 & \frac{10}{7} & 0 & 1 & 50 \end{array}\right]$$

All indicators are ≥ 0. The maximum is 50 when $y = 5$ and $x = 0$.

29. $$\left[\begin{array}{ccccc|c} 1 & \mathbf{2} & 1 & 0 & 0 & 10 \\ 1 & 1 & 0 & 1 & 0 & 7 \\ \hline -2 & -3 & 0 & 0 & 1 & 0 \end{array}\right] \tfrac{1}{2}R_1\to R_1 \left[\begin{array}{ccccc|c} \frac{1}{2} & 1 & \frac{1}{2} & 0 & 0 & 5 \\ 1 & 1 & 0 & 1 & 0 & 7 \\ \hline -2 & -3 & 0 & 0 & 1 & 0 \end{array}\right] \begin{array}{c} \to \\ -R_1+R_2\to R_2 \\ 3R_1+R_3\to R_3 \end{array}$$

$$\left[\begin{array}{ccccc|c} \frac{1}{2} & 1 & \frac{1}{2} & 0 & 0 & 5 \\ \frac{1}{2} & 0 & -\frac{1}{2} & 1 & 0 & 2 \\ \hline -\frac{1}{2} & 0 & \frac{3}{2} & 0 & 1 & 15 \end{array}\right] 2R_2\to R_2 \left[\begin{array}{ccccc|c} \frac{1}{2} & 1 & \frac{1}{2} & 0 & 0 & 5 \\ 1 & 0 & -1 & 2 & 0 & 4 \\ \hline -\frac{1}{2} & 0 & \frac{3}{2} & 0 & 1 & 15 \end{array}\right] \begin{array}{c} -\frac{1}{2}R_2+R_1\to R_1 \\ \to \\ \frac{1}{2}R_2+R_3\to R_3 \end{array} \left[\begin{array}{ccccc|c} 0 & 1 & 1 & -1 & 0 & 3 \\ 1 & 0 & -1 & 2 & 0 & 4 \\ \hline 0 & 0 & 1 & 1 & 1 & 17 \end{array}\right]$$

Maximum is 17 at $x = 4, y = 3$.

31. $$\left[\begin{array}{cccccc|c} 1 & 2 & 1 & 0 & 0 & 0 & 12 \\ 0 & \mathbf{1} & 0 & 1 & 0 & 0 & 5 \\ 1 & 1 & 0 & 0 & 1 & 0 & 9 \\ \hline -1 & -3 & 0 & 0 & 0 & 1 & 0 \end{array}\right] \begin{array}{c} -2R_2+R_1\to R_1 \\ \to \\ -R_2+R_3\to R_3 \\ 3R_2+R_4\to R_4 \end{array}$$

$$\left[\begin{array}{cccccc|c} \mathbf{1} & 0 & 1 & -2 & 0 & 0 & 2 \\ 0 & 1 & 0 & 1 & 0 & 0 & 5 \\ 1 & 0 & 0 & -1 & 1 & 0 & 4 \\ \hline -1 & 0 & 0 & 3 & 0 & 1 & 15 \end{array}\right] \begin{array}{c} \to \\ \\ -R_1+R_3\to R_3 \\ R_1+R_4\to R_4 \end{array} \left[\begin{array}{cccccc|c} 1 & 0 & 1 & -2 & 0 & 0 & 2 \\ 0 & 1 & 0 & 1 & 0 & 0 & 5 \\ 0 & 0 & -1 & 1 & 1 & 0 & 2 \\ \hline 0 & 0 & 1 & 1 & 0 & 1 & 17 \end{array}\right]$$

Maximum is 17 at $x = 2, y = 5$.

33. $$\left[\begin{array}{cccccc|c} -1 & 1 & 1 & 0 & 0 & 0 & 2 \\ 1 & 2 & 0 & 1 & 0 & 0 & 10 \\ \mathbf{3} & 1 & 0 & 0 & 1 & 0 & 15 \\ \hline -2 & -1 & 0 & 0 & 0 & 1 & 0 \end{array}\right] \begin{array}{l} \frac{2}{(-1)}<0 \\ \frac{10}{1}=10 \quad \to \\ \frac{15}{3}=5 \quad \frac{1}{3}R_3\to R_3 \\ \end{array} \left[\begin{array}{cccccc|c} -1 & 1 & 1 & 0 & 0 & 0 & 2 \\ 1 & 2 & 0 & 1 & 0 & 0 & 10 \\ 1 & \frac{1}{3} & 0 & 0 & \frac{1}{3} & 0 & 5 \\ \hline -2 & -1 & 0 & 0 & 0 & 1 & 0 \end{array}\right] \begin{array}{c} R_3+R_1\to R_1 \\ -R_3+R_2\to R_2 \\ \to \\ 2R_3+R_4\to R_4 \end{array}$$

$$\left[\begin{array}{cccccc|c} 0 & \frac{4}{3} & 1 & 0 & \frac{1}{3} & 0 & 7 \\ 0 & \frac{5}{3} & 0 & 1 & -\frac{1}{3} & 0 & 5 \\ 1 & \frac{1}{3} & 0 & 0 & \frac{1}{3} & 0 & 5 \\ \hline 0 & -\frac{1}{3} & 0 & 0 & \frac{2}{3} & 1 & 10 \end{array}\right] \begin{array}{l} 7\div\left(\frac{4}{3}\right)=\frac{21}{4} \quad \to \\ 5\div\left(\frac{5}{3}\right)=3 \quad \frac{3}{5}R_2\to R_2 \\ 5\div\left(\frac{1}{3}\right)=15 \\ \end{array} \left[\begin{array}{cccccc|c} 0 & \frac{4}{3} & 1 & 0 & \frac{1}{3} & 0 & 7 \\ 0 & 1 & 0 & \frac{3}{5} & -\frac{1}{5} & 0 & 3 \\ 1 & \frac{1}{3} & 0 & 0 & \frac{1}{3} & 0 & 5 \\ \hline 0 & -\frac{1}{3} & 0 & 0 & \frac{2}{3} & 1 & 10 \end{array}\right] \begin{array}{c} -\frac{4}{3}R_2+R_1\to R_1 \\ \to \\ -\frac{1}{3}R_2+R_3\to R_3 \\ \frac{1}{3}R_2+R_4\to R_4 \end{array}$$

$$\left[\begin{array}{cccccc|c} 0 & 0 & 1 & -\frac{4}{5} & \frac{3}{5} & 0 & 3 \\ 0 & 1 & 0 & \frac{3}{5} & -\frac{1}{5} & 0 & 3 \\ 1 & 0 & 0 & -\frac{1}{5} & \frac{2}{5} & 0 & 4 \\ \hline 0 & 0 & 0 & \frac{1}{5} & \frac{3}{5} & 1 & 11 \end{array}\right]$$

Maximum is 11 at $x = 4, y = 3$.

35. $\left[\begin{array}{ccccccc|c} 3 & 5 & 4 & 1 & 0 & 0 & 0 & 30 \\ 3 & \mathbf{2} & 0 & 0 & 1 & 0 & 0 & 4 \\ 1 & 2 & 0 & 0 & 0 & 1 & 0 & 8 \\ \hline -7 & -10 & -4 & 0 & 0 & 0 & 1 & 0 \end{array}\right] \begin{array}{c} \frac{1}{2}R_2 \to R_2 \\ \to \end{array} \left[\begin{array}{ccccccc|c} 3 & 5 & 4 & 1 & 0 & 0 & 0 & 30 \\ \frac{3}{2} & 1 & 0 & 0 & \frac{1}{2} & 0 & 0 & 2 \\ 1 & 2 & 0 & 0 & 0 & 1 & 0 & 8 \\ \hline -7 & -10 & -4 & 0 & 0 & 0 & 1 & 0 \end{array}\right] \begin{array}{c} -5R_2 + R_1 \to R_1 \\ \to \\ -2R_2 + R_3 \to R_3 \\ 10R_2 + R_4 \to R_4 \end{array}$

$\left[\begin{array}{ccccccc|c} -\frac{9}{2} & 0 & \mathbf{4} & 1 & -\frac{5}{2} & 0 & 0 & 20 \\ \frac{3}{2} & 1 & 0 & 0 & \frac{1}{2} & 0 & 0 & 2 \\ -2 & 0 & 0 & 0 & -1 & 1 & 0 & 4 \\ \hline 8 & 0 & -4 & 0 & 5 & 0 & 1 & 20 \end{array}\right] \begin{array}{c} \frac{1}{4}R_1 \to R_1 \\ \to \end{array} \left[\begin{array}{ccccccc|c} -\frac{9}{8} & 0 & 1 & \frac{1}{4} & -\frac{5}{8} & 0 & 0 & 5 \\ \frac{3}{2} & 1 & 0 & 0 & \frac{1}{2} & 0 & 0 & 2 \\ -2 & 0 & 0 & 0 & -1 & 1 & 0 & 4 \\ \hline 8 & 0 & -4 & 0 & 5 & 0 & 1 & 20 \end{array}\right] \begin{array}{c} \to \\ 4R_1 + R_4 \to R_4 \end{array}$

$\left[\begin{array}{ccccccc|c} -\frac{9}{8} & 0 & 1 & \frac{1}{4} & -\frac{5}{8} & 0 & 0 & 5 \\ \frac{3}{2} & 1 & 0 & 0 & \frac{1}{2} & 0 & 0 & 2 \\ -2 & 0 & 0 & 0 & -1 & 1 & 0 & 4 \\ \hline \frac{7}{2} & 0 & 0 & 1 & \frac{5}{2} & 0 & 1 & 40 \end{array}\right]$

Maximum is 40 at $x = 0$, $y = 2$, $z = 5$.

37. $\left[\begin{array}{ccccccc|c} 1 & 0 & 1 & 1 & 0 & 0 & 0 & 40 \\ \mathbf{1} & 1 & 0 & 0 & 1 & 0 & 0 & 30 \\ 0 & 1 & 1 & 0 & 0 & 1 & 0 & 40 \\ \hline -20 & -12 & -12 & 0 & 0 & 0 & 1 & 0 \end{array}\right] \begin{array}{c} -R_2 + R_1 \to R_1 \\ \to \\ \\ 20R_2 + R_4 \to R_4 \end{array} \left[\begin{array}{ccccccc|c} 0 & -1 & \mathbf{1} & 1 & -1 & 0 & 0 & 10 \\ 1 & 1 & 0 & 0 & 1 & 0 & 0 & 30 \\ 0 & 1 & 1 & 0 & 0 & 1 & 0 & 40 \\ \hline 0 & 8 & -12 & 0 & 20 & 0 & 1 & 600 \end{array}\right] \begin{array}{c} \\ \to \\ -R_1 + R_3 \to R_3 \\ 12R_1 + R_4 \to R_4 \end{array}$

$\left[\begin{array}{ccccccc|c} 0 & -1 & 1 & 1 & -1 & 0 & 0 & 10 \\ 1 & 1 & 0 & 0 & 1 & 0 & 0 & 30 \\ 0 & \mathbf{2} & 0 & -1 & 1 & 1 & 0 & 30 \\ \hline 0 & -4 & 0 & 12 & 8 & 0 & 1 & 720 \end{array}\right] \begin{array}{c} \to \\ \\ \frac{1}{2}R_3 \to R_3 \end{array} \left[\begin{array}{ccccccc|c} 0 & -1 & 1 & 1 & -1 & 0 & 0 & 10 \\ 1 & 1 & 0 & 0 & 1 & 0 & 0 & 30 \\ 0 & 1 & 0 & -\frac{1}{2} & \frac{1}{2} & \frac{1}{2} & 0 & 15 \\ \hline 0 & -4 & 0 & 12 & 8 & 0 & 1 & 720 \end{array}\right] \begin{array}{c} R_3 + R_1 \to R_1 \\ -R_3 + R_2 \to R_2 \\ \to \\ 4R_3 + R_4 \to R_4 \end{array}$

$\left[\begin{array}{ccccccc|c} 0 & 0 & 1 & \frac{1}{2} & -\frac{1}{2} & \frac{1}{2} & 0 & 25 \\ 1 & 0 & 0 & \frac{1}{2} & \frac{1}{2} & -\frac{1}{2} & 0 & 15 \\ 0 & 1 & 0 & -\frac{1}{2} & \frac{1}{2} & \frac{1}{2} & 0 & 15 \\ \hline 0 & 0 & 0 & 10 & 10 & 2 & 1 & 780 \end{array}\right]$

Maximum is 780 at $x = 15$, $y = 15$, $z = 25$.

39. $\left[\begin{array}{ccccccc|c} 2 & 1 & 1 & 1 & 0 & 0 & 0 & 40 \\ \mathbf{1} & 2 & 0 & 0 & 1 & 0 & 0 & 10 \\ 0 & 1 & 3 & 0 & 0 & 1 & 0 & 80 \\ \hline -10 & -8 & -5 & 0 & 0 & 0 & 1 & 0 \end{array}\right] \begin{array}{c} -2R_2 + R_1 \to R_1 \\ \to \\ \\ 10R_2 + R_4 \to R_4 \end{array} \left[\begin{array}{ccccccc|c} 0 & -3 & \mathbf{1} & 1 & -2 & 0 & 0 & 20 \\ 1 & 2 & 0 & 0 & 1 & 0 & 0 & 10 \\ 0 & 1 & 3 & 0 & 0 & 1 & 0 & 80 \\ \hline 0 & 12 & -5 & 0 & 10 & 0 & 1 & 100 \end{array}\right] \begin{array}{c} \to \\ \\ -3R_1 + R_3 \to R_3 \\ 5R_1 + R_4 \to R_4 \end{array}$

$\left[\begin{array}{ccccccc|c} 0 & -3 & 1 & 1 & -2 & 0 & 0 & 20 \\ 1 & 2 & 0 & 0 & 1 & 0 & 0 & 10 \\ 0 & \mathbf{10} & 0 & -3 & 6 & 1 & 0 & 20 \\ \hline 0 & -3 & 0 & 5 & 0 & 0 & 1 & 200 \end{array}\right] \begin{array}{c} \to \\ \\ \frac{1}{10}R_3 \to R_3 \end{array} \left[\begin{array}{ccccccc|c} 0 & -3 & 1 & 1 & -2 & 0 & 0 & 20 \\ 1 & 2 & 0 & 0 & 1 & 0 & 0 & 10 \\ 0 & 1 & 0 & -\frac{3}{10} & \frac{3}{5} & \frac{1}{10} & 0 & 2 \\ \hline 0 & -3 & 0 & 5 & 0 & 0 & 1 & 200 \end{array}\right] \begin{array}{c} 3R_3 + R_1 \to R_1 \\ -2R_3 + R_2 \to R_2 \\ \to \\ 3R_3 + R_4 \to R_4 \end{array}$

$\left[\begin{array}{ccccccc|c} 0 & 0 & 1 & \frac{1}{10} & -\frac{1}{5} & \frac{3}{10} & 0 & 26 \\ 1 & 0 & 0 & \frac{3}{5} & -\frac{1}{5} & -\frac{1}{5} & 0 & 6 \\ 0 & 1 & 0 & -\frac{3}{10} & \frac{3}{5} & \frac{1}{10} & 0 & 2 \\ \hline 0 & 0 & 0 & \frac{41}{10} & \frac{9}{5} & \frac{3}{10} & 1 & 206 \end{array}\right]$

Maximum is 206 at $x = 6$, $y = 2$, $z = 26$.

41. $\left[\begin{array}{ccccc|c} 1 & \frac{1}{2} & 1 & 0 & 0 & 16 \\ 0 & \frac{1}{2} & -1 & 1 & 0 & 8 \\ \hline 0 & 0 & 2 & 0 & 1 & 32 \end{array}\right] \xrightarrow{2R_2 \to R_2} \left[\begin{array}{ccccc|c} 1 & \frac{1}{2} & 1 & 0 & 0 & 16 \\ 0 & 1 & -2 & 2 & 0 & 16 \\ \hline 0 & 0 & 2 & 0 & 1 & 32 \end{array}\right] \xrightarrow{-\frac{1}{2}R_2 + R_1 \to R_1} \left[\begin{array}{ccccc|c} 1 & 0 & 2 & -1 & 0 & 8 \\ 0 & 1 & -2 & 2 & 0 & 16 \\ \hline 0 & 0 & 2 & 0 & 1 & 32 \end{array}\right]$

The maximum is 32 at $x = 8, y = 16$.

43. $\left[\begin{array}{ccccc|c} \mathbf{1} & -10 & 1 & 0 & 0 & 10 \\ -1 & 1 & 0 & 1 & 0 & 40 \\ \hline -3 & -2 & 0 & 0 & 1 & 0 \end{array}\right] \xrightarrow[3R_1 + R_3 \to R_3]{R_1 + R_2 \to R_2} \left[\begin{array}{ccccc|c} 1 & -10 & 1 & 0 & 0 & 10 \\ 0 & -9 & 1 & 1 & 0 & 50 \\ \hline 0 & -32 & 3 & 0 & 1 & 30 \end{array}\right]$

No nonnegative quotient exists. There is no solution.

45. $\left[\begin{array}{ccccc|c} 2 & 1 & 1 & 0 & 0 & 120 \\ 1 & \mathbf{4} & 0 & 1 & 0 & 200 \\ \hline -3 & -12 & 0 & 0 & 1 & 0 \end{array}\right] \xrightarrow{\frac{1}{4}R_2 \to R_2} \left[\begin{array}{ccccc|c} 2 & 1 & 1 & 0 & 0 & 120 \\ \frac{1}{4} & 1 & 0 & \frac{1}{4} & 0 & 50 \\ \hline -3 & -12 & 0 & 0 & 1 & 0 \end{array}\right] \xrightarrow[12R_2 + R_3 \to R_3]{-R_2 + R_1 \to R_1}$

$\left[\begin{array}{ccccc|c} \frac{7}{4} & 0 & 1 & -\frac{1}{4} & 0 & 70 \\ \frac{1}{4} & 1 & 0 & \frac{1}{4} & 0 & 50 \\ \hline 0 & 0 & 0 & 3 & 1 & 600 \end{array}\right] \; \frac{4}{7}R_1 \to R_1$

Maximum is 600 at $x = 0, y = 50$. Continuing to find a second solution we have:

$\left[\begin{array}{ccccc|c} 1 & 0 & \frac{4}{7} & -\frac{1}{7} & 0 & 40 \\ \frac{1}{4} & 1 & 0 & \frac{1}{4} & 0 & 50 \\ \hline 0 & 0 & 0 & 3 & 1 & 600 \end{array}\right] \xrightarrow{-\frac{1}{4}R_1 + R_2 \to R_2} \left[\begin{array}{ccccc|c} 1 & 0 & \frac{4}{7} & -\frac{1}{7} & 0 & 40 \\ 0 & 1 & -\frac{1}{7} & \frac{2}{7} & 0 & 40 \\ \hline 0 & 0 & 0 & 3 & 1 & 600 \end{array}\right]$

Maximum is 600 at $x = 40, y = 40$.

47. Let x = number of inkjet printers, y = number of laser printers.
Maximize $f = 40x + 60y$
Constraints: $x + y \le 60, \; x + 3y \le 120$

a. $\left[\begin{array}{ccccc|c} 1 & 1 & 1 & 0 & 0 & 60 \\ 1 & \mathbf{3} & 0 & 1 & 0 & 120 \\ \hline -40 & -60 & 0 & 0 & 1 & 0 \end{array}\right] \xrightarrow{\frac{1}{3}R_2 \to R_2} \left[\begin{array}{ccccc|c} 1 & 1 & 1 & 0 & 0 & 60 \\ \frac{1}{3} & 1 & 0 & \frac{1}{3} & 0 & 40 \\ \hline -40 & -60 & 0 & 0 & 1 & 0 \end{array}\right] \xrightarrow[60R_2 + R_3 \to R_3]{-R_2 + R_1 \to R_1}$

$\left[\begin{array}{ccccc|c} \frac{2}{3} & 0 & 1 & -\frac{1}{3} & 0 & 20 \\ \frac{1}{3} & 1 & 0 & \frac{1}{3} & 0 & 40 \\ \hline -20 & 0 & 0 & 20 & 1 & 2400 \end{array}\right] \xrightarrow{\frac{3}{2}R_1 \to R_1} \left[\begin{array}{ccccc|c} 1 & 0 & \frac{3}{2} & -\frac{1}{2} & 0 & 30 \\ \frac{1}{3} & 1 & 0 & \frac{1}{3} & 0 & 40 \\ \hline -20 & 0 & 0 & 20 & 1 & 2400 \end{array}\right] \xrightarrow[20R_1 + R_3 \to R_3]{-\frac{1}{3}R_1 + R_2 \to R_2}$

$\left[\begin{array}{ccccc|c} 1 & 0 & \frac{3}{2} & -\frac{1}{2} & 0 & 30 \\ 0 & 1 & -\frac{1}{2} & \frac{1}{2} & 0 & 30 \\ \hline 0 & 0 & 30 & 10 & 1 & 3000 \end{array}\right]$

b. Maximum profit is \$3000 with 30 ink jet and 30 laser printers.

49. Let x = number of axles, y = number of wheels.
Maximize: $f = 300x + 300y$
Constraints: $4x + 3y \le 50,\ 3x + 5y \le 43$

a.
$$\left[\begin{array}{ccccc|c} 4 & 3 & 1 & 0 & 0 & 50 \\ 3 & 5 & 0 & 1 & 0 & 43 \\ \hline -300 & -300 & 0 & 0 & 1 & 0 \end{array}\right] \xrightarrow{\frac{1}{5}R_2 \to R_2} \left[\begin{array}{ccccc|c} 4 & 3 & 1 & 0 & 0 & 50 \\ \frac{3}{5} & 1 & 0 & \frac{1}{5} & 0 & \frac{43}{5} \\ \hline -300 & -300 & 0 & 0 & 1 & 0 \end{array}\right] \begin{array}{c} -3R_2 + R_1 \to R_1 \\ \to \\ 300R_2 + R_3 \to R_3 \end{array}$$

$$\left[\begin{array}{ccccc|c} \frac{11}{5} & 0 & 1 & -\frac{3}{5} & 0 & \frac{121}{5} \\ \frac{3}{5} & 1 & 0 & \frac{1}{5} & 0 & \frac{43}{5} \\ \hline -120 & 0 & 0 & 60 & 1 & 2580 \end{array}\right] \xrightarrow{\frac{5}{11}R_1 \to R_1} \left[\begin{array}{ccccc|c} 1 & 0 & \frac{5}{11} & -\frac{3}{11} & 0 & 11 \\ \frac{3}{5} & 1 & 0 & \frac{1}{5} & 0 & \frac{43}{5} \\ \hline -120 & 0 & 0 & 60 & 1 & 2580 \end{array}\right] \begin{array}{c} \to \\ -\frac{3}{5}R_1 + R_2 \to R_2 \\ 120R_1 + R_3 \to R_3 \end{array}$$

$$\left[\begin{array}{ccccc|c} 1 & 0 & \frac{5}{11} & -\frac{3}{11} & 0 & 11 \\ 0 & 1 & -\frac{3}{11} & \frac{4}{11} & 0 & 2 \\ \hline 0 & 0 & \frac{600}{11} & \frac{300}{11} & 1 & 3900 \end{array}\right]$$

b. Maximum profit is \$3900 with 11 axles and 2 wheels.

51. Let t = crates of tomatoes and p = crates of peaches. Objective function: $f = t + 2p$
Constraints: $\frac{1}{2}t + \frac{5}{4}p \le 2500,\ 60t + 50p \le 120{,}000$

$$\left[\begin{array}{ccccc|c} \frac{1}{2} & \frac{5}{4} & 1 & 0 & 0 & 2500 \\ 60 & 50 & 0 & 1 & 0 & 120{,}000 \\ \hline -1 & -2 & 0 & 0 & 1 & 0 \end{array}\right] \xrightarrow{\frac{4}{5}R_1 \to R_1} \left[\begin{array}{ccccc|c} \frac{2}{5} & 1 & \frac{4}{5} & 0 & 0 & 2000 \\ 60 & 50 & 0 & 1 & 0 & 120{,}000 \\ \hline -1 & -2 & 0 & 0 & 1 & 0 \end{array}\right] \begin{array}{c} \to \\ -50R_1 + R_2 \to R_2 \\ 2R_1 + R_3 \to R_3 \end{array}$$

$$\left[\begin{array}{ccccc|c} \frac{2}{5} & 1 & \frac{4}{5} & 0 & 0 & 2000 \\ \mathbf{40} & 0 & -40 & 1 & 0 & 20{,}000 \\ \hline -\frac{1}{5} & 0 & \frac{8}{5} & 0 & 1 & 4000 \end{array}\right] \xrightarrow{\frac{1}{40}R_2 \to R_2} \left[\begin{array}{ccccc|c} \frac{2}{5} & 1 & \frac{4}{5} & 0 & 0 & 2000 \\ 1 & 0 & -1 & \frac{1}{40} & 0 & 500 \\ \hline -\frac{1}{5} & 0 & \frac{8}{5} & 0 & 1 & 4000 \end{array}\right] \begin{array}{c} -\frac{2}{5}R_2 + R_1 \to R_1 \\ \to \\ \frac{1}{5}R_2 + R_3 \to R_3 \end{array}$$

$$\left[\begin{array}{ccccc|c} 0 & 1 & \frac{6}{5} & -0.01 & 0 & 1800 \\ 1 & 0 & -1 & \frac{1}{40} & 0 & 500 \\ \hline 0 & 0 & \frac{7}{5} & 0.005 & 1 & 4100 \end{array}\right]$$

There is a maximum profit of \$4100 obtained from 500 crates of tomatoes and 1800 crates of peaches.

53. Let x = newspaper ads and y = radio ads. Objective function: $f = 6000x + 8000y$
Constraints: $100x + 300y \le 6000,\ x \le 21,\ y \le 28$

$$\left[\begin{array}{cccccc|c} 1 & \mathbf{3} & 1 & 0 & 0 & 0 & 60 \\ 1 & 0 & 0 & 1 & 0 & 0 & 21 \\ 0 & 1 & 0 & 0 & 1 & 0 & 28 \\ \hline -6000 & -8000 & 0 & 0 & 0 & 1 & 0 \end{array}\right] \xrightarrow{\frac{1}{3}R_1 \to R_1} \left[\begin{array}{cccccc|c} \frac{1}{3} & 1 & \frac{1}{3} & 0 & 0 & 0 & 20 \\ 1 & 0 & 0 & 1 & 0 & 0 & 21 \\ 0 & 1 & 0 & 0 & 1 & 0 & 28 \\ \hline -6000 & -8000 & 0 & 0 & 0 & 1 & 0 \end{array}\right] \begin{array}{c} \to \\ \\ -R_1 + R_3 \to R_3 \\ 8000R_1 + R_4 \to R_4 \end{array}$$

$$\left[\begin{array}{cccccc|c} \frac{1}{3} & 1 & \frac{1}{3} & 0 & 0 & 0 & 20 \\ \mathbf{1} & 0 & 0 & 1 & 0 & 0 & 21 \\ -\frac{1}{3} & 0 & -\frac{1}{3} & 0 & 1 & 0 & 8 \\ \hline -\frac{10{,}000}{3} & 0 & \frac{8{,}000}{3} & 0 & 0 & 1 & 160{,}000 \end{array}\right] \begin{array}{c} -\frac{1}{3}R_2 + R_1 \to R_1 \\ \to \\ \frac{1}{3}R_2 + R_3 \to R_3 \\ \frac{10{,}000}{3}R_2 + R_4 \to R_4 \end{array} \left[\begin{array}{cccccc|c} 0 & 1 & \frac{1}{3} & -\frac{1}{3} & 0 & 0 & 13 \\ 1 & 0 & 0 & 1 & 0 & 0 & 21 \\ 0 & 0 & -\frac{1}{3} & \frac{1}{3} & 1 & 0 & 15 \\ \hline 0 & 0 & -\frac{8000}{3} & \frac{10{,}000}{3} & 0 & 1 & 230{,}000 \end{array}\right]$$

Maximum exposure occurs with 21 newspaper ads and 13 radio ads. The maximum number of exposures is 230,000.

55. Let x = number of ad packages in medium 1
y = number of ad packages in medium 2
z = number of ad packages in medium 3
Maximize: $f = 3100x + 2000y + 2400z$
Constraints: $10x + 4y + 5z \le 200,\ x \le 18,\ y \le 10,\ z \le 12$

$$\left[\begin{array}{cccccccc|c} 10 & 4 & 5 & 1 & 0 & 0 & 0 & 0 & 200 \\ \mathbf{1} & 0 & 0 & 0 & 1 & 0 & 0 & 0 & 18 \\ 0 & 1 & 0 & 0 & 0 & 1 & 0 & 0 & 10 \\ 0 & 0 & 1 & 0 & 0 & 0 & 1 & 0 & 12 \\ \hline -3100 & -2000 & -2400 & 0 & 0 & 0 & 0 & 1 & 0 \end{array}\right] \begin{array}{l} -10R_2 + R_1 \to R_1 \\ \\ \to \\ \\ 3100R_2 + R_5 \to R_5 \end{array}$$

$$\left[\begin{array}{cccccccc|c} 0 & 4 & \mathbf{5} & 1 & -10 & 0 & 0 & 0 & 20 \\ 1 & 0 & 0 & 0 & 1 & 0 & 0 & 0 & 18 \\ 0 & 1 & 0 & 0 & 0 & 1 & 0 & 0 & 10 \\ 0 & 0 & 1 & 0 & 0 & 0 & 1 & 0 & 12 \\ \hline 0 & -2000 & -2400 & 0 & 3100 & 0 & 0 & 1 & 55800 \end{array}\right] \begin{array}{l} \frac{1}{5}R_1 \to R_1 \\ \\ \to \\ \\ \\ \end{array}$$

$$\left[\begin{array}{cccccccc|c} 0 & \frac{4}{5} & 1 & \frac{1}{5} & -2 & 0 & 0 & 0 & 4 \\ 1 & 0 & 0 & 0 & 1 & 0 & 0 & 0 & 18 \\ 0 & 1 & 0 & 0 & 0 & 1 & 0 & 0 & 10 \\ 0 & 0 & 1 & 0 & 0 & 0 & 1 & 0 & 12 \\ \hline 0 & -2000 & -2400 & 0 & 3100 & 0 & 0 & 1 & 55800 \end{array}\right] \begin{array}{l} \\ \to \\ \\ -R_1 + R_4 \to R_4 \\ 2400R_1 + R_5 \to R_5 \end{array}$$

$$\left[\begin{array}{cccccccc|c} 0 & \frac{4}{5} & 1 & \frac{1}{5} & -2 & 0 & 0 & 0 & 4 \\ 1 & 0 & 0 & 0 & 1 & 0 & 0 & 0 & 18 \\ 0 & 1 & 0 & 0 & 0 & 1 & 0 & 0 & 10 \\ 0 & -\frac{4}{5} & 0 & -\frac{1}{5} & 2 & 0 & 1 & 0 & 8 \\ \hline 0 & -80 & 0 & 480 & -1700 & 0 & 0 & 1 & 65400 \end{array}\right] \begin{array}{l} \\ \to \\ \\ \frac{1}{2}R_4 \to R_4 \\ \\ \end{array}$$

$$\left[\begin{array}{cccccccc|c} 0 & \frac{4}{5} & 1 & \frac{1}{5} & -2 & 0 & 0 & 0 & 4 \\ 1 & 0 & 0 & 0 & 1 & 0 & 0 & 0 & 18 \\ 0 & 1 & 0 & 0 & 0 & 1 & 0 & 0 & 10 \\ 0 & -\frac{2}{5} & 0 & -\frac{1}{10} & 1 & 0 & \frac{1}{2} & 0 & 4 \\ \hline 0 & -80 & 0 & 480 & -1700 & 0 & 0 & 1 & 65,400 \end{array}\right] \begin{array}{l} 2R_4 + R_1 \to R_1 \\ -R_4 + R_2 \to R_2 \\ \to \\ \\ 1700R_4 + R_5 \to R_5 \end{array}$$

$$\left[\begin{array}{cccccccc|c} 0 & 0 & 1 & 0 & 0 & 0 & 1 & 0 & 12 \\ 1 & \frac{2}{5} & 0 & \frac{1}{10} & 0 & 0 & -\frac{1}{2} & 0 & 14 \\ 0 & 1 & 0 & 0 & 0 & 1 & 0 & 0 & 10 \\ 0 & -\frac{2}{5} & 0 & -\frac{1}{10} & 1 & 0 & \frac{1}{2} & 0 & 4 \\ \hline 0 & -760 & 0 & 310 & 0 & 0 & 850 & 1 & 72,200 \end{array}\right] \begin{array}{l} \\ -\frac{2}{5}R_3 + R_2 \to R_2 \\ \to \\ \frac{2}{5}R_3 + R_4 \to R_4 \\ 760R_3 + R_5 \to R_5 \end{array}$$

$$\left[\begin{array}{cccccccc|c} 0 & 0 & 1 & 0 & 0 & 0 & 1 & 0 & 12 \\ 1 & 0 & 0 & \frac{1}{10} & 0 & -\frac{2}{5} & -\frac{1}{2} & 0 & 10 \\ 0 & 1 & 0 & 0 & 0 & 1 & 0 & 0 & 10 \\ 0 & 0 & 0 & -\frac{1}{10} & 1 & \frac{2}{5} & \frac{1}{2} & 0 & 8 \\ \hline 0 & 0 & 0 & 310 & 0 & 760 & 850 & 1 & 79,800 \end{array}\right]$$

Purchase 10 each of medium I and II. Purchase 12 units of medium III.

57. Objective function: $f = 30A + 9B + 15C$

Constraints: $9A + 3B + 0.5C \le 477$, $5A + 4B \le 350$, $3A + 2C \le 150$, $B \le 20$

$$\left[\begin{array}{cccccccc|c} 9 & 3 & \frac{1}{2} & 1 & 0 & 0 & 0 & 0 & 477 \\ 5 & 4 & 0 & 0 & 1 & 0 & 0 & 0 & 350 \\ 3 & 0 & 2 & 0 & 0 & 1 & 0 & 0 & 150 \\ 0 & 1 & 0 & 0 & 0 & 0 & 1 & 0 & 20 \\ \hline -30 & -9 & -15 & 0 & 0 & 0 & 0 & 1 & 0 \end{array}\right] \begin{array}{l} \rightarrow \\ \\ \frac{1}{3}R_3 \rightarrow R_3 \\ \\ \\ \end{array} \left[\begin{array}{cccccccc|c} 9 & 3 & \frac{1}{2} & 1 & 0 & 0 & 0 & 0 & 477 \\ 5 & 4 & 0 & 0 & 1 & 0 & 0 & 0 & 350 \\ 1 & 0 & \frac{2}{3} & 0 & 0 & \frac{1}{3} & 0 & 0 & 50 \\ 0 & 1 & 0 & 0 & 0 & 0 & 1 & 0 & 20 \\ \hline -30 & -9 & -15 & 0 & 0 & 0 & 0 & 1 & 0 \end{array}\right] \begin{array}{l} -9R_3 + R_1 \rightarrow R_1 \\ -5R_3 + R_2 \rightarrow R_2 \\ \\ \\ 30R_3 + R_5 \rightarrow R_5 \end{array}$$

$$\left[\begin{array}{cccccccc|c} 0 & 3 & -\frac{11}{2} & 1 & 0 & -3 & 0 & 0 & 27 \\ 0 & 4 & -\frac{10}{3} & 0 & 1 & -\frac{5}{3} & 0 & 0 & 100 \\ 1 & 0 & \frac{2}{3} & 0 & 0 & \frac{1}{3} & 0 & 0 & 50 \\ 0 & 1 & 0 & 0 & 0 & 0 & 1 & 0 & 20 \\ \hline 0 & -9 & 5 & 0 & 0 & 10 & 0 & 1 & 1500 \end{array}\right] \begin{array}{l} \frac{1}{3}R_1 \rightarrow R_1 \\ \\ \rightarrow \\ \\ \\ \end{array} \left[\begin{array}{cccccccc|c} 0 & 1 & -\frac{11}{6} & \frac{1}{3} & 0 & -1 & 0 & 0 & 9 \\ 0 & 4 & -\frac{10}{3} & 0 & 1 & -\frac{5}{3} & 0 & 0 & 100 \\ 1 & 0 & \frac{2}{3} & 0 & 0 & \frac{1}{3} & 0 & 0 & 50 \\ 0 & 1 & 0 & 0 & 0 & 0 & 1 & 0 & 20 \\ \hline 0 & -9 & 5 & 0 & 0 & 10 & 0 & 1 & 1500 \end{array}\right] \begin{array}{l} \\ -4R_1 + R_2 \rightarrow R_2 \\ \rightarrow \\ -R_1 + R_4 \rightarrow R_4 \\ 9R_1 + R_5 \rightarrow R_5 \end{array}$$

$$\left[\begin{array}{cccccccc|c} 0 & 1 & -\frac{11}{6} & \frac{1}{3} & 0 & -1 & 0 & 0 & 9 \\ 0 & 0 & 4 & -\frac{4}{3} & 1 & \frac{7}{3} & 0 & 0 & 64 \\ 1 & 0 & \frac{2}{3} & 0 & 0 & \frac{1}{3} & 0 & 0 & 50 \\ 0 & 0 & \frac{11}{6} & -\frac{1}{3} & 0 & 1 & 1 & 0 & 11 \\ \hline 0 & 0 & -\frac{23}{6} & 3 & 0 & 1 & 0 & 1 & 1581 \end{array}\right] \begin{array}{l} \\ \rightarrow \\ \\ \frac{6}{11}R_4 \rightarrow R_4 \\ \\ \end{array}$$

$$\left[\begin{array}{cccccccc|c} 0 & 1 & -\frac{11}{6} & \frac{1}{3} & 0 & -1 & 0 & 0 & 9 \\ 0 & 0 & 4 & -\frac{4}{3} & 1 & \frac{7}{3} & 0 & 0 & 64 \\ 1 & 0 & \frac{2}{3} & 0 & 0 & \frac{1}{3} & 0 & 0 & 50 \\ 0 & 0 & 1 & -\frac{2}{11} & 0 & \frac{6}{11} & \frac{6}{11} & 0 & 6 \\ \hline 0 & 0 & -\frac{23}{2} & 3 & 0 & 1 & 0 & 1 & 1581 \end{array}\right] \begin{array}{l} \frac{11}{6}R_4 + R_1 \rightarrow R_1 \\ -4R_4 + R_2 \rightarrow R_2 \\ -\frac{2}{3}R_4 + R_3 \rightarrow R_3 \\ \rightarrow \\ \frac{23}{2}R_4 + R_5 \rightarrow R_5 \end{array} \left[\begin{array}{cccccccc|c} 0 & 1 & 0 & 0 & 0 & 0 & 1 & 0 & 20 \\ 0 & 0 & 0 & -\frac{20}{33} & 1 & \frac{5}{33} & -\frac{24}{11} & 0 & 40 \\ 1 & 0 & 0 & \frac{4}{33} & 0 & -\frac{1}{33} & -\frac{4}{33} & 0 & 46 \\ 0 & 0 & 1 & -\frac{2}{11} & 0 & \frac{6}{11} & \frac{6}{11} & 0 & 6 \\ \hline 0 & 0 & 0 & \frac{10}{11} & 0 & \frac{80}{11} & \frac{69}{11} & 1 & 1650 \end{array}\right]$$

Maximum profit is \$1650 with 46A, 20B, and 6C.

59. Objective function: $f = 3x_1 + 4x_2 + 8x_3$

Constraints: $5x_1 + 12x_2 + 24x_3 \le 90{,}000$, $1x_1 + 2x_2 + 4x_3 \le 12{,}000$, $1x_1 + 1x_2 + 1x_3 \le 9{,}000$

$$\left[\begin{array}{ccccccc|c} 5 & 12 & 24 & 1 & 0 & 0 & 0 & 90{,}000 \\ 1 & 2 & 4 & 0 & 1 & 0 & 0 & 12{,}000 \\ 1 & 1 & 1 & 0 & 0 & 1 & 0 & 9{,}000 \\ \hline -3 & -4 & -8 & 0 & 0 & 0 & 1 & 0 \end{array}\right] \begin{array}{l} \\ \frac{1}{4}R_2 \rightarrow R_2 \\ \rightarrow \\ \\ \end{array} \left[\begin{array}{ccccccc|c} 5 & 12 & 24 & 1 & 0 & 0 & 0 & 90{,}000 \\ \frac{1}{4} & \frac{1}{2} & 1 & 0 & \frac{1}{4} & 0 & 0 & 3{,}000 \\ 1 & 1 & 1 & 0 & 0 & 1 & 0 & 9{,}000 \\ \hline -3 & -4 & -8 & 0 & 0 & 0 & 1 & 0 \end{array}\right] \begin{array}{l} -24R_2 + R_1 \rightarrow R_1 \\ \rightarrow \\ -1R_2 + R_3 \rightarrow R_3 \\ 8R_2 + R_4 \rightarrow R_4 \end{array}$$

$$\left[\begin{array}{ccccccc|c} -1 & 0 & 0 & 1 & -6 & 0 & 0 & 18{,}000 \\ \frac{1}{4} & \frac{1}{2} & 1 & 0 & \frac{1}{4} & 0 & 0 & 3{,}000 \\ \frac{3}{4} & \frac{1}{2} & 0 & 0 & -\frac{1}{4} & 1 & 0 & 6{,}000 \\ \hline -1 & 0 & 0 & 0 & 2 & 0 & 1 & 24{,}000 \end{array}\right] \begin{array}{l} \rightarrow \\ \\ \frac{4}{3}R_3 \rightarrow R_3 \\ \\ \end{array}$$

$$\left[\begin{array}{ccccccc|c} -1 & 0 & 0 & 1 & -6 & 0 & 0 & 18{,}000 \\ \frac{1}{4} & \frac{1}{2} & 1 & 0 & \frac{1}{4} & 0 & 0 & 3{,}000 \\ 1 & \frac{2}{3} & 0 & 0 & -\frac{1}{3} & \frac{4}{3} & 0 & 8{,}000 \\ \hline -1 & 0 & 0 & 0 & 2 & 0 & 1 & 24{,}000 \end{array}\right] \begin{array}{l} R_3 + R_1 \rightarrow R_1 \\ -\frac{1}{4}R_3 + R_2 \rightarrow R_2 \\ \rightarrow \\ R_3 + R_4 \rightarrow R_4 \end{array} \left[\begin{array}{ccccccc|c} 0 & \frac{2}{3} & 0 & 1 & -\frac{19}{3} & \frac{4}{3} & 0 & 26{,}000 \\ 0 & \frac{1}{3} & 1 & 0 & \frac{1}{3} & -\frac{1}{3} & 0 & 1{,}000 \\ 1 & \frac{2}{3} & 0 & 0 & -\frac{1}{3} & \frac{4}{3} & 0 & 8{,}000 \\ \hline 0 & \frac{2}{3} & 0 & 0 & \frac{5}{3} & \frac{4}{3} & 1 & 32{,}000 \end{array}\right]$$

Maximum profit is \$32,000 with 8000 x_1 units and 1000 x_3 units.

61. Objective function: $f = 500x_1 + 800x_2 + 1150x_3$

Constraints: $x_1 + 2x_2 + 3x_3 \le 250,\ x_1 \le 26,\ x_2 \le 40,\ x_3 \le 60$

$$\left[\begin{array}{cccccccc|c} 1 & 2 & 3 & 1 & 0 & 0 & 0 & 0 & 250 \\ 1 & 0 & 0 & 0 & 1 & 0 & 0 & 0 & 26 \\ 0 & 1 & 0 & 0 & 0 & 1 & 0 & 0 & 40 \\ 0 & 0 & \mathbf{1} & 0 & 0 & 0 & 1 & 0 & 60 \\ \hline -500 & -800 & -1150 & 0 & 0 & 0 & 0 & 1 & 0 \end{array}\right] \begin{array}{l} -3R_4 + R_1 \to R_1 \\ \\ \to \\ \\ 1150R_4 + R_5 \to R_5 \end{array}$$

$$\left[\begin{array}{cccccccc|c} 1 & 2 & 0 & 1 & 0 & 0 & -\frac{3}{2} & 0 & 35 \\ 1 & 0 & 0 & 0 & 1 & 0 & 0 & 0 & 26 \\ 0 & 1 & 0 & 0 & 0 & 1 & 0 & 0 & 40 \\ 0 & 0 & 1 & 0 & 0 & 0 & 1 & 0 & 60 \\ \hline -500 & -800 & 0 & 0 & 0 & 0 & 1150 & 1 & 69{,}000 \end{array}\right] \begin{array}{l} \frac{1}{2}R_1 \to R_1 \\ \\ \to \\ \\ \\ \end{array}$$

$$\left[\begin{array}{cccccccc|c} \frac{1}{2} & \mathbf{1} & 0 & \frac{1}{2} & 0 & 0 & -\frac{3}{2} & 0 & 35 \\ 1 & 0 & 0 & 0 & 1 & 0 & 0 & 0 & 26 \\ 0 & 1 & 0 & 0 & 0 & 1 & 0 & 0 & 40 \\ 0 & 0 & 1 & 0 & 0 & 0 & 1 & 0 & 60 \\ \hline -500 & -800 & 0 & 0 & 0 & 0 & 1150 & 1 & 69{,}000 \end{array}\right] \begin{array}{l} \to \\ \\ -R_1 + R_3 \to R_3 \\ \\ 800R_1 + R_5 \to R_5 \end{array}$$

$$\left[\begin{array}{cccccccc|c} \frac{1}{2} & 1 & 0 & \frac{1}{2} & 0 & 0 & -\frac{3}{2} & 0 & 35 \\ \mathbf{1} & 0 & 0 & 0 & 1 & 0 & 0 & 0 & 26 \\ -\frac{1}{2} & 0 & 0 & -\frac{1}{2} & 0 & 1 & \frac{3}{2} & 0 & 5 \\ 0 & 0 & 1 & 0 & 0 & 0 & 1 & 0 & 60 \\ \hline -100 & 0 & 0 & 400 & 0 & 0 & -50 & 1 & 97{,}000 \end{array}\right] \begin{array}{l} -\frac{1}{2}R_2 + R_1 \to R_1 \\ \to \\ \frac{1}{2}R_2 + R_3 \to R_3 \\ \\ 100R_2 + R_5 \to R_5 \end{array}$$

$$\left[\begin{array}{cccccccc|c} 0 & 1 & 0 & \frac{1}{2} & -\frac{1}{2} & 0 & -\frac{3}{2} & 0 & 22 \\ 1 & 0 & 0 & 0 & 1 & 0 & 0 & 0 & 26 \\ 0 & 0 & 0 & -\frac{1}{2} & \frac{1}{2} & 1 & \frac{3}{2} & 0 & 18 \\ 0 & 0 & 1 & 0 & 0 & 0 & 1 & 0 & 60 \\ \hline 0 & 0 & 0 & 400 & 100 & 0 & -50 & 1 & 99{,}600 \end{array}\right] \begin{array}{l} \\ \to \\ \frac{2}{3}R_3 \to R_3 \\ \\ \\ \end{array}$$

$$\left[\begin{array}{cccccccc|c} 0 & 1 & 0 & \frac{1}{2} & -\frac{1}{2} & 0 & -\frac{3}{2} & 0 & 22 \\ 1 & 0 & 0 & 0 & 1 & 0 & 0 & 0 & 26 \\ 0 & 0 & 0 & -\frac{1}{3} & \frac{1}{3} & \frac{2}{3} & \mathbf{1} & 0 & 12 \\ 0 & 0 & 1 & 0 & 0 & 0 & 1 & 0 & 60 \\ \hline 0 & 0 & 0 & 400 & 100 & 0 & -50 & 1 & 99{,}600 \end{array}\right] \begin{array}{l} \frac{3}{2}R_3 + R_1 \to R_1 \\ \to \\ \\ -R_3 + R_4 \to R_4 \\ 50R_3 + R_5 \to R_5 \end{array}$$

$$\left[\begin{array}{cccccccc|c} 0 & 1 & 0 & 0 & 0 & 1 & 0 & 0 & 40 \\ 1 & 0 & 0 & 0 & 1 & 0 & 0 & 0 & 26 \\ 0 & 0 & 0 & -\frac{1}{3} & \frac{1}{3} & \frac{2}{3} & 1 & 0 & 12 \\ 0 & 0 & 1 & \frac{1}{3} & -\frac{1}{3} & -\frac{2}{3} & 0 & 0 & 48 \\ \hline 0 & 0 & 0 & \frac{1150}{3} & \frac{350}{3} & \frac{100}{3} & 0 & 1 & 100{,}200 \end{array}\right]$$

Maximum revenue is \$100,200 by renting 26 1-BR, 40 2-BR, and 48 3-BR apartments.

Exercise 4.5

1. a. $\left[\begin{array}{cc|c} 5 & 2 & 10 \\ 1 & 2 & 6 \\ \hline 3 & 1 & g \end{array}\right]$ transpose: $\left[\begin{array}{cc|c} 5 & 1 & 3 \\ 2 & 2 & 1 \\ \hline 10 & 6 & g \end{array}\right]$

b. From the transpose in (a):
Maximize $f = 10x + 6y$
Constraints: $5x + y \le 3$
$2x + 2y \le 1$
$x, y > 0$

3. a. $\left[\begin{array}{cc|c} 1 & 1 & 9 \\ 1 & 3 & 15 \\ \hline 5 & 2 & g \end{array}\right]$ transpose: $\left[\begin{array}{cc|c} 1 & 1 & 5 \\ 1 & 3 & 2 \\ \hline 9 & 15 & g \end{array}\right]$

b. From the transpose in (a):
Maximize $f = 9x + 15y$
Constraints: $x + y \le 5$
$x + 3y \le 2$
$x, y \ge 0$

5. a. For the minimization problem, the last entries in the columns corresponding to the slack variables are the solution.
Minimum: $g = 252$; $y_1 = 8, y_2 = 2, y_3 = 0$

b. Maximum: $f = 252$; $x_1 = 5, x_2 = 0, x_3 = 9$

7. From problem 1 we obtain

$$\left[\begin{array}{ccccc|c} 5 & 1 & 1 & 0 & 0 & 3 \\ \mathbf{2} & 2 & 0 & 1 & 0 & 1 \\ \hline -10 & -6 & 0 & 0 & 1 & 0 \end{array}\right]$$

Use this matrix to solve problem 7.

$$\begin{array}{c} \\ \frac{1}{2}R_2 \to R_2 \\ \\ \end{array} \left[\begin{array}{ccccc|c} 5 & 1 & 1 & 0 & 0 & 3 \\ 1 & 1 & 0 & \frac{1}{2} & 0 & \frac{1}{2} \\ \hline -10 & -6 & 0 & 0 & 1 & 0 \end{array}\right]$$

$$\begin{array}{c} -5R_2 + R_1 \to R_1 \\ \to \\ 10R_2 + R_3 \to R_3 \end{array} \left[\begin{array}{ccccc|c} 0 & -4 & 1 & -\frac{5}{2} & 0 & \frac{1}{2} \\ 1 & 1 & 0 & \frac{1}{2} & 0 & \frac{1}{2} \\ \hline 0 & 4 & 0 & 5 & 1 & 5 \end{array}\right]$$

Minimum: $g = 5$ at $y_1 = 0$, $y_2 = 5$.

Maximum: $f = 5$ at $x_1 = \dfrac{1}{2}$, $x_2 = 0$.

9. From problem 3 we obtain:

$$\left[\begin{array}{ccccc|c} 1 & 1 & 1 & 0 & 0 & 5 \\ 1 & \mathbf{3} & 0 & 1 & 0 & 2 \\ \hline -9 & -15 & 0 & 0 & 1 & 0 \end{array}\right] \begin{array}{c} \\ \frac{1}{3}R_2 \to R_2 \\ \to \end{array} \left[\begin{array}{ccccc|c} 1 & 1 & 1 & 0 & 0 & 5 \\ \frac{1}{3} & 1 & 0 & \frac{1}{3} & 0 & \frac{2}{3} \\ \hline -9 & -15 & 0 & 0 & 1 & 0 \end{array}\right] \begin{array}{c} -R_2 + R_1 \to R_1 \\ \to \\ 15R_2 + R_3 \to R_3 \end{array}$$

$$\left[\begin{array}{ccccc|c} \frac{2}{3} & 0 & 1 & -\frac{1}{3} & 0 & \frac{13}{3} \\ \frac{1}{3} & 1 & 0 & \frac{1}{3} & 0 & \frac{2}{3} \\ \hline -4 & 0 & 0 & 5 & 1 & 10 \end{array}\right] \begin{array}{c} \\ 3R_2 \to R_2 \\ \to \end{array} \left[\begin{array}{ccccc|c} \frac{2}{3} & 0 & 1 & -\frac{1}{3} & 0 & \frac{13}{3} \\ \mathbf{1} & 3 & 0 & 1 & 0 & 2 \\ \hline -4 & 0 & 0 & 5 & 1 & 10 \end{array}\right] \begin{array}{c} -\frac{2}{3}R_2 + R_1 \to R_1 \\ \to \\ 4R_2 + R_3 \to R_3 \end{array}$$

$$\left[\begin{array}{ccccc|c} 0 & -2 & 1 & -1 & 0 & 3 \\ 1 & 3 & 0 & 1 & 0 & 2 \\ \hline 0 & 12 & 0 & 9 & 1 & 18 \end{array}\right]$$

Minimum: $g = 18$ at $y_1 = 0, y_2 = 9$. Maximum: $f = 18$ at $x_1 = 2, x_2 = 0$

11. Maximize $f = 11x_1 + 12x_2 + 6x_3$

Constraints: $4x_1 + 3x_2 + 3x_3 \le 3,\ \ x_1 + 2x_2 + x_3 \le 1,\ \ x_1, x_2, x_3 \ge 0$

$$\left[\begin{array}{cccccc|c} 4 & 3 & 3 & 1 & 0 & 0 & 3 \\ 1 & \mathbf{2} & 1 & 0 & 1 & 0 & 1 \\ \hline -11 & -12 & -6 & 0 & 0 & 1 & 0 \end{array}\right] \begin{array}{c} \to \\ \frac{1}{2}R_2 \to R_2 \\ \ \end{array} \left[\begin{array}{cccccc|c} 4 & 3 & 3 & 1 & 0 & 0 & 3 \\ \frac{1}{2} & 1 & \frac{1}{2} & 0 & \frac{1}{2} & 0 & \frac{1}{2} \\ \hline -11 & -12 & -6 & 0 & 0 & 1 & 0 \end{array}\right] \begin{array}{c} -3R_2 + R_1 \to R_1 \\ \to \\ 12R_2 + R_3 \to R_3 \end{array}$$

$$\left[\begin{array}{cccccc|c} \frac{5}{2} & 0 & \frac{3}{2} & 1 & -\frac{3}{2} & 0 & \frac{3}{2} \\ \frac{1}{2} & 1 & \frac{1}{2} & 0 & \frac{1}{2} & 0 & \frac{1}{2} \\ \hline -5 & 0 & 0 & 0 & 6 & 1 & 6 \end{array}\right] \begin{array}{c} \frac{2}{5}R_1 \to R_1 \\ \to \\ \ \end{array} \left[\begin{array}{cccccc|c} 1 & 0 & \frac{3}{5} & \frac{2}{5} & -\frac{3}{5} & 0 & \frac{3}{5} \\ \frac{1}{2} & 1 & \frac{1}{2} & 0 & \frac{1}{2} & 0 & \frac{1}{2} \\ \hline -5 & 0 & 0 & 0 & 6 & 1 & 6 \end{array}\right] \begin{array}{c} \to \\ -\frac{1}{2}R_1 + R_2 \to R_2 \\ 5R_1 + R_3 \to R_3 \end{array}$$

$$\left[\begin{array}{cccccc|c} 1 & 0 & \frac{3}{5} & \frac{2}{5} & -\frac{3}{5} & 0 & \frac{3}{5} \\ 0 & 1 & \frac{1}{5} & -\frac{1}{5} & \frac{4}{5} & 0 & \frac{1}{5} \\ \hline 0 & 0 & 3 & 2 & 3 & 1 & 9 \end{array}\right]$$

Maximum; $f = 9;\ x_1 = \dfrac{3}{5},\ x_2 = \dfrac{1}{5},\ x_3 = 0$

Minimum: $g = 9,\ y_1 = 2,\ y_2 = 3$

13. Maximize: $f = x_1 + 3x_2 + x_3$

Constraints: $x_1 + 4x_2 \le 12,\ \ 3x_1 + 6x_2 + 4x_3 \le 48,\ \ x_2 + x_3 \le 8,\ \ x_1, x_2, x_3 \ge 0$

$$\left[\begin{array}{ccccccc|c} 1 & \mathbf{4} & 0 & 1 & 0 & 0 & 0 & 12 \\ 3 & 6 & 4 & 0 & 1 & 0 & 0 & 48 \\ 0 & 1 & 1 & 0 & 0 & 1 & 0 & 8 \\ \hline -1 & -3 & -1 & 0 & 0 & 0 & 1 & 0 \end{array}\right] \begin{array}{c} \frac{1}{4}R_1 \to R_1 \\ \to \\ \ \\ \ \end{array} \left[\begin{array}{ccccccc|c} \frac{1}{4} & 1 & 0 & \frac{1}{4} & 0 & 0 & 0 & 3 \\ 3 & 6 & 4 & 0 & 1 & 0 & 0 & 48 \\ 0 & 1 & 1 & 0 & 0 & 1 & 0 & 8 \\ \hline -1 & -3 & -1 & 0 & 0 & 0 & 1 & 0 \end{array}\right] \begin{array}{c} \to \\ -6R_1 + R_2 \to R_2 \\ -R_1 + R_3 \to R_3 \\ 3R_1 + R_4 \to R_4 \end{array}$$

$$\left[\begin{array}{ccccccc|c} \frac{1}{4} & 1 & 0 & \frac{1}{4} & 0 & 0 & 0 & 3 \\ \frac{3}{2} & 0 & 4 & -\frac{3}{2} & 1 & 0 & 0 & 30 \\ -\frac{1}{4} & 0 & 1 & -\frac{1}{4} & 0 & 1 & 0 & 5 \\ \hline -\frac{1}{4} & 0 & -1 & \frac{3}{4} & 0 & 0 & 1 & 9 \end{array}\right] \begin{array}{c} \ \\ -4R_3 + R_2 \to R_2 \\ \to \\ R_3 + R_4 \to R_4 \end{array} \left[\begin{array}{ccccccc|c} \frac{1}{4} & 1 & 0 & \frac{1}{4} & 0 & 0 & 0 & 3 \\ \frac{5}{2} & 0 & 0 & -\frac{1}{2} & 1 & -4 & 0 & 10 \\ -\frac{1}{4} & 0 & 1 & -\frac{1}{4} & 0 & 1 & 0 & 5 \\ \hline -\frac{1}{2} & 0 & 0 & \frac{1}{2} & 0 & 1 & 1 & 14 \end{array}\right] \begin{array}{c} \ \\ \frac{2}{5}R_2 \to R_2 \\ \to \\ \ \end{array}$$

$$\left[\begin{array}{ccccccc|c} \frac{1}{4} & 1 & 0 & \frac{1}{4} & 0 & 0 & 0 & 3 \\ 1 & 0 & 0 & -\frac{1}{5} & \frac{2}{5} & -\frac{8}{5} & 0 & 4 \\ -\frac{1}{4} & 0 & 1 & -\frac{1}{4} & 0 & 1 & 0 & 5 \\ \hline -\frac{1}{2} & 0 & 0 & \frac{1}{2} & 0 & 1 & 1 & 14 \end{array}\right] \begin{array}{c} -\frac{1}{4}R_2 + R_1 \to R_1 \\ \to \\ \frac{1}{4}R_2 + R_3 \to R_3 \\ \frac{1}{2}R_2 + R_4 \to R_4 \end{array} \left[\begin{array}{ccccccc|c} 0 & 1 & 0 & \frac{3}{10} & -\frac{1}{10} & \frac{2}{5} & 0 & 2 \\ 1 & 0 & 0 & -\frac{1}{5} & \frac{2}{5} & -\frac{8}{5} & 0 & 4 \\ 0 & 0 & 1 & -\frac{3}{10} & \frac{1}{10} & \frac{3}{5} & 0 & 6 \\ \hline 0 & 0 & 0 & \frac{2}{5} & \frac{1}{5} & \frac{1}{5} & 1 & 16 \end{array}\right]$$

Minimum: $g = 16$ at $y_1 = \dfrac{2}{5},\ y_2 = \dfrac{1}{5}$, and $y_3 = \dfrac{1}{5}$. Maximum: $f = 16$ at $x_1 = 4,\ x_2 = 2,\ x_3 = 6$.

15. $\left[\begin{array}{cc|c} 2 & 1 & 11 \\ 1 & 3 & 11 \\ 1 & 4 & 16 \\ \hline 2 & 10 & g \end{array}\right]$ transpose: $\left[\begin{array}{ccc|c} 2 & 1 & 1 & 2 \\ 1 & 3 & 4 & 10 \\ \hline 11 & 11 & 16 & g \end{array}\right]$

$$\left[\begin{array}{cccccc|c} 2 & 1 & \mathbf{1} & 1 & 0 & 0 & 2 \\ 1 & 3 & 4 & 0 & 1 & 0 & 10 \\ \hline -11 & -11 & -16 & 0 & 0 & 1 & 0 \end{array}\right] \begin{array}{c} \to \\ -4R_1 + R_2 \to R_2 \\ 16R_1 + R_3 \to R_3 \end{array} \left[\begin{array}{cccccc|c} 2 & 1 & 1 & 1 & 0 & 0 & 2 \\ -7 & -1 & 0 & -4 & 1 & 0 & 2 \\ \hline 21 & 5 & 0 & 16 & 0 & 1 & 32 \end{array}\right]$$

Minimum is 32 at $x = 16,\ y = 0$.

17. $\left[\begin{array}{ccc|c} 1 & 1 & 1 & 3 \\ 0 & 1 & 2 & 2 \\ 1 & 0 & 0 & 2 \\ \hline 8 & 7 & 12 & g \end{array}\right]$ transpose: $\left[\begin{array}{ccc|c} 1 & 0 & 1 & 8 \\ 1 & 1 & 0 & 7 \\ 1 & 2 & 0 & 12 \\ \hline 3 & 2 & 2 & g \end{array}\right]$

$$\left[\begin{array}{ccccccc|c} 1 & 0 & 1 & 1 & 0 & 0 & 0 & 8 \\ \mathbf{1} & 1 & 0 & 0 & 1 & 0 & 0 & 7 \\ 1 & 2 & 0 & 0 & 0 & 1 & 0 & 12 \\ \hline -3 & -2 & -2 & 0 & 0 & 0 & 1 & 0 \end{array}\right] \begin{array}{l} -R_2+R_1 \to R_1 \\ \to \\ -R_2+R_3 \to R_3 \\ 3R_2+R_4 \to R_4 \end{array} \left[\begin{array}{ccccccc|c} 0 & -1 & \mathbf{1} & 1 & -1 & 0 & 0 & 1 \\ 1 & 1 & 0 & 0 & 1 & 0 & 0 & 7 \\ 0 & 1 & 0 & 0 & -1 & 1 & 0 & 5 \\ \hline 0 & 1 & -2 & 0 & 3 & 0 & 1 & 21 \end{array}\right] \begin{array}{l} \\ \to \\ \\ 2R_1+R_4 \to R_4 \end{array}$$

$$\left[\begin{array}{ccccccc|c} 0 & -1 & 1 & 1 & -1 & 0 & 0 & 1 \\ 1 & 1 & 0 & 0 & 1 & 0 & 0 & 7 \\ 0 & \mathbf{1} & 0 & 0 & -1 & 1 & 0 & 5 \\ \hline 0 & -1 & 0 & 2 & 1 & 0 & 1 & 23 \end{array}\right] \begin{array}{l} R_3+R_1 \to R_1 \\ -R_3+R_2 \to R_2 \\ \to \\ R_3+R_4 \to R_4 \end{array} \left[\begin{array}{ccccccc|c} 0 & 0 & 1 & 1 & -2 & 1 & 0 & 6 \\ 1 & 0 & 0 & 0 & 2 & -1 & 0 & 2 \\ 0 & 1 & 0 & 0 & -1 & 1 & 0 & 5 \\ \hline 0 & 0 & 0 & 2 & 0 & 1 & 1 & 28 \end{array}\right]$$

Minimum: $g = 28$ at $x = 2, y = 0, z = 1$

19. $\left[\begin{array}{ccc|c} 1 & 2 & 1 & 16 \\ 1 & 5 & 2 & 18 \\ 2 & 5 & 3 & 38 \\ \hline 40 & 90 & 30 & g \end{array}\right]$ transpose: $\left[\begin{array}{ccc|c} 1 & 1 & 2 & 40 \\ 2 & 5 & 5 & 90 \\ 1 & 2 & 3 & 30 \\ \hline 16 & 18 & 38 & g \end{array}\right]$

$$\left[\begin{array}{ccccccc|c} 1 & 1 & 2 & 1 & 0 & 0 & 0 & 40 \\ 2 & 5 & 5 & 0 & 1 & 0 & 0 & 90 \\ 1 & 2 & \mathbf{3} & 0 & 0 & 1 & 0 & 30 \\ \hline -16 & -18 & -38 & 0 & 0 & 0 & 1 & 0 \end{array}\right] \begin{array}{l} \to \\ \\ \frac{1}{3}R_3 \to R_3 \\ \\ \end{array} \left[\begin{array}{ccccccc|c} 1 & 1 & 2 & 1 & 0 & 0 & 0 & 40 \\ 2 & 5 & 5 & 0 & 1 & 0 & 0 & 90 \\ \frac{1}{3} & \frac{2}{3} & 1 & 0 & 0 & \frac{1}{3} & 0 & 10 \\ \hline -16 & -18 & -38 & 0 & 0 & 0 & 1 & 0 \end{array}\right] \begin{array}{l} -2R_3+R_1 \to R_1 \\ -5R_3+R_2 \to R_2 \\ \to \\ 38R_3+R_4 \to R_4 \end{array}$$

$$\left[\begin{array}{ccccccc|c} \frac{1}{3} & -\frac{1}{3} & 0 & 1 & 0 & -\frac{2}{3} & 0 & 20 \\ \frac{1}{3} & \frac{5}{3} & 0 & 0 & 1 & -\frac{5}{3} & 0 & 40 \\ \frac{1}{3} & \frac{2}{3} & 1 & 0 & 0 & \frac{1}{3} & 0 & 10 \\ \hline -\frac{10}{3} & \frac{22}{3} & 0 & 0 & 0 & \frac{38}{3} & 1 & 380 \end{array}\right] \begin{array}{l} \to \\ \\ 3R_3 \to R_3 \\ \\ \end{array} \left[\begin{array}{ccccccc|c} \frac{1}{3} & -\frac{1}{3} & 0 & 1 & 0 & -\frac{2}{3} & 0 & 20 \\ \frac{1}{3} & \frac{5}{3} & 0 & 0 & 1 & -\frac{5}{3} & 0 & 40 \\ 1 & 2 & 3 & 0 & 0 & 1 & 0 & 30 \\ \hline -\frac{10}{3} & \frac{22}{3} & 0 & 0 & 0 & \frac{38}{3} & 1 & 380 \end{array}\right] \begin{array}{l} -\frac{1}{3}R_3+R_1 \to R_1 \\ -\frac{1}{3}R_3+R_2 \to R_2 \\ \to \\ \frac{10}{3}R_3+R_4 \to R_4 \end{array}$$

$$\left[\begin{array}{ccccccc|c} 0 & -1 & -1 & 1 & 0 & -1 & 0 & 10 \\ 0 & 1 & -1 & 0 & 1 & -2 & 0 & 30 \\ 1 & 2 & 3 & 0 & 0 & 1 & 0 & 30 \\ \hline 0 & 14 & 10 & 0 & 0 & 16 & 1 & 480 \end{array}\right]$$

Minimum is 480 at $y_1 = 0, y_2 = 0, y_3 = 16$.

21. **a.** Minimize: $g = 120y_1 + 50y_2$

Constraints: $3y_1 + y_2 \geq 40,\ \ 2y_1 + y_2 \geq 20$

b. $$\left[\begin{array}{ccccc|c} \mathbf{3} & 2 & 1 & 0 & 0 & 120 \\ 1 & 1 & 0 & 1 & 0 & 50 \\ \hline -40 & -20 & 0 & 0 & 1 & 0 \end{array}\right] \begin{array}{l} \frac{1}{3}R_1 \to R_1 \\ \to \\ \\ \end{array}$$

$$\left[\begin{array}{ccccc|c} \mathbf{1} & \frac{2}{3} & \frac{1}{3} & 0 & 0 & 40 \\ 1 & 1 & 0 & 1 & 0 & 50 \\ \hline -40 & -20 & 0 & 0 & 1 & 0 \end{array}\right] \begin{array}{l} \to \\ -R_1+R_2 \to R_2 \\ 40R_1+R_3 \to R_3 \end{array} \left[\begin{array}{ccccc|c} 1 & \frac{2}{3} & \frac{1}{3} & 0 & 0 & 40 \\ 0 & \frac{1}{3} & -\frac{1}{3} & 1 & 0 & 10 \\ \hline 0 & \frac{20}{3} & \frac{40}{3} & 0 & 1 & 1600 \end{array}\right]$$

Primal: Maximum $f = 1600$ at $x_1 = 40, x_2 = 0$, Dual: Minimum $g = 1600$ at $y_1 = \frac{40}{3}, y_2 = 0$

23. x = hours for Line 1, y = hours for Line 2
Minimize: $C = 200x + 400y$
Constraints: $30x + 150y \geq 270 \rightarrow x + 5y \geq 9$
$40x + 40y \geq 200 \rightarrow x + y \geq 5$
We must maximize for:

$$\left[\begin{array}{ccccc|c} 1 & 1 & 1 & 0 & 0 & 200 \\ 5 & 1 & 0 & 1 & 0 & 400 \\ \hline -9 & -5 & 0 & 0 & 1 & 0 \end{array}\right] \begin{array}{c} \\ \frac{1}{5}R_2 \rightarrow R_2 \\ \\ \end{array} \left[\begin{array}{ccccc|c} 1 & 1 & 1 & 0 & 0 & 200 \\ 1 & \frac{1}{5} & 0 & \frac{1}{5} & 0 & 80 \\ \hline -9 & -5 & 0 & 0 & 1 & 0 \end{array}\right] \begin{array}{c} -R_2 + R_1 \rightarrow R_1 \\ \rightarrow \\ 9R_2 + R_3 \rightarrow R_3 \end{array}$$

$$\left[\begin{array}{ccccc|c} 0 & \frac{4}{5} & 1 & -\frac{1}{5} & 0 & 120 \\ 1 & \frac{1}{5} & 0 & \frac{1}{5} & 0 & 80 \\ \hline 0 & -\frac{16}{5} & 0 & \frac{9}{5} & 1 & 720 \end{array}\right] \begin{array}{c} -\frac{5}{4}R_1 \rightarrow R_1 \\ \rightarrow \\ \\ \end{array} \left[\begin{array}{ccccc|c} 0 & 1 & \frac{5}{4} & -\frac{1}{4} & 0 & 150 \\ 1 & \frac{1}{5} & 0 & \frac{1}{5} & 0 & 80 \\ \hline 0 & -\frac{16}{5} & 0 & \frac{9}{5} & 1 & 720 \end{array}\right] \begin{array}{c} \rightarrow \\ -\frac{1}{5}R_1 + R_2 \rightarrow R_2 \\ \frac{16}{5}R_1 + R_3 \rightarrow R_3 \end{array}$$

$$\left[\begin{array}{ccccc|c} 0 & 1 & \frac{5}{4} & -\frac{1}{4} & 0 & 150 \\ 1 & 0 & -\frac{1}{4} & \frac{1}{4} & 0 & 50 \\ \hline 0 & 0 & 4 & 1 & 1 & 1200 \end{array}\right]$$

The minimum cost is \$1200 for 4 hours on Line 1, and 1 hour on Line 2.

25. Minimize: $\text{Cost} = 1000A + 3000B + 4000C$
Constraints: $200A + 200B + 400C \geq 2000 \rightarrow A + B + 2C \geq 10$
$100A + 200B + 100C \geq 1200 \rightarrow A + 2B + C \geq 12$
We must maximize for:

$$\left[\begin{array}{cccccc|c} 1 & 1 & 1 & 0 & 0 & 0 & 1000 \\ 1 & 2 & 0 & 1 & 0 & 0 & 3000 \\ 2 & 1 & 0 & 0 & 1 & 0 & 4000 \\ \hline -10 & -12 & 0 & 0 & 0 & 1 & 0 \end{array}\right] \begin{array}{c} \rightarrow \\ -2R_1 + R_2 \rightarrow R_2 \\ -R_1 + R_3 \rightarrow R_3 \\ 12R_1 + R_4 \rightarrow R_4 \end{array} \left[\begin{array}{cccccc|c} 1 & 1 & 1 & 0 & 0 & 0 & 1000 \\ -1 & 0 & -2 & 1 & 0 & 0 & 1000 \\ 1 & 0 & -1 & 0 & 1 & 0 & 3000 \\ \hline 2 & 0 & 12 & 0 & 0 & 1 & 12000 \end{array}\right]$$

The minimum cost of \$12,000 is obtained by using facility A for 12 weeks, and 0 weeks for B and C.

27. x = days for Factory 1; y = days for Factory 2
Minimize: $C = 10{,}000x + 20{,}000y$
Constraints: $80x + 20y \geq 1600 \rightarrow 4x + y \geq 80$
$10x + 10y \geq 500 \rightarrow x + y \geq 50$
$50x + 20y \geq 1900 \rightarrow 5x + 2y \geq 190$
We must maximize for:

$$\left[\begin{array}{cccccc|c} 4 & 1 & 5 & 1 & 0 & 0 & 10{,}000 \\ 1 & 1 & 2 & 0 & 1 & 0 & 20{,}000 \\ \hline -80 & -50 & -190 & 0 & 0 & 1 & 0 \end{array}\right] \begin{array}{c} \frac{1}{5}R_1 \rightarrow R_1 \\ \rightarrow \\ \rightarrow \end{array} \left[\begin{array}{cccccc|c} \frac{4}{5} & \frac{1}{5} & 1 & \frac{1}{5} & 0 & 0 & 2000 \\ 1 & 1 & 2 & 0 & 1 & 0 & 20{,}000 \\ \hline -80 & -50 & -190 & 0 & 0 & 1 & 0 \end{array}\right] \begin{array}{c} \rightarrow \\ -2R_1 + R_2 \rightarrow R_2 \\ 190R_1 + R_3 \rightarrow R_3 \end{array}$$

$$\left[\begin{array}{cccccc|c} \frac{4}{5} & \frac{1}{5} & 1 & \frac{1}{5} & 0 & 0 & 2000 \\ -\frac{3}{5} & \frac{3}{5} & 0 & -\frac{2}{5} & 1 & 0 & 16{,}000 \\ \hline 72 & -12 & 0 & 38 & 0 & 1 & 380{,}000 \end{array}\right] \begin{array}{c} 5R_1 \rightarrow R_1 \\ \\ \\ \end{array} \left[\begin{array}{cccccc|c} 4 & 1 & 5 & 1 & 0 & 0 & 10{,}000 \\ -\frac{3}{5} & \frac{3}{5} & 0 & -\frac{2}{5} & 1 & 0 & 16{,}000 \\ \hline 72 & -12 & 0 & 38 & 0 & 1 & 380{,}000 \end{array}\right] \begin{array}{c} -\frac{3}{5}R_1 + R_2 \rightarrow R_2 \\ \rightarrow \\ 12R_1 + R_3 \rightarrow R_3 \end{array}$$

$$\left[\begin{array}{cccccc|c} 4 & 1 & 5 & 1 & 0 & 0 & 10{,}000 \\ -3 & 0 & -3 & -1 & 1 & 0 & 10{,}000 \\ \hline 120 & 0 & 60 & 50 & 0 & 1 & 500{,}000 \end{array}\right]$$

Minimum cost is \$500,000 using 50 days at Factory 1 and 0 days at Factory 2.

29. Minimize: $C = 500T + 100R$

Constraints: $0.9T + 0.6R \ge 63$

$T + R \ge 90$

We must maximize for:

$$\left[\begin{array}{ccccc|c} 0.9 & 1 & 1 & 0 & 0 & 500 \\ 0.6 & \mathbf{1} & 0 & 1 & 0 & 100 \\ \hline -63 & -90 & 0 & 0 & 1 & 0 \end{array}\right] \begin{array}{c} -R_2 + R_1 \to R_1 \\ \to \\ 90R_2 + R_3 \to R_3 \end{array} \left[\begin{array}{ccccc|c} 0.3 & 0 & 1 & -1 & 0 & 400 \\ \mathbf{0.6} & 1 & 0 & 1 & 0 & 100 \\ \hline -9 & 0 & 0 & 90 & 1 & 9000 \end{array}\right] \begin{array}{c} \to \\ \tfrac{5}{3}R_2 \to R_2 \\ \end{array}$$

$$\left[\begin{array}{ccccc|c} 0.3 & 0 & 1 & -1 & 0 & 400 \\ \mathbf{1} & \frac{5}{3} & 0 & \frac{5}{3} & 0 & \frac{500}{3} \\ \hline -9 & 0 & 0 & 90 & 1 & 9000 \end{array}\right] \begin{array}{c} -\frac{3}{10}R_2 + R_1 \to R_1 \\ \to \\ 9R_2 + R_3 \to R_3 \end{array} \left[\begin{array}{ccccc|c} 0 & -\frac{1}{2} & 1 & -\frac{3}{2} & 0 & 350 \\ 1 & \frac{5}{3} & 0 & \frac{5}{3} & 0 & \frac{500}{3} \\ \hline 0 & 15 & 0 & 105 & 1 & 10{,}500 \end{array}\right]$$

Minimum cost is \$10,500 using 0 minutes of TV time and 105 minutes of radio time.

31. Minimize: $\text{Cost} = 120x_1 + 140x_2 + 126x_3$

Constraints: $10x_1 + 5x_2 + 6x_3 \ge 230$

$5x_1 + 8x_2 + 6x_3 \ge 240$

$6x_1 + 6x_2 + 6x_3 \ge 210 \quad \to \quad x_1 + x_2 + x_3 \ge 35$

$$\left[\begin{array}{ccccccc|c} 10 & 5 & 1 & 1 & 0 & 0 & 0 & 120 \\ 5 & \mathbf{8} & 1 & 0 & 1 & 0 & 0 & 140 \\ 6 & 6 & 1 & 0 & 0 & 1 & 0 & 126 \\ \hline -230 & -240 & -35 & 0 & 0 & 0 & 1 & 0 \end{array}\right] \begin{array}{c} \to \\ \frac{1}{8}R_2 \to R_2 \\ \\ \end{array} \left[\begin{array}{ccccccc|c} 10 & 5 & 1 & 1 & 0 & 0 & 0 & 120 \\ 0.625 & \mathbf{1} & 0.125 & 0 & 0.125 & 0 & 0 & 17.5 \\ 6 & 6 & 1 & 0 & 0 & 1 & 0 & 126 \\ \hline -230 & -240 & -35 & 0 & 0 & 0 & 1 & 0 \end{array}\right]$$

$$\begin{array}{c} -5R_2 + R_1 \to R_1 \\ \to \\ -6R_2 + R_3 \to R_3 \\ 240R_2 + R_4 \to R_4 \end{array} \left[\begin{array}{ccccccc|c} \mathbf{6.875} & 0 & 0.375 & 1 & -0.625 & 0 & 0 & 32.5 \\ 0.625 & 1 & 0.125 & 0 & 0.125 & 0 & 0 & 17.5 \\ 2.250 & 0 & 0.250 & 0 & -0.750 & 1 & 0 & 21 \\ \hline -80 & 0 & -5 & 0 & 30 & 0 & 1 & 4200 \end{array}\right] \begin{array}{c} \frac{1}{6.875}R_1 \to R_1 \\ \\ \to \\ \end{array}$$

$$\left[\begin{array}{ccccccc|c} \mathbf{1} & 0 & 0.0545 & 0.1455 & -0.0909 & 0 & 0 & 4.7273 \\ 0.625 & 1 & 0.125 & 0 & 0.125 & 0 & 0 & 17.5 \\ 2.250 & 0 & 0.250 & 0 & -0.750 & 1 & 0 & 21 \\ \hline -80 & 0 & -5 & 0 & 30 & 0 & 1 & 4200 \end{array}\right] \begin{array}{c} \to \\ -0.625R_1 + R_2 \to R_2 \\ -2.25R_1 + R_3 \to R_3 \\ 80R_1 + R_4 \to R_4 \end{array}$$

$$\left[\begin{array}{ccccccc|c} 1 & 0 & 0.0545 & 0.1455 & -0.0909 & 0 & 0 & 4.7273 \\ 0 & 1 & 0.0909 & -0.0909 & 0.1818 & 0 & 0 & 14.5455 \\ 0 & 0 & \mathbf{0.1273} & -0.3273 & -0.5455 & 1 & 0 & 10.3636 \\ \hline 0 & 0 & -0.6363 & 11.6364 & 22.7273 & 0 & 1 & 4578.1818 \end{array}\right] \begin{array}{c} \to \\ \\ \frac{1}{0.1273}R_3 \to R_3 \\ \end{array}$$

$$\left[\begin{array}{ccccccc|c} 1 & 0 & 0.0545 & 0.1455 & -0.0909 & 0 & 0 & 4.7273 \\ 0 & 1 & 0.0909 & -0.0909 & 0.1818 & 0 & 0 & 14.5455 \\ 0 & 0 & \mathbf{1} & -2.5709 & -4.2848 & 7.8558 & 0 & 81.4111 \\ \hline 0 & 0 & -0.6363 & 11.6364 & 22.7273 & 0 & 1 & 4578.1818 \end{array}\right] \begin{array}{c} -0.0545R_3 + R_1 \to R_1 \\ -0.0909R_3 + R_2 \to R_2 \\ \to \\ 0.6363R_3 + R_4 \to R_4 \end{array}$$

$$\left[\begin{array}{ccccccc|c} 1 & 0 & 0 & 0.2856 & 0.1426 & -0.4281 & 0 & 0.2904 \\ 0 & 1 & 0 & 0.1428 & 0.5713 & -0.7141 & 0 & 7.1452 \\ 0 & 0 & 1 & -2.5709 & -4.2848 & 7.8555 & 0 & 81.4111 \\ \hline 0 & 0 & 0 & 10.0005 & 20.0009 & 4.9984 & 1 & 4629.98 \end{array}\right]$$

Since we used 4-decimal accuracy, we round off to obtain a minimum cost of \$4630 for 10 Georgia packages, 20 Union packages, and 5 Pacific packages.

33. x = ounces of Food I, y = ounces of Food II, z = ounces of Food III
Minimize: $g = x + 5y + 2z$
Constraints: $2x + 2y + 2z \geq 24 \quad \rightarrow \quad x + y + z \geq 12$
$x + y + z \geq 12$
$x + 5y + z \geq 16$
We must maximize for:

$$\left[\begin{array}{cccccc|c} 1 & 1 & 1 & 0 & 0 & 0 & 1 \\ 1 & 5 & 0 & 1 & 0 & 0 & 5 \\ 1 & 1 & 0 & 0 & 1 & 0 & 2 \\ \hline -12 & -16 & 0 & 0 & 0 & 1 & 0 \end{array}\right] \begin{array}{c} \rightarrow \\ -5R_1 + R_2 \rightarrow R_2 \\ -R_1 + R_3 \rightarrow R_3 \\ 16R_1 + R_4 \rightarrow R_4 \end{array} \left[\begin{array}{cccccc|c} 1 & 1 & 1 & 0 & 0 & 0 & 1 \\ -4 & 0 & -5 & 1 & 0 & 0 & 0 \\ 0 & 0 & -1 & 0 & 1 & 0 & 1 \\ \hline 4 & 0 & 16 & 0 & 0 & 1 & 16 \end{array}\right]$$

Minimum cost is \$16 with 16 ounces of Food I, and 0 ounces of Food II and Food III.

Exercise 4.6

1. $3x - y \geq 5$
$-3x + y \leq -5$

3. $y \geq 40 - 6x$
$-6x - y \leq -40$

5.
$$\left[\begin{array}{ccccc|c} 1 & 2 & 1 & 0 & 0 & 6 \\ -4 & -2 & 0 & 1 & 0 & -12 \\ \hline -4 & -5 & 0 & 0 & 1 & 0 \end{array}\right]$$

7.
$$\left[\begin{array}{ccccc|c} 6 & 4 & 1 & 0 & 0 & 24 \\ -5 & -2 & 0 & 1 & 0 & -16 \\ \hline 2 & 2 & 0 & 0 & 1 & 0 \end{array}\right]$$

9. Maximize: $f = 2x + 3y$
Constraints: $7x + 4y \leq 28$
$-3x + y \geq 2$
$x,\ y \geq 0$

a. $f = 2x + 3y$
$7x + 4y \leq 28$
$3x - y \leq -2$
$x,\ y \geq 0$

b.
$$\begin{array}{cccccc|c} x & y & s_1 & s_2 & s_3 & f & \end{array}$$
$$\left[\begin{array}{ccccc|c} 7 & 4 & 1 & 0 & 0 & 28 \\ 3 & \mathbf{-1} & 0 & 1 & 0 & -2 \\ \hline -2 & -3 & 0 & 0 & 1 & 0 \end{array}\right]$$

11. Minimize: $g = 3x + 8y$
Constraints: $4x - 5y \leq 50$
$x + y \leq 80$
$-x + 2y \geq 4$
$x,\ y \geq 0$

a. $-g = -3x - 8y$ (to maximize)
$4x - 5y \leq 50$
$x + y \leq 80$
$x - 2y \leq -4$
$x,\ y \geq 0$

b.
$$\begin{array}{ccccccc} x & y & s_1 & s_2 & s_3 & -g & \end{array}$$
$$\left[\begin{array}{cccccc|c} 4 & -5 & 1 & 0 & 0 & 0 & 50 \\ 1 & 1 & 0 & 1 & 0 & 0 & 80 \\ 1 & \mathbf{-2} & 0 & 0 & 1 & 0 & -4 \\ \hline 3 & 8 & 0 & 0 & 0 & 1 & 0 \end{array}\right]$$

13. The minimum is $f = 120$ at $x = 6$, $y = 8$, $z = 12$.

15. Maximize: $f = 4x + y$

Constraints: $5x + 2y \le 84,\ 3x - 2y \le -4,\ x,\ y \ge 0$

$$\left[\begin{array}{ccccc|c} 5 & 2 & 1 & 0 & 0 & 84 \\ 3 & \mathbf{-2} & 0 & 1 & 0 & -4 \\ \hline -4 & -1 & 0 & 0 & 1 & 0 \end{array}\right] -\tfrac{1}{2}R_2 \to R_2 \qquad \left[\begin{array}{ccccc|c} 5 & 2 & 1 & 0 & 0 & 84 \\ -\frac{3}{2} & \mathbf{1} & 0 & -\frac{1}{2} & 0 & 2 \\ \hline -4 & -1 & 0 & 0 & 1 & 0 \end{array}\right] \begin{array}{c} -2R_2 + R_1 \to R_1 \\ \to \\ R_2 + R_3 \to R_3 \end{array}$$

$$\left[\begin{array}{ccccc|c} \mathbf{8} & 0 & 1 & 1 & 0 & 80 \\ -\frac{3}{2} & 1 & 0 & -\frac{1}{2} & 0 & 2 \\ \hline -\frac{11}{2} & 0 & 0 & -\frac{1}{2} & 1 & 2 \end{array}\right] \tfrac{1}{8}R_1 \to R_1 \qquad \left[\begin{array}{ccccc|c} \mathbf{1} & 0 & \frac{1}{8} & \frac{1}{8} & 0 & 10 \\ -\frac{3}{2} & 1 & 0 & -\frac{1}{2} & 0 & 2 \\ \hline -\frac{11}{2} & 0 & 0 & -\frac{1}{2} & 1 & 2 \end{array}\right] \begin{array}{c} \to \\ \frac{3}{2}R_1 + R_2 \to R_2 \\ \frac{11}{2}R_1 + R_3 \to R_3 \end{array}$$

$$\left[\begin{array}{ccccc|c} 1 & 0 & \frac{1}{8} & \frac{1}{8} & 0 & 10 \\ 0 & 1 & \frac{3}{16} & -\frac{5}{16} & 0 & 17 \\ \hline 0 & 0 & \frac{11}{16} & \frac{3}{16} & 1 & 57 \end{array}\right]$$

A maximum of 57 occurs when $x = 10$ and $y = 17$.

17. Maximize: $-f = -2x - 3y$

Constraints:
$$\begin{aligned} -x &\le -5 \\ y &\le 13 \\ x - y &\le -2 \\ x,\ y &\ge 0 \end{aligned}$$

$$\left[\begin{array}{cccccc|c} \mathbf{-1} & 0 & 1 & 0 & 0 & 0 & -5 \\ 0 & 1 & 0 & 1 & 0 & 0 & 13 \\ 1 & -1 & 0 & 0 & 1 & 0 & -2 \\ \hline 2 & 3 & 0 & 0 & 0 & 1 & 0 \end{array}\right] \begin{array}{c} -R_1 \to R_1 \\ \to \\ \\ \\ \end{array} \qquad \left[\begin{array}{cccccc|c} 1 & 0 & -1 & 0 & 0 & 0 & 5 \\ 0 & 1 & 0 & 1 & 0 & 0 & 13 \\ 1 & -1 & 0 & 0 & 1 & 0 & -2 \\ \hline 2 & 3 & 0 & 0 & 0 & 1 & 0 \end{array}\right] \begin{array}{c} \to \\ \\ -R_1 + R_3 \to R_3 \\ -2R_1 + R_4 \to R_4 \end{array}$$

$$\left[\begin{array}{cccccc|c} 1 & 0 & -1 & 0 & 0 & 0 & 5 \\ 0 & 1 & 0 & 1 & 0 & 0 & 13 \\ 0 & \mathbf{-1} & 1 & 0 & 1 & 0 & -7 \\ \hline 0 & 3 & 2 & 0 & 0 & 1 & -10 \end{array}\right] \begin{array}{c} \to \\ \\ -R_3 \to R_3 \\ \\ \end{array} \qquad \left[\begin{array}{cccccc|c} 1 & 0 & -1 & 0 & 0 & 0 & 5 \\ 0 & 1 & 0 & 1 & 0 & 0 & 13 \\ 0 & 1 & -1 & 0 & -1 & 0 & 7 \\ \hline 0 & 3 & 2 & 0 & 0 & 1 & -10 \end{array}\right] \begin{array}{c} -R_3 + R_2 \to R_2 \\ \to \\ -3R_3 + R_4 \to R_4 \end{array}$$

$$\left[\begin{array}{cccccc|c} 1 & 0 & -1 & 0 & 0 & 0 & 5 \\ 0 & 0 & 1 & 1 & 1 & 0 & 6 \\ 0 & 1 & -1 & 0 & -1 & 0 & 7 \\ \hline 0 & 0 & 5 & 0 & 3 & 1 & -31 \end{array}\right]$$

Minimum is 31 at $x = 5$ and $y = 7$.

19. Maximize: $-f = -2x - y$

Constraints:
$$\begin{aligned} x &\le 12 \\ -x - 2y &\le -20 \\ -3x + 2y &\le 4 \\ x, y &\ge 0 \end{aligned}$$

$$\left[\begin{array}{cccccc|c} \mathbf{1} & 0 & 1 & 0 & 0 & 0 & 12 \\ -1 & -2 & 0 & 1 & 0 & 0 & -20 \\ -3 & 2 & 0 & 0 & 1 & 0 & 4 \\ \hline 2 & 1 & 0 & 0 & 0 & 1 & 0 \end{array}\right] \begin{array}{c} \to \\ R_1 + R_2 \to R_2 \\ 3R_1 + R_3 \to R_3 \\ -2R_1 + R_4 \to R_4 \end{array} \left[\begin{array}{cccccc|c} 1 & 0 & 1 & 0 & 0 & 0 & 12 \\ 0 & \mathbf{-2} & 1 & 1 & 0 & 0 & -8 \\ 0 & 2 & 3 & 0 & 1 & 0 & 40 \\ \hline 0 & 1 & -2 & 0 & 0 & 1 & -24 \end{array}\right] \begin{array}{c} \to \\ -\tfrac{1}{2}R_2 \to R_2 \\ \\ \\ \end{array}$$

$$\left[\begin{array}{cccccc|c} 1 & 0 & 1 & 0 & 0 & 0 & 12 \\ 0 & \mathbf{1} & -\tfrac{1}{2} & -\tfrac{1}{2} & 0 & 0 & 4 \\ 0 & 2 & 3 & 0 & 1 & 0 & 40 \\ \hline 0 & 1 & -2 & 0 & 0 & 1 & -24 \end{array}\right] \begin{array}{c} \to \\ \\ -2R_2 + R_3 \to R_3 \\ -R_2 + R_4 \to R_4 \end{array} \left[\begin{array}{cccccc|c} 1 & 0 & 1 & 0 & 0 & 0 & 12 \\ 0 & 1 & -\tfrac{1}{2} & -\tfrac{1}{2} & 0 & 0 & 4 \\ 0 & 0 & \mathbf{4} & 1 & 1 & 0 & 32 \\ \hline 0 & 0 & -\tfrac{3}{2} & \tfrac{1}{2} & 0 & 1 & -28 \end{array}\right] \begin{array}{c} \\ \\ \tfrac{1}{4}R_3 \to R_3 \\ \to \end{array}$$

$$\left[\begin{array}{cccccc|c} 1 & 0 & 1 & 0 & 0 & 0 & 12 \\ 0 & 1 & -\tfrac{1}{2} & -\tfrac{1}{2} & 0 & 0 & 4 \\ 0 & 0 & \mathbf{1} & \tfrac{1}{4} & \tfrac{1}{4} & 0 & 8 \\ \hline 0 & 0 & -\tfrac{3}{2} & \tfrac{1}{2} & 0 & 1 & -28 \end{array}\right] \begin{array}{c} -R_3 + R_1 \to R_1 \\ \to \\ \tfrac{1}{2}R_3 + R_2 \to R_2 \\ \tfrac{3}{2}R_3 + R_4 \to R_4 \end{array} \left[\begin{array}{cccccc|c} 1 & 0 & 0 & -\tfrac{1}{4} & -\tfrac{1}{4} & 0 & 4 \\ 0 & 1 & 0 & -\tfrac{3}{8} & \tfrac{1}{8} & 0 & 8 \\ 0 & 0 & 1 & \tfrac{1}{4} & \tfrac{1}{4} & 0 & 8 \\ \hline 0 & 0 & 0 & \tfrac{7}{8} & \tfrac{3}{8} & 1 & -16 \end{array}\right]$$

Minimum is 16 at $x = 4, y = 8$.

21. Maximize: $f = 5x + 2y$

Constraints: $y \le 20,\ 2x + y \le 32,\ x - 2y \le -4,\ x, y \ge 0$

$$\left[\begin{array}{cccccc|c} 0 & 1 & 1 & 0 & 0 & 0 & 20 \\ 2 & 1 & 0 & 1 & 0 & 0 & 32 \\ 1 & \mathbf{-2} & 0 & 0 & 1 & 0 & -4 \\ \hline -5 & -2 & 0 & 0 & 0 & 1 & 0 \end{array}\right] \begin{array}{c} \to \\ \\ -\tfrac{1}{2}R_3 \to R_3 \\ \\ \end{array} \left[\begin{array}{cccccc|c} 0 & 1 & 1 & 0 & 0 & 0 & 20 \\ 2 & 1 & 0 & 1 & 0 & 0 & 32 \\ -\tfrac{1}{2} & 1 & 0 & 0 & -\tfrac{1}{2} & 0 & 2 \\ \hline -5 & -2 & 0 & 0 & 0 & 1 & 0 \end{array}\right] \begin{array}{c} -R_3 + R_1 \to R_1 \\ -R_3 + R_2 \to R_2 \\ \to \\ 2R_3 + R_4 \to R_4 \end{array}$$

$$\left[\begin{array}{cccccc|c} \tfrac{1}{2} & 0 & 1 & 0 & \tfrac{1}{2} & 0 & 18 \\ \tfrac{5}{2} & 0 & 0 & 1 & \tfrac{1}{2} & 0 & 30 \\ -\tfrac{1}{2} & 1 & 0 & 0 & -\tfrac{1}{2} & 0 & 2 \\ \hline -6 & 0 & 0 & 0 & -1 & 1 & 4 \end{array}\right] \begin{array}{c} \to \\ \tfrac{2}{5}R_2 \to R_2 \\ \\ \\ \end{array} \left[\begin{array}{cccccc|c} \tfrac{1}{2} & 0 & 1 & 0 & \tfrac{1}{2} & 0 & 18 \\ 1 & 0 & 0 & \tfrac{2}{3} & \tfrac{1}{5} & 0 & 12 \\ -\tfrac{1}{2} & 1 & 0 & 0 & -\tfrac{1}{2} & 0 & 2 \\ \hline -6 & 0 & 0 & 0 & -1 & 1 & 4 \end{array}\right] \begin{array}{c} -\tfrac{1}{2}R_2 + R_1 \to R_1 \\ \to \\ \tfrac{1}{2}R_2 + R_3 \to R_3 \\ 6R_2 + R_4 \to R_4 \end{array}$$

$$\left[\begin{array}{cccccc|c} 0 & 0 & 1 & -\tfrac{1}{5} & \tfrac{2}{5} & 0 & 12 \\ 1 & 0 & 0 & \tfrac{2}{5} & \tfrac{1}{5} & 0 & 12 \\ 0 & 1 & 0 & \tfrac{1}{5} & -\tfrac{2}{5} & 0 & 8 \\ \hline 0 & 0 & 0 & \tfrac{12}{5} & \tfrac{1}{5} & 1 & 76 \end{array}\right]$$

Maximum is 76 at $x = 12, y = 8$.

23. Minimize: $f = 3x + 2y$
Subject to: $-x + y \leq 10$
$x + y \geq 20$
$x + y \leq 35$
$x,\ y \geq 0$

Maximize: $-f = -3x - 2y$
Subject to: $-x + y \leq 10$
$-x - y \leq -20$
$x + y \leq 35$
$x,\ y \geq 0$

$$\left[\begin{array}{rrrrrr|r} -1 & 1 & 1 & 0 & 0 & 0 & 10 \\ \mathbf{-1} & -1 & 0 & 1 & 0 & 0 & -20 \\ 1 & 1 & 0 & 0 & 1 & 0 & 35 \\ \hline 3 & 2 & 0 & 0 & 0 & 1 & 0 \end{array}\right] \xrightarrow{} -R_2 \to R_2 \left[\begin{array}{rrrrrr|r} -1 & 1 & 1 & 0 & 0 & 0 & 10 \\ \mathbf{1} & 1 & 0 & -1 & 0 & 0 & 20 \\ 1 & 1 & 0 & 0 & 1 & 0 & 35 \\ \hline -3 & -2 & 0 & 0 & 0 & 1 & 0 \end{array}\right] \begin{array}{l} R_2 + R_1 \to R_1 \\ \to \\ -R_2 + R_3 \to R_3 \\ -3R_2 + R_4 \to R_4 \end{array}$$

$$\left[\begin{array}{rrrrrr|r} 0 & \mathbf{2} & 1 & -1 & 0 & 0 & 30 \\ 1 & 1 & 0 & -1 & 0 & 0 & 20 \\ 0 & 0 & 0 & 1 & 1 & 0 & 15 \\ \hline 0 & -1 & 0 & 3 & 0 & 1 & -60 \end{array}\right] \begin{array}{l} \frac{1}{2}R_1 \to R_1 \\ \\ \to \end{array} \left[\begin{array}{rrrrrr|r} 0 & 1 & \frac{1}{2} & -\frac{1}{2} & 0 & 0 & 15 \\ 1 & 1 & 0 & -1 & 0 & 0 & 20 \\ 0 & 0 & 0 & 1 & 1 & 0 & 15 \\ \hline 0 & -1 & 0 & 3 & 0 & 1 & -60 \end{array}\right] \begin{array}{l} \to \\ -R_1 + R_2 \to R_2 \\ \\ R_1 + R_4 \to R_4 \end{array}$$

$$\left[\begin{array}{rrrrrr|r} 0 & 1 & \frac{1}{2} & -\frac{1}{2} & 0 & 0 & 15 \\ 1 & 0 & -\frac{1}{2} & -\frac{1}{2} & 0 & 0 & 5 \\ 0 & 0 & 0 & 1 & 1 & 0 & 15 \\ \hline 0 & 0 & \frac{1}{2} & \frac{5}{2} & 0 & 1 & -45 \end{array}\right]$$

Maximum of 45 occurs when $x = 5$ and $y = 15$.

25. Maximize: $f = 2x + 5y$
Constraints: $-x + y \leq 10,\ x + y \leq 30,\ 2x - y \leq 24,\ x, y \geq 0$

$$\left[\begin{array}{rrrrrr|r} -1 & \mathbf{1} & 1 & 0 & 0 & 0 & 10 \\ 1 & 1 & 0 & 1 & 0 & 0 & 30 \\ 2 & -1 & 0 & 0 & 1 & 0 & 24 \\ \hline -2 & -5 & 0 & 0 & 0 & 1 & 0 \end{array}\right] \begin{array}{l} \to \\ -R_1 + R_2 \to R_2 \\ R_1 + R_3 \to R_3 \\ 5R_1 + R_4 \to R_4 \end{array} \left[\begin{array}{rrrrrr|r} -1 & 1 & 1 & 0 & 0 & 0 & 10 \\ \mathbf{2} & 0 & -1 & 1 & 0 & 0 & 20 \\ 1 & 0 & 1 & 0 & 1 & 0 & 34 \\ \hline -7 & 0 & 5 & 0 & 0 & 1 & 50 \end{array}\right] \begin{array}{l} \\ \frac{1}{2}R_2 \to R_2 \\ \to \end{array}$$

$$\left[\begin{array}{rrrrrr|r} -1 & 1 & 1 & 0 & 0 & 0 & 10 \\ 1 & 0 & -\frac{1}{2} & \frac{1}{2} & 0 & 0 & 10 \\ 1 & 0 & 1 & 0 & 1 & 0 & 34 \\ \hline -7 & 0 & 5 & 0 & 0 & 1 & 50 \end{array}\right] \begin{array}{l} R_2 + R_1 \to R_1 \\ \to \\ -R_2 + R_3 \to R_3 \\ 7R_2 + R_4 \to R_4 \end{array} \left[\begin{array}{rrrrrr|r} 0 & 1 & \frac{1}{2} & \frac{1}{2} & 0 & 0 & 20 \\ 1 & 0 & -\frac{1}{2} & \frac{1}{2} & 0 & 0 & 10 \\ 0 & 0 & \frac{3}{2} & -\frac{1}{2} & 1 & 0 & 24 \\ \hline 0 & 0 & \frac{3}{2} & \frac{7}{2} & 0 & 1 & 120 \end{array}\right]$$

Maximum is 120 at $x = 10$, $y = 20$.

27. Maximize: $-f = -x - 2y - 3z$

Constraints: $x + z \le 20$

$-x - y \le -30$

$y + z \le 20$

$x,\ y,\ z \ge 0$

$$\left[\begin{array}{ccccccc|c} \mathbf{1} & 0 & 1 & 1 & 0 & 0 & 0 & 20 \\ -1 & -1 & 0 & 0 & 1 & 0 & 0 & -30 \\ 0 & 1 & 1 & 0 & 0 & 1 & 0 & 20 \\ \hline 1 & 2 & 3 & 0 & 0 & 0 & 1 & 0 \end{array}\right] \begin{array}{c} \to \\ R_1 + R_2 \to R_2 \\ \to \\ -R_1 + R_4 \to R_4 \end{array} \left[\begin{array}{ccccccc|c} 1 & 0 & 1 & 1 & 0 & 0 & 0 & 20 \\ 0 & \mathbf{-1} & 1 & 1 & 1 & 0 & 0 & -10 \\ 0 & 1 & 1 & 0 & 0 & 1 & 0 & 20 \\ \hline 0 & 2 & 2 & -1 & 0 & 0 & 1 & -20 \end{array}\right] \begin{array}{c} \\ -R_2 \to R_2 \\ \to \\ \\ \end{array}$$

$$\left[\begin{array}{ccccccc|c} 1 & 0 & 1 & 1 & 0 & 0 & 0 & 20 \\ 0 & 1 & -1 & -1 & -1 & 0 & 0 & 10 \\ 0 & 1 & 1 & 0 & 0 & 1 & 0 & 20 \\ \hline 0 & 2 & 2 & -1 & 0 & 0 & 1 & -20 \end{array}\right] \begin{array}{c} \to \\ \\ -R_2 + R_3 \to R_3 \\ -2R_2 + R_4 \to R_4 \end{array} \left[\begin{array}{ccccccc|c} 1 & 0 & 1 & 1 & 0 & 0 & 0 & 20 \\ 0 & 1 & -1 & -1 & -1 & 0 & 0 & 10 \\ 0 & 0 & 2 & 1 & 1 & 1 & 0 & 10 \\ \hline 0 & 0 & 4 & 1 & 2 & 0 & 1 & -40 \end{array}\right]$$

Minimum is 40 at $x = 20, y = 10, z = 0$.

Note: recall basic variables have columns with a single entry of 1 and other entries, 0.

29. Maximize: $f = 2x - y + 4z$

Constraints: $x + y + z \le 8,\ -x + y - z \le -4,\ -x - y + z \le -2,\ x,\ y,\ z \ge 0$

$$\left[\begin{array}{ccccccc|c} 1 & 1 & 1 & 1 & 0 & 0 & 0 & 8 \\ -1 & 1 & -1 & 0 & 1 & 0 & 0 & -4 \\ \mathbf{-1} & -1 & 1 & 0 & 0 & 1 & 0 & -2 \\ \hline -2 & 1 & -4 & 0 & 0 & 0 & 1 & 0 \end{array}\right] \begin{array}{c} \\ \\ -R_3 \to R_3 \\ \\ \end{array} \left[\begin{array}{ccccccc|c} 1 & 1 & 1 & 1 & 0 & 0 & 0 & 8 \\ -1 & 1 & -1 & 0 & 1 & 0 & 0 & -4 \\ 1 & 1 & -1 & 0 & 0 & -1 & 0 & 2 \\ \hline -2 & 1 & -4 & 0 & 0 & 0 & 1 & 0 \end{array}\right] \begin{array}{c} -R_3 + R_1 \to R_1 \\ R_3 + R_2 \to R_2 \\ \to \\ 2R_3 + R_4 \to R_4 \end{array}$$

$$\left[\begin{array}{ccccccc|c} 0 & 0 & 2 & 1 & 0 & 1 & 0 & 6 \\ 0 & 2 & \mathbf{-2} & 0 & 1 & -1 & 0 & -2 \\ 1 & 1 & -1 & 0 & 0 & -1 & 0 & 2 \\ \hline 0 & 3 & -6 & 0 & 0 & -2 & 1 & 4 \end{array}\right] \begin{array}{c} \\ -\frac{1}{2}R_2 \to R_2 \\ \\ \\ \end{array} \left[\begin{array}{ccccccc|c} 0 & 0 & 2 & 1 & 0 & 1 & 0 & 6 \\ 0 & -1 & 1 & 0 & -\frac{1}{2} & \frac{1}{2} & 0 & 1 \\ 1 & 1 & -1 & 0 & 0 & -1 & 0 & 2 \\ \hline 0 & 3 & -6 & 0 & 0 & -2 & 1 & 4 \end{array}\right] \begin{array}{c} -2R_2 + R_1 \to R_1 \\ \to \\ R_2 + R_3 \to R_3 \\ 6R_2 + R_4 \to R_4 \end{array}$$

$$\left[\begin{array}{ccccccc|c} 0 & \mathbf{2} & 0 & 1 & 1 & 0 & 0 & 4 \\ 0 & -1 & 1 & 0 & -\frac{1}{2} & \frac{1}{2} & 0 & 1 \\ 1 & 0 & 0 & 0 & -\frac{1}{2} & -\frac{1}{2} & 0 & 3 \\ \hline 0 & -3 & 0 & 0 & -3 & 1 & 1 & 10 \end{array}\right] \begin{array}{c} \frac{1}{2}R_1 \to R_1 \\ \\ \\ \\ \end{array} \left[\begin{array}{ccccccc|c} 0 & 1 & 0 & \frac{1}{2} & \frac{1}{2} & 0 & 0 & 2 \\ 0 & -1 & 1 & 0 & -\frac{1}{2} & \frac{1}{2} & 0 & 1 \\ 1 & 0 & 0 & 0 & -\frac{1}{2} & -\frac{1}{2} & 0 & 3 \\ \hline 0 & -3 & 0 & 0 & -3 & 1 & 1 & 10 \end{array}\right] \begin{array}{c} \\ R_1 + R_2 \to R_2 \\ \to \\ 3R_1 + R_4 \to R_4 \end{array}$$

$$\left[\begin{array}{ccccccc|c} 0 & 1 & 0 & \frac{1}{2} & \frac{1}{2} & 0 & 0 & 2 \\ 0 & 0 & 1 & \frac{1}{2} & 0 & \frac{1}{2} & 0 & 3 \\ 1 & 0 & 0 & 0 & -\frac{1}{2} & -\frac{1}{2} & 0 & 3 \\ \hline 0 & 0 & 0 & \frac{3}{2} & -\frac{3}{2} & 1 & 1 & 16 \end{array}\right] \begin{array}{c} 2R_1 \to R_1 \\ \\ \\ \\ \end{array} \left[\begin{array}{ccccccc|c} 0 & 2 & 0 & 1 & 1 & 0 & 0 & 4 \\ 0 & 0 & 1 & \frac{1}{2} & 0 & \frac{1}{2} & 0 & 3 \\ 1 & 0 & 0 & 0 & -\frac{1}{2} & -\frac{1}{2} & 0 & 3 \\ \hline 0 & 0 & 0 & \frac{3}{2} & -\frac{3}{2} & 1 & 1 & 16 \end{array}\right] \begin{array}{c} \to \\ \\ \frac{1}{2}R_1 + R_3 \to R_3 \\ \frac{3}{2}R_1 + R_4 \to R_4 \end{array}$$

$$\left[\begin{array}{ccccccc|c} 0 & 2 & 0 & 1 & 1 & 0 & 0 & 4 \\ 0 & 0 & 1 & \frac{1}{2} & 0 & \frac{1}{2} & 0 & 3 \\ 1 & 1 & 0 & \frac{1}{2} & 0 & -\frac{1}{2} & 0 & 5 \\ \hline 0 & 3 & 0 & 3 & 0 & 1 & 1 & 22 \end{array}\right]$$

Maximum is 22 at $x = 5, y = 0, z = 3$. See note in problem 27.

31. Maximize: $f = 2x + 3y - z$

Constraints: $3x + y + 2z \le 240$

$2x + 2y - z \le -10$

$x, y, z \ge 0$

$$\left[\begin{array}{cccccc|c} 3 & 1 & 2 & 1 & 0 & 0 & 240 \\ 2 & 2 & \mathbf{-1} & 0 & 1 & 0 & -10 \\ \hline -2 & -3 & 1 & 0 & 0 & 1 & 0 \end{array}\right] \begin{array}{c} \to \\ -R_2 \to R_2 \\ \end{array} \left[\begin{array}{cccccc|c} 3 & 1 & 2 & 1 & 0 & 0 & 240 \\ -2 & -2 & 1 & 0 & -1 & 0 & 10 \\ \hline -2 & -3 & 1 & 0 & 0 & 1 & 0 \end{array}\right] \begin{array}{c} -2R_2 + R_1 \to R_1 \\ \to \\ -R_2 + R_3 \to R_3 \end{array}$$

$$\left[\begin{array}{cccccc|c} 7 & \mathbf{5} & 0 & 1 & 2 & 1 & 220 \\ -2 & -2 & 1 & 0 & -1 & 0 & 10 \\ \hline 0 & -1 & 0 & 0 & 1 & 1 & -10 \end{array}\right] \begin{array}{c} \frac{1}{5}R_1 \to R_1 \\ \to \\ \end{array} \left[\begin{array}{cccccc|c} \frac{7}{5} & 1 & 0 & \frac{1}{5} & \frac{2}{5} & 0 & 44 \\ -2 & -2 & 1 & 0 & -1 & 0 & 10 \\ \hline 0 & -1 & 0 & 0 & 1 & 1 & -10 \end{array}\right] \begin{array}{c} \to \\ 2R_1 + R_2 \to R_2 \\ R_1 + R_3 \to R_3 \end{array}$$

$$\left[\begin{array}{cccccc|c} \frac{7}{5} & 1 & 0 & \frac{1}{5} & \frac{2}{5} & 0 & 44 \\ \frac{4}{5} & 0 & 1 & \frac{2}{5} & -\frac{1}{5} & 0 & 98 \\ \hline \frac{7}{5} & 0 & 0 & \frac{1}{5} & \frac{7}{5} & 1 & 34 \end{array}\right]$$

Maximum is 34 at $x = 0$, $y = 44$, $z = 98$.

33. Maximize: $-f = -10x - 30y - 35z$

Constraints: $x + y + z \le 250$, $-x - y - 2z \le -150$, $2x + y + z \le 180$

$$\left[\begin{array}{ccccccc|c} 1 & 1 & 1 & 1 & 0 & 0 & 0 & 250 \\ -1 & -1 & \mathbf{-2} & 0 & 1 & 0 & 0 & -150 \\ 2 & 1 & 1 & 0 & 0 & 1 & 0 & 180 \\ \hline 10 & 30 & 35 & 0 & 0 & 0 & 1 & 0 \end{array}\right] \begin{array}{c} \\ -\frac{1}{2}R_2 \to R_2 \\ \to \\ \end{array} \left[\begin{array}{ccccccc|c} 1 & 1 & 1 & 1 & 0 & 0 & 0 & 250 \\ \frac{1}{2} & \frac{1}{2} & \mathbf{1} & 0 & -\frac{1}{2} & 0 & 0 & 75 \\ 2 & 1 & 1 & 0 & 0 & 1 & 0 & 180 \\ \hline 10 & 30 & 35 & 0 & 0 & 0 & 1 & 0 \end{array}\right] \begin{array}{c} -R_2 + R_1 \to R_1 \\ \to \\ -R_2 + R_3 \to R_3 \\ -35R_2 + R_4 \to R_4 \end{array}$$

$$\left[\begin{array}{ccccccc|c} \frac{1}{2} & \frac{1}{2} & 0 & 1 & \frac{1}{2} & 0 & 0 & 175 \\ \frac{1}{2} & \frac{1}{2} & 1 & 0 & -\frac{1}{2} & 0 & 0 & 75 \\ \mathbf{\frac{3}{2}} & \frac{1}{2} & 0 & 0 & \frac{1}{2} & 1 & 0 & 105 \\ \hline -\frac{15}{2} & \frac{25}{2} & 0 & 0 & \frac{35}{2} & 0 & 1 & -2625 \end{array}\right] \begin{array}{c} \to \\ \\ \frac{2}{3}R_3 \to R_3 \\ \end{array} \left[\begin{array}{ccccccc|c} \frac{1}{2} & \frac{1}{2} & 0 & 1 & \frac{1}{2} & 0 & 0 & 175 \\ \frac{1}{2} & \frac{1}{2} & 1 & 0 & -\frac{1}{2} & 0 & 0 & 75 \\ \mathbf{1} & \frac{1}{3} & 0 & 0 & \frac{1}{3} & \frac{2}{3} & 0 & 70 \\ \hline -\frac{15}{2} & \frac{25}{2} & 0 & 0 & \frac{35}{2} & 0 & 1 & -2625 \end{array}\right] \begin{array}{c} -\frac{1}{2}R_3 + R_1 \to R_1 \\ -\frac{1}{2}R_3 + R_2 \to R_2 \\ \to \\ \frac{15}{2}R_3 + R_4 \to R_4 \end{array}$$

$$\left[\begin{array}{ccccccc|c} 0 & \frac{1}{3} & 0 & 1 & \frac{1}{3} & -\frac{1}{3} & 0 & 140 \\ 0 & \frac{1}{3} & 1 & 0 & -\frac{2}{3} & -\frac{1}{3} & 0 & 40 \\ 1 & \frac{1}{3} & 0 & 0 & \frac{1}{3} & \frac{2}{3} & 0 & 70 \\ \hline 0 & 15 & 0 & 0 & 20 & 5 & 1 & -2100 \end{array}\right]$$

At $x = 70$, $y = 0$, $z = 40$, $-f = -2100$. f has a minimum value of 2100.

35.

	Total Needed	Monaca	Hamburg
Heating Components	500	--	--
Produced	--	x	$500-x$
Profit	--	$400x$	$390(500-x)$
Domestic Furnaces	750	--	--
Produced	--	y	$750-y$
Profit	--	$200y$	$215(750-y)$

Profit = objective function = $P = 400x + 390(500-x) + 200y + 215(750-y)$

$= 10x - 15y + 356,250$

Constraints: $x, y \geq 0$

$x + y \leq 1000$ Capacity at Monaca

$(500-x)+(750-y) \leq 850$ or $x + y \geq 400$ Capacity at Hamburg

$x \leq 100 + \frac{1}{2}y$ or $2x - y \leq 200$ Monaca capacity restriction

$$\left[\begin{array}{cccccc|c} 1 & 1 & 1 & 0 & 0 & 0 & 1000 \\ -1 & \mathbf{-1} & 0 & 1 & 0 & 0 & -400 \\ 2 & -1 & 0 & 0 & 1 & 0 & 200 \\ \hline -10 & 15 & 0 & 0 & 0 & 1 & 356,250 \end{array}\right] \xrightarrow{-R_2 \to R_2} \left[\begin{array}{cccccc|c} 1 & 1 & 1 & 0 & 0 & 0 & 1000 \\ 1 & 1 & 0 & -1 & 0 & 0 & 400 \\ 2 & -1 & 0 & 0 & 1 & 0 & 200 \\ \hline -10 & 15 & 0 & 0 & 0 & 1 & 356,250 \end{array}\right] \begin{array}{c} -R_2 + R_1 \to R_1 \\ \to \\ R_2 + R_3 \to R_3 \\ -15R_2 + R_4 \to R_4 \end{array}$$

$$\left[\begin{array}{cccccc|c} 0 & 0 & 1 & 1 & 0 & 0 & 600 \\ 1 & 1 & 0 & -1 & 0 & 0 & 400 \\ \mathbf{3} & 0 & 0 & -1 & 1 & 0 & 600 \\ \hline -25 & 0 & 0 & 15 & 0 & 1 & 350,250 \end{array}\right] \xrightarrow{\frac{1}{3}R_3 \to R_3} \left[\begin{array}{cccccc|c} 0 & 0 & 1 & 1 & 0 & 0 & 600 \\ 1 & 1 & 0 & -1 & 0 & 0 & 400 \\ 1 & 0 & 0 & -\frac{1}{3} & \frac{1}{3} & 0 & 200 \\ \hline -25 & 0 & 0 & 15 & 0 & 1 & 350,250 \end{array}\right] \begin{array}{c} -R_3 + R_2 \to R_2 \\ \to \\ 25R_3 + R_4 \to R_4 \end{array}$$

$$\left[\begin{array}{cccccc|c} 0 & 0 & 1 & 1 & 0 & 0 & 600 \\ 0 & 1 & 0 & -\frac{2}{3} & -\frac{1}{3} & 0 & 200 \\ 1 & 0 & 0 & -\frac{1}{3} & \frac{1}{3} & 0 & 200 \\ \hline 0 & 0 & 0 & \frac{20}{3} & \frac{25}{3} & 1 & 355,250 \end{array}\right]$$

Maximum profit is $355,250.
$x = 200$ heating components at Monaca and 500 – 200 = 300 at Hamburg.
$y = 200$ domestic furnaces at Monaca and 750 – 200 = 550 at Hamburg.

37.

	Total Needed	Monaca	Hamburg
Heating Components	500	--	--
Produced	--	x	$500-x$
Cost	--	$380x$	$400(500-x)$
Domestic Furnaces	750	--	--
Produced	--	y	$750-y$
Cost	--	$200y$	$185(750-y)$

Cost = objective function = $C = 380x + 400(500-x) + 200y + 185(750-y)$

$= -20x + 15y + 338,750$

Constraints: $x, y \ge 0$

$x + y \le 1000$ Capacity at Monaca

$(500-x)+(750y) \le 850$ or $x + y \ge 400$ Capacity at Hamburg

$x \le 100 + \frac{1}{2}y$ or $2x - y \le 200$ Monaca capacity restriction

Need to maximize $-C = 20x - 15y - 338,750$

$$\left[\begin{array}{cccccc|c} 1 & 1 & 1 & 0 & 0 & 0 & 1000 \\ -1 & \mathbf{-1} & 0 & 1 & 0 & 0 & -400 \\ 2 & -1 & 0 & 0 & 1 & 0 & 200 \\ \hline -20 & 15 & 0 & 0 & 0 & 1 & -338,750 \end{array}\right] \begin{array}{c} \to \\ -R_2 \to R_2 \\ \\ \\ \end{array} \left[\begin{array}{cccccc|c} 1 & 1 & 1 & 0 & 0 & 0 & 1000 \\ 1 & \mathbf{1} & 0 & -1 & 0 & 0 & 400 \\ 2 & -1 & 0 & 0 & 1 & 0 & 200 \\ \hline -20 & 15 & 0 & 0 & 0 & 1 & -338,750 \end{array}\right] \begin{array}{c} -R_2 + R_1 \to R_1 \\ \to \\ R_2 + R_3 \to R_3 \\ -15R_2 + R_4 \to R_4 \end{array}$$

$$\left[\begin{array}{cccccc|c} 0 & 0 & 1 & 1 & 0 & 0 & 600 \\ 1 & 1 & 0 & -1 & 0 & 0 & 400 \\ \mathbf{3} & 0 & 0 & -1 & 1 & 0 & 600 \\ \hline -35 & 0 & 0 & 15 & 0 & 1 & -344,750 \end{array}\right] \begin{array}{c} \to \\ \\ \frac{1}{3}R_3 \to R_3 \\ \\ \end{array} \left[\begin{array}{cccccc|c} 0 & 0 & 1 & 1 & 0 & 0 & 600 \\ 1 & 1 & 0 & -1 & 0 & 0 & 400 \\ 1 & 0 & 0 & -\frac{1}{3} & \frac{1}{3} & 0 & 200 \\ \hline -35 & 0 & 0 & 15 & 0 & 1 & -344,750 \end{array}\right] \begin{array}{c} \\ -R_3 + R_2 \to R_2 \\ \to \\ 35R_3 + R_4 \to R_4 \end{array}$$

$$\left[\begin{array}{cccccc|c} 0 & 0 & 1 & 1 & 0 & 0 & 600 \\ 0 & 1 & 0 & -\frac{2}{3} & -\frac{1}{3} & 0 & 200 \\ 1 & 0 & 0 & -\frac{1}{3} & \frac{1}{3} & 0 & 200 \\ \hline 0 & 0 & 0 & \frac{10}{3} & \frac{35}{3} & 1 & -337,750 \end{array}\right]$$

Minimum cost is $337,750.

$x = 200$ heating components at Monaca and 500 – 200 = 300 at Hamburg.

$y = 200$ domestic furnaces at Monaca and 750 – 200 = 550 at Hamburg.

39.

	All Beef	Regular	Available
Beef	$0.75x$	$0.18y$	≤ 1020
Spices	$0.2x$	$0.2y$	≥ 500
Pork	--	$0.3y$	≤ 600

Objective function: $P = 0.6x + 0.4y$

$$\left[\begin{array}{cccccc|c} \mathbf{0.75} & 0.18 & 1 & 0 & 0 & 0 & 1020 \\ -0.20 & -0.20 & 0 & 1 & 0 & 0 & -500 \\ 0 & 0.3 & 0 & 0 & 1 & 0 & 600 \\ \hline -0.60 & -0.40 & 0 & 0 & 0 & 1 & 0 \end{array}\right] \quad \begin{array}{l} \frac{1020}{0.75} = 1360 \quad \frac{4}{3}R_1 \to R_1 \\ \frac{-500}{(-0.2)} = 2500 \\ \\ \\ \end{array}$$

Convert to fractions.

$$\left[\begin{array}{cccccc|c} 1 & \frac{6}{25} & \frac{4}{3} & 0 & 0 & 0 & 1360 \\ -\frac{1}{5} & -\frac{1}{5} & 0 & 1 & 0 & 0 & -500 \\ 0 & \frac{3}{10} & 0 & 0 & 1 & 0 & 600 \\ \hline -\frac{3}{5} & -\frac{2}{5} & 0 & 0 & 0 & 1 & 0 \end{array}\right] \begin{array}{l} \\ \frac{1}{5}R_1 + R_2 \to R_2 \\ \to \\ \frac{3}{5}R_1 + R_4 \to R_4 \end{array}$$

$$\left[\begin{array}{cccccc|c} 1 & \frac{6}{25} & \frac{4}{3} & 0 & 0 & 0 & 1360 \\ 0 & -\frac{19}{125} & \frac{4}{15} & 1 & 0 & 0 & -228 \\ 0 & \frac{3}{10} & 0 & 0 & 1 & 0 & 600 \\ \hline 0 & -\frac{32}{125} & \frac{4}{5} & 0 & 0 & 1 & 816 \end{array}\right] \begin{array}{l} \\ -\frac{125}{19}R_2 \to R_2 \\ \\ \\ \end{array}$$

$$\left[\begin{array}{cccccc|c} 1 & \frac{6}{25} & \frac{4}{3} & 0 & 0 & 0 & 1360 \\ 0 & 1 & -\frac{100}{57} & -\frac{125}{19} & 0 & 0 & 1500 \\ 0 & \frac{3}{10} & 0 & 0 & 1 & 0 & 600 \\ \hline 0 & -\frac{32}{125} & \frac{4}{5} & 0 & 0 & 1 & 816 \end{array}\right] \begin{array}{l} -\frac{6}{25}R_2 + R_1 \to R_1 \\ -\frac{3}{10}R_2 + R_3 \to R_2 \\ \to \\ \frac{32}{125}R_2 + R_4 \to R_4 \end{array}$$

$$\left[\begin{array}{cccccc|c} 1 & 0 & \frac{100}{57} & \frac{30}{19} & 0 & 0 & 1000 \\ 0 & 1 & -\frac{100}{57} & -\frac{125}{19} & 0 & 0 & 1500 \\ 0 & 0 & \frac{10}{19} & \frac{75}{38} & 1 & 0 & 150 \\ \hline 0 & 0 & \frac{20}{57} & -\frac{32}{19} & 0 & 1 & 1200 \end{array}\right] \begin{array}{l} \to \\ \\ \frac{38}{75}R_3 \to R_3 \\ \\ \end{array}$$

$$\left[\begin{array}{cccccc|c} 1 & 0 & \frac{100}{57} & \frac{30}{19} & 0 & 0 & 1000 \\ 0 & 1 & -\frac{100}{57} & -\frac{125}{19} & 0 & 0 & 1500 \\ 0 & 0 & \frac{4}{15} & 1 & \frac{38}{75} & 0 & 76 \\ \hline 0 & 0 & \frac{20}{57} & -\frac{32}{19} & 0 & 1 & 1200 \end{array}\right] \begin{array}{l} -\frac{30}{19}R_3 + R_1 \to R_1 \\ \frac{125}{19}R_3 + R_2 \to R_2 \\ \to \\ \frac{32}{19}R_3 + R_4 \to R_4 \end{array} \left[\begin{array}{cccccc|c} 1 & 0 & \frac{4}{3} & 0 & -\frac{4}{5} & 0 & 880 \\ 0 & 1 & 0 & 0 & \frac{10}{3} & 0 & 2000 \\ 0 & 0 & \frac{4}{15} & 1 & \frac{38}{75} & 0 & 76 \\ \hline 0 & 0 & \frac{4}{5} & 0 & \frac{64}{75} & 1 & 1328 \end{array}\right]$$

Maximum profit is \$1328 with 880 lbs of all beef and 2000 lbs of regular.

41.

	Plant 1	Plant 2	Plant 3	Totals
Million gallons	x	y	z	≤ 10
Impurities	$0.20x$	$0.15y$	$0.10z$	$\le 0.15(x+y+z) \Rightarrow 0.05x - 0.05z \le 0$
Cost (Thousands)	$20x$	$30y$	$40z$	$= C$
Other constraints	x		z	≥ 6

$$\left[\begin{array}{ccccccc|c} 1 & 1 & 1 & 1 & 0 & 0 & 0 & 10 \\ .05 & 0 & -.05 & 0 & 1 & 0 & 0 & 0 \\ \mathbf{-1} & 0 & -1 & 0 & 0 & 1 & 0 & -6 \\ \hline 20 & 30 & 40 & 0 & 0 & 0 & 1 & 0 \end{array}\right] \begin{array}{c} \to \\ 100R_2 \to R_2 \\ -R_3 \to R_3 \\ \end{array} \left[\begin{array}{ccccccc|c} 1 & 1 & 1 & 1 & 0 & 0 & 0 & 10 \\ 5 & 0 & -5 & 0 & 100 & 0 & 0 & 0 \\ 1 & 0 & 1 & 0 & 0 & -1 & 0 & 6 \\ \hline 20 & 30 & 40 & 0 & 0 & 0 & 1 & 0 \end{array}\right] \begin{array}{c} -R_3 + R_1 \to R_1 \\ -5R_3 + R_2 \to R_2 \\ \to \\ -20R_3 + R_4 \to R_4 \end{array}$$

$$\left[\begin{array}{ccccccc|c} 0 & 1 & 0 & 1 & 0 & 1 & 0 & 4 \\ 0 & 0 & \mathbf{-10} & 0 & 100 & 5 & 0 & -30 \\ 1 & 0 & 1 & 0 & 0 & -1 & 0 & 6 \\ \hline 0 & 30 & 20 & 0 & 0 & 20 & 1 & -120 \end{array}\right] \begin{array}{c} \\ -\tfrac{1}{10}R_2 \to R_2 \\ \to \\ \end{array} \left[\begin{array}{ccccccc|c} 0 & 1 & 0 & 1 & 0 & 1 & 0 & 4 \\ 0 & 0 & 1 & 0 & -10 & -\tfrac{1}{2} & 0 & 3 \\ 1 & 0 & 1 & 0 & 0 & -1 & 0 & 6 \\ \hline 0 & 30 & 20 & 0 & 0 & 20 & 1 & -120 \end{array}\right]$$

$$\begin{array}{c} \to \\ \\ -R_2 + R_3 \to R_3 \\ -20R_2 + R_4 \to R_4 \end{array} \left[\begin{array}{ccccccc|c} 0 & 1 & 0 & 1 & 0 & 1 & 0 & 4 \\ 0 & 0 & 1 & 0 & -10 & -\tfrac{1}{2} & 0 & 3 \\ 1 & 0 & 0 & 0 & 10 & -\tfrac{1}{2} & 0 & 3 \\ \hline 0 & 30 & 0 & 0 & 200 & 30 & 1 & -180 \end{array}\right]$$

Minimum cost is \$180,000 with plants 1 and 3 each handling 3,000,000 gallons. Plant 2 is zero, not 4, because the y-column still represents a nonbasic variable.

43. x = number of footballs, y = number of soccer balls, z = number of volleyballs

Maximize: $f = 30x + 25y + 20z$

Subject to: $10x + 8y + 8z \ge 10{,}000 \quad \to \quad -10x - 8y - 8z \le -10{,}000$

$10x + 12y + 8z \le 20{,}000$

$$\left[\begin{array}{cccccc|c} \mathbf{-10} & -8 & -8 & 1 & 0 & 0 & -10{,}000 \\ 10 & 12 & 8 & 0 & 1 & 0 & 20{,}000 \\ \hline -30 & -25 & -20 & 0 & 0 & 1 & 0 \end{array}\right] \begin{array}{c} -\tfrac{1}{10}R_1 \to R_1 \\ \to \\ \end{array}$$

$$\left[\begin{array}{cccccc|c} 1 & 0.8 & 0.8 & -0.1 & 0 & 0 & 1000 \\ 10 & 12 & 8 & 0 & 1 & 0 & 20{,}000 \\ \hline -30 & -25 & -20 & 0 & 0 & 1 & 0 \end{array}\right] \begin{array}{c} \to \\ -10R_1 + R_2 \to R_2 \\ 30R_1 + R_3 \to R_3 \end{array}$$

$$\left[\begin{array}{cccccc|c} 1 & 0.8 & 0.8 & -0.1 & 0 & 0 & 1000 \\ 0 & 4 & 0 & \mathbf{1} & 1 & 0 & 10{,}000 \\ \hline 0 & -1 & 4 & -3 & 0 & 1 & 30{,}000 \end{array}\right] \begin{array}{c} 0.1R_2 + R_1 \to R_1 \\ \to \\ 3R_2 + R_3 \to R_3 \end{array} \left[\begin{array}{cccccc|c} 1 & 1.2 & 0.8 & 0 & 0.1 & 0 & 2000 \\ 0 & 4 & 0 & 1 & 1 & 0 & 10{,}000 \\ \hline 0 & 11 & 4 & 0 & 3 & 1 & 60{,}000 \end{array}\right]$$

The maximum revenue is \$60,000 when 2000 footballs and no soccer balls or volleyballs are produced.

Review Exercises

1. $3x-9 \le 4(3-x)$

$3x-9 \le 12-4x$

$7x \le 21$

$x \le 3$

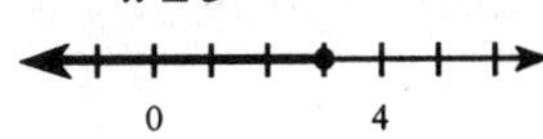

2. $\frac{2}{5}x \le x+4$

$-4 \le \frac{3}{5}x$

$-\frac{20}{3} \le x$

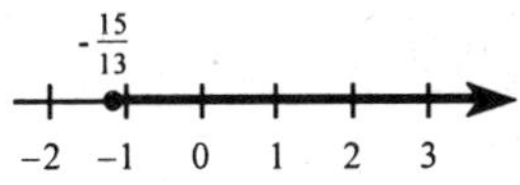

3. $5x+1 \ge \frac{2}{3}(x-6)$

$15x+3 \ge 2x-12$

$13x \ge -15$

$x \ge -\frac{15}{13}$

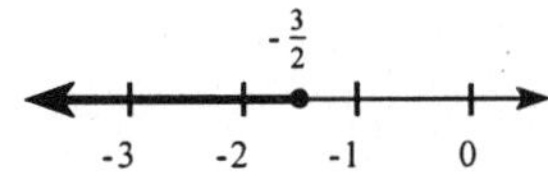

4. $\frac{4(x-2)}{3} \ge 3x-\frac{1}{6}$

$6\left(\frac{4(x-2)}{3}\right) \ge 6\left(3x-\frac{1}{6}\right)$

$8x-16 \ge 18x-1$

$-15 \ge 10x$

$-\frac{3}{2} \ge x$

-3/2

-3 -2 -1 0

5. **a.** $0 \le x \le 5$ closed [0, 5]
b. $3 \le x < 7$ half open [3, 7)
c. $-3 < x < 2$ open (– 3, 2)

6. **a.** $-1 < x < 16$
b. $-12 \le x \le -8$
c. $x < -1$

7.

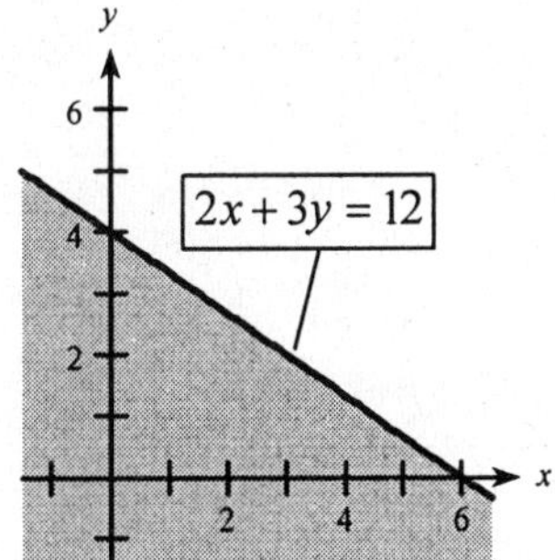

8.

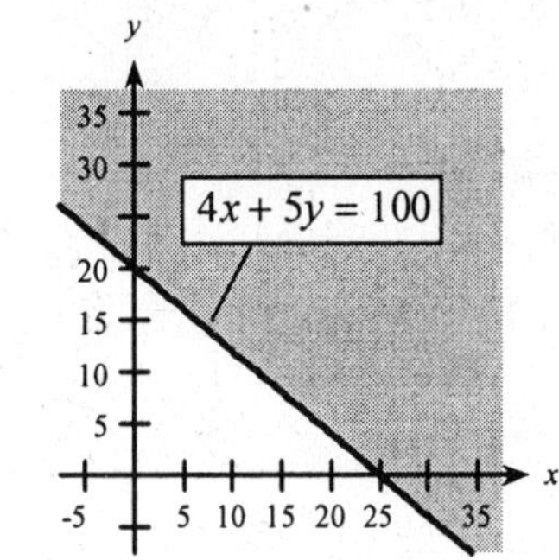

9.

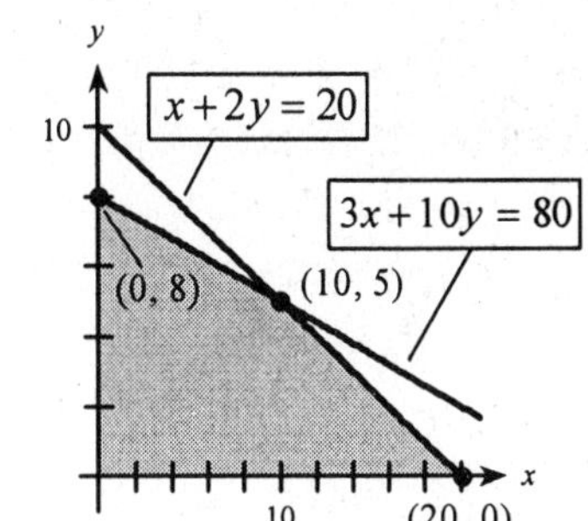

10.

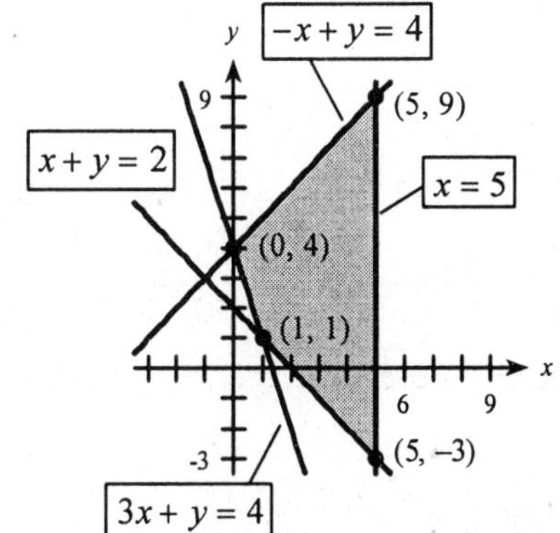

11.

Corners	$f = -x + 3y$
$(0,5)$	15
$(5,10)$	25
$(10,6)$	8
$(12,0)$	-12
$(0,0)$	0

Maximum is 25 at (5,10).
Minimum is -12 at (12,0).

12.

Corners	$f = 6x + 4y$
$(0,30)$	120
$(0,40)$	160
$(17,23)$	194
$(8,14)$	104

Maximum is 194 at (17,23).
Minimum is 104 at (8,14).

13. Maximize $f = 5x + 6y$

Subject to: $x + 3y \le 24$

$4x + 3y \le 42$

$2x + y \le 20$

Solving $\begin{cases} 4x + 3y = 42 \\ x + 3y = 24 \end{cases}$ we get $A(6, 6)$.

Solving $\begin{cases} 2x + y = 20 \\ 4x + 3y = 42 \end{cases}$ we get $B(9, 2)$.

Feasible Corners	$f = 5x + 6y$
$(0,8)$	48
$(6,6)$	66
$(9,2)$	57
$(10,0)$	50

Maximum is 66 at (6,6).

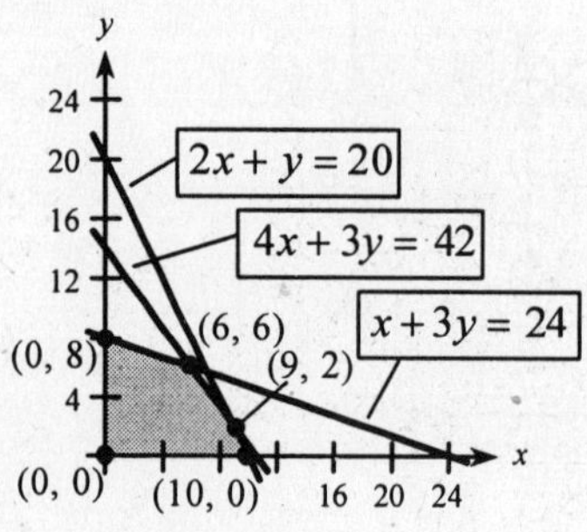

14. Maximize $f = 9x + 5y$

Subject to: $x + 3y \le 24$

$4x + 3y \le 42$

$2x + y \le 20$

Using the information from problem 13, the feasible corners are (0, 8), (6, 6), (9, 2), and (10, 0).

At (0, 8), $f = 40$
At (6, 6), $f = 84$
At (9, 2), $f = 91$ ← Maximum
At (10, 0), $f = 90$

15. Maximize $f = x + 4y$

Subject to: $7x + 3y \le 105$

$2x + 5y \le 59$

$x + 7y \le 70$

Solving $\begin{cases} 7x + 3y = 105 \\ 2x + 5y = 59 \end{cases}$ we get $B(12, 7)$.

Solving $\begin{cases} x + 7y = 70 \\ 2x + 5y = 59 \end{cases}$ we get $A\ (7, 9)$.

Feasible Corners	$f = x + 4y$
$(0,10)$	40
$(7,9)$	43
$(12,7)$	40
$(15,0)$	15

Maximum is 43 at (7,9).

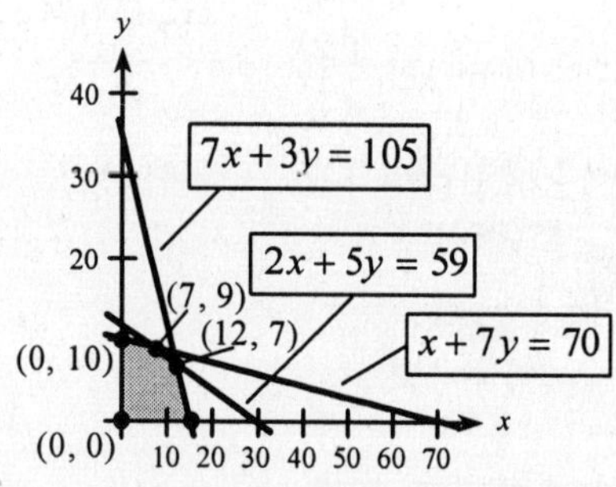

16. Maximize $f = 2x + y$

Subject to: $7x + 3y \le 105$

$2x + 5y \le 59$

$x + 7y \le 70$

Using the information from problem 15, the feasible corners are (0, 10), (7, 9), (12, 7), and (15, 0).

At (0, 10), $f = 10$
At (7, 9), $f = 23$
At (12, 7), $f = 31$ ← Maximum
At (15, 0), $f = 30$

17. Minimize $g = 5x + 3y$

Subject to: $3x + y \ge 12$

$x + y \ge 6$

$x + 6y \ge 11$

Solving $\begin{cases} 3x + y = 12 \\ x + y = 6 \end{cases}$ we get $A(3, 3)$.

Solving $\begin{cases} x + 6y = 11 \\ x + y = 6 \end{cases}$ we get $B(5, 1)$.

Feasible Corners	$g = 5x + 3y$
$(0,12)$	36
$(3,3)$	24
$(5,1)$	28
$(11,0)$	55

Minimum is 24 at (3,3).

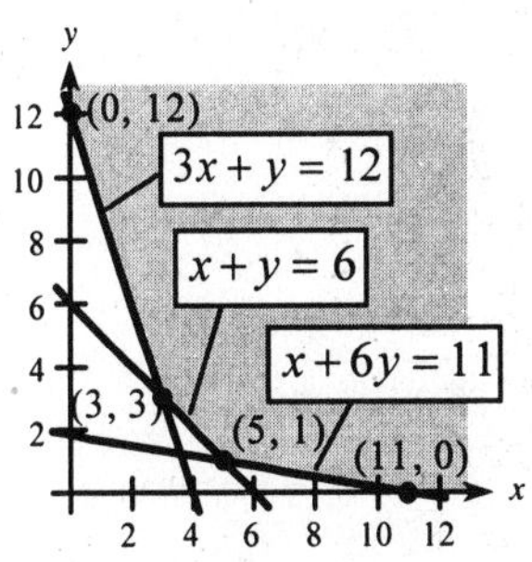

18. Minimize $g = 3x + 5y$

Subject to: $3x + y \ge 12$

$x + y \ge 6$

$x + 6y \ge 11$

Using the information from problem 17, the feasible corners are (11, 0), (5, 1), (3, 3), and $(0,12)$.

At (11, 0), $g = 33$
At (5, 1), $g = 20$ ← Minimum
At (3, 3), $g = 24$
At (0, 12), $g = 60$

19. Minimize $g = x + 5y$

Subject to: $8x + y \ge 85$

$x + y \ge 50$

$x + 4y \ge 80$

$x + 10y \ge 94$

Solving $\begin{cases} 8x + y = 85 \\ x + y = 50 \end{cases}$ we get $A(5, 45)$.

Solving $\begin{cases} x + y = 50 \\ x + 4y = 80 \end{cases}$ we get $B(40, 10)$.

Solving $\begin{cases} x + 4y = 80 \\ x + 10y = 94 \end{cases}$ we get $C\left(\frac{212}{3}, \frac{7}{3}\right)$.

Feasible Corners	$g = x + 5y$
$(0,85)$	425
$(5,45)$	230
$(40,10)$	90
$\left(\frac{212}{3}, \frac{7}{3}\right)$	$\frac{247}{3}$
$(94,0)$	94

Minimum is $\frac{247}{3}$ at $\left(\frac{212}{3}, \frac{7}{3}\right)$.

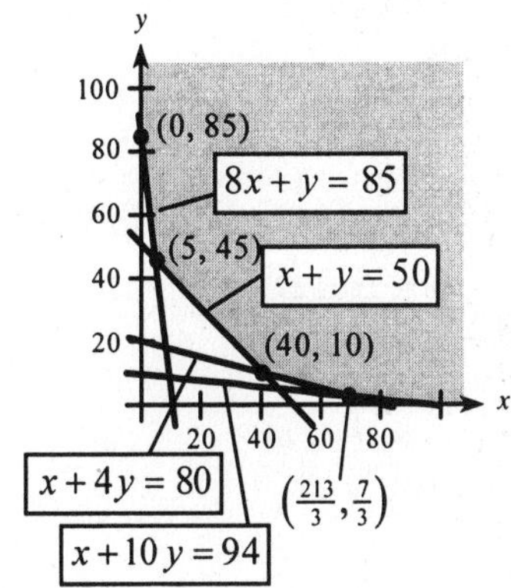

20. Minimize $g = 7x + y$

Subject to: $8x + y \geq 85$

$x + y \geq 50$

$x + 4y \geq 80$

$x + 10y \geq 94$

Using the information from problem 19, the feasible corners are (0, 85), (5, 45), (40, 10), $\left(\frac{212}{3}, \frac{7}{3}\right)$, and (94, 0).

At (0, 85), $g = 85$

At (5, 45), $g = 80 \leftarrow$ Minimum

At (40, 10), $g = 290$

At $\left(\frac{212}{3}, \frac{7}{3}\right)$, $g = 497$

At (94, 0), $g = 658$

21. $\left[\begin{array}{cccccc|c} 7 & 3 & 1 & 0 & 0 & 0 & 105 \\ 2 & 5 & 0 & 1 & 0 & 0 & 59 \\ 1 & 7 & 0 & 0 & 1 & 0 & 70 \\ \hline -7 & -12 & 0 & 0 & 0 & 1 & 0 \end{array}\right] \begin{array}{c} \to \\ \\ \frac{1}{7}R_3 \to R_3 \\ \to \end{array} \left[\begin{array}{cccccc|c} 7 & 3 & 1 & 0 & 0 & 0 & 105 \\ 2 & 5 & 0 & 1 & 0 & 0 & 59 \\ \frac{1}{7} & 1 & 0 & 0 & \frac{1}{7} & 0 & 10 \\ \hline -7 & -12 & 0 & 0 & 0 & 1 & 0 \end{array}\right] \begin{array}{l} -3R_3 + R_1 \to R_1 \\ -5R_3 + R_2 \to R_2 \\ \to \\ 12R_3 + R_4 \to R_4 \end{array}$

$\left[\begin{array}{cccccc|c} \frac{46}{7} & 0 & 1 & 0 & -\frac{3}{7} & 0 & 75 \\ \frac{9}{7} & 0 & 0 & 1 & -\frac{5}{7} & 0 & 9 \\ \frac{1}{7} & 1 & 0 & 0 & \frac{1}{7} & 0 & 10 \\ \hline -\frac{37}{7} & 0 & 0 & 0 & \frac{12}{7} & 1 & 120 \end{array}\right] \frac{7}{9}R_2 \to R_2 \text{ then } \left\{\begin{array}{l} -\frac{46}{7}R_2 + R_1 \to R_1 \\ \\ -\frac{1}{7}R_2 + R_3 \to R_3 \\ \frac{37}{7}R_2 + R_4 \to R_4 \end{array}\right. \left[\begin{array}{cccccc|c} 0 & 0 & 1 & -\frac{46}{9} & \frac{29}{9} & 0 & 29 \\ 1 & 0 & 0 & \frac{7}{9} & -\frac{5}{9} & 0 & 7 \\ 0 & 1 & 0 & -\frac{1}{9} & \frac{2}{9} & 0 & 9 \\ \hline 0 & 0 & 0 & \frac{37}{9} & -\frac{11}{9} & 1 & 157 \end{array}\right]$

$\begin{array}{c} \frac{9}{29}R_1 \to R_1 \\ \to \end{array} \left[\begin{array}{cccccc|c} 0 & 0 & \frac{9}{29} & -\frac{46}{29} & 1 & 0 & 9 \\ 1 & 0 & 0 & \frac{7}{9} & -\frac{5}{9} & 0 & 7 \\ 0 & 1 & 0 & -\frac{1}{9} & \frac{2}{9} & 0 & 9 \\ \hline 0 & 0 & 0 & \frac{37}{9} & -\frac{11}{9} & 1 & 157 \end{array}\right] \begin{array}{l} \to \\ \frac{5}{9}R_1 + R_2 \to R_2 \\ -\frac{2}{9}R_1 + R_3 \to R_3 \\ \frac{11}{9}R_1 + R_4 \to R_4 \end{array} \left[\begin{array}{cccccc|c} 0 & 0 & \frac{9}{29} & -\frac{46}{29} & 1 & 0 & 9 \\ 1 & 0 & \frac{5}{29} & -\frac{3}{29} & 0 & 0 & 12 \\ 0 & 1 & -\frac{2}{29} & \frac{7}{29} & 0 & 0 & 7 \\ \hline 0 & 0 & \frac{11}{29} & \frac{63}{29} & 0 & 1 & 168 \end{array}\right]$

Maximum is 168 at $x = 12, y = 7$.

22.

$$\left[\begin{array}{cccccc|c} 1 & 4 & 1 & 0 & 0 & 0 & 160 \\ 1 & 2 & 0 & 1 & 0 & 0 & 100 \\ 4 & 3 & 0 & 0 & 1 & 0 & 300 \\ \hline -3 & -4 & 0 & 0 & 0 & 1 & 0 \end{array}\right] \begin{array}{l} \frac{1}{4}R_1 \to R_1 \\ \to \\ \\ \\ \end{array} \left[\begin{array}{cccccc|c} \frac{1}{4} & 1 & \frac{1}{4} & 0 & 0 & 0 & 40 \\ 1 & 2 & 0 & 1 & 0 & 0 & 100 \\ 4 & 3 & 0 & 0 & 1 & 0 & 300 \\ \hline -3 & -4 & 0 & 0 & 0 & 1 & 0 \end{array}\right] \begin{array}{l} \to \\ -2R_1 + R_2 \to R_2 \\ -3R_1 + R_3 \to R_3 \\ 4R_1 + R_4 \to R_4 \end{array}$$

$$\left[\begin{array}{cccccc|c} \frac{1}{4} & 1 & \frac{1}{4} & 0 & 0 & 0 & 40 \\ \frac{1}{2} & 0 & -\frac{1}{2} & 1 & 0 & 0 & 20 \\ \frac{13}{4} & 0 & -\frac{3}{4} & 0 & 1 & 0 & 180 \\ \hline -2 & 0 & 1 & 0 & 0 & 1 & 160 \end{array}\right] \begin{array}{l} \\ 2R_2 \to R_2 \\ \to \\ \\ \end{array} \left[\begin{array}{cccccc|c} \frac{1}{4} & 1 & \frac{1}{4} & 0 & 0 & 0 & 40 \\ 1 & 0 & -1 & 2 & 0 & 0 & 40 \\ \frac{13}{4} & 0 & -\frac{3}{4} & 0 & 1 & 0 & 180 \\ \hline -2 & 0 & 1 & 0 & 0 & 1 & 160 \end{array}\right] \begin{array}{l} -\frac{1}{4}R_2 + R_1 \to R_1 \\ \to \\ -\frac{13}{4}R_2 + R_3 \to R_3 \\ 2R_2 + R_4 \to R_4 \end{array}$$

$$\left[\begin{array}{cccccc|c} 0 & 1 & \frac{1}{2} & -\frac{1}{2} & 0 & 0 & 30 \\ 1 & 0 & -1 & 2 & 0 & 0 & 40 \\ 0 & 0 & \frac{5}{2} & -\frac{13}{2} & 1 & 0 & 50 \\ \hline 0 & 0 & -1 & 4 & 0 & 1 & 240 \end{array}\right] \begin{array}{l} \to \\ \\ \frac{2}{5}R_3 \to R_3 \\ \\ \end{array} \left[\begin{array}{cccccc|c} 0 & 1 & \frac{1}{2} & -\frac{1}{2} & 0 & 0 & 30 \\ 1 & 0 & -1 & 2 & 0 & 0 & 40 \\ 0 & 0 & 1 & -\frac{13}{5} & \frac{2}{5} & 0 & 20 \\ \hline 0 & 0 & -1 & 4 & 0 & 1 & 240 \end{array}\right] \begin{array}{l} -\frac{1}{2}R_3 + R_1 \to R_1 \\ R_3 + R_2 \to R_2 \\ \to \\ R_3 + R_4 \to R_4 \end{array}$$

$$\left[\begin{array}{cccccc|c} 0 & 1 & 0 & \frac{4}{5} & -\frac{1}{5} & 0 & 20 \\ 1 & 0 & 0 & -\frac{3}{5} & \frac{2}{5} & 0 & 60 \\ 0 & 0 & 1 & -\frac{13}{5} & \frac{2}{5} & 0 & 20 \\ \hline 0 & 0 & 0 & \frac{7}{5} & \frac{2}{5} & 1 & 260 \end{array}\right]$$

Solution: $x = 60$, $y = 20$; $f = 260$

23.

$$\left[\begin{array}{cccccc|c} 1 & 4 & 1 & 0 & 0 & 0 & 160 \\ 1 & 2 & 0 & 1 & 0 & 0 & 100 \\ 4 & 3 & 0 & 0 & 1 & 0 & 300 \\ \hline -3 & -8 & 0 & 0 & 0 & 1 & 0 \end{array}\right] \begin{array}{l} \frac{1}{4}R_1 \to R_1 \\ \to \\ \\ \\ \end{array} \left[\begin{array}{cccccc|c} \frac{1}{4} & 1 & \frac{1}{4} & 0 & 0 & 0 & 40 \\ 1 & 2 & 0 & 1 & 0 & 0 & 100 \\ 4 & 3 & 0 & 0 & 1 & 0 & 300 \\ \hline -3 & -8 & 0 & 0 & 0 & 1 & 0 \end{array}\right] \begin{array}{l} \to \\ -2R_1 + R_2 \to R_2 \\ -3R_1 + R_3 \to R_3 \\ 8R_1 + R_4 \to R_4 \end{array}$$

$$\left[\begin{array}{cccccc|c} \frac{1}{4} & 1 & \frac{1}{4} & 0 & 0 & 0 & 40 \\ \frac{1}{2} & 0 & -\frac{1}{2} & 1 & 0 & 0 & 20 \\ \frac{13}{4} & 0 & -\frac{3}{4} & 0 & 1 & 0 & 180 \\ \hline -1 & 0 & 2 & 0 & 0 & 1 & 320 \end{array}\right] \begin{array}{l} \to \\ 2R_2 \to R_2 \\ \\ \\ \end{array} \left[\begin{array}{cccccc|c} \frac{1}{4} & 1 & \frac{1}{4} & 0 & 0 & 0 & 40 \\ 1 & 0 & -1 & 2 & 0 & 0 & 40 \\ \frac{13}{4} & 0 & -\frac{3}{4} & 0 & 1 & 0 & 180 \\ \hline -1 & 0 & 2 & 0 & 0 & 1 & 320 \end{array}\right] \begin{array}{l} -\frac{1}{4}R_2 + R_1 \to R_1 \\ \to \\ -\frac{13}{4}R_2 + R_3 \to R_3 \\ R_2 + R_4 \to R_4 \end{array}$$

$$\left[\begin{array}{cccccc|c} 0 & 1 & \frac{1}{2} & -\frac{1}{2} & 0 & 0 & 30 \\ 1 & 0 & -1 & 2 & 0 & 0 & 40 \\ 0 & 0 & \frac{5}{2} & -\frac{13}{2} & 1 & 0 & 50 \\ \hline 0 & 0 & 1 & 2 & 0 & 1 & 360 \end{array}\right]$$

Maximum is 360 at $x = 40$, $y = 30$.

24.

$$\left[\begin{array}{ccccccc|c} 1 & 0 & 1 & 1 & 0 & 0 & 0 & 7 \\ 3 & 5 & 0 & 0 & 1 & 0 & 0 & 30 \\ 3 & 1 & 0 & 0 & 0 & 1 & 0 & 18 \\ \hline -39 & -5 & -30 & 0 & 0 & 0 & 1 & 0 \end{array}\right] \xrightarrow{\frac{1}{3}R_3 \to R_3} \left[\begin{array}{ccccccc|c} 1 & 0 & 1 & 1 & 0 & 0 & 0 & 7 \\ 3 & 5 & 0 & 0 & 1 & 0 & 0 & 30 \\ 1 & \frac{1}{3} & 0 & 0 & 0 & \frac{1}{3} & 0 & 6 \\ \hline -39 & -5 & -30 & 0 & 0 & 0 & 1 & 0 \end{array}\right] \begin{array}{l} -R_3 + R_1 \to R_1 \\ -3R_3 + R_2 \to R_2 \\ \to \\ 39R_3 + R_4 \to R_4 \end{array}$$

$$\left[\begin{array}{ccccccc|c} 0 & -\frac{1}{3} & 1 & 1 & 0 & -\frac{1}{3} & 0 & 1 \\ 0 & 4 & 0 & 0 & 1 & -1 & 0 & 12 \\ 1 & \frac{1}{3} & 0 & 0 & 0 & \frac{1}{3} & 0 & 6 \\ \hline 0 & 8 & -30 & 0 & 0 & 13 & 1 & 234 \end{array}\right] \xrightarrow{30R_1 + R_4 \to R_4} \left[\begin{array}{ccccccc|c} 0 & -\frac{1}{3} & 1 & 1 & 0 & -\frac{1}{3} & 0 & 1 \\ 0 & 4 & 0 & 0 & 1 & -1 & 0 & 12 \\ 1 & \frac{1}{3} & 0 & 0 & 0 & \frac{1}{3} & 0 & 6 \\ \hline 0 & -2 & 0 & 30 & 0 & 3 & 1 & 264 \end{array}\right] \frac{1}{4}R_2 \to R_2$$

$$\left[\begin{array}{ccccccc|c} 0 & -\frac{1}{3} & 1 & 1 & 0 & -\frac{1}{3} & 0 & 1 \\ 0 & 1 & 0 & 0 & \frac{1}{4} & -\frac{1}{4} & 0 & 3 \\ 1 & \frac{1}{3} & 0 & 0 & 0 & \frac{1}{3} & 0 & 6 \\ \hline 0 & -2 & 0 & 30 & 0 & 3 & 1 & 264 \end{array}\right] \begin{array}{l} \frac{1}{3}R_2 + R_1 \to R_1 \\ \to \\ -\frac{1}{3}R_2 + R_3 \to R_3 \\ 2R_2 + R_4 \to R_4 \end{array} \left[\begin{array}{ccccccc|c} 0 & 0 & 1 & 1 & \frac{1}{12} & -\frac{5}{12} & 0 & 2 \\ 0 & 1 & 0 & 0 & \frac{1}{4} & -\frac{1}{4} & 0 & 3 \\ 1 & 0 & 0 & 0 & -\frac{1}{12} & \frac{5}{12} & 0 & 5 \\ \hline 0 & 0 & 0 & 30 & \frac{1}{2} & \frac{5}{2} & 1 & 270 \end{array}\right]$$

Solution: $x = 5, y = 3, z = 2; f = 270$

25.

$$\left[\begin{array}{cccccc|c} 1 & 5 & 1 & 0 & 0 & 0 & 500 \\ 1 & 2 & 0 & 1 & 0 & 0 & 230 \\ 1 & 1 & 0 & 0 & 1 & 0 & 160 \\ \hline -4 & -4 & 0 & 0 & 0 & 1 & 0 \end{array}\right] \xrightarrow{\frac{1}{5}R_1 \to R_1} \left[\begin{array}{cccccc|c} \frac{1}{5} & 1 & \frac{1}{5} & 0 & 0 & 0 & 100 \\ 1 & 2 & 0 & 1 & 0 & 0 & 230 \\ 1 & 1 & 0 & 0 & 1 & 0 & 160 \\ \hline -4 & -4 & 0 & 0 & 0 & 1 & 0 \end{array}\right] \begin{array}{l} \to \\ -2R_1 + R_2 \to R_2 \\ -R_1 + R_3 \to R_3 \\ 4R_1 + R_4 \to R_4 \end{array}$$

$$\left[\begin{array}{cccccc|c} \frac{1}{5} & 1 & \frac{1}{5} & 0 & 0 & 0 & 100 \\ \frac{3}{5} & 0 & -\frac{2}{5} & 1 & 0 & 0 & 30 \\ \frac{4}{5} & 0 & -\frac{1}{5} & 0 & 1 & 0 & 60 \\ \hline -\frac{16}{5} & 0 & \frac{4}{5} & 0 & 0 & 1 & 400 \end{array}\right] \xrightarrow{\frac{5}{3}R_2 \to R_2} \left[\begin{array}{cccccc|c} \frac{1}{5} & 1 & \frac{1}{5} & 0 & 0 & 0 & 100 \\ 1 & 0 & -\frac{2}{3} & \frac{5}{3} & 0 & 0 & 50 \\ \frac{4}{5} & 0 & -\frac{1}{5} & 0 & 1 & 0 & 60 \\ \hline -\frac{16}{5} & 0 & \frac{4}{5} & 0 & 0 & 1 & 400 \end{array}\right] \begin{array}{l} -\frac{1}{5}R_2 + R_1 \to R_1 \\ \to \\ -\frac{4}{5}R_2 + R_3 \to R_3 \\ \frac{16}{5}R_2 + R_4 \to R_4 \end{array}$$

$$\left[\begin{array}{cccccc|c} 0 & 1 & \frac{1}{3} & -\frac{1}{3} & 0 & 0 & 90 \\ 1 & 0 & -\frac{2}{3} & \frac{5}{3} & 0 & 0 & 50 \\ 0 & 0 & \frac{1}{3} & -\frac{4}{3} & 1 & 0 & 20 \\ \hline 0 & 0 & -\frac{4}{3} & \frac{16}{3} & 0 & 1 & 560 \end{array}\right] \xrightarrow{3R_3 \to R_3} \left[\begin{array}{cccccc|c} 0 & 1 & \frac{1}{3} & -\frac{1}{3} & 0 & 0 & 90 \\ 1 & 0 & -\frac{2}{3} & \frac{5}{3} & 0 & 0 & 50 \\ 0 & 0 & 1 & -4 & 3 & 0 & 60 \\ \hline 0 & 0 & -\frac{4}{3} & \frac{16}{3} & 0 & 1 & 560 \end{array}\right] \begin{array}{l} -\frac{1}{3}R_3 + R_1 \to R_1 \\ \frac{2}{3}R_3 + R_2 \to R_2 \\ \to \\ \frac{4}{3}R_3 + R_4 \to R_4 \end{array}$$

$$\left[\begin{array}{cccccc|c} 0 & 1 & 0 & 1 & -1 & 0 & 70 \\ 1 & 0 & 0 & -1 & 2 & 0 & 90 \\ 0 & 0 & 1 & -4 & 3 & 0 & 60 \\ \hline 0 & 0 & 0 & 0 & 4 & 1 & 640 \end{array}\right]$$

There are multiple solutions. Maximum is 640 at $x = 90$, $y = 70$.

$$\begin{array}{l} \to \\ R_1 + R_2 \to R_2 \\ 4R_1 + R_3 \to R_3 \end{array} \left[\begin{array}{cccccc|c} 0 & 1 & 0 & 1 & -1 & 0 & 70 \\ 1 & 1 & 0 & 0 & 1 & 0 & 160 \\ 0 & 4 & 1 & 0 & -1 & 0 & 340 \\ \hline 0 & 0 & 0 & 0 & 4 & 1 & 640 \end{array}\right]$$

Second maximum occurs at $x = 160$, $y = 0$.

26.

$$\left[\begin{array}{ccccc|c} -4 & 1 & 1 & 0 & 0 & 40 \\ 1 & -7 & 0 & 1 & 0 & 70 \\ \hline -2 & -5 & 0 & 0 & 1 & 0 \end{array}\right] \begin{array}{l} \to \\ 7R_1 + R_2 \to R_2 \\ 5R_1 + R_3 \to R_3 \end{array} \left[\begin{array}{ccccc|c} -4 & 1 & 1 & 0 & 0 & 40 \\ -27 & 0 & 7 & 1 & 0 & 350 \\ \hline -22 & 0 & 5 & 0 & 1 & 200 \end{array}\right]$$

No solution

27. $\left[\begin{array}{cc|c} 5 & 2 & 16 \\ 3 & 7 & 27 \\ \hline 7 & 6 & g \end{array}\right]$ Dual: $\left[\begin{array}{cc|c} 5 & 3 & 7 \\ 2 & 7 & 6 \\ \hline 16 & 27 & g \end{array}\right]$

$$\left[\begin{array}{ccccc|c} 5 & 3 & 1 & 0 & 0 & 7 \\ 2 & \mathbf{7} & 0 & 1 & 0 & 6 \\ \hline -16 & -27 & 0 & 0 & 1 & 0 \end{array}\right] \begin{array}{c} \rightarrow \\ \frac{1}{7}R_2 \rightarrow R_2 \\ \\ \end{array} \left[\begin{array}{ccccc|c} 5 & 3 & 1 & 0 & 0 & 7 \\ \frac{2}{7} & 1 & 0 & \frac{1}{7} & 0 & \frac{6}{7} \\ \hline -16 & -27 & 0 & 0 & 1 & 0 \end{array}\right] \begin{array}{c} -3R_2 + R_1 \rightarrow R_1 \\ \rightarrow \\ 27R_2 + R_3 \rightarrow R_3 \end{array}$$

$$\left[\begin{array}{ccccc|c} \frac{29}{7} & 0 & 1 & -\frac{3}{7} & 0 & \frac{31}{7} \\ \frac{2}{7} & 1 & 0 & \frac{1}{7} & 0 & \frac{6}{7} \\ \hline -\frac{58}{7} & 0 & 0 & \frac{27}{7} & 1 & \frac{162}{7} \end{array}\right] \begin{array}{c} \frac{7}{29}R_1 \rightarrow R_1 \\ \\ \\ \end{array} \left[\begin{array}{ccccc|c} 1 & 0 & \frac{7}{29} & -\frac{3}{29} & 0 & \frac{31}{29} \\ \frac{2}{7} & 1 & 0 & \frac{1}{7} & 0 & \frac{6}{7} \\ \hline -\frac{58}{7} & 0 & 0 & \frac{27}{7} & 1 & \frac{162}{7} \end{array}\right] \begin{array}{c} \\ -\frac{2}{7}R_1 + R_2 \rightarrow R_2 \\ \frac{58}{7}R_1 + R_3 \rightarrow R_3 \end{array}$$

$$\left[\begin{array}{ccccc|c} 1 & 0 & \frac{7}{29} & -\frac{3}{29} & 0 & \frac{31}{29} \\ 0 & 1 & -\frac{2}{29} & \frac{5}{29} & 0 & \frac{16}{29} \\ \hline 0 & 0 & 2 & 3 & 1 & 32 \end{array}\right]$$

Minimum is 32 at $y_1 = 2, y_2 = 3$.

28. $\left[\begin{array}{cc|c} 3 & 1 & 8 \\ 1 & 1 & 6 \\ 2 & 5 & 18 \\ \hline 3 & 4 & g \end{array}\right] \rightarrow \left[\begin{array}{ccc|c} 3 & 1 & 2 & 3 \\ 1 & 1 & 5 & 4 \\ \hline 8 & 6 & 18 & g \end{array}\right]$

Maximize $g = 8x_1 + 6x_2 + 18x_3$

Constraints: $3x_1 + x_2 + 2x_3 \le 3$

$x_1 + x_2 + 5x_3 \le 4$

$x_1, x_2, x_3 \ge 0$

$$\left[\begin{array}{cccccc|c} 3 & 1 & 2 & 1 & 0 & 0 & 3 \\ 1 & 1 & \mathbf{5} & 0 & 1 & 0 & 4 \\ \hline -8 & -6 & -18 & 0 & 0 & 1 & 0 \end{array}\right] \begin{array}{c} \\ \frac{1}{5}R_2 \rightarrow R_2 \\ \rightarrow \end{array} \left[\begin{array}{cccccc|c} 3 & 1 & 2 & 1 & 0 & 0 & 3 \\ \frac{1}{5} & \frac{1}{5} & 1 & 0 & \frac{1}{5} & 0 & \frac{4}{5} \\ \hline -8 & -6 & -18 & 0 & 0 & 1 & 0 \end{array}\right] \begin{array}{c} -2R_2 + R_1 \rightarrow R_1 \\ \rightarrow \\ 18R_2 + R_3 \rightarrow R_3 \end{array}$$

$$\left[\begin{array}{cccccc|c} \mathbf{\frac{13}{5}} & \frac{3}{5} & 0 & 1 & -\frac{2}{5} & 0 & \frac{7}{5} \\ \frac{1}{5} & \frac{1}{5} & 1 & 0 & \frac{1}{5} & 0 & \frac{4}{5} \\ \hline -\frac{22}{5} & -\frac{12}{5} & 0 & 0 & \frac{18}{5} & 1 & \frac{72}{5} \end{array}\right] \begin{array}{c} \frac{5}{13}R_1 \rightarrow R_1 \\ \rightarrow \\ \\ \end{array} \left[\begin{array}{cccccc|c} 1 & \frac{3}{13} & 0 & \frac{5}{13} & -\frac{2}{13} & 0 & \frac{7}{13} \\ \frac{1}{5} & \frac{1}{5} & 1 & 0 & \frac{1}{5} & 0 & \frac{4}{5} \\ \hline -\frac{22}{5} & -\frac{12}{5} & 0 & 0 & \frac{18}{5} & 1 & \frac{72}{5} \end{array}\right] \begin{array}{c} \rightarrow \\ -\frac{1}{5}R_1 + R_2 \rightarrow R_2 \\ \frac{22}{5}R_1 + R_3 \rightarrow R_3 \end{array}$$

$$\left[\begin{array}{cccccc|c} 1 & \mathbf{\frac{3}{13}} & 0 & \frac{5}{13} & -\frac{2}{13} & 0 & \frac{7}{13} \\ 0 & \frac{2}{13} & 1 & -\frac{1}{13} & \frac{3}{13} & 0 & \frac{9}{13} \\ \hline 0 & -\frac{18}{13} & 0 & \frac{22}{13} & \frac{38}{13} & 1 & \frac{218}{13} \end{array}\right] \begin{array}{c} \frac{13}{3}R_1 \rightarrow R_1 \\ \rightarrow \\ \\ \end{array} \left[\begin{array}{cccccc|c} \frac{13}{3} & 1 & 0 & \frac{5}{3} & -\frac{2}{3} & 0 & \frac{7}{3} \\ 0 & \frac{2}{13} & 1 & -\frac{1}{13} & \frac{3}{13} & 0 & \frac{9}{13} \\ \hline 0 & -\frac{18}{13} & 0 & \frac{22}{13} & \frac{38}{13} & 1 & \frac{218}{13} \end{array}\right] \begin{array}{c} \rightarrow \\ -\frac{2}{13}R_1 + R_2 \rightarrow R_2 \\ \frac{18}{13}R_1 + R_3 \rightarrow R_3 \end{array}$$

$$\left[\begin{array}{cccccc|c} \frac{13}{3} & 1 & 0 & \frac{5}{3} & -\frac{2}{3} & 0 & \frac{7}{3} \\ -\frac{2}{3} & 0 & 1 & -\frac{1}{3} & \frac{1}{3} & 0 & \frac{1}{3} \\ \hline 6 & 0 & 0 & 4 & 2 & 1 & 20 \end{array}\right]$$

Solution: $y_1 = 4, y_2 = 2; g = 20$

29. $\left[\begin{array}{cc|c} 3 & 1 & 8 \\ 1 & 1 & 6 \\ 2 & 5 & 18 \\ \hline 2 & 1 & g \end{array}\right]$ Dual: $\left[\begin{array}{ccc|c} 3 & 1 & 1 & 2 \\ 1 & 1 & 5 & 1 \\ \hline 8 & 6 & 18 & g \end{array}\right]$

$$\left[\begin{array}{cccccc|c} 3 & 1 & 2 & 1 & 0 & 0 & 2 \\ 1 & 1 & 5 & 0 & 1 & 0 & 1 \\ \hline -8 & -6 & -18 & 0 & 0 & 1 & 0 \end{array}\right] \begin{array}{c} \tfrac{1}{5}R_2 \to R_2 \\ \to \end{array} \left[\begin{array}{cccccc|c} 3 & 1 & 2 & 1 & 0 & 0 & 2 \\ \tfrac{1}{5} & \tfrac{1}{5} & 1 & 0 & \tfrac{1}{5} & 0 & \tfrac{1}{5} \\ \hline -8 & -6 & -18 & 0 & 0 & 1 & 0 \end{array}\right] \begin{array}{c} -2R_2 + R_1 \to R_1 \\ \to \\ 18R_2 + R_3 \to R_3 \end{array}$$

$$\left[\begin{array}{cccccc|c} \tfrac{13}{5} & \tfrac{3}{5} & 0 & 1 & -\tfrac{2}{5} & 0 & \tfrac{8}{5} \\ \tfrac{1}{5} & \tfrac{1}{5} & 1 & 0 & \tfrac{1}{5} & 0 & \tfrac{1}{5} \\ \hline -\tfrac{22}{5} & -\tfrac{12}{5} & 0 & 0 & \tfrac{18}{5} & 1 & \tfrac{18}{5} \end{array}\right] \tfrac{5}{13}R_1 \to R_1 \left[\begin{array}{cccccc|c} 1 & \tfrac{3}{13} & 0 & \tfrac{5}{13} & -\tfrac{2}{13} & 0 & \tfrac{8}{13} \\ \tfrac{1}{5} & \tfrac{1}{5} & 1 & 0 & \tfrac{1}{5} & 0 & \tfrac{1}{5} \\ \hline -\tfrac{22}{5} & -\tfrac{12}{5} & 0 & 0 & \tfrac{18}{5} & 1 & \tfrac{18}{5} \end{array}\right] \begin{array}{c} \to \\ -\tfrac{1}{5}R_1 + R_2 \to R_2 \\ \tfrac{22}{5}R_1 + R_3 \to R_3 \end{array}$$

$$\left[\begin{array}{cccccc|c} 1 & \tfrac{3}{13} & 0 & \tfrac{5}{13} & -\tfrac{2}{13} & 0 & \tfrac{8}{13} \\ 0 & \tfrac{2}{13} & 1 & -\tfrac{1}{13} & \tfrac{3}{13} & 0 & \tfrac{1}{13} \\ \hline 0 & -\tfrac{18}{13} & 0 & \tfrac{22}{13} & \tfrac{38}{13} & 1 & \tfrac{82}{13} \end{array}\right] \begin{array}{c} \to \\ \tfrac{13}{2}R_2 \to R_2 \end{array} \left[\begin{array}{cccccc|c} 1 & \tfrac{3}{13} & 0 & \tfrac{5}{13} & -\tfrac{2}{13} & 0 & \tfrac{8}{13} \\ 0 & 1 & \tfrac{13}{2} & -\tfrac{1}{2} & \tfrac{3}{2} & 0 & \tfrac{1}{2} \\ \hline 0 & -\tfrac{18}{13} & 0 & \tfrac{22}{13} & \tfrac{38}{13} & 1 & \tfrac{82}{13} \end{array}\right] \begin{array}{c} -\tfrac{3}{13}R_2 + R_1 \to R_1 \\ \to \\ \tfrac{18}{13}R_2 + R_3 \to R_3 \end{array}$$

$$\left[\begin{array}{cccccc|c} 1 & 0 & -\tfrac{3}{2} & \tfrac{1}{2} & -\tfrac{1}{2} & 0 & \tfrac{1}{2} \\ 0 & 1 & \tfrac{13}{2} & -\tfrac{1}{2} & \tfrac{3}{2} & 0 & \tfrac{1}{2} \\ \hline 0 & 0 & 9 & 1 & 5 & 1 & 7 \end{array}\right]$$

Minimum is 7 at $y_1 = 1, y_2 = 5$.

30. $\left[\begin{array}{ccc|c} 1 & 3 & 0 & 1 \\ 4 & 6 & 1 & 3 \\ 0 & 4 & 1 & 1 \\ \hline 12 & 48 & 8 & g \end{array}\right] \to \left[\begin{array}{ccc|c} 1 & 4 & 0 & 12 \\ 3 & 6 & 4 & 48 \\ 0 & 1 & 1 & 8 \\ \hline 1 & 3 & 1 & g \end{array}\right]$

$$\left[\begin{array}{ccccccc|c} 1 & 4 & 0 & 1 & 0 & 0 & 0 & 12 \\ 3 & 6 & 4 & 0 & 1 & 0 & 0 & 48 \\ 0 & 1 & 1 & 0 & 0 & 1 & 0 & 8 \\ \hline -1 & -3 & -1 & 0 & 0 & 0 & 1 & 0 \end{array}\right] \begin{array}{c} \tfrac{1}{4}R_1 \to R_1 \\ \to \end{array} \left[\begin{array}{ccccccc|c} \tfrac{1}{4} & 1 & 0 & \tfrac{1}{4} & 0 & 0 & 0 & 3 \\ 3 & 6 & 4 & 0 & 1 & 0 & 0 & 48 \\ 0 & 1 & 1 & 0 & 0 & 1 & 0 & 8 \\ \hline -1 & -3 & -1 & 0 & 0 & 0 & 1 & 0 \end{array}\right] \begin{array}{c} \to \\ -6R_1 + R_2 \to R_2 \\ -R_1 + R_3 \to R_3 \\ 3R_1 + R_4 \to R_4 \end{array}$$

$$\left[\begin{array}{ccccccc|c} \tfrac{1}{4} & 1 & 0 & \tfrac{1}{4} & 0 & 0 & 0 & 3 \\ \tfrac{3}{2} & 0 & 4 & -\tfrac{3}{2} & 1 & 0 & 0 & 30 \\ -\tfrac{1}{4} & 0 & 1 & -\tfrac{1}{4} & 0 & 1 & 0 & 5 \\ \hline -\tfrac{1}{4} & 0 & -1 & \tfrac{3}{4} & 0 & 0 & 1 & 9 \end{array}\right] \begin{array}{c} -4R_3 + R_2 \to R_2 \\ \to \\ R_3 + R_4 \to R_4 \end{array} \left[\begin{array}{ccccccc|c} \tfrac{1}{4} & 1 & 0 & \tfrac{1}{4} & 0 & 0 & 0 & 3 \\ \tfrac{5}{2} & 0 & 0 & -\tfrac{1}{2} & 1 & -4 & 0 & 10 \\ -\tfrac{1}{4} & 0 & 1 & -\tfrac{1}{4} & 0 & 1 & 0 & 5 \\ \hline -\tfrac{1}{2} & 0 & 0 & \tfrac{1}{2} & 0 & 1 & 1 & 14 \end{array}\right] \begin{array}{c} \to \\ \tfrac{2}{5}R_2 \to R_2 \end{array}$$

$$\left[\begin{array}{ccccccc|c} \tfrac{1}{4} & 1 & 0 & \tfrac{1}{4} & 0 & 0 & 0 & 3 \\ 1 & 0 & 0 & -\tfrac{1}{5} & \tfrac{2}{5} & -\tfrac{8}{5} & 0 & 4 \\ -\tfrac{1}{4} & 0 & 1 & -\tfrac{1}{4} & 0 & 1 & 0 & 5 \\ \hline -\tfrac{1}{2} & 0 & 0 & \tfrac{1}{2} & 0 & 1 & 1 & 14 \end{array}\right] \begin{array}{c} -\tfrac{1}{4}R_2 + R_1 \to R_1 \\ \to \\ \tfrac{1}{4}R_2 + R_3 \to R_3 \\ \tfrac{1}{2}R_2 + R_4 \to R_4 \end{array} \left[\begin{array}{ccccccc|c} 0 & 1 & 0 & \tfrac{3}{10} & -\tfrac{1}{10} & \tfrac{2}{5} & 0 & 2 \\ 1 & 0 & 0 & -\tfrac{1}{5} & \tfrac{2}{5} & -\tfrac{8}{5} & 0 & 4 \\ 0 & 0 & 1 & -\tfrac{3}{10} & \tfrac{1}{10} & \tfrac{3}{5} & 0 & 6 \\ \hline 0 & 0 & 0 & \tfrac{2}{5} & \tfrac{1}{5} & \tfrac{1}{5} & 1 & 16 \end{array}\right]$$

Solution $y_1 = \dfrac{2}{5}, y_2 = \dfrac{1}{5}, y_3 = \dfrac{1}{5}; g = 16$

31. Maximize $f = 3x + 5y$

Constraints: $-x - y \le -19,\ x - y \le -1,\ -x + 10y \le 190,\ x, y \ge 0$

$$\left[\begin{array}{cccccc|c} -1 & -1 & 1 & 0 & 0 & 0 & -19 \\ 1 & \mathbf{-1} & 0 & 1 & 0 & 0 & -1 \\ -1 & 10 & 0 & 0 & 1 & 0 & 190 \\ \hline -3 & -5 & 0 & 0 & 0 & 1 & 0 \end{array}\right] \begin{array}{c} \\ -R_2 \to R_2 \\ \\ \\ \end{array} \left[\begin{array}{cccccc|c} -1 & -1 & 1 & 0 & 0 & 0 & -19 \\ -1 & 1 & 0 & -1 & 0 & 0 & 1 \\ -1 & 10 & 0 & 0 & 1 & 0 & 190 \\ \hline -3 & -5 & 0 & 0 & 0 & 1 & 0 \end{array}\right] \begin{array}{c} R_2 + R_1 \to R_1 \\ \to \\ -10R_2 + R_3 \to R_3 \\ 5R_2 + R_4 \to R_4 \end{array}$$

$$\left[\begin{array}{cccccc|c} \mathbf{-2} & -0 & 1 & -1 & 0 & 0 & -18 \\ -1 & 1 & 0 & -1 & 0 & 0 & 1 \\ 9 & 0 & 0 & 10 & 1 & 0 & 180 \\ \hline -8 & 0 & 0 & -5 & 0 & 1 & 5 \end{array}\right] \begin{array}{c} -\frac{1}{2}R_1 \to R_1 \\ \to \\ \\ \\ \end{array} \left[\begin{array}{cccccc|c} \mathbf{1} & 0 & -\frac{1}{2} & \frac{1}{2} & 0 & 0 & 9 \\ -1 & 1 & 0 & -1 & 0 & 0 & 1 \\ 9 & 0 & 0 & 10 & 1 & 0 & 180 \\ \hline -8 & 0 & 0 & -5 & 0 & 1 & 5 \end{array}\right] \begin{array}{c} \to \\ R_1 + R_2 \to R_2 \\ -9R_1 + R_3 \to R_3 \\ 8R_1 + R_4 \to R_4 \end{array}$$

$$\left[\begin{array}{cccccc|c} 1 & 0 & -\frac{1}{2} & \frac{1}{2} & 0 & 0 & 9 \\ 0 & 1 & -\frac{1}{2} & -\frac{1}{2} & 0 & 0 & 10 \\ 0 & 0 & \frac{9}{2} & \frac{11}{2} & 1 & 0 & 99 \\ \hline 0 & 0 & -4 & -1 & 0 & 1 & 77 \end{array}\right] \begin{array}{c} \to \\ \\ \frac{2}{9}R_3 \to R_3 \\ \\ \end{array} \left[\begin{array}{cccccc|c} 1 & 0 & -\frac{1}{2} & \frac{1}{2} & 0 & 0 & 9 \\ 0 & 1 & -\frac{1}{2} & -\frac{1}{2} & 0 & 0 & 10 \\ 0 & 0 & 1 & \frac{11}{9} & \frac{2}{9} & 0 & 22 \\ \hline 0 & 0 & -4 & -1 & 0 & 1 & 77 \end{array}\right] \begin{array}{c} \frac{1}{2}R_3 + R_1 \to R_1 \\ \frac{1}{2}R_3 + R_2 \to R_2 \\ \to \\ 4R_3 + R_4 \to R_4 \end{array}$$

$$\left[\begin{array}{cccccc|c} 1 & 0 & 0 & \frac{10}{9} & \frac{1}{9} & 0 & 20 \\ 0 & 1 & 0 & \frac{1}{9} & \frac{1}{9} & 0 & 21 \\ 0 & 0 & 1 & \frac{11}{9} & \frac{2}{9} & 0 & 22 \\ \hline 0 & 0 & 0 & \frac{35}{9} & \frac{8}{9} & 1 & 165 \end{array}\right]$$

Maximum is 165 at $x = 20, y = 21$.

32.

$$\left[\begin{array}{cccccc|c} 2 & 5 & 1 & 0 & 0 & 0 & 37 \\ 5 & -1 & 0 & 1 & 0 & 0 & 34 \\ 1 & \mathbf{-2} & 0 & 0 & 1 & 0 & -4 \\ \hline -4 & -6 & 0 & 0 & 0 & 1 & 0 \end{array}\right] \begin{array}{c} \to \\ \\ -\frac{1}{2}R_3 \to R_3 \\ \to \end{array} \left[\begin{array}{cccccc|c} 2 & 5 & 1 & 0 & 0 & 0 & 37 \\ 5 & -1 & 0 & 1 & 0 & 0 & 34 \\ -\frac{1}{2} & 1 & 0 & 0 & -\frac{1}{2} & 0 & 2 \\ \hline -4 & -6 & 0 & 0 & 0 & 1 & 0 \end{array}\right] \begin{array}{c} -5R_3 + R_1 \to R_1 \\ R_3 + R_2 \to R_2 \\ \to \\ 6R_3 + R_4 \to R_4 \end{array}$$

$$\left[\begin{array}{cccccc|c} \frac{9}{2} & 0 & 1 & 0 & \frac{5}{2} & 0 & 27 \\ \frac{9}{2} & 0 & 0 & 1 & -\frac{1}{2} & 0 & 36 \\ -\frac{1}{2} & 1 & 0 & 0 & -\frac{1}{2} & 0 & 2 \\ \hline -7 & 0 & 0 & 0 & -3 & 1 & 12 \end{array}\right] \begin{array}{c} \frac{2}{9}R_1 \to R_1 \\ \to \\ \\ \\ \end{array} \left[\begin{array}{cccccc|c} 1 & 0 & \frac{2}{9} & 0 & \frac{5}{9} & 0 & 6 \\ \frac{9}{2} & 0 & 0 & 1 & -\frac{1}{2} & 0 & 36 \\ -\frac{1}{2} & 1 & 0 & 0 & -\frac{1}{2} & 0 & 2 \\ \hline -7 & 0 & 0 & 0 & -3 & 1 & 12 \end{array}\right] \begin{array}{c} \to \\ -\frac{9}{2}R_1 + R_2 \to R_2 \\ \frac{1}{2}R_1 + R_3 \to R_3 \\ 7R_1 + R_4 \to R_4 \end{array}$$

$$\left[\begin{array}{cccccc|c} 1 & 0 & \frac{2}{9} & 0 & \frac{5}{9} & 0 & 6 \\ 0 & 0 & -1 & 1 & -3 & 0 & 9 \\ 0 & 1 & \frac{1}{9} & 0 & -\frac{2}{9} & 0 & 5 \\ \hline 0 & 0 & \frac{14}{9} & 0 & \frac{8}{9} & 1 & 54 \end{array}\right]$$

Solution: $x = 6, y = 5; f = 54$

33. Maximize : $-f=-10x-3y$

Constraints: $x-10y\le -5,\ -4x-y\le -62,\ x+y\le 50,\ x, y\ge 0$

$$\left[\begin{array}{cccccc|c} 1 & -10 & 1 & 0 & 0 & 0 & -5\\ -4 & -1 & 0 & 1 & 0 & 0 & -62\\ 1 & 1 & 0 & 0 & 1 & 0 & 50\\ \hline 10 & 3 & 0 & 0 & 0 & 1 & 0 \end{array}\right] -\tfrac{1}{10}R_1\to R_1 \quad \left[\begin{array}{cccccc|c} -\frac{1}{10} & 1 & -\frac{1}{10} & 0 & 0 & 0 & \frac{1}{2}\\ -4 & -1 & 0 & 1 & 0 & 0 & -62\\ 1 & 1 & 0 & 0 & 1 & 0 & 50\\ \hline 10 & 3 & 0 & 0 & 0 & 1 & 0 \end{array}\right] \begin{array}{c}\to\\ R_1+R_2\to R_2\\ -R_1+R_3\to R_3\\ -3R_1+R_4\to R_4\end{array}$$

$$\left[\begin{array}{cccccc|c} -\frac{1}{10} & 1 & -\frac{1}{10} & 0 & 0 & 0 & \frac{1}{2}\\ -\frac{41}{10} & 0 & -\frac{1}{10} & 1 & 0 & 0 & -\frac{123}{2}\\ \frac{11}{10} & 0 & \frac{1}{10} & 0 & 1 & 0 & \frac{99}{2}\\ \hline \frac{103}{10} & 0 & \frac{3}{10} & 0 & 0 & 1 & -\frac{3}{2} \end{array}\right] \begin{array}{c} -\frac{10}{41}R_2\to R_2\\ \to \end{array} \quad \left[\begin{array}{cccccc|c} -\frac{1}{10} & 1 & -\frac{1}{10} & 0 & 0 & 0 & \frac{1}{2}\\ 1 & 0 & \frac{1}{41} & -\frac{10}{41} & 0 & 0 & 15\\ \frac{11}{10} & 0 & \frac{1}{10} & 0 & 1 & 0 & \frac{99}{2}\\ \hline \frac{103}{10} & 0 & \frac{3}{10} & 0 & 0 & 1 & -\frac{3}{2} \end{array}\right] \begin{array}{c}\frac{1}{10}R_2+R_1\to R_1\\ \to\\ -\frac{11}{10}R_2+R_3\to R_3\\ -\frac{103}{10}R_2+R_4\to R_4\end{array}$$

$$\left[\begin{array}{cccccc|c} 0 & 1 & -\frac{4}{41} & -\frac{1}{41} & 0 & 0 & 2\\ 1 & 0 & \frac{1}{41} & -\frac{10}{41} & 0 & 0 & 15\\ 0 & 0 & \frac{3}{41} & \frac{11}{41} & 1 & 0 & 33\\ \hline 0 & 0 & \frac{2}{41} & \frac{103}{41} & 0 & 1 & -156 \end{array}\right]$$

Minimum is 156 at $x = 15, y = 2$.

34. Maximize $-f=-4x-3y$

Subject to: $x-y\le -1,\ x+y\le 45,\ -10x-y\le -45$

$$\left[\begin{array}{cccccc|c} 1 & -1 & 1 & 0 & 0 & 0 & -1\\ 1 & 1 & 0 & 1 & 0 & 0 & 45\\ -10 & -1 & 0 & 0 & 1 & 0 & -45\\ \hline 4 & 3 & 0 & 0 & 0 & 1 & 0 \end{array}\right] -R_1\to R_1 \quad \left[\begin{array}{cccccc|c} -1 & 1 & -1 & 0 & 0 & 0 & 1\\ 1 & 1 & 0 & 1 & 0 & 0 & 45\\ -10 & -1 & 0 & 0 & 1 & 0 & -45\\ \hline 4 & 3 & 0 & 0 & 0 & 1 & 0 \end{array}\right] \begin{array}{c}\to\\ -R_1+R_2\to R_2\\ R_1+R_3\to R_3\\ -3R_1+R_4\to R_4\end{array}$$

$$\left[\begin{array}{cccccc|c} -1 & 1 & -1 & 0 & 0 & 0 & 1\\ 2 & 0 & 1 & 1 & 0 & 0 & 44\\ -11 & 0 & -1 & 0 & 1 & 0 & -44\\ \hline 7 & 0 & 3 & 0 & 0 & 1 & -3 \end{array}\right] -\tfrac{1}{11}R_3\to R_3 \quad \left[\begin{array}{cccccc|c} -1 & 1 & -1 & 0 & 0 & 0 & 1\\ 2 & 0 & 1 & 1 & 0 & 0 & 44\\ 1 & 0 & \frac{1}{11} & 0 & -\frac{1}{11} & 0 & 4\\ \hline 7 & 0 & 3 & 0 & 0 & 1 & -3 \end{array}\right] \begin{array}{c}R_3+R_1\to R_1\\ -2R_3+R_2\to R_2\\ \to\\ -7R_3+R_4\to R_4\end{array}$$

$$\left[\begin{array}{cccccc|c} 0 & 1 & -\frac{10}{11} & 0 & -\frac{1}{11} & 0 & 5\\ 0 & 0 & \frac{9}{11} & 1 & \frac{2}{11} & 0 & 36\\ 1 & 0 & \frac{1}{11} & 0 & -\frac{1}{11} & 0 & 4\\ \hline 0 & 0 & \frac{26}{11} & 0 & \frac{7}{11} & 1 & -31 \end{array}\right]$$

Minimum is 31 at $x = 4, y = 5$.

35. Let x = number of large sets and y = number of small sets.

Maximize $f=100x+50y$

Constraints: $5x+2y\le 700,\ x+y\le 185$

Solving the two equations we get $A(110,75)$.

Feasible Corners	$f=100x+50y$
$(0,185)$	9250
$(110,75)$	14,750
$(140,0)$	14,000

Maximum profit is \$14,750 with 110 large sets and 75 small sets.

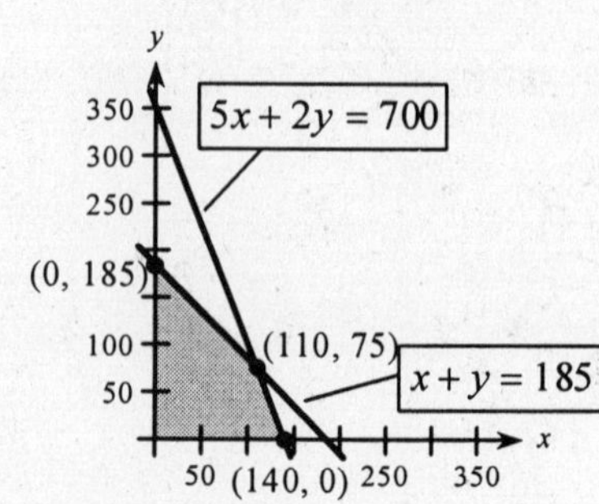

36. x_1 = days for factory 1, x_2 = days for factory 2

Minimize: $g = 5000x_1 + 6000x_2$

Subject to: $x_1 + 2x_2 \geq 80$

$3x_1 + 2x_2 \geq 140$

$x_1, x_2 \geq 0$

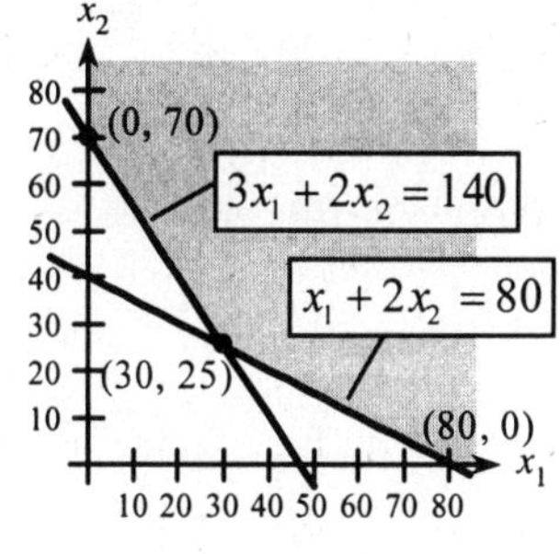

$x_1 + 2x_2 = 80$ and $3x_1 + 2x_2 = 140$ intersect at $x_1 = 30, x_2 = 25$.

The feasible corners are (80, 0), (30, 25), and (0, 70).

At (80, 0), $g = 400{,}000$

At (30, 25), $g = 300{,}000$ ← Minimum

At (0, 70), $g = 420{,}000$

The minimum cost is \$300,000 when factory 1 operates 30 days and factory 2 operates 25 days.

37. Let x = number of Is and y = number of IIs.

Maximize $P = 6x + 4y$

Constraints: $2x + y \leq 100$, $x + y \leq 60$

$$\left[\begin{array}{ccccc|c} 2 & 1 & 1 & 0 & 0 & 100 \\ 1 & 1 & 0 & 1 & 0 & 60 \\ \hline -6 & -4 & 0 & 0 & 1 & 0 \end{array}\right] \begin{array}{c} \tfrac{1}{2}R_1 \to R_1 \\ \to \\ \end{array} \left[\begin{array}{ccccc|c} 1 & \tfrac{1}{2} & \tfrac{1}{2} & 0 & 0 & 50 \\ 1 & 1 & 0 & 1 & 0 & 60 \\ \hline -6 & -4 & 0 & 0 & 1 & 0 \end{array}\right] \begin{array}{c} \to \\ -R_1 + R_2 \to R_2 \\ 6R_1 + R_3 \to R_3 \end{array}$$

$$\left[\begin{array}{ccccc|c} 1 & \tfrac{1}{2} & \tfrac{1}{2} & 0 & 0 & 50 \\ 0 & \tfrac{1}{2} & -\tfrac{1}{2} & 1 & 0 & 10 \\ \hline 0 & -1 & 3 & 0 & 1 & 300 \end{array}\right] \begin{array}{c} \to \\ 2R_2 \to R_2 \\ \end{array} \left[\begin{array}{ccccc|c} 1 & \tfrac{1}{2} & \tfrac{1}{2} & 0 & 0 & 50 \\ 0 & 1 & -1 & 2 & 0 & 20 \\ \hline 0 & -1 & 3 & 0 & 1 & 300 \end{array}\right] \begin{array}{c} -\tfrac{1}{2}R_2 + R_1 \to R_1 \\ \to \\ R_2 + R_3 \to R_3 \end{array}$$

$$\left[\begin{array}{ccccc|c} 1 & 0 & 1 & -1 & 0 & 40 \\ 0 & 1 & -1 & 2 & 0 & 20 \\ \hline 0 & 0 & 2 & 2 & 1 & 320 \end{array}\right]$$

Maximum profit is \$320 with 40 of the Is and 20 of the IIs.

38. Let x = number of Jacob's Ladders, let y = number of locomotive engines.

Maximize $P = 3x + 5y$

Constraints: $x + y \leq 120$ finishing hours

$\tfrac{1}{2}x + y \leq 75$ carpentry hours

$x \leq 100$

$x, y \geq 0$

$$\left[\begin{array}{cccccc|c} 1 & 1 & 1 & 0 & 0 & 0 & 120 \\ \tfrac{1}{2} & 1 & 0 & 1 & 0 & 0 & 75 \\ 1 & 0 & 0 & 0 & 1 & 0 & 100 \\ \hline -3 & -5 & 0 & 0 & 0 & 1 & 0 \end{array}\right] \begin{array}{c} -R_2 + R_1 \to R_1 \\ \to \\ \\ 5R_2 + R_4 \to R_4 \end{array} \left[\begin{array}{cccccc|c} \tfrac{1}{2} & 0 & 1 & -1 & 0 & 0 & 45 \\ \tfrac{1}{2} & 1 & 0 & 1 & 0 & 0 & 75 \\ 1 & 0 & 0 & 0 & 1 & 0 & 100 \\ \hline -\tfrac{1}{2} & 0 & 0 & 5 & 0 & 1 & 375 \end{array}\right] \begin{array}{c} 2R_1 \to R_1 \\ \\ \\ \end{array}$$

$$\left[\begin{array}{cccccc|c} 1 & 0 & 2 & -2 & 0 & 0 & 90 \\ \tfrac{1}{2} & 1 & 0 & 1 & 0 & 0 & 75 \\ 1 & 0 & 0 & 0 & 1 & 0 & 100 \\ \hline -\tfrac{1}{2} & 0 & 0 & 5 & 0 & 1 & 375 \end{array}\right] \begin{array}{c} \\ -\tfrac{1}{2}R_1 + R_2 \to R_2 \\ -R_1 + R_3 \to R_3 \\ \tfrac{1}{2}R_1 + R_4 \to R_4 \end{array} \left[\begin{array}{cccccc|c} 1 & 0 & 2 & -2 & 0 & 0 & 90 \\ 0 & 1 & -1 & 2 & 0 & 0 & 30 \\ 0 & 0 & -2 & 2 & 1 & 0 & 10 \\ \hline 0 & 0 & 1 & 4 & 0 & 1 & 420 \end{array}\right]$$

Maximum profit is \$420, obtained by producing 90 Jacob's Ladders and 30 locomotive engines.

39. We display the information in a table.

	x food I	y food II	Requirement
A	2	10	5
B	1	10	30
cost	30 cents/ ounce	20 cents/ ounce	

Minimize: $g = 30x + 20y$

Subject to: $2x + 10y \geq 5$

$x + 10y \geq 30$

$x,\ y \geq 0$

primal: $\left[\begin{array}{cc|c} 2 & 10 & 5 \\ 1 & 10 & 30 \\ \hline 30 & 20 & g \end{array}\right]$ dual: $\left[\begin{array}{cc|c} 2 & 1 & 30 \\ 10 & 10 & 20 \\ \hline 5 & 30 & g \end{array}\right]$

Maximization dual problem:

$$\left[\begin{array}{ccccc|c} 2 & 1 & 1 & 0 & 0 & 30 \\ 10 & \mathbf{10} & 0 & 1 & 0 & 20 \\ \hline -5 & -30 & 0 & 0 & 1 & 0 \end{array}\right] \begin{array}{c} \\ \frac{1}{10}R_2 \to R_2 \\ \to \end{array} \left[\begin{array}{ccccc|c} 2 & 1 & 1 & 0 & 0 & 30 \\ 1 & 1 & 0 & \frac{1}{10} & 0 & 2 \\ \hline -5 & -30 & 0 & 0 & 1 & 0 \end{array}\right] \begin{array}{c} -R_2 + R_1 \to R_1 \\ \to \\ 30R_2 + R_3 \to R_3 \end{array}$$

$$\left[\begin{array}{ccccc|c} 1 & 0 & 1 & -\frac{1}{10} & 0 & 28 \\ 1 & 1 & 0 & \frac{1}{10} & 0 & 2 \\ \hline 25 & 0 & 0 & 3 & 1 & 60 \end{array}\right]$$

Use none of food I and 3 ounces of food II for a minimum cost of 60 cents.

40. Minimize $C = 14A + 16B$

Constraints: $A + 4B \geq 40$

$2A + B \geq 80$

$\left[\begin{array}{cc|c} 1 & 4 & 40 \\ 2 & 1 & 80 \\ \hline 14 & 16 & C \end{array}\right]$ Dual: $\left[\begin{array}{cc|c} 1 & 2 & 14 \\ 4 & 1 & 16 \\ \hline 40 & 80 & C \end{array}\right]$

$$\left[\begin{array}{ccccc|c} 1 & \mathbf{2} & 1 & 0 & 0 & 14 \\ 4 & 1 & 0 & 1 & 0 & 16 \\ \hline -40 & -80 & 0 & 0 & 1 & 0 \end{array}\right] \begin{array}{c} \frac{1}{2}R_1 \to R_1 \\ \to \\ \ \end{array}$$

$$\left[\begin{array}{ccccc|c} \frac{1}{2} & 1 & \frac{1}{2} & 0 & 0 & 7 \\ 4 & 1 & 0 & 1 & 0 & 16 \\ \hline -40 & -80 & 0 & 0 & 1 & 0 \end{array}\right] \begin{array}{c} -R_1 + R_2 \to R_2 \\ 80R_1 + R_3 \to R_3 \\ \ \end{array} \left[\begin{array}{ccccc|c} \frac{1}{2} & 1 & \frac{1}{2} & 0 & 0 & 7 \\ \frac{7}{2} & 0 & -\frac{1}{2} & 1 & 0 & 9 \\ \hline 0 & 0 & 40 & 0 & 1 & 560 \end{array}\right]$$

Minimum cost is $C = \$5.60$ with 40 units of A and 0 units of B.

41. x = days at factory A, y = days of factory B, z = days at factory C
Minimize $g = 200x + 300y + 500z$

Subject to:
$$\begin{aligned} 10x + 20z &\ge 200 &\rightarrow\quad x + 2z &\ge 20 \\ 10x + 20y + 20z &\ge 500 & x + 2y + 2z &\ge 50 \\ 10x + 20y + 10z &\ge 300 & x + 2y + z &\ge 30 \\ x, y, z &\ge 0 \end{aligned}$$

$$\left[\begin{array}{ccc|c} 1 & 0 & 2 & 20 \\ 1 & 2 & 2 & 50 \\ 1 & 2 & 1 & 30 \\ \hline 200 & 300 & 500 & g \end{array}\right] \qquad \text{Dual: } \left[\begin{array}{ccc|c} 1 & 1 & 1 & 200 \\ 0 & 2 & 2 & 300 \\ 2 & 2 & 1 & 500 \\ \hline 20 & 50 & 30 & g \end{array}\right]$$

$$\left[\begin{array}{ccccccc|c} 1 & 1 & 1 & 1 & 0 & 0 & 0 & 200 \\ 0 & \mathbf{2} & 2 & 0 & 1 & 0 & 0 & 300 \\ 2 & 2 & 1 & 0 & 0 & 1 & 0 & 500 \\ \hline -20 & -50 & -30 & 0 & 0 & 0 & 1 & 0 \end{array}\right] \begin{array}{c} \\ \frac{1}{2}R_2 \rightarrow R_2 \\ \\ \\ \end{array} \left[\begin{array}{ccccccc|c} 1 & 1 & 1 & 1 & 0 & 0 & 0 & 200 \\ 0 & 1 & 1 & 0 & \frac{1}{2} & 0 & 0 & 150 \\ 2 & 2 & 1 & 0 & 0 & 1 & 0 & 500 \\ \hline -20 & -50 & -30 & 0 & 0 & 0 & 1 & 0 \end{array}\right] \begin{array}{c} -R_2 + R_1 \rightarrow R_1 \\ \rightarrow \\ -2R_2 + R_3 \rightarrow R_3 \\ 50R_2 + R_4 \rightarrow R_4 \end{array}$$

$$\left[\begin{array}{ccccccc|c} \mathbf{1} & 0 & 0 & 1 & -\frac{1}{2} & 0 & 0 & 50 \\ 0 & 1 & 1 & 0 & \frac{1}{2} & 0 & 0 & 150 \\ 2 & 0 & -1 & 0 & -1 & 1 & 0 & 200 \\ \hline -20 & 0 & 20 & 0 & 25 & 0 & 1 & 7500 \end{array}\right] \begin{array}{c} \rightarrow \\ \\ -2R_1 + R_3 \rightarrow R_3 \\ 20R_1 + R_4 \rightarrow R_4 \end{array} \left[\begin{array}{ccccccc|c} 1 & 0 & 0 & 1 & -\frac{1}{2} & 0 & 0 & 50 \\ 0 & 1 & 1 & 0 & \frac{1}{2} & 0 & 0 & 150 \\ 0 & 0 & -1 & -2 & 0 & 1 & 0 & 100 \\ \hline 0 & 0 & 20 & 20 & 15 & 0 & 1 & 8500 \end{array}\right]$$

Solution: $x = 20, y = 15, z = 0$; minimum = \$8500

42.

	Flour	Shortening	Sugar	Objective Function
Pancake: x	$0.6x$	$0.1x$	--	$P = 0.35x + 0.25y$
Cake: y	$0.4y$	$0.1y$	$0.4y$	

Constraints: $0.6x + 0.4y \le 6000$ Flour

$0.1x + 0.1y \ge 500$ Shortening

$0.4y \le 1200$ Sugar

$x, y \ge 0$

$$\left[\begin{array}{cccccc|c} 0.6 & 0.4 & 1 & 0 & 0 & 0 & 6000 \\ \mathbf{-0.1} & -0.1 & 0 & 1 & 0 & 0 & -500 \\ 0 & 0.4 & 0 & 0 & 1 & 0 & 1200 \\ \hline -0.35 & -0.25 & 0 & 0 & 0 & 1 & 0 \end{array}\right] \begin{array}{l} -10R_2 \to R_2 \\ \to \end{array}$$

$$\left[\begin{array}{cccccc|c} 0.6 & 0.4 & 1 & 0 & 0 & 0 & 6000 \\ 1 & 1 & 0 & -10 & 0 & 0 & 5000 \\ 0 & 0.4 & 0 & 0 & 1 & 0 & 1200 \\ \hline -0.35 & -0.25 & 0 & 0 & 0 & 1 & 0 \end{array}\right] \begin{array}{l} -.6R_2 + R_1 \to R_1 \\ \to \\ 0.35R_2 + R_4 \to R_4 \end{array}$$

$$\left[\begin{array}{cccccc|c} 0 & -0.2 & 1 & \mathbf{6} & 0 & 0 & 3000 \\ 1 & 1 & 0 & -10 & 0 & 0 & 5000 \\ 0 & 0.4 & 0 & 0 & 1 & 0 & 1200 \\ \hline 0 & 0.1 & 0 & -3.5 & 0 & 1 & 1750 \end{array}\right] \tfrac{1}{6}R_1 \to R_1$$

$$\left[\begin{array}{cccccc|c} 0 & -0.0\overline{3} & 0.1\overline{6} & 1 & 0 & 0 & 500 \\ 1 & 1 & 0 & -10 & 0 & 0 & 5000 \\ 0 & 0.4 & 0 & 0 & 1 & 0 & 1200 \\ \hline 0 & 0.1 & 0 & -3.5 & 0 & 1 & 1750 \end{array}\right] \begin{array}{l} 10R_1 + R_2 \to R_2 \\ \to \\ 3.5R_1 + R_4 \to R_4 \end{array}$$

$$\left[\begin{array}{cccccc|c} 0 & -0.0\overline{3} & 0.1\overline{6} & 1 & 0 & 0 & 500 \\ 1 & 0.\overline{6} & 1.\overline{6} & 0 & 0 & 0 & 10,000 \\ 0 & \mathbf{0.4} & 0 & 0 & 1 & 0 & 1200 \\ \hline 0 & -0.01\overline{6} & 0.58\overline{3} & 0 & 0 & 1 & 3500 \end{array}\right] \begin{array}{l} 2.5R_3 \to R_3 \\ \to \end{array}$$

$$\left[\begin{array}{cccccc|c} 0 & -0.0\overline{3} & 0.1\overline{6} & 1 & 0 & 0 & 500 \\ 1 & 0.\overline{6} & 1.\overline{6} & 0 & 0 & 0 & 10,000 \\ 0 & 1 & 0 & 0 & 2.5 & 0 & 3000 \\ \hline 0 & -0.01\overline{6} & 0.58\overline{3} & 0 & 0 & 1 & 3500 \end{array}\right] \begin{array}{l} 0.0\overline{3}R_3 + R_1 \to R_1 \\ \to \\ -0.\overline{6}R_3 + R_2 \to R_2 \\ 0.01\overline{6}R_3 + R_4 \to R_4 \end{array}$$

$$\left[\begin{array}{cccccc|c} 0 & 0 & 0.1\overline{6} & 1 & 0.08\overline{3} & 0 & 600 \\ 1 & 0 & 1.\overline{6} & 0 & -1.\overline{6} & 0 & 8000 \\ 0 & 1 & 0 & 0 & 2.5 & 0 & 3000 \\ \hline 0 & 0 & 0.58\overline{3} & 0 & 0.041\overline{6} & 1 & 3550 \end{array}\right]$$

Maximum profit is \$3550 with 8000 lbs of pancake mix and 3000 lbs of cake mix.

43. x_1 = number of desks at Texas, x_2 = number of tables at Texas,

x_3 = number of desks at Louisiana, x_4 = number of tables at Louisiana

Minimize $g = 12x_1 + 20x_2 + 14x_3 + 19x_4$

Subject to: $x_1 + x_2 \le 120,\ x_3 + x_4 \le 150,\ x_1 + x_3 \ge 130,\ x_2 + x_4 \ge 130,\ -x_1 + x_2 \ge 10,\ x_1, x_2, x_3, x_4 \ge 0$

$$\left[\begin{array}{cccccccccc|c} 1 & 1 & 0 & 0 & 1 & 0 & 0 & 0 & 0 & 0 & 120 \\ 0 & 0 & 1 & 1 & 0 & 1 & 0 & 0 & 0 & 0 & 150 \\ \mathbf{-1} & 0 & -1 & 0 & 0 & 0 & 1 & 0 & 0 & 0 & -130 \\ 0 & -1 & 0 & -1 & 0 & 0 & 0 & 1 & 0 & 0 & -130 \\ 1 & -1 & 0 & 0 & 0 & 0 & 0 & 0 & 1 & 0 & -10 \\ \hline 12 & 20 & 14 & 19 & 0 & 0 & 0 & 0 & 0 & 1 & 0 \end{array}\right] \begin{array}{l} \to \\ \\ -R_3 \to R_3 \\ \to \\ \\ \\ \end{array}$$

$$\left[\begin{array}{cccccccccc|c} 1 & 1 & 0 & 0 & 1 & 0 & 0 & 0 & 0 & 0 & 120 \\ 0 & 0 & 1 & 1 & 0 & 1 & 0 & 0 & 0 & 0 & 150 \\ 1 & 0 & 1 & 0 & 0 & 0 & -1 & 0 & 0 & 0 & 130 \\ 0 & -1 & 0 & -1 & 0 & 0 & 0 & 1 & 0 & 0 & -130 \\ 1 & -1 & 0 & 0 & 0 & 0 & 0 & 0 & 1 & 0 & -10 \\ \hline 12 & 20 & 14 & 19 & 0 & 0 & 0 & 0 & 0 & 1 & 0 \end{array}\right] \begin{array}{l} -R_3 + R_1 \to R_1 \\ \\ \to \\ \\ -R_3 + R_5 \to R_5 \\ -12R_3 + R_6 \to R_6 \end{array}$$

$$\left[\begin{array}{cccccccccc|c} 0 & 1 & -1 & 0 & 1 & 0 & 1 & 0 & 0 & 0 & -10 \\ 0 & 0 & 1 & 1 & 0 & 1 & 0 & 0 & 0 & 0 & 150 \\ 1 & 0 & 1 & 0 & 0 & 0 & -1 & 0 & 0 & 0 & 130 \\ 0 & -1 & 0 & -1 & 0 & 0 & 0 & 1 & 0 & 0 & -130 \\ 0 & \mathbf{-1} & -1 & 0 & 0 & 0 & 1 & 0 & 1 & 0 & -140 \\ \hline 0 & 20 & 2 & 19 & 0 & 0 & 12 & 0 & 0 & 1 & -1560 \end{array}\right] \left.\begin{array}{l} R_5 + R_1 \to R_1 \\ \to \\ \\ -R_5 + R_4 \to R_4 \\ \\ 20R_5 + R_6 \to R_6 \end{array}\right\} \text{then } -R_5 \to R_5$$

$$\left[\begin{array}{cccccccccc|c} 0 & 0 & -2 & 0 & 1 & 0 & 2 & 0 & 1 & 0 & -150 \\ 0 & 0 & 1 & 1 & 0 & 1 & 0 & 0 & 0 & 0 & 150 \\ 1 & 0 & 1 & 0 & 0 & 0 & -1 & 0 & 0 & 0 & 130 \\ 0 & 0 & \mathbf{1} & -1 & 0 & 0 & -1 & 1 & -1 & 0 & 10 \\ 0 & 1 & 1 & 0 & 0 & 0 & -1 & 0 & -1 & 0 & 140 \\ \hline 0 & 0 & -18 & 19 & 0 & 0 & 32 & 0 & 20 & 1 & -4360 \end{array}\right] \begin{array}{l} 2R_4 + R_1 \to R_1 \\ -R_4 + R_2 \to R_2 \\ -R_4 + R_3 \to R_3 \\ \to \\ -R_4 + R_5 \to R_5 \\ 18R_4 + R_6 \to R_6 \end{array}$$

$$\left[\begin{array}{cccccccccc|c} 0 & 0 & 0 & \mathbf{-2} & 1 & 0 & 0 & 2 & -1 & 0 & -130 \\ 0 & 0 & 0 & 2 & 0 & 1 & 1 & -1 & 1 & 0 & 140 \\ 1 & 0 & 0 & 1 & 0 & 0 & 0 & -1 & 1 & 0 & 120 \\ 0 & 0 & 1 & -1 & 0 & 0 & -1 & 1 & -1 & 0 & 10 \\ 0 & 1 & 0 & 1 & 0 & 0 & 0 & -1 & 0 & 0 & 130 \\ \hline 0 & 0 & 0 & 1 & 0 & 0 & 14 & 18 & 2 & 1 & -4180 \end{array}\right] \left.\begin{array}{l} \\ R_1 + R_2 \to R_2 \\ \frac{1}{2}R_1 + R_3 \to R_3 \\ -\frac{1}{2}R_1 + R_4 \to R_4 \\ \frac{1}{2}R_1 + R_5 \to R_5 \\ \frac{1}{2}R_1 + R_6 \to R_6 \end{array}\right\} \text{then } -\tfrac{1}{2}R_1 \to R_1$$

$$\left[\begin{array}{cccccccccc|c} 0 & 0 & 0 & 1 & -\frac{1}{2} & 0 & 0 & -1 & \frac{1}{2} & 0 & 65 \\ 0 & 0 & 0 & 0 & 1 & 1 & 1 & 1 & 0 & 0 & 10 \\ 1 & 0 & 0 & 0 & \frac{1}{2} & 0 & 0 & 0 & \frac{1}{2} & 0 & 55 \\ 0 & 0 & 1 & 0 & -\frac{1}{2} & 0 & -1 & 0 & -\frac{1}{2} & 0 & 75 \\ 0 & 1 & 0 & 0 & \frac{1}{2} & 0 & 0 & 0 & -\frac{1}{2} & 0 & 65 \\ \hline 0 & 0 & 0 & 0 & \frac{1}{2} & 0 & 14 & 19 & \frac{3}{2} & 1 & -4245 \end{array}\right]$$

Texas: 55 desks, 65 tables; Louisiana: 75 desks, 65 tables; cost = \$4245

Chapter Test

1. $-\frac{7}{3}t \le 21$

$t \ge 21 \cdot \left(-\frac{3}{7}\right)$

$t \ge -9$

2.

Corners	$f = 3x + 5y$
$(0,24)$	120
$(8,18)$	114
$(20,0)$	60
$(0,0)$	0

Maximum is 120 at (0,24).

3. **a.** *C* Multiple solution is possible since there is a zero indicator in a nonbasic variable column. Pivot element is R_2, C_2.
Steps needed :
$-2R_2 + R_1 \to R_1; -3R_2 + R_3 \to R_3$

b. *A* Pivot column is C_3 but no pivot element is defined.

4. $(-1, 4]: -1 < x \le 4$

5. $\left[\begin{array}{ccc|c} 3 & 5 & 1 & 100 \\ 4 & 6 & 3 & 120 \\ \hline 2 & 3 & 5 & g \end{array}\right]$

Take the transpose to obtain:
Maximize $f = 100x_1 + 120x_2$
Subject to: $3x_1 + 4x_2 \le 2$
$5x_1 + 6x_2 \le 3$
$x_1 + 3x_2 \le 5$
$x_1, x_2 \ge 0$

6. **a.**

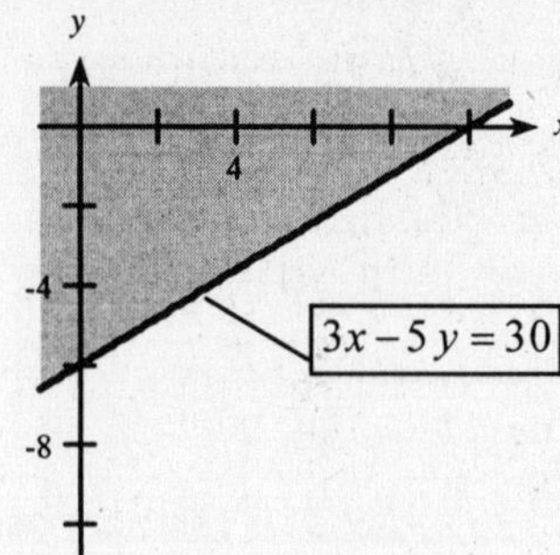

b.

7. Maximize (change signs) $-g = -7x - 3y$
Subject to: $x - 4y \le -4$
$x - y \le 5$
$2x + 3y \le 30$

8. $x > 2$

a.

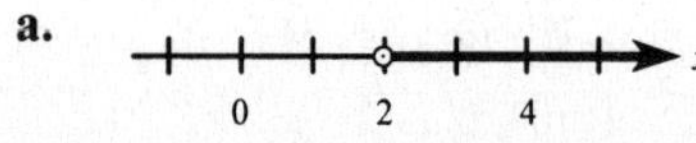

b.

9. Maximum is $f = 658$ at $x_1 = 17, x_2 = 15, x_3 = 0.$
Minimum is $g = 658$ at $y_1 = 4, y_2 = 18, y_3 = 0.$

10. 

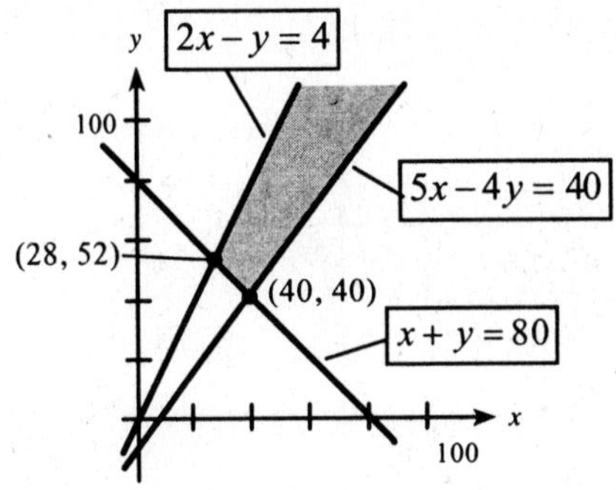

Minimum is 136 at (28, 52). There are only two feasible corners.

11. Maximize: $f = 70x + 5y$

Subject to: $x + 1.5y \le 150$

$x + 0.5y \le 90$

$x, y \ge 0$

Clear of decimals

$$\left[\begin{array}{ccccc|c} 2 & 3 & 1 & 0 & 0 & 300 \\ 2 & 1 & 0 & 1 & 0 & 180 \\ \hline -70 & -5 & 0 & 0 & 1 & 0 \end{array}\right] \begin{array}{l} \\ \frac{1}{2}R_2 \to R_2 \\ \to \end{array}$$

$$\left[\begin{array}{ccccc|c} 2 & 3 & 1 & 0 & 0 & 300 \\ 1 & \frac{1}{2} & 0 & \frac{1}{2} & 0 & 90 \\ \hline -70 & -5 & 0 & 0 & 1 & 0 \end{array}\right] \begin{array}{l} -2R_2 + R_1 \to R_1 \\ \to \\ 70R_2 + R_3 \to R_3 \end{array}$$

$$\left[\begin{array}{ccccc|c} 0 & 2 & 1 & -1 & 0 & 120 \\ 1 & \frac{1}{2} & 0 & \frac{1}{2} & 0 & 90 \\ \hline 0 & 30 & 0 & 35 & 1 & 6300 \end{array}\right]$$

Maximum is 6300 at $x = 90$, $y = 0$.

12. Let x = barrels of lager, y = barrels of ale.

Maximize: $P = 35x + 30y$

Subject to: $3x + 2y \le 1200$

$2x + 2y \le 1000$

$$\left[\begin{array}{ccccc|c} 3 & 2 & 1 & 0 & 0 & 1200 \\ 2 & 2 & 0 & 1 & 0 & 1000 \\ \hline -35 & -30 & 0 & 0 & 1 & 0 \end{array}\right] \begin{array}{l} \frac{1}{3}R_1 \to R_1 \\ \\ \\ \end{array}$$

$$\left[\begin{array}{ccccc|c} 1 & \frac{2}{3} & \frac{1}{3} & 0 & 0 & 400 \\ 2 & 2 & 0 & 1 & 0 & 1000 \\ \hline -35 & -30 & 0 & 0 & 1 & 0 \end{array}\right] \begin{array}{l} \to \\ -2R_1 + R_2 \to R_2 \\ 35R_1 + R_3 \to R_3 \end{array}$$

$$\left[\begin{array}{ccccc|c} 1 & \frac{2}{3} & \frac{1}{3} & 0 & 0 & 400 \\ 0 & \frac{2}{3} & -\frac{2}{3} & 1 & 0 & 200 \\ \hline 0 & -\frac{20}{3} & \frac{35}{3} & 0 & 1 & 14{,}000 \end{array}\right] \begin{array}{l} -R_2 + R_1 \to R_1 \\ \to \\ 10R_2 + R_3 \to R_3 \end{array}$$

$$\left[\begin{array}{ccccc|c} 1 & 0 & 1 & -1 & 0 & 200 \\ 0 & \frac{2}{3} & -\frac{2}{3} & 1 & 0 & 200 \\ \hline 0 & 0 & 5 & 10 & 1 & 16{,}000 \end{array}\right] \begin{array}{l} \\ \frac{3}{2}R_2 \to R_2 \\ \\ \end{array}$$

$$\left[\begin{array}{ccccc|c} 1 & 0 & 1 & -1 & 0 & 200 \\ 0 & 1 & -1 & \frac{3}{2} & 0 & 300 \\ \hline 0 & 0 & 5 & 10 & 1 & 16{,}000 \end{array}\right]$$

Maximum is \$16,000 at $x = 200$, $y = 300$.

13. Let x = number of day calls,
let y = number of evening calls.

Minimize $C = 3x + 4y$

Subject to: $0.3x + 0.3y \ge 150$

$0.1x + 0.3y \ge 120$

$x, y \ge 0$

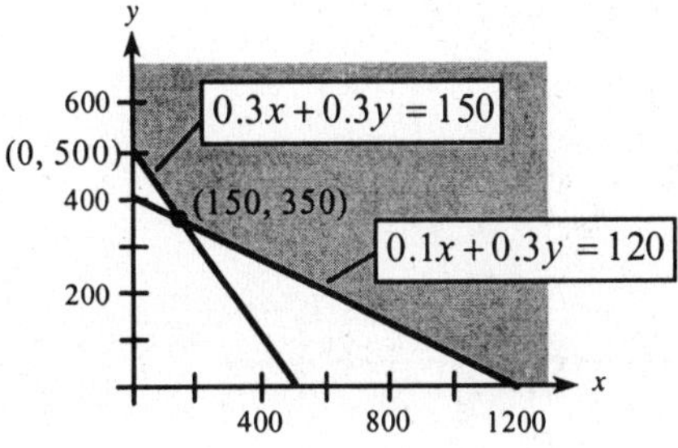

Evaluating the objective function at (0, 500), (150, 350) and (1200, 0) the minimum is \$1850 at $x = 150$, $y = 350$.

Chapter 5: Exponential and Logarithmic Functions

Exercise 5.1

1. $10^{0.5}$ 10 [y^x] 0.5 [=] 3.1623

3. $5^{-2.7}$ 5 [y^x] 2.7 [±] [=] 0.012965

5. $3^{1/3}$ 3 [y^x] 0.3333 [=] 1.44225

7. e^2 e [y^x] 2 [=] 7.3891

9.

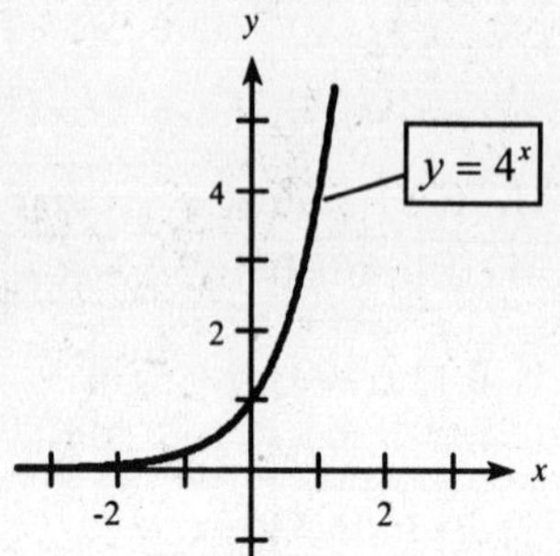

11.

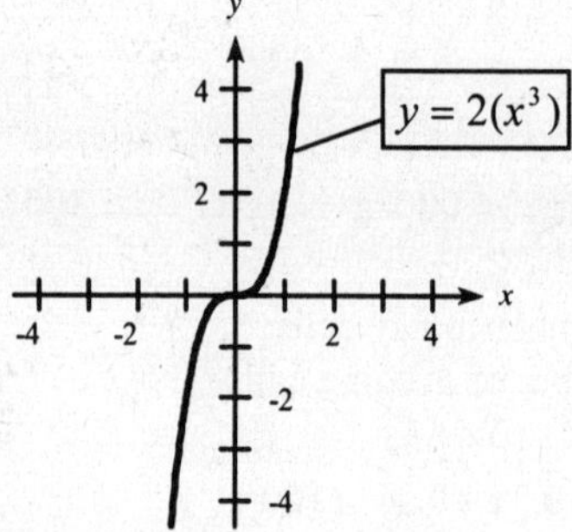

13.

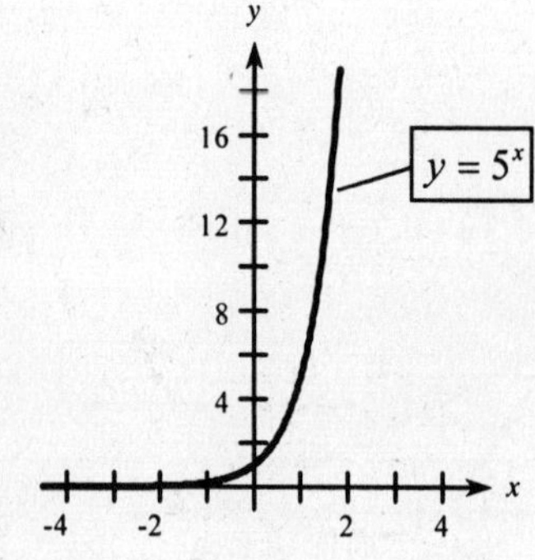

15.

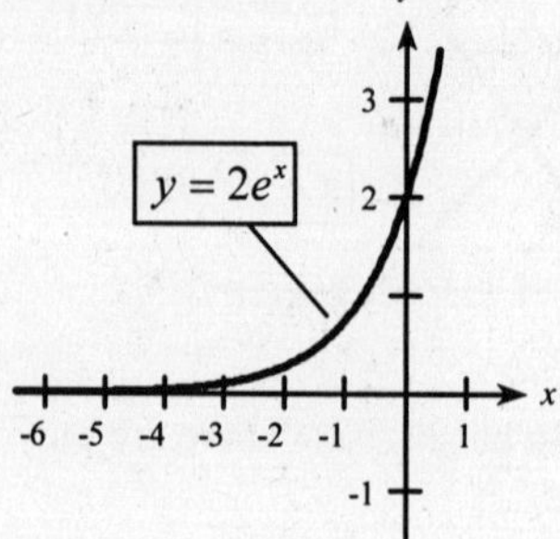

17.

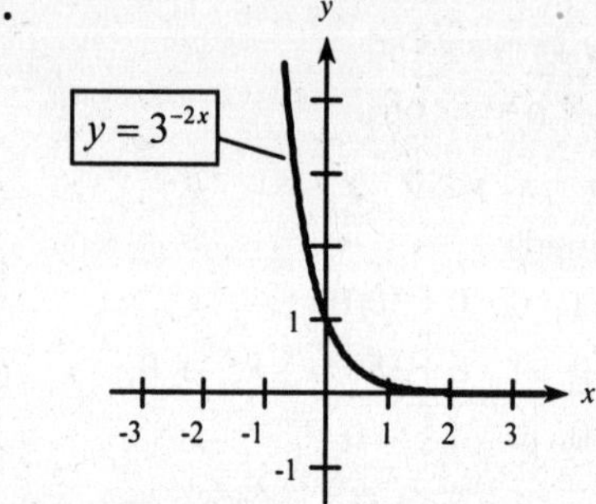

19. a. $y = 3\left(\frac{2}{5}\right)^x = 3\left(\frac{5}{2}\right)^{-x}$

b. These functions are decay exponentials. They are algebraically equivalent and an exponential decay function is one of the form $f(x) = cb^{-x}$ where $b > 1$. In this form $b = 5/2$ which is greater than 1.

c. Graphs are identical and falling.

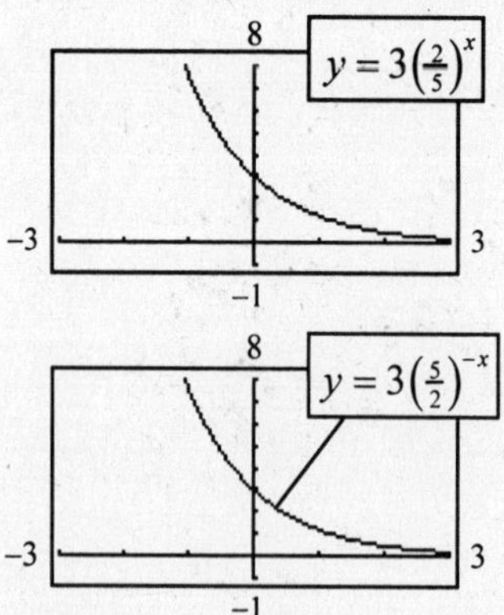

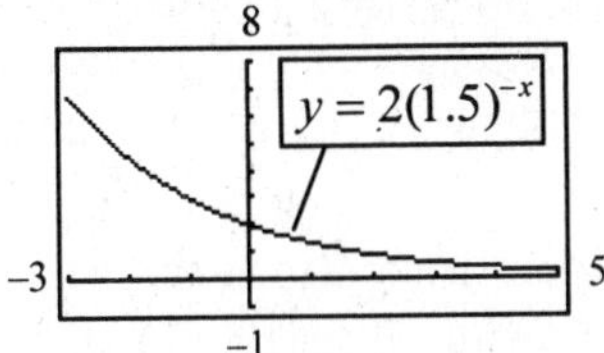

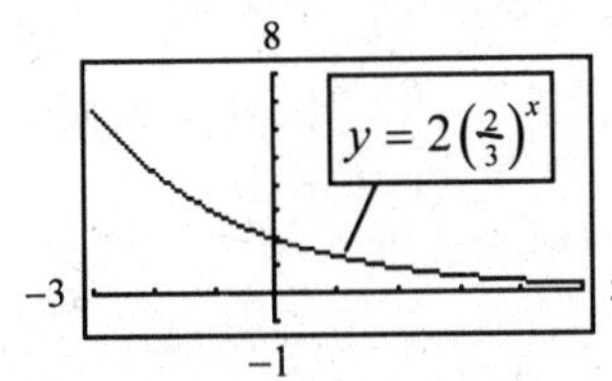

c. $y = 2(1.5)^{-x} = 2\left(\frac{3}{2}\right)^{-x} = 2\left(\frac{2}{3}\right)^{x}$

23. $f(x) = e^{-x}$, $f(kx) = e^{-kx}$

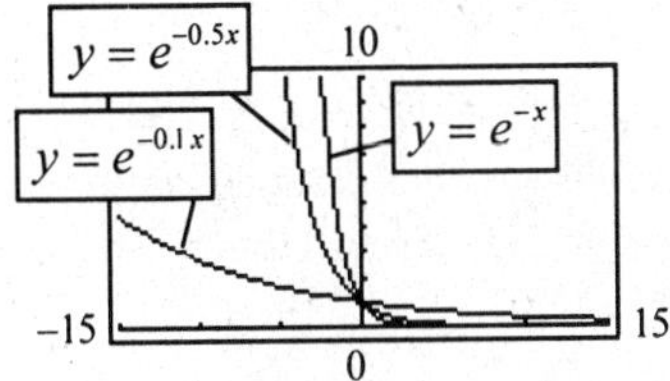

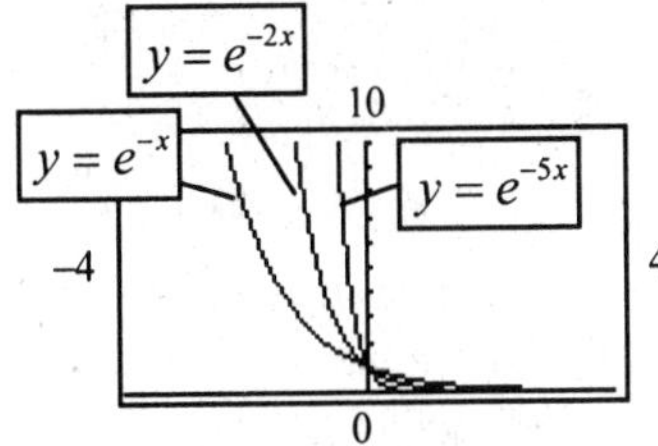

For $k > 1$ the graphs rise and fall more sharply than $y = e^{-x}$. For $k < 1$ the graphs rise and fall more slowly than $y = e^{-x}$.

25. $f(x) = 4^x$ $\quad f(x) + C = 4^x + C$

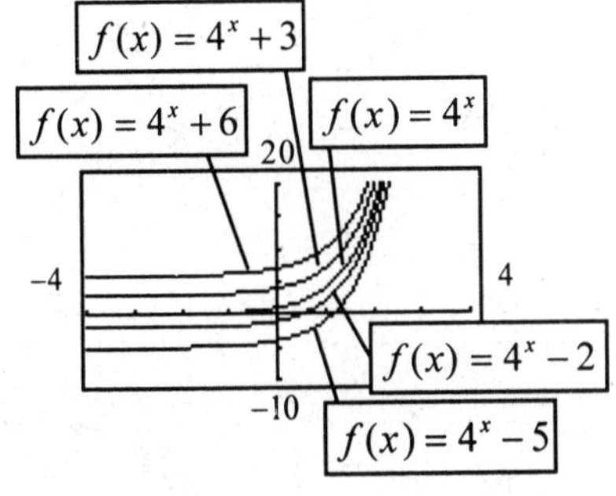

The graphs are the same but $f(x) + C$ is shifted C units on the y-axis.

27. a.

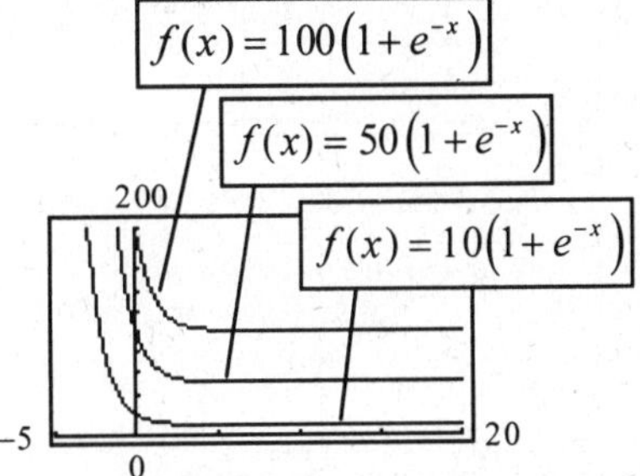

b. As c changes, the y-intercept and the horizontal asymptote change.

29. a.

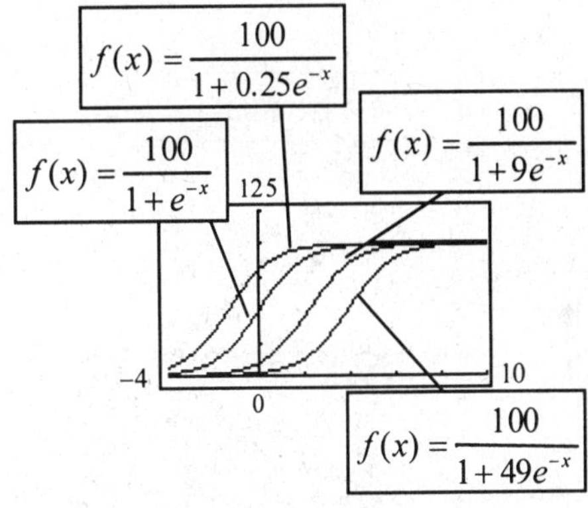

b. Different c values change only the y-intercept.

31. $S(x) = 1000(1.02)^{4x}$

$S(8) = 1.02$ [y^x] 32 [×] 1000 [=] \$1884.54

33.

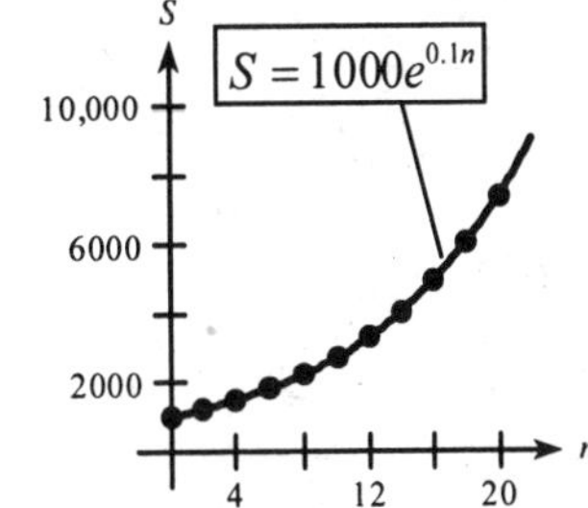

35.

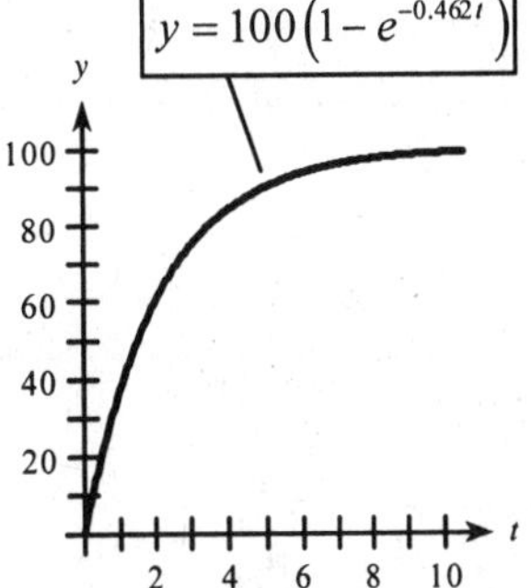

After 10 hours the drug is almost completely in the bloodstream.

37.

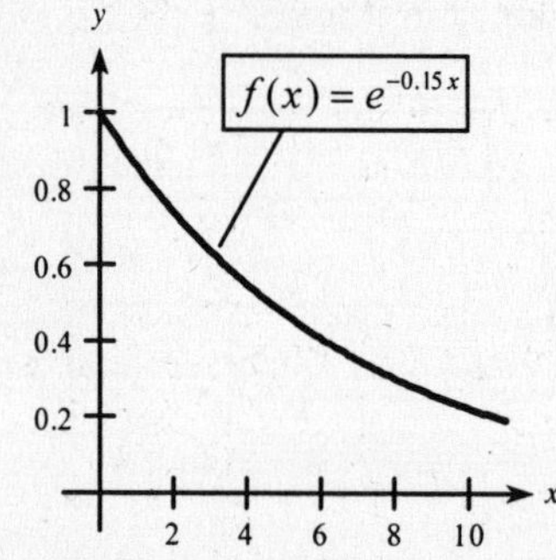

After 10 years about 20% of the television sets are still in service.

39.

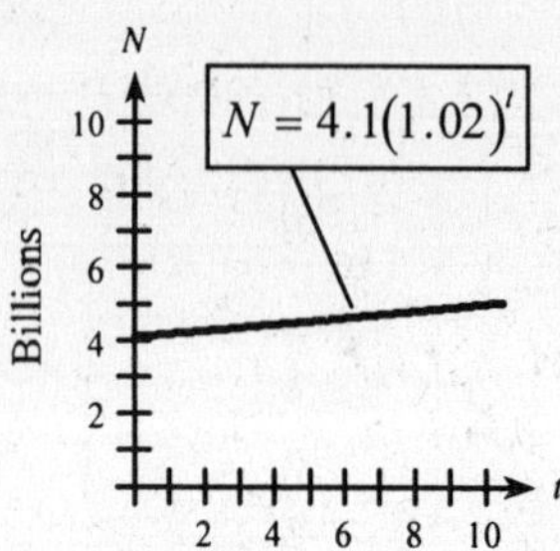

41.

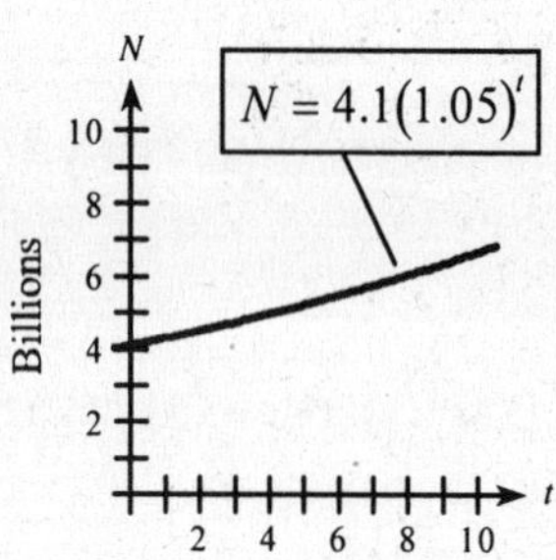

43. a. $N=630.2e^{0.3739t}$ is an exponential growth model since $e>1$. Graph rises.

b.

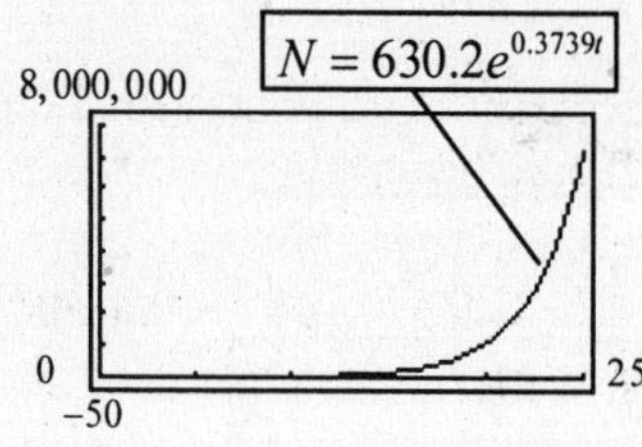

45. Compare the two graphs together. The linear model gives a negative number of processors in 2007. The exponential model is always non-negative.

47. a. Letting $x=0$ correspond to the year 1960, enter the data values $(0,409.4)$, $(10,831.4)$, etc. and obtain the exponential regression equation $y=423.6498(1.0793)^x$.

b. $x=50$ in 2010, so $y=423.6498(1.0793)^{50}\approx 19236.0$. If model is accurate, the total U.S. personal income in 2010 will be $19236.0 billion.

49. From the scatter plot of points, an exponential function of the form $a\cdot b^x$ is the best fit.

a. $y=2.4476(1.0440)^x$

b. $f(110)=2.4476(1.0440)^{110}\approx 279.1$

c. Graph $y_1=300$ and $y_2=2.4476(1.0440)^x$. The point of intersection is the solution. In 112.55 years the CPI will reach 300. So, the year is 2012-2013.

51. a. $x=0$ in 1980, $x=5$ in 1985, etc. Enter the data values $(5,75)$, $(6,50)$, $(7,37)$, etc., and obtain the exponential regression equation $y=128.2209(0.8487)^x$.

b. This is an exponential decay problem. Since $b=0.8487$ is less than one, the y values will decrease as x increases.

c. In 2010 $x=30$, so $y=128.2209(0.8487)^{30}\approx 0.9$. This model predicts an average of 0.9 students per computer in public schools in 2010.

Exercise 5.2

NOTE: $a^y=x$ means $y=\log_a x$.

1. $4=\log_2 16$

$y=4, a=2, x=16$

$2^4=16$

3. $\frac{1}{2}=\log_4 2$

$y=\frac{1}{2}, a=4, x=2$

$4^{1/2}=2$

5. $\log_2 x = 3$

$2^3 = x$

$x = 8$

7. $\log_8 x = -\dfrac{1}{3}$

$8^{-1/3} = x$

$x = \dfrac{1}{8^{1/3}} = \dfrac{1}{2}$

9. $2^5 = 32$

$y = 5, a = 2, x = 32$

$5 = \log_2 32$

11. $4^{-1} = \dfrac{1}{4}$

$y = -1, a = 4, x = \dfrac{1}{4}$

$-1 = \log_4\left(\dfrac{1}{4}\right)$

13.

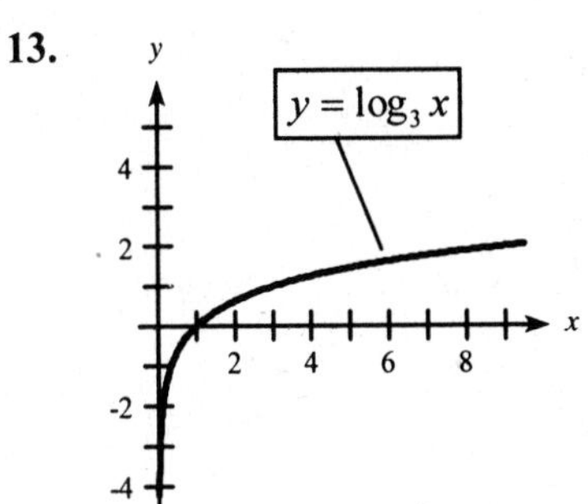

15.

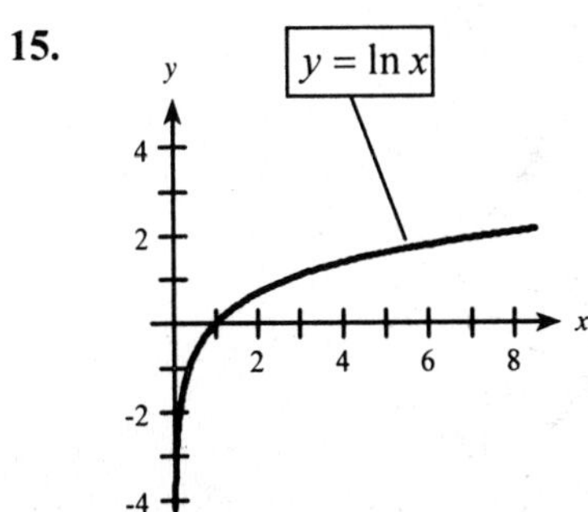

17.

$y = \log_2(-x)$

19. a. $\log_3 27 = \log_3 3^3 = 3\log_3 3 = 3 \cdot 1 = 3$

b. $\log_5\left(\dfrac{1}{5}\right) = \log_5 5^{-1} = -1\log_5 5 = -1 \cdot 1 = -1$

21. $f(x) = \ln(x)$

$f(e^x) = \ln(e^x) = x \ln e = x \cdot 1 = x$

23. $f(x) = e^x$

$f(\ln 3) = e^{\ln 3} = 3 \quad (e^{\ln 3} = e^{\log_e 3})$

25. a. $\log_a(xy) = \log_a x + \log_a y = 3.1 + 1.8 = 4.9$

b. $\log_a\left(\dfrac{x}{z}\right) = \log_a x - \log_a z = 3.1 - 2.7 = 0.4$

c. $\log_a x^4 = 4\log_a x = 4(3.1) = 12.4$

d. $\log_a \sqrt{y} = \dfrac{1}{2}\log_a y = \dfrac{1}{2}(1.8) = 0.9$

27. $\log\left(\dfrac{x}{x+1}\right) = \log x - \log(x+1)$

29. $\log_7 x\sqrt[3]{x+4} = \log_7 x + \log_7 (x+4)^{1/3}$

$= \log_7 x + \dfrac{1}{3}\log_7 (x+4)$

31. $\ln x - \ln y = \ln\left(\dfrac{x}{y}\right)$

33. $\log_5(x+1) + \dfrac{1}{2}\log_5 x = \log_5(x+1) + \log_5 x^{1/2}$

$= \log_5\left[x^{1/2}(x+1)\right]$

35. a. $\ln(4 \cdot 6)^{1/2} = 1.5890$

b. $\dfrac{1}{2}(\ln 4 + \ln 6)$

The expressions are equivalent. Properties V and III are illustrated.

37. a. $\log\left(\dfrac{8}{5}\right)^{1/3} = \log(8 \div 5)^{1/3} = 0.06804$

b. $\dfrac{1}{3}\log 8 - \log 5 = \dfrac{1}{3}(0.9031) - 0.6990$

$\neq 0.06804$

The expressions are not equivalent.

Change (a) to $\log\dfrac{\sqrt[3]{8}}{5}$ to get (a) = (b).

39. a.

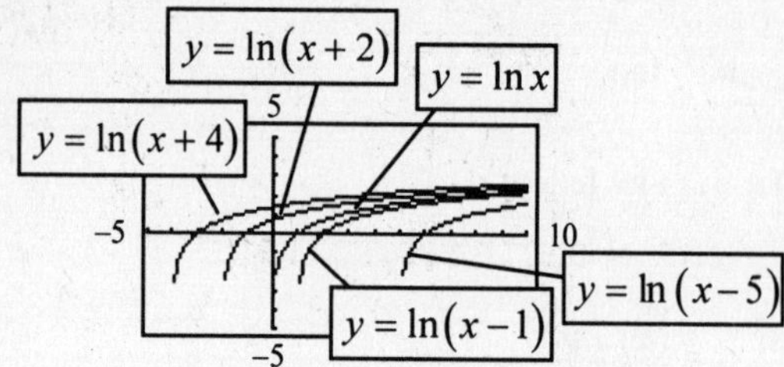

b. For each c, the domain is $x > c$ and the vertical asymptote is at $x = c$.

c. Each x-intercept is at $x = c + 1$.
$\ln(x-c) = \ln(c+1-c) = \ln 1 = 0$.

d. The graph of $f(x-c)$ is the graph of $f(x)$ shifted c units on the x-axis.

41. a. $\log_2 17 = \dfrac{\ln 17}{\ln 2} = \dfrac{2.8332}{0.6931} = 4.0875$

b. $\log_5(0.78) = \dfrac{\ln(0.78)}{\ln 5} = \dfrac{-0.2485}{1.6094} = -0.1544$

43. $y = \log_5 x \qquad y = \dfrac{\ln x}{\ln 5} = \dfrac{1}{1.6094}\ln x$

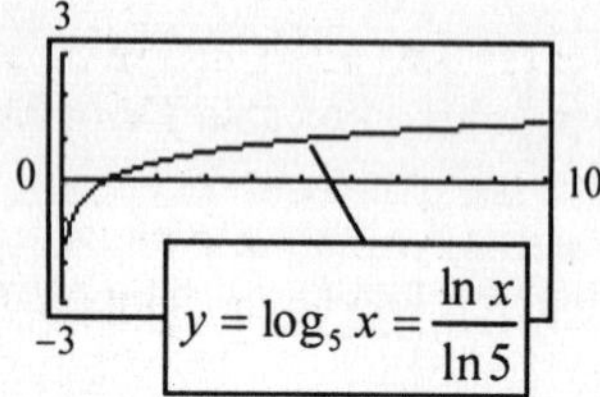

45. $y = \log_{13} x \qquad y = \dfrac{\ln x}{\ln 13} = \dfrac{1}{2.5649}\ln x$

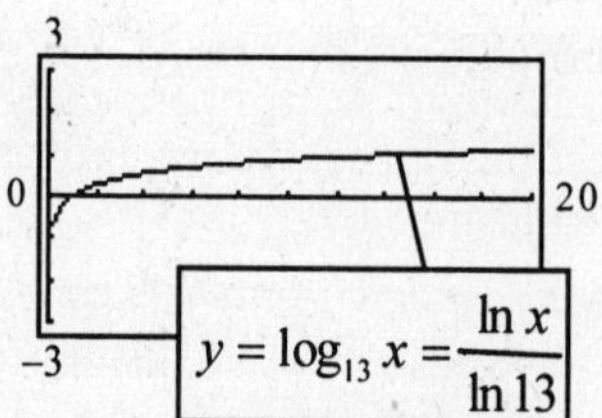

47. Let $u = \log_a M$ and $v = \log_a N$. Then, $M = a^u$ and $N = a^v$. So,

$$\log_a\left(\frac{M}{N}\right) = \log_a\left(\frac{a^u}{a^v}\right) = \log_a(a^{u-v})$$
$$= (u-v)\log_a a$$
$$= u - v = \log_a M - \log_a N.$$

49. $y = 3.974(1.05414)^{-x}$

To change to base e we have

$(1.05414)^{-x} = e^{-kx} = \left(e^k\right)^{-x}$ so

$1.05414 = e^k$

$\ln 1.05414 = k$

$k \approx 0.05273$ thus $y = 3.974e^{-0.05273x}$.

51. $7.7 = \log\left(\dfrac{I}{I_0}\right) \qquad 7.3 = \log\left(\dfrac{I}{I_0}\right)$

$10^{7.7} = \dfrac{I}{I_0} \qquad 10^{7.3} = \dfrac{I}{I_0}$

$\dfrac{10^{7.7}}{10^{7.3}} = 10^{0.4} \approx 2.5$ So the earthquake was approximately 2.5 times as severe.

53. Intensity of the 1906 San Francisco quake:
$10^{8.25} I_0$
Intensity of the 1989 San Francisco quake:
$10^{7.1} I_0$
The 1906 quake was
$$\frac{10^{8.25} I_0}{10^{7.1} I_0} = 10^{8.25-7.1} = 10^{1.15} \approx 14$$
times as severe.

55. $L = 10\log(10{,}000) = 10 \cdot 4 = 40$

57. $L = 10\log\left(\dfrac{I}{I_0}\right)$

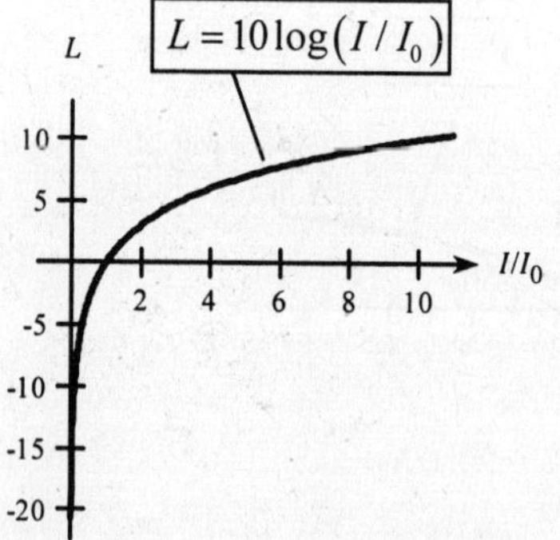

59. When pH = 1, $\quad 1 = -\log[H^+]$
$$-1 = \log[H^+]$$
$$10^{-1} = [H^+]$$
$$0.1 = \frac{1}{10} = [H^+]$$
When pH = 14, $[H^+] = 10^{-14}$

61. $\text{pH} = \log\left(\frac{1}{[\text{H}^+]}\right)$

Since $\log\left(\frac{1}{[\text{H}^+]}\right) = \log 1 - \log[\text{H}^+]$ and $\log 1 = 0$, we have $\text{pH} = -\log[\text{H}^+]$.

63. In 1999,

$L(99) = 11.485 + 14.180\ln(99) \approx 76.644$ and

$\ell(99) = 11.6164 + 14.1442\ln(99) = 76.61$.

In 2010,

$L(110) = 11.485 + 14.180\ln 110 \approx 78.138$

$\ell(110) = 11.6164 + 14.1442\ln 110 \approx 78.101$

The additional data gave predictions that were very similar to predictions based on the earlier model.

65. a.

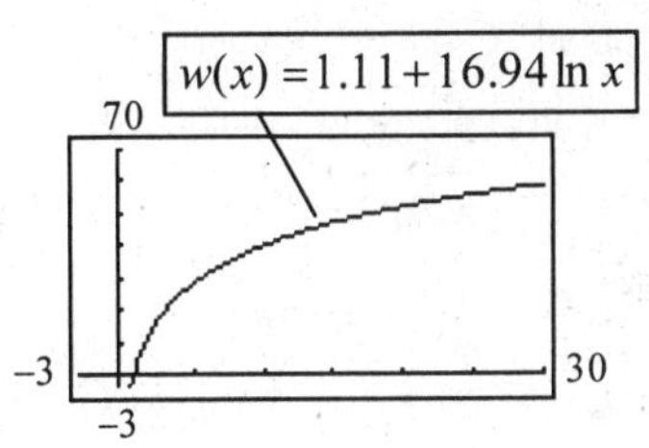

b. In 1994 $x = 24$, so $w(24) = 1.11 + 16.94\ln 24 \approx 54.95$. In 1994, 54.95 percent of mothers returned to the work force within one year after they had a child.

c. Graph $y_1 = 1.11 + 16.94\ln x$ and $y_2 = 40$ and find where they intersect. The point of intersection is $(9.93, 40)$. The percent of mothers who returned to the work force within one year after they had a child reached 40% in 1980.

67. a. For x the number of years after 1970, enter the data values $(10, 22.6)$, $(11, 28.3)$, etc., and obtain the logarithmic regression equation $y = -64.8867 + 40.39228\ln x$.

b. $y = -64.8867 + 40.39228\ln 40 \approx 84.1$

If this model remains valid, the percentage of U.S. households with cable TV in the year 2010 is predicted to be 84.1.

c. For this model to remain valid the percentage of households cannot exceed 100.

Exercise 5.3

1. $S = 50{,}000e^{-0.8x}$

a. $x = 4$

$S = 50{,}000e^{-3.2}$

$= 50{,}000(0.0408) = \$2038.11$

b. $S = 1000$

$1000 = 50{,}000e^{-0.8x}$

$e^{-0.8x} = 0.02$

$-0.8x = \ln 0.02$

$x = \frac{\ln 0.02}{-0.8} = \frac{-3.912}{-0.8} = 4.89$

In the fifth month sales will drop below \$1000.

3. a. $30{,}000 = 60{,}000e^{-0.05t}$

$e^{-0.05t} = \frac{30{,}000}{60{,}000}$

$e^{-0.05t} = 0.5$ take ln of both sides

$-0.05t \approx -0.6931$

$t = \frac{-0.6931}{-0.05} \approx 13.9$ years

b. $50{,}000 = A \cdot e^{-0.05(15)}$

$50{,}000 = Ae^{-0.75}$

$A = \frac{50{,}000}{e^{-0.75}} = \frac{50{,}000}{0.47237} \approx \$105{,}849$

5. $Q(t) = 100e^{-0.02828t}$

At $t = 0$ the initial amount is 100 gm.

Thus,

$$50 = 100e^{-0.02828t}$$

$$0.5 = e^{-0.02828t}$$

$\ln 0.5 = -0.02828t$ take ln of both sides

$$t = \frac{\ln 0.5}{-0.02828} = \frac{-0.6931}{-0.02828} = 24.5 \text{ years}$$

7. $y = P_0 e^{ht}$

$$110{,}517 = 100{,}000e^{10h}$$

$$e^{10h} = 1.10517$$

$10h = \ln 1.10517$ (Take ln of both sides.)

$$10h = 0.09999$$

$$h = 0.009999 = 0.01$$

2015 gives $t = 15$ and $P = 110{,}517$.

So, $y = 110{,}517e^{0.01(15)} = 110{,}517e^{0.15}$

$$= 110{,}517(1.161834) = 128{,}402$$

9. $N = 3000(0.2)^{0.6^t}$

a. $t = 0$ $\quad N = 3000(0.2)^1$

$$= 600$$

b. $t = 3$ $\quad N = 3000(0.2)^{0.6^t}$

$$= 3000(0.2)^{0.216}$$

$$= 3000(0.706354)$$

$$= 2119$$

c. The largest value of $(0.2)^{0.6^t}$ occurs when $0.6^t = 0$. Then, $(0.2)^0 = 1$ and the maximum sales are 3000.

d.

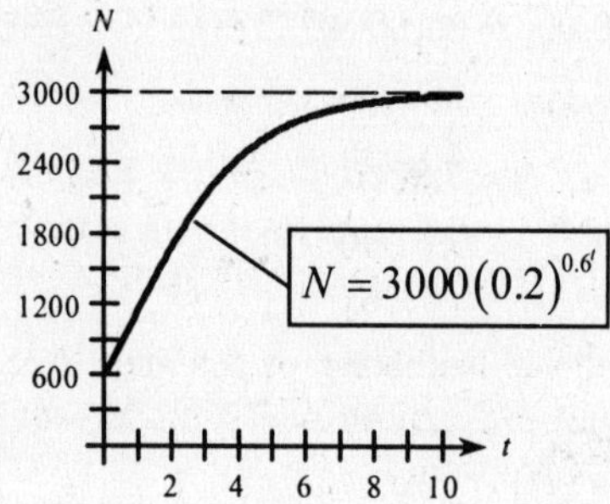

11. $N = 500(0.02)^{0.7^t}$

a. If $t = 0$, $N = 500(0.02) = 10$ employees when company opens.

b. $N = 100$

$$100 = 500(0.02)^{0.7^t}$$

$$(0.02)^{0.7^t} = 0.2$$

$$0.7^t \log 0.02 = \log 0.2$$

$$0.7^t = \frac{\log 0.2}{\log 0.02} = \frac{-0.69897}{-1.69897} = 0.41141$$

$$t \log 0.7 = \log 0.41141$$

$$t = \frac{\log 0.41141}{\log 0.7} = \frac{-0.38573}{-0.15490} = 2.49$$

At least 100 employees are working $2\frac{1}{2}$ years later.

13. $y = 100(1 - e^{-0.462t})$

a. $t = 1$

$$y = 100(1 - e^{-0.462}) = 100(1 - 0.6300) = 37$$

b. $y = 50$

$$50 = 100(1 - e^{-0.462t})$$

$$1 - e^{-0.462t} = 0.5$$

$$e^{-0.462t} = 0.5$$

$$-0.462t = \ln 0.5 = -0.6931$$

$$t = \frac{-0.6931}{-0.462} = 1.5 \text{ hours}$$

15. $y = \dfrac{10{,}000}{1 + 9999e^{-0.99t}}$

a. $t = 4$

$$y = \frac{10{,}000}{1 + 9999e^{-3.96}} = \frac{10{,}000}{1 + 190.61} \approx 52$$

b. $y = 5000$

$$5000 = \frac{10{,}000}{1 + 9999e^{-0.99t}}$$

$$1 + 9999e^{-0.99t} = 2$$

$$e^{-0.99t} = \frac{1}{9999}$$

$$-0.99t = \ln 1 - \ln 9999$$

$$t = \frac{-\ln 9999}{-0.99} = \frac{9.2102}{0.99} = 9.30$$

School will close during the tenth day.

17. $p = 100e^{-q/2}$

a. $q = 6$

$p = 100e^{-3} = 100(0.04978) = \$4.98$

b. $p = \$1.83$

$1.83 = 100e^{-q/2}$

$0.0183 = e^{-q/2}$

$-q/2 = \ln(0.0183)$

$q = -2(\ln 0.0183)$

$= -2(-4.000) = 8$ units

19. $p = \dfrac{100e^q}{q+1}$

$q = \dfrac{300}{100} = 3$

$p = \dfrac{100e^3}{4} = 25e^3 = \$502.14$

21. $C(x) = e^{0.1x} + 400$

$C(30) = e^3 + 400 = \$420.09$

23. $p = 200e^{-2} = \$27.07$

100 units will give a revenue of \$2707.

25. $S(t) = 8500e^{0.115t}$

a. $S(1.5) = 8500e^{0.1725} = 8500(1.1883)$

$= \$10,100.31$

b. $17,000 = 8500e^{0.115t}$

$2 = e^{0.115t}$

$0.115t = \ln 2$

$t = \dfrac{\ln 2}{0.115} = \dfrac{0.6931}{0.115} = 6.03$

The account will double in 6.03 years.

27. $S = 5000(1.0075)^t$

a. $t = 12$

$S = 5000(1.0075)^{12} = 5000(1.0938)$

$= \$5469.03$

b. $S = 10,000$

$10,000 = 5000(1.0075)^t$

$1.0075^t = 2$

$t \ln 1.0075 = \ln 2$

$t = \dfrac{\ln 2}{\ln 1.0075} = \dfrac{0.6931}{0.00747}$

$= 92.8$ months

Thus, $t = 7$ years, 9 months.

29. $y = 40 - 40e^{-0.05t}$

a. $t = 1$

$y = 40 - 40e^{-0.05} = 40 - 40(0.95123)$

$\approx 1.95\% \approx 2\%$

b. $y = 25$

$25 = 40 - 40e^{-0.05t}$

$40e^{-0.05t} = 15$

$e^{-0.05t} = \dfrac{15}{40} = 0.375$

$t = \dfrac{\ln 0.375}{-0.05} = \dfrac{-0.9808}{-0.05}$

$t = 19.6$ or 20 months

31. a. In 2000 $t = 20$

$P(20) = 21.4e^{0.131(20)} - 18.6e^{0.131(20)}$

$= (21.4 - 18.6)e^{0.131(20)}$

$= 2.8e^{0.131(20)}$

≈ 38.46

Predicted profit in 2000 is \$38.5 billion.

b. $2 = e^{0.131t}$

$\ln 2 = \ln e^{0.131t}$

$\ln 2 = 0.131t$

$\dfrac{\ln 2}{0.131} = t$

$5.29 \approx t$

Profit will double in 5.29 years.

c. September 11, 2001 might affect the prediction in part b.

33. a. $2500 = 28.8e^{0.0994t}$

$e^{0.0994t} = \dfrac{2500}{28.8}$ take ln of both sides

$0.0994t = \ln 2500 - \ln 28.8$

$\approx 7.8240 - 3.3604 = 4.4636$

$t = \dfrac{4.4636}{0.0994} \approx 44.9$ years

b. National health care is designed to reduce future expenditures. This model would not be appropriate.

35. **a.** $P(10) = 3.97e^{-0.0527(10)}$

$= 3.97e^{-0.527} \approx 2.34$

The purchasing power of a 1983 dollar was \$2.34 in 1970.

$P(70) = 3.97e^{-0.0527(70)}$

$= 3.97e^{-3.689} \approx 0.10$

The purchasing power of a 1983 dollar will be \$0.10 in 2030.

b. $0.25 = 3.974e^{-0.0527t}$

$e^{-0.0527t} = \dfrac{0.25}{3.974}$

$e^{-0.0527t} \approx 0.06291$ take ln of both sides

$-0.0527t = \ln(0.06291)$

$-0.0527t = -2.7661$

$t = \dfrac{-2.7661}{-0.0527} = 52.488$

It will take 52.5 years before it will cost \$1 to purchase goods that cost \$0.25 in 1983. The year is 2013.

37. **a.** $2 = (1.005)^t$

$\log_{1.005}(2) = t\log_{1.005} 1.005 = t$

b. $\ln 2 = t\ln 1.005$

$t = \dfrac{\ln 2}{\ln 1.005} = \dfrac{0.6931}{0.0050} \approx 139$ month

39. $x = 0.05 + 0.18e^{-0.38t}$

a. $t = 0 \quad x = 0.05 + 0.18(1) = 0.23\text{km}^3$

b. 30% of 0.23 = 0.069.

So, for $x = 0.069$ we have

$0.069 = 0.05 + 0.18e^{-0.38t}$

$e^{-0.38t} = \dfrac{0.069 - 0.05}{0.18} = 0.10556$

$t = \dfrac{\ln 0.10556}{-0.38} = \dfrac{-2.2485}{-0.38} = 5.9$ years

41. **a.** $t = 0$

$x = \dfrac{120[1-1]}{4-1} = 40(0)$

$= 0$ pounds are present.

b. $t = 4$

$x = \dfrac{120[1-0.6^{12}]}{4-0.6^{12}} = \dfrac{120[1-0.0022]}{4-0.0022}$

$= \dfrac{120(0.9978)}{3.9978} = 29.95$ lbs.

c. $x = 10$

$10 = \dfrac{120[1-(0.6)^{3t}]}{4-(0.6)^{3t}}$

$40 - 10(0.6^{3t}) = 120 - 120(0.6^{3t})$

$110(0.6^{3t}) = 80$

$3t(\ln 0.6) = \ln 80 - \ln 110$

$t = \dfrac{\ln 80 - \ln 110}{3\ln 0.6}$

$= \dfrac{4.3820 - 4.7004}{3(-0.5108)}$

$= 0.21$ min

43. $x = 4.8 + 11.2e^{-t/4}$

a. $t = 0 \quad x = 4.8 + 11.2(1) = 16$ cu ft of CO_2.

$16 = R\%$ of 8000 gives $R = 16/8000 = 0.2\%$.

b. 0.07% concentration means $0.0007(8000) = 5.6$ cu ft of CO_2 in the room.

$5.6 = 4.8 + 11.2e^{-t/4}$

$11.2e^{-t/4} = 0.8$

$-t/4 = \ln 8 - \ln 112$

$-t = 4(\ln 8 - \ln 112)$

$t = 4(\ln 112 - \ln 8)$

$t = 4(4.718 - 2.079)$

$t = 10.56$ minutes

c. As t gets large $e^{-t/4}$ gets close to 0. Thus, there would be 4.8 cu ft of CO_2 always present. We have 4.8 cu ft is 0.06% concentration.

45. a. Age is $x+15$, so at 15 years old $x=0$. Input data pairs $(0,16.6)$, $(1,28.7)$, etc., and calculate logistic model $y=\dfrac{89.7857}{1+4.6531e^{-0.8256x}}$

b.

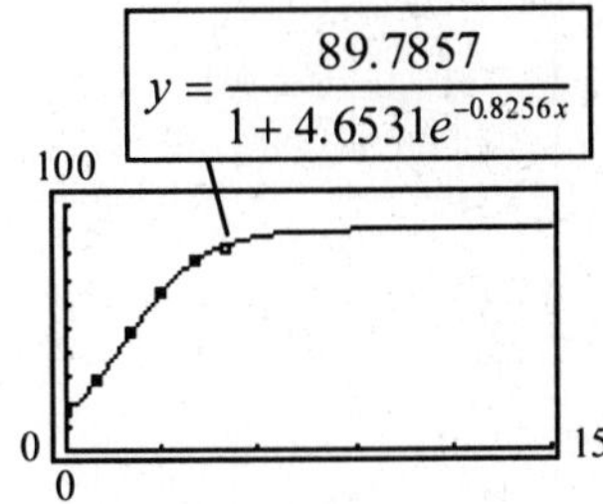

c. The model estimates the cumulative percent to be 29.6 for boys whose age is 16.

d. The model estimates the cumulative percent to be 86.9 for boys of age 21.

e.
$$64.55=\frac{89.7857}{1+4.6531e^{-0.8256x}}$$
$$64.55\left(1+4.6531e^{-0.8256x}\right)=89.7857$$
$$64.55+300.3576e^{-0.8256x}=89.7857$$
$$300.3576e^{-0.8256x}=25.2357$$
$$e^{-0.8256x}=\frac{25.2357}{300.3576}\approx 0.0840 \text{ take ln of both sides}$$
$$-0.8256x=\ln(0.0840)$$
$$-0.8256x=-2.4769$$
$$x=\frac{-2.4769}{-0.8256}=3.0$$

Age is $x+15$, so $3+15=18$ years of age and under. So model estimates the cumulative percent to be 64.55 for boys 18 years of age and under.

Review Exercises

1. a. $2^x = y \rightarrow x = \log_2 y$

   b. $3^y = 2x \rightarrow y = \log_3(2x)$

2. a. $\log_7\left(\frac{1}{49}\right) = -2 \rightarrow 7^{-2} = \frac{1}{49}$

   b. $\log_4 x = -1 \rightarrow 4^{-1} = x$

3.

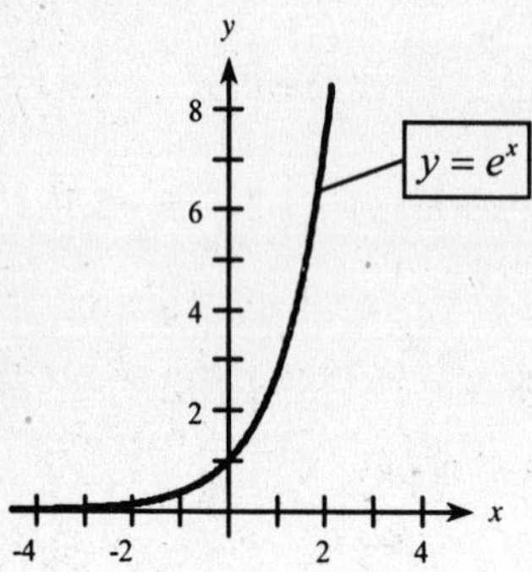

4.

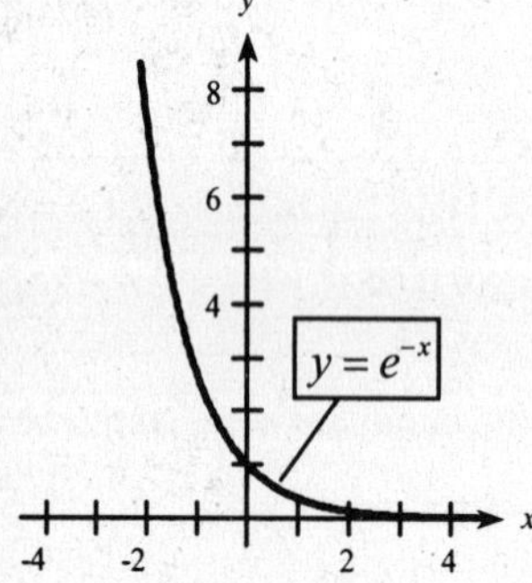

5.

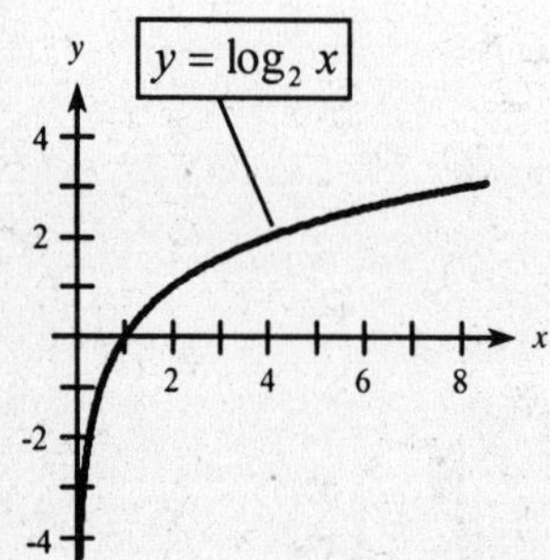

6.

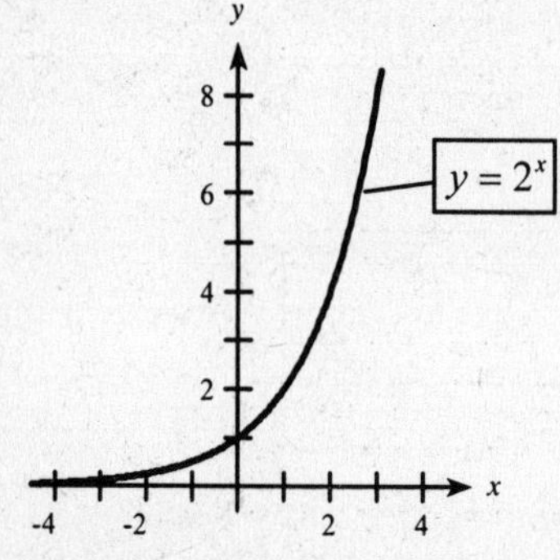

7.

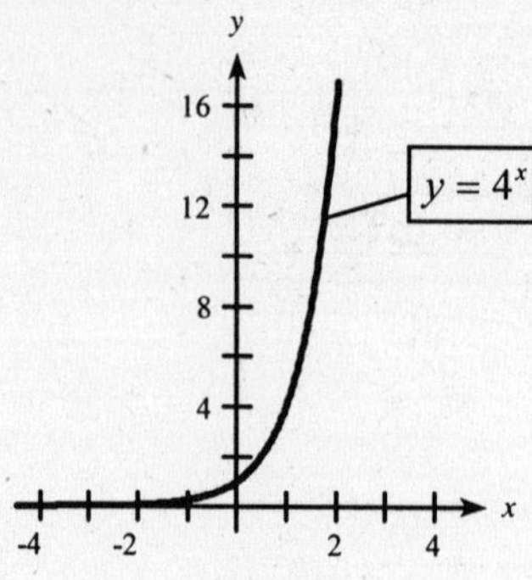

8.

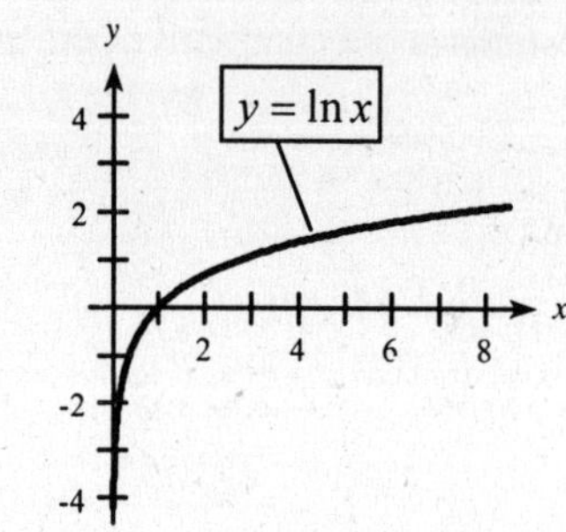

9.

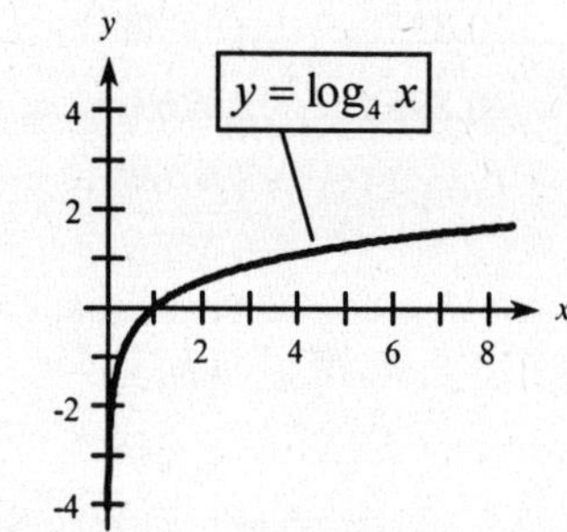

10.

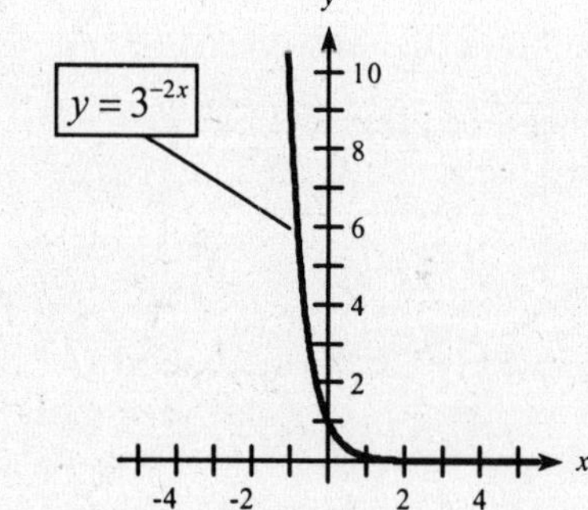

11.

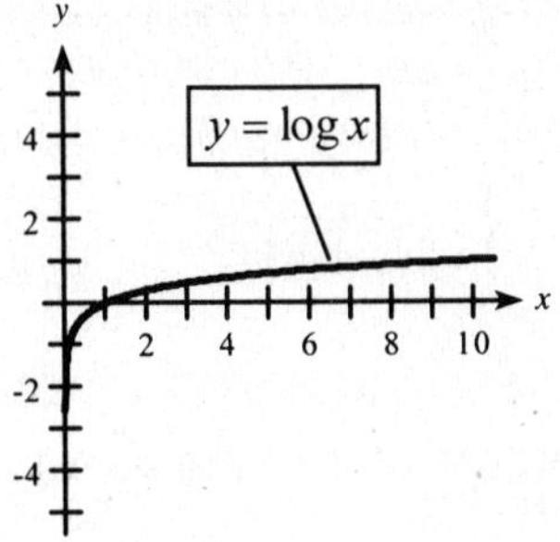

12.

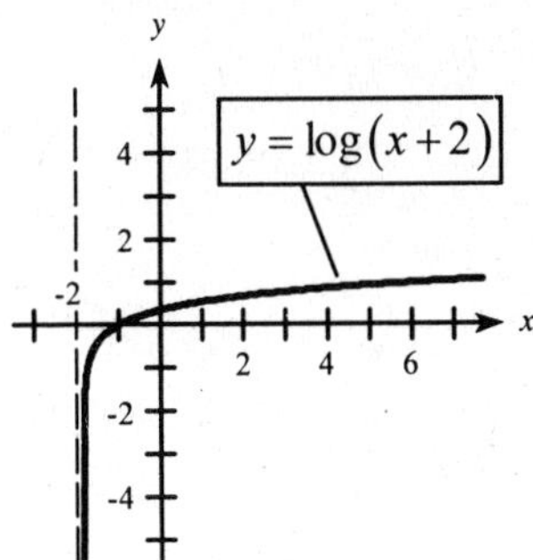

13.

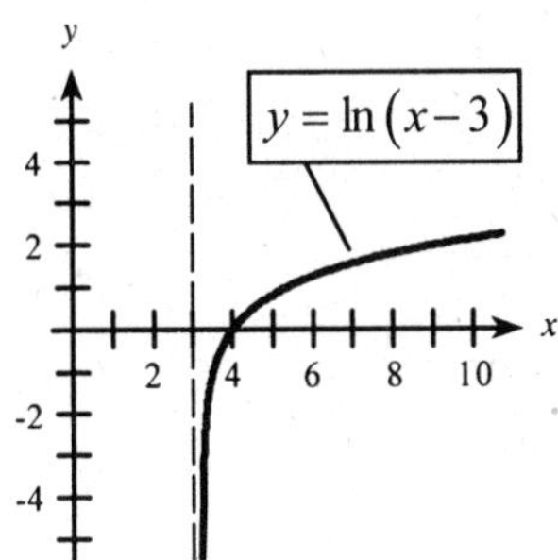

14.

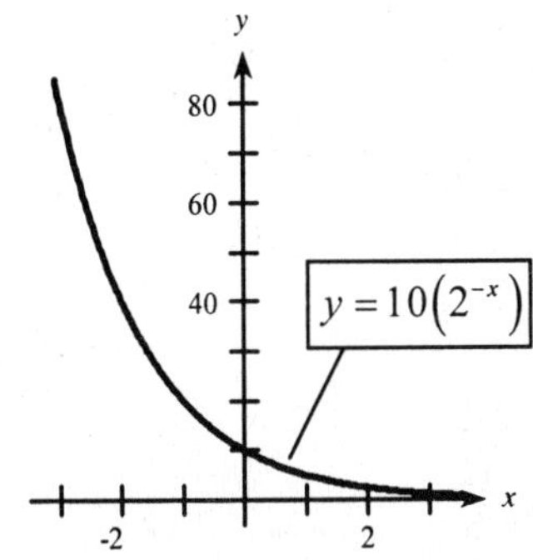

15. $\log_5 1 = 0$

16. $\log_8 64 = \log_8 8^2 = 2$

17. $\log_{25} 5 = \log_{25}(25)^{1/2} = \dfrac{1}{2}$

18. $\log_3\left(\dfrac{1}{3}\right) = \log_3 3^{-1} = -1$

19. $\log_3 3^8 = 8\log_3 3 = 8\cdot 1 = 8$

20. $\ln e = \ln e^1 = 1$

21. $e^{\ln 5} = 5$

22. $10^{\log 3.15} = 3.15$

23. $\log_a\left(\dfrac{x}{y}\right) = \log_a x - \log_a y = 1.2 - 3.9 = -2.7$

24. $\log_a x = 1.2$

$\log_a \sqrt{x} = \dfrac{1}{2}\log_a x = \dfrac{1}{2}(1.2) = 0.6$

25. $\log_a(xy) = \log_a x + \log_a y = 1.2 + 3.9 = 5.1$

26. $\log_a y = 3.9$

$\log_a(y^4) = 4\log_a y = 4(3.9) = 15.6$

27. $\log(yz) = \log y + \log z$

28. $\ln\sqrt{\dfrac{x+1}{x}} = \dfrac{1}{2}\ln\dfrac{x+1}{x} = \dfrac{1}{2}[\ln(x+1) - \ln x]$

$= \dfrac{1}{2}\ln(x+1) - \dfrac{1}{2}\ln x$

29. No, Let $x = y = 1$. Then $\ln 1 + \ln 1 = 0 + 0 = 0$ and $\ln(1+1) = \ln 2 \approx 0.6931$.

30. $f(e^{-2}) = \ln(e^{-2}) = -2$

31. $f(x) = 2^x + \log(7x-4)$

$f(2) = 2^2 + \log(14-4) = 4 + 1 = 5$

32. $f(0) = e^0 + \ln 1 = 1 + 0 = 1$

33. $f(x) = \ln(3e^x - 5)$

$f(\ln 2) = \ln(3e^{\ln 2} - 5)$

$= \ln(3\cdot 2 - 5)$

$= \ln 1 = 0$

34. $\log_9 2158 = \dfrac{\log 2158}{\log 9} \approx 3.494$

35. $\log_{12} 0.0195 = \dfrac{\ln 0.0195}{\ln 12} = \dfrac{-3.9373}{2.4849} = -1.5845$

36.

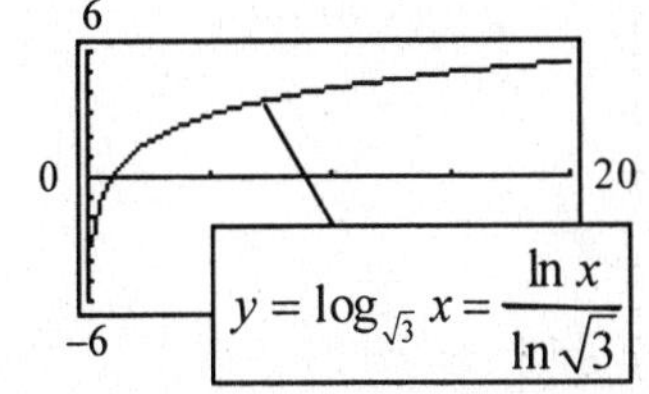

37. $f(x) = \log_{11}(2x-5)$

$$= \frac{\ln(2x-5)}{\ln 11} = \frac{\ln(2x-5)}{2.3979}$$

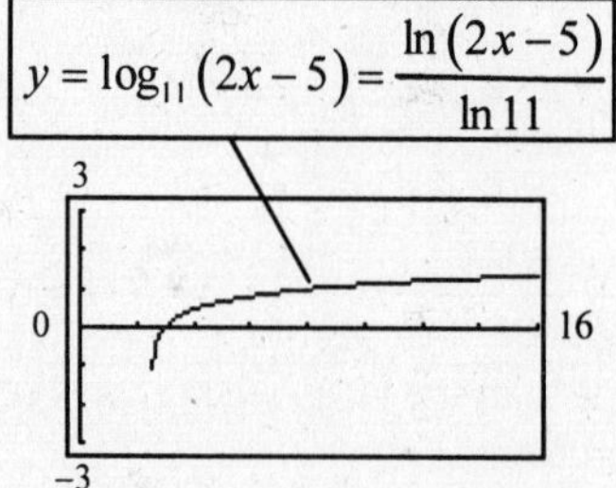

38. For the given graph, growth exponential would be the best model because the general outline has the same shape as a growth exponential.

39. a.

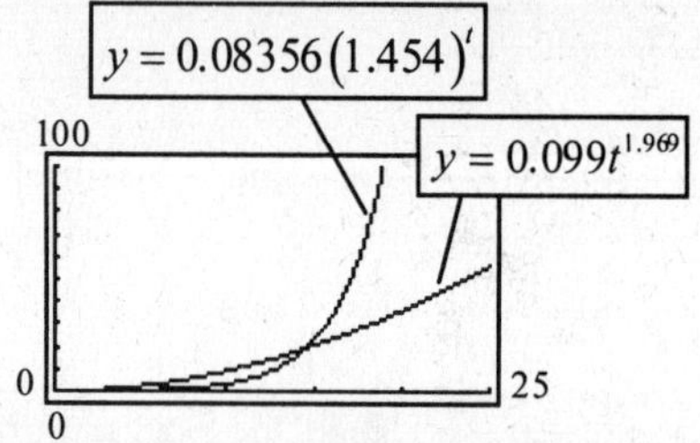

b. Is AIDS growing exponentially or more slowly, such as a power function?

40. Exponential because the general shape is similar to the graph of a decay exponential.

41. a. Letting x be the number of years since 1980, enter the data values $(0, 329821)$, $(5, 502507)$, etc., and obtain the exponential regression equation $y = 348275(1.076005)^x$

b. The model predicts a prison population of 1,507,329 in 2000 (using $x = 20$). This is 125,437 more than actual data reflects.

c. No. From 1999 to 2000 the curve appears to rise at a much slower rate.

42. a. Letting x be the number of years since 1900, enter the data values $(50, 23.5)$, $(60, 34.7)$, etc., and obtain the logarithmic regression equation $y = -197.0261 + 56.6496 \ln x$.

b. In 2010, $x = 110$.
$y = -197.0261 + 56.6496 \ln 110 = 69.2542$.
So the model predicts 69.3 percent paved roads in 2010.

c.
$$75 = -197.0261 + 56.6496 \ln x$$
$$272.0261 = 56.6496 \ln x$$
$$\ln x = \frac{272.0261}{56.6496} \approx 4.8019$$
$$x = e^{4.8019}$$
$$x \approx 121.7 \text{ years past 1900}$$
The model predicts 75 percent paved roads in the year 2022.

43. a. $M = -\frac{5}{2}\log\left(\frac{36.3B_0}{B_0}\right) = -\frac{5}{2}\log 36.3$

$$= -\frac{5}{2}(1.5599) = -3.8998$$

b.
$$2.1 = -\frac{5}{2}\log\left(\frac{B}{B_0}\right)$$
$$\log\frac{B}{B_0} = \frac{-4.2}{5} = -0.84$$
$$\frac{B}{B_0} = 10^{-0.84}$$
$$B = 10^{-0.84}B_0 = 0.14B_0$$

c.
$$6 = -\frac{5}{2}\log\left(\frac{B}{B_0}\right)$$
$$\log\frac{B}{B_0} = \frac{-12}{5} = -2.4$$
$$\frac{B}{B_0} = 10^{-2.4}$$
$$B = 10^{-2.4}B_0 = 0.004B_0$$

d. Yes

44. $N = 10{,}000(0.3)^{0.5^t}$

a. $t = 0$

$N = 10{,}000(0.3)^1 = \$3000$

b. $t = 3$

$N = 10{,}000(0.3)^{0.125} = 10{,}000(0.86028)$

$\approx \$8603$

c. Sales are a maximum when $(0.3)^{0.5^t}$ is a maximum. Since 0.3 is between 0 and 1 we know that 0.3 to a positive power cannot be greater than 1. Thus, maximum sales will be \$10,000.

45. a. $S = 50{,}000e^{-0.1(6)} = 50{,}000(0.5488)$

$= \$27{,}440.58$

b. $15{,}000 = 50{,}000e^{-0.1x}$

$e^{-0.1x} = \dfrac{15{,}000}{50{,}000}$ Take ln of each side.

$-0.1x = \ln 15{,}000 - \ln 50{,}000$

$x = \dfrac{9.6158 - 10.8198}{-0.1} = 12$ weeks

46. $S = 50{,}000e^{-0.6x}$

$x = 6$

$S = 50{,}000e^{-3.6} = 50{,}000(0.02732) = \$1366.19$

47. $2000 = 1000(1.01)^{12t}$

$(1.01)^{12t} = 2$

$12t = \dfrac{\ln 2}{\ln 1.01} = 69.66$

$t = \dfrac{69.66}{12} = 5.8$ years

48. $S = 5000e^{0.135t}$

a. $t = 0.75$

$S = 5000e^{0.10125} = 5000(1.10655)$

$= \$5532.77$

b. $S = 10{,}000$

$10{,}000 = 5000e^{0.135t}$

$2 = e^{0.135t}$

$t = \dfrac{\ln 2}{0.135} = \dfrac{0.6931}{0.135} = 5.13$ years

49. a.

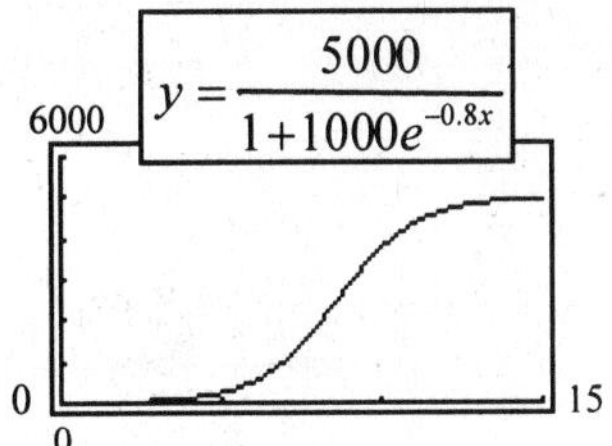

b. $x = 0$ when first discovered.

$y = \dfrac{5000}{1+1000e^{-0.8(0)}} = \dfrac{5000}{1001} = 4.995 \approx 5$

5 students had the virus when it was first discovered.

c. When $x = 15$,

$y = \dfrac{5000}{1+1000e^{-0.8(15)}} = 4969.4665$.

The total number infected by the virus during the first 15 days is 4969.

d. $3744 = \dfrac{5000}{1+1000e^{-0.8x}}$

$3744\left(1+1000e^{-0.8x}\right) = 5000$

$3744 + 3744000e^{-0.8x} = 5000$

$3744000e^{-0.8x} = 1256$

$e^{-0.8x} = \dfrac{1256}{3744000}$

$e^{-0.8x} \approx 0.0003547$

take ln of both sides

$-0.8x = \ln 0.0003547$

$-0.8x = -7.9442$

$x = \dfrac{-7.9442}{-0.8} \approx 9.9$

$= 10$ to the nearest whole number

In 10 days the number infected will reach 3744.

Chapter Test

1.

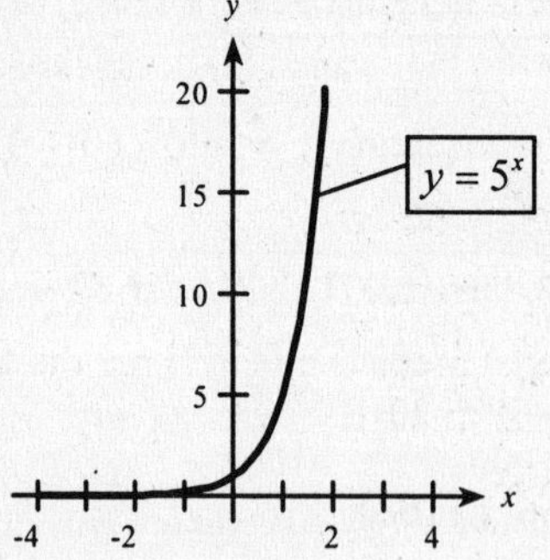

2.

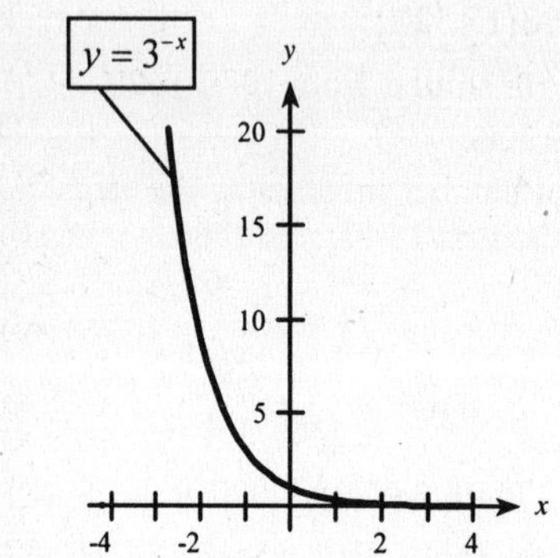

3.

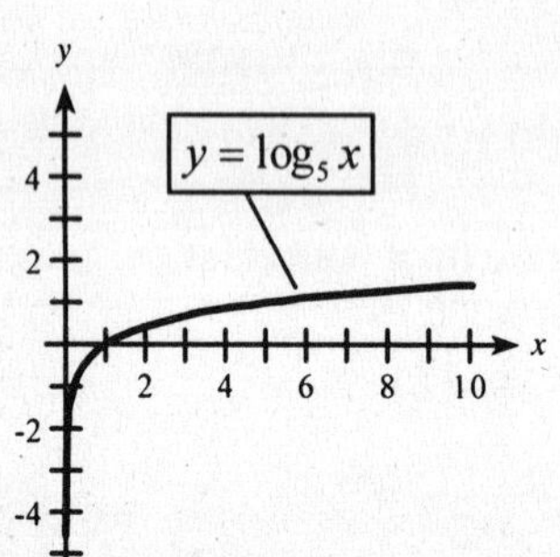

4.

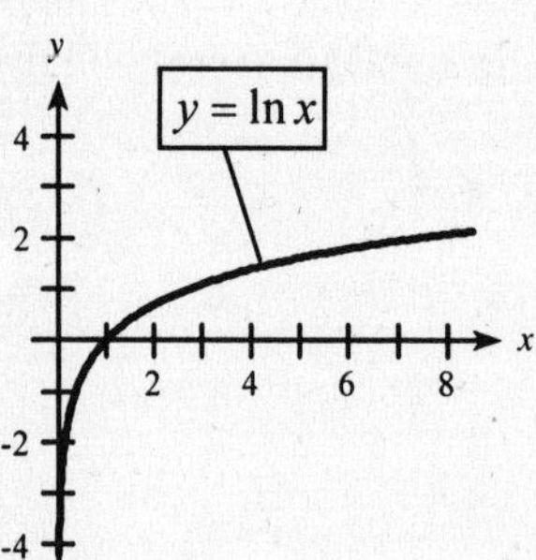

5.

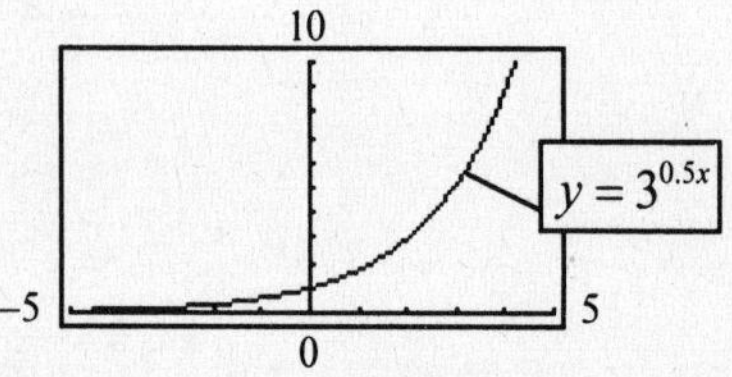

6.

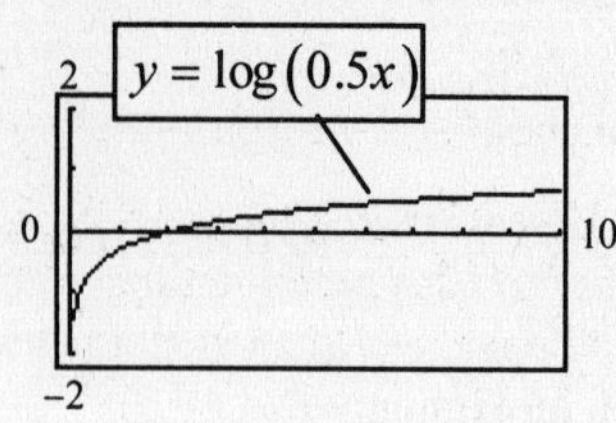

7.

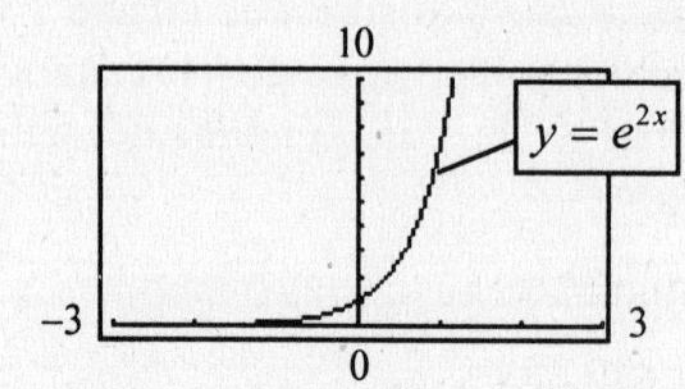

8.

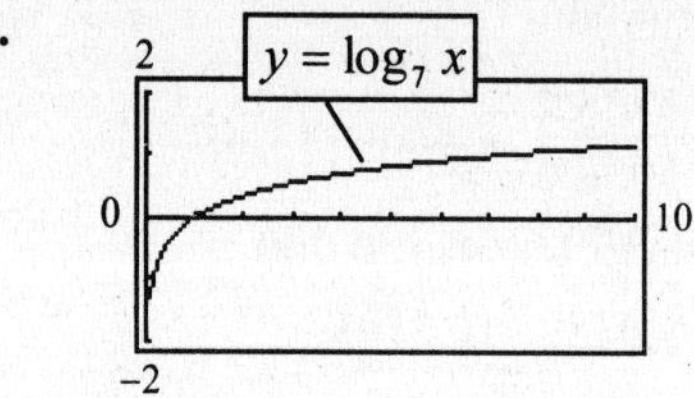

9. [2nd] [ln] 4 [=] 54.5982

10. [2nd] [ln] [(−)] 2.1 [=] 0.1225

11. [ln] 4 [=] 1.3863

12. [ln] 21 [=] 3.0445

13. $\log_{17} x = 3.1$

$$x = 17^{3.1}$$

$$x = 6522.163$$

14. $3^{2x} = 27$

$$2x = \log_3 27$$

$$x = \frac{1}{2}\log_3(3^3) = \frac{1}{2}\cdot 3 = \frac{3}{2}$$

15. $\log_2 8 = \log_2 2^3 = 3\log_2 2 = 3\cdot 1 = 3$

16. $e^{\ln x^4} = x^4$ (Property II)

17. $\log_7 7^3 = 3$ (Property I)

18. $\ln e^{x^2} = x^2$ (Property I)

19. $\ln(M \cdot N) = \ln M + \ln N$

20. $\ln\left(\frac{x^3 - 1}{x + 2}\right) = \ln\left(x^3 - 1\right) - \ln\left(x + 2\right)$

21. $\log_4\left(x^3+1\right)=\dfrac{\ln(x^3+1)}{\ln 4}$ or $0.721\ln\left(x^3+1\right)$

22. A decay exponential is more appropriate.

23. Exponential. If years were placed on horizontal axis and number of subscribers on the vertical axis, you would have a shape approximated by an exponential growth curve.

24. $y = 4155e^{0.0242x}$

a. $8310 = 4155e^{0.0242x}$

$2 = e^{0.0242x}$

$\ln 2 = 0.0242x \ln e$

$x = \dfrac{\ln 2}{0.0242} = 28.6$ years

b. $f(28.6) = 4155e^{0.0242(28.6)}$

$= 4155(1.9979)$

$= 8301$

Yes. Rounding causes the slight difference.

25. a. Letting x be the number of years since 1950, enter the data values $(10, 507.95)$, $(15, 588.95)$, etc., and obtain the exponential regression equation $y = 246.4660\left(1.0713^x\right)$.

b. In 2010, $x = 60$ so

$y = 246.4660\left(1.0713^{60}\right) = 15361.14$.

This model predicts a federal per capita tax of $15,361 in 2010.

26. $y = 317.4366(1.1162)^x$

27. $y = 254.9138(1.1327)^x$

The graphs are similar from 1980-1990.

Chapter 6: Mathematics of Finance

Exercise 6.1

1. $P = 10{,}000,\ t = 6,\ r = 0.16$

a. $I = Prt = 10{,}000(0.16)(6) = \$9600$

b. $S = P + I = 10{,}000 + 9600 = \$19{,}600$

3. $P = 1000,\ t = \dfrac{1}{4},\ r = 0.12$

a. $I = Prt = 1000(0.12)\left(\dfrac{1}{4}\right) = \$30$

b. $S = P + I = 1000 + 30 = \$1030$

5. $P = 800,\ t = \dfrac{1}{2},\ r = 0.16$

$I = Prt = 800(0.16)\left(\dfrac{1}{2}\right) = \$64$

$S = P + I = \$864$

7. $P = 3500,\ t = \dfrac{15}{12} = \dfrac{5}{4},\ r = 0.08$

$I = 3500(0.08)\left(\dfrac{5}{4}\right) = \$350$

$S = P + I = 3500 + 350 = \$3850$

9. $P = 30,\ t = 1,\ I = 0.90$

From $I = Prt$ we have

$r = \dfrac{I}{Pt} = \dfrac{0.90}{30(1)} = 0.03$. Thus, the couple earned 3% from the dividend.

$P = 30,\ t = 1,\ I = 3.90$

So, $r = \dfrac{3.90}{30(1)} = 0.13$. The total rate earned on the investment is 13%.

11. $S = 10{,}000,\ P = 9685.23,\ t = \dfrac{1}{2},$

a. $I = S - P = 314.77$

So, $r = \dfrac{I}{Pt} = \dfrac{314.77}{(9685.23)(\frac{1}{2})} = 0.065$. The interest rate is 6.5%.

b. $P = 9685.23 + 40 = 9725.23$

$I = S - P = 274.77$

$r = \dfrac{274.77}{(9725.23)(\frac{1}{2})} = 0.0565$

So the interest rate is 5.65%.

13. $S = 12(140) = 1680,\ t = \dfrac{1}{4},\ r = 0.12$, find P

$S = P + I = P + Prt = P(1 + rt)$

Thus,

$$P = \frac{S}{1+rt} = \frac{1680}{1+(0.12)(\frac{1}{4})} = \frac{1680}{1.03} = \$1631.07$$

The firm must deposit \$1631.07 now in order to pay the bill in 90 days.

15. $S = 13{,}500,\ P = ?\ \ t = \dfrac{10}{12} = \dfrac{5}{6},\ r = 0.15$

$S = P + Prt = P + P(0.15)\left(\dfrac{5}{6}\right) = 1.125P$

$P = 13500 \div 1.125 = \$12{,}000$

17. $S = 9000\ \ P = \$5000\ \ t = ??\ r = 0.08$

$I = S - P = \$4000$

$4000 = 5000\,(0.08)(t)$

$t = \dfrac{4000}{5000(0.08)} = 10$ years

19. 45 day note: $I = 500{,}000(0.06)\left(\dfrac{45}{360}\right) = \$3750$

60 day note:

$I = 500{,}000(0.07)\left(\dfrac{60}{360}\right) = \$5833.33$

Penalty is \$5000. By paying on time the retailer will earn \$3750 versus \$833.33 if he waits.

21. a. $P = 2000,\ t = \dfrac{9}{12},\ r = 0.08$

$S = P(1 + rt) =$

$2000\left[1+(0.08)\left(\dfrac{9}{12}\right)\right] = \$2120$

b. $S = 2120,\ t = \dfrac{3}{12},\ r = 0.10$

$P = \dfrac{S}{1+rt} = \dfrac{2120}{1+(\frac{1}{10})(\frac{3}{12})} = \$2068.29$

23.

n	1	2	3	4	5	6	7	8	9	10
a_n	3	6	9	12	15	18	21	24	27	30

25. $a_n = \dfrac{(-1)^n}{2n+1}$

n	1	2	3	4	5	6
a_n	$\frac{-1}{3}$	$\frac{1}{5}$	$\frac{-1}{7}$	$\frac{1}{9}$	$-\frac{1}{11}$	$\frac{1}{13}$

27. $a_n = \dfrac{n-4}{n(n+2)}$

$a_1 = \dfrac{1-4}{1(1+2)} = -1 \quad a_2 = \dfrac{2-4}{2(2+2)} = -\dfrac{1}{4}$

$a_3 = \dfrac{3-4}{3(5)} = -\dfrac{1}{15} \quad a_4 = \dfrac{0}{4(6)} = 0$

$a_{10} = \dfrac{10-4}{10(12)} = \dfrac{1}{20}$

29. **a.** $d = 3,\ a_1 = 2$
b. 11, 14, 17

31. **a.** $d = \dfrac{9}{2} - \dfrac{6}{2} = \dfrac{3}{2},\ a_1 = 3$
b. $\dfrac{15}{2}, 9, \dfrac{21}{2}$

33. $a_1 = 6,\ d = -\dfrac{1}{2},\ n = 83$

$a_{83} = 6 + (83-1)\left(-\dfrac{1}{2}\right) = -35$

35. $a_1 = 5,\ a_8 = 19,\ n = 100$

$19 = 5 + (8-1)d$ or $14 = 7d$ or $d = 2$

$a_{100} = 5 + (100-1)2 = 203$

37. $a_1 = 2,\ d = 3,\ n = 38,\ a_{38} = 113$

$s_{38} = \dfrac{38}{2}(2+113) = 2185$

39. $a_1 = 10, d = \dfrac{1}{2},\ n = 70$

$a_{70} = 10 + (70-1)\left(\dfrac{1}{2}\right) = \dfrac{89}{2}$

$s_{70} = \dfrac{70}{2}\left(10 + \dfrac{89}{2}\right) = \dfrac{3815}{2}$

41. $a_1 = 6, d = -\dfrac{3}{2}, n = 150$

$a_{150} = 6 + (150-1)\left(-\dfrac{3}{2}\right) = -\dfrac{435}{2}$

$s_{150} = \dfrac{150}{2}\left(6 - \dfrac{435}{2}\right) = -\dfrac{31,725}{2}$

43. Beginning with the third term and following terms: The value of the term is the sum of the values of the preceding two terms. 21, 34, 55

45. $a_1 = -2000, d = 400, n = 12$, Find a_{12}

$a_{12} = -2000 + (12-1)(400) = \$2400$

47. $a_1 = 20,000, d = 1000, n = 10$, Find a_{10}

$a_{10} = 20,000 + (10-1)1000 = \$29,000$

$a_1 = 18,000, d = 1200, n = 10$, Find a_{10}

$a_{10} = 18,000 + (10-1)1200 = \$28,800$

Through 10 years of work, the \$20,000 beginning salary is better.

49. **a.** The total base salary for 3 years is \$60,000. The total salary received for the 3 years is \$63,000. Thus, the sum of the raises is \$3000.
b. The total salary received for 3 years is \$64,500. Thus, the sum of the raises is \$4500.
c. Plan II is better by a total of \$1500.
d. Total salary for Plan I for 5 years is
$63,000 + 11,500 + 11,500$
$+12,000 + 12,000 = \$110,000$
Total base salary is 5(20,000) = \$100,000.
Sum of raises is \$10,000.
e. Total salary for Plan II for 5 years is:
$64,500 + 11,800 + 12,100 + 12,400$
$+12,700 = \$113,500$
Sum of the raises is \$13,500.
f. Plan II is better by \$3500.
g. Choose Plan II. As the years increase, its pay advantage also increases.

Exercise 6.2

1. **a.** 8%
b. 7
c. $\dfrac{8\%}{4} = 2\% = 0.02$
d. $4 \cdot 7 = 28$

3. **a.** 9%
b. 5
c. $\dfrac{9\%}{12} = \dfrac{3}{4}\% = 0.0075$
d. $12 \cdot 5 = 60$

5. $P = 8000,\ \frac{r}{m} = \frac{0.12}{1} = 0.12,\ mt = 1(10) = 10$

$S = 8000(1+0.12)^{10}$

$= 8,000(3.105848) = \$24,846.79$

7. $P = 10,000,\ \frac{r}{m} = \frac{0.09}{12} = 0.0075,\ mt = (12)3 = 36$

$S = 10,000(1+0.0075)^{36}$

$= 10,000(1.308645) = \$13,086.45$

Interest = \$3086.45 ($I = S - P$)

9. $P = 3200,\ \frac{r}{m} = \frac{0.08}{4} = 0.02,\ mt = 4(5) = 20$

$S = 3200(1+0.02)^{20} = 3200(1.485947)$

$= \$4,755.03$

11. $P = 8600,\ \frac{r}{m} = \frac{0.10}{2} = 0.05,\ mt = 2(6) = 12$

$S = 8600(1+0.05)^{12}$

$= 8600(1.795856) = \$15,444.36$

$I = S - P = \$6844.36$

13. $S = 40,000,\ P = ??\ \frac{r}{m} = \frac{0.10}{12} = 0.008333$

$mt = 12(18) = 216$

$P = \frac{40,000}{(1+0.008333)^{216}} = \$6661.50$

15. $S = 10,000\ \ P = ??$

$\frac{r}{m} = \frac{0.06}{1} = 0.06$

$mt = 1(10) = 10$

$P = \frac{10,000}{(1+0.06)^{10}} = \$5583.95$

17. $P = 5100,\ r = 0.09,\ t = 4,\ rt = 0.36$

$S = 5100e^{0.36}$

$= 5100(1.433329) = \$7309.98$

19. $P = 410, r = 0.08, t = 10, rt = 0.8$

$S = 410e^{0.8} = 410(2.225541) = \$912.47$

$I = S - P = \$502.47$

21. $S = 100,000,\ P = ??,\ t = 20$

a. $r = 0.105,\ rt = 2.1$

$P = \frac{S}{e^{rt}} = \frac{100,000}{e^{2.1}} = \frac{100,000}{8.16617} = \$12,245.64$

b. $r = 0.11,\ rt = 2.2$

$P = \frac{S}{e^{rt}} = \frac{100,000}{e^{2.2}} = \frac{100,000}{9.02501} = \$11,080.32$

c. A $\frac{1}{2}$% increase in the interest rate reduces the investment by \$1165.32

23. $P = 1000,\ \frac{r}{m} = 0.08,\ mt = 5$

$S = 1000(1+0.08)^5$

$= 1000(1.469328) = \$1469.33$

$P = 1000,\ r = 0.07,\ t = 5,\ rt = 0.35$

$S = 1000e^{0.35} = 1000(1.419068) = \$1419.07$

The 8% compounded annually yields \$50.26 more interest.

25. $APY = \left(1 + \frac{0.073}{12}\right)^{12} - 1$

$= 1.0755 - 1 = 7.55\%$

27. $S = Pe^r = Pe^{0.06} = P(1.0618)$

$APY = 1.0618 - 1 = 0.0618$

Effective annual rate is 6.18%.

29. The highest yield occurs with the most compounding periods. Rank: 8% compounded monthly, 8% compounded quarterly, 8% compounded annually.

31. $S_1 = Pe^r = Pe^{0.082} - P(1.08546)$

$S_2 = P(1 + \frac{0.084}{4})^4 = P(1.08668)$

$APY_1 = 1.08546 - 1 = 0.08546 = 8.546\%$

$APY_2 = 1.08668 - 1 = 0.08668 = 8.668\%$

The 8.4% investment is better.

33. The higher graph represents the continuous compounding since it has the higher yield.

35. $P = 10{,}000$, $S = 133{,}675$, $n = 10$, $i = ?$

$$S = P(1+i)^n$$
$$133{,}675 = 10{,}000(1+i)^{10}$$
$$(1+i)^{10} = 13.3675$$
$$10\ln(1+i) = \ln(13.3675)$$
$$\ln(1+i) = \frac{2.5928}{10} = 0.25928$$
$$1+i = e^{0.25928} = 1.296$$
$$i = 1.296 - 1 = 0.296 = 29.6\%$$

37. $P = 700$, $S = 700 + 300 = 1000$, $r = 0.119$, $t = ?$

$$1000 = 700e^{0.119t}$$
$$e^{0.119t} = \frac{1000}{700} \quad \text{Take ln of both sides.}$$
$$0.119t = \ln 1000 - \ln 700 = 6.9078 - 6.5511$$
$t = 2.997$ or approximately 3 years.

39. $P = 20{,}000$, $S = 26{,}425.82$, $\frac{r}{m} = \frac{r}{4}$, $t = 7$

$$26{,}425.82 = 20{,}000\left(1+\frac{r}{4}\right)^{28}$$
$$\left(1+\frac{r}{4}\right) = (1.32129)^{1/28}$$
$$= 1.00999 = 1.01$$
$$\frac{r}{4} = 0.01$$
$$r = 0.04 = 4\%$$

41. $P = 1000$, $\frac{r}{m} = \frac{0.08}{1} = 0.08$, $mt = 1(18) = 18$

$$S = 1000(1+0.08)^{18} = \$3996.02$$

43. **a.** $P = 12{,}860$, $\frac{r}{m} = \frac{0.075}{1} = 0.075$,

$$mt = 1(22) = 22$$
$$S = 12{,}860(1+0.075)^{22}$$
$$= 12{,}860(4.908923) = \$63{,}128.75$$

b. Change the interest rate to 0.085.

$$S = 12{,}860(1+0.085)^{22}$$
$$= 12{,}860(6.0180285) = \$77{,}391.85$$

At the higher rate of interest there will be an additional \$14,263.10 available for retirement.

45. $P = 10{,}000$, $S = 15{,}000$,

$$\frac{r}{m} = \frac{0.08}{4} = 0.02,\ mt = 4t$$
$$15{,}000 = 10{,}000(1+0.02)^{4t}$$
$$(1+0.02)^{4t} = 1.5000$$
$$4t \cdot \ln(1+0.02) = \ln(1.500)$$
$$t = \frac{\ln 1.5}{4\ln 1.02} \approx 5.1 \text{ years}$$

47. $P_1 = 2500$, $t = 3$, $r = 0.085$

$$S_1 = 2500[1+3(0.085)] = \$3137.50$$
$$S_1 = P_2 = 3137.50,\ n = 9,\ r = 0.18$$
$$S_2 = 3137.50(1+0.18)^9 = \$13{,}916.24$$

49. This problem is more meaningful if a spreadsheet is used.

a. From a table, the investments will be worth more than \$7500 during the sixth year.

b. $P = 5000$

	$i = \frac{0.063}{4}$	$i = \frac{0.063}{12}$
3 years	$5000\left(1+\frac{.063}{4}\right)^{12} = \$6031.31$	$5000\left(1+\frac{.063}{12}\right)^{36} = \$6037.22$
7 years	$5000\left(1+\frac{.063}{4}\right)^{28} = \$7744.634$	$5000\left(1+\frac{.063}{12}\right)^{84} = 7762.34$
10 years	$5000\left(1+\frac{.063}{4}\right)^{40} = 9342.07$	$5000\left(1+\frac{.063}{12}\right)^{120} = \$9372.59$

51. a. 3, 6, 12 Ratio is 2.
Next three terms are 24, 48, 96.

b. 81, 54, 36

$81 \cdot r = 54$ or $r = \dfrac{54}{81} = \dfrac{2}{3}$

Next three terms are 24, 16, $\dfrac{32}{3}$.

53. $a_1 = 10, r = 2, n = 13$

$a_{13} = 10 \cdot 2^{13-1} = 10 \cdot 4096 = 40{,}960$

55. $a_1 = 4, r = \dfrac{3}{2}, n = 16$

$a_{16} = 4\left(\dfrac{3}{2}\right)^{16-1} = 4\left(\dfrac{3}{2}\right)^{15}$

57. $a_1 = 6, r = 3, n = 17$

$s_{17} = \dfrac{6\left(1-3^{17}\right)}{1-3} = -3\left(1-3^{17}\right)$

59. $a_1 = 4,\ r = -\dfrac{1}{2},\ n = 21$

$s_{21} = \dfrac{4\left[1-\left(-\frac{1}{2}\right)^{21}\right]}{1-\left(-\frac{1}{2}\right)} = \dfrac{8}{3}\left[1+\left(\dfrac{1}{2}\right)^{21}\right]$

61. $a_1 = 1,\ r = 3, n = 35$

$s_{35} = \dfrac{1-3^{35}}{1-3} = \dfrac{3^{35}-1}{2}$

63. $a_1 = 6, r = \dfrac{2}{3}, n = 18$

$s_{18} = \dfrac{6\left[1-\left(\frac{2}{3}\right)^{18}\right]}{1-\frac{2}{3}} = 18\left[1-\left(\frac{2}{3}\right)^{18}\right]$

65. $a_1 = 160{,}000, r = 1.04,$
$n = 21$ (at the end of 20 years)

$a_{21} = 160{,}000(1.04)^{20} = \$350{,}580$

67. $a_1 = 20(\text{million}), r = 1.02, n = 11$

$a_{11} = 20(1.02)^{10}$
$= 20(1.218994) = 24.38$ million

Note: As in problem 65, n is 11 since 20 million is the value at the beginning of the year.

69. $a_1 = P, a_n = 2P, r = 1.02$

$2P = P(1.02)^n$

$(1.02)^n = 2$

$n \ln 1.02 = \ln 2$

$n = \dfrac{\ln 2}{\ln 1.02} = \dfrac{0.6931}{0.0198} = 35$ years

71. Down: 128 96 72 54

Up: 96 72 54 $\dfrac{81}{2}$

Note that there are two separate sequences.

Ball will bounce up $\dfrac{81}{2}$ or 40.5 ft.

73. Value now: 10,000

Value end of year	1	2	3	4
	8000	6400	5120	4096

75. Number now: 5000

End of Hour	1	2	3	4	5
Number	10,000	20,000	40,000	80,000	160,000

At end of 6th hour 320,000 bacteria are present. $N = 10{,}000 \cdot 2^{6-1}$

77. Profit now: $8,000,000

End of Year	2003	2004	2005	2006	2007
Profit	$7,840,000	$7,683,200	$7,529,536	$7,378,945	$7,231,366

Note: Problems 65 - 77 are all equivalent problems. Selecting the correct value of *n* is the hard part of solving these geometric sequence problems.

79. Payment #1 $0.1(12{,}000)=1200$ Payment #2 $0.1(10{,}800)=1080$
Payment #3 $0.1(9720)=972$ Payment #4 $0.1(8748)=874.80$
The sequence of outstanding balances is 10,800, 9720, 8748, $\cdots$. The 18th term is
$a_{18}=10{,}800(0.9)^{17}=\$1801.14$.

81. Five people receive your letter with your name in position 6. They cross out the top name and each sends out 5 letters. Your name is now in place 5 on $25=5^2$ letters. Each of these sends out 5 letters and now your name is in position 4 on $25\cdot 5=5^3$ letters. Continuing this pattern we have: position 3 5^4 letters
position 2 5^5 letters
position 1 5^6 letters

You will receive $5^6=15{,}625$ dimes.

83. Total number of letters mailed: $5+5^2+5^3+\cdots+5^{12}$ $s_{12}=\dfrac{5(1-5^{12})}{1-5}=305{,}175{,}780$ letters

Exercise 6.3

1. **a.** The higher graph is the \$1120 per year annuity.

b. $S=1000\left[\dfrac{(1+0.08)^{25}-1}{0.08}\right]=\$73{,}105.94$

$S=1120\left[\dfrac{(1+0.08)^{25}-1}{0.08}\right]=\$81{,}878.65$

The difference is
$\$81{,}878.65-73{,}105.94=\$8772.71$

3. $R=1300,\quad i=0.06,\quad n=5$

$S=1300\left[\dfrac{(1+0.06)^{5}-1}{0.06}\right]=\$7328.22$

5. $R=80, n=4\cdot 3=12, i=\dfrac{0.08}{4}=0.02$

$S=80\left[\dfrac{(1+0.02)^{12}-1}{0.02}\right]$

$=80(13.4121)=\$1072.97$

7. $R=500, n=4\cdot 2=8, i=\dfrac{0.10}{2}=0.05$

$S=500\left[\dfrac{(1+0.05)^{8}-1}{0.05}\right]$

$=500(9.5491)=\$4774.55$

9. $S=\$50{,}000, n=8, i=0.10$

$R=50{,}000\left(\dfrac{0.10}{(1+0.10)^{8}-1}\right)$

$=50{,}000(0.0874)=\$4372.20$

11. $S=\$200{,}000, n=20\cdot 4=80,$

$i=\dfrac{0.076}{4}=0.019$

$R=200{,}000\left(\dfrac{0.019}{(1+0.019)^{80}-1}\right)$

$=200{,}000(0.005417)=\$1083.40$

13. $S=\$20{,}000, n=2\cdot 12=24, i=\dfrac{0.12}{12}=0.01$

$R=20{,}000\left(\dfrac{0.01}{(1+0.01)^{24}-1}\right)$

$=20{,}000(0.03707)=\$741.47$

15. The 10% rate is better. A sinking fund is an account that accumulates money.

17. Twin 1: $R=2000, n=10, i=0.08$

$S_1=2000\left[\dfrac{(1+0.08)^{10}-1}{0.08}\right]=\$28{,}973.12$

$P_2=28{,}973.12, i=0.08, n=32$ (age 65)

$S_2=28{,}973.12(1+0.08)^{32}=340{,}060$ (rounded)

Twin 2: $S=340{,}060, i=0.08, n=25$ (age 65)

$R=340{,}060\left[\dfrac{0.08}{(1+0.08)^{25}-1}\right]=\$4651.61$

19. $R=100, n=\dfrac{5}{2}\cdot 4=10, i=\dfrac{0.12}{4}=0.03$

$S_{due}=100\left[\dfrac{(1+0.3)^{10}-1}{0.03}\right](1+0.03)=\$1180.78$

21. $R = 200, n = 8 \cdot 2 = 16, i = \frac{0.06}{2} = 0.03$

$$S_{due} = 200\left[\frac{(1+.03)^{16} - 1}{0.03}\right](1+0.03) = \$4152.32$$

23. $S_{due} = 24000, n = 5, i = 0.08$

$$R = 24000\left[\frac{0.08}{(1+.08)^5 - 1}\right]\left(\frac{1}{1+0.08}\right) = \$3787.92$$

25. $S = \$300{,}000$, $n = 25 \cdot 12 = 300$, $i = \frac{0.10}{12}$

$$R = 300{,}000\left(\frac{0.10/12}{(1+0.10/12)^{300} - 1}\right)$$
$$= 300{,}000(0.00075) = \$226.10$$

27. $R = 300$, $n = 12 \cdot 12 = 144$, $i = \frac{0.09}{12} = 0.0075$

$$S = 300\left[\frac{(1+0.0075)^{144} - 1}{0.0075}\right]$$
$$= 300(257.7116) = 77{,}313.47$$

29. $S_{due} = 180{,}000$, $n = 18 \cdot 12 = 216$,
$i = \frac{0.12}{12} = 0.01$

$$R = 180{,}000\left[\frac{0.01}{(1+0.1)^{216} - 1}\right]\left(\frac{1}{1+0.01}\right)$$
$$= \$235.16$$

31. $R = 500$, $n = 9 \cdot 4 = 36$, $i = \frac{0.08}{4} = 0.02$

$$S_{due} = 500\left[\frac{(1+0.02)^{36} - 1}{0.02}\right](1+0.02)$$
$$= \$26{,}517.13$$

33. $S = \$20{,}000$, $n = 10 \cdot 4 = 40$, $i = \frac{0.12}{4} = 0.03$

$$R = 20{,}000\left(\frac{0.03}{(1+0.03)^{40} - 1}\right)$$
$$= 20{,}000(0.01326) = \$265.25$$

35. $R = 100$, $n = 8 \cdot 12 = 96$, $i = \frac{0.09}{12} = 0.0075$

$$S_1 = 100\left[\frac{(1+0.0075)^{96} - 1}{0.0075}\right]$$
$$= 100(139.8562) = \$13{,}985.62$$

Now: $P = 13{,}985.62$, $n = 15 \cdot 12 = 180$,
$i = 0.0075$

$$S_2 = 13{,}985.62(1+0.0075)^{180} = 53{,}677.41$$

37. $S = 100\left[\frac{(1.006)^{120} - 1}{0.006}\right] = \$17{,}500.30$

and $S_{due} = S(1.006) = \$17605.30$

a. $n = 10 \cdot 12 = 120$ for each annuity. Total contributed: $100(120) = \$12{,}000$

b. The annuity due has more money. Each payment earns one month's interest more than is earned by the payments for the ordinary annuity.

Exercise 6.4

1. $R = 100, n = 17, i = 0.07$

$$A = 100\left[\frac{1-(1+.07)^{-17}}{0.07}\right] = \$976.32$$

3. $R = 6000, n = 8 \cdot 2 = 16, i = \frac{0.08}{2} = 0.04$

$$A = 6000\left[\frac{1-(1+.04)^{-16}}{0.04}\right] = \$69{,}913.77$$

5. $A = 135{,}000, n = 10 \cdot 4 = 40, i = \frac{0.064}{4} = 0.016$

$$R = 135{,}000\left[\frac{0.016}{1-(1+.016)^{-40}}\right] = \$4595.46$$

7. $R = 250{,}000\, n = 20, i = 0.10$

$$A = 250{,}000\left[\frac{1-(1+.10)^{-20}}{0.10}\right] = \$2{,}128{,}390.93$$

9. $A = 500{,}000, n = 10, i = 0.065$

a. $R = 500{,}000\left[\frac{0.065}{1-(1+.065)^{-10}}\right]$

$= \$69{,}552.35$

b. $A = 500{,}000, n = 10, i = 0.0682$

$R = 500{,}000\left[\frac{0.0682}{1-(1+.0682)^{-10}}\right]$

$= \$70{,}597.58$

HMO can save $70,597.58 – 69,552.35 = $1045.23 on each payment or $10,452.30 over 10 years.

11. a. The higher graph corresponds to 8%.

b. $1500 (approximately)

c. A 10% interest rate has a present value of $9000. An 8% interest rate has a present value of $10,500.

13. The payments are at the end of each period for an ordinary annuity. The payments are at the beginning of each period for an annuity due.

15. $R = 3000,\ n = 7\cdot 4 = 28,\ i = \frac{0.058}{4} = 0.0145$

$A_{(n,due)} = 3000\left[\frac{1-(1+.0145)^{-28}}{0.0145}\right](1+.0145)$

$= \$69{,}632.02$

17. $R = 50{,}000,\ n = 12,\ i = 0.0592$

$A_{(n,due)} = 50{,}000\left[\frac{1-(1+.0592)^{-12}}{0.0592}\right](1+.0592)$

$= \$445{,}962.23$

19. $A_{(n,due)} = 25{,}000,\ n = 12,\ i = \frac{0.0648}{12} = 0.0054$

$R = 25{,}000\left[\frac{0.0054}{1-(1+0.0054)^{-12}}\right]\left(\frac{1}{1+.0054}\right)$

$= \$2145.59$

21. $A_{(n,due)} = 800{,}000,\ n = 3\cdot 2 = 6,$

$i = \frac{0.077}{2} = 0.0385$

$R = 800{,}000\left[\frac{0.0385}{1-(1+.0385)^{-6}}\right]\left(\frac{1}{1+.0385}\right)$

$= \$146{,}235.06$

23. $R = 800,\ n = \frac{5}{2}\cdot 12 = 30,\ i = \frac{0.048}{12} = 0.004$

$A_{(n,due)} = 800\left[\frac{1-(1+.004)^{-30}}{0.004}\right](1+.004)$

$= \$22{,}663.74$

25. $A_n = 750{,}000$, $n = 40\cdot 12 = 480$,

$i = \frac{0.084}{12} = 0.007$

$R = 750{,}000\left[\frac{0.007}{1-(1+0.007)^{-480}}\right] = \$5441.23$

27. $R = 40{,}000,\ n = \frac{9}{2}\cdot 2 = 9,$

$i = \frac{0.0668}{2} = 0.0334$

$A_{(n,due)} = 40{,}000\left[\frac{1-(1+.0334)^{-9}}{0.0334}\right](1+.0334)$

$= \$316{,}803.61$

29. $R = 2000$, $n = 16$, $i = \frac{0.072}{4} = 0.018$

$A_n = 2000\left[\frac{1-(1+0.018)^{-16}}{0.018}\right] = \$27{,}590.62$

31. $P = 2500$, $n = 12\cdot 12 = 144$,

$i = \frac{0.078}{12} = 0.0065$

$S_1 = 2500(1+0.0065)^{144} = \$6355.13$

$S_2 = 100\left[\frac{(1+0.0065)^{144}-1}{0.0065}\right] = \$23{,}723.86$

a. After the last deposit, $S_1 + S_2 = \$30{,}078.99$ is in the account.

b. $\$2500 + \$14{,}400 = \$16{,}900$ was deposited.

c. $A_n = 30{,}078.99$, $n = 5\cdot 12 = 60$, $i = 0.0065$

$R = 30{,}078.99\left[\frac{0.0065}{1-(1+0.0065)^{-60}}\right]$

$= \$607.02$

d. The total amount withdrawn is $60(607.02) = \$36{,}421.20$.

33. $R = 2000,\ n = 8,\ k = 6,\ i = 0.07$

$= 2000\left[\frac{1-(1+0.07)^{-8}}{0.07}\right](1+0.07)^{-6}$

$= \$7957.86$

35. $R = 16{,}000,\ n = 7\text{(difficult part)},$
$k = 3,\ i = 0.06$

$$A_{(n,k)} = 16{,}000\left[\frac{1-(1+.06)^{-7}}{0.06}\right](1+.06)^{-3}$$

$$= \$74{,}993.20$$

37. $R = 10{,}000,\ n = 8,\ k = 36,$

$$i = \frac{0.07}{2} = 0.035$$

$$A_{(8,36)} = 10{,}000\left[\frac{1-(1+.035)^{-8}}{0.035}\right](1+.035)^{-36}$$

$$= \$19{,}922.97$$

39. $A_{(4,18)} = 1600,\ n = 4,\ k = 18,\ i = 0.06$

$$1600 = R\left[\frac{1-(1+.06)^{-4}}{0.06}\right](1+.06)^{-18}$$

$$= R(1.21398)$$

$$R = \frac{1600}{1.21398} = \$1317.98\text{ (Alternate method)}$$

41. $A_{(480,180)} = 2{,}500{,}000,\ n = 480,\ k = 180,$
$i = 0.008$

$$2{,}500{,}000 = R\left[\frac{1-(1+0.008)^{-480}}{0.008}\right](1+.008)^{-180}$$

$$= R(29.1361)$$

$$R = \frac{2{,}500{,}000}{29.1361} = \$85{,}804.29$$

43. a.

	A	B	C	D
1	End of Month	Acct. Value	Payment	New Balance
2	0	$100,000.00	$0.00	$100,000.00
3	1	$100,650.00	$1,000.00	$99,650.00
4	2	$100,297.73	$1,000.00	$99,297.73
5	3	$99,943.16	$1,000.00	$98,943.16
6	4	$99,586.29	$1,000.00	$98,586.29
7	5	$99,227.10	$1,000.00	$98,227.10
8	6	$98,865.58	$1,000.00	$97,865.58
9	7	$98,501.70	$1,000.00	$97,501.70
10	8	$98,135.47	$1,000.00	$97,135.47
11	9	$97,766.85	$1,000.00	$96,766.85
12	10	$97,395.83	$1,000.00	$96,395.83
⋮	⋮	⋮	⋮	⋮
159	157	$5,938.16	$1,000.00	$4,938.16
160	158	$4,970.26	$1,000.00	$3,970.26
161	159	$3,996.06	$1,000.00	$2,996.06
162	160	$3,015.54	$1,000.00	$2,015.54
163	161	$2,028.64	$1,000.00	$1,028.64
164	162	$1,035.32	$1,000.00	$35.32
164	163	$35.55	$35.55	$0.00

b.

	A	B	C	D
1	End of Month	Acct. Value	Payment	New Balance
2	0	$100,000.00	$0.00	$100,000.00
3	1	$100,650.00	$2,500.00	$98,150.00
4	2	$98,787.98	$2,500.00	$96,287.98
5	3	$96,913.85	$2,500.00	$94,413.85
6	4	$95,027.54	$2,500.00	$92,527.54
7	5	$93,128.97	$2,500.00	$90,628.97
8	6	$91,218.05	$2,500.00	$88,718.05
9	7	$89,294.72	$2,500.00	$86,794.72
10	8	$87,358.89	$2,500.00	$84,858.89
11	9	$85,410.47	$2,500.00	$82,910.47
12	10	$83,449.39	$2,500.00	$80,949.39
⋮	⋮	⋮	⋮	⋮
43	41	$15,902.26	$2,500.00	$13,402.26
44	42	$13,489.37	$2,500.00	$10,989.37
45	43	$11,060.81	$2,500.00	$8,560.81
46	44	$8,616.45	$2,500.00	$6,116.45
47	45	$6,156.21	$2,500.00	$3,656.21
48	46	$3,679.98	$2,500.00	$1,179.98
49	47	$1,187.65	$1,187.65	$0.00

Exercise 6.5

1. **a.** The 10 year loan requires more payment to principal since the loan must be paid more quickly.
b. The 25 year loan requires lower payment period since the loan is paid more slowly.

3. $A_n = 10,000,\ n = 5,\ i = 0.08,$

$$a_{\overline{5}|0.08} = \frac{1-(1+0.08)^{-5}}{0.08}$$

$$R = \frac{A_n}{a_{\overline{5}|0.08}} = A_n \cdot \frac{1}{a_{\overline{5}|0.08}} = \frac{10,000}{3.992710}$$

$$= 10,000(0.25045646) = \$2504.56$$

5. $A_n = 8000,\ n = 8,\ i = \frac{0.12}{2} = 0.06,$

$$a_{\overline{8}|0.06} = \frac{1-(1+0.06)^{-8}}{0.06}$$

$$R = \frac{8000}{a_{\overline{8}|0.06}} = 8000 \cdot \frac{1}{a_{\overline{8}|0.06}} = \frac{8000}{6.209794}$$

$$= 8000(0.161036) = \$1288.29$$

7. $A_n = 100,000,\ n = 3, i = 0.09,\ a_{\overline{3}|0.09} = \frac{1-(1+0.09)^{-3}}{0.09}$ $\quad R = \frac{100,000}{a_{\overline{3}|0.09}} = \frac{100,000}{2.531295} = \$39,505.48$

Payment – Interest = Balance Reduction

Period	Payment	Interest	Balance Reduction	Unpaid Balance
				100,000.00
1	39,505.48	9000.00	30,505.48	69,494.52
2	39,505.48	6254.51	33,250.98	36,243.55
3	39,505.47	3261.92	36,243.55	0.00

Last payment usually leaves a few cents on balance. Also, method used can affect answer by 2 or 3 cents.

9. $A_n = 20,000,\ n = 4,\ i = \dfrac{0.12}{4} = 0.03,\ a_{\overline{4}|0.03} = \dfrac{1-(1+0.03)^{-4}}{0.03}$

$R = \dfrac{20,000}{a_{\overline{4}|0.03}} = \dfrac{20,000}{3.717098} = \$5380.54$

Period	Payment	Interest	Balance Reduction	Unpaid Balance
				20,000.00
1	5380.54	600.00	4780.54	15,219.46
2	5380.54	456.58	4923.96	10,295.50
3	5380.54	308.87	5071.67	5223.83
4	5380.54	156.71	5223.83	0.00

11. $R = 334.27,\ \ n = 10\cdot 4,\ \ k = 6,\ \ i = \dfrac{0.06}{4} = 0.015$

Unpaid balance: $334.27\left[\dfrac{1-(1+.015)^{-(40-6)}}{0.015}\right] = \$8,852.05$

13. $R = 342.44,\ n = 48,\ k = 30,\ i = \dfrac{0.081}{12} = 0.00675$

Unpaid balance: $342.44\left[\dfrac{1-(1+.00675)^{-(48-30)}}{0.00675}\right] = \$5785.83$

15. $A_n = 100,000\,,\ n = 10\cdot 2 = 20\,,\ i = \dfrac{0.12}{2} = 0.06$

a. $R = 100,000\left[\dfrac{0.06}{1-(1+.06)^{-20}}\right] = \$8718.46$

b. Total paid = 20(8718.46) = \$174,369.20

c. Interest paid = 174,869.20 – 100,000 = \$74,369.20

17. $A_n = 8000,\ n = 4\cdot 2 = 8,\ i = \dfrac{0.10}{2} = 0.05$

a. $R = 8000\left[\dfrac{0.05}{1-(1+.05)^{-8}}\right] = \$1237.78$

b. Total paid = 8(1237.78) = \$9902.24

c. Interest paid = \$9902.24 – 8000 = \$1902.24

19. $R = 300, n = 36, i = 0.01$

$A_n = 300\left[\dfrac{1-(1+.01)^{-36}}{0.01}\right] = \$9032.25$

Down payment is \$12,000 – 9032.25 = \$2967.75.

21. $R = 500, n = 300, i = 0.01$

$$A_n = 500\left[\frac{1-(1+.01)^{-300}}{0.01}\right] = \$47,473.28$$

Amount available:
$\$15,000 + 47,473.28 = \$62,473.28$

23. $R = 45,000, n = 40$

a. $i = 0.0196$

$$A_n = 45,000\left[\frac{1-(1+.0196)^{-40}}{0.0196}\right]$$

$$= \$1,239,676.52$$

b. $i = 0.0182$

$$A_n = 45,000\left[\frac{1-(1+.0182)^{-40}}{0.0182}\right]$$

$$= \$1,270,768.38$$

25. $R = 610.91, n = 360, k = 12, i = 0.006$

a. Unpaid balance:

$$610.91\left[\frac{1-(1+.006)^{-(360-12)}}{0.006}\right] = \$89,120.53$$

b. Interest paid: 12(610.91) – 879.47
= $6451.45

27. The line is the total amount paid (amount per month x months). The other curve is the amount paid toward the principal.

29. The length of the vertical line segment from the lower curve to the line at $x = 250$ represents the total interest paid after 250 months.

31. a. $A_n = 15,000, \quad n = 12 \cdot 4 = 48$

$$i = \frac{0.08}{12} \rightarrow R = 15,000\left[\frac{\left(\frac{.08}{12}\right)}{1-\left(1+\frac{.08}{12}\right)^{-48}}\right] = \$366.19$$

$$i = \frac{0.085}{12} \rightarrow R = 15,000\left[\frac{\left(\frac{.085}{12}\right)}{1-\left(1+\frac{.085}{12}\right)^{-48}}\right] = \$369.72$$

Total interest at:

$8\% \quad : 48(366.19) - 15,000 = \$2577.12$

$8.5\% \quad : 48(369.72) - 15,000 = \$2746.56$

b. $A_n = 80,000, \quad n = 12 \cdot 30 = 360$

$$i = \frac{0.0675}{12} \rightarrow R = 80,000\left[\frac{\frac{.0675}{12}}{1-\left(1+\frac{.0675}{12}\right)^{-360}}\right] = \$518.88$$

$$i = \frac{0.0725}{12} \rightarrow R = 80000\left[\frac{\left(\frac{.0725}{12}\right)}{1-\left(1+\frac{.0725}{12}\right)^{-360}}\right] = \$545.74$$

Total interest at:

$6.75\% \quad : 360(518.88) - 80,000 = \$106,796.80$

$7.25\% \quad : 360(545.74) - 80,000 = \$116.466.40$

c. The duration of the loan has the greatest effect on the borrower. It affects R and the total interest paid.

33. $A_n = 100,000, n = 25 \cdot 12 = 300$

i. $$R = 100,000\left[\frac{\left(\frac{.075}{12}\right)}{1-\left(1+\frac{.075}{12}\right)^{-300}}\right] = \$738.99$$

ii. $$R = 100,000\left[\frac{\left(\frac{.0725}{12}\right)}{1-\left(1+\frac{.0725}{12}\right)^{-300}}\right] = \$722.81$$

iii. $$R = 100,000\left[\frac{\left(\frac{.07}{12}\right)}{1-\left(1+\frac{.07}{12}\right)^{-300}}\right] = \$706.78$$

a.

Payment	Points	Total Paid $[(R \cdot 300) + \text{Points}]$
$738.99	0	$221,697
722.81	$1000	217,843
706.78	$2000	214,034

b. The lower total cost is the 7%, 2 points loan.

35. $A_n = 16,700,\ i = \dfrac{0.082}{12},\ n = (4)(12) = 48$ $$R = 16,700\left[\frac{\frac{0.082}{12}}{1-\left(1+\frac{0.082}{12}\right)^{-48}}\right] = \$409.27$$

	A	B	C	D	E
1	**Period**	**Payment**	**Interest**	**Bal. Reduction**	**Unpaid Bal.**
2	0	$0.00	$0.00	$0.00	$16,700.00
3	1	$409.27	$114.12	$295.15	$16,404.85
4	2	$409.27	$112.10	$297.17	$16,107.68
5	3	$409.27	$110.07	$299.20	$15,808.48
6	4	$409.27	$108.02	$301.25	$15,507.23
7	5	$409.27	$105.97	$303.30	$15,203.93
8	6	$409.27	$103.89	$305.38	$14,898.55
9	7	$409.27	$101.81	$307.46	$14,591.09
10	8	$409.27	$99.71	$309.56	$14,281.52
⋮	⋮	⋮	⋮	⋮	⋮
45	43	$409.27	$16.38	$392.89	$2,004.81
46	44	$409.27	$13.70	$395.57	$1,609.24
47	45	$409.27	$11.00	$398.27	$1,210.97
48	46	$409.27	$8.27	$401.00	$809.98
49	47	$409.27	$5.53	$403.74	$406.24
50	48	$409.02	$2.78	$406.69	$0.00

37. Scheduled payment: $R = 18000\left[\dfrac{0.007}{1-(1.007)^{-60}}\right] = \$368.43$

	A	B	C	D	E
1	**Period**	**Payment**	**Interest**	**Bal. Reduction**	**Unpaid Bal.**
2	0	$0.00	$0.00	$0.00	$18,000.00
3	1	$383.43	$126.00	$257.43	$17,742.57
4	2	$383.43	$124.20	$259.23	$17,483.34
5	3	$383.43	$122.38	$261.05	$17,222.29
6	4	$383.43	$120.56	$262.87	$16,959.42
7	5	$383.43	$118.72	$264.71	$16,694.70
8	6	$383.43	$116.86	$266.57	$16,428.14
9	7	$383.43	$115.00	$268.43	$16,159.70
10	8	$383.43	$113.12	$270.31	$15,889.39
⋮	⋮	⋮	⋮	⋮	⋮
55	53	$383.43	$13.44	$369.99	$1,549.81
56	54	$383.43	$10.85	$372.58	$1,177.23
57	55	$383.43	$8.24	$375.19	$802.04
58	56	$383.43	$5.61	$377.82	$424.22
59	57	$383.43	$2.97	$380.46	$43.76
60	58	$44.07	$0.31	$43.76	$0.00

Following the 57th payment of $44.07, the loan is paid off.

Review Exercises

1. $a_n = \frac{1}{n^2}$ $a_1 = \frac{1}{1^2} = 1$ $a_2 = \frac{1}{2^2} = \frac{1}{4}$

   $a_3 = \frac{1}{3^2} = \frac{1}{9}$ $a_4 = \frac{1}{4^2} = \frac{1}{16}$

2. 12, 7, 2, −3, … arithmetic with $d = -5$.

   $\frac{1}{6}, \frac{2}{6}, \frac{3}{6}, \frac{4}{6}$, … arithmetic with $d = \frac{1}{6}$.

3. $a_1 = -2, d = 3, n = 80, a_{80} = -2 + (80-1) = 235$

4. $a_3 = 10, a_8 = 25, n = 36$

   $\begin{Bmatrix} 25 = a_1 + 7d \\ 10 = a_1 + 2d \end{Bmatrix}$ $\begin{matrix} d = 3 \\ a_1 = 4 \end{matrix}$

   $a_{36} = 4 + (36-1)3 = 109$

5. $a_1 = \frac{1}{3}, d = \frac{1}{6}, n = 60$

   $a_{60} = \frac{1}{3} + (60-1)\frac{1}{6} = \frac{61}{6}$

   $s_{60} = \frac{60}{2}\left(\frac{1}{3} + \frac{61}{6}\right) = 315$

6. $\frac{1}{4}$, 2, 16, 128, … geometric with $r = 8$.

   16, −12, 9, $-\frac{27}{4}$, … geometric with $r = -\frac{3}{4}$.

   $\left(16r = -12, r = -\frac{3}{4}\right)$

7. $a_1 = 64, a_8 = \frac{1}{2}, n = 4$

   $\frac{1}{2} = 64r^7, r^7 = \frac{1}{128}$

   or $r = \frac{1}{2}, a_4 = 64\left(\frac{1}{2}\right)^3 = 8$

8. $a_1 = \frac{1}{9}, r = 3, n = 16,$

   $s_{16} = \left(\frac{1}{9}\right)\frac{\left(1-3^{16}\right)}{1-3} = 2,391,484\frac{4}{9}$

9. $P = 8000, r = 0.12, t = 3$

   $S = P + Prt$

   $= 8000 + 8000(0.12)(3) = \$10,880$

10. $P = 2000, S = 2100, t = 0.75, r = ?$

    $I = S - P = 100, 100 = 2000 \cdot r(0.75)$

    $r = \frac{100}{2000(0.75)} = 6\frac{2}{3}\%$

11. $S = 3000, r = 0.06, t = \frac{4}{12} = \frac{1}{3}$

    $3000 = P\left(1 + (.06)\frac{1}{3}\right),$

    $P = \frac{3000}{1.02} = \$2941.18$

12. $a_1 = 10, a_2 = 20, a_3 = 30, \ldots, a_{30} = 300$

    $s_{30} = \frac{30}{2}(10 + 300) = \$4650$

13. Job #1

    $n = 10,\ a_1 = 20,000,$

    $d = 1000,\ a_{10} = 29,000$

    $s_{10} = \frac{10}{2}(20,000 + 29,000) = \$245,000$

    Job #2

    $n = 10,\ a_1 = 18,000,$

    $d = 1200,\ a_{10} = 28,800$

    $s_{10} = \frac{10}{2}(18,000 + 28,800) = \$234,000.$

14. **a.** $n = 4 \cdot 10 = 40$

    b. $i = \frac{0.08}{4} = 0.02 = 2\%$

15. **a.** $S = P\left(1 + \frac{r}{m}\right)^{mt} = P(1+i)^n$

    b. $S = Pe^{rt}$

16. Monthly compounding will earn more.

17. Interest = $1000(1.02)^{16} - 1000 = \$372.79$

18. $S = 18000, n = 4 \cdot 12 = 48, i = \frac{0.054}{12} = 0.0045$

    $P = \frac{18000}{(1 + 0.0045)^{48}} = \$14,510.26$

19. $S = 1000e^{(0.08)(6)} = \$1616.07$

20. $S = 100,000, t = 15, r = 0.1031$

    $P = \frac{100,000}{e^{(.1031)(15)}} = \$21,299.21$

Note: In 21–22 we can use either log or ln.

21. $25{,}000 = 15{,}000\left(1+\dfrac{0.06}{4}\right)^{4t}$

$\dfrac{5}{3} = (1.015)^{4t}$

$\log\left(\dfrac{5}{3}\right) = 4t\log(1.015)$

$t = \dfrac{\log 5 - \log 3}{4\log(1.015)} \approx 8.577$ years

22. a. $257{,}000 = 35{,}000e^{15r}$

$\ln\left(\dfrac{257{,}000}{35{,}000}\right) = 15r$

$\dfrac{\ln 257{,}000 - \ln 35{,}000}{15} = r$

$r \approx 0.1329 = 13.29\%$

b. $APY = e^{0.1329} - 1 = 0.1421 = 14.21\%$

23. a. $APY = \left(1+\dfrac{r}{m}\right)^m - 1,\ m = 4,\ r = 0.072$

$APY = \left(1+\dfrac{0.072}{4}\right)^4 - 1$

$= 1.0740 - 1 = 0.0740 = 7.40\%$

b. $APY = e^r - 1 = e^{.072} - 1$

$= 0.07466 = 7.47\%$

24. $a_1 = 1, a_2 = 2, \ldots, r = 2;$

$a_{64} = 1\cdot 2^{64-1} = 2^{63}$

25. $a_1 = 1, r = 2, a_{32} = 1\cdot 2^{32-1} = 2^{31}$

$s_{32} = \dfrac{1(1-2^{32})}{1-2} = 2^{32} - 1 \approx 4.295\times 10^9$

26. $R = 800, n = 20, \dfrac{0.12}{2} = 0.06$

$S = 800\cdot s_{\overline{20}|0.06} = 800(36.785592)$

$= \$29{,}428.47$

27. $S = 80{,}000, n = 10, i = 0.06;$

$R = S\cdot\dfrac{1}{s_{\overline{10}|0.06}} = 80{,}000(0.075868) = \$6096.44$

28. $R = 800, n = 10\cdot 2 = 20, i = \dfrac{0.12}{2} = 0.06$

$S_{due} = 800\left[\dfrac{(1+.06)^{20}-1}{0.06}\right](1+.06) = \$31{,}194.18$

29. $S = 250{,}000, n = \dfrac{9}{2}\cdot 4 = 18, i = \dfrac{0.102}{4} = 0.0255$

$R = 250{,}000\left[\dfrac{0.0255}{(1+.0255)^{18}-1}\right]\left(\dfrac{1}{1+.0255}\right)$

$= \$10{,}841.24$

30. $R = 10{,}000,\ n = 10\cdot 2 = 20, i = \dfrac{0.09}{2} = 0.045$

$A_n = 10{,}000\left[\dfrac{1-(1+.045)^{-20}}{0.045}\right] = \$130{,}079.36$

31. $R = 3000, n = 6, k = 5\cdot 12 - 1 = 59,$

$i = \dfrac{0.078}{12} = 0.0065$

$A_{(n,k)} = 3000\left[\dfrac{1-(1+.0065)^{-6}}{0.0065}\right](1+.0065)^{-59}$

$= \$12{,}007.09$

First payment is before they leave.

32. a. $R = \dfrac{295.7}{25}$ million $= \$11{,}828{,}000$

b. $R = 11{,}828{,}000, n = 24, i = 0.0591$

$A_n = 11{,}828{,}000\left[\dfrac{1-(1+.0591)^{-24}}{0.0591}\right]$

$= \$149{,}688{,}218$

Now add first payment at beginning: $A = \$161{,}516{,}218$. This is also an annuity due with $n = 25$.

33. $A_n = 20{,}000,\ n = 12,\ i = \dfrac{0.066}{12} = 0.0055$

$R = 20{,}000\left[\dfrac{0.0055}{1-(1+.0055)^{-12}}\right] = \$1{,}726.85$

34. $A_{due} = 250{,}000,\ n = 20\cdot 4 = 80,$

$i = \dfrac{0.062}{4} = 0.0155$

$A_{due} : R = 250{,}000\left[\dfrac{0.0155}{1-(1+.0155)^{-80}}\right]\left(\dfrac{1}{1+.0155}\right)$

$= \$5390.77$

35. $A_n = 1000,\ n = 12,\ i = 0.01$

$R = 1000\left[\dfrac{0.01}{1-(1+.01)^{-12}}\right] = \$88.85$

36. $R = 1288.29, n = 8, k = 5, i = 0.06$

$$A_{n-k} = 1288.29\left[\frac{1-(1+.06)^{-(8-5)}}{0.06}\right] = \$3443.61$$

37. Present value of the 18 quarterly payment is

$$A_n = 1500 \cdot a_{\overline{18}|1\%} = 1500 \cdot \frac{1-(1.01)^{-18}}{0.01}$$

or \$24,597.40

Total cost: 24,597.40 + 10,000 = \$34,597.40

38. $i = 0.00625$

Interest = (Unpaid Balance)(.00625)(1)

Balance Reduction = Payment – Interest

57: Interest = (95042.20)(.00625) = 594.01

Payment	Amount	Interest	Balance Reduction	Unpaid Balance
57	\$699.22	\$594.01	\$105.21	\$94,936.99
58	\$699.22	\$593.36	\$105.86	\$94,831.13

39. $P = 2500, S = 38,000, n = 18$

$$38,000 = 2500(1+i)^{18}$$

$$(1+i)^{18} = \frac{38,000}{2500}$$

$$18\ln(1+i) = \ln 38,000 - \ln 2500$$

$$18\ln(1+i) = 10.5453 - 7.8240 = 2.7213$$

$$\ln(1+i) = \frac{2.7213}{18} = 0.1512$$

$$1+i = e^{0.1512} = 1.1632$$

$$i = 0.1632 = 16.32\%$$

40. $S = 40,000, n = 10 \cdot 12 = 120, i = \dfrac{0.084}{12} = 0.007$

$$R = 40,000\left[\frac{0.007}{(1+.007)^{120}-1}\right] = \$213.81$$

41. a. $P = 1000,\ r = 8\%,\ t = 6$

$$S = P + Prt = 1000 + 1000(0.08)(6) = \$1,480$$

b. $P = 1000,\ i = \dfrac{0.08}{2} = 0.04,\ n = 6 \cdot 2 = 12$

$$S = P(1+i)^n = 1000(1+0.04)^{12} = \$1601.03$$

42. $R = 500,\ i = \dfrac{0.08}{4} = 0.02,\ n = 4 \cdot 4 = 16$

$$S = R\left[\frac{(1+i)^n - 1}{i}\right] = 500\left[\frac{(1+0.02)^{16}-1}{0.02}\right] = 500(18.6393) = \$9319.64$$

43. $P = 8000, S = 22,000, r = 0.07$

$$22,000 = 8000e^{0.07t}$$
$$e^{0.07t} = \frac{22,000}{8000}$$
$$0.07t = \ln 22,000 - \ln 8000$$
$$0.07t = 9.9988 - 8.9872 = 1.0116$$
$$t = 14.45 \text{ years}$$

44. $P = 87.89$, $S = 105.34$, $t = 1/4$

$$I = S - P = 105.34 - 87.89 = 17.45$$
$$17.45 = 87.89 \cdot r \cdot \tfrac{1}{4}$$
$$r = \frac{4(17.45)}{87.89} = 0.794 = 79.4\%$$

45. $R = 400$, $n = 4 \cdot 12 = 48$, $i = \dfrac{0.054}{12} = 0.0045$

$$S_{due} = R\left[\frac{(1+i)^n - 1}{i}\right](1+i) = 400\left[\frac{(1+0.0045)^{48} - 1}{0.0045}\right](1+0.0045) = \$21,474.08$$

46. $A_{(n,k)} = 72,000,\ n = 9 \cdot 2 = 18,\ k = 11 \cdot 2 = 22,\ i = 0.0365$

$$A_{(n,k)} : R = 72,000\left[\frac{0.0365}{1-(1+.0365)^{-18}}\right](1+.0365)^{22} = \$12,162.06$$

47. $APY = \left(1 + \dfrac{r}{m}\right)^m - 1 = \left(1 + \dfrac{0.0652}{4}\right)^4 - 1 = 0.066811$

$APY = e^r - 1 = e^{0.0648} - 1 = 0.066946$

The compounded continuous offer is the higher rate.

48. $r = 1000, n = 40, i = \dfrac{0.12}{12} = 0.01$

$$A_n = 1000\left[\frac{1-(1+.01)^{-40}}{0.01}\right] = \$32,834.69$$

49. $P = 8000, i = 0.12, n = 3$

$$S = Pe^{rt} = 8000e^{0.36} = 8000(1.433329415) = \$11,466.64$$
$$S - P = \$3,466.64$$

50. $A_n = 92,000,\ n = 25(12) = 300,\ i = \dfrac{6\%}{12} = 0.005$

a. $R = A_n\left(\dfrac{1}{a_{\overline{n}|i\%}}\right) = 92,000\left(\dfrac{0.005}{1-(1+0.005)^{-300}}\right) = 92,000(0.006443014) = \$592.76$

b. Total payment: 300(592.76) = \$177,828

c. Total interest: \$177,828 – 92,000 = \$85,828

d. Find the present value for $n = 300 - 84 = 216$.

$$A_{n-k} = 592.76\left[\frac{1-(1+0.005)^{-216}}{0.005}\right] = \$78,183.79$$

51. Bonus: $P = 12{,}500$, $n = 150$, $i = \dfrac{10.8\%}{12} = 0.009$

$S_1 = 12{,}500(1+0.009)^{150} = \$47{,}927.52$

Annuity: $S_2 = 150\left[\dfrac{(1+0.009)^{150}-1}{0.009}\right] = \$47{,}236.69$

a. $S = S_1 + S_2 = \$95{,}164.21$

b. $R = \$95{,}164.21\left[\dfrac{0.009}{1-(1+0.009)^{-120}}\right] = \$1300.14$

52. The withdrawals begin one month after the last deposit. Look at the problem this way. Amount needed in account at first withdrawal is $A = 5000 + 5000\left[\dfrac{1-(1+0.0055)^{-3}}{0.0055}\right] = \$19{,}836.50$. Now we have

$19{,}836.50 = R\left[\dfrac{(1+0.0055)^{36}-1}{0.0055}\right](1+0.0055)$. Solving we have $R = \$497.04$.

Chapter Test

1. $S = 47{,}000, P = 8000, r = 0.07,$

$S = Pe^{rt}$

$47{,}000 = 8000e^{0.07t}$

$e^{0.07t} = \dfrac{47{,}000}{8000} = 5.875$

$0.07t \ln e = \ln 5.875$

$t = \dfrac{\ln 5.875}{0.07} = 25.3$ years

2. $S = 12{,}000, n = 6\cdot 2 = 12, i = 0.031$

$R = 12{,}000\left[\dfrac{0.031}{(1+.031)^{12}-1}\right] = \$840.75$

3. $S = 5000, P = 4756.69, t = 0.75$

$I = S - P = 243.31$

$r = \dfrac{I}{Pt} = \dfrac{243.31}{4756.69(.75)} = 0.0682 = 6.82\%$

4. $R = 14{,}357.78, n = 40, k = 25, i = .041$

Unpaid balance;

$= 14{,}357.78\left[\dfrac{1-(1+.041)^{-(40-25)}}{0.041}\right]$

$= \$158{,}524.90$

5. $P = 1000, S = 13{,}500, n = 9$

$13{,}500 = 1000(1+i)^9$

$1+i = \left(\dfrac{13{,}500}{1000}\right)^{1/9} = 1.3353$

$i = 0.3353 = 33.53\%$

This problem can also be solved by taking ln of both sides.

6. $A_n = 97{,}000,\ n = 25\cdot 12 = 300,\ i = 0.006$

a. $R = 97{,}000\left[\dfrac{0.006}{1-(1+.006)^{-300}}\right] = \$698$

b. Interest paid $= 300(698) - 97{,}000 =$ \$112,400

7. $P = 2500, r = 0.04, t = \dfrac{15}{12} = 1.25$

$S = 2500 + 2500(.04)(1.25) = \$2625$

8. $R = 100, n = \dfrac{11}{2}\cdot 12 = 66, i = \dfrac{0.069}{12} = 0.00575$

$S_n = 100\left[\dfrac{(1+.00575)^{66}-1}{0.00575}\right] = \$7999.41$

9. $i = \dfrac{0.084}{12} = 0.007,\ APY = (1+.007)^{12} - 1$

$= 0.0873 = 8.73\%$

10. $R = 3000, n = 15 \cdot 4 = 60, i = 0.015$

$$A_{due} = 3000\left[\frac{1-(1+.015)^{-60}}{0.015}\right](1+.015)$$

$$= \$119,912.92$$

11. $P = 10,000, r = 0.07, t = 20$

$$S = 10,000e^{(.07)(20)} = \$40,552$$

12. $R = 1500, n = 18, i = 0.02$

$$A_n = 1500\left[\frac{1-(1+.02)^{-18}}{0.02}\right] = \$22,488 \text{ (rounded)}$$

Cost of car = \$22,488 + 10,000 = \$32,488

13. $S = 9,500, n = 5 \cdot 4 = 20, i = 0.017$

$$P = \frac{9500}{(1+.017)^{20}} = \$6781.17$$

14. $P = 10,000, n = \frac{9}{2} \cdot 4 = 18, i = 0.01925$

$$S_1 = 10,000(1+.01925)^{18} = 14,094.61$$

$$S_2 = 50,000 - S_1 = 35,905.39$$

$$R = 35,905.39\left[\frac{0.01925}{(1+.01925)^{18}-1}\right] = \$1688.02$$

15. a. $R = 3000, n = 8 \cdot 2 = 16, i = 0.0375$

$$S_1 = 3000\left[\frac{(1+.0375)^{16}-1}{0.0375}\right] = 64,178.22$$

$$S = 64,178.22(1+0.0375)^{40} = \$279,841.35$$

b. $A = 279,841.35, n = 20 \cdot 2 = 40, i = 0.0375$

$$A_{due}: R = 279,841.35\left[\frac{0.0375}{1-(1+.0375)^{-40}}\right](1+.0375)^{-1}$$

$$= \$13,124.75$$

16. $R = 80, n = 15 \cdot 12 = 180, i = 0.005$

$$S_{due} = 80\left[\frac{(1+.005)^{180}-1}{0.005}\right](1+.005)$$

$$= \$23,381.82$$

17. Deferred Annuity: $R = 4000$, $n = 16$, $k = 40$, $i = 0.016$

$$A = 4000\left[\frac{1-(1+.016)^{-16}}{0.016}\right](1+.016)^{-40}$$

$$= \$29,716.47$$

18. a. The difference between successive terms is -5.5.

b. $a_{51} = 298.8 + (51-1)(-5.5) = 23.8$

c. $s_{51} = \frac{51}{2}(298.8 + 23.8) = 8226.3$

19. $a_1 = 400, r = 0.6, n = 31$ (Look at sequence.)

$$s_{31} = 400\frac{(1-0.6^{31})}{1-0.6} = 1000 \text{ mg}$$

Chapter 7: Introduction to Probability

Exercise 7.1

1. a. $\Pr(R) = \frac{4}{10} = \frac{2}{5}$

b. $\Pr(G) = \frac{0}{10} = 0$

c. $\Pr(R \text{ or } W) = \frac{10}{10} = 1$

3. $\Pr(4,8,12) = \frac{3}{12} = \frac{1}{4}$

5. $\Pr(\text{greater than } 0) = \frac{6}{6} = 1$

7. a. $\Pr(\text{Red}) = \frac{3}{10}$

b. $\Pr(\text{Odd}) = \frac{5}{10} = \frac{1}{2}$

c. $\Pr(\text{Red and Odd}) = \Pr(1 \text{ or } 3) = \frac{2}{10} = \frac{1}{5}$

d. $\Pr(\text{Red or Odd})$

$= \Pr(1,2,3,5,7,\text{or } 9) = \frac{6}{10} = \frac{3}{5}$

e. $\Pr(\text{Not Black})$

$= 1 - \Pr(\text{Black}) = 1 - \frac{3}{10} = \frac{7}{10}$

9. a. $\Pr(\text{Queen}) = \frac{4}{52} = \frac{1}{13}$

b. $\Pr(\text{Heart or Diamond}) = \frac{1}{2}$

c. $\Pr(\text{Spade}) = \frac{1}{4}$

11. Sample space = {HH, HT, TH, TT}

a. $\Pr(0H) = \frac{1}{4}$

b. $\Pr(1H) = \frac{2}{4} = \frac{1}{2}$

c. $\Pr(2H) = \frac{1}{4}$

13. a. $\Pr(\text{Sum} = 4) = \frac{3}{36} = \frac{1}{12}$

b. $\Pr(\text{Sum} = 10) = \frac{3}{36} = \frac{1}{12}$

c. $\Pr(\text{Sum} = 12) = \frac{1}{36}$

15. a. $\Pr(4 \le S \le 7) = \frac{3+4+5+6}{36} = \frac{1}{2}$

(3 ways to roll a 4, etc.)

b. $\Pr(8 \le S \le 12) = \frac{5+4+3+2+1}{36} = \frac{15}{36} = \frac{5}{12}$

(5 ways to roll 8, etc.)

17. a. $\Pr(6) = \frac{431}{1200}$

b. The die is biased since Pr(6) should be about $1/6$.

19. a. $2:3$ or $\frac{2}{3}$

b. $3:2$ or $\frac{3}{2}$

21. a. $\Pr(\text{Win}) = \frac{1}{20+1} = \frac{1}{21}$

b. $\Pr(\text{Lose}) = \frac{20}{20+1} = \frac{20}{21}$

23. $\Pr(\text{In city}) = 0.46$

25. a. $\Pr(\text{defective turn signal}) = \frac{63}{425} = 0.1482$

b. $\Pr(\text{defective tires}) = \frac{32}{425} = 0.753$

27. a. $\Pr(\text{Republican will vote})$

$= \frac{2835}{4500} = \frac{63}{100} = 0.63$

$\Pr(\text{Democrat will vote})$

$= \frac{2501}{6100} = \frac{41}{100} = 0.41$

$\Pr(\text{Independent will vote})$

$= \frac{1122}{2200} = \frac{51}{100} = 0.51$

b. Probability is highest that a Republican will vote in the next election.

29. a. $\Pr(100\%\text{ discount}) = \dfrac{1}{3601}$

b. $\Pr(50\%\text{ discount}) = \dfrac{100}{3601}$

c. $\Pr(< 50\%\text{ discount}) = \dfrac{3000+500}{3601} = \dfrac{3500}{3601}$

d. $\Pr(30\%\text{ discount}) = \frac{500}{3601}$;

$\Pr(> 30\%\text{ discount}) = \frac{101}{3601}$.

So, more likely to get a 30% discount than a higher discount.

31. a. $\Pr(\text{from developed nation})$

$= \dfrac{1188}{6157} = 0.1930$

b. $\Pr(\text{age 20-39}) = \dfrac{1946}{6157} = 0.3161$

c. $\Pr(\text{age 60-79}) = \dfrac{545}{6157} = 0.0885$

33. $S = \{A^+, A^-, B^+, B^-, AB^+, AB^-, O^+, O^-\}$

35. a. $\Pr(AB) = 0.04$
b. $\Pr(\text{not }AB) = 0.96$

37. a. $\Pr(A) = 0.10$
b. $\Pr(\text{not }A) = 0.90$

39. $\Pr(\text{Empty}) = \dfrac{60{,}000}{2{,}000{,}000} = \dfrac{3}{100}$

41. Pr(has lactose intolerance) = 0.75

43. $\Pr(\text{Woman}) = \dfrac{3}{9} = \dfrac{1}{3}$

45. $\Pr(\text{grade}) = \dfrac{2}{6} = \dfrac{1}{3}$

47. Pr(No V and DS) = 0.22; yes

49. $\Pr(\text{win}) = \dfrac{1}{3+1} = \dfrac{1}{4}$

51. Sample space = {GGG, GGB, GBG, BGG, BBG, BGB, GBB, BBB}; $\Pr(2G) = \dfrac{3}{8}$

53. a. No
b. Sample space = {GG, GB, BG, BB}
c. $\Pr(1B\text{ and }1G) = \dfrac{2}{4} = \dfrac{1}{2}$

55. $\Pr(\text{Q1 is correct}) = \dfrac{1}{5}$

57. $\Pr(\text{Defective}) = \dfrac{6}{250} = \dfrac{3}{125}$

59. $\Pr(A) = \dfrac{182}{25{,}500(365)} \approx 0.0000196$

$\Pr(B) = \dfrac{51}{3890(365)} \approx 0.0000359$

$\Pr(C) = \dfrac{118}{8580(365)} \approx 0.0000377$

Intersection C is the most dangerous.

61. a. $\Pr(A) = \dfrac{557}{1200}$

b. $\Pr(\text{No Vote}) = \dfrac{110}{1200} = \dfrac{11}{120}$

63. a. $\Pr(B) = \dfrac{2}{10} = \dfrac{1}{5}$

$\Pr(G) = \dfrac{8}{10} = \dfrac{4}{5}$

b. $\Pr(B) = \dfrac{1194}{1220+1194} \approx 0.4946$

$\Pr(G) = \dfrac{1220}{2414} \approx 0.5054$

c. $\Pr(B) = \Pr(G) = \dfrac{1}{2}$; part b

Exercise 7.2

1. $\Pr(6, 12) = \dfrac{2}{12} = \dfrac{1}{6}$

3. $\Pr(1, 3, 5) = \dfrac{1}{2}$

5. $\Pr(E') = 1 - \Pr(E) = 1 - \dfrac{3}{5} = \dfrac{2}{5}$

7. a. $\Pr(R \cap \text{Even}) = \Pr(2, 4) = \dfrac{2}{14} = \dfrac{1}{7}$

b. $\Pr(R \cup \text{Even})$

$= \Pr(R) + \Pr(\text{Even}) - \Pr(R \cap \text{Even})$

$= \dfrac{5}{14} + \dfrac{7}{14} - \dfrac{2}{14} = \dfrac{10}{14} = \dfrac{5}{7}$

9. Pr(odd or {4, 8, 12})

$$=\frac{6}{12}+\frac{3}{12}-0=\frac{9}{12}=\frac{3}{4}$$

11. $\text{Pr(odd or }\{3, 6, 9, 12\}) = \frac{6}{12}+\frac{4}{12}-\frac{2}{12}$

$$=\frac{8}{12}=\frac{2}{3}$$

13. $\Pr(W \cup G) = \frac{4}{17}+\frac{6}{17}-0=\frac{10}{17}$

15. $\Pr(R \cup W) = \frac{2}{6}+\frac{2}{6}-0=\frac{4}{6}=\frac{2}{3}$

17. a. $\Pr(\text{not } E) = \frac{9}{18}=\frac{1}{2}$

b. $\Pr(R \cap E) = \frac{6}{18}=\frac{1}{3}$

c. $\Pr(R \cup E) = \frac{13}{18}+\frac{9}{18}-\frac{6}{18}=\frac{16}{18}=\frac{8}{9}$

d. $\Pr(R' \cap E') = \Pr(15, 17) = \frac{2}{18}=\frac{1}{9}$

19. $\Pr(IC') = 1-\Pr(IC) = 1-0.46 = 0.54$

21. a. $\Pr(DT') = 1-\Pr(DT) = 1-\frac{63}{425}=\frac{362}{425}$

b. $\Pr(\text{No Defects}) = 1-\frac{63+32}{425}=\frac{330}{425}=\frac{66}{85}$

23. $\Pr(F \cup G) = \frac{24}{100}+\frac{18}{100}-\frac{8}{100}=\frac{34}{100}=\frac{17}{50}$

25. a. $\Pr(A \cup B) = \frac{1}{3}+\frac{1}{5}-0=\frac{8}{15}$

b. $\Pr(A' \cap B') = 1-\frac{8}{15}=\frac{7}{15}$

27. a. $\Pr(\text{female} \cap \text{less than } \$30{,}000) = 0.35$

b. $\Pr(\text{at least } \$50{,}000) = 0.08$

c. $\Pr(\text{male} \cup \text{ less than } \$30{,}000)$

$$= \Pr(\text{male}) + \Pr(\text{less than } \$30{,}000) - \Pr(\text{male} \cap \text{less than } \$30{,}000) = 0.48+0.60-0.25 = 0.83$$

29. a. $\Pr(\text{developed nation or 40-59}) = \frac{1188}{6157}+\frac{1200}{6157}-\frac{318}{6157}=\frac{2070}{6157}=0.336$

b. $\Pr(\text{20-39} \cap \text{less developed nation}) = \frac{1605}{6157}=0.261$

c. $\Pr(\text{age 40 or more}) = \Pr(\text{40-59}) + \Pr(\text{60-79}) + \Pr(\text{80+}) = \frac{318+882}{6157}+\frac{196+349}{6157}+\frac{38+35}{6157}=\frac{1818}{6157}=0.295$

31. a. $\Pr(\text{Hispanic}) = \frac{\text{\# of Hispanics}}{\text{Total}}=\frac{2674+763}{15{,}383+5074}=\frac{3437}{20{,}457}=0.1680$

b. $\Pr(\text{female}) = \frac{\text{\# of females}}{\text{Total}}=\frac{5074}{20{,}457}=0.2480$

c. $\Pr(\text{female} \cup \text{Hispanic}) = \Pr(\text{female}) + \Pr(\text{Hispanic}) - \Pr(\text{female} \cap \text{Hispanic})$

$$=\frac{5074}{20{,}457}+\frac{3437}{20{,}457}-\frac{763}{20{,}457}=\frac{7748}{20{,}457}=0.3787$$

d. $\Pr(\text{male or Black}) = \Pr(\text{male}) + \Pr(\text{Black}) - \Pr(\text{male} \cap \text{Black})$

$$=\frac{15{,}383}{20{,}457}+\frac{9945}{20{,}457}-\frac{6667}{20{,}457}=\frac{18{,}661}{20{,}457}=0.9122$$

33. a. $\Pr(A \cup B) = \frac{7}{12} + \frac{7}{12} - \frac{3}{12} = \frac{11}{12}$

b. $\Pr(B \cup C) = \frac{7}{12} + \frac{8}{12} - \frac{5}{12} = \frac{10}{12} = \frac{5}{6}$

35. a. $\Pr(S \cup LA) = \frac{12}{32} + \frac{4}{32} - 0 = \frac{16}{32} = \frac{1}{2}$

b. $\Pr(E \cup S) = \frac{16}{32} + \frac{12}{32} - 0 = \frac{28}{32} = \frac{7}{8}$

c. $\Pr(E \cup F) = \frac{16}{32} + \frac{13}{32} - \frac{5}{32} = \frac{24}{32} = \frac{3}{4}$

37. $\Pr(R \cup {>}S) = \frac{400+50}{1000} + \frac{435}{1000} - \frac{300+25}{1000} = \frac{560}{1000} = 0.56$

39. $\Pr(W \cup S') = \frac{400+350}{1000} + \frac{565}{1000} - \frac{100+250}{1000} = \frac{965}{1000} = 0.965$

41. a. $\Pr(F \cup RA) = \frac{390+160}{1000} + \frac{560}{1000} - \frac{390}{1000} = \frac{720}{1000} = 0.72$

b. $\Pr(M \cup RA) = \frac{450}{1000} + \frac{560}{1000} - \frac{170}{1000} = \frac{840}{1000} = 0.84$

c. $\Pr(M \cup RA') = \frac{450}{1000} + \frac{440}{1000} - \frac{280}{1000} = \frac{610}{1000} = 0.61$

43. $\Pr(M \cup BO) = \frac{30}{65} + \frac{29}{65} - \frac{8}{65} = \frac{51}{65}$

45. $\Pr(PC') = 1 - 0.87 = 0.13$

Exercise 7.3

1. a. $\Pr(H \mid R) = \frac{13 \text{ Hearts}}{26 \text{ Red}} = \frac{1}{2}$

b. $\Pr(K \mid R) = \frac{2 \text{ Kings}}{26 \text{ Red}} = \frac{1}{13}$

3. a. $\Pr(6 \mid \text{even})$

$= \frac{\Pr(6 \text{ and even})}{\Pr(\text{even})} = \frac{\frac{2}{9}}{\frac{6}{9}} = \frac{2}{6} = \frac{1}{3}$

b. $\Pr(3 \mid 3 \text{ or } 6)$

$= \frac{\Pr(3 \text{ and } (3 \text{ or } 6))}{\Pr(3 \text{ or } 6)} = \frac{\frac{1}{9}}{\frac{3}{9}} = \frac{1}{3}$

5. $\Pr(R \mid E) = \frac{\Pr(R \text{ and } E)}{\Pr(E)} = \frac{\frac{4}{15}}{\frac{7}{15}} = \frac{4}{7}$

7. a. $\Pr(W \mid R) = \frac{6(W)}{9(\text{Total})} = \frac{2}{3}$

b. $\Pr(R \mid W) = \frac{4(R)}{9(\text{Total})} = \frac{4}{9}$

c. First draw has no effect on second draw.

$\Pr(W) = \frac{6}{10} = \frac{3}{5}$

9. a. $\Pr(H2 \text{ and } H3 \mid H1)$

$= \frac{\Pr(H1H2H3)}{\Pr(H1)} = \frac{\frac{1}{8}}{\frac{1}{2}} = \frac{1}{4}$

b. $\Pr(H3 \mid H1 \text{ and } H2)$

$= \frac{\Pr(H1H2H3)}{\Pr(H1 \text{ and } H2)} = \frac{\frac{1}{8}}{\frac{1}{4}} = \frac{1}{2}$

11. $\Pr(3 \text{ first} \cap 6 \text{ second}) = \frac{1}{6} \cdot \frac{1}{6} = \frac{1}{36}$

13. a. $\Pr(HHH) = \frac{1}{2} \cdot \frac{1}{2} \cdot \frac{1}{2} = \frac{1}{8}$

b. $\Pr(\text{at least 1 tail})$

$= 1 - \Pr(\text{no tails}) = 1 - \frac{1}{8} = \frac{7}{8}$

15. a. $\Pr(\text{R1 and W2}) = \Pr(\text{R})\cdot\Pr(\text{W}) = \frac{3}{10}\cdot\frac{2}{10} = \frac{3}{50}$

b. $\Pr(\text{R1 and W2}) = \Pr(\text{R})\cdot\Pr(\text{W} \mid \text{R}) = \frac{3}{10}\cdot\frac{2}{9} = \frac{1}{15}$

c. The events in (a) are independent. Since the ball was replaced it did not affect the probability of getting a white ball the second time. The events in (b) were dependent. Not replacing the first ball changed the sample space which affected the probability of getting a white ball the second time.

17. a. $\Pr(\text{RR}) = \Pr(\text{R})\cdot\Pr(\text{R}) = \frac{2}{5}\cdot\frac{2}{5} = \frac{4}{25}$

b. $\Pr(\text{WW}) = \Pr(\text{W})\cdot\Pr(\text{W}) = \frac{3}{5}\cdot\frac{3}{5} = \frac{9}{25}$

c. $\Pr(\text{R1 and W2}) = \Pr(\text{R})\cdot\Pr(\text{W}) = \frac{2}{5}\cdot\frac{3}{5} = \frac{6}{25}$

d. $\Pr(\text{Black}) = 0$

19. $\Pr(\text{N1Q2N3}) = \Pr(\text{N})\cdot\Pr(\text{Q} \mid \text{N})\cdot\Pr(\text{N} \mid \text{N} \cap \text{Q}) = \frac{9}{18}\cdot\frac{5}{17}\cdot\frac{8}{16} = \frac{5}{68}$

21. a. $\Pr(\text{R1 and W2}) = \Pr(\text{R})\cdot\Pr(\text{W} \mid \text{R}) = \frac{1}{5}\cdot\frac{4}{4} = \frac{1}{5}$

b. $\Pr(\text{W1 and W2}) = \Pr(\text{W})\cdot\Pr(\text{W} \mid \text{W}) = \frac{4}{5}\cdot\frac{3}{4} = \frac{3}{5}$

c. $\Pr(\text{R1 and R2}) = 0$ (Only one red ball in box)

23. a. $\Pr(\text{S1 and S2}) = \Pr(\text{S})\cdot\Pr(\text{S} \mid \text{S}) = \frac{13}{52}\cdot\frac{12}{51} = \frac{1}{17}$

b. $\Pr(\text{H1 and C2}) = \Pr(\text{H})\cdot\Pr(\text{C} \mid \text{H}) = \frac{13}{52}\cdot\frac{13}{51} = \frac{13}{204}$

25. a. $\Pr(\text{R} \mid \text{W}) = \frac{13}{17}$

b. $\Pr(\text{E1 and E2}) = \Pr(\text{E})\cdot\Pr(\text{E} \mid \text{E}) = \frac{9}{18}\cdot\frac{8}{17} = \frac{4}{17}$

c. $\Pr(\text{ER1 and E2}) = \Pr(\text{E and R})\cdot\Pr(\text{E} \mid \text{E} \cap \text{R}) = \frac{6}{18}\cdot\frac{8}{17} = \frac{8}{51}$

27. $\Pr(\text{F} \mid \text{D}) = \frac{310}{520} = \frac{31}{52}$; $\Pr(\text{F} \mid \text{D}) = \frac{\Pr(\text{F} \cap \text{D})}{\Pr(\text{D})}$

29. $\Pr(\text{F} \mid \text{R}) = \frac{125}{480} = \frac{25}{96}$; $\Pr(\text{F} \mid \text{R}) = \frac{\Pr(\text{F} \cap \text{R})}{\Pr(\text{R})}$

31. $\Pr(\text{O} \mid \text{NW}) = \frac{25+190}{50+200} = \frac{215}{250} = \frac{43}{50}$

33. $\Pr(\text{R} \mid \text{F}) = \frac{300+25}{435} = \frac{325}{435} = \frac{65}{87}$

35. $\Pr(\text{NW} \mid \text{F}) = \frac{25+10}{435} = \frac{35}{435} = \frac{7}{87}$

37. $\Pr(\text{WR and O}) = \dfrac{100}{1000} = \dfrac{1}{10}$ (No "given that" in this problem)

39. Events are independent. $\Pr(\text{H1 and H1}) = \dfrac{1}{12{,}000} \cdot \dfrac{1}{12{,}000} = \dfrac{1}{144{,}000{,}000}$

41. $\Pr(\text{A and SD and PGM}) = (0.337)(0.796)(0.016) = 0.004292032$

43. $\Pr(\text{not identified on test 1} \mid \text{defective}) = 0.3$

$\Pr(\text{not identified on test 2} \mid \text{defective}) = 0.2$

$\Pr(\text{not identified on either test} \mid \text{defective}) = (0.3)(0.2) = 0.06$

45. $\Pr(\text{H and LI}) = \Pr(\text{H}) \cdot \Pr(\text{LI} \mid \text{H}) = (0.09)(0.5) = 0.045$

47. $\Pr(\text{Good}) = 0.95$; $\Pr(\text{GGGGG}) = (0.95)^5$ (Independent events)

49. $\Pr(\text{See and Buy}) = \Pr(\text{See}) \cdot \Pr(\text{Buy} \mid \text{See}) = (0.3)(0.2) = 0.06$

51. Events (each woman) are independent.

a. $\Pr(\text{0 pregnancies}) = (0.99)^{100} \approx 0.366$

b. $\Pr(\text{at least 1 pregnancy}) \approx 1 - 0.366 = 0.634$

53. a. $\Pr(\text{0 defects}) = (0.77)^3 \approx 0.456$

b. Pr(at least 1 defect) $1 - 0.456 = 0.544$

55. Events are independent.

a. $\Pr(\text{all correct}) = \dfrac{1}{3} \cdot \dfrac{1}{3} \cdot \dfrac{1}{3} \cdot \dfrac{1}{5} \cdot \dfrac{1}{5} \cdot \dfrac{1}{5} \cdot \dfrac{1}{5} = \dfrac{1}{16{,}875}$

b. $\Pr(\text{no correct}) = \dfrac{2}{3} \cdot \dfrac{2}{3} \cdot \dfrac{2}{3} \cdot \dfrac{4}{5} \cdot \dfrac{4}{5} \cdot \dfrac{4}{5} \cdot \dfrac{4}{5} = \dfrac{2048}{16{,}875}$

c. $\Pr(\text{at least 1 correct}) = 1 - \dfrac{2048}{16{,}875} = \dfrac{14{,}827}{16{,}875}$

57. $\Pr(\text{Portia}) = \dfrac{3}{11}$ $\qquad \Pr(\text{Trinka}) = \dfrac{1}{11}$

$\Pr(\text{P or T}) = \Pr(\text{P}) + \Pr(\text{T}) = \dfrac{3}{11} + \dfrac{1}{11} = \dfrac{4}{11}$

Odds for Portia or Trinka $= \dfrac{4}{7}$ or $= 4 : 7$

59. Choose one person. Now choose the other person.

a. $\Pr(\text{different from birthday \#1}) = \dfrac{364}{365}$

b. $\Pr(\text{same birthday as \#1}) = \dfrac{1}{365}$

61. a. $\Pr(\text{Different days}) = \underset{\substack{\updownarrow \\ 1}}{1} \; \underset{\substack{\updownarrow \\ 2}}{\left(\dfrac{364}{365}\right)} \; \underset{\substack{\updownarrow \\ 3}}{\left(\dfrac{363}{365}\right)} \cdots \underset{\substack{\updownarrow \\ 20}}{\left(\dfrac{346}{365}\right)}$

$\approx 0.588 \approx 0.59$

b. Pr(At least 2 same) = 1 – Pr(Different days) $\quad 1 - 0.59 = 0.41$

Exercise 7.4

1.

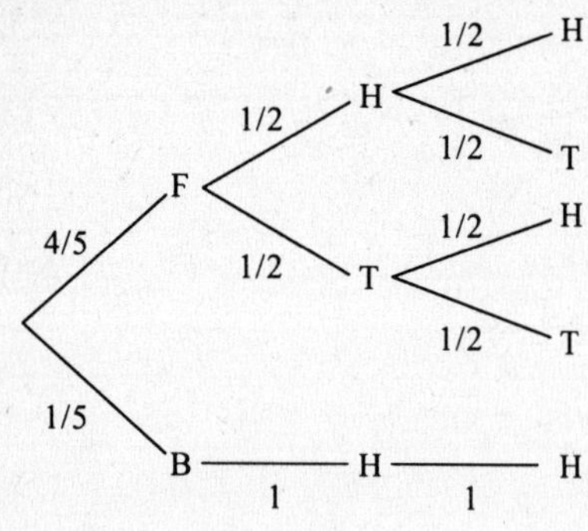

$$\Pr(2H) = \frac{4}{5}\cdot\frac{1}{2}\cdot\frac{1}{2} + \frac{1}{5}\cdot 1\cdot 1 = \frac{2}{5}$$

3.

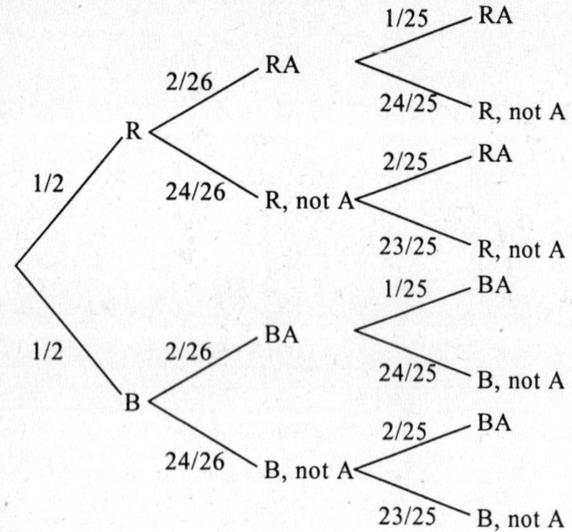

$\Pr(2A \mid \text{same color})$

$$= \frac{1}{2}\cdot\frac{2}{26}\cdot\frac{1}{25} + \frac{1}{2}\cdot\frac{2}{26}\cdot\frac{1}{25} = \frac{1}{325}$$

5.

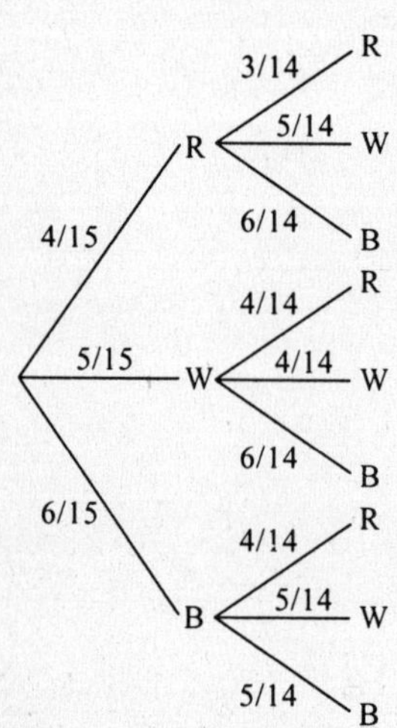

a. $\Pr(WW) = \frac{5}{15}\cdot\frac{4}{14} = \frac{2}{21}$

b. $\Pr(RW \text{ or } WR) = \frac{4}{15}\cdot\frac{5}{14} + \frac{5}{15}\cdot\frac{4}{14} = \frac{4}{21}$

c. Pr(at least one is black)

$= \Pr(RB \text{ or } WB \text{ or } BR \text{ or } BW \text{ or } BB)$

$$= \frac{4}{15}\cdot\frac{6}{14} + \frac{5}{15}\cdot\frac{6}{14} + \frac{6}{15}\cdot\left(\frac{4}{14} + \frac{5}{14} + \frac{5}{14}\right) = \frac{23}{35}$$

7.

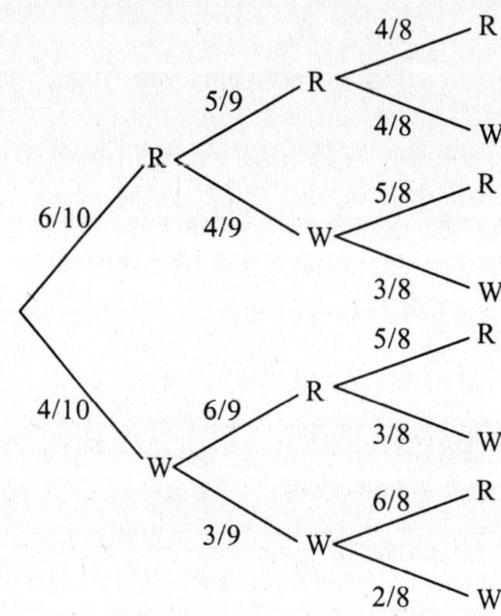

a. $\Pr(WWW) = \frac{4}{10}\cdot\frac{3}{9}\cdot\frac{2}{8} = \frac{1}{30}$

b. $\Pr(WRR \text{ or } RWR \text{ or } RRW)$

$$= \frac{4}{10}\cdot\frac{6}{9}\cdot\frac{5}{8} + \frac{6}{10}\cdot\frac{4}{9}\cdot\frac{5}{8} + \frac{6}{10}\cdot\frac{5}{9}\cdot\frac{4}{8}$$

$$= \frac{3(4\cdot 5\cdot 6)}{8\cdot 9\cdot 10} = \frac{1}{2}$$

c. $\Pr(\text{at least } 1W) = 1 - \Pr(\text{No } W)$

$$= 1 - \frac{6}{10}\cdot\frac{5}{9}\cdot\frac{4}{8} = \frac{5}{6}$$

9. Referring to the probability tree in exercise 8, we have

Pr(HHT or HTH or THH or HHH)

$$= 3\left(\frac{4}{5}\cdot\frac{1}{2}\cdot\frac{1}{2}\cdot\frac{1}{2}\right) + \left(\frac{4}{5}\cdot\frac{1}{2}\cdot\frac{1}{2}\cdot\frac{1}{2} + \frac{1}{5}\cdot 1\cdot 1\cdot 1\right) = \frac{3}{5}$$

11. a. $\Pr(\text{R1 and W2}) = \frac{6}{10}\cdot\frac{4}{10} = \frac{24}{100} = \frac{6}{25}$

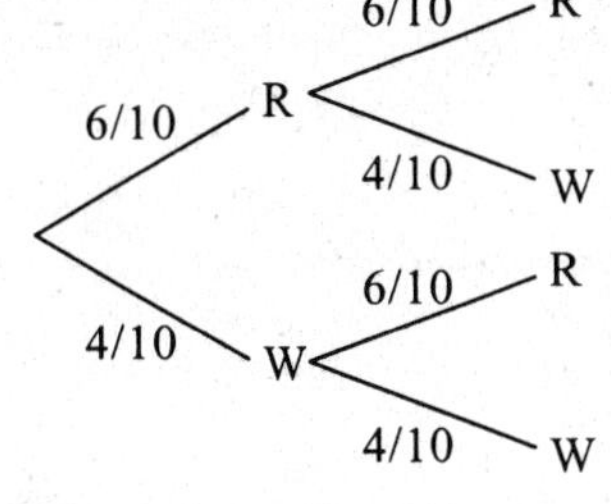

b. $\Pr(\text{RR}) = \frac{6}{10}\cdot\frac{6}{10} = \frac{36}{100} = \frac{9}{25}$

c. $\Pr(\text{R1W2 or W1R2}) = \frac{6}{10}\cdot\frac{4}{10} + \frac{4}{10}\cdot\frac{6}{10}$

$= \frac{48}{100} = \frac{12}{25}$

d. $\Pr(\text{RR or RW or WW})$

$= \frac{6}{10}\cdot\frac{6}{10} + \frac{6}{10}\cdot\frac{4}{10} + \frac{4}{10}\cdot\frac{4}{10} = \frac{76}{100} = \frac{19}{25}$

13.

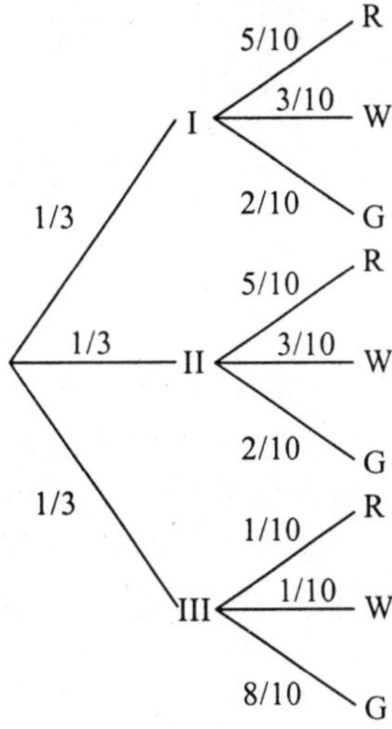

a. $\Pr(\text{III} \mid \text{G})$

$= \dfrac{\text{product of branch probabilities on III-G path}}{\text{sum of all branch products leading to G}}$

$= \dfrac{\frac{1}{3}\cdot\frac{8}{10}}{\frac{2}{30}+\frac{2}{30}+\frac{8}{30}} = \dfrac{\frac{8}{30}}{\frac{12}{30}} = \dfrac{2}{3}$

b. $\Pr(\text{III})|\text{G})$

$= \dfrac{\Pr(\text{III})\cdot\Pr(\text{G} \mid \text{III})}{\Pr(\text{III})\cdot\Pr(\text{G} \mid \text{III}) + \Pr(\text{II})\cdot\Pr(\text{G} \mid \text{II}) + \Pr(\text{I})\cdot\Pr(\text{G} \mid \text{I})}$

$= \dfrac{\frac{1}{3}\cdot\frac{8}{10}}{\frac{1}{3}\cdot\frac{8}{10}+\frac{1}{3}\cdot\frac{2}{10}+\frac{1}{3}\cdot\frac{2}{10}} = \dfrac{\frac{8}{30}}{\frac{12}{30}} = \dfrac{2}{3}$

15. a. $\Pr(\text{I} \mid \text{G})$

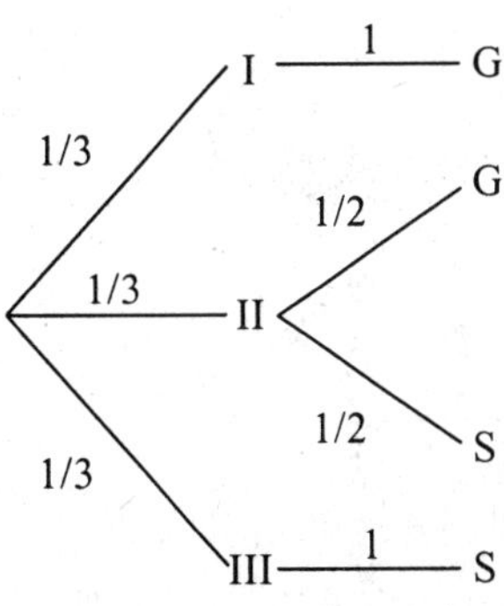

$= \dfrac{\text{product of branch probabilities on I-G path}}{\text{sum of all branch probabilities leading to G}}$

$= \dfrac{\frac{1}{3}\cdot 1}{\frac{1}{3}+\frac{1}{6}} = \dfrac{\frac{1}{3}}{\frac{3}{6}} = \dfrac{2}{3}$

b. $\Pr(\text{I} \mid \text{G})$

$= \dfrac{\Pr(\text{I})\cdot\Pr(\text{G} \mid \text{I})}{\Pr(\text{I})\cdot\Pr(\text{G} \mid \text{I}) + \Pr(\text{II})\cdot\Pr(\text{G} \mid \text{II}) + \Pr(\text{III})\cdot\Pr(\text{G} \mid \text{III})}$

$= \dfrac{\frac{1}{3}\cdot 1}{\frac{1}{3}\cdot 1+\frac{1}{3}\cdot\frac{1}{2}+\frac{1}{3}\cdot 0} = \dfrac{\frac{1}{3}}{\frac{3}{6}} = \dfrac{2}{3}$

17. $\Pr(W) = 0.76$ $\Pr(H) = 0.09$ $\Pr(AANA) = 0.15$

$\Pr(I \mid W) = 0.20$ $\Pr(I \mid H) = 0.50$ $\Pr(I \mid AANA) = 0.75$

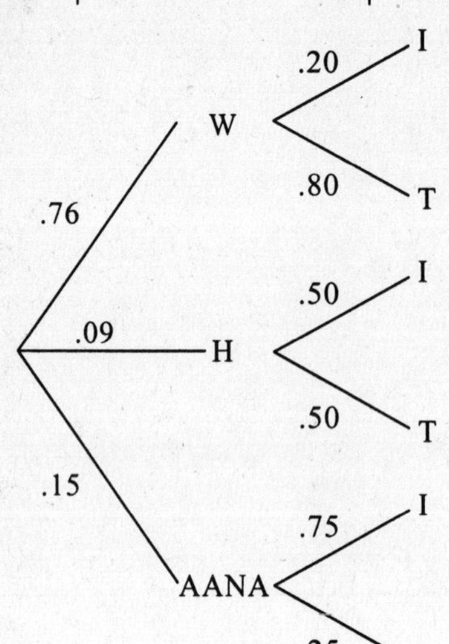

$\Pr(I) = .76(.20) + .09(0.50) + .15(.75) = 0.3095$

19. a. $\Pr(HHHH) = \frac{15}{50}\cdot\frac{15}{50}\cdot\frac{15}{50}\cdot\frac{15}{50} = \frac{81}{10{,}000}$

b. There are 6 successful events, each with the same probability.
Success: HHMM, HMHM, HMMH, MHMH, MMHH, MHHM

$\Pr(\text{Success}) = 6\left(\frac{15}{50}\cdot\frac{15}{50}\cdot\frac{35}{50}\cdot\frac{35}{50}\right) = \frac{1323}{5000}$

21. a. $\Pr(G1D2) = \frac{12}{15}\cdot\frac{3}{14} = \frac{6}{35}$

b. $\Pr(D1G2) = \frac{3}{15}\cdot\frac{12}{14} = \frac{6}{35}$

c. $\Pr(G1D2 \text{ or } D1G2) = \frac{6}{35} + \frac{6}{35} = \frac{12}{35}$

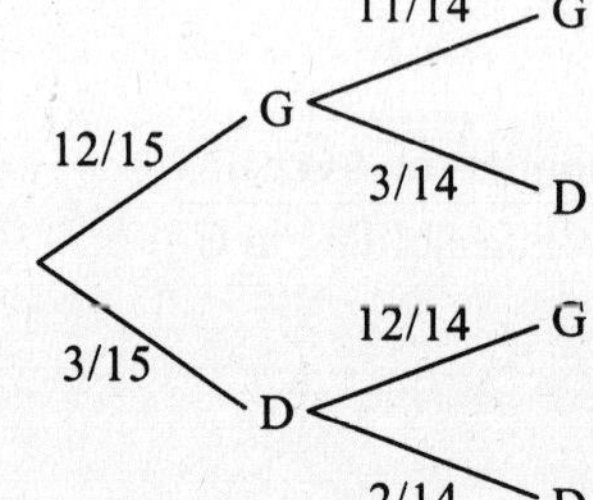

23. a. $\Pr(\text{Male}) = \frac{8}{14} = \frac{4}{7}$

b. 3200 males and 1800 females drink heavily.

$\Pr(DH) = \frac{3200 + 1800}{14{,}000} = \frac{5}{14}$

c. $\Pr(M \text{ or } DH) = \frac{8}{14} + \frac{5}{14} - \frac{8}{14}\cdot(.4) = \frac{7}{10}$

d. Using Bayes' formula and the probability tree,

we have $\Pr(M \mid DH) = \frac{(.4)\left(\frac{8}{14}\right)}{(.4)\left(\frac{8}{14}\right) + .3\left(\frac{6}{14}\right)} = \frac{16}{25}$

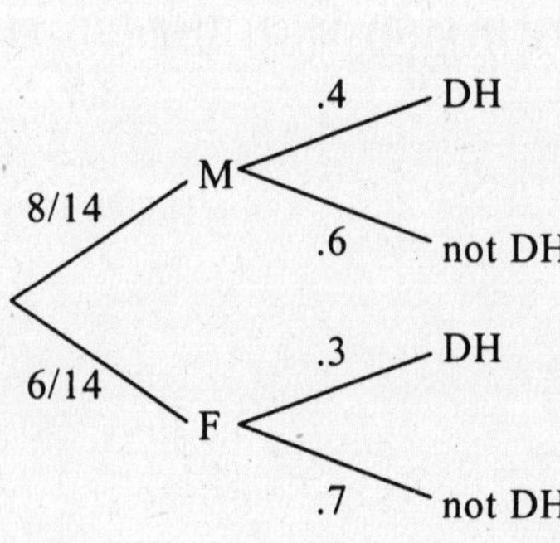

25. $\Pr(\text{F} \mid \text{M}) = \dfrac{170}{170+280} = \dfrac{170}{450} = \dfrac{17}{45}$

27. T = Test indicates pregnant (P)
not T = Test shows not pregnant

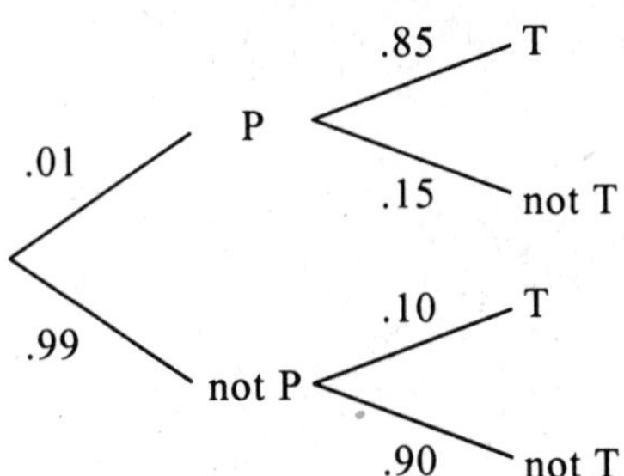

Using Bayes' formula and the probability tree, we have $\Pr(\text{P} \mid \text{T}) = \dfrac{(0.01)(0.85)}{(0.01)(0.85)+(0.99)(0.1)} \approx 0.079$

Alternate Method:

	T	not T	
P	.0085		.01
not P	.099		0.99
			1

	T	not T	
P	.0085	.0015	0.01
not P	.099	.891	0.99
	.1075	.8925	1

$\Pr(\text{P} \mid \text{T}) = \dfrac{\Pr(\text{P and T})}{\Pr(\text{T})} = \dfrac{0.0085}{0.1075} \approx 0.079$

29. **a.** $\Pr(\text{Female}) = \dfrac{10+12+6+13+8}{30+25+20+15+10} = \dfrac{49}{100}$

b. $\Pr(\text{S} \mid \text{F}) = \dfrac{12 \text{ Science Females}}{49 \text{ Females}} = \dfrac{12}{49}$

Exercise 7.5

1. ${}_6P_4 = \dfrac{6!}{(6-4)!} = \dfrac{6\cdot5\cdot4\cdot3\cdot2!}{2!} = 360$

3. ${}_{10}P_6 = \dfrac{10!}{(10-6)!} = \dfrac{10\cdot9\cdot8\cdot7\cdot6\cdot5\cdot4!}{4!} = 151{,}200$

5. ${}_5P_0 = \dfrac{5!}{(5-0)!} = \dfrac{5!}{5!} = 1$

7. **a.** By Fundamental Counting Principle:
$6\cdot5\cdot4\cdot3 = 360$

b. By Fundamental Counting Principle:
$6\cdot6\cdot6\cdot6 = 1296$

9. ${}_nP_n = \dfrac{n!}{(n-n)!} = \dfrac{n!}{1} = n!$

11. $\dfrac{(n+1)!}{n!} = \dfrac{(n+1)n!}{n!} = n+1$

13. $(n+1)! = 17n!$
$(n+1)n! = 17n!$
$n+1 = 17$
$n = 16$

15. ${}_{100}C_{98} = \dfrac{100!}{98!2!} = \dfrac{100\cdot99}{2} = 4950$

17. ${}_4C_4 = \frac{4!}{0!4!} = \frac{4!}{1\cdot 4!} = 1$

19. $\binom{5}{0} = \frac{5!}{0!5!} = 1$

21.

$$ {}_nC_6 = {}_nC_4 $$
$$ \frac{n!}{6!(n-6)!} = \frac{n!}{4!(n-4)!} $$
$$ 4!(n-4)! = 6!(n-6)! $$
$$ 4!(n-4)(n-5)(n-6)! = 6\cdot 5\cdot 4!(n-6)! $$
$$ n^2 - 9n + 20 = 30 $$
$$ n^2 - 9n - 10 = 0 $$
$$ (n-10)(n+1) = 0 $$

$n = 10$ is the only possible solution.

23. a. By FCP: $2\cdot 3\cdot 2 = 12$
b. By FCP: $2\cdot 3\cdot 2\cdot 6 = 72$

25. By FCP: $10\cdot 9\cdot 8\cdot 7\cdot 6\cdot 5\cdot 4 = 604{,}800$

27. By FCP: $6\cdot 5\cdot 4 = 120$

29. By FCP: $4\cdot 3\cdot 2\cdot 1 = 24$

31. By FCP: $4\cdot 4\cdot 4 = 64$

33. By FCP: $6\cdot 5\cdot 4\cdot 3\cdot 2\cdot 1 = 720$

35. By FCP: $2\cdot 2\cdot 2\cdots 2 = 2^{10} = 1024$
Question | | | |
1 2 3 10

37. By FCP: $10\cdot 9\cdot 8\cdots 1 = 10! = 3{,}628{,}800$
| | | |
Question 1 2 3 10

39. ${}_{10}C_5 = \frac{10!}{5!5!} = \frac{10\cdot 9\cdot 8\cdot 7\cdot 6\cdot 5!}{5!\cdot 5\cdot 4\cdot 3\cdot 2\cdot 1} = 252$

41. ${}_{30}C_{20} = \frac{30!}{20!10!} = 30{,}045{,}015$

43. ${}_{12}C_5 = \frac{12!}{5!7!} = \frac{12\cdot 11\cdot 10\cdot 9\cdot 8\cdot 7!}{5\cdot 4\cdot 3\cdot 2\cdot 1\cdot 7!} = 792$

45. ${}_{10}C_4 = \frac{10!}{4!6!} = \frac{10\cdot 9\cdot 8\cdot 7\cdot 6!}{4\cdot 3\cdot 2\cdot 1\cdot 6!} = 210$

47. "AND" means multiply.
${}_{20}C_6 \cdot {}_{22}C_6 = \frac{20!}{6!14!}\cdot\frac{22!}{6!16!} = 2{,}891{,}999{,}880$

49. By FCP: $10\cdot 370{,}000 = 3{,}700{,}000$

Exercise 7.6

1. $n(\text{E}) = 1$ DOG is the only way for a success.
$n(\text{S}) = \underline{6}\cdot\underline{5}\cdot\underline{4} = 120$

So, $\Pr(\text{E}) = \frac{1}{120}$. Also, $\Pr(\text{E}) = \frac{1}{{}_6P_3} = \frac{1}{120}$

3. a. By FCP: $5\cdot 4\cdot 3\cdot 2\cdot 1 = 120$
b. Pr(alphabetical order) $= \frac{1}{120}$

5. Total License Plates:
$26\cdot 26\cdot 26\cdot 10\cdot 10\cdot 10 = 17{,}576{,}000$
Letters and numbers different:
$26\cdot 25\cdot 24\cdot 10\cdot 9\cdot 8 = 11{,}232{,}000$

Pr(Success) $= \frac{11{,}232{,}000}{17{,}576{,}000} \approx 0.639$

7. a. Pr(Success) $= \frac{1}{10\cdot 10\cdot 10\cdot 10} = \frac{1}{10{,}000}$

b. Pr(Success) $= \frac{1}{10\cdot 9\cdot 8\cdot 7} = \frac{1}{5040}$

9. Total Phone Numbers:
$8\cdot 10\cdot 10\cdot 10\cdot 10\cdot 10\cdot 10 = 8{,}000{,}000$
There are 8 phone numbers that have the same digits.

Pr(Success) $= \frac{8}{8{,}000{,}000} = \frac{1}{1{,}000{,}000}$

11. $\Pr(\text{John and Jill}) = \frac{\binom{4}{2}\cdot\binom{496}{2}}{\binom{500}{4}} \approx 2.8\times 10^{-4}$

13. Total ways to answer test $= \underline{10}\cdot\underline{9}\cdot\underline{8}\cdots\underline{1} = 10!$
Question: 1 2 3⋯10

Also, order is important, thus ${}_{10}P_{10} = 10!$ is another way to obtain the total.

Pr(All correct) $= \frac{1}{10!} = \frac{1}{3{,}628{,}800}$

15. **a.** $\Pr(DD) = \dfrac{\binom{3}{2}}{\binom{12}{2}} = \dfrac{3}{66} = \dfrac{1}{22} \approx 0.0455$

b. $\Pr(GG) = \dfrac{\binom{9}{2}}{\binom{12}{2}} = \dfrac{36}{66} = \dfrac{6}{11} \approx 0.5455$

c. $\Pr(1D) = 1 - [\Pr(DD) + \Pr(GG)]$

$\approx 1 - 0.5910 = 0.409$

Also,

$\Pr(1G1D) = \dfrac{\binom{9}{1}\binom{3}{1}}{\binom{12}{2}} = \dfrac{9 \cdot 3}{66} = \dfrac{9}{22} \approx 0.409$

17. Method 1:

Def Good Def Good

$$\Pr(E) = \frac{\binom{2}{1}\cdot\binom{98}{4} + \binom{2}{2}\cdot\binom{98}{3}}{\binom{100}{5}}$$

$$= \frac{7{,}224{,}560 + 152{,}096}{75{,}287{,}520} \approx 0.098$$

Method 2:

$$n(1 \text{ or } 2 \text{ Def}) = \text{Total} - \text{All Good} = \binom{100}{5} - \binom{98}{5}$$

$$= 75{,}287{,}520 - 67{,}910{,}864$$

$$= 7{,}376{,}656$$

$$\Pr(E) = \frac{7{,}376{,}656}{\binom{100}{5}} \approx 0.098$$

19. In the sample of 30, there are 2 minority loans. Thus, ${}_{10}C_2$ is the numbers of ways to obtain 2 minority loans from the 10 minority loans. Also, ${}_{90}C_{28}$ is the number of ways of getting 28 non minority loans from the total of non minority loans.

$$\Pr(E) = \frac{\binom{90}{28}\cdot\binom{10}{2}}{\binom{100}{30}}$$

21. **a.** $\Pr(\text{No minority}) = \dfrac{\binom{6}{4}}{\binom{9}{4}} = \dfrac{15}{126} = \dfrac{5}{42} \approx 0.119$

b. $\Pr(\text{All minority}) = \dfrac{\binom{6}{1}\cdot\binom{3}{3}}{\binom{9}{4}} = \dfrac{6}{126}$

$= \dfrac{1}{21} \approx 0.0476$

c. $\Pr(1 \text{ minority}) = \dfrac{\binom{6}{3}\cdot\binom{3}{1}}{\binom{9}{4}} = \dfrac{20\cdot 3}{126} = \dfrac{10}{21}$

≈ 0.476

23. $\Pr(E) = \dfrac{\binom{6}{5}}{\binom{10}{5}} = \dfrac{6}{252} = \dfrac{1}{42} \approx 0.0238$

25. **a.** $\Pr(2M) = \dfrac{\binom{23}{2}}{\binom{27}{2}} = \dfrac{253}{351} \approx 0.721$

b. $\Pr(MW) = \dfrac{\binom{23}{1}\cdot\binom{4}{1}}{\binom{27}{2}} = \dfrac{23\cdot 4}{351} = \dfrac{92}{351} \approx 0.262$

c. $\Pr(\text{at least } 1W) = 1 - \Pr(2M) \quad 1 - 0.721$
$= 0.279$

Also,

$\Pr(\text{at least } 1W) = \dfrac{\binom{23}{1}\binom{4}{1} + \binom{4}{2}}{\binom{27}{2}} = \dfrac{92+6}{351}$

$= \dfrac{98}{351} \approx 0.279$

27. **a.** $\Pr(E) = \dfrac{1 \text{ most able}}{3 \text{ men}} = \dfrac{1}{3}$

b. $\Pr(E) = \dfrac{1}{{}_3P_3} = \dfrac{1}{3\cdot 2\cdot 1} = \dfrac{1}{6}$

29. $\Pr(E) = \dfrac{\text{ways to get 10 winning numbers}}{\text{total ways to get 10 numbers}}$

$= \dfrac{{}_{20}C_{10}}{{}_{80}C_{10}} \approx 1.12 \times 10^{-7}$

31. **a.** $\Pr(E) = \dfrac{\binom{2}{2}\binom{23}{3}}{\binom{25}{5}} = \dfrac{1\cdot 1771}{53{,}130} \approx 0.033$

b. $\Pr(E) = \dfrac{\binom{23}{5}}{\binom{25}{5}} = \dfrac{33{,}649}{53{,}130} \approx 0.633$

33. Order is not important.

a. $\Pr(5\text{ spades}) = \dfrac{\binom{13}{5}}{\binom{52}{5}} = \dfrac{1287}{2{,}598{,}960} \approx 0.0005$

b. Pr(5 of same suit)
= Pr(All S or All C or All D or All H)
$\approx 4(0.0005) = 0.002$

35. $\Pr(4S4H4D1C) = \dfrac{\binom{13}{4}\cdot\binom{13}{4}\cdot\binom{13}{4}\cdot\binom{13}{1}}{\binom{52}{13}}$
$\approx \dfrac{715\cdot 715\cdot 715\cdot 13}{6.35\times 10^{11}}$
≈ 0.00748

Exercise 7.7

1. Can be since $\frac{1}{4}+\frac{3}{4}=1$.

3. Cannot be since $\frac{1}{4}+\frac{3}{5}+\frac{1}{6}\neq 1$.

5. Cannot be since matrix is not square.

7. Can be since the sum of entries in each row is 1, and matrix is square.

9. $[.2 \quad .8]\begin{bmatrix} .1 & .9 \\ .3 & .7 \end{bmatrix} = [.26 \quad .74];\ [.26 \quad .74]\begin{bmatrix} .1 & .9 \\ .3 & .7 \end{bmatrix} = [.248 \quad .752]$

11. $[.1 \quad .3 \quad .6]\begin{bmatrix} .5 & .3 & .2 \\ .3 & .5 & .2 \\ .1 & .1 & .8 \end{bmatrix} = [.20 \quad .24 \quad .56];\ [.20 \quad .24 \quad .56]\begin{bmatrix} .5 & .3 & .2 \\ .3 & .5 & .2 \\ .1 & .1 & .8 \end{bmatrix} = [.228 \quad .236 \quad .536]$

13. $[V_1 \quad V_2]\begin{bmatrix} .1 & .9 \\ .3 & .7 \end{bmatrix} = [V_1 \quad V_2]$ or $\begin{matrix} .1V_1 + .3V_2 = V_1 \\ .9V_1 + .7V_2 = V_2 \end{matrix}$ or $\begin{matrix} -.9V_1 + .3V_2 = 0 \\ .9V_1 - .3V_2 = 0 \end{matrix}$

Thus, $V_1 = \frac{0.3}{0.9}V_2 = \frac{1}{3}V_2$. $V_1 + V_2 = 1$ gives $\frac{1}{3}V_2 + V_2 = 1$ or $V_2 = \frac{3}{4}$.

Then, $V_1 = \frac{1}{4}$. The steady-state vector is $\begin{bmatrix} \frac{1}{4} & \frac{3}{4} \end{bmatrix}$.

15. $\begin{bmatrix} V_1 & V_2 & V_3 \end{bmatrix}\begin{bmatrix} .5 & .3 & .2 \\ .3 & .5 & .2 \\ .1 & .1 & .8 \end{bmatrix} = \begin{bmatrix} V_1 & V_2 & V_3 \end{bmatrix}$

$$\begin{aligned} .5V_1 + .3V_2 + .1V_3 &= V_1 & \quad -.5V_1 + .3V_2 + .1V_3 &= 0 \\ .3V_1 + .5V_2 + .1V_3 &= V_2 \text{ or} & .3V_1 - .5V_2 + .1V_3 &= 0 \\ .2V_1 + .2V_2 + .8V_3 &= V_3 & .2V_1 + .2V_2 - .2V_3 &= 0 \end{aligned}$$

$$\left[\begin{array}{ccc|c} -.5 & .3 & .1 & 0 \\ .3 & -.5 & .1 & 0 \\ .2 & .2 & -.2 & 0 \end{array}\right] \to \left[\begin{array}{ccc|c} 1 & -.6 & -.2 & 0 \\ .3 & -.5 & .1 & 0 \\ .2 & .2 & -.2 & 0 \end{array}\right] \to \left[\begin{array}{ccc|c} 1 & -.6 & -.2 & 0 \\ 0 & -.32 & .16 & 0 \\ 0 & .32 & -.16 & 0 \end{array}\right] \to \left[\begin{array}{ccc|c} 1 & -.6 & -.2 & 0 \\ 0 & -.32 & .16 & 0 \\ 0 & 0 & 0 & 0 \end{array}\right] \to$$

$$\left[\begin{array}{ccc|c} 1 & -.6 & -.2 & 0 \\ 0 & 1 & -.5 & 0 \\ 0 & 0 & 0 & 0 \end{array}\right] \to \left[\begin{array}{ccc|c} 1 & 0 & -.5 & 0 \\ 0 & 1 & -.5 & 0 \\ 0 & 0 & 0 & 0 \end{array}\right]$$

So, $V_1 = 0.5V_3$

$V_2 = 0.5V_3$

$V_1 + V_2 + V_3 = 0.5V_3 + 0.5V_3 + V_3 = 1$

Thus, $V_3 = \frac{1}{2}$ and the steady-state vector is $\begin{bmatrix} \frac{1}{4} & \frac{1}{4} & \frac{1}{2} \end{bmatrix}$.

17. $\begin{bmatrix} 1 & 0 & 0 \end{bmatrix}\begin{bmatrix} .5 & .4 & .1 \\ .4 & .5 & .1 \\ .3 & .3 & .4 \end{bmatrix} = \begin{bmatrix} .5 & .4 & .1 \end{bmatrix}$; $\begin{bmatrix} .5 & .4 & .1 \end{bmatrix}\begin{bmatrix} .5 & .4 & .1 \\ .4 & .5 & .1 \\ .3 & .3 & .4 \end{bmatrix} = \begin{bmatrix} .44 & .43 & .13 \end{bmatrix}$

$\begin{bmatrix} .44 & .43 & .13 \end{bmatrix}\begin{bmatrix} .5 & .4 & .1 \\ .4 & .5 & .1 \\ .3 & .3 & .4 \end{bmatrix} = \begin{bmatrix} .431 & .430 & .139 \end{bmatrix}$

$\begin{bmatrix} .431 & .430 & .139 \end{bmatrix}\begin{bmatrix} .5 & .4 & .1 \\ .4 & .5 & .1 \\ .3 & .3 & .4 \end{bmatrix} = \begin{bmatrix} .4292 & .4291 & .1417 \end{bmatrix}$

19.

		Daughter R	Daughter N
Mother	R	.8	.2
	N	.3	.7

21. $\begin{bmatrix} 0 & 1 \end{bmatrix}\begin{bmatrix} .8 & .2 \\ .3 & .7 \end{bmatrix} = \begin{bmatrix} .3 & .7 \end{bmatrix}$

$\begin{bmatrix} .3 & .7 \end{bmatrix}\begin{bmatrix} .8 & .2 \\ .3 & .7 \end{bmatrix} = \begin{bmatrix} .45 & .55 \end{bmatrix}$

Pr(Granddaughter regular) = 0.45

23.

	A	F	VW
A	0	.7	.3
F	.6	0	.4
VW	.8	.2	0

25. $\begin{bmatrix}0 & 1 & 0\end{bmatrix}\begin{bmatrix}0 & 0.7 & 0.3\\0.6 & 0 & 0.4\\0.8 & 0.2 & 0\end{bmatrix}=\begin{bmatrix}0.6 & 0 & 0.4\end{bmatrix}$

$\begin{bmatrix}.6 & 0 & .4\end{bmatrix}\begin{bmatrix}0 & .7 & .3\\.6 & 0 & .4\\.8 & .2 & 0\end{bmatrix}=\begin{bmatrix}.32 & .50 & .18\end{bmatrix}$; $\begin{bmatrix}.32 & .50 & .18\end{bmatrix}\begin{bmatrix}0 & .7 & .3\\.6 & 0 & .4\\.8 & .2 & 0\end{bmatrix}=\begin{bmatrix}.444 & .260 & .296\end{bmatrix}$

$\begin{bmatrix}.444 & .260 & .296\end{bmatrix}\begin{bmatrix}0 & .7 & .3\\.6 & 0 & .4\\.8 & .2 & 0\end{bmatrix}=\begin{matrix}[.3928 & .37 & .2372]\\ \text{A} & \text{F} & \text{VW}\end{matrix}$

27. $\begin{bmatrix}V_1 & V_2 & V_3\end{bmatrix}\begin{bmatrix}0 & .7 & .3\\.6 & 0 & .4\\.8 & .2 & 0\end{bmatrix}=\begin{bmatrix}V_1 & V_2 & V_3\end{bmatrix}$

$$\begin{matrix} & .6V_2+.8V_3=V_1 & & -V_1+.6V_2+.8V_3=0\\ .7V_1 & +.2V_3=V_2 & \text{or} & .7V_1-\ V_2+.2V_3=0\\ .3V_1+.4V_2 & =V_3 & & .3V_1+.4V_2\ -V_3=0\end{matrix}$$

$$\left[\begin{array}{ccc|c}1 & -.6 & -.8 & 0\\.7 & -1 & .2 & 0\\.3 & .4 & -1 & 0\end{array}\right]\to\left[\begin{array}{ccc|c}1 & -.60 & -.80 & 0\\0 & -.58 & .76 & 0\\0 & .58 & -.76 & 0\end{array}\right]\to\left[\begin{array}{ccc|c}1 & -.60 & -.80 & 0\\0 & -.58 & .76 & 0\\0 & 0 & 0 & 0\end{array}\right]\to\left[\begin{array}{ccc|c}1 & -.60 & -.80 & 0\\0 & 1 & -1.31 & 0\\0 & 0 & 0 & 0\end{array}\right]\to$$

$$\left[\begin{array}{ccc|c}1 & 0 & -1.586 & 0\\0 & 1 & -1.310 & 0\\0 & 0 & 0 & 0\end{array}\right]\quad\begin{matrix}V_1=1.586V_3\\V_2=1.31V_3\\V_1+V_2+V_3=1\end{matrix}$$

So, $1.586V_3+1.31V_3+V_3=1$ or $V_3=\dfrac{1}{3.896}\approx 0.257$.

$V_1\approx 1.586(0.257)\approx 0.407$ and $V_2\approx 1.31(0.257)\approx 0.336$

Steady-state vector $\approx\begin{bmatrix}.407 & .336 & .257\end{bmatrix}$.

Actual steady-state vector is $\begin{bmatrix}\frac{46}{113} & \frac{38}{113} & \frac{29}{113}\end{bmatrix}$.

29. $\begin{matrix} & \text{R} & \text{U}\\ \text{R} & .7 & .3\\ \text{U} & .1 & .9\end{matrix}$

$\begin{bmatrix}V_1 & V_2\end{bmatrix}\begin{bmatrix}.7 & .3\\.1 & .9\end{bmatrix}=\begin{bmatrix}V_1 & V_2\end{bmatrix}$

$\begin{matrix}.7V_1+.1V_2=V_1\\.3V_1+.9V_2=V_2\end{matrix}$ or $\begin{matrix}-.3V_1+.1V_2=0\\.3V_1-.1V_2=0\end{matrix}$;

$V_1=\frac{1}{3}V_2$, $V_1+V_2=1$ gives $V_1=\frac{1}{4}$, $V_2=\frac{3}{4}$.

Steady-state vector is $\begin{bmatrix}\frac{1}{4} & \frac{3}{4}\end{bmatrix}$.

31. $\begin{bmatrix} V_1 & V_2 & V_3 \end{bmatrix}\begin{bmatrix} .7 & .2 & .1 \\ .1 & .6 & .3 \\ 0 & .1 & .9 \end{bmatrix} = \begin{bmatrix} V_1 & V_2 & V_3 \end{bmatrix}$ or $\begin{aligned} -.3V_1 + .1V_2 \quad &= 0 \\ .2V_1 - .4V_2 + .1V_3 &= 0 \\ .1V_1 + .3V_2 - .1V_3 &= 0 \end{aligned}$

$$\left[\begin{array}{ccc|c} 1 & 3 & -1 & 0 \\ -.3 & .1 & 0 & 0 \\ .2 & -.4 & .1 & 0 \end{array}\right] \to \left[\begin{array}{ccc|c} 1 & 3 & -1 & 0 \\ 0 & 1 & -.3 & 0 \\ 0 & -1 & .3 & 0 \end{array}\right] \to \left[\begin{array}{ccc|c} 1 & 0 & -.1 & 0 \\ 0 & 1 & -.3 & 0 \\ 0 & 0 & 0 & 0 \end{array}\right] \begin{array}{l} V_1 = \frac{1}{10}V_3 \\ V_2 = \frac{3}{10}V_3 \end{array}$$

$V_1 + V_2 + V_3 = 1$ So, $V_3 = \frac{10}{14} = \frac{5}{7}$.

This gives $V_1 = \frac{1}{14}$ and $V_2 = \frac{3}{14}$.

Thus, the steady-state vector is $\begin{bmatrix} \frac{1}{14} & \frac{3}{14} & \frac{5}{7} \end{bmatrix}$.

33. $\begin{bmatrix} V_1 & V_2 & V_3 \end{bmatrix}\begin{bmatrix} .7 & .2 & .1 \\ .4 & .4 & .2 \\ .4 & .4 & .2 \end{bmatrix} = \begin{bmatrix} V_1 & V_2 & V_3 \end{bmatrix}$ or $\begin{aligned} -.3V_1 + .4V_2 + .4V_3 &= 0 \\ .2V_1 - .6V_2 + .4V_3 &= 0 \\ .1V_1 + .2V_2 - .8V_3 &= 0 \end{aligned}$

$$\left[\begin{array}{ccc|c} 1 & 2 & -8 & 0 \\ -.3 & .4 & .4 & 0 \\ .2 & -.6 & .4 & 0 \end{array}\right] \to \left[\begin{array}{ccc|c} 1 & 2 & -8 & 0 \\ 0 & 1 & -2 & 0 \\ 0 & -1 & 2 & 0 \end{array}\right] \to \left[\begin{array}{ccc|c} 1 & 0 & -4 & 0 \\ 0 & 1 & -2 & 0 \\ 0 & 0 & 0 & 0 \end{array}\right] \begin{array}{l} V_1 = 4V_3 \\ V_2 = 2V_3 \end{array}$$

$4V_3 + 2V_3 + V_3 = 1$ or $V_3 = \frac{1}{7}$, $V_1 = \frac{4}{7}$, $V_2 = \frac{2}{7}$

Steady-state vector is $\begin{bmatrix} \frac{4}{7} & \frac{2}{7} & \frac{1}{7} \end{bmatrix}$.

35. From exercise 34, it is clear that the vector $\begin{bmatrix} 0.49 & 0.42 & 0.09 \end{bmatrix}$ is the steady-state vector for the given transition matrix.

Review Exercises

1. **a.** $\Pr(\text{odd}) = \frac{5}{9}$

b. $\Pr(3, 6, 9) = \frac{3}{9} = \frac{1}{3}$

c. $\Pr(3, 9) = \frac{2}{9}$

2. **a.** $\Pr(R) = \frac{9}{12} = \frac{3}{4}$

b. $\Pr(\text{odd}) = \frac{6}{12} = \frac{1}{2}$

c. $\Pr(10, 12) = \frac{1}{6}$

d. $\Pr(W \text{ or } 1, 3, 5, 7, 9) = \frac{3+5}{12} = \frac{2}{3}$

3. $\Pr(E) = \frac{3}{7}$ means Win 3, Lose 4.

a. Odds for E: 3:4
b. Odds against E: 4:3

4. Equiprobable sample space: {(HH), (HT), (TH), (TT)}

a. $\Pr(0 \text{ heads}) = \frac{1}{4}$

b. $\Pr(1 \text{ head}) = \frac{2}{4} = \frac{1}{2}$

c. $\Pr(2 \text{ heads}) = \frac{1}{4}$

5. {HHH, HHT, HTH, HTT, TTT, TTH, THT, THH}

a. $\Pr(2H) = \frac{3}{8}$

b. $\Pr(3H) = \frac{1}{8}$

c. $\Pr(1H) = \frac{3}{8}$

6. $\Pr(\text{Queen or Jack}) = \frac{8}{52} = \frac{2}{13}$

7. $\Pr(\text{Success}) = \frac{16}{52} \cdot \frac{16}{52} = \frac{16}{169}$

8. $\Pr(L) = 1 - \Pr(W) = \frac{3}{4}$

9. $\Pr(A \text{ or } 10) = \frac{8}{52} = \frac{2}{13}$

10. $\Pr(\text{King or Red}) = \Pr(\text{King}) + \Pr(\text{Red}) - \Pr(\text{King and Red}) = \frac{4}{52} + \frac{26}{52} - \frac{2}{52} = \frac{28}{52} = \frac{7}{13}$

11. **a.** There are 2 even numbered R. $\Pr(\text{Even and R}) = \frac{2}{9}$

b. $\Pr(R \text{ or } E) = \Pr(R) + \Pr(E) - \Pr(R \text{ and } E) = \frac{4}{9} + \frac{4}{9} - \frac{2}{9} = \frac{2}{3}$

c. $\Pr(W \text{ or } O) = \Pr(W) + \Pr(O) - \Pr(W \text{ and } O) = \frac{5}{9} + \frac{5}{9} - \frac{3}{9} = \frac{7}{9}$

12. $\Pr(\text{both balls red}) = \Pr(\text{1st ball red}) \cdot \Pr(\text{2nd ball red} \mid \text{1st ball red}) = \left(\frac{4}{7}\right)\left(\frac{3}{6}\right) = \frac{12}{42} = \frac{2}{7}$

13. There are 4 balls in box, given that the first ball drawn is black.

$S = \{R, R, B, B\}$ Thus, $\Pr(R \mid B) = \frac{2}{4} = \frac{1}{2}$.

14. $\Pr(W \text{ and } R) = \Pr(WR \text{ or } RW) = \Pr(W) \cdot \Pr(R \mid W) + \Pr(R) \cdot \Pr(W \mid R) = \left(\frac{6}{10}\right) \cdot \left(\frac{4}{9}\right) + \left(\frac{4}{10}\right) \cdot \left(\frac{6}{9}\right) = \frac{48}{90} = \frac{8}{15}$

15. a. $\Pr(\text{U1 and R}) = \Pr(\text{U1})\cdot\Pr(\text{R}\mid\text{U1}) = \frac{1}{2}\cdot\frac{3}{7} = \frac{3}{14}$

b. $\Pr(\text{R}) = \Pr((\text{U1 and R}) \text{ or } (\text{U2 and R})) = \Pr(\text{U1})\cdot\Pr(\text{R}\mid\text{U1}) + \Pr(\text{U2})\cdot\Pr(\text{R}\mid\text{U2}) = \frac{1}{2}\cdot\frac{3}{7}+\frac{1}{2}\cdot\frac{5}{7} = \frac{8}{14} = \frac{4}{7}$

c. Using Bayes' formula, we have

$$\Pr(\text{U1}\mid\text{R}) = \frac{\Pr(\text{U1})\cdot\Pr(\text{R}\mid\text{U1})}{\Pr(\text{U1})\cdot\Pr(\text{R}\mid\text{U1})+\Pr(\text{U2})\cdot\Pr(\text{R}\mid\text{U2})} = \frac{\frac{1}{2}\cdot\frac{3}{7}}{\frac{1}{2}\cdot\frac{3}{7}+\frac{1}{2}\cdot\frac{5}{7}} = \frac{3}{3+5} = \frac{3}{8}$$

16. ${}_6P_2 = \frac{6!}{4!} = 30$

17. ${}_7C_3 = \frac{7!}{3!4!} = \frac{7\cdot6\cdot5\cdot4!}{6\cdot4!} = 35$

18. $26^3 = 17{,}576$

19. $\Pr(\le 50) = \frac{50{,}000}{80{,}000} = \frac{5}{8}$

20. Pr(French or German)

$= \Pr(\text{Fr}) + \Pr(\text{Ger}) - \Pr(\text{Fr and Ger})$

$= \frac{30}{100} + \frac{40}{100} - \frac{12}{100} = 0.58$

21. $\Pr(\text{Demo and Med. Pro}) = \frac{25}{280} = \frac{5}{56}$

22. $\Pr(\text{HP or A}) = \frac{75}{280}+\frac{120}{280}-\frac{30}{280} = \frac{165}{280} = \frac{33}{56}$

23. No formulas are needed. Read table.

$\Pr(\text{Auth}\mid\text{Med Pro}) = \frac{75}{110} = \frac{15}{22}$

24. By FCP: $4\cdot3\cdot2\cdot1 = 24$

25. ${}_8P_4 = \frac{8!}{4!} = 1680$

26. $\binom{12}{4} = \frac{12!}{4!8!} = 495$

27. ${}_8C_4 = \frac{8!}{4!4!} = 70$

28. a. ${}_{12}C_2 = \binom{12}{2} = \frac{12!}{2!10!} = 66$

b. ${}_{12}C_3 = \binom{12}{3} = \frac{12!}{3!9!} = 220$

29. There are 36 choices for each position on the plate and repetitions are allowed.
Total plates $= 36 + 36\cdot36 + 36\cdot36\cdot36 + 36^4 + 36^5 = 62{,}193{,}780$

30. (# of blood groups) · (pos or neg) · (# blood types) · (#Rh types) $= 4\cdot2\cdot4\cdot8 = 256$
So, the conclusion that there are 288 unique groups is incorrect if the given information is correct.

31. a. $\Pr(\text{Color Blind}) = \frac{63}{2000}$

b. $\Pr(\text{M}\mid\text{CB}) = \frac{60}{63}$

32.

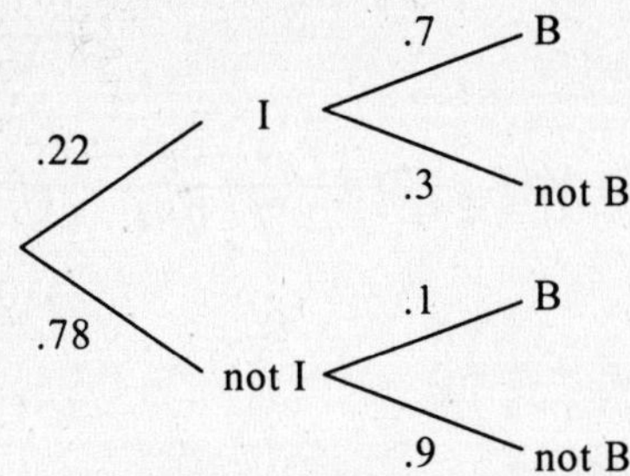

Using Bayes' formula and the probability tree, we have

$$\Pr(\text{No intention} \mid \text{Bought}) = \frac{(.78)(.10)}{(.78)(.10)+(.22)(.7)} = \frac{39}{116}$$

33. Number of ways to enter courses = 4! = 24.

$\Pr(\text{alphabetical}) = \frac{1}{24}$.

34. There are 24 ways to win. Assume 0 can be in any position and digits can be repeated.

$$\Pr(\text{Win}) = \frac{24}{10 \cdot 10 \cdot 10 \cdot 10} = \frac{3}{1250}$$

35. Number of three digit possibilities $= 10^3 = 1000$.

Number of permutations of 3 different digits = 6. $\Pr(\text{winner}) = \frac{6}{1000} = \frac{3}{500}$

36. a. $\Pr(0\ \text{def}) = \dfrac{\binom{180}{10}}{\binom{200}{10}} \approx \dfrac{7.6283 \times 10^{15}}{2.2451 \times 10^{16}} \approx 0.3398$

b. $\Pr(2\ \text{def}) = \dfrac{\binom{20}{2} \cdot \binom{180}{8}}{\binom{200}{10}} \approx \dfrac{(190)(2.3342 \times 10^{13})}{2.2451 \times 10^{16}} \approx 0.1975$

37. $\Pr(3\ \text{best}) = \dfrac{1}{{}_5C_3} = \dfrac{1}{\binom{5}{3}} = \dfrac{1}{10}$

38. a. $\Pr(1\ \text{def}) = \dfrac{\binom{2}{1} \cdot \binom{10}{5}}{\binom{12}{6}} = \dfrac{2(252)}{924} = \dfrac{6}{11} \approx 0.545$

b. $\Pr(\text{At least 1 def}) = 1 - \Pr(\text{No def}) = 1 - \dfrac{\binom{10}{6}}{\binom{12}{6}} = 1 - \dfrac{210}{924} = 1 - \dfrac{5}{22} = \dfrac{17}{22} \approx 0.773$

Alternate Method: $\Pr(E) = \dfrac{\binom{2}{1} \cdot \binom{10}{5} + \binom{2}{2} \cdot \binom{10}{4}}{\binom{12}{6}} = \dfrac{504 + 210}{924} \approx 0.773$

39. $\begin{bmatrix} .7 & .2 & .1 \end{bmatrix} \begin{bmatrix} .15 & .60 & .25 \\ .15 & .35 & .50 \\ 0 & .20 & .80 \end{bmatrix} = \begin{bmatrix} .135 & .51 & .355 \end{bmatrix}$

$\begin{bmatrix} .135 & .51 & .355 \end{bmatrix} \begin{bmatrix} .15 & .60 & .25 \\ .15 & .35 & .50 \\ 0 & .20 & .80 \end{bmatrix} = \begin{bmatrix} .09675 & .3305 & .57275 \end{bmatrix}$

$\begin{bmatrix} .09675 & .3305 & .57275 \end{bmatrix} \begin{bmatrix} .15 & .60 & .25 \\ .15 & .35 & .50 \\ 0 & .20 & .80 \end{bmatrix} = \begin{bmatrix} .0640875 & .288275 & .6476375 \end{bmatrix}$

40. $\begin{bmatrix} V_1 & V_2 & V_3 \end{bmatrix} \begin{bmatrix} .15 & .60 & .25 \\ .15 & .35 & .50 \\ 0 & .20 & .80 \end{bmatrix} = \begin{bmatrix} V_1 & V_2 & V_3 \end{bmatrix}$

$$\begin{aligned} .15V_1 + .15V_2 &= V_1 & \quad -.85V_1 + .15V_2 &= 0 \\ .60V_1 + .35V_2 + .20V_3 &= V_2 \quad \text{or} & \quad .60V_1 - .65V_2 + .20V_3 &= 0 \\ .25V_1 + .50V_2 + .80V_3 &= V_3 & \quad .25V_1 + .50V_2 - .20V_3 &= 0 \end{aligned}$$

$$\left[\begin{array}{ccc|c} 1 & 2 & -.80 & 0 \\ -.85 & .15 & 0 & 0 \\ .60 & -.65 & .20 & 0 \end{array}\right] \to \left[\begin{array}{ccc|c} 1 & 2 & -.80 & 0 \\ 0 & 1.85 & -.68 & .0 \\ 0 & -1.85 & .68 & 0 \end{array}\right] \to \left[\begin{array}{ccc|c} 1 & 2 & -.80 & 0 \\ 0 & 1 & -\frac{68}{185} & 0 \\ 0 & 0 & 0 & 0 \end{array}\right] \to \left[\begin{array}{ccc|c} 1 & 0 & -\frac{12}{185} & 0 \\ 0 & 1 & -\frac{68}{185} & 0 \\ 0 & 0 & 0 & 0 \end{array}\right]$$

$V_1 = \frac{12}{185}V_3,\ V_2 = \frac{68}{185}V_3$; So $\frac{12}{185}V_3 + \frac{68}{185}V_3 + V_3 = 1$ gives $V_3 = \frac{37}{53}$.

$V_1 = \frac{12}{185} \cdot \frac{37}{53} = \frac{12}{265}$ $\qquad V_2 = \frac{68}{185} \cdot \frac{37}{53} = \frac{68}{265}$

The steady-state vector is $\begin{bmatrix} \frac{12}{265} & \frac{68}{265} & \frac{37}{53} \end{bmatrix}$.

Chapter Test

1. **a.** $\Pr(\text{odd}) = \frac{4}{7}$

b. $\Pr(\text{white}) = \frac{3}{7}$

2. **a.** $\Pr(\text{black and even}) = \frac{2}{7}$

b. $\Pr(\text{white or even}) = \frac{5}{7}$

3. **a.** $\Pr(\text{red}) = 0$
b. $\Pr(< 8) = 1$

4. $\Pr(\text{sum} = 7)$
$= \Pr((1, 6), (2, 5), (3, 4))$
$= \frac{3}{\binom{7}{2}} = \frac{3}{21} = \frac{1}{7}$

5. $\Pr(\text{W and W}) = \frac{3}{7} \cdot \frac{2}{6} = \frac{1}{7}$

6. **a.** $\Pr(\text{W1 and B2}) = \frac{3}{7} \cdot \frac{4}{6} = \frac{2}{7}$

b. $\Pr(\text{W1B2 or B1W2}) = \frac{3}{7} \cdot \frac{4}{6} + \frac{4}{7} \cdot \frac{3}{6} = \frac{4}{7}$

7. $\Pr(\text{both odd}) = \frac{4}{7} \cdot \frac{3}{6} = \frac{2}{7}$

8.

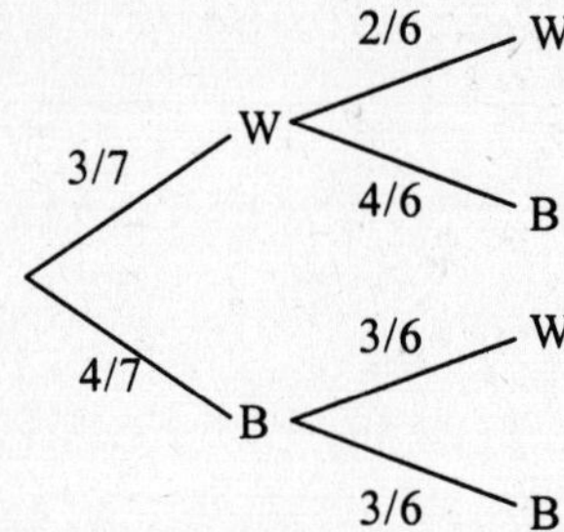

$$\Pr(W1W2 \text{ or } B1W2) = \frac{3}{7}\cdot\frac{2}{6}+\frac{4}{7}\cdot\frac{3}{6}=\frac{3}{7}$$

9. Using Bayes' formula and the probability tree in exercise 8, we have

$$\Pr(B1 \mid W2) = \frac{\frac{4}{7}\cdot\frac{3}{6}}{\frac{4}{7}\cdot\frac{3}{6}+\frac{3}{7}\cdot\frac{2}{6}} = \frac{12}{18} = \frac{2}{3}$$

10. Total outcomes: $26\cdot 26\cdot 26$

$$\Pr(RAT) = \frac{1}{26^3} = \frac{1}{17{,}576}$$

11. Assume a large sample space and events are independent.

Pr(2 Am in 3 draws)

$$= \binom{3}{2}(.35)^2(.65) = 0.238875$$

12. **a.** $\Pr(\text{left handed}) = \dfrac{4}{20} = \dfrac{1}{5}$

b. $\Pr(\text{ambidextrous}) = \dfrac{1}{20}$

13. **a.** $\Pr(LL) = \dfrac{4}{20}\cdot\dfrac{3}{19} = \dfrac{3}{95}$

b. There are other ways to solve this problem.

$$\Pr(1R \text{ and } 1L) = \frac{\binom{15}{1}\binom{4}{1}}{\binom{20}{2}} = \frac{6}{19}$$

c. $\Pr(2R) = \dfrac{15}{20}\cdot\dfrac{14}{19} = \dfrac{21}{38}$

d. Pr(2 ambidextrous) = 0

14. **a.** $\binom{42}{6} = 5{,}245{,}786$

b. $\Pr(\text{Win}) = \dfrac{1}{5{,}245{,}786}$

15. **a.** $\binom{50}{5} = 2{,}118{,}760$

b. $\Pr(\text{Win}) = \dfrac{1}{2{,}118{,}760}$

16. Pr(Def) = Pr(#1 and Def or #2 and Def)
= (.8)(.06) + (.2)(.08) = 0.064

17.

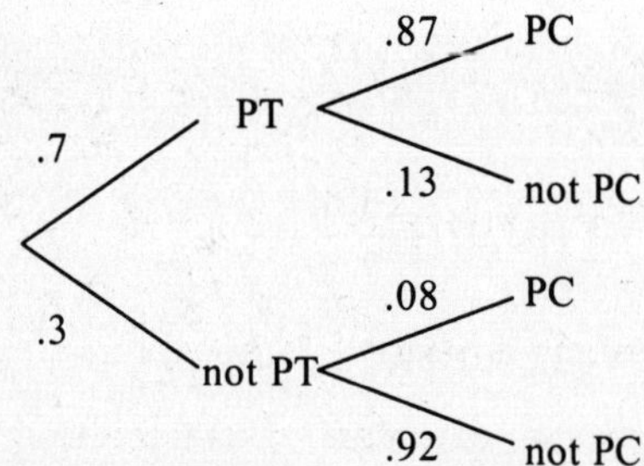

a. Pr(Pass T and Pass C or Fail T and Pass C)

$$= (.7)(.87) + (.3)(.08) = 0.633$$

b. Using Bayes' formula and the probability tree, we have

$$\Pr(\text{Pass T} \mid \text{Pass C}) = \frac{(.7)(.87)}{(.7)(.87)+(.3)(.08)} \approx 0.96$$

18.

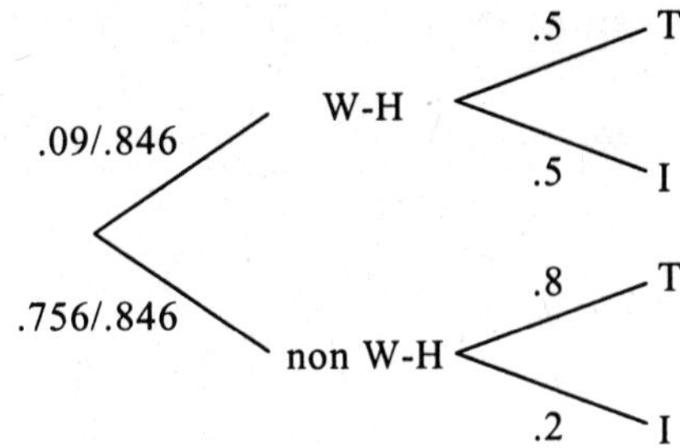

Using Bayes' formula and the probability tree, we have $\Pr(H \mid I) = \dfrac{(\frac{0.09}{0.846})(0.5)}{(\frac{0.09}{0.846})(0.5) + (\frac{0.756}{0.846})(0.2)} \approx 0.229$

19. a. $\Pr(\text{Def WW}) = \dfrac{70}{350} = \dfrac{1}{5}$

b. $\Pr(\text{Def TL}) = \dfrac{25}{350} = \dfrac{1}{14}$

c. $\Pr(\text{No Def TL}) = 1 - \Pr(\text{Def TL}) = \dfrac{13}{14}$

20. $\Pr(\text{Paint}) = 0.14$
$\Pr(\text{BW and Paint}) = 0.03$
$\Pr(\text{BW} \mid \text{Paint})$
$= \dfrac{\Pr(\text{BW and Paint})}{\Pr(\text{Paint})} = \dfrac{0.03}{0.14} = \dfrac{3}{14}$

21. a. off/on $2 \cdot 2 \cdot 2 \cdot \ldots \cdot 2 = 2^{10} = 1024$
Switch 1 2 3 … 10

b. $\Pr(\text{open}) = \dfrac{1}{1024}$

c. $\Pr(\text{open}) = \dfrac{1}{3}$

d. Change the code (sequence).

22. a.

$$\begin{array}{cc} & \begin{matrix} T & O \end{matrix} \\ A = & \begin{bmatrix} 0.80 & 0.20 \\ 0.07 & 0.93 \end{bmatrix} \begin{matrix} \text{Text} \\ \text{Other} \end{matrix} \end{array}$$

b. $\begin{bmatrix} 0.25 & 0.75 \end{bmatrix} \cdot A = \begin{bmatrix} 0.2525 & 0.7475 \end{bmatrix}$
$\begin{bmatrix} 0.2525 & 0.7475 \end{bmatrix} \cdot A$
$= \begin{bmatrix} 0.254325 & 0.745675 \end{bmatrix}$
$\begin{bmatrix} 0.2543 & 0.7457 \end{bmatrix} \cdot A$
$= \begin{bmatrix} 0.25565725 & 0.74434275 \end{bmatrix}$
Percent of market share 3 editions later is 25.6%.

c. $\begin{bmatrix} V_1 & V_2 \end{bmatrix} \begin{bmatrix} .80 & .20 \\ .07 & .93 \end{bmatrix} = \begin{bmatrix} V_1 & V_2 \end{bmatrix}$

$.8V_1 + .07V_2 = V_1$ $\quad -.2V_1 + .07V_2 = 0$
or ;
$.2V_1 + .93V_2 = V_2$ $\quad .2V_1 - .07V_2 = 0$

$V_1 = \dfrac{7}{20}V_2$, $V_1 + V_2 = 1$ gives

$V_1 = \dfrac{7}{27}$, $V_2 = \dfrac{20}{27}$.

Steady-state vector is $\begin{bmatrix} \dfrac{7}{27}, \dfrac{20}{27} \end{bmatrix}$.

Hence the percent of market this text will have is $\dfrac{7}{27} \approx 25.9\%$.

Chapter 8: Further Topics in Probability: Data Description

Exercise 8.1

1. $p = 0.3,\ q = 0.7,\ n = 6,\ x = 4$

$$\binom{6}{4}(0.3)^4(0.7)^2 = 15(0.0081)(0.49) \approx 0.0595$$

3. **a.** $p = \frac{1}{6}$ **b.** $q = \frac{5}{6}$ **c.** $n = 18$

d. $\Pr(6\ 4\text{'s}) = \binom{18}{6}\left(\frac{1}{6}\right)^6\left(\frac{5}{6}\right)^{12} = 18564\left(\frac{1}{46656}\right)\left(\frac{244140625}{2176782336}\right) \approx 0.04463$

5. $p = \frac{1}{2},\ q = \frac{1}{2},\ n = 6$

a. $x = 6$ $\binom{6}{6}\left(\frac{1}{2}\right)^6\left(\frac{1}{2}\right)^0 = 1\left(\frac{1}{64}\right)(1) = \frac{1}{64}$

b. $x = 3$ $\binom{6}{3}\left(\frac{1}{2}\right)^3\left(\frac{1}{2}\right)^3 = 20\left(\frac{1}{8}\right)\left(\frac{1}{8}\right) = \frac{5}{16}$

c. $x = 2$ $\binom{6}{2}\left(\frac{1}{2}\right)^2\left(\frac{1}{2}\right)^4 = 15\left(\frac{1}{4}\right)\left(\frac{1}{16}\right) = \frac{15}{64}$

7. $p = \frac{1}{6},\ q = \frac{5}{6},\ n = 12,\ x = 5$

$$\binom{12}{5}\left(\frac{1}{6}\right)^5\left(\frac{5}{6}\right)^7 = 792\left(\frac{1}{7776}\right)\left(\frac{78{,}125}{279{,}936}\right) \approx 0.0284$$

9. **a.** $p = \frac{3}{5},\ q = \frac{2}{5},\ n = 5,\ x = 2$

$$\binom{5}{2}\left(\frac{3}{5}\right)^2\left(\frac{2}{5}\right)^3 = 10\left(\frac{9}{25}\right)\left(\frac{8}{125}\right) = 0.2304$$

b. $p = \frac{2}{5},\ q = \frac{3}{5},\ n = 5,\ x = 5$

$$\binom{5}{5}\left(\frac{2}{5}\right)^5\left(\frac{3}{5}\right)^0 = 1\left(\frac{32}{3125}\right)(1) = 0.01024$$

c. $\Pr(\text{At least 3B}) = \binom{5}{3}\left(\frac{2}{5}\right)^3\left(\frac{3}{5}\right)^2 + \binom{5}{4}\left(\frac{2}{5}\right)^4\left(\frac{3}{5}\right) + \binom{5}{5}\left(\frac{2}{5}\right)^5\left(\frac{3}{5}\right)^0$

$$= 10\left(\frac{8}{125}\right)\left(\frac{9}{25}\right) + 5\left(\frac{16}{625}\right)\left(\frac{3}{5}\right) + 0.01024 = 0.2304 + 0.0768 + 0.01024 = 0.31744$$

11. $p = \frac{4}{36} = \frac{1}{9},\ q = \frac{8}{9},\ n = 4,\ x = 2;$

$$\binom{4}{2}\left(\frac{1}{9}\right)^2\left(\frac{8}{9}\right)^2 = 6\left(\frac{1}{81}\right)\left(\frac{64}{81}\right) \approx 0.0585$$

13. $p = 0.85$, $q = 0.15$, $n = 10$, $x = 8$

$$\binom{10}{8}(0.85)^8(0.15)^2 \approx 45(0.2725)(0.0225) \approx 0.2759$$

15. $p = \frac{1}{2}$, $q = \frac{1}{2}$, $n = 4$

a. $x = 2$ $\quad \binom{4}{2}\left(\frac{1}{2}\right)^2\left(\frac{1}{2}\right)^2 = 6\cdot\frac{1}{4}\cdot\frac{1}{4} = \frac{3}{8} = 0.375$

b. $x = 4$ $\quad \binom{4}{4}\left(\frac{1}{2}\right)^4\left(\frac{1}{2}\right)^0 = 1\cdot\frac{1}{16}\cdot 1 = 0.0625$

17. Pr(Def) $= \frac{1}{6}$, $n = 4$

a. Pr(2Def) $= \binom{4}{2}\left(\frac{1}{6}\right)^2\left(\frac{5}{6}\right)^2 = 6\cdot\frac{1}{36}\cdot\frac{25}{36} = \frac{25}{216} \approx 0.1157$

b. Pr(0Def) $= \binom{4}{0}\left(\frac{1}{6}\right)^0\left(\frac{5}{6}\right)^4 = \frac{625}{1296} \approx 0.4823$

19. Pr(Blue) $= \frac{1}{4}$, $n = 4$

a. Pr(1 Blue) $= \binom{4}{1}\left(\frac{1}{4}\right)^1\left(\frac{3}{4}\right)^3 = 4\cdot\frac{1}{4}\cdot\frac{27}{64} = \frac{27}{64}$

b. Pr(2 Blue) $= \binom{4}{2}\left(\frac{1}{4}\right)^2\left(\frac{3}{4}\right)^2 = 6\cdot\frac{1}{16}\cdot\frac{9}{16} = \frac{27}{128}$

c. Pr(0 Blue) $= \binom{4}{0}\left(\frac{1}{4}\right)^0\left(\frac{3}{4}\right)^4 = 1\cdot 1\cdot\frac{81}{256} = \frac{81}{256}$

21. Pr(Death) = 0.1, $n = 5$

a. Pr(2 Deaths) $= \binom{5}{2}(.1)^2(.9)^3 = 10(.01)(.729) = 0.0729$

b. Pr(0 Deaths) $= \binom{5}{0}(.1)^0(.9)^5 = 1(1)(.59049) = 0.59049$

c. Pr(0 or 1 or 2 Deaths) $= 0.59049 + \binom{5}{1}(.1)^1(.9)^4 + 0.0729 = .59049 + .32805 + .0729 = 0.9914$

23. Pr(Boy) $= \frac{105}{205} \approx 0.5122$, $n = 6$

Pr(4 Boys) $= \binom{6}{4}(.5122)^4(.4878)^2 \approx 15(.06883)(.2379) \approx 0.2457$

25. Pr(Fire) = 0.004, $n = 10$

Pr(2 Fires) $= \binom{10}{2}(.004)^2(.996)^8 \approx 45(.000016)(.96844) \approx 0.0007$

27. $\Pr(\text{Hit}) = 0.3,\ n = 5$

a. $\Pr(3 \text{ Hits}) = \binom{5}{3}(.3)^3(.7)^2 = 10(.027)(.49) = 0.1323$

b. $\Pr(4 \text{ or } 5 \text{ Hits}) = \binom{5}{4}(.3)^4(.7)^1 + \binom{5}{5}(.3)^5(.7)^0 = 5(.0081)(.7) + 1(.00243)(1)$

$= 0.02835 + 0.00243 = 0.03078$

29. $\Pr(\text{Def}) = 0.01,\ n = 10$

a. $\Pr(0 \text{ Def}) = \binom{10}{0}(.01)^0(.99)^{10} = (.99)^{10} \approx 0.9044$

b. $\Pr(1 \text{ Def}) = \binom{10}{1}(.01)^1(.99)^9 \approx 10(.01)(.9135) = 0.09135$

c. $\Pr(>1 \text{ Def}) = 1 - \Pr(0 \text{ or } 1 \text{ Def}) \approx 1 - (.9044 + .09135) = 0.00425$

31. $\Pr(\text{correct}) = \frac{1}{3}$

a. To get 60%, student needs to get at least 1 of the last 5 correct.

$\Pr(\ \ 60\%) = 1 - \Pr(0 \text{ correct}) = 1 - \binom{5}{0}\left(\frac{1}{3}\right)^0\left(\frac{2}{3}\right)^5 = 1 - \frac{32}{243} = \frac{211}{243} \approx 0.8683$

b. To get 80%, student needs to get at least 3 of the last 5 correct.

$\Pr(\ \ 80\%) = \Pr(3 \text{ or } 4 \text{ or } 5 \text{ correct}) = \binom{5}{3}\left(\frac{1}{3}\right)^3\left(\frac{2}{3}\right)^2 + \binom{5}{4}\left(\frac{1}{3}\right)^4\left(\frac{2}{3}\right) + \binom{5}{5}\left(\frac{1}{3}\right)^5\left(\frac{2}{3}\right)^0$

$= 10\cdot\frac{1}{27}\cdot\frac{4}{9} + 5\cdot\frac{1}{81}\cdot\frac{2}{3} + \frac{1}{243} \approx 0.1646 + 0.0412 + 0.0041 = 0.2099$

33. $\Pr(\text{No suicide}) = 0.997,\ n = 100$

$\Pr(\text{No suicide}) = \binom{100}{100}(.997)^{100}(.003)^0 \approx 0.7405$

Exercise 8.2

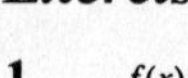

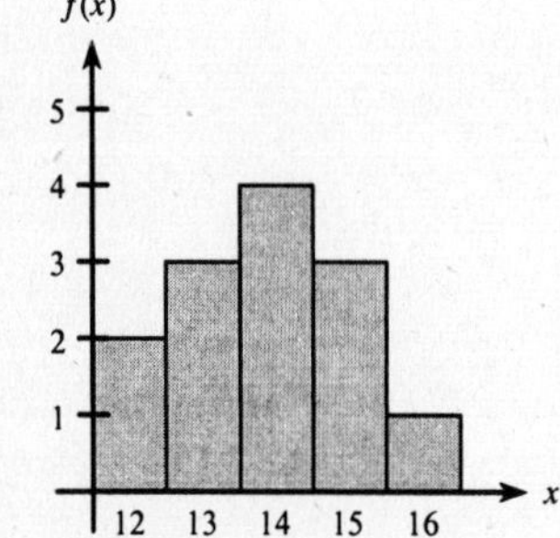

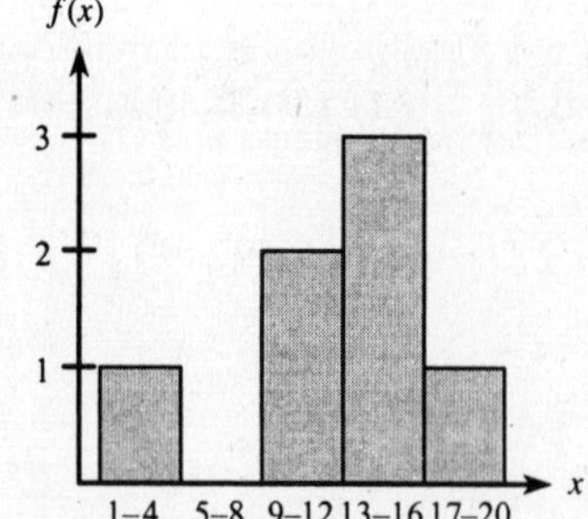

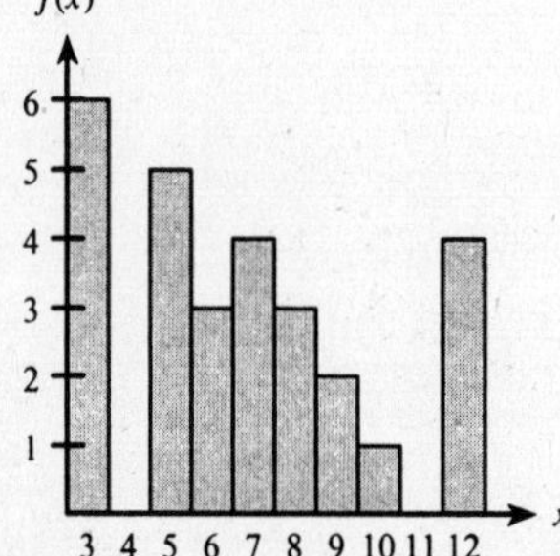

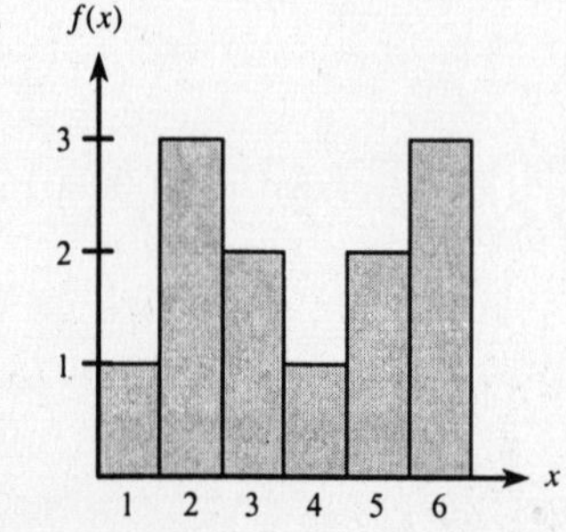

9. The mode is the score that occurs most frequently. 3 (4 times)

11. 13 (3 times)

13. Arrange scores in ascending order.
1, 1, 2, **2**, 2, 3, 4
The median is 2.

15. Arrange scores in ascending order.
0, 0, 1, 1, **1**, 2, 2, 14, 37
The median is 1.

17. Arranged in order: 1, 2, 2, **3**, **6**, 8, 12, 14
The mode is 2. The median is $(3+6)\div 2 = 4.5$.
The mean is $\bar{x} = (1+2+2+3+6+8+12+14)\div 8 = 6$.

19. Arranged in order: 14, 17, **17**, **20**, 31, 42
The mode is 17 since 17 occurs most often.
The median is $(17+20)\div 2 = 18.5$.
The mean is $\bar{x} = (14+17+17+20+31+42)\div 6 = 141\div 6 = 23.5$.

21. Arranged in order: 2.8, 5.3, **5.3**, 6.4, 6.8
The mode is 5.3. The median is 5.3.
The mean is $\bar{x} = (2.8+5.3+5.3+6.4+6.8)\div 5 = (26.6)\div 5 = 5.32$.

23.

Scores	Class marks	Frequencies
1–4	2.5	1
5–8	6.5	0
9–12	10.5	2
13–16	14.5	3
17–20	18.5	1

mean: $\bar{x} = \dfrac{2.5(1)+6.5(0)+10.5(2)+14.5(3)+18.5(1)}{1+0+2+3+1} \approx 12.21$
mode : 14.5 (most frequent score)
median: 14.5 (middle (4th) score)

25. Range: $11 - 2 = 9$

27. Range: $11 - (-3) = 14$

29. $\bar{x} = (5+7+1+3+0+8+6+2)\div 8 = 4$

$$s^2 = \left[1^2 + 3^2 + (-3)^2 + (-1)^2 + (-4)^2 + 4^2 + 2^2 + (-2)^2\right]\div 7 = \frac{60}{7} \approx 8.57$$

$s = \sqrt{8.57} \approx 2.93$

31. $\bar{x} = (11+12+13+14+15+16+17)\div 7 = 14$

$$s^2 = \left[(-3)^2 + (-2)^2 + (-1)^2 + 0^2 + 1^2 + 2^2 + 3^2\right]\div 6 = \frac{28}{6} \approx 4.67$$

$s = \sqrt{4.67} \approx 2.16$

33. $\bar{x} = (3\cdot 1 + 1\cdot 2 + 4\cdot 3 + 2\cdot 4 + 1\cdot 5)\div 11 = \dfrac{30}{11} \approx 2.73$

$$s^2 = \left[3(-1.73)^2 + 1(-0.73)^2 + 4(0.27)^2 + 2(1.27)^2 + 1(2.27)^2\right]\div 10 \approx 1.82$$

$s = \sqrt{1.82} \approx 1.35$

35. $\bar{x} = (6\cdot3+0\cdot4+5\cdot5+3\cdot6+4\cdot7+3\cdot8+2\cdot9+1\cdot10+0\cdot11+4\cdot12)\div28 = \frac{189}{28} = 6.75$

$$s^2 = \frac{6(-3.75)^2+5(-1.75)^2+3(-.75)^2+4(.25)^2+3(1.25)^2+2(2.25)^2+1(3.25)^2+4(5.25)^2}{27}$$

$$= \frac{237.25}{27} \approx 8.787$$

$$s = \sqrt{8.787} \approx 2.96$$

37.

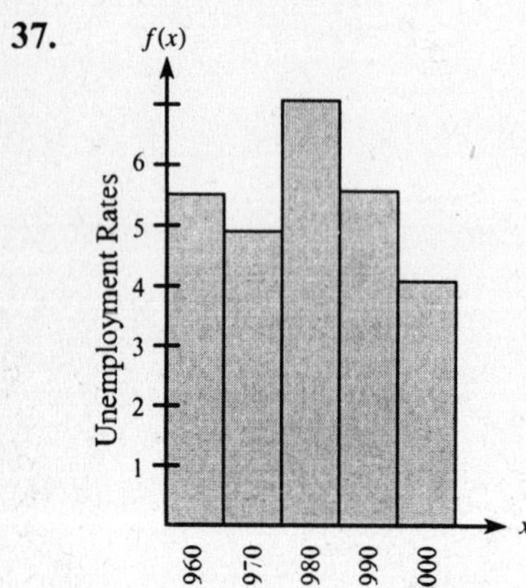

39. a. $\bar{x} = \frac{1(80,000)+1(60,000)+2(30,000)+1(20,000)+5(16,000)}{1+1+2+1+5} = \$30,000$

b. Median salary is $\frac{20,000+16,000}{2} = \$18,000$

c. Mode: $16,000

41. Using the mean will give the highest measure.

43. Using the median would give the most representative average.

45. $\bar{x} = \frac{\Sigma xf}{\Sigma f} = \frac{(30,000)(4)+(32,000)(2)+(34,000)(3)+(36,000)(2)+(38,000)(1)}{12} = 33,000$

47. $\bar{x} = \frac{\Sigma xf}{\Sigma f} = \frac{531.8}{160} = 3.32375 \qquad s^2 = \frac{72.970}{159} \approx 0.4589 \qquad s = 0.6774$

49. a. $\bar{x} = \frac{63.03}{8} = 7.88\%$

b. $\bar{x} = \frac{49.03}{8} = 6.13\%$

c. $\bar{x} = \frac{112.06}{16} = 7.00\%$

d. $s = 1.2480$

51. **a.**

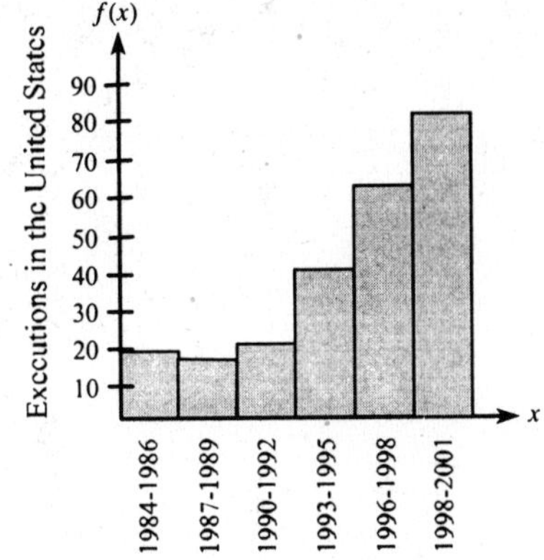

b. $\bar{x} = \dfrac{19+17.3+22.7+41.7+62.3+83}{6} = 41$

c. $s = 26.975$

d. No

53. Gold Prod: 1249.1 Gold Sales: 1233.9 Rev: 423
Avg Price: 355.7 Oper Costs: 179.7 Net I: 138.3

Exercise 8.3

1. No. $\Pr(x)$ must be ≥ 0.

3. Yes. $0 \leq \Pr(x) \leq 1$ and $\sum \Pr(x) = 1$.

5. Yes. $0 \leq \Pr(x) \leq 1$ and $\sum \Pr(x) = 1$.

7. No. $\sum \Pr(x) \neq 1$.

9. $E(x) = \sum x\Pr(x) = 0 \cdot \frac{1}{8} + 1 \cdot \frac{1}{4} + 2 \cdot \frac{1}{4} + 3 \cdot \frac{3}{8} = \frac{15}{8}$

11. $E(x) = \sum x\Pr(x) = 4 \cdot \frac{1}{3} + 5 \cdot \frac{1}{3} + 6 \cdot \frac{1}{3} + 7 \cdot 0 = 5$

13. $\mu = \sum x\Pr(x) = 0 \cdot \frac{1}{4} + 1 \cdot \frac{1}{4} + 2 \cdot \frac{1}{8} + 3 \cdot \frac{3}{8} = \frac{13}{8}$

$$\sigma^2 = \sum (x-\mu)^2 \Pr(x) = \left(0 - \frac{13}{8}\right)^2 \cdot \frac{1}{4} + \left(1 - \frac{13}{8}\right)^2 \cdot \frac{1}{4} + \left(2 - \frac{13}{8}\right)^2 \cdot \frac{1}{8} + \left(3 - \frac{13}{8}\right)^2 \cdot \frac{3}{8}$$

$$= \frac{169}{256} + \frac{25}{256} + \frac{9}{512} + \frac{363}{512} = \frac{760}{512} = \frac{95}{64} \approx 1.48$$

$\sigma \approx \sqrt{1.48} \approx 1.22$

15. $\mu = \sum x\Pr(x) = 0 \cdot \frac{1}{5} + 2 \cdot \frac{1}{4} + 4 \cdot \frac{1}{10} + 6 \cdot \frac{3}{20} + 8 \cdot \frac{3}{10} = \frac{42}{10} = 4.2$

$$\sigma^2 = \sum (x-\mu)^2 \Pr(x) = (-4.2)^2 \cdot \frac{1}{5} + (-2.2)^2 \cdot \frac{1}{4} + (-0.2)^2 \cdot \frac{1}{10} + (1.8)^2 \cdot \frac{3}{20} + (3.8)^2 \cdot \frac{3}{10}$$

$$= 3.528 + 1.21 + 0.004 + 0.486 + 4.332 = 9.56$$

$\sigma = \sqrt{9.56} \approx 3.09$

17. $\mu = 0\cdot\frac{0}{21}+1\cdot\frac{1}{21}+2\cdot\frac{2}{21}+3\cdot\frac{3}{21}+4\cdot\frac{4}{21}+5\cdot\frac{5}{21}+6\cdot\frac{6}{21}=\frac{91}{21}=\frac{13}{3}$

$\sigma^2 = \left(-\frac{13}{3}\right)^2\cdot 0+\left(-\frac{10}{3}\right)^2\cdot\frac{1}{21}+\left(-\frac{7}{3}\right)^2\cdot\frac{2}{21}+\left(-\frac{4}{3}\right)^2\cdot\frac{3}{21}+\left(-\frac{1}{3}\right)^2\cdot\frac{4}{21}+\left(\frac{2}{3}\right)^2\cdot\frac{5}{21}+\left(\frac{5}{3}\right)^2\cdot\frac{6}{21}$

$=\frac{1}{21}\left(\frac{100}{9}+\frac{98}{9}+\frac{48}{9}+\frac{4}{9}+\frac{20}{9}+\frac{150}{9}\right)=\frac{420}{9\cdot 21}=\frac{20}{9}\approx 2.22$

$\sigma = \sqrt{2.22}\approx 1.49$

19. $E(x)=\sum x\Pr(x)=0\cdot\frac{0}{10}+1\cdot\frac{1}{10}+2\cdot\frac{2}{10}+3\cdot\frac{3}{10}+4\cdot\frac{4}{10}=3$

21. $E(x)=0\cdot\frac{1}{5}+1\cdot\frac{1}{5}+2\cdot\frac{1}{5}+3\cdot\frac{1}{5}+4\cdot\frac{1}{5}=\frac{10}{5}=2$

23. $E(x)=0(.04)+1(.35)+2(.38)+3(.18)+4(.05)=1.85$

25. TV: $E(x)=100,000(0.01)+50,000(0.47)+25,000(0.52)=37,500$

PA: $E(x)=80,000(0.02)+50,000(0.47)+20,000(0.51)=35,300$

27. $E(x)=39,900\left(\frac{1}{1500}\right)+4900\left(\frac{1}{1500}\right)+2400\left(\frac{1}{1500}\right)+1400\left(\frac{1}{1500}\right)-100\left(\frac{1496}{1500}\right)$

$=26.60+3.27+1.60+0.93-99.73=-67.33$

Expected loss \$67.33

29. Questions Correct x: 0, 1, 2, 3, 4

Questions Correct x:	0	1	2	3	4
$\Pr(x)$	$\binom{4}{0}\left(\frac{1}{4}\right)^0\left(\frac{3}{4}\right)^4$	$\binom{4}{1}\left(\frac{1}{4}\right)\left(\frac{3}{4}\right)^3$	$\binom{4}{2}\left(\frac{1}{4}\right)^2\left(\frac{3}{4}\right)^2$	$\binom{4}{3}\left(\frac{1}{4}\right)^3\left(\frac{3}{4}\right)$	$\binom{4}{4}\left(\frac{1}{4}\right)^4\left(\frac{3}{4}\right)^0$

$E(x)=0\cdot\left(\frac{3}{4}\right)^4+1\cdot\frac{108}{4^4}+2\cdot\frac{54}{4^4}+3\cdot\frac{12}{4^4}+4\cdot\frac{1}{4^4}=\frac{256}{4^4}=1$

Expect to get credit for $E(x) - 1 = 0$.

31. $E(x)=500\cdot\frac{1}{1000}+100\cdot\frac{1}{1000}+10\cdot\frac{1}{1000}-1=-\$0.39$

33. $E(x)=15\cdot\frac{1}{13}+10\cdot\frac{1}{13}+1\cdot\frac{1}{13}-4(1)=-\$2.00$

35. Buy 0: There is no profit.

Buy 100: $E(x)=3(100)(.25)+3(50)(.20)+3(10)(.55)-1(50)(.20)-1(90)(.55)=\$62$

Buy 200: $E(x)=3(180)(.25)+3(50)(.20)+3(10)(.55)-1(20)(.25)-1(150)(.20)-1(190)(.55)=\$42$

Buy 100 for the best profit.

37. Radio: $E(x)=80,000(0.01)+40,000(0.47)+25,000(0.52)=32,600$

TM: $E(x)=70,000(0.04)+50,000(0.38)+30,000(0.58)=39,200$

Telemarketing will reach more people.

39. No. The variance is not known. They could be accurate only at altitudes close to 1000 ft. or only at very high and very low altitude.

41. $E(x)=20\cdot\frac{8}{30}+10\cdot\frac{14}{30}=\frac{16}{3}+\frac{14}{3}=\$10$

Exercise 8.4

1. Let x = number of rolls where 5 occurs.

a.

x	$\Pr(x)$
0	$\binom{3}{0}\left(\frac{1}{6}\right)^0\left(\frac{5}{6}\right)^3 = \frac{125}{216}$
1	$\binom{3}{1}\left(\frac{1}{6}\right)^1\left(\frac{5}{6}\right)^2 = \frac{75}{216} = \frac{25}{72}$
2	$\binom{3}{2}\left(\frac{1}{6}\right)^2\left(\frac{5}{6}\right)^1 = \frac{15}{216} = \frac{5}{72}$
3	$\binom{3}{3}\left(\frac{1}{6}\right)^3\left(\frac{5}{6}\right)^0 = \frac{1}{216}$

b. $\mu = np = 3\left(\frac{1}{6}\right) = \frac{1}{2}$

c. $\sigma = \sqrt{npq} = \sqrt{3\left(\frac{1}{6}\right)\left(\frac{5}{6}\right)} = \frac{1}{6}\sqrt{15} \approx 0.645$

3. **a.** $\mu = np = 60(0.7) = 42$

b. $\sigma = \sqrt{npq} = \sqrt{60(0.7)(0.3)} \approx 3.550$

5. $\mu = np = 12 \cdot \frac{1}{6} = 2$

$\sigma = \sqrt{npq} = \sqrt{12 \cdot \frac{1}{6} \cdot \frac{5}{6}} = \sqrt{\frac{5}{3}} \approx 1.291$

7. Expected number of successes is np.

$\mu = 6 \cdot \frac{2}{3} = 4$

9. $\mu = 12 \cdot \frac{6}{36} = 2$

11. $\mu = 24 \cdot \frac{1}{12} = 2$

13. By the binomial formula the coefficient of a^4b^2 is $\binom{6}{4} = 15$.

15. $(a+b)^6 = \binom{6}{6}a^6b^0 + \binom{6}{5}a^5b + \binom{6}{4}a^4b^2 + \binom{6}{3}a^3b^3 + \binom{6}{2}a^2b^4 + \binom{6}{1}ab^5 + \binom{6}{0}a^0b^6$

$= a^6 + 6a^5b + 15a^4b^2 + 20a^3b^3 + 15a^2b^4 + 6ab^5 + b^6$

17. $(x+h)^4 = \binom{4}{4}x^4h^0 + \binom{4}{3}x^3h + \binom{4}{2}x^2h^2 + \binom{4}{1}xh^3 + \binom{4}{0}x^0h^4$

$= x^4 + 4x^3h + 6x^2h^2 + 4xh^3 + h^4$

19. $n = 100$, $p = 0.1$

a. $\mu = np = 100(0.1) = 10$

b. $\sigma = \sqrt{npq} = \sqrt{100(0.1)(0.9)} = 3$

21. $n = 100{,}000$, $p = 0.6$

a. $\mu = np = 100{,}000(0.6) = 60{,}000$

b. $\sigma = \sqrt{npq} = \sqrt{100{,}000(0.6)(0.4)} = \sqrt{24{,}000} \approx 155$

23. In problem 21 we have $\sigma \approx 155$. Thus, 2 standard deviations is 2(155) = 310 votes. The candidate actually received 60,000 – 310 = 59,690 votes.

25. $n = 20,\ p = \frac{1}{5},\ q = \frac{4}{5}$

a. $\mu = np = 20 \cdot \frac{1}{5} = 4$

b. $\sigma = \sqrt{npq} = \sqrt{20 \cdot \frac{1}{5} \cdot \frac{4}{5}} = \sqrt{\frac{16}{5}} \approx 1.79$

27. $n = 200$, $\Pr(\text{Def}) = 0.01 = p$, $q = 0.99$

$\mu = np = 200(0.01) = 2$

$\sigma = \sqrt{200(0.01)(0.99)} \approx 1.41$

29. Pr(Germinate) = 0.85, Pr(Fail to germinate) = 0.15
For this question $n = 2000$, $p = 0.15$
$\mu = 2000(0.15) = 300$

Exercise 8.5

1. Problems 1 and 3 are the same from a probability standpoint. $\Pr(0 \le z \le 1.8) = 0.4641$

3. Problems 3 and 1 are the same from a probability standpoint. $\Pr(-1.8 \le z \le 0) = 0.4641$

5. $\Pr(-1.5 \le z \le 0) = 0.4332 \qquad \Pr(0 \le z \le 2.1) = 0.4821$
Thus, $\Pr(-1.5 \le z \le 2.1) = 0.4332 + 0.4821 = 0.9153$

7. $\Pr(-1.9 \le z \le 0) = 0.4713 \qquad \Pr(-1.1 \le z \le 0) = 0.3643$
Thus, $\Pr(-1.9 \le z \le -1.1) = 0.4713 - 0.3643 = 0.1070$

9. $\Pr(0 \le z \le 3) = 0.4987 \qquad \Pr(0 \le z \le 2.1) = 0.4821$
Thus, $\Pr(2.1 \le z \le 3) = 0.4987 - 0.4821 = 0.0166$

11. $\Pr(0 \le z \le 2) = 0.4773$
$\Pr(z > 2) = 0.5 - 0.4773 = 0.0227$

13. $\Pr(0 \le x \le 1.2) = 0.3849$
$\Pr(z < 1.2) = 0.5 + 0.3849 = 0.8849$

15. 20: $z = \dfrac{20-20}{5} = 0 \qquad$ 22.5: $z = \dfrac{22.5-20}{5} = 0.5$
$\Pr(20 \le x \le 22.5) = \Pr(0 \le z \le 0.5) = 0.1915$

17. 13.75: $z = \dfrac{13.75-20}{5} = -1.25 \qquad$ 20: $z = \dfrac{20-20}{5} = 0$
$\Pr(13.75 \le x \le 20) = \Pr(0 \le z \le 1.25) = 0.3944$

19. 45: $z = \dfrac{45-50}{10} = -0.5 \qquad$ 55: $z = \dfrac{55-50}{10} = 0.5$
$\Pr(45 \le x \le 55) = 2\Pr(0 \le z \le 0.5) = 2(0.1915) = 0.3830$

21. 35: $z = \dfrac{35-50}{10} = -1.5 \qquad$ 60: $z = \dfrac{60-50}{10} = 1$
$\Pr(35 \le x \le 60) = \Pr(0 \le z \le 1.5) + \Pr(0 \le z \le 1) = 0.4332 + 0.3413 = 0.7745$

23. 134: $z = \dfrac{134-110}{12} = 2$
$\Pr(x < 134) = 0.5 + \Pr(0 \le z \le 2) = 0.5 + 0.4773 = 0.9773$

25. 134: $z = \dfrac{134-110}{12} = 2$

$\Pr(x > 134) = 0.5 - \Pr(0 \le z \le 2) = 0.5 - 0.4773 = 0.0227$

27. a. 15: $z = 0$ 19: $z = \dfrac{19-15}{4} = 1$ $\Pr(0 \le z \le 1) = 0.3413$

b. 10: $z = \dfrac{10-15}{4} = -1.25$ $\Pr(-1.25 \le z \le 0) = 0.3944$

29. 10: $z = \dfrac{10-20}{4} = -2.5$ $z = \dfrac{30-20}{4} = 2.5$

$\Pr(10 \le x \le 30) = \Pr(-2.5 \le z \le 0) + \Pr(0 \le z \le 2.5) = 0.4938 + 0.4938 = 0.9876$

31. a. 160: $z = 0$ 181: $z = \dfrac{181-160}{15} = 1.4$ $\Pr(0 \le z \le 1.4) = 0.4192$

b. 190: $z = \dfrac{190-160}{15} = 2$ $\Pr(z \ge 2) = 0.5000 - 0.4773 = 0.0227$

c. $\Pr(1.4 \le z \le 2) = 0.4773 - 0.4192 = 0.0581$

d. 130: $z = \dfrac{130-160}{15} = -2$

$\Pr(-2 \le z \le 1.4) = \Pr(-2 \le z \le 0) + \Pr(0 \le z \le 1.4) = 0.4773 + 0.4192 = 0.8965$

33. a. 22: $z = \dfrac{22-28}{4} = -1.5$ $\Pr(z \le -1.5) = 0.5000 - 0.4332 = 0.0668$

b. 30: $z = \dfrac{30-28}{4} = 0.5$ $\Pr(z \ge 0.5) = 0.5000 - 0.1915 = 0.3085$

c. 26: $z = \dfrac{26-28}{4} = -0.5$

$\Pr(-0.5 \le z \le 0.5) = \Pr(-0.5 \le z \le 0) + \Pr(0 \le z \le 0.5) = 0.1915 + 0.1915 = 0.3830$

35. a. 140: $z = \dfrac{140-120}{12} = 1.67$ $\Pr(z \ge 1.67) = 0.5000 - 0.4525 = 0.0475$

b. 110: $z = \dfrac{110-120}{12} = -0.83$ $\Pr(z \le -0.83) = 0.5000 - 0.2967 = 0.2033$

c. 130: $z = 0.83$ $\Pr(-0.83 \le z \le 0.83) = 0.2967 + 0.2967 = 0.5934$

37. a. 0.9: $z = \dfrac{0.9-0.7}{0.1} = 2$ $\Pr(z \ge 2) = 0.5000 - 0.4773 = 0.0227$

b. 0.6: $z = \dfrac{0.6-0.7}{0.1} = -1$ $\Pr(z \le -1) = 0.5000 - 0.3413 = 0.1587$

c. $\Pr(-1 \le z \le 2) = 0.3413 + 0.4773 = 0.8186$

Review Exercises

1. $\Pr(5\text{ successes}) = \binom{7}{5}(0.4)^5(0.6)^2 = 21(0.01024)(0.36) = 0.07741$

2. a. $\Pr(2\text{ black out of }4) = \binom{4}{2}\left(\frac{5}{12}\right)^2\left(\frac{7}{12}\right)^2 \approx 0.354$

   b. $\Pr(\text{at least }2) = \Pr(2)+\Pr(3)+\Pr(4) = \binom{4}{2}\left(\frac{5}{12}\right)^2\left(\frac{7}{12}\right)^2 + \binom{4}{3}\left(\frac{5}{12}\right)^3\left(\frac{7}{12}\right)^1 + \binom{4}{4}\left(\frac{5}{12}\right)^4\left(\frac{7}{12}\right)^0$

   $= 0.3545 + 0.1688 + 0.03014 = 0.5534$

3. Pr(At least 2 rolls are greater than 4) = Pr(2 rolls) + Pr(3 rolls) + Pr(4 rolls)

   $= \binom{4}{2}\left(\frac{1}{3}\right)^2\left(\frac{2}{3}\right)^2 + \binom{4}{3}\left(\frac{1}{3}\right)^3\left(\frac{2}{3}\right) + \binom{4}{4}\left(\frac{1}{3}\right)^4\left(\frac{2}{3}\right)^0 = 6\cdot\frac{1}{9}\cdot\frac{4}{9} + 4\cdot\frac{1}{27}\cdot\frac{2}{3} + 1\cdot\frac{1}{81}\cdot 1 = \frac{11}{27} \approx 0.407$

4.

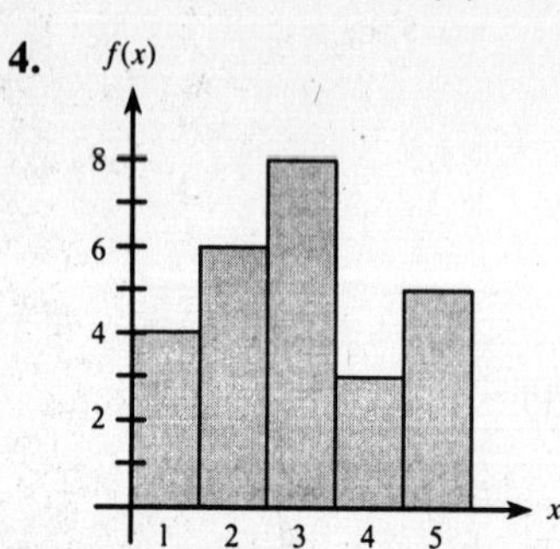

5. mode = 3

6. $\bar{x} = \dfrac{4\cdot1+6\cdot2+8\cdot3+3\cdot4+5\cdot5}{4+6+8+3+5} \approx 2.96$

7. Median is between the 13th and 14th score. It is $\dfrac{3+3}{2} = 3$.

8.

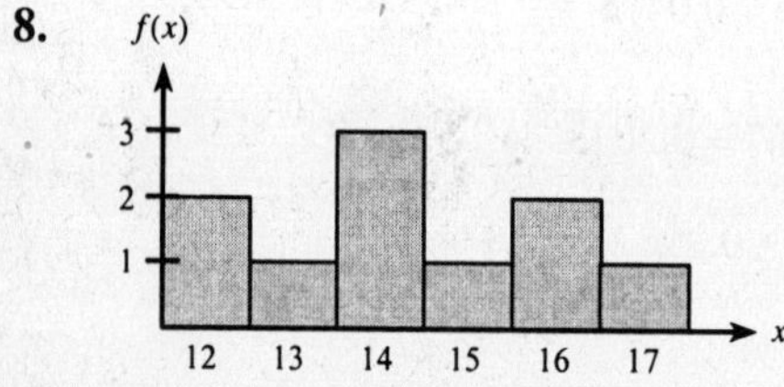

9. Median is between the 5th and 6th score. It is 14.

10. Mode is 14 (3 times).

11. $\text{mean} = \dfrac{\text{sum of scores}}{10} = \dfrac{143}{10} = 14.3$

12. Mean: $\bar{x} = \dfrac{4+3+4+6+8+0+2}{7} = \dfrac{27}{7} \approx 3.86$

    Variance: $s^2 = \dfrac{2(0.14)^2 + (-0.86)^2 + (2.14)^2 + (4.14)^2 + (-3.86)^2 + (-1.86)^2}{6} \approx 6.81$

    Standard deviation: $s = \sqrt{6.81} \approx 2.61$

13. $\bar{x} = \frac{\Sigma x}{n} = \frac{20}{10} = 2$

$s^2 = \frac{\Sigma(x-\bar{x})^2}{n-1} = \frac{1+0+1+9+1+4+4+0+1+1}{9} = \frac{22}{9} = 2.4\overline{4}$

$s \approx 1.56$

14. $E(x) = 1(0.2) + 2(0.3) + 3(0.4) + 4(0.1) = 2.4$

15. $\sum \Pr(x) = \frac{1}{15} + \frac{2}{15} + \frac{3}{15} + \frac{4}{15} + \frac{5}{15} = 1$

Yes.

16. No, because $\sum \Pr(x) \neq 1$.

17. $\sum \Pr(x) = \frac{1}{6} + \frac{1}{3} + \frac{1}{3} + \frac{1}{6} = 1$ Yes.

18. No, because $\Pr(x)$ cannot be negative.

19. $E(x) = 1(0.4) + 2(0.3) + 3(0.2) + 4(0.1) = 2.0$

20. **a.** $\mu = \sum x \Pr(x) = 1\left(\frac{1}{16}\right) + 2\left(\frac{2}{16}\right) + 3\left(\frac{3}{16}\right) + 4\left(\frac{4}{16}\right) + 6\left(\frac{6}{16}\right) = \frac{66}{16} = 4.125$

b. $\sigma^2 = \sum (x-\mu)^2 \Pr(x)$

$= \left(1 - \frac{33}{8}\right)^2 \cdot \frac{1}{16} + \left(2 - \frac{33}{8}\right)^2 \cdot \frac{2}{16} + \left(3 - \frac{33}{8}\right)^2 \cdot \frac{3}{16} + \left(4 - \frac{33}{8}\right)^2 \cdot \frac{4}{16} + \left(6 - \frac{33}{8}\right)^2 \cdot \frac{6}{16} = \frac{1}{16} \cdot \frac{175}{4} = 2.734375$

c. $\sigma = \sqrt{2.734375} \approx 1.6536$

21. **a.** $\mu = \sum x \Pr(x) = 1 \cdot \frac{1}{12} + 2 \cdot \frac{1}{6} + 3 \cdot \frac{1}{3} + 4 \cdot \frac{5}{12} = \frac{37}{12}$

b. $\sigma^2 = \sum (x-\mu)^2 \cdot \Pr(x) = \left(-\frac{25}{12}\right)^2 \cdot \frac{1}{12} + \left(-\frac{13}{12}\right)^2 \cdot \frac{1}{6} + \left(-\frac{1}{12}\right)^2 \cdot \frac{1}{3} + \left(\frac{11}{12}\right)^2 \cdot \frac{5}{12} = \frac{131}{144} \approx 0.9097$

c. $\sigma = \sqrt{0.9097} \approx 0.9538$

22. $\mu = np = 6\left(\frac{2}{3}\right) = 4$ $\sigma = \sqrt{npq} = \sqrt{6\left(\frac{2}{3}\right)\left(\frac{1}{3}\right)} = \frac{\sqrt{12}}{3} \approx 1.15$

23. $\mu = np = 18\left(\frac{1}{6}\right) = 3$

24. $(x+y)^5 = x^5 + 5x^4y + 10x^3y^2 + 10x^2y^3 + 5xy^4 + y^5$

25. $\Pr(-1.6 \le z \le 1.9) = 0.4452 + 0.4713 = 0.9165$

26. $\Pr(-1 \le z \le -0.5) = 0.3413 - 0.1915 = 0.1498$

27. $\Pr(1.23 \le z \le 2.55) = 0.4946 - 0.3907 = 0.1039$

28. $\Pr(25 \le x \le 30) = \Pr\left(\frac{25-25}{5} \le \frac{x-\mu}{\sigma} \le \frac{30-25}{5}\right) = \Pr(0 \le z \le 1) = 0.3413$

29. z scores are -1 and 1 respectively.

$\Pr(-1 \le z \le 1) = 0.3413 + 0.3413 = 0.6826$

30. $\Pr(30 \le x \le 35) = \Pr\left(\frac{30-25}{5} \le \frac{x-\mu}{\sigma} \le \frac{35-25}{5}\right) = \Pr(1 \le z \le 2) = 0.4773 - 0.3413 = 0.1360$

31. $\Pr(2 \text{ blond}) = \binom{6}{2}\left(\frac{1}{4}\right)^2\left(\frac{3}{4}\right)^4 = 15 \cdot \frac{1}{16} \cdot \frac{81}{256} \approx 0.297$

32. $\Pr(x \ge 3) = \binom{5}{3}(0.3)^3(0.7)^2 + \binom{5}{4}(0.3)^4(0.7)^1 + \binom{5}{5}(0.3)^5(0.7)^0 = 0.1323 + 0.02835 + 0.00243 = 0.16308$

33. $\Pr(\text{victim}) = \frac{1}{5}$ $\quad\Pr(\text{not a victim}) = \frac{4}{5}$

$\Pr(2 \text{ victims}) = \binom{5}{2}\left(\frac{1}{5}\right)^2\left(\frac{4}{5}\right)^3 = 10\left(\frac{1}{25}\right)\left(\frac{64}{125}\right) = \frac{640}{3125} = 0.2048$

34. a. $\Pr(\text{exactly } 1) = \binom{100{,}000}{1}\left(\frac{1}{100{,}000}\right)^1\left(\frac{99{,}999}{100{,}000}\right)^{99{,}999} \approx 100{,}000\left(\frac{1}{100{,}000}\right)(0.36788) \approx 0.37$

b. $\Pr(\text{at least } 1) = 1 - \Pr(0 \text{ get disease})$

$= 1 - \binom{100{,}000}{0}\left(\frac{1}{100{,}000}\right)^0\left(\frac{99{,}999}{100{,}000}\right)^{100{,}000} \approx 1 - 0.3679 \approx 0.63$

35.

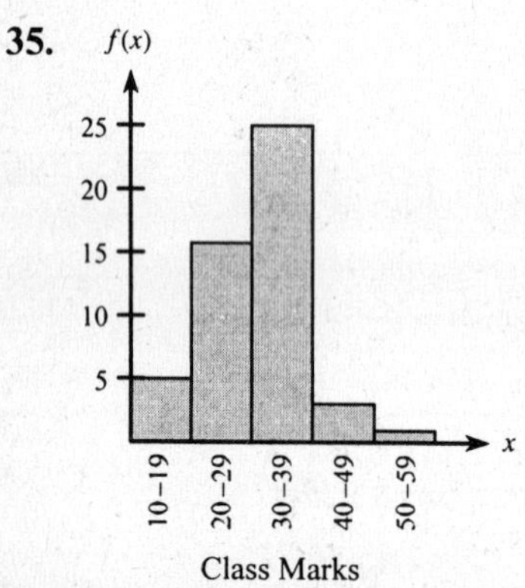

36.

Scores	Class marks	Frequencies
10–19	14.5	5
20–29	24.5	16
30–39	34.5	25
40–49	44.5	3
50–59	54.5	1

$\overline{x} = \frac{5(14.5)+16(24.5)+25(34.5)+3(44.5)+1(54.5)}{50} = \frac{1515}{50} = 30.3$

37. $s^2 = \frac{5(14.5-30.3)^2 + 16(24.5-30.3)^2 + 25(34.5-30.3)^2 + 3(44.5-30.3)^2 + 1(54.5-30.3)^2}{49}$

≈ 69.76

$s = \sqrt{69.76} \approx 8.35$

38. Pr(testing positive) = 0.91
Thus, 500(0.91) = 455 of the 500 are expected to test positively.

39. Each capsule is an independent event.

Expected number of empty capsules $= np = 100\left(\dfrac{60{,}000}{2{,}000{,}000}\right) = 3$.

40. $\Pr(\text{Win}) = \dfrac{3}{3+12} = \dfrac{1}{5}$ Expected value $= 98\left(\dfrac{1}{5}\right) + \dfrac{4}{5}(-2) = \$18.00$

41. $E(x) = (-1)(0.999) + 499(0.001) = -\$0.50$

42.

x	0	1	2	3	4	5
$\Pr(x)$	$\frac{1024}{5^5}$	$\frac{1280}{5^5}$	$\frac{640}{5^5}$	$\frac{160}{5^5}$	$\frac{20}{5^5}$	$\frac{1}{5^5}$

a. $E(x) = 0\cdot\dfrac{1024}{5^5} + 1\cdot\dfrac{1280}{5^5} + 2\cdot\dfrac{640}{5^5} + 3\cdot\dfrac{160}{5^5} + 4\cdot\dfrac{20}{5^5} + 5\cdot\dfrac{1}{5^5} = \dfrac{3125}{5^5} = 1$

b. $\Pr(1\text{ victim}) = \binom{5}{1}\left(\dfrac{1}{5}\right)\left(\dfrac{4}{5}\right)^4 = \left(\dfrac{4}{5}\right)^4 = 0.4096$

43. a. $\Pr(1000 \le x \le 1400) = \Pr\left(\dfrac{1000-1000}{200} \le \dfrac{x-\mu}{\sigma} \le \dfrac{1400-1000}{200}\right) = \Pr(0 \le z \le 2) = 0.4773$

b. $\Pr(1200 \le x \le 1400) = \Pr(1 \le z \le 2) = 0.4773 - 0.3413 = 0.1360$

c. $\Pr(x > 1400) = \Pr(z > 2) = 0.5000 - 0.4773 = 0.0227$

44. $\sigma = 96{,}000,\ \mu = 611{,}000$

$z_1 = \dfrac{700{,}000 - 611{,}000}{96{,}000} \approx 0.93$ $z_2 = \dfrac{800{,}000 - 611{,}000}{96{,}000} \approx 1.97$

$\Pr(0.93 \le z \le 1.97) = 0.4756 - 0.3238 = 0.1518 \approx 15\%$

Chapter Test

1. a. $\Pr(3\text{H}) = \binom{5}{3}\left(\dfrac{1}{3}\right)^3\left(\dfrac{2}{3}\right)^2 = \dfrac{40}{243}$

b. $\Pr(3\text{H or 4H or 5H}) = \dfrac{40}{243} + \binom{5}{4}\left(\dfrac{1}{3}\right)^4\left(\dfrac{2}{3}\right) + \binom{5}{5}\left(\dfrac{1}{3}\right)^5\left(\dfrac{2}{3}\right)^0 = \dfrac{51}{243}$

2. For a binomial distribution, the expected number of successes is np. The expected number of heads is $12\left(\dfrac{1}{3}\right) = 4$.

3. The mean $\mu = np$ or $\mu = 12\left(\dfrac{1}{3}\right) = 4$.

$\sigma = \sqrt{npq} = \sqrt{12\left(\dfrac{1}{3}\right)\left(\dfrac{2}{3}\right)} = \sqrt{\dfrac{8}{3}} = \sqrt{\dfrac{24}{9}} = \dfrac{2}{3}\sqrt{6} \approx 1.6$ The variance $\sigma^2 = \left(\sqrt{\dfrac{8}{3}}\right)^2 = \dfrac{8}{3}$.

4. $E(x) = \sum x\Pr(x) = 0.6 + 1.2 + 0.5 + 0.6 + 1.4 + 0.8 = 5.1$

5. $\mu = E(x) = \sum x\Pr(x) = 1 + 3.6 + 1.5 + 3.6 + 2 + 5 = 16.7$

$\sigma^2 = \sum (x-\mu)^2\Pr(x) = 26.61$ $\sigma = \sqrt{26.61} \approx 5.16$

6. mean $\mu = \dfrac{100+147+66+92+48}{5+7+3+4+2} \approx 21.57$

median = 21 (11th score), mode = 21

7. $z = \dfrac{x-\mu}{\sigma}$

a. $z_1 = \dfrac{14-16}{6} \approx -.33$, $z_2 = \dfrac{22-16}{6} = 1$

$\Pr(14 \le x \le 22) = \Pr(-0.33 \le z \le 1) = 0.1293 + 0.3413 = 0.4706$

b. $\Pr(x \le 22) = \Pr(z \le 1) = 0.5000 + 0.3413 = 0.8413$

c. $\Pr(22 \le x \le 24) = \Pr(1 \le z \le 1.33) = 0.4082 - 0.3413 = 0.0669$

8. $z = \dfrac{x-\mu}{\sigma}$

a. $\Pr(73 \le x \le 97) = \Pr(0.25 \le z \le 2.25) = 0.4878 - 0.0987 = 0.3891$

b. $\Pr(65 \le x \le 84) = \Pr(-0.42 \le z \le 1.17) = 0.1628 + 0.3790 = 0.5418$

c. $\Pr(x \ge 84) = \Pr(z \ge 1.17) = 0.5000 - 0.3790 = 0.1210$

9.

$f(x)$

20,000
15,000
10,000
5000

15 24
25-34
35-44
45-54
55-64
65+

x

Age of Householder (years)

10. Class marks: 19.5, 29.5, …, 59.5, 72.5

$\bar{x} = \dfrac{\Sigma \text{ (class mark)} \cdot \text{number}}{\Sigma \text{number}} = \dfrac{3,221,572}{69,312} \approx 46.48$ years

$s^2 = \dfrac{\Sigma(x-\bar{x})^2 \cdot \text{ number}}{n-1} = \dfrac{16,432,190.25}{69,311} \approx 237.079 \qquad s = \sqrt{237.079} \approx 15.40$

11. Class marks: 2, 7, 12, 17, … , 92, 97

$\bar{x} = \dfrac{\Sigma\text{(class mark)(number)}}{\Sigma \text{ number}} = \dfrac{8,882,963}{257,674} \approx 34.47$

$s^2 = \dfrac{\Sigma(x-\bar{x})^2 \cdot \text{ number}}{n-1} \approx \dfrac{122,383,833.6}{257,673} \approx 474.96 \qquad s = \sqrt{474.96} \approx 21.8$

12. $\bar{x} = \dfrac{666.8}{104.6} \approx 6.4$ In 1980 the average age is 6.4 years.

$\bar{x} = \dfrac{1052.1}{129.6} \approx 8.1$ In 1994 the average age is 8.1 years.

The average age is higher. Two reasons for this are better quality and much higher prices.

13. a. $\Pr(10 \text{ of } 100 \text{ are defective}) = \binom{100}{10}(0.02)^{10}(0.98)^{90} \approx 0.00003$

b. If 1500 chips are shipped, the expected number of defective chips is (0.02)(1500) = 30.

14. In this group of 30, 0.06(30) = 1.8 or 2 could expect to become pregnant.

15. In this group of 30, 0.18(30) = 5.4 or 5 could expect to become pregnant.

16. In this group of 30, 0.06(0.03)(30) = 0.054 or 0 could expect to become pregnant.

17. Use $z = \dfrac{x-\mu}{\sigma}$, $\mu = 18620$, and $\sigma = 4012$

a. $z = \dfrac{10000-18620}{4012} \approx -2.15$

$\Pr(x < 10000) = \Pr(z < -2.15) = 0.5000 - 0.4842 = 0.0158$

b. $z = \dfrac{24000-18620}{4012} \approx 1.34$

$\Pr(x > 24000) = \Pr(z > 1.34) = 0.5000 - 0.4099 = 0.0901$

c. $\Pr(15000 < x < 21000) = \Pr(-0.90 < z < 0) + \Pr(0 < z < 0.59) = 0.3159 + 0.2224 = 0.5383$

Chapter 9: Derivatives

Exercise 9.1

1. **a.** $\lim_{x\to c} f(x)=1$

b. $f(c)=1$

3. **a.** $\lim_{x\to c} f(x)=-8$

b. $f(c)=-8$

5. **a.** $\lim_{x\to c} f(x)=10$

b. $f(x)$ is not defined at $x=c$.

7. **a.** $\lim_{x\to c} f(x)=0$

b. $f(c)=-6$

9. **a.** $\lim_{x\to c^-} f(x)=+\infty$

b. $\lim_{x\to c^+} f(x)=+\infty$

c. $\lim_{x\to c} f(x)=+\infty$

d. $f(c)$ is not defined.

11. **a.** $\lim_{x\to c^-} f(x)=3$

b. $\lim_{x\to c^+} f(x)=-6$

c. $\lim_{x\to c} f(x)$ does not exist.

d. $f(c)=-6$

13.

x	0.9	0.99	0.999	$\to 1 \leftarrow$	1.001	1.01	1.1
$f(x)$	−2.9	−2.99	−2.999		−3.001	−3.01	−3.1

$\lim_{x\to 1} f(x)=-3$

15.

x	0.9	0.99	0.999	$\to 1 \leftarrow$	1.001	1.01	1.1
$f(x)$	3.5	3.95	3.995		4.995999	4.9599	4.59

$\lim_{x\to 1} f(x)$ does not exist since $4\neq 5$.

17. $\lim_{x\to -35}(34+x)=34+(-35)=-1$

19. $\lim_{x\to -1}(4x^3-2x^2+2)=4(-1)-2(1)+2=-4$

21. $\lim_{x\to -\frac{1}{2}} \dfrac{4x-2}{4x^2+1}=\dfrac{-2-2}{1+1}=-2$

23. $\lim_{x\to 3}\dfrac{x^2-9}{x-3}=\lim_{x\to 3}\dfrac{(x-3)(x+3)}{x-3}=\lim_{x\to 3}(x+3)=6$

25. $\lim_{x\to 7}\dfrac{(x^2-8x+7)}{(x^2-6x-7)}=\lim_{x\to 7}\dfrac{(x-7)(x-1)}{(x-7)(x+1)}=\dfrac{7-1}{7+1}=\dfrac{3}{4}$

27. $\lim_{x\to -2}\dfrac{x^2+4x+4}{x^2+3x+2}=\lim_{x\to -2}\dfrac{(x+2)(x+2)}{(x+2)(x+1)}=\dfrac{0}{-1}=0$

29. $\lim_{x\to 3^-} f(x)=4$ $\lim_{x\to 3^+} f(x)=6$ $\lim_{x\to 3} f(x)$ does not exist.

31. $\lim_{x\to (-1)^-} f(x)=-3$ $\lim_{x\to (-1)^+} f(x)=-3$ $\lim_{x\to -1} f(x)=-3$

33. $\lim_{x\to 2}\dfrac{x^2+6x+9}{x-2}=\dfrac{\lim_{x\to 2}(x^2+6x+9)}{\lim_{x\to 2}(x-2)}=\dfrac{25}{0}=$ does not exist (unbounded at $x=2$)

35. $\lim_{x\to -1}\frac{x^2+5x+6}{x+1}=\frac{\lim_{x\to -1}(x^2+5x+6)}{\lim_{x\to -1}(x+1)}=\frac{2}{0}=$ does not exist (unbounded at $x=-1$)

37. $\lim_{h\to 0}\frac{(x+h)^3-x^3}{h}=\lim_{h\to 0}\frac{x^3+3x^2h+3xh^2+h^3-x^3}{h}=\lim_{h\to 0}(3x^2+3xh+h^2)=3x^2$

39. $\lim_{x\to 2}\frac{x^3-4x}{2x^2-x^3}=\lim_{x\to 2}\frac{x(x+2)(x-2)}{x^2(2-x)}=\lim_{x\to 2}\frac{x+2}{-x}=\frac{4}{-2}=-2$

41. $\lim_{x\to 10}\frac{x^2-19x+90}{3x^2-30x}=\lim_{x\to 10}\frac{(x-10)(x-9)}{3x(x-10)}=\lim_{x\to 10}\frac{x-9}{3x}=\frac{1}{30}$

43. $\lim_{x\to -1}\frac{x^3-x}{x^2+2x+1}=\lim_{x\to -1}\frac{x(x+1)(x-1)}{(x+1)(x+1)}=\lim_{x\to -1}\frac{x(x-1)}{x+1}=\frac{2}{0}$ Limit does not exist.

45. $\lim_{x\to 6}\frac{x^2-2x-24}{x^2+2x-48}=\lim_{x\to 6}\frac{(x-6)(x+4)}{(x-6)(x+8)}=\frac{10}{14}=\frac{5}{7}$

47. $\lim_{x\to -2}\frac{x^4-4x^2}{x^2+8x+12}=\lim_{x\to -2}\frac{x^2(x-2)(x+2)}{(x+6)(x+2)}=\frac{4(-4)}{4}=-4$

49. $\lim_{x\to 4^-}f(x)=9 \quad \lim_{x\to 4^+}f(x)=9 \quad \lim_{x\to 4}f(x)=9$

51.

a	0.1	0.01	0.001	0.0001	0.00001	$\to 0$
$(1+a)^{1/a}$	2.5937	2.7048	2.7169	2.7181	2.71827	$e\approx 2.71828$

53. **a.** $\lim_{x\to 3}[f(x)+g(x)]=4+(-2)=2$

b. $\lim_{x\to 3}[f(x)-g(x)]=4-(-2)=6$

c. $\lim_{x\to 3}[f(x)\cdot g(x)]=4(-2)=-8$

d. $\lim_{x\to 3}\left[\frac{g(x)}{f(x)}\right]=-\frac{2}{4}=-\frac{1}{2}$

55. $\lim_{x\to 100}(1600x-x^2)=160{,}000-10{,}000=\$150{,}000$

57. **a.** $\lim_{x\to 4^+}\left(\frac{4}{x}+30+\frac{x}{4}\right)=1+30+1=\$32$ (thousands)

b. $\lim_{x\to 100^-}\left(\frac{4}{x}+30+\frac{x}{4}\right)=0.04+30+25=\$55.04$ (thousands)

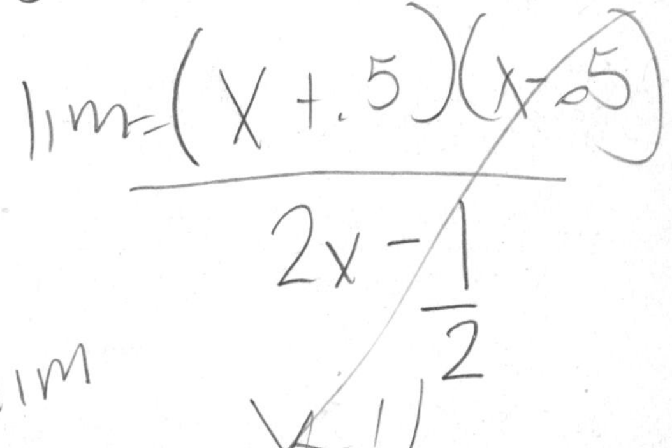

59. **a.** $S(0)=400+\frac{2400}{1}=\$2800$

b. $\lim_{t\to 7}\left(400+\frac{2400}{t+1}\right)=400+\frac{2400}{8}=\$700$

c. $\lim_{t\to 14}\left(400+\frac{2400}{t+1}\right)=400+\frac{2400}{15}=\$560$

61. a. $\lim_{t\to 4}\frac{128t(t+6)}{(t^2+6t+18)^2}=\frac{128\cdot 4\cdot 10}{(16+24+18)^2}\approx 1.52$ units/hr

b. $\lim_{t\to 8^-}\frac{128t(t+6)}{(t^2+6t+18)^2}=\frac{128\cdot 8\cdot 14}{(64+48+18)^2}\approx 0.85$ units/hr

c. lunch time

63. a. $\lim_{p\to 100^-} C(p)=0$ This is basically untreated water.

b. $\lim_{p\to 0^+} C(p)=\infty$

c. No. The cost would be extremely large since $C(0)$ is undefined.

65. a. $\lim_{x\to 27{,}050^-} T(x)=(0.15)(27{,}050)=\$4057.50$

b. $\lim_{x\to 27{,}050^+} T(x)=4057.50+0.275(0)=\$4057.50$

c. $\lim_{x\to 27{,}050} T(x)=\$4057.50$

67. $C(x)=\begin{cases}1.557x & \text{if } 0\le x\le 100\\ 155.70+1.040(x-100) & \text{if } 100<x\le 1000\\ 155.70+936.00+0.689(x-1000) & \text{if } x>1000\end{cases}$

$\lim_{x\to 1000} C(x)=155.70+936.00=\$1091.70$

69. $\lim_{t\to 9{:}30\text{AM}^+} D(t)\approx 9924.15$

This is the opening average on January 16, 2002.

71. a. $\lim_{t\to 200} f(t)=f(200)=1000\cdot\frac{-59.34}{40962.4}\approx -1.449$

b. This is the percentage of US workers in farm occupations in the year 2000.

c. The model is inaccurate. The percentage cannot be negative.

Exercise 9.2

1. a. continuous

b. discontinuous; $f(1)$ does not exist.

c. discontinuous; $\lim_{x\to 3} f(x)$ does not exist.

d. discontinuous; $\lim_{x\to 0} f(x)$ does not exist.
$f(0)$ does not exist.

3. $f(x)$ is continuous at $x=0$.

5. $f(x)$ is continuous at $x=-2$.

7. $f(x)$ is discontinuous at $x=-3$.
$f(x)$ is not defined at $x=-3$.

9. $f(x)$ is discontinuous at $x=-1$.
$f(x)$ is not defined at $x=-1$.
$\lim_{x\to 1} f(x)$ does not exist.

11. $f(x)$ is continuous at $x=0$.

13. $f(x)$ is discontinuous at $x=1$.
$\lim_{x\to 1} f(x)$ does not exist.

15. $f(x)$ is continuous everywhere.

17. $g(x)$ is discontinuous at $x=-2$ since $g(-2)$ is not defined and $\lim_{x\to -2} g(x)$ does not exist.

19. $f(x)$ is continuous everywhere.

21. $f(x)$ is continuous everywhere.

23. $y=\frac{x^2-5x-6}{x+1}$. $f(-1)$ is not defined. $f(x)$ is discontinuous at $x=-1$.

25. $f(x)$ is discontinuous at $x=3$ since $\lim_{x\to 3} f(x)$ does not exist.

27. a. VA: $x = -2$; $\lim_{x\to+\infty} f(x) = 0$; $\lim_{x\to-\infty} f(x) = 0$

b. The denominator is 0 and the numerator is not 0 at $x = -2$, so f has a vertical asymptote at $x = -2$.

$$\lim_{x\to+\infty} \frac{8}{x+2} = \lim_{x\to+\infty} \frac{\frac{8}{x}}{1+\frac{2}{x}} = \frac{0}{1} = 0$$

$$\lim_{x\to-\infty} \frac{8}{x+2} = \lim_{x\to-\infty} \frac{\frac{8}{x}}{1+\frac{2}{x}} = \frac{0}{1} = 0$$

29. a. VA: $x = -2$, $x = 3$;
$\lim_{x\to-\infty} f(x) = 2$; $\lim_{x\to+\infty} f(x) = 2$

b. The denominator is 0 and the numerator is not 0 at $x = -2$ and $x = 3$, so f has vertical asymptotes at $x = -2$ and $x = 3$.

$$\lim_{x\to+\infty} \frac{2(x+1)^3(x+5)}{(x-3)^2(x+2)^2} = \lim_{x\to+\infty} \frac{2(x+1)^3(x+5)\cdot\frac{1}{x^4}}{(x-3)^2(x+2)^2\cdot\frac{1}{x^4}}$$

$$= \lim_{x\to+\infty} \frac{2\left(1+\frac{1}{x}\right)^3\left(1+\frac{5}{x}\right)}{\left(1-\frac{3}{x}\right)^2\left(1+\frac{2}{x}\right)^2} = \frac{2(1)(1)}{(1)(1)} = 2$$

Similarly, $\lim_{x\to-\infty} f(x) = 2$

31. $\lim_{x\to+\infty} \frac{3}{x+1} = \lim_{x\to+\infty} \frac{\frac{3}{x}}{1+\frac{1}{x}} = \frac{0}{1+0} = 0$

33. $\lim_{x\to+\infty} \frac{x^3-1}{x^3+4} = \lim_{x\to+\infty} \frac{1-\frac{1}{x^3}}{1+\frac{4}{x^3}} = \frac{1-0}{1+0} = 1$

35. $\lim_{x\to-\infty} \frac{5x^3-4x}{3x^3-2} = \lim_{x\to-\infty} = \frac{5-\frac{4}{x^2}}{3-\frac{2}{x^3}} = \frac{5-0}{3-0} = \frac{5}{3}$

37. $\lim_{x\to+\infty} \frac{3x^2+5x}{6x+1} = \lim_{x\to+\infty} \frac{3+\frac{5}{x}}{\frac{6}{x}+\frac{1}{x^2}} \to +\infty$

Because the limit of the denominator is 0, this limit does not exist.

39. a.

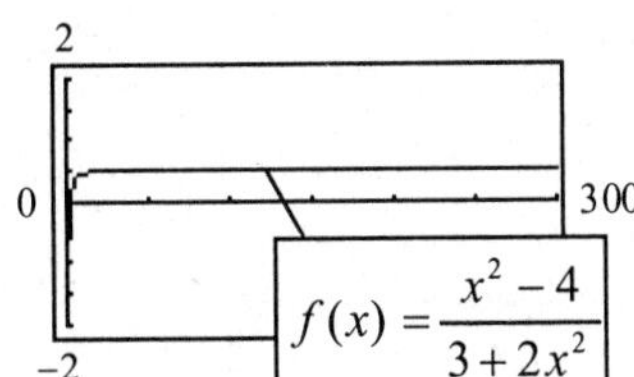

$\lim_{x\to+\infty} f(x) = 0.5$

b. $f(100{,}000) = 0.49999$

$f(1{,}000{,}000) = 0.5$

The table supports the conclusion.

41. $f(x) = \frac{1000(x-1)}{x+1000}$

a. Discontinuous at $x = -1000$

b. Since $f(x) = \frac{1000\left(1-\frac{1}{x}\right)}{1+\frac{1000}{x}}$,

$y = 1000$ is a horizontal asymptote.

c. For large values, an attempt to find the appropriate window is difficult. The asymptote may never be located.

43. $y = \frac{32}{(p+8)^{2/5}}$

a. If all values are allowed, then y is not continuous at $p = -8$.

b. Yes

c. Yes

d. $p > 0$

45. a. Discontinuous at $q = -1$.

b. Yes

47. a. $\lim_{n\to\infty} A_n = \lim_{n\to\infty} R\left(\frac{1-\frac{1}{(1+i)^n}}{i}\right) = R\left[\frac{1-0}{i}\right] = \frac{R}{i}$

b. $i = 0.01$, $R = 100$; $A = \frac{100}{0.01} = \$10{,}000$

49. $p \le 100$ for problem to have meaning. Thus, the function is continuous for the domain of the function.

51. $p = \frac{100C}{7300+C} = \frac{100}{\frac{7300}{C}+1}$ $\lim_{C\to\infty} p = \frac{100}{0+1} = 100\%$

100% of pollution cannot be removed. Cost would be impossible to afford.

53. $R(x)$ is discontinuous at each point that the rate changes. $x = 45{,}200,\ 109{,}250,\ 166{,}500,\ 297{,}350$

55. a. $C(1100) = 49.40 + 0.05(1100-500)$
$= \$79.40$

b. $\lim_{x\to100^-} C(x) = 19.40$

$\lim_{x\to100^+} C(x) = 19.40 + 0.075(100-100)$
$= 19.40$

Thus, $\lim_{x\to100} C(x) = 19.40$.

By similar procedure, $\lim_{x\to500} C(x) = 49.40$.

c. Yes, since $\lim_{x\to a} C(x) = C(a)$.

57. a. $d(t) = -0.00001719t^4 + 0.004697t^3 - 0.4579t^2 + 18.9049t - 268.314$

b. $d(110) = 5.55$

c. $\lim_{t \to +\infty} d(t) = -\infty$

d. No. $d(t) < 0$ for $t > 112$.

e. $t < 30$ and $t > 112$ since $d(t) < 0$.

Exercise 9.3

1. a. average rate of change $= \dfrac{f(5)-f(0)}{5-0} = \dfrac{18-(-12)}{5} = \dfrac{30}{5} = 6$

b. average rate of change $= \dfrac{f(10)-f(-3)}{10-(-3)} = \dfrac{98-(-6)}{13} = \dfrac{104}{13} = 8$

3. a. average rate of change $= \dfrac{f(5)-f(2)}{5-2} = \dfrac{30-20}{3} = \dfrac{10}{3} = 3.\overline{3}$

b. average rate of change $= \dfrac{f(4)-f(3.8)}{4-3.8} = \dfrac{16-17}{0.2} = \dfrac{-1}{0.2} = -5$

5. a. average rate of change over $[2.9,3] = \dfrac{f(3)-f(2.9)}{3-2.9} = \dfrac{-3-(-2.61)}{3-2.9} = \dfrac{-0.39}{0.1} = -3.9$

average rate of change over $[2.99,3] = \dfrac{f(3)-f(2.99)}{3-2.99} = \dfrac{-3-(-2.96)}{3-2.99} = \dfrac{-0.04}{0.01} = -4$

b. average rate of change over $[3,3.1] = \dfrac{f(3.1)-f(3)}{3.1-3} = \dfrac{-3.41-(-3)}{3.1-3} = \dfrac{-0.41}{0.1} = -4.1$

average rate of change over $[3,3.01] = \dfrac{f(3.01)-f(3)}{3.01-3} = \dfrac{-3.04-(-3)}{3.01-3} = \dfrac{-0.04}{0.01} = 4$

c. The calculations suggest that the instantaneous rate of change of $f(x)$ at $x = 3$ might be -4.

7. a. Instantaneous rate of change means $f'(x)$. $f'(4) = 8 \cdot 4 = 32$.

b. Slope of tangent to the graph means $f'(x)$. $f'(4) = 8 \cdot 4 = 32$.

c. $f(4) = 4(4)^2 = 64$ Point on graph: (4, 64)

9. $f(x) = 2x^2 - x$

a. $f(x+h) - f(x) = 2(x+h)^2 - (x+h) - [2x^2 - x] = 4xh + 2h^2 - h$

$\dfrac{f(x+h)-f(x)}{h} = 4x + 2h - 1$

$f'(x) = \lim_{h \to 0}(4x + 2h - 1) = 4x - 1$

b. $f'(-1) = 4(-1) - 1 = -5$

c. $f'(-1) = 4(-1) - 1 = -5$

d. $f(-1) = 2(-1)^2 - (-1) = 3$ Point on graph: (–1, 3)

11. a. $P(x,y) = P(1,1) \quad A(x,y) = A(3,0)$

b. $m_{\tan} = \dfrac{y_2 - y_1}{x_2 - x_1} = \dfrac{0-1}{3-1} = -\dfrac{1}{2}$

c. $f'(1) = -1/2$

d. $f'(1) = -1/2$

13. a. $P(x, y) = P(1, 3)$ $A(x, y) = A(0, 3)$

b. $m_{\tan} = \dfrac{y_2 - y_1}{x_2 - x_1} = \dfrac{3-3}{0-1} = 0$

c. $f'(1) = 0$

d. $f'(1) = 0$

15. $f(x) = 4x^2 - 2x + 1$

Follow the step-by-step procedure of the text.

a. $f(x+h) - f(x) = [4(x+h)^2 - 2(x+h) + 1] - (4x^2 - 2x + 1) = 8xh + 4h^2 - 2h$

$\dfrac{f(x+h) - f(x)}{h} = \dfrac{8xh + 4h^2 - 2h}{h} = 8x + 4h - 2$

$f'(x) = \lim\limits_{h \to 0} \dfrac{f(x+h) - f(x)}{h} = \lim\limits_{h \to 0}(8x + 4h - 2) = 8x - 2$

b. $f'(x) = 8x - 2,\ f'(-3) = -26$

c. $f'(-3) = -26$

17. a. $p(q) = q^2 + 4q + 1$

$p(q+h) - p(q) = [(q+h)^2 + 4(q+h) + 1] - (q^2 + 4q + 1) = 2qh + h^2 + 4h$

$\dfrac{p(q+h) - p(q)}{h} = \dfrac{2qh + h^2 + 4h}{h} = 2q + h + 4$

$p'(q) = \lim\limits_{h \to 0}(2q + h + 4) = 2q + 4$

b. $p'(q) = 2q + 4,\ p'(5) = 14$

c. $p'(5) = 14$

19. a. $nDeriv(3x^4 - 7x - 5,\ x,\ 2) = 89.000024$

b. $\dfrac{f(2.0001) - f(2)}{0.0001} \approx \dfrac{29.0089 - 29}{0.0001} \approx 89.0072$

c. $f'(2) \approx 89$

21. a. $nDeriv((2x-1)^3,\ x,\ 4) = 294.000008$

b. $\dfrac{f(4.0001) - f(4)}{0.0001} \approx \dfrac{343.0294 - 343}{0.0001} \approx 294.0084$

c. $f'(4) \approx 294$

23. $f'(13) = \dfrac{f(12.99) - f(13)}{-0.01} = \dfrac{17.42 - 17.11}{-0.01} = -31$

25.

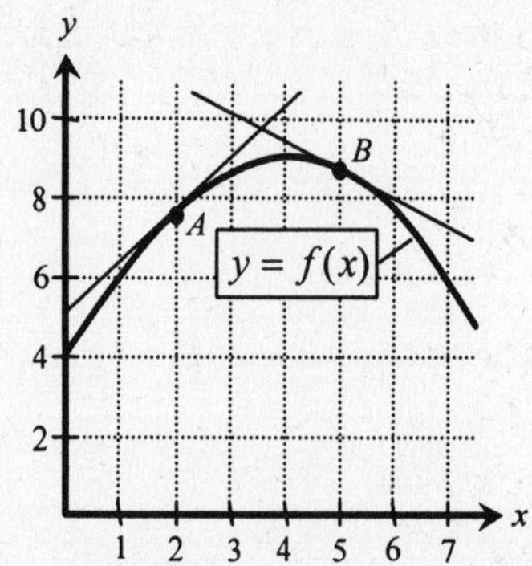

a. $f'(x)$ is greater at point A. The slope of the tangent line is positive and rate of change is greater.

b. Answers may vary. $f'(x)$ at $B \approx -\frac{2}{3}$.

27. (a, b) is on the graph of $f(x)$.
If $(a, b) = (-3, -9)$, then $f(a) = -9$.
(a, b) is on the graph of the tangent line at the point of tangency.
$f'(-3)$ is the slope of the tangent line. If

$5x - 2y = 3$, then $y = \frac{5}{2}x - \frac{3}{2}$ and $f'(-3) = \frac{5}{2}$.

29. $f'(1) = 3$

$y - y_1 = m(x - x_1)$

$y - (-1) = 3(x - 1)$

yields $y = 3x - 4$

31. **a.** $f'(x) > 0$ means $f(x)$ is increasing.
Answer: *a, b, d*

b. $f'(x) < 0$ means $f(x)$ is decreasing.
Answer: *c*

c. $f'(x) = 0$ at *A, C,* and *E*.

33. **a.** Function is continuous at *A, B, C,* and *D*.

b. Function is differentiable at *A* and *D*.
($f'(B)$ and $f'(C)$ are undefined.)

35. **a.** $f(x+h) - f(x)$

$= [(x+h)^2 + (x+h)] - (x^2 + x)$

$= 2xh + h^2 + h$

$$\frac{f(x+h) - f(x)}{h} = \frac{2xh + h^2 + h}{h} = 2x + h + 1$$

$$f'(x) = \lim_{h \to 0}(2x + h + 1) = 2x + 1$$

b. $f'(2) = 2(2) + 1 = 5$

c. $y - y_1 = m(x - x_1)$

$y - 6 = 5(x - 2)$ yields $y = 5x - 4$

d.

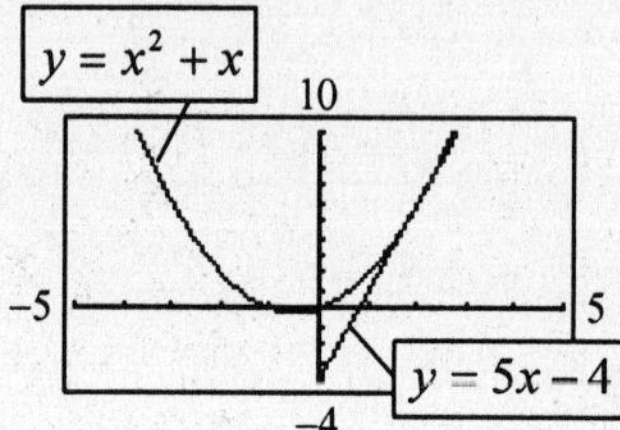

37. **a.** $f(x+h) - f(x)$

$= [(x+h)^3 + 3] - (x^3 + 3)$

$= 3x^2h + 3xh^2 + h^3$

$$\frac{f(x+h) - f(x)}{h} = \frac{3x^2h + 3xh^2 + h^3}{h}$$

$$= 3x^2 + 3xh + h^2$$

$$f'(x) = \lim_{h \to 0}(3x^2 + 3xh + h^2) = 3x^2$$

b. $f'(1) = 3(1)^2 = 3$

c. $y - y_1 = m(x - x_1)$

$y - 4 = 3(x - 1)$ yields $y = 3x + 1$

d.

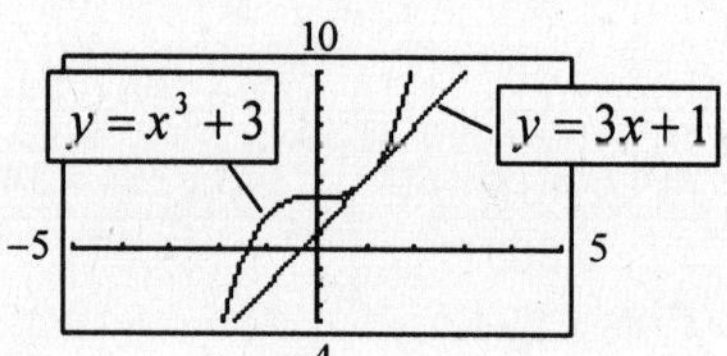

39. **a.** $\dfrac{f(300) - f(100)}{300 - 100} = \dfrac{14550 - 5950}{200} = \dfrac{8600}{200} = 43$

b. $\dfrac{f(600) - f(300)}{600 - 300} = \dfrac{43200 - 14550}{300} = \dfrac{28650}{300} = 95.50$

c. The average rate of change of total cost of production is greater when 300 to 600 printers are produced.

41. **a.** $\dfrac{D(25) - D(1)}{25 - 1} = \dfrac{\left(\frac{1000}{\sqrt{25}} - 1\right) - \left(\frac{1000}{\sqrt{1}} - 1\right)}{24} = \dfrac{199 - 999}{24} = -\dfrac{100}{3}$

b. $\dfrac{D(100) - D(25)}{100 - 25} = \dfrac{(100 - 1) - (199)}{75} = -\dfrac{4}{3}$

43. Smallest to largest: A to B, A to C, B to C. The slope of the line from A to B was less than the slope from A to C and the slope of the line from A to C was less than the slope of the line from B to C.

45. $R(x) = 300x - x^2$

a. $R'(x) = \lim_{h\to 0} \dfrac{[300(x+h)-(x+h)^2]-(300x-x^2)}{h} = 300-2x$

b. $R'(50) = 300 - 2(50) = 200$ A positive value for $R'(x)$ means that $R(x)$ is increasing.

c. $R'(200) = 300 - 2(200) = -100$ A negative value for $R'(x)$ means that $R(x)$ is decreasing.

d. $R'(150) = 300 - 2(150) = 0$ A zero value for $R'(x)$ means that $R(x)$ is stationary.

e. The revenue changes from increasing to decreasing at $x = 150$.

47. $Q(x) = 15{,}000 + 2x^2$

$Q'(x) = \lim_{h\to 0} \dfrac{[15{,}000+2(x+h)^2]-(15{,}000+2x^2)}{h} = \lim_{h\to 0}(4x+2h) = 4x$

$Q'(50) = 4(50) = 200$

49. $P(x) = 500x - x^2 - 100$

a. *nDeriv*$(500x - x^2 - 100,\ x,\ 200) = 100$ Profit is increasing.

b. *nDeriv*$(500x - x^2 - 100,\ x,\ 300) = -100$ Profit is decreasing.

51. $f(x) = 0.009x^2 + 0.139x + 91.875$

a. *nDeriv*$(0.009x^2 + 0.139x + 91.875,\ x,\ 50) = 1.039$

b. If humidity changes 1%, the heat index will change by about 1.039°F.

53. a. The $\overline{MR}$ is greater at 300. The slope of the tangent line to the curve is greater at 300 than it is at 700.

b. The sale of the 301st cell phone brings in more revenue since the rate of change in revenue is greater at 300 and it indicates the predicted increase in total revenue when one more unit is sold.

Exercise 9.4

1. $y = 4$

$y'(x) = 0$

Derivative of a constant is zero.

3. $y = x$

$\dfrac{dy}{dx} = 1 \cdot x^{1-1} = 1$

5. $f(x) = 2x^3 - x^5$

$f'(x) = 2 \cdot 3x^2 - 5 \cdot x^4$

$= 6x^2 - 5x^4$

7. $y = 6x^4 - 5x^2 + x - 2$

$\dfrac{dy}{dx} = 6 \cdot 4x^3 - 5 \cdot 2x + 1 \cdot x^0 - 0$

$= 24x^3 - 10x + 1$

9. $g(x) = 10x^9 - 5x^5 + 7x^3 + 5x - 6$

$g'(x) = 10 \cdot 9x^8 - 5 \cdot 5x^4 + 7 \cdot 3x^2 + 5x^0 - 0$

$= 90x^8 - 25x^4 + 21x^2 + 5$

11. $y = 4x^2 + 3x$

$y' = 8x + 3$

a. At $x = 2$ we have $y' = 8(2) + 3 = 19$.

b. At $x = 2$ we have $y' = 19$.

13. $P(x) = x^2 - 4x$

$P'(x) = 2x - 4$

a. $P'(2) = 2(2) - 4 = 0$

b. $P'(2) = 0$

15. $y = x^{-5} + x^{-8} - 3$

$y' = -5x^{-5-1} + (-8)x^{-8-1} - 0$

$= -5x^{-6} - 8x^{-9}$

17. $y = 3x^{11/3} - 2x^{7/4} - x^{1/2} + 8$

$y' = 3\left(\frac{11}{3}x^{8/3}\right) - 2\left(\frac{7}{4}x^{3/4}\right) - \frac{1}{2}x^{-1/2} + 0$

$= 11x^{8/3} - \frac{7}{2}x^{3/4} - \frac{1}{2}x^{-1/2}$

19. $f(x) = 5x^{-4/5} + 2x^{-4/3}$

$f'(x) = 5\left(-\frac{4}{5}x^{-9/5}\right) + 2\left(-\frac{4}{3}x^{-7/3}\right)$

$= -4x^{-9/5} - \frac{8}{3}x^{-7/3}$

21. $g(x) = 3x^{-4} + 2x^{-5} + 5x^{1/3}$

$g'(x) = 3\left(-4x^{-5}\right) + 2\left(-5x^{-6}\right) + 5\left(\frac{1}{3}x^{-2/3}\right)$

$= -12x^{-5} - 10x^{-6} + \frac{5}{3}x^{-2/3}$

23. $y = x^3 - 3x^2 + 5$ $y - y_1 = m(x - x_1)$

$y' = 3x^2 - 6x$ $y - 3 = -3(x - 1)$

$y = -3x + 6$

At $x = 1$, $y' = 3(1)^2 - 6(1) = -3$

At $x = 1$, $y = 1 - 3 + 5 = 3$

25. $f(x) = 4x^2 - x^{-1}$

$f'(x) = 8x + x^{-2} = 8x + \frac{1}{x^2}$

$f'\left(-\frac{1}{2}\right) = 8\left(-\frac{1}{2}\right) + \frac{1}{\left(-\frac{1}{2}\right)^2} = -4 + 4 = 0$

$f\left(-\frac{1}{2}\right) = 4\left(-\frac{1}{2}\right)^2 - \frac{1}{-\frac{1}{2}} = 1 + 2 = 3$

$y - y_1 = m(x - x_1)$

$y - 3 = 0\left(x - \frac{1}{2}\right)$

$y = 3$

27–29 A horizontal tangent means $f'(x) = 0$

27. $f(x) = -x^3 + 9x^2 - 15x + 6$

$f'(x) = -3x^2 + 18x - 15$

$= -3(x^2 - 6x + 5)$

$= -3(x - 5)(x - 1)$

$f'(x) = 0$ if $x = 1$ or 5

Answer: (1, –1), (5, 31)

29. $f(x) = x^4 - 4x^3 + 9$

$f'(x) = 4x^3 - 12x^2$

$= 4x^2(x - 3)$

$f'(x) = 0$ if $x = 0$ or 3

Answer: (0, 9), (3, –18)

31. $y = 5 - 2x^{1/2}$

a. $y' = -x^{-1/2} = \frac{-1}{\sqrt{x}}$ At $x = 4$, $y' = \frac{-1}{\sqrt{4}} = -\frac{1}{2}$

b. $nDeriv\left(5 - 2x^{1/2},\ x,\ 4\right) = -0.500000004$

33. $f(x) = 2x^3 + 5x - \pi^4 + 8$

a. $f'(x) = 6x^2 + 5$

b.

$y = 6x^2 + 5$

60

–3   3

0

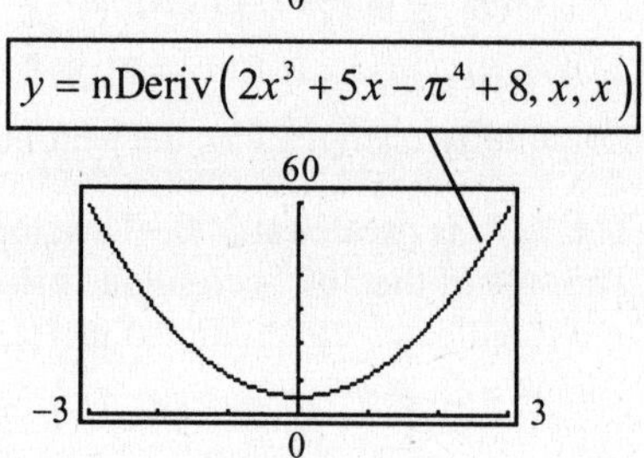

35. $h(x) = 10x^{-3} - 10x^{-2/5} + x^2 + 1$

a. $h'(x) = -30x^{-4} + 4x^{-7/5} + 2x$

$= \frac{-30}{x^4} + \frac{4}{\sqrt[5]{x^7}} + 2x$

b.

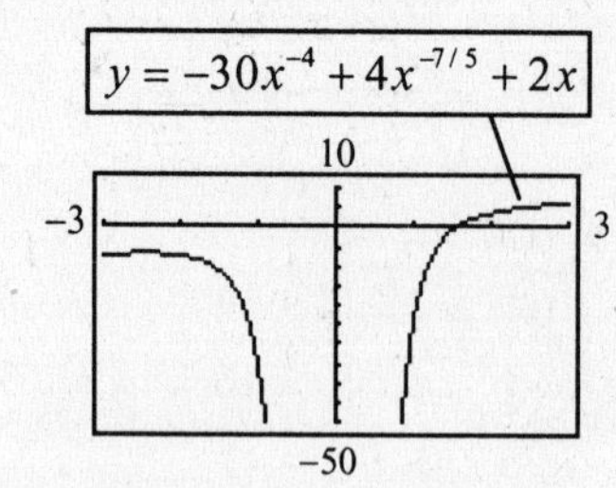

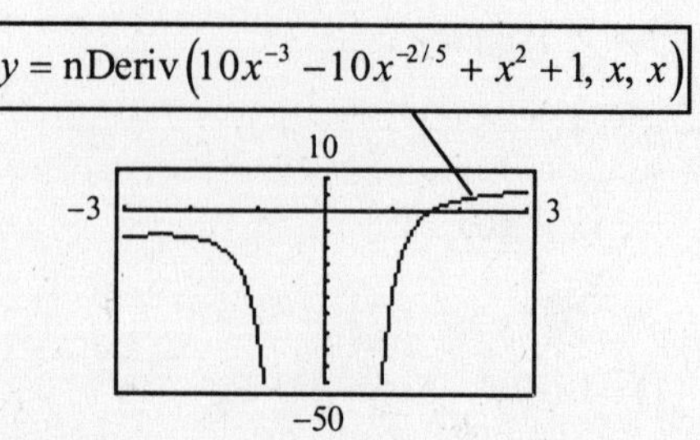

37. a. $f(x) = 3x^2 + 2x$ $\quad y - y_1 = m(x - x_1)$

$f'(x) = 6x + 2$ $\quad y - 5 = 8(x - 1)$

$f'(1) = 6 + 2 = 8$ $\quad y = 8x - 3$

b.

$y = 3x^2 + 2x$

15

$y = 8x - 3$

−3 3

−5

c. Answers may vary.
Starting with the window $[0, 2] \times [0, 10]$ and zooming in four times, we find that, in the window
$[0.99609375, 1.00390625] \times$
$[4.98046875, 5.01953125]$,
the function and tangent line cannot be distinguished.

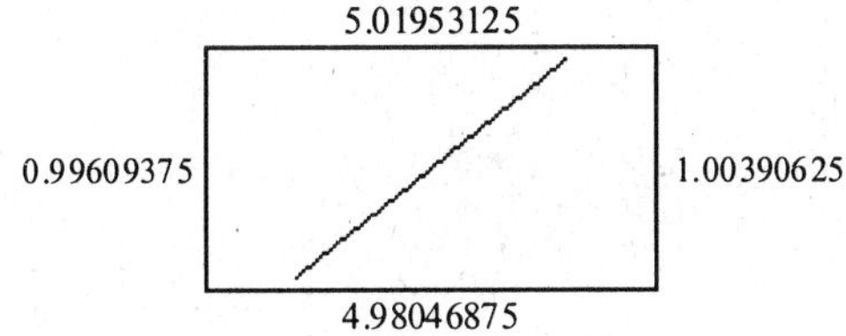

39. $f(x) = 8 - 2x - x^2$

a. $f'(x) = -2 - 2x$

b.

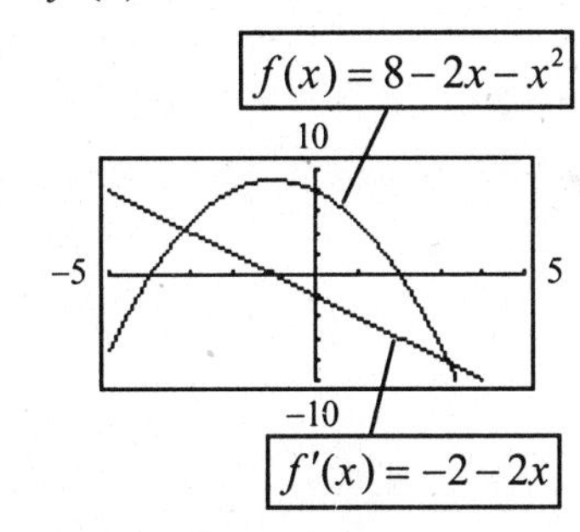

c. $f'(x) = 0$ at $x = -1$.
$f'(x) > 0$ if $x < -1$.
$f'(x) < 0$ if $x > -1$.

d. $f(x)$ has a max if $x = -1$.
$f(x)$ rises if $x < -1$.
$f(x)$ falls if $x > -1$.

41. $f(x) = x^3 - 12x - 5$

a. $f'(x) = 3x^2 - 12$

b.

$f'(x) = 3x^2 - 12$

25

−5 5

−25

$f(x) = x^3 - 12x - 5$

c. $f'(x) = 0$ if $x = \pm 2$.
$f'(x) > 0$ if $x < -2$ and if $x > 2$.
$f'(x) < 0$ if $-2 < x < 2$.

d. $f(x)$ has a max if $x = -2$.
$f(x)$ has a min if $x = 2$.
$f(x)$ rises if $x < -2$ and if $x > 2$.
$f(x)$ falls if $-2 < x < 2$.

43. $R(x) = 100x - 0.1x^2$

$R'(x) = 100 - 0.2x$

a. $R'(300) = 100 - 0.2(300) = 40$
The additional revenue from the 301st unit is about \$40.

b. $R'(600) = 100 - 0.2(600) = -20$
The sale of the 601st unit results in a loss of revenue of about \$20.

45. a. $Q'(60) = 200 + 12(60) = 920$

b. $Q(61) - Q(60)$

$= [200(61) + 6(61)^2] - [200(60) + 6(60)^2]$

$= 34{,}526 - 33{,}600 = 926$

47. $q = 1000p^{-1/2} - 1$

$$q'(p) = -500p^{-3/2} = \frac{-500}{\left(\sqrt{p}\right)^3}$$

a. $q'(25) = \dfrac{-500}{\left(\sqrt{25}\right)^3} = \dfrac{-500}{125} = -4$

If price increases by \$1, the demand will drop approximately 4 units.

b. $q'(100) = \dfrac{-500}{\left(\sqrt{100}\right)^3} = \dfrac{-500}{1000} = -\dfrac{1}{2}$

If price increases to \$101, the demand will drop approximately $\frac{1}{2}$ unit.

49. $\overline{C}(x)=\dfrac{4000}{x}+55+0.1x$

a. $\overline{C}'(x)=\dfrac{-4000}{x^2}+0.1$

b. $\dfrac{-4000}{x^2}+0.1=0$

$0.1x^2=4000$

$x^2=40,000$

$x=200$

c. $C'(x)=55+0.2x$. So $C'(200)=95$

$\overline{C}(200)=\frac{4000}{200}+55+0.1(200)=95$.

So $C'(200)=\overline{C}(200)$.

51. $C=\dfrac{120,000}{p}-1200$

$C'(p)=-\dfrac{120,000}{p^2}$

a. $C'(1)=-120,000$

b. If impurities increase 1%, the cost will decrease $120,000.

53. a. $WC=48.064+0.474(15)-0.020(15s)-1.85s+0.304\left(15\sqrt{s}\right)-27.74\sqrt{s}=55.17-2.15s-23.18\sqrt{s}$

b. $WC'(s)=-2.15-\dfrac{11.59}{\sqrt{s}}$

$WC'(25)=-2.15-\dfrac{11.59}{\sqrt{25}}=-4.468$

c. If the wind speed increases 1 mph, then the wind-chill will decrease approximately 4.468°.

55. a. $u(x)=0.00036999x^3-0.0861x^2+6.0827x-103.0527$

b. Average change $=\dfrac{13.5-16.1}{100-90}=-0.26\%$

c. $u'(x)=0.0011x^2-0.1722x+6.0827$

$u'(100)\approx-0.14\%$ per year.

57. a. $f(t)=0.0060315t^4-0.121931t^3+0.838258t^2-2.44124t+5.55315$

b. $f'(t)=0.024126t^3-0.365793t^2+1.67652t-2.44124$

c. $f'(1)\approx-1.11$

$f'(10)\approx1.90$

d. The -1.11 means that from '90 to '91 the CPI fell about 1.11%.
The 1.90 means that from '00 to '01 the CPI rose about 1.90%

Exercise 9.5

1. $y=(x+3)(x^2-2x) \quad u'=1 \quad v'=2x-2$

$y'=(x+3)(2x-2)+(x^2-2x)\cdot 1=2x^2+4x-6+x^2-2x=3x^2+2x-6$

3. $p=(3q-1)(q^2+2) \quad u'=3 \quad v'=2q$

$\frac{dp}{dq}=(3q-1)(2q)+(q^2+2)\cdot 3=9q^2-2q+6$

5. $f(x)=(x^{12}+3x^4+4)(4x^3-1)$

$f'(x)=(x^{12}+3x^4+4)(12x^2)+(4x^3-1)(12x^{11}+12x^3)$

$=12x^{14}+36x^6+48x^2+48x^{14}+48x^6-12x^{11}-12x^3$

$=60x^{14}-12x^{11}+84x^6-12x^3+48x^2$

7. $y=(7x^6-5x^4+2x^2-1)(4x^9+3x^7-5x^2+3x)$

$\frac{dy}{dx}=(7x^6-5x^4+2x^2-1)(36x^8+21x^6-10x+3)+(4x^9+3x^7-5x^2+3x)(42x^5-20x^3+4x)$

9. $y=(x^2+x+1)(x^{1/3}-2x^{1/2}+5)$

$\frac{dy}{dx}=(x^2+x+1)\left(\frac{1}{3}x^{-2/3}-x^{-1/2}\right)+(x^{1/3}-2x^{1/2}+5)(2x+1)$

11. $f(x)=(x^2+1)(x^3-4x)$

$f'(x)=(x^2+1)(3x^2-4)+(x^3-4x)(2x)=5x^4-9x^2-4$

$f'(-2)=5(-2)^4-9(-2)^2-4=40$

a. Slope of tangent line at (–2, 0) is $f'(-2)=40$.

b. Instantaneous rate of change at (–2, 0) is $f'(-2)=40$.

13. $y=\frac{x}{x^2-1},\ u'=1,\ v'=2x \qquad y'=\frac{(x^2-1)(1)-x(2x)}{(x^2-1)^2}=\frac{-x^2-1}{(x^2-1)^2}$

15. $p=\frac{q^2+1}{q-2},\ u'=2q,\ v'=1 \qquad \frac{dp}{dq}=\frac{(q-2)(2q)-(q^2+1)(1)}{(q-2)^2}=\frac{q^2-4q-1}{(q-2)^2}$

17. $y=\frac{1-2x^2}{x^4-2x^2+5} \qquad \frac{dy}{dx}=\frac{(x^4-2x^2+5)(-4x)-(1-2x^2)(4x^3-4x)}{(x^4-2x^2+5)^2}=\frac{4x^5-4x^3-16x}{(x^4-2x^2+5)^2}$

19. $z=x^2+\frac{x^2}{1-x-2x^2} \qquad \frac{dz}{dx}=2x+\frac{(1-x-2x^2)(2x)-x^2(-1-4x)}{(1-x-2x^2)^2}=2x+\frac{-x^2+2x}{(1-x-2x^2)^2}$

21. $p=\frac{3\sqrt[3]{q}}{1-q},\ u'=\frac{1}{q^{2/3}},\ v'=-1 \qquad \frac{dp}{dq}=\frac{(1-q)\left(\frac{1}{q^{2/3}}\right)-3q^{1/3}(-1)}{(1-q)^2}=\frac{(1-q)-3q(-1)}{q^{2/3}(1-q)^2}=\frac{1+2q}{q^{2/3}(1-q)^2}$

23. $y=\frac{x(x^2+4)}{x-2}=\frac{x^3+4x}{x-2} \qquad y'=\frac{(x-2)(3x^2+4)-(x^3+4x)(1)}{(x-2)^2}=\frac{2x^3-6x^2-8}{(x-2)^2}$

25. $f(x)=\dfrac{x^2+1}{x+3}$ $\quad f'(x)=\dfrac{(x+3)(2x)-(x^2+1)(1)}{(x+3)^2}=\dfrac{x^2+6x-1}{(x+3)^2}$ $\quad f'(2)=\dfrac{4+12-1}{25}=\dfrac{15}{25}=\dfrac{3}{5}$

a. The slope of the tangent line at (2, 1) is $\dfrac{3}{5}$.

b. The instantaneous rate of change is also $\dfrac{3}{5}$ at this point.

27. $y=(9x^2-6x+1)(1+2x)$ $\quad \dfrac{dy}{dx}=(9x^2-6x+1)(2)+(1+2x)(18x-6)$

At $x=1$, $\dfrac{dy}{dx}=4(2)+3(12)=44$. At $x=1$, $y=4(3)=12$.

$y-y_1=m(x-x_1)$
$y-12=44(x-1)$
$y=44x-32$

29. $f(x)=\dfrac{3x^4-2x-1}{4-x^2}$ $\quad f'(x)=\dfrac{(4-x^2)(12x^3-2)-(3x^4-2x-1)(-2x)}{(4-x^2)^2}=\dfrac{-6x^5+48x^3-2x^2-2x-8}{(4-x^2)^2}$

At $x=1$ we have $f(x)=\dfrac{3-2-1}{3}=0$. At $x=1$ we have $f'(x)=\dfrac{-6+48-2-2-8}{9}=\dfrac{30}{9}=\dfrac{10}{3}$.

$y-y_1=m(x-x_1) \;\rightarrow\; y-0=\dfrac{10}{3}(x-1) \;\rightarrow\; y=\dfrac{10}{3}x-\dfrac{10}{3}$

31. $nDeriv\left(\left(4\sqrt{x}+3x^{-1}\right)\cdot\left(3x^{1/3}-5x^{-2}-25\right),\ x,\ 1\right)=104.0002584$

33. $nDeriv\left(\left(4x-4\right)/\left(3x^{2/3}\right),\ x,\ 1\right)=1.333334074$

35. $f(x)=(x^2+4x+4)(x-7)$

a. $f'(x)=(x^2+4x+4)(1)+(x-7)(2x+4)$
$=3x^2-6x-24$
$=3(x^2-2x-8)$
$=3(x-4)(x+2)$

b.

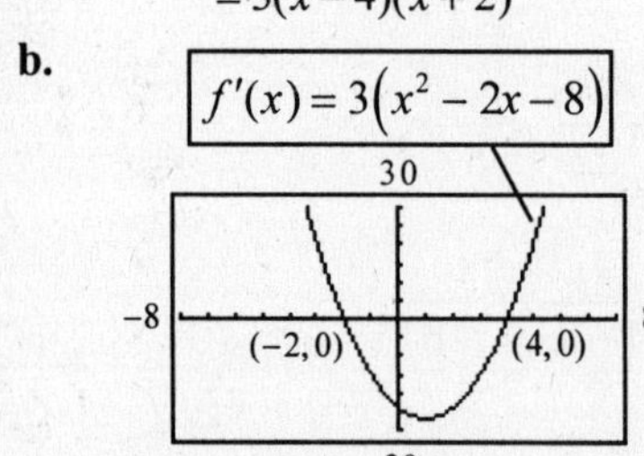

The slope of the tangent is 0 at $x=-2$ and at $x=4$.

c.

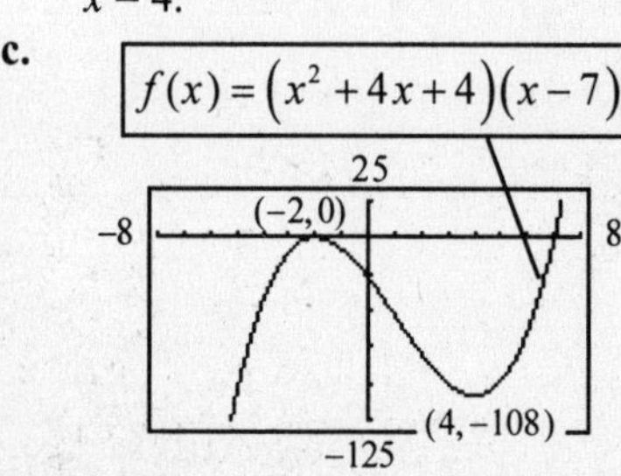

37. $y=\dfrac{x^2}{x-2}$

a. $\dfrac{dy}{dx}=\dfrac{(x-2)(2x)-x^2(1)}{(x-2)^2}$
$=\dfrac{x^2-4x}{(x-2)^2}=\dfrac{x(x-4)}{(x-2)^2}$

b.

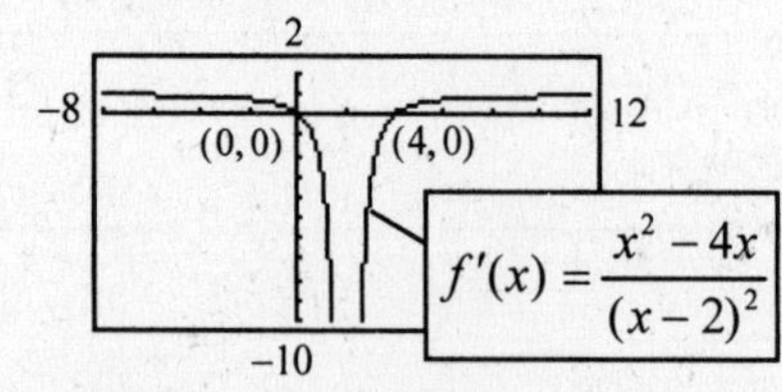

The slope of the tangent is 0 at $x=0$ and at $x=4$.

c.

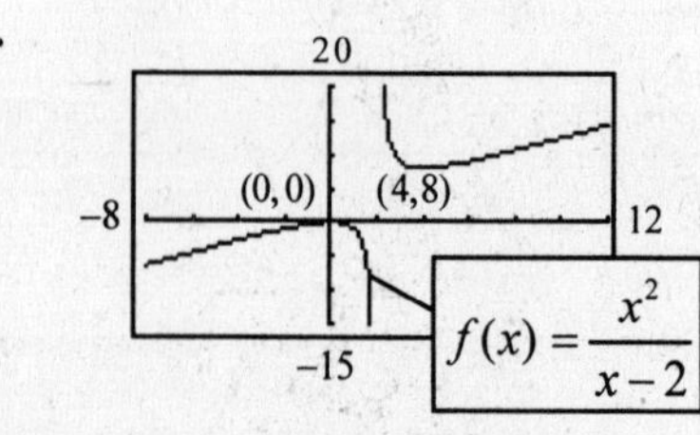

39. $f(x)=\dfrac{10x^2}{x^2+1}$

a. $f'(x)=\dfrac{(x^2+1)(20x)-10x^2(2x)}{(x^2+1)^2}=\dfrac{20x}{(x^2+1)^2}$

b.

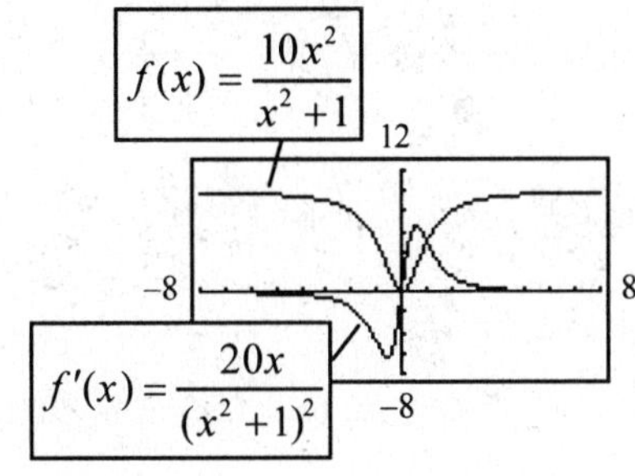

c. $f'(x)=0$ if $x=0$. $f'(x)>0$ if $x>0$. $f'(x)<0$ if $x<0$.
Read graph of $f'(x)$.

d. f has a min at $x=0$. f is increasing for $x>0$. f is decreasing for $x<0$.

41. $f'(x)=\lim\limits_{h\to 0}\dfrac{\frac{u(x+h)}{v(x+h)}-\frac{u(x)}{v(x)}}{h}=\lim\limits_{h\to 0}\dfrac{u(x+h)v(x)-u(x)v(x+h)}{h\cdot v(x)v(x+h)}$

$=\lim\limits_{h\to 0}\dfrac{u(x+h)v(x)-u(x)v(x)+u(x)v(x)-u(x)v(x+h)}{h\cdot v(x)v(x+h)}$

$=\lim\limits_{h\to 0}\dfrac{v(x)\left[\frac{u(x+h)-u(x)}{h}\right]-u(x)\left[\frac{v(x+h)-v(x)}{h}\right]}{v(x)v(x+h)}=\dfrac{v(x)u'(x)-u(x)v'(x)}{[v(x)]^2}$

43. $C(p)=\dfrac{8100p}{100-p}$ $\quad C'(p)=\dfrac{(100-p)(8100)-8100p(-1)}{(100-p)^2}=\dfrac{810{,}000}{(100-p)^2}$

45. $R(x)=\dfrac{60x^2+74x}{2x+2}$

$R'(x)=\dfrac{(2x+2)(120x+74)-(60x^2+74x)2}{(2x+2)^2}=\dfrac{120x^2+240x+148}{(2x+2)^2}$ $\quad R'(49)=\dfrac{300{,}028}{(100)^2}\approx \$30$

The revenue from the sale of the next unit is approximately \$30.

47. $R(x)=(25+x)(300-10x)=7500+50x-10x^2$ $\quad R'(x)=50-20x$ $\quad R'(5)=-50$
The revenue will decrease approximately \$50 if the group adds one person.

49. $R(x)=500x^2-\dfrac{1}{3}x^3$ $\quad R'(x)=1000x-x^2$

51. $R(n)=\dfrac{nr}{1+nr-r}$, r is a constant. $\quad R'(n)=\dfrac{(1+nr-r)r-nr(r)}{(1+nr-r)^2}=\dfrac{r(1-r)}{[1+(n-1)r]^2}$

53. $P(t)=\dfrac{13t}{t^2+100}+0.18$ $\quad P'(t)=\dfrac{(t^2+100)13-13t(2t)}{(t^2+100)^2}=\dfrac{-13t^2+1300}{(t^2+100)^2}$

a. $P'(6)\approx 0.045$ In the next month the recognition will increase about 4.5%.

b. $P'(12)\approx -0.010$ In the next month the recognition will decrease about 1%.

c. Positive means increasing recognition.

55. $f(x)=\dfrac{289.173-58.5731x}{x+1}$

$f'(x)=\dfrac{(x+1)(-58.5731)-(289.173-58.5731x)(1)}{(x+1)^2}=\dfrac{-347.7461}{(x+1)^2}$

a. $f'(20)\approx -0.79$

b. At 0°F, if the windspeed increases 1 mph, the wind-chill will decrease about 0.79°F.

57. **a.** $B'(t)=\left(0.01t+3\right)\left(0.04766t-9.79\right)+\left(0.01\right)\left(0.02383t^2-9.79t+3097.19\right)$

$=0.0004766t^2+0.04508t-29.37+0.0002383t^2-0.0979t+30.9719$

$=0.0007149t^2-0.05282t+1.6019$

b. Instantaneous rate of change in 2010 $=B'(60)=1.006$

This indicates the number of beneficiaries (in millions) is predicted to increase by 1.006 from 2010 to 2011.

c. From 2000 to 2010 $=\dfrac{53.3-44.8}{10}=0.85$ From 2010 to 2020 $=\dfrac{68.8-53.3}{10}=1.55$

From 2000 to 2020 $=\dfrac{68.8-44.8}{20}=1.2$

The average rate of change from 2000 to 2020 best approximates the instantaneous rate of change in 2010.

d. The instantaneous rate of change appears to be the smallest around 1987. The graph of the derivative is the smallest when t is around 37.

59. $f(t)=1000\cdot\dfrac{-8.0912t+1558.9}{1.09816t^2-122.183t+21,472.6}$

a. $f'(t)=1000\left[\dfrac{(1.09816t^2-122.183t+21,472.6)(-8.0912)-(-8.0912t+1558.9)(2.19632t-122.183)}{(1.09816t^2-122.183t+21,472.6)^2}\right]$

$=1000\left[\dfrac{8.885432192t^2-3423.843248t+16,731.97758}{(1.09816t^2-122.183t+21,472.6)^2}\right]$

b. $f'(70)\approx -0.5356\%$ $f'(170)\approx -0.2932\%$

c. $f'(70)$ means that from 1870 to 1871, the model predicts a change of about –0.5356% in U.S. workers in farm occupations. $f'(170)$ means that from 1970 to 1971, the model predicts a change of about –0.2932% in U.S. workers in farm occupations.

Exercise 9.6

Unless otherwise specified the Power Rule will be used in calculating the derivatives.

$y=u^n\quad \dfrac{dy}{dx}=nu^{n-1}\cdot\dfrac{du}{dx}$

1. $y=u^3$ and $u=x^2+1$

$\dfrac{dy}{du}=3u^2;\quad \dfrac{du}{dx}=2x;$

$\dfrac{dy}{dx}=\dfrac{dy}{du}\cdot\dfrac{du}{dx}=3u^2\cdot 2x=6x\left(x^2+1\right)^2$

3. $y=u^4$ and $u=4x^2-x+8$

$\dfrac{dy}{du}=4x^3;\quad \dfrac{du}{dx}=8x-1;$

$\dfrac{dy}{dx}=\dfrac{dy}{du}\cdot\dfrac{du}{dx}=4u^3(8x-1)$

$=4(8x-1)(4x^2-x+8)^3$

5. $f'(x)=25\left(2x^4-5\right)^{24}\left(8x^3\right)$

$=200x^3\left(2x^4-5\right)^{24}$

7. $h'(x) = \frac{2}{3} \cdot 8\left(x^6 + 3x^2 - 11\right)^7 \left(6x^5 + 6x\right)$

$= \frac{16\left(6x^5 + 6x\right)\left(x^6 + 3x^2 - 11\right)^7}{3}$

$= \frac{96x\left(x^4 + 1\right)\left(x^6 + 3x^2 - 11\right)^7}{3}$

$= 32x\left(x^4 + 1\right)\left(x^6 + 3x^2 - 11\right)^7$

9. $g(x) = (x^2 + 4x)^{-2}$

$g'(x) = -2(x^2 + 4x)^{-3}(2x + 4)$

$= -4(x + 2)(x^2 + 4x)^{-3}$

11. $f(x) = \frac{1}{(x^2 + 2)^3} = (x^2 + 2)^{-3}$

Preferred: $f'(x) = -3(x^2 + 2)^{-4}(2x) = \frac{-6x}{(x^2 + 2)^4}$

Acceptable:

$f'(x) = \frac{(x^2 + 2)^3(0) - 1 \cdot 3(x^2 + 2)^2(2x)}{(x^2 + 2)^6}$

$= \frac{-6x(x^2 + 2)^2}{(x^2 + 2)^6} = \frac{-6x}{(x^2 + 2)^4}$

13. $g(x) = \left(2x^3 + 3x + 5\right)^{-3/4}$

$g'(x) = -\frac{3}{4}\left(2x^3 + 3x + 5\right)^{-7/4}\left(6x^2 + 3\right)$

$= \frac{-18x^2 - 9}{4(2x^3 + 3x + 5)^{7/4}}$

15. $y = \left(x^2 + 4x + 5\right)^{1/2}$

$y' = \frac{1}{2}\left(x^2 + 4x + 5\right)^{-1/2}(2x + 4) = \frac{x + 2}{\sqrt{x^2 + 4x + 5}}$

17. $y = \frac{8}{5}\left(x^2 - 3\right)^5$

$y' = \frac{8}{5} \cdot 5\left(x^2 - 3\right)^4(2x) = 16x\left(x^2 - 3\right)^4$

19. $y = \frac{1}{7}\left[(3x + 1)^5 - 3x\right]$

$y' = \frac{1}{7}\left[5(3x + 1)^4(3) - 3\right] = \frac{3}{7}\left[5(3x + 1)^4 - 1\right]$

21. $y = f(x) = \left(x^3 + 2x\right)^4$

$f'(x) = 4\left(x^3 + 2x\right)^3\left(3x^2 + 2\right)$

$f'(2) = 4(8 + 4)^3(12 + 2) = 96,768$

a. The slope of the tangent line at $x = 2$ is 96,768.

b. The instantaneous rate of change of the function at $x = 2$ is also 96,768.

$nDeriv\left(\left(x^3 + 2x\right)^4, x, 2\right) = 96,768.28378$

23. $f(x) = \left(x^3 + 1\right)^{1/2}$

$f'(x) = \frac{1}{2}\left(x^3 + 1\right)^{-1/2}\left(3x^2\right) = \frac{3x^2}{2\sqrt{x^3 + 1}}$

$f'(2) = \frac{12}{2\sqrt{8 + 1}} = \frac{6}{\sqrt{9}} = 2$

a. The slope of the tangent line at (2, 3) is 2.

b. The instantaneous rate of change of the function at (2, 3) is also 2.

$nDeriv\left(\left(x^3 + 1\right)^{1/2}, x, 2\right) = 2$

25. $f(x) = \left(x^2 - 3x + 3\right)^3$

$f'(x) = 3\left(x^2 - 3x + 3\right)^2(2x - 3)$

$f'(1) = 3(1)(-1) = -3$

$y - y_1 = m(x - x_1)$

$y - 1 = -3(x - 1)$

$y = -3x + 4$

27. $f(x) = \left(3x^2 - 2\right)^{1/2}$

$f'(x) = \frac{1}{2}\left(3x^2 - 2\right)^{-1/2}(6x) = \frac{3x}{\sqrt{3x^2 - 2}}$

$f(3) = \sqrt{27 - 2} = 5$

$f'(3) = \frac{9}{\sqrt{27 - 2}} = \frac{9}{5}$

$y - y_1 = m(x - x_1)$

$y - 5 = \frac{9}{5}(x - 3)$

$9x - 5y = 2$

29. $f(x) = (x^2 - 4)^3 + 12$

a. $f'(x) = 3(x^2 - 4)^2 (2x) = 6x(x-2)^2 (x+2)^2$

b.

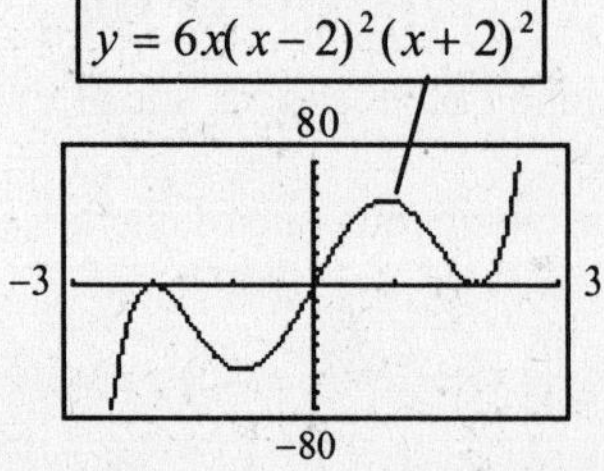

$y = \text{nDeriv}((x^2 - 4)^3 + 12, x, x)$

80

−3 3

−80

c. $f'(x) = 0$ at $x = 0, 2, -2$

d. At $\begin{cases} x = & 0 & 2 & -2 \\ y = & -52 & 12 & 12 \end{cases}$

Points: (0, −52), (2, 12), (−2, 12)

e.

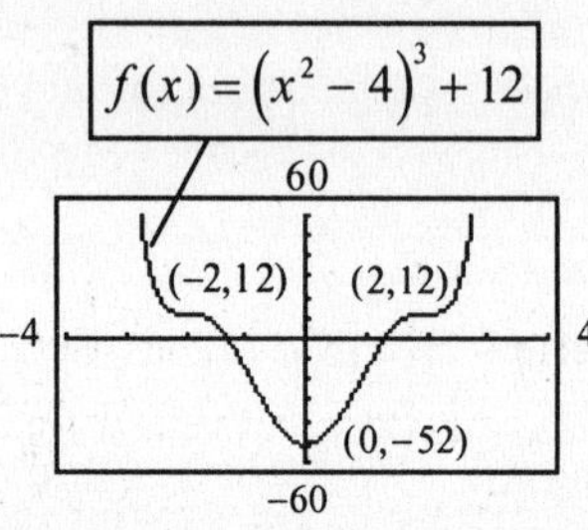

31. $f(x) = 5 - 3(1 - x^2)^{4/3}$

a. $f'(x) = -4(1-x^2)^{1/3}(-2x) = 8x(1-x^2)^{1/3}$

b.

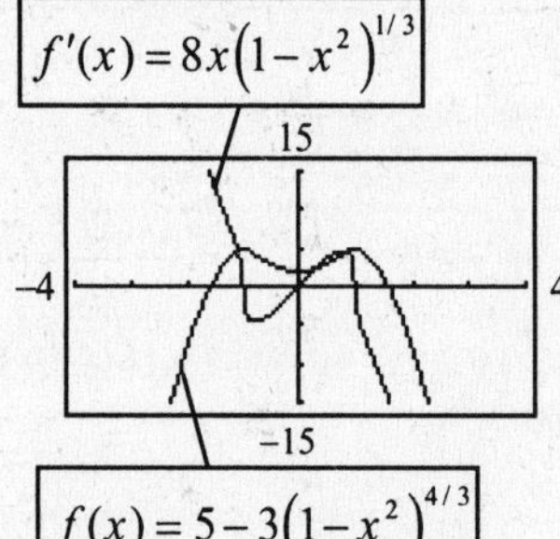

c.

x		−1		0		1	
$f'(x)$	+	0	−	0	+	0	−

d. max at $x = -1$ and $x = 1$. min at $x = 0$.
$f(x)$ increasing for $x < -1$ and $0 < x < 1$.
$f(x)$ decreasing for $x > 1$ and $-1 < x < 0$.

33. a. $y = \frac{2}{3}x^3 \quad \frac{dy}{dx} = \frac{2}{3} \cdot 3x^2 = 2x^2$

b. $y = \frac{2}{3}x^{-3} \quad \frac{dy}{dx} = \frac{2}{3}(-3x^{-4}) = -2x^{-4} = -\frac{2}{x^4}$

c. $y = \frac{1}{3}(2x)^3 \quad \frac{dy}{dx} = \frac{1}{3} \cdot 3(2x)^2 \cdot 2 = 2(2x)^2$

d. $y = 2(3x)^{-3}$

$$\frac{dy}{dx} = 2(-3)(3x)^{-4}(3) = -\frac{18}{(3x)^4}$$

35. $s = 27 - (3 - 10t)^3$

$s'(t) = 0 - 3(3-10t)^2(0 - 10) = 30(3-10t)^2$

$s'\left(\frac{1}{10}\right) = 30(3-1)^2 = 120 \text{ in/sec} = 10 \text{ ft/sec}$

37. $R = 1500x + 3000(2x+3)^{1} - 1000$

$R'(x) = 1500 - 3000(2x+3)^{-2}(2)$

$= 1500 - \frac{6000}{(2x+3)^2}$

$R'(100) = 1500 - \frac{6000}{203^2} = \$1499.85$

So if the sales go from 100 units sold to 101 units sold, the revenue will increase by about \$1499.85.

39. $y = 32(3p+1)^{-2/5}$

$y'(p) = 32\left(-\frac{2}{5}\right)(3p+1)^{-7/5}(3)$

$= \frac{-192}{5(3p+1)^{7/5}}$

a. $y'(21) = \frac{-192}{5(64)^{7/5}} \approx \frac{-192}{5(337.79)} \approx -0.114$

b. If the price increases \$1, the sales volume will decrease by 114 units.

41. $p = 200{,}000(q+1)^{-2}$

$p'(q) = -400{,}000(q+1)^{-3}(1)$

b. $p'(49) = \frac{-400{,}000}{50^3} = -\$3.20$

c. If the quantity demanded increases one unit, the price will decrease about \$3.20.

43. $y = k(x - x_0)^{8/5}$ k and x_0 are constants.

$$\frac{dy}{dx} = k \cdot \frac{8}{5}(x - x_0)^{3/5} = \frac{8k}{5}(x - x_0)^{3/5}$$

45. $p = 100(2q+1)^{-1/2}$

$p'(q) = -50(2q+1)^{-3/2}(2)$

$= \dfrac{-100}{(2q+1)^{3/2}}$

47. $K_c = 4\sqrt{4v+1}$

$K_c' = 4\cdot\dfrac{1}{2}(4v+1)^{-1/2}\cdot 4 = \dfrac{8}{\sqrt{4v+1}}$

49. $S = 1000\left[1+\dfrac{0.01r}{12}\right]^{240}$

$S'(r) = 240{,}000\left[1+\dfrac{0.01r}{12}\right]^{239}\left(\dfrac{0.01}{12}\right) = 200\left[1+\dfrac{0.01r}{12}\right]^{239}$

a. $S'(6) = 200[1+0.005]^{239} = 200(3.2937) = \$658.75$

b. $S'(12) = 200[1+0.01]^{239} = 200(10.7847) = \$2156.94$

In each case the future value of the investment would have an increase of the above values.

51. a. $d'(t) = -0.17189(4)(0.1t+3)^3(0.1) + 4.698(3)(0.1t+3)^2(0.1) - 45.791(2)(0.1t+3)(0.1) + 189.049(0.1)$

$= -0.068756(0.1t+3)^3 + 1.4094(0.1t+3)^2 - 9.1582(0.1t+3) + 18.9049$

$d'(50) = 0.6378$ means from 1980 to 1981 the model predicts the interest paid as a percent of federal expenditures would increase by 0.6378%.

$d'(70) = -0.4931$ means from 2000 to 2001 the model predicts the interest paid as a percent of federal expenditures would decrease by 0.4931%.

b. From 1975 to 1985 average interest paid is $\dfrac{18.9-9.8}{10} = 0.91$

53. a. $p'(t) = -0.002211(3)(t+60)^2 + 0.5503(2)(t+60) - 44.93$

$= -0.006633(t+60)^2 + 1.1006(t+60) - 44.93$

$p'(10) = -0.3897$ means that during the next year the number of persons below the poverty level would decrease by 389,700 (0.3897 million). $p'(38) = -0.7745$ means that during the next year the number of persons below the poverty level would decrease by 774,500 (0.7745 million).

b. From 1965 to 1975 the average is -0.73 million per year. From 1997 to 1999 the average is -1.65 million per year.

Exercise 9.7

1. $f(x) = \pi^4 \quad f'(x) = 0$

Derivative of a constant.

3. $g(x) = 4x^{-4} \quad g'(x) = 4(-4)x^{-5} = \dfrac{-16}{x^5}$

5. $g(x) = 5x^3 + 4x^{-1}$

$g'(x) = 15x^2 - 4x^{-2} = 15x^2 - \dfrac{4}{x^2}$

7. $y = (x^2-2)(x+4)$

$y' = (x^2-2)(1) + (x+4)(2x) = 3x^2 + 8x - 2$

9. $f(x) = \dfrac{x^3+1}{x^2} = \dfrac{x^3}{x^2} + \dfrac{1}{x^2} = x + x^{-2}$

Preferred: $f'(x) = 1 - 2x^{-3} = 1 - \dfrac{2}{x^3} = \dfrac{x^3-2}{x^3}$

Acceptable: $f'(x) = \dfrac{x^2(3x^2) - (x^3+1)(2x)}{x^4}$

$= \dfrac{x^4-2x}{x^4} = \dfrac{x^3-2}{x^3}$

11. $y = \frac{1}{10}\left(x^3 - 4x\right)^{10} \qquad y' = \frac{1}{10} \cdot 10\left(x^3 - 4x\right)^9\left(3x^2 - 4\right) = \left(3x^2 - 4\right)\left(x^3 - 4x\right)^9$

13. $y = \frac{5}{3}x^3\left(4x^5 - 5\right)^3$

$$y' = \frac{5}{3}x^3 \cdot 3\left(4x^5 - 5\right)^2\left(20x^4\right) + \left(4x^5 - 5\right)^3 \cdot \frac{5}{3} \cdot 3x^2 = \left(4x^5 - 5\right)^2\left[\left(5x^3\right)\left(20x^4\right) + 5x^2\left(4x^5 - 5\right)\right]$$
$$= 5x^2\left(4x^5 - 5\right)^2\left[20x^5 + \left(4x^5 - 5\right)\right] = 5x^2\left(4x^5 - 5\right)^2\left(24x^5 - 5\right)$$

15. $f(x) = \left(x - 1\right)^2\left(x^2 + 1\right)$

$$f'(x) = \left(x - 1\right)^2(2x) + \left(x^2 + 1\right)(2)\left(x - 1\right)(1) = (x - 1)\left[\left(x - 1\right)2x + 2\left(x^2 + 1\right)\right]$$
$$= (x - 1)\left(4x^2 - 2x + 2\right) = 2(x - 1)\left(2x^2 - x + 1\right)$$

17. $y = \frac{\left(x^2 - 4\right)^3}{x^2 + 1}$

$$y' = \frac{\left(x^2 + 1\right) \cdot 3\left(x^2 - 4\right)^2(2x) - \left(x^2 - 4\right)^3(2x)}{\left(x^2 + 1\right)^2} = \frac{2x\left(x^2 - 4\right)^2\left[3\left(x^2 + 1\right) - \left(x^2 - 4\right)\right]}{\left(x^2 + 1\right)^2} = \frac{2x\left(x^2 - 4\right)^2\left(2x^2 + 7\right)}{\left(x^2 + 1\right)^2}$$

19. $p = \left(q + 1\right)^3\left(q^3 - 3\right)^3$

$$p' = \left(q + 1\right)^3 \cdot 3\left(q^3 - 3\right)^2\left(3q^2\right) + \left(q^3 - 3\right)^3 \cdot 3\left(q + 1\right)^2(1) = 3\left(q^3 - 3\right)^2\left[\left(q + 1\right)^3\left(3q^2\right) + \left(q^3 - 3\right)\left(q + 1\right)^2\right]$$
$$= 3\left(q^3 - 3\right)^2\left(q + 1\right)^2\left[\left(q + 1\right)\left(3q^2\right) + \left(q^3 - 3\right)\right] = 3\left(q^3 - 3\right)^2\left(q + 1\right)^2\left(4q^3 + 3q^2 - 3\right)$$

21. $R(x) = x^8(x^2 + 3x)^4$

$$R'(x) = x^8 \cdot 4\left(x^2 + 3x\right)^3(2x + 3) + \left(x^2 + 3x\right)^4 \cdot 8x^7 = 4\left(x^2 + 3x\right)^3\left[x^8(2x + 3) + \left(x^2 + 3x\right)2x^7\right]$$
$$= 4x^7\left(x^2 + 3x\right)^3\left[x\left(2x + 3\right) + 2\left(x^2 + 3x\right)\right] = 4x^7\left(x^2 + 3x\right)^3\left[4x^2 + 9x\right] = 4x^8\left(x^2 + 3x\right)^3(4x + 9)$$

23. $y = \frac{(2x - 1)^4}{\left(x^2 + x\right)^4}$

$$y' = \frac{\left(x^2 + x\right)^4 \cdot 4(2x - 1)^3(2) - (2x - 1)^4 \cdot 4\left(x^2 + x\right)^3(2x + 1)}{\left(x^2 + x\right)^8}$$
$$= \frac{4\left(x^2 + x\right)^3\left[\left(x^2 + x\right)(2x - 1)^3(2) - (2x - 1)^4(2x + 1)\right]}{\left(x^2 + x\right)^8} = \frac{4(2x - 1)^3\left(-2x^2 + 2x + 1\right)}{\left(x^2 + x\right)^5}$$

25. $g(x) = (8x^4 + 3)^2(x^3 - 4x)^3$

$$g'(x) = (8x^4 + 3)^2 \cdot 3(x^3 - 4x)^2(3x^2 - 4) + (x^3 - 4x)^3 \cdot 2(8x^4 + 3)(32x^3)$$
$$= (x^3 - 4x)^2[3(8x^4 + 3)^2(3x^2 - 4) + 2(x^3 - 4x)(8x^4 + 3)(32x^3)]$$
$$= (x^3 - 4x)^2(8x^4 + 3)[3(8x^4 + 3)(3x^2 - 4) + 64x^3(x^3 - 4x)]$$
$$= (x^3 - 4x)^2(8x^4 + 3)(136x^6 - 352x^4 + 27x^2 - 36)$$

27. $f(x)=\dfrac{(x^2+5)^{1/3}}{4-x^2}$

$$f'(x)=\frac{(4-x^2)\cdot\frac{1}{3}(x^2+5)^{-2/3}(2x)-(x^2+5)^{1/3}(-2x)}{(4-x^2)^2}$$

$$=\frac{2[x(4-x^2)-(x^2+5)(-3x)]}{3(x^2+5)^{2/3}(4-x^2)^2}=\frac{2(2x^3+19x)}{3(x^2+5)^{2/3}(4-x^2)^2}=\frac{2x(2x^2+19)}{3(x^2+5)^{2/3}(4-x^2)^2}$$

29. $y=x^2(4x-3)^{1/4}$

$$y'=x^2\cdot\frac{1}{4}(4x-3)^{-3/4}(4)+(4x-3)^{1/4}(2x)=\frac{x^2}{(4x-3)^{3/4}}+\frac{2x(4x-3)^{1/4}}{1}=\frac{x^2+2x(4x-3)}{(4x-3)^{3/4}}=\frac{9x^2-6x}{(4x-3)^{3/4}}$$

31. $c(x)=2x(x^3+1)^{1/2}$

$$c'(x)=(2x)\cdot\frac{1}{2}(x^3+1)^{-1/2}(3x^2)+(x^3+1)^{1/2}\cdot(2)=\frac{3x^3}{(x^3+1)^{1/2}}+\frac{2(x^3+1)^{1/2}}{1}=\frac{3x^3+2(x^3+1)}{(x^3+1)^{1/2}}=\frac{5x^3+2}{(x^3+1)^{1/2}}$$

33. **a.** $F_1(x)=\frac{3}{5}(x^4+1)^5$ $\quad F_1'(x)=\frac{3}{5}\cdot5(x^4+1)^4(4x^3)=12x^3(x^4+1)^4$

b. $F_2(x)=\frac{3}{5}(x^4+1)^{-5}$ $\quad F_2'(x)=\frac{3}{5}\cdot(-5)(x^4+1)^{-6}(4x^3)=\dfrac{-12x^3}{(x^4+1)^6}$

c. $F_3(x)=\frac{1}{5}(3x^4+1)^5$ $\quad F_3'(x)=\frac{1}{5}\cdot5(3x^4+1)^4(12x^3)=12x^3(3x^4+1)^4$

d. $F_4(x)=3\left(5x^4+1\right)^{-5}$ $\quad F_4'(x)=3\cdot(-5)\left(5x^4+1\right)^{-6}\left(20x^3\right)=\dfrac{-300x^3}{\left(5x^4+1\right)^6}$

35. $P=10(3x+1)^3-10$ $\quad \dfrac{dP}{dx}=30(3x+1)^2(3)=90(3x+1)^2$

37. $R(x)=60{,}000x+40{,}000(10+x)^{-1}-4000$ $\quad R'(x)=60{,}000-40{,}000(10+x)^{-2}(1)$

a. $R'(10)=60{,}000-\dfrac{40{,}000}{20^2}=\$59{,}900$

b. The revenue is increasing.

39. $C(y)=2(y+1)^{1/2}+0.4y+4$ $\quad C'(y)=2\cdot\frac{1}{2}(y+1)^{-1/2}(1)+0.4=\dfrac{1}{(y+1)^{1/2}}+0.4$

41. $V=x(12-2x)^2$

$$V'(x)=x\cdot2(12-2x)(0-2)+(12-2x)^2\cdot1=(12-2x)[-4x+(12-2x)]$$

$$=(12-2x)(12-6x)=144-96x+12x^2$$

43. $S(t)=\dfrac{200t}{(t+1)^2}$

$$S'(t)=\frac{(t+1)^2(200)-200t\cdot2(t+1)\cdot1}{(t+1)^4}=\frac{200(t+1)[(t+1)-2t]}{(t+1)^4}=\frac{200(1-t)}{(t+1)^3}\qquad S'(9)=\frac{200(-8)}{10^3}=-1.6$$

From the 9th to the 10th week, sales will decrease approximately \$1.60 thousand or \$1,600.

45. $f(t)=\dfrac{1000(-8.0912(t+20)+1558.9)}{1.09816(t+20)^2-122.183(t+20)+21472.6}=\dfrac{-8091.2(t+20)+1,558,900}{1.09816(t+20)^2-122.183(t+20)+21472.6}$

$t=0 \leftrightarrow 1820$

a. Let $d(t)=1.09816(t+20)^2-122.183(t+20)+21472.6$

Using the quotient rule gives

$$f'(t)=\frac{d(t)(-8091.2)-(-8091.2(t+20)+1,558,900)(2.19632(t+20)-122.183)}{[d(t)]^2}$$

Numerator of $f'(t)$ $=[-8885.432192(t+20)^2+988,607.0896(t+20)-173,739,101.12]$

$-[-17,770.864384(t+20)^2+4,412,450.3376(t+20)-190,471,078.7]$

Hence, $f'(t)=\dfrac{8885.432192(t+20)^2-3,423,843.248(t+20)+16,731,977.58}{[d(t)]^2}$.

b. 1850: $f'(30)\approx -0.403$ 1950: $f'(130)\approx -0.383$

c. $f'(30)$ means that from 1850 to 1851, the model predicts a change of -0.403% in U.S. workers in farm occupations. $f'(130)$ means that from 1950 to 1951, the model predicts a change of -0.383% in U.S. workers in farm occupations.

Exercise 9.8

1. $f(x)=4x^3-15x^2+3x+2$

$f'(x)=12x^2-30x+3$

$f''(x)=24x-30$

3. $y=10x^3-x^2+14x+3$

$y'=30x^2-2x+14$

$y''=60x-2$

5. $g(x)=x^3-x^{-1}$

$g'(x)=3x^2+x^{-2}$

$g''(x)=6x-2x^{-3}=6x-\dfrac{2}{x^3}$

7. $y=x^3-x^{1/2}$

$y'=3x^2-\dfrac{1}{2}x^{-1/2}$

$y''=6x+\dfrac{1}{4}x^{-3/2}=6x+\dfrac{1}{4x^{3/2}}$

9. $y=x^5-16x^3+12$

$y'=5x^4-48x^2$

$y''=20x^3-96x$

$y'''=60x^2-96$

11. $f(x)=2x^9-6x^6$

$f'(x)=18x^8-36x^5$

$f''(x)=144x^7-180x^4$

$f'''(x)=1008x^6-720x^3$

13. $y=x^{-1}$

$y'=-1x^{-2}$

$y''=2x^{-3}$

$y'''=-6x^{-4}$

15. $y=x^{1/2}$

$y'=\dfrac{1}{2}x^{-1/2}$

$y''=-\dfrac{1}{4}x^{-3/2}$

$y'''=\dfrac{3}{8}x^{-5/2}$

17. $y=x^5-x^{1/2}$

$\dfrac{dy}{dx}=5x^4-\dfrac{1}{2}x^{-1/2}$

$\dfrac{d^2y}{dx^2}=20x^3+\dfrac{1}{4}x^{-3/2}$

19. $f(x)=(x+1)^{1/2}$

$f'(x)=\dfrac{1}{2}(x+1)^{-1/2}(1)$

$f''(x)=-\dfrac{1}{4}(x+1)^{-3/2}(1)$

$f'''(x)=\dfrac{3}{8}(x+1)^{-5/2}(1)=\dfrac{3}{8}(x+1)^{-5/2}$

21. $y = 4x^3 - 16x$

$y' = 12x^2 - 16$

$y'' = 24x$

$y''' = 24$

$y^{(4)} = 0$

23. $f(x) = x^{1/2}$

From problem 15 we have $f^{(3)}(x) = \frac{3}{8}x^{-5/2}$.

$f^{(4)}(x) = -\frac{15}{16}x^{-7/2}$

25. $y' = (4x-1)^{1/2}$

$y'' = \frac{1}{2}(4x-1)^{-1/2}(4) = 2(4x-1)^{-1/2}$

$y''' = -1(x-1)^{-3/2}(4) = -4(4x-1)^{-3/2}$

$y^{(4)} = \frac{12}{2}(4x-1)^{-5/2}(4) = 24(4x-1)^{-5/2}$

27. $f^{(4)}(x) = \dfrac{x}{x+1}$

$f^{(5)}(x) = \dfrac{(x+1)(1) - x(1)}{(x+1)^2} = (x+1)^{-2}$

$f^{(6)}(x) = -2(x+1)^{-3}(1) = -\dfrac{2}{(x+1)^3}$

29. $f(x) = 16x^2 - x^3$

$f'(x) = 32x - 3x^2$

$f''(x) = 32 - 6x$

At $x = 1$, $f''(1) = 32 - 6 = 26$.

31. $f''(3) \approx nDeriv(nDeriv(x^3 - 27/x, x, x,), x, 3) = 15.99999925$

33. $f''(21) \approx nDeriv(nDeriv(\sqrt{\ }(x^2+4), x, x,), x, 21) = 0.000426$

35. $f(x) = x^3 - 3x^2 + 5$

a. $f'(x) = 3x^2 - 6x$

$f''(x) = 6x - 6$

b.

$f'(x) = 3x^2 - 6x$

15

−3

5

−10

$f(x) = x^3 - 3x^2 + 5$

$f''(x) = 6x - 6$

c. $f''(x) = 0$ at $x = 1$.

$f''(x) > 0$ if $x > 1$;

$f''(x) < 0$ if $x < 1$.

d. $f'(x)$ [NOT $f(x)$] min at $x = 1$.

$f'(x)$ increasing if $x > 1$. ($f''(x) > 0$)

$f'(x)$ decreasing if $x < 1$. ($f''(x) < 0$)

e. $f''(x) < 0$

f. $f''(x) > 0$ [For now look at graphs.]

37. $f(x) = -\frac{1}{3}x^3 - x^2 + 3x + 7$

a. $f'(x) = -x^2 - 2x + 3$ $\quad f''(x) = -2x - 2$

b.

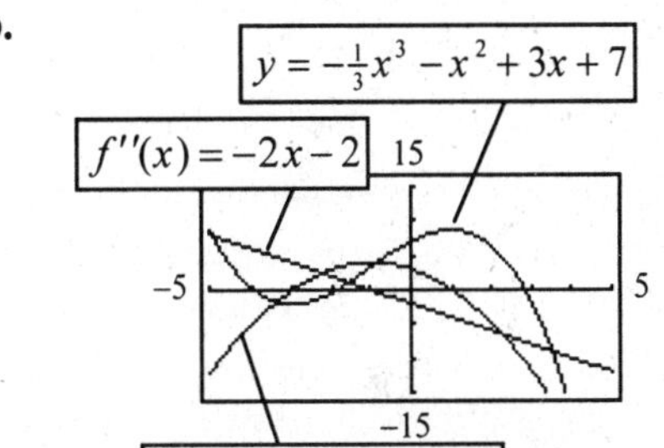

c. $f''(x) = 0$ at $x = -1$.

$f''(x) > 0$ if $x < -1$;

$f''(x) < 0$ if $x > -1$.

d. $f'(x)$ [NOT $f(x)$] max at $x = -1$.

$f'(x)$ increasing if $x < -1$. ($f''(x) > 0$)

$f'(x)$ decreasing if $x > -1$. ($f''(x) < 0$)

e. $f''(x) < 0$

f. $f''(x) > 0$

[For now look at graphs.]

39. $s(t) = 100 + 10t + 0.01t^3$ distance

$s'(t) = 10 + 0.03t^2$ velocity

$s''(t) = 0.06t$ acceleration

$s''(2) = 0.12$

41. $R(x) = 100x - 0.01x^2$

$R'(x) = 100 - 0.02x$

$R''(x) = -0.02$

43. $R = m^2\left(\frac{c}{2} - \frac{m}{3}\right) = \frac{c}{2}m^2 - \frac{1}{3}m^3$

a. $R' = cm - m^2$

b. $R'' = c - 2m$

c. Second derivative

45. $R(x) = 15x + 30(4x+1)^{-1} - 30$

$\overline{MR} = R'(x) = 15 - 30(4x+1)^{-2}(4)$

$= 15 - 120(4x+1)^{-2}$

$\overline{MR}' = R''(x) = 0 + 240(4x+1)^{-3}(4)$

$= \frac{960}{(4x+1)^3}$

a. $R''(25) = \frac{960}{101^3} \approx 0.0009$

b. When the next unit is sold the marginal revenue will increase about $0.90 per unit.

47. $S(t) = 1 + 3(t+3)^{-1} - 18(t+3)^{-2}$

a. $S'(t) = -3(t+3)^{-2} + 36(t+3)^{-3}$

$= \frac{-3}{(t+3)^2} + \frac{36}{(t+3)^3}$

b. $S''(t) = 6(t+3)^{-3} - 108(t+3)^{-4}$

$= \frac{6}{(t+3)^3} - \frac{108}{(t+3)^4}$

$S''(15) = \frac{6}{18^3} - \frac{108}{18^4}$

$= 0$

c. The rate of change of the rate of change in sales (after 15 weeks) is zero.

49. $p(t) = -0.0022605t^3 + 0.154805t^2 - 2.78785t + 41.038$

a. $p'(t) = -0.0067815t^2 + 0.309610t - 2.78785$

b. $p''(t) = -0.0135630t + 0.309610$

1980: $p''(20) = 0.0384$ 1998: $p''(38) = -0.2058$

c. $p'(20) = 0.6918$ means that in the next year (1981), the number of people who lived below the poverty level was expected to increase by about 691,800 people.

$p''(20) = 0.0384$ means that in the next year (1981), the rate of change of the number of people who lived below the poverty level was expected to increase by about 38,400 people. From 1980 to 1981 the number of people who lived below the poverty level was increasing at an increasing rate.

51. a. $f(t) = 0.0060315t^4 - 0.24256t^3 + 3.57194t^2 - 22.9844t + 57.7268$

b. $f'(t) = 0.0241260t^3 - 0.72768t^2 + 7.14388t - 22.9844$

c. $f''(t) = 0.072378t^2 - 1.45536t + 7.14388$

1994: $f''(9) = -0.0917$ 2000: $f''(15) = 1.5985$

d. $f'(15) \approx 1.8711$ means that in 2000, the CPI was increasing 1.8711% per year.

$f''(15) = 1.5985$ means that in 2000, the rate of change of the CPI was increasing 1.5985%. This means that the CPI was increasing at an increasing rate.

Exercise 9.9

1. $C(x) = 40 + 8x \quad \overline{MC} = C'(x) = 8$

3. $C(x) = 500 + 13x + x^2 \quad \overline{MC} = C'(x) = 13 + 2x$

5. $C = x^3 - 6x^2 + 24x + 10$

   $\overline{MC} = C' = 3x^2 - 12x + 24$

7. $C = 400 + 27x + x^3 \quad \overline{MC} = C' = 27 + 3x^2$

9. **a.** $C(x) = 40 + x^2$

   $\overline{MC} = C'(x) = 2x$

   $C'(5) = 2(5) = 10$

   The cost to produce the 6th unit is predicted to be \$10.

   b. $C(6) - C(5) = 76 - 65 = \$11$

11. $C(x) = x^3 - 4x^2 + 30x + 20$

   $\overline{MC} = C'(x) = 3x^2 - 8x + 30$

   $C'(4) = 48 - 32 + 30 = 46$

   The cost will increase by \$46.

13. $C(x) = 300 + 4x + x^2 \quad \overline{MC} = C'(x) = 4 + 2x$

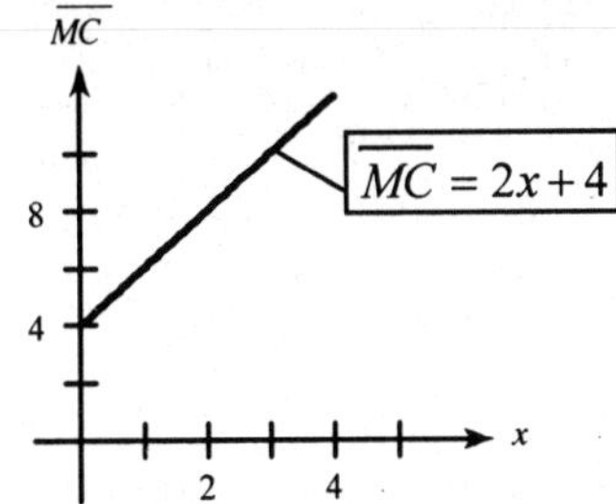

15. **a.** Remember that the slope gives the cost of the next unit. $C'(100)$ is greater than $C'(500)$. Thus the 101st unit will cost more than the 501st unit.

   b. Since the slope is decreasing, the cost of each additional unit is less than the previous unit. The manufacturing process is more efficient.

17. **a.** $\overline{MR} = R'(x) = 4$

   b. Each unit sold brings in \$4 revenue.

19. $R(x) = 36x - 0.01x^2$

   a. $R(100) = 3600 - 0.01(10{,}000)$

   $R(100) = \$3500$

   100 units produce \$3500 of revenue.

   b. $\overline{MR} = R'(x) = 36 - 0.02x$

   c. $R'(100) = 36 - 2 = 34$

   The sale of the next unit will increase the revenue by about \$34.

   d. $R(101) - R(100) = 3533.99 - 3500$

   $= \$33.99$

   (Actual revenue from 101st unit.)

21. **a.** $R(x) = px = (80 - 0.4x)x = 80x - 0.4x^2$

   (in hundreds of dollars)

   b. $50 = 80 - 0.4x$

   $0.4x = 80 - 50 = 30$

   $x = \dfrac{30}{0.4} = 75$ hundreds

   (or 7500 subscribers)

   $R = 50 \cdot 7500 = \$375{,}000$

   c. More subscribers are attracted by lower prices.

   d. $R'(x) = 80 - 0.8x$

   $p = 50 \rightarrow x = 75$

   $R'(75) = 80 - 0.8(75) = \$20$

   If the number of subscribers increases from 75 to 76 (hundred) the revenue increases $\$20(100) = \$2{,}000$. Increasing the number of subscribers will occur by lowering the monthly charges.

23. **a.** $\overline{MR} = 36 - 0.02x$

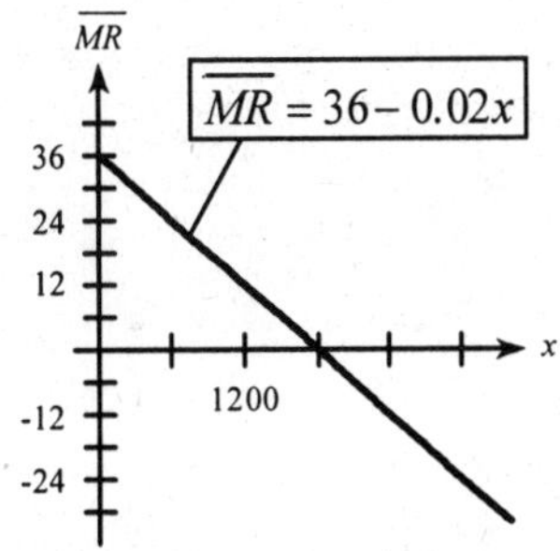

   b. Maximum revenue at $x = 1800$ ($\overline{MR} = 0$)

   c. $R(1800) = \$32{,}400$

25. $P(x) = 5x - 25 \qquad \overline{MP} = P'(x) = 5$

   The next unit sold earns a \$5 profit.

27. $P(x) = R(x) - C(x) = 30x - x^2 - 200$

a. $P(20) = 600 - 400 - 200 = 0$

b. $\overline{MP} = P'(x) = 30 - 2x$

c. $P'(20) = 30 - 40 = -10$

The total profit will decrease by approximately \$10 on the sale of the next (21st) unit.

d. $P(21) - P(20) = (-11) - (0) = -\$11$ (The actual loss on the sale of the 21st unit.)

29. a. From $P = R - C$ we obtain the following profits in ascending order: 100, 700, and 400 units. There is a loss at 100 units since the cost curve is above the revenue curve.

b. $MP = MR - MC$ Look at the slopes of each curve at $x = 100$, 400, and 700. Ranking from high to low:

$\overline{MP}(100) > \overline{MP}(400) > \overline{MP}(700)$.

$\overline{MP}(700) < 0$ since $R' - C' < 0$.

31. a. From low to high: $A < B < C$.

The graph shows a loss at A.

b. $P' > 0$ at each point.

Evaluate the slope at each point to obtain $P'(C) < P'(B) < P'(A)$.

33. $\overline{MP} = 30 - 2x$

a.

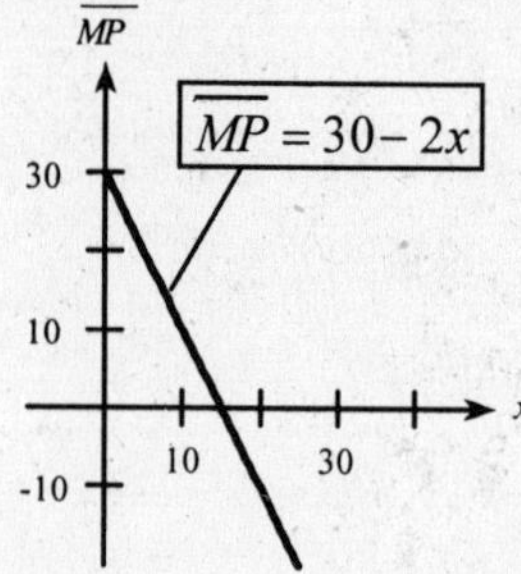

b. $\overline{MP} = 0$ if $30 - 2x = 0$. Thus, $x = 15$.

c. Profit is a maximum at the vertex of the parabola $P(x) = 30x - x^2 - 200$.

Vertex is at $x = \dfrac{-b}{2a} = \dfrac{-30}{2(-1)} = 15$.

d. $P(15) = 450 - 225 - 200 = \$25$

35. $P(x) = R(x) - C(x) = 300x - (160 + x)x$

$= 140x - x^2$

$P'(x) = 140 - 2x$

$P'(x) = 0$ at $140 - 2x = 0$ or $x = 70$.

Note that cost is per unit. Total cost is the cost per unit times the number of units.
Production of 70 units will maximize profit.

37. $P(x) = R(x) - C(x) = 50x - (10 + 2x)x$

$= 40x - 2x^2$

$P'(x) = 40 - 4x$

$P'(x) = 0$ at $x = 10$.

Maximum profit $= P(10) = 400 - 200 = \$200$

Review Exercises

1. **a.** $f(-2)=2$
b. $\lim_{x\to-2} f(x)=2$

2. **a.** From the graph, $f(-1)=0$.
b. $\lim_{x\to-1} f(x)=0$

3. **a.** $f(4)=2$
b. $\lim_{x\to4^-} f(x)=1$

4. **a.** From the graph, $\lim_{x\to4^+} f(x)=2$.
b. $\lim_{x\to4} f(x)$ does not exist.

5. **a.** $f(1)$ is not defined
b. $\lim_{x\to1} f(x)=2$

6. **a.** From the graph, $f(2)$ does not exist.
b. $\lim_{x\to2} f(x)$ does not exist.

7. $\lim_{x\to4}(3x^2+x+3)=48+4+3=55$

8. $\lim_{x\to4}\frac{x^2-16}{x+4}=\frac{4^2-16}{4+4}=0$

9. $\lim_{x\to-1}\frac{x^2-1}{x+1}=\lim_{x\to-1}\frac{(x+1)(x-1)}{x+1}$
$=\lim_{x\to-1}(x-1)=-2$

10. $\lim_{x\to3}\frac{x^2-9}{x-3}=\lim_{x\to3}\frac{(x-3)(x+3)}{x-3}$
$=\lim_{x\to3}(x+3)=3+3=6$

11. $\lim_{x\to2}\frac{4x^3-8x^2}{4x^3-16x}=\lim_{x\to2}\frac{4x^2(x-2)}{4x(x-2)(x+2)}$
$=\lim_{x\to2}\frac{x}{x+2}=\frac{2}{4}=\frac{1}{2}$

12. $\lim_{x\to-\frac{1}{2}}\frac{x^2-\frac{1}{4}}{6x^2+x-1}=\lim_{x\to-\frac{1}{2}}\frac{\left(x+\frac{1}{2}\right)\left(x-\frac{1}{2}\right)}{(3x-1)(2)\left(x+\frac{1}{2}\right)}$
$=\lim_{x\to-\frac{1}{2}}\frac{x-\frac{1}{2}}{2(3x-1)}=\frac{-1}{-5}=\frac{1}{5}$

13. $\lim_{x\to3}\frac{x^2-16}{x-3}=\frac{-7}{0}$
Thus, the limit does not exist.

14. $\lim_{x\to-3}\frac{x^2-9}{x-3}=\frac{(-3)^2-9}{-3-3}=\frac{0}{-6}=0$

15. $\lim_{x\to1}\frac{x^2-9}{x-3}=\frac{-8}{-2}=4$

16. $\lim_{x\to2}\frac{x^2-8}{x-2}$ does not exist.

17. $\lim_{x\to1^-} f(x)=4-1=3$
$\lim_{x\to1^+} f(x)=2(1)+1=3$
Thus, $\lim_{x\to1} f(x)=3$

18. $\lim_{x\to-2^-} f(x)=-8+2=-6$
$\lim_{x\to-2^+} f(x)=2-4=-2$
Since these two limits are not equal, $\lim_{x\to-2} f(x)$ does not exist.

19. $\lim_{h\to0}\frac{3(x+h)^2-3x^2}{h}$
$=\lim_{h\to0}\frac{3x^2+6xh+3h^2-3x^2}{h}$
$=\lim_{h\to0}\frac{6xh+3h^2}{h}=\lim_{h\to0}(6x+3h)=6x$

20. $\lim_{h\to0}\frac{[(x+h)-2(x+h)^2]-(x-2x^2)}{h}$
$=\lim_{h\to0}\frac{x+h-2x^2-4xh-2h^2-x+2x^2}{h}$
$=\lim_{h\to0}\frac{h-4xh-2h^2}{h}$
$=\lim_{h\to0}(1-4x-2h)=1-4x$

21. $\lim_{x\to2}\frac{(x+12)(x-2)}{(x-3)(x-2)}$
$=\lim_{x\to2}\frac{x+12}{x-3}=\frac{14}{-1}=-14$

22. $\lim_{x\to-\frac{1}{2}}\frac{\left(x+\frac{1}{2}\right)\left(x-\frac{1}{3}\right)}{\left(x+\frac{1}{2}\right)\left(x+\frac{1}{3}\right)}$
$=\lim_{x\to-\frac{1}{2}}\frac{x-\frac{1}{3}}{x+\frac{1}{3}}=\frac{-\frac{1}{2}-\frac{1}{3}}{-\frac{1}{2}+\frac{1}{3}}=5$

23. **a.** From the graph, $f(-1)=0=\lim_{x\to-1} f(x)$, so $f(x)$ is continuous at $x=-1$.
b. From the graph, $f(1)$ does not exist, so $f(x)$ is not continuous at $x=1$.

24. a. From the graph, $f(-2) = 2 = \lim_{x\to -2} f(x)$, so $f(x)$ is continuous at $x = -2$.

b. From the graph, $f(2)$ and $\lim_{x\to 2} f(x)$ do not exist, so $f(x)$ is not continuous at $x = 2$.

25. $\lim_{x\to -1^+} f(x) = (-1)^2 + 1 = 2$

$\lim_{x\to -1^-} f(x) = (-1)^2 + 1 = 2$

Thus, $\lim_{x\to -1} f(x) = 2$

26. $\lim_{x\to 0^-} f(x) = 0^2 + 1 = 1$

$\lim_{x\to 0^+} f(x) = 0$

Thus, $\lim_{x\to 0} f(x)$ does not exist.

27. $\lim_{x\to 1^+} f(x) = 2(1)^2 - 1 = 1$

$\lim_{x\to 1^-} f(x) = 1$

Thus, $\lim_{x\to 1} f(x) = 1$

28. $f(x)$ is not continuous at $x = 0$ since the $\lim_{x\to 0} f(x)$ does not exist.

29. $f(1) = 2(1)^2 - 1 = 1 = \lim_{x\to 1} f(x)$

Yes, $f(x)$ is continuous at $x = 1$.

30. Since $f(-1) = 2 = \lim_{x\to -1} f(x)$, $f(x)$ is continuous at $x = -1$.

31. Function is discontinuous at $x = 5$.
(Function is not defined.)

32. $f(x)$ is not continuous at $x = 2$ since $f(2)$ does not exist.

33. $f(2) = 4$

$\lim_{x\to 2} f(x) = 4$

Function is continuous everywhere.

34. $f(x)$ is not continuous at $x = 1$ since $\lim_{x\to 1} f(x)$ does not exist.

35. a. $f(x)$ is discontinuous at $x = 0$ and $x = 1$.

b. $\lim_{x\to +\infty} f(x) = 0$

c. $\lim_{x\to -\infty} f(x) = 0$

36. a. From the graph, $f(x)$ is discontinuous at $x = 0$ and $x = -1$.

b. $\lim_{x\to +\infty} f(x) = \frac{1}{2}$

c. $\lim_{x\to -\infty} f(x) = \frac{1}{2}$

37. $\lim_{x\to -\infty} \frac{2x^2}{1-x^2} = \lim_{x\to -\infty} \frac{2}{\frac{1}{x^2}-1} = \frac{2}{0-1} = -2$

38. $\lim_{x\to +\infty} \frac{3x^{2/3}}{x+1} = \lim_{x\to +\infty} \frac{\frac{3}{x^{1/3}}}{1+\frac{1}{x}} = \frac{0}{1} = 0$

39. $\text{Avg} = \frac{f(2)-f(-1)}{2-(-1)} = \frac{33-12}{3} = 7$

40. True.

41. False. The given expression gives the slope of the tangent line at $x = c$.

42. $f(x) = 3x^2 + 2x - 1$

$$f'(x) = \lim_{h\to 0} \frac{[3(x+h)^2 + 2(x+h) - 1] - (3x^2 + 2x - 1)}{h} = \lim_{h\to 0} \frac{6xh + 3h^2 + 2h}{h} = \lim_{h\to 0}(6x + 3h + 2) = 6x + 2$$

43. $$f'(x) = \lim_{h\to 0} \frac{x+h-(x+h)^2-(x-x^2)}{h} = \lim_{h\to 0} \frac{h-2xh-h^2}{h} = \lim_{h\to 0}(1-2x-h) = 1-2x$$

44. $[-3,0]$: $\text{Avg} = \frac{f(0)-f(-3)}{0-(-3)} = \frac{1-0}{3} = \frac{1}{3}$

$[-1,0]$: $\text{Avg} = \frac{f(0)-f(-1)}{0-(-1)} = \frac{1-0}{1} = 1$

From $[-1,0]$ the change is greater.

45. a. No **b.** No

46. a. From the graph, $f(x)$ is differentiable at $x = -2$.

b. $f(x)$ is not differentiable at $x = 2$.

47. a. $nDeriv\left((4x)^{1/3}/(3x^2-10)^2,\ x,\ 2\right)$

$= -5.917062766$

b. $\dfrac{f(2+.0001)-f(2)}{0.0001}$

$= \dfrac{0.49941-0.5}{0.0001} = -5.91$

48. a. $\text{Avg} = \dfrac{g(5)-g(2)}{5-2} = \dfrac{18.1-13.2}{3} = 1.633$

b. $g'(4) = \dfrac{g(4.3)-g(4)}{4.3-4} = \dfrac{2.1}{0.3} = 7$

49. Approximate the tangent through $(0,2)$ and $(8,0)$. $f'(4) \approx \dfrac{0-2}{8-0} = -\dfrac{1}{4}$

50. $A: f'(2) \approx 0$

$B: f'(6) \approx \dfrac{-1-1}{8-4} = -\dfrac{1}{2}$

$C: \text{Avg} = \dfrac{-1-1.2}{10-2} = -0.275$

Rank: $B < C < A$

51. $c = 4x^5 - 6x^3$

$c' = 4(5x^4) - 6(3x^2) = 20x^4 - 18x^2$

52. $f(x) = 4x^2 - 1 \qquad f'(x) = 8x$

53. $p = 3q + \sqrt{7} \qquad \dfrac{dp}{dq} = 3 + 0 = 3$

54. $y = \sqrt{x} = x^{1/2} \qquad y' = \dfrac{1}{2}x^{-1/2} = \dfrac{1}{2\sqrt{x}}$

55. $f(z) = 2^{4/3} \qquad f'(z) = 0$

56. $v(x) = \dfrac{4}{\sqrt[3]{x}} = 4x^{-1/3}$

$v'(x) = -\dfrac{4}{3}x^{-4/3} = -\dfrac{4}{3\sqrt[3]{x^4}}$

57. $y = x^{-1} - x^{-1/2}$

$y' = -1(x^{-2}) - \left(-\dfrac{1}{2}x^{-3/2}\right) = -\dfrac{1}{x^2} + \dfrac{1}{2x^{3/2}}$

58. $f(x) = \dfrac{3}{2x^2} - \sqrt[3]{x} + 4^5 = \dfrac{3}{2}x^{-2} - x^{1/3} + 4^5$

$f'(x) = -3x^{-3} - \dfrac{1}{3}x^{-2/3} = -\dfrac{3}{x^3} - \dfrac{1}{3\sqrt[3]{x^2}}$

59. $f(x) = 3x^5 - 6 \qquad f'(x) = 15x^4$

$f(1) = 3 - 6 = -3 \qquad f'(1) = 15$

So, $y - (-3) = 15(x-1)$ or $y = 15x - 18$.

60. $f(x) = 3x^3 - 2x \qquad f(2) = 20$

$f'(x) = 9x^2 - 2 \qquad f'(2) = 34$

So, $y - 20 = 34(x-2)$ or $y = 34x - 48$.

61. $f'(x) = 3x^2 - 6x = 3x(x-2)$

a. $f'(x) = 0$ if $x = 0, 2$.

b. $f(0) = 0 - 0 + 1 = 1$ Point: $(0, 1)$

$f(2) = 8 - 12 + 1 = -3$ Point: $(2, -3)$

c.

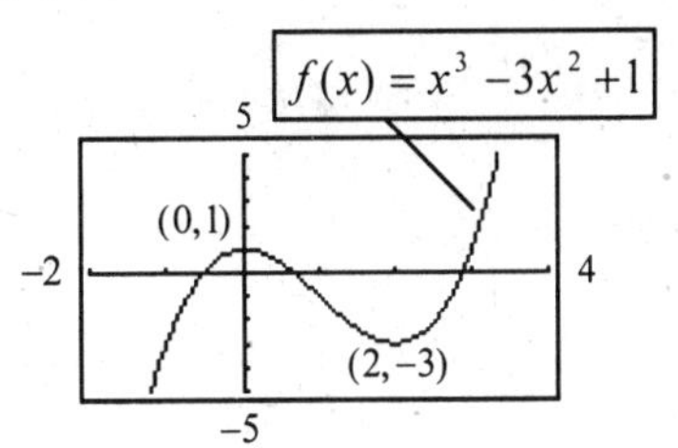

62. $f(x) = x^6 - 6x^4 + 8$

$f'(x) = 6x^5 - 24x^3 = 6x^3(x^2-4)$

$= 6x^3(x+2)(x-2)$

a. $f'(x) = 0$ when $x = 0$, $x = -2$, and when $x = 2$.

b. The associated points where the slope equals zero are (0, 8), (–2, –24) and (2, –24).

c.

$f(x) = x^6 - 6x^4 + 8$

28

−3 3

(0,8)

(−2,−24) −28 (2,−24)

63. $f(x) = (3x-1)(x^2-4x)$

$f'(x) = (3x-1)(2x-4) + (x^2-4x)(3)$

$= 9x^2 - 26x + 4$

64. $y = (x^2+1)(3x^3+1)$

$y' = (x^2+1)(9x^2) + (3x^3+1)(2x)$

$= 15x^4 + 9x^2 + 2x$

65. $p = \dfrac{2q-1}{q^2}$

$\dfrac{dp}{dq} = \dfrac{q^2(2) - (2q-1)2q}{q^4} = \dfrac{-2q^2+2q}{q^4} = \dfrac{2-2q}{q^3}$

66. $s = \dfrac{\sqrt{t}}{3t+1} = \dfrac{t^{1/2}}{3t+1}$ $\quad \dfrac{ds}{dt} = \dfrac{(3t+1)\cdot\frac{1}{2}t^{-1/2} - t^{1/2}\cdot 3}{(3t+1)^2} = \dfrac{\frac{1}{2}t^{-1/2}(3t+1-6t)}{(3t+1)^2} = \dfrac{1-3t}{2\sqrt{t}(3t+1)^2}$

67. $y = \sqrt{x}(3x+2) = 3x^{3/2} + 2x^{1/2}$ $\quad \dfrac{dy}{dx} = \dfrac{9}{2}x^{1/2} + x^{-1/2} = \dfrac{9x+2}{2\sqrt{x}}$

68. $C = \dfrac{5x^4 - 2x^2 + 1}{x^3 + 1}$

$$\frac{dC}{dx} = \frac{(x^3+1)(20x^3-4x)-(5x^4-2x^2+1)(3x^2)}{(x^3+1)^2}$$

$$= \frac{20x^6 - 4x^4 + 20x^3 - 4x - 15x^6 + 6x^4 - 3x^2}{(x^3+1)^2} = \frac{5x^6 + 2x^4 + 20x^3 - 3x^2 - 4x}{(x^3+1)^2}$$

69. $y = \left(x^3 - 4x^2\right)^3$ $\quad y' = 3\left(x^3-4x^2\right)^2\left(3x^2-8x\right) = \left(x^3-4x^2\right)^2\left(9x^2-24x\right)$

70. $y = \left(5x^6 + 6x^4 + 5\right)^6$ $\quad y' = 6\left(5x^6+6x^4+5\right)^5\left(30x^5+24x^3\right)$

71. $y = \left(2x^4 - 9\right)^9$ $\quad \dfrac{dy}{dx} = 9\left(2x^4-9\right)^8 \cdot 8x^3 = 72x^3\left(2x^4-9\right)^8$

72. $g(x) = \dfrac{1}{\sqrt{x^3-4x}} = \left(x^3-4x\right)^{-1/2}$ $\quad g'(x) = -\dfrac{1}{2}\left(x^3-4x\right)^{-3/2}\left(3x^2-4\right) = -\dfrac{3x^2-4}{2\sqrt{(x^3-4x)^3}}$

73. $f(x) = x^2\left(2x^4+5\right)^8$

$$f'(x) = x^2\cdot 8\left(2x^4+5\right)^7\left(8x^3\right) + \left(2x^4+5\right)^8(2x) = 2x\left(2x^4+5\right)^7\left[32x^4 + \left(2x^4+5\right)\right] = 2x\left(2x^4+5\right)^7\left(34x^4+5\right)$$

74. $S = \dfrac{(3x+1)^2}{x^2-4}$

$$S' = \frac{(x^2-4)(2)(3x+1)(3)-(3x+1)^2(2x)}{(x^2-4)^2} = \frac{(3x+1)[(6x^2-24)-(6x^2+2x)]}{(x^2-4)^2} = -\frac{2(3x+1)(x+12)}{(x^2-4)^2}$$

75. $y = (3x+1)^{12}(2x^3-1)^{12}$

$$\frac{dy}{dx} = (3x+1)^{12}\cdot 12\left(2x^3-1\right)^{11}\left(6x^2\right) + \left(2x^3-1\right)^{12}\cdot 12(3x+1)^{11}(3)$$

$$= 12\left(2x^3-1\right)^{11}\left[6x^2(3x+1)^{12} + 3\left(2x^3-1\right)(3x+1)^{11}\right]$$

$$= 36\left(2x^3-1\right)^{11}(3x+1)^{11}\left[2x^2(3x+1) + \left(2x^3-1\right)\right]$$

$$= 36\left(2x^3-1\right)^{11}(3x+1)^{11}\left(8x^3+2x^2-1\right)$$

76. $y = \left(\dfrac{x+1}{1-x^2}\right)^3 = \left(\dfrac{1}{1-x}\right)^3 = (1-x)^{-3}$ $\quad y' = -3(1-x)^{-4}(-1) = \dfrac{3}{(1-x)^4}$

77. $y = x(x^2-4)^{1/2}$

$$y' = x\cdot\frac{1}{2}(x^2-4)^{-1/2}(2x) + (x^2-4)^{1/2}\cdot 1 = \frac{x^2}{\sqrt{x^2-4}} + \frac{\sqrt{x^2-4}}{1} = \frac{x^2+(x^2-4)}{\sqrt{x^2-4}} = \frac{2x^2-4}{\sqrt{x^2-4}}$$

78. $y=\dfrac{x}{\sqrt[3]{3x-1}}=\dfrac{x}{(3x-1)^{1/3}}$ $\quad \dfrac{dy}{dx}=\dfrac{(3x-1)^{1/3}\cdot 1-x\cdot\frac{1}{3}(3x-1)^{-2/3}\cdot 3}{(3x-1)^{2/3}}=\dfrac{(3x-1)^{-2/3}[(3x-1)-x]}{(3x-1)^{2/3}}=\dfrac{2x-1}{(3x-1)^{4/3}}$

79. $y=x^{1/2}-x^2$ $\quad y'=\dfrac{1}{2}x^{-1/2}-2x$ $\quad y''=-\dfrac{1}{4}x^{-3/2}-2$

80. $y=x^4-x^{-1}$ $\quad y'=4x^3+x^{-2}$ $\quad y''=12x^2-2x^{-3}=12x^2-\dfrac{2}{x^3}$

81. $y=(2x+1)^4$ Since the largest power in any term is 4, the fifth derivative equals 0.

82. $y=\dfrac{1}{24}(1-x)^6$ $\quad y'=-\dfrac{1}{4}(1-x)^5$ $\quad y''=\dfrac{5}{4}(1-x)^4$

$y'''=-5(1-x)^3$ $\quad y^{(4)}=15(1-x)^2$ $\quad y^{(5)}=-30(1-x)$

83. $\dfrac{dy}{dx}=(x^2-4)^{1/2}$

$\dfrac{d^2y}{dx^2}=\dfrac{1}{2}(x^2-4)^{-1/2}(2x)=\dfrac{x}{(x^2-4)^{1/2}}$

$\dfrac{d^3y}{dx^3}=\dfrac{(x^2-4)^{1/2}(1)-x\left[\frac{x}{(x^2-4)^{1/2}}\right]}{(x^2-4)}\cdot\dfrac{(x^2-4)^{1/2}}{(x^2-4)^{1/2}}=\dfrac{x^2-4-x^2}{(x^2-4)^{3/2}}=\dfrac{-4}{(x^2-4)^{3/2}}$

84. $\dfrac{d^2y}{dx^2}=\dfrac{x}{x^2+1}$

$\dfrac{d^3y}{dx^3}=\dfrac{x^2+1-x(2x)}{(x^2+1)^2}=\dfrac{-x^2+1}{(x^2+1)^2}$

$\dfrac{d^4y}{dx^4}=\dfrac{(x^2+1)^2(-2x)-(-x^2+1)(2)(x^2+1)(2x)}{(x^2+1)^4}=\dfrac{(x^2+1)[-2x(x^2+1)-4x(-x^2+1)]}{(x^2+1)^4}=\dfrac{2x^3-6x}{(x^2+1)^3}=\dfrac{2x(x^2-3)}{(x^2+1)^3}$

85. **a.** Men: $\text{Avg}=\dfrac{18.6-63.1}{100}=-0.445\%$

b. Women: $\text{Avg}=\dfrac{10-8.3}{100}=0.017\%$

86. **a.** 1990-2000: $\text{Avg}=\dfrac{18.6-17.6}{10}=0.1\%$

Increased possibly because men chose to work past age 65.

b. $1950-1960$: $\text{Avg}=\dfrac{10.3-7.8}{10}=0.25\%$

Increased possibly because women were needed because of so many men killed in WW II.

87. $x(p)=\dfrac{100}{p}-1$ $\quad x'(p)=\dfrac{-100}{p^2}$

a. $x'(10)=\dfrac{-100}{10^2}=-1$

As the price per unit increases from \$10 to \$11, the demand is expected to drop by 1 unit.

b. $x'(20)=\dfrac{-100}{20^2}=-\dfrac{1}{4}$

As the price per unit increases from \$20 to \$21, the demand is expected to drop by $\dfrac{1}{4}$ units.

88. $P(x) = 64.37x^{-0.0917}$ $\quad P'(x) = -5.9027x^{-1.0917}$

a. $P(12) = 51.25$

b. $P'(12) = -0.3917$

c. In 2000 the country was divided on their choice for president. Also, there was no incumbent president.

89. The $(A+1)$st item will produce more revenue. Reason: $R'(A) > R'(B)$.

90. $q = 10{,}000 - 50(0.02p^2 + 500)^{1/2}$ $\quad q' = -25(0.02p^2 + 500)^{-1/2}(0.04p) = -\dfrac{p}{\sqrt{0.02p^2 + 500}}$

91. $x(p) = \sqrt{p-1}$ $\quad x'(p) = \dfrac{1}{2\sqrt{p-1}}$ $\quad x'(10) = \dfrac{1}{2\sqrt{9}} = \dfrac{1}{6}$

If the price increases \$1, the number of units supplied will increase about 1/6.

92. $C(x) = 3x^2 + 6x + 600$

a. $C'(x) = 6x + 6 = \overline{MC}$

b. When $x = 30$, $C'(30) = 186$.

c. If a 31st unit is produced, costs will increase by about \$186.

93. $C(x) = 400 + 5x + x^3$ $\quad \overline{MC} = C'(x) = 5 + 3x^2$ $\quad C'(4) = 5 + 48 = 53$

It will cost about \$53 for the 5th unit. Also, the 5th unit produced will increase total costs by \$53.

94. $R = 40x - 0.02x^2$

a. $R' = 40 - 0.04x$

b. $\overline{MR} = 0$ when $40 - 0.04x = 0$ or when $x = 1000$.

95. $P(x) = 60x - (200 + 10x + 0.1x^2)$ $\quad P'(x) = 50 - 0.2x$ $\quad P'(10) = 50 - 2 = 48$

The next unit sold will have a profit of about \$48.

96. $R = 80x - 0.04x^2$

a. $R' = 80 - 0.08x$

b. At $x = 100$, $R' = 80 - 0.08(100) = 72$.

c. Selling the 101st unit is expected to result in \$72 extra revenue.

97. $R(x) = \dfrac{60x^2}{2x+1}$ $\quad R'(x) = \dfrac{(2x+1)(120x) - 60x^2(2)}{(2x+1)^2} = \dfrac{120x^2 + 120x}{(2x+1)^2}$

98. $C(x) = 45{,}000 + 100x + x^3$ $\quad R(x) = 4600x$

$P(x) = R(x) - C(x) = 4600x - (45{,}000 + 100x + x^3) = -x^3 + 4500x - 45{,}000$

$P'(x) = \overline{MP} = -3x^2 + 4500$

99. $P(x) = 46x - \left(100 + 30x + \dfrac{x^2}{10}\right) = 16x - 100 - \dfrac{x^2}{10}$ $\quad P'(x) = 16 - \dfrac{x}{5} = 16 - 0.2x$

100. Answers are decided from C', R', and $R' - C' = P'$

a. A, $C'(A) < C'(B) < C'(C)$

b. Profit is greatest at B since $R - C$ is a maximum distance at B.

c. $R'(A) > R'(B) > R'(C)$ $\quad C'(A) < C'(B) < C'(C)$

The marginal profit is greatest at A since $R'(A) - C'(A)$ has its largest value.

d. Profit is reduced at C since $R'(C) - C'(C) < 0$.

Chapter Test

1. a. $\lim_{x\to -2}\frac{4x-x^2}{4x-8}=\lim_{x\to -2}\frac{x(4-x)}{4(x-2)}=\frac{(-2)(6)}{4(-4)}=\frac{3}{4}$

b. $\lim_{x\to\infty}\frac{8x^2-4x+1}{2+x-5x^2}=\lim_{x\to\infty}\frac{8-\frac{4}{x}+\frac{1}{x^2}}{-5+\frac{1}{x}+\frac{2}{x^2}}=-\frac{8}{5}$

c. $\lim_{x\to 7}\frac{x^2-5x-14}{x^2-6x-7}=\lim_{x\to 7}\frac{(x-7)(x+2)}{(x-7)(x+1)}=\frac{7+2}{7+1}=\frac{9}{8}$

d. $\lim_{x\to -5}\frac{5x-25}{x+5}$ does not exist. (division by zero)

2. a. $f'(x)=\lim_{h\to 0}\frac{f(x+h)-f(x)}{h}$

b. $f'(x)=\lim_{h\to 0}\frac{3(x+h)^2-(x+h)+9-(3x^2-x+9)}{h}=\lim_{h\to 0}\frac{6xh+3h^2-h}{h}=\lim_{h\to 0}(6x+3h-1)=6x-1$

3. $f(x)=\frac{4x}{x(x-8)}$ $f(x)$ is not continuous at $x=0$ and $x=8$ as there is division by zero.

4. a. $y=\frac{3x^3}{2x^7+11}$ $\frac{dy}{dx}=\frac{(2x^7+11)(9x^2)-3x^3(14x^6)}{(2x^7+11)^2}=\frac{-24x^9+99x^2}{(2x^7+11)^2}$

b. $f(x)=(3x^5-2x+3)(4x^{10}+10x^4-17)$

$f'(x)=(3x^5-2x+3)(40x^9+40x^3)+(4x^{10}+10x^4-17)(15x^4-2)$

c. $g(x)=\frac{3}{4}(2x^5+7x^3-5)^{12}$ $g'(x)=9(2x^5+7x^3-5)^{11}(10x^4+21x^2)$

d. $y=(x^2+3)(2x+5)^6$

$y'=(x^2+3)\cdot 6(2x+5)^5(2)+(2x+5)^6\cdot 2x$

$=2(2x+5)^5[6(x^2+3)+x(2x+5)]$

$=2(2x+5)^5(8x^2+5x+18)$

e. $f(x)=12x^{1/2}-10x^{-2}+17$ $f'(x)=6x^{-1/2}+20x^{-3}=\frac{6}{\sqrt{x}}+\frac{20}{x^3}$

5. $y=x^3-x^{-3}$ $y'=3x^2+3x^{-4}$ $y''=6x-12x^{-5}$ $y'''=\frac{d^3y}{dx^3}=6+60x^{-6}$

6. $f(x)=x^3-3x^2-24x-10$

a. $f'(x)=3x^2-6x-24=3(x-4)(x+2)$

$f(-1)=(-1)^3-3(-1)^2-24(-1)-10=10$

$f'(-1)=3(-1)^2-6(-1)-24=-15$

$y-y_1=m(x-x_1)$

$y-10=-15(x+1)$ or $y=-15x-5$

b. $f'(x)=0$ at $x=4$ and $x=-2$ (from **a**).

$f(4)=-90$, $f(-2)=18$

Points: (4, –90), (–2, 18)

7. $f(x)=4-x-2x^2$ over $[1,6]$. Avg rate $=\frac{f(6)-f(1)}{6-1}=\frac{-74-1}{5}=-15$

8. **a.** $\lim_{x \to 5} f(x) = 2$

b. $\lim_{x \to 5} g(x) =$ DNE

c. $\lim_{x \to 5^-} g(x) = -4$

9. $g(x) = \begin{cases} 6 - x \text{ if } x \le -2 \\ x^3 \text{ if } x > -2 \end{cases}$

$g(-2) = 6 - (-2) = 8$

$\lim_{x \to -2^-} g(x) = 8 \qquad \lim_{x \to -2^+} g(x) = -8$

$g(x)$ is not continuous at $x = -2$ since the left hand limit does not equal the right hand limit.

10. $C(x) = 200x + 10000 \qquad R(x) = 250x - 0.01x^2$

a. $P(x) = R(x) - C(x) = -0.01x^2 + 50x - 10000$

b. $\overline{MP} = P'(x) = -0.02x + 50$

c. $P'(1000) = 30$. The company will make approximately $30 on the sale of the next unit.

11. $f'(3) \approx \dfrac{f(3) - f(2.999)}{3 - 2.999} = \dfrac{0.104}{0.001} = 104$

12. **a.** $f(1) = -5$ **b.** $\lim_{x \to 6} f(x) = -1$ **c.** $\lim_{x \to 3^-} f(x) = 4$

d. $\lim_{x \to -4} f(x) =$ DNE **e.** $\lim_{x \to -\infty} f(x) = 2$ **f.** $f'(x) =$ DNE at $x = -4$, 1, 3, and 6.

g. $f(x)$ is not continuous at $x = -4$, 3, and 6.

h. $f'(4) \approx \dfrac{3}{2}$ **i.** $f'(-2) <$ average rate $< f'(2)$

13. $y = \dfrac{2}{3}x - 8$ is tangent to $f(x)$ at $x = 6$.

a. $f'(6) = \dfrac{2}{3}$ **b.** $f(6) = \dfrac{2}{3}(6) - 8 = -4$

c. At $x = 6$, the instantaneous rate of change is $\dfrac{2}{3}$.

14. **a.** *B* $R(x) - C(x)$ is greatest at this point.

b. *A* $C(x) - R(x) > 0$

c. *A* and *B* $R'(x) - C'(x) > 0$ at each of these points.

d. C $R'(x) - C'(x) < 0$ at this point.

Chapter 10: Applications of Derivatives

Exercise 10.1

1. a. (1, 5)
   b. (4, 1)
   c. (–1, 2)

3. a. (1, 5)
   b. (4, 1)
   c. (–1, 2)

5. a. Critical values: $x = 3$ and $x = 7$
   b. Increasing: $3 < x < 7$
   c. Decreasing: $x < 3$, $x > 7$
   d. Max at $x = 7$
   e. Min at $x = 3$

7. $y = 2x^3 - 12x^2 + 6$
   $y' = 6x^2 - 24x = 6x(x-4)$
   $y' = 0$ if $x = 0$ or 4

9. $y = 2x^3 - 12x^2 + 6$
   $y' = 6x^2 - 24x = 6x(x-4)$

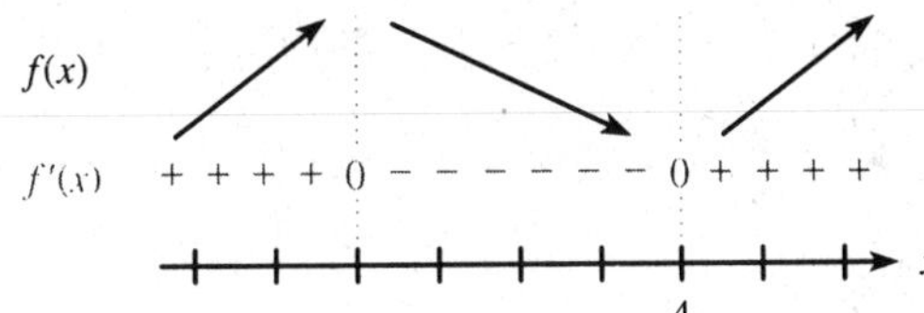

11. $y = x^3 - 3x + 4$
   a. Rel. max: $(-1, 6)$
      Rel. min: $(1, 2)$
   b. $f'(x) = 3x^2 - 3 = 3(x^2 - 1)$
      $f'(x) = 0$ if $3(x+1)(x-1) = 0$ or $x = -1, 1$.
   c. At $x = -1$ we have $y = (-1)^3 - 3(-1) + 4 = 6$.
      At $x = 1$ we have $y = 1^3 - 3 \cdot 1 + 4 = 2$
      Critical points (–1, 6), (1, 2)
   d. Yes.

13. a. HPI: $(-1, -3)$
   b. $y = x^3 + 3x^2 + 3x - 2$
      $f'(x) = 3x^2 + 6x + 3$
      Critical values: $3(x^2 + 2x + 1) = 0$ or $x = -1$
   c. Critical point: (–1, –3)
   d. Yes. From the graph, (–1, –3) is a horizontal point of inflection.

15. $y = \frac{1}{2}x^2 - x$
   a. $f'(x) = x - 1$
   b. $f'(x) = 0$ if $x = 1$
   c. At $x = 1$ we have $y = \frac{1}{2}(1)^2 - 1 = -\frac{1}{2}$.
      Critical point: $\left(1, -\frac{1}{2}\right)$
   d.

x	0	1	2
$f'(x)$	–	0	+

   Decreasing if $x < 1$
   Increasing if $x > 1$
   e.

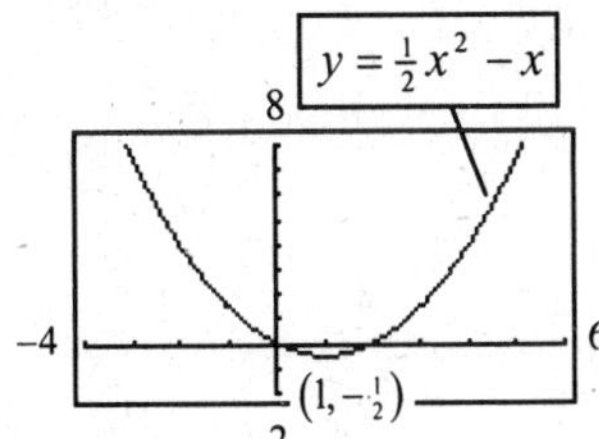

17. $y = \frac{1}{3}x^3 + \frac{1}{2}x^2 - 2x + 1$
   a. $f'(x) = x^2 + x - 2$
   b. $x^2 + x - 2 = 0$
      $(x+2)(x-1) = 0$
      Critical values are $x = -2, 1$.
   c. $f(-2) = -\frac{8}{3} + 2 + 4 + 1 = \frac{13}{3}$
      $f(1) = \frac{1}{3} + \frac{1}{2} - 2 + 1 = -\frac{1}{6}$
      Critical points: $\left(-2, \frac{13}{3}\right), \left(1, -\frac{1}{6}\right)$
   d.

x		-2		1	
$f'(x)$	+	0	–	0	+

   Increasing: $x < -2$ and $x > 1$
   Decreasing: $-2 < x < 1$
   e.

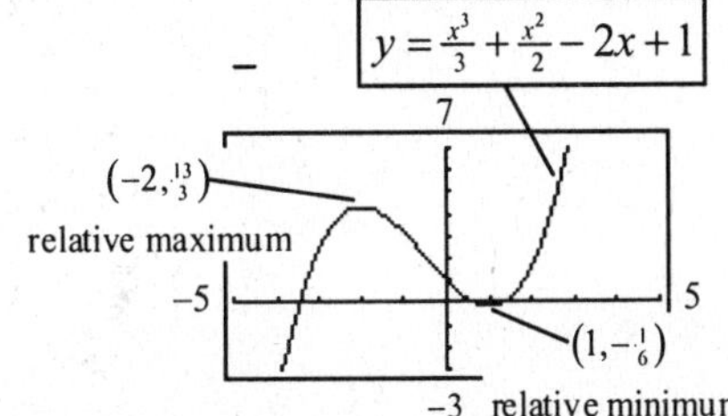

19. $y = x^{2/3}$

a. $f'(x) = \dfrac{2}{3x^{1/3}}$

b. Critical value is $x = 0$ since $f'(x)$ does not exist at $x = 0$.

c. Critical point: (0, 0)

d. Decreasing: $x < 0$ ($f'(x) < 0$)

Increasing: $x > 0$

e.

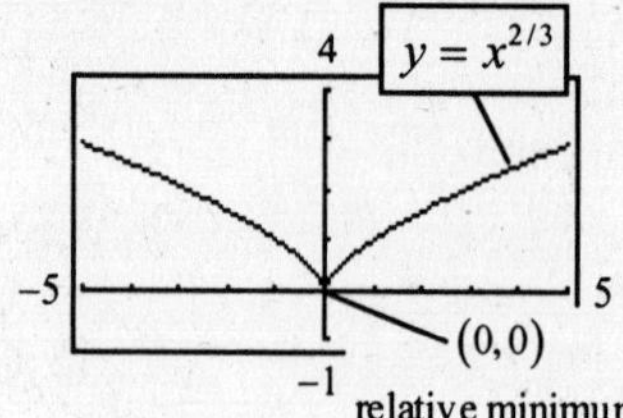

21. a. $y' = 0$ if $x = -\dfrac{1}{2}$; $y'(x) > 0$ if $x < -\dfrac{1}{2}$;

$y'(x) < 0$ if $x > -\dfrac{1}{2}$

b. $y'(x) = -1 - 2x$

$-1 - 2x = 0$ if $x = -\dfrac{1}{2}$

Substitute −1 and +1 to verify conclusion.

23. a. $y'(x) = 0$ at $x = 0, 3, -3$

$y'(x) > 0$ for $-3 < x < 3,\ x \neq 0$

$y'(x) < 0$ for $x < -3$ and for $x > 3$

b. $y'(x) = 3x^2 - \dfrac{1}{3}x^4 = 3x^2\left(1 - \dfrac{1}{9}x^2\right)$

Substitute −4, −3, −1, 0, 1, 3, and 4 to verify conclusion.

25.

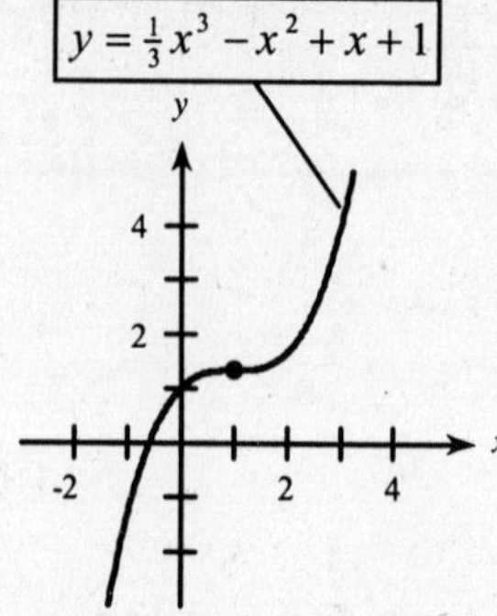

$$y = \frac{1}{3}x^3 - x^2 + x + 1$$

$$\frac{dy}{dx} = x^2 - 2x + 1 = (x-1)^2$$

Critical value is $x = 1$.

$$f(1) = \frac{1}{3} - 1 + 1 + 1 = \frac{4}{3}$$

Critical point: $\left(1, \dfrac{4}{3}\right)$

x	1
$f'(x)$	+ 0 +

No relative maximum or minimum.
Horizontal point of inflection at $x = 1$.

27.

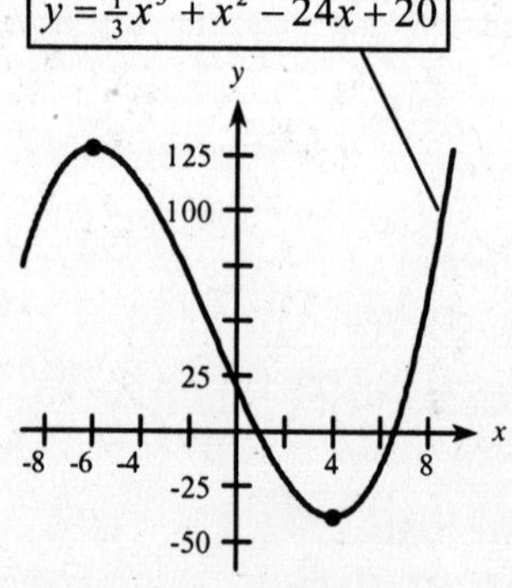

$$y = \frac{1}{3}x^3 + x^2 - 24x + 20$$

$$\frac{dy}{dx} = x^2 + 2x - 24 = (x+6)(x-4)$$

Critical values are $x = -6, 4$.

$$f(-6) = 128;\ f(4) = -\frac{116}{3}$$

Critical points: (−6, 128), $\left(4, -\frac{116}{3}\right)$

x	−6 4
$f'(x)$	+ 0 − 0 +

Relative maximum at $x = -6$.
Relative minimum at $x = 4$.

29.

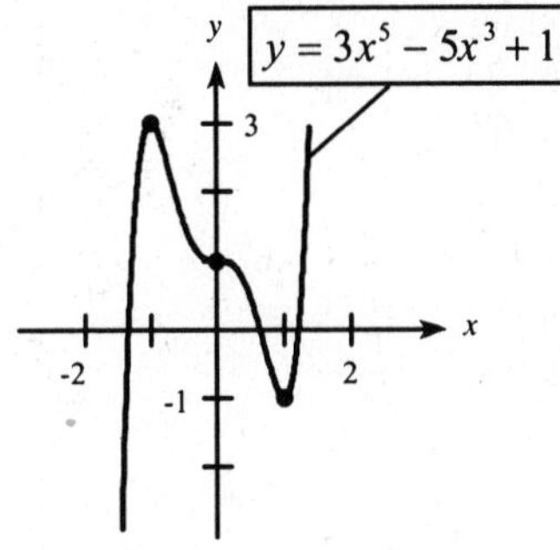

$y = 3x^5 - 5x^3 + 1$

$y' = 15x^4 - 15x^2 = 15x^2(x^2 - 1)$

Critical values are $x = 0, 1, -1$.

Critical points are (0, 1), (1, −1), (−1, 3)

x		−1		0		1	
$f'(x)$	+	0	−	0	−	0	+

HPI at $x = 0$.

Relative maximum at $x = -1$.

Relative minimum at $x = 1$.

31.

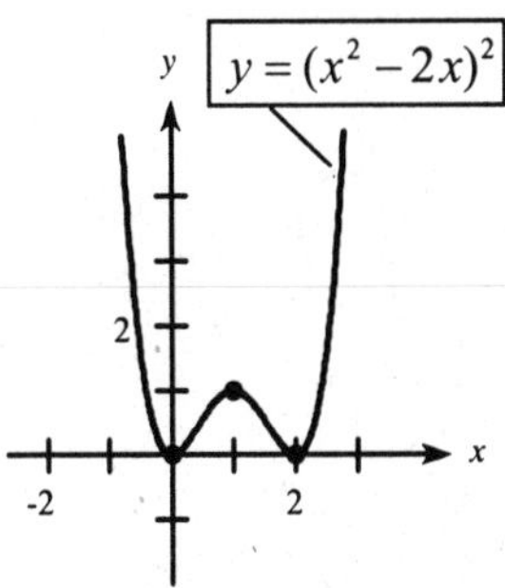

$y = (x^2 - 2x)^2$

$y' = 4x(x-1)(x-2)$

Critical values at $x = 0, 1, 2$.

$f(0) = 0$; $f(1) = 1$; $f(2) = 0$

Critical points: (0, 0), (1, 1), (2, 0)

x		0		1		2	
$f'(x)$	−	0	+	0	−	0	+

Relative minimum at $x = 0, 2$.

Relative maximum at $x = 1$.

33.

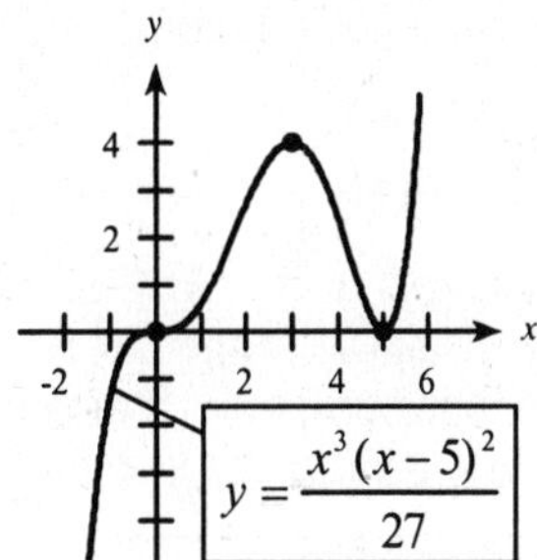

$y = \frac{1}{27}x^3(x-5)^2$

$y' = \frac{5x^2(x-3)(x-5)}{27}$

Critical values are $x = 0, 5, 3$.

Critical points are (0, 0), (5, 0), (3, 4).

x		0		3		5	
$f'(x)$	+	0	+	0	−	0	+

Relative maximum at $x = 3$.

Relative minimum at $x = 5$.

HPI at $x = 0$.

35.

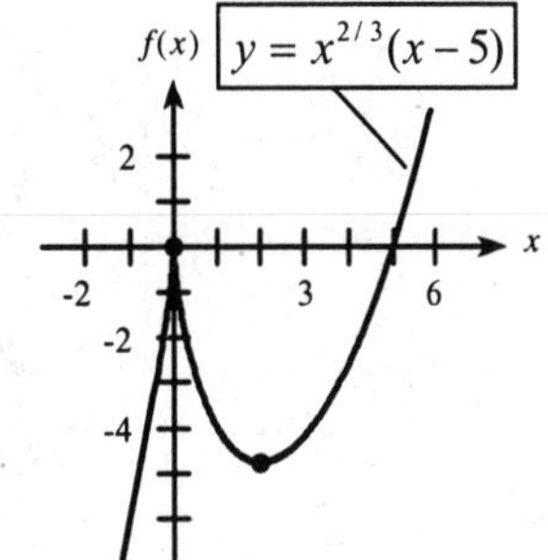

$f(x) = x^{2/3}(x-5)$

$f'(x) = \frac{5(x-2)}{3x^{1/3}}$

Critical values are $x = 0, 2$.

Critical points are (0, 0) and $\left(2, -3\sqrt[3]{4}\right)$.

Note that $-3\sqrt[3]{4} \approx -4.8$.

x	−1	0	1	2	3
$f'(x)$	+	*	−	0	+

* means $f'(0)$ is undefined.

Relative maximum at $x = 0$.

Relative minimum at $x = 2$.

37. $f(x) = x^3 - 225x^2 + 15000x - 12000$

$f'(x) = 3x^2 - 450x + 15000$

$= 3(x^2 - 150x + 5000)$

$= 3(x - 100)(x - 50)$

Critical values: $x = 50$ and $x = 100$

Critical points: (50, 300,500), (100, 238,000)

Viewing window: $0 \le x \le 150,\ 0 \le y \le 400{,}000$

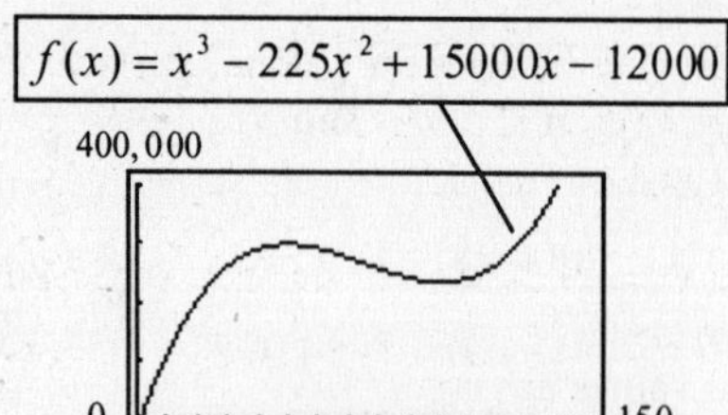

39. $f(x) = x^4 - 160x^3 + 7200x^2 - 40000$

$f'(x) = 4x^3 - 480x^2 + 14400x$

$= 4x(x^2 - 120x + 3600)$

$= 4x(x - 60)(x - 60)$

Critical values: $x = 0$ and $x = 60$

Critical points: (0, –40,000) and (60, 4,280,000)

Viewing window:

$-30 \le x \le 90,\ -500{,}000 \le y \le 5{,}000{,}000$

(There are other windows.)

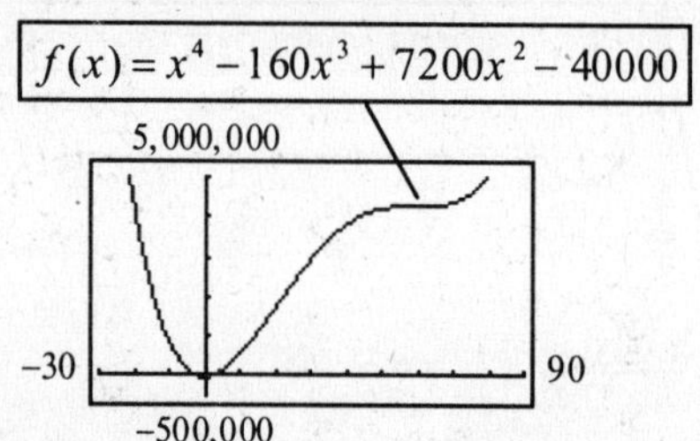

41. $y = 7.5x^4 - x^3 + 2$

$y' = 30x^3 - 3x^2 = 3x^2(10x - 1)$

Critical values: $x = 0,\ 0.1$

Critical points: (0, 2) and (0.1, 1.99975)

Viewing window:

$-0.1 \le x \le 0.2,\ \ 1.9997 \le y \le 2.0007$

Viewing windows can vary.

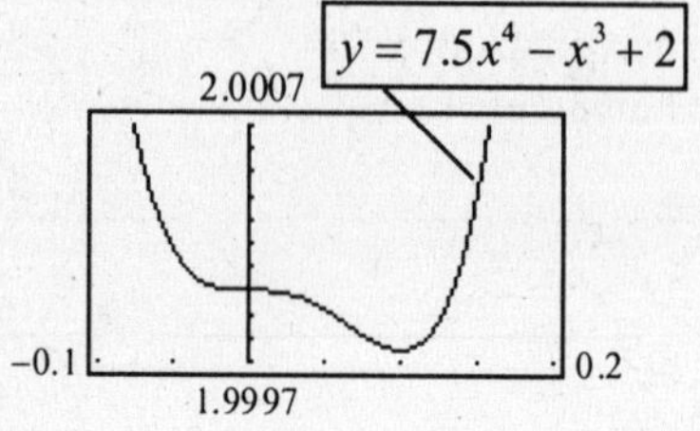

43. Possible graph for $f(x)$.

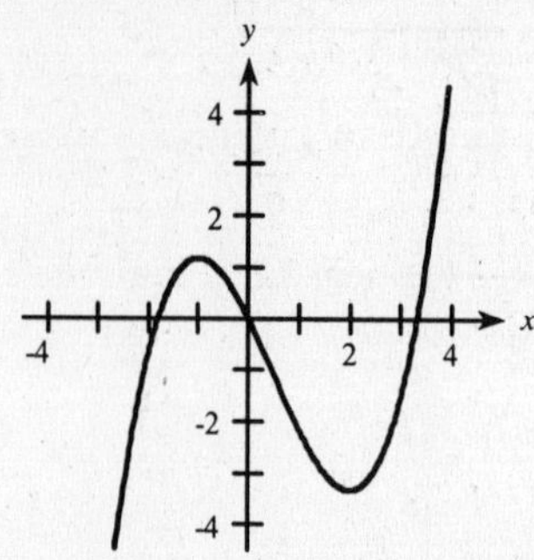

$f'(x) = x^2 - x - 2 = (x - 2)(x + 1)$

Note: The given graph is $f'(x)$ and NOT $f(x)$.

Critical values: $x = -1,\ x = 2$

$f(x)$ is increasing if $x < -1$ and $x > 2$.

$f(x)$ is decreasing if $-1 < x < 2$.

Thus, rel max at $x = -1$; rel min at $x = 2$.

45. Possible graph of $f(x)$.

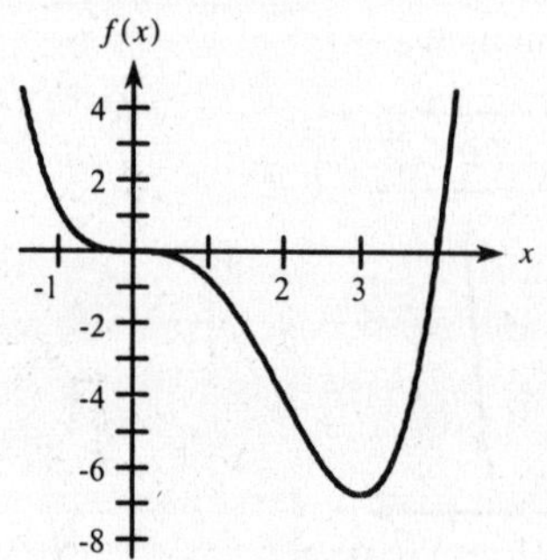

$f'(x) = x^3 - 3x^2 = x^2(x - 3)$

Note: The given graph is $f'(x)$ and NOT $f(x)$.

Critical values: $x = 0,\ x = 3$

$f(x)$ is increasing if $x > 3$.

$f(x)$ is decreasing if $x < 3,\ x \ne 0$.

Thus, relative min at $x = 3$; HPI at $x = 0$.

47. Graph on the left is $f(x)$.

Graph on the right is $f'(x)$.

If $f'(x) > 0$, then $f(x)$ is increasing (rising).

If $f'(x) < 0$, then $f(x)$ is decreasing (falling).

49. $S(t) = 1000 + 400(t + 1)^{-1}$ $\quad S'(t) = \dfrac{-400}{(t + 1)^2}$

$S'(t) < 0$ for all t. Thus, $S(t)$ is always decreasing for $t \ge 0$.

51. $P(t)=27t+6t^2-t^3$

$P'(t)=27+12t-3t^2=3(9+4t-t^2)$

a. $t=\frac{-4\pm\sqrt{16+36}}{-2}=\frac{4\pm2\sqrt{13}}{2}=2\pm\sqrt{13}$

b. $t=2+\sqrt{13}$ is the only value in the domain of the function.

c.

t		$2+\sqrt{13}$	
$P'(t)$	+	0	−

$P(t)$ increasing if $0\le t<2+\sqrt{13}$.

d.

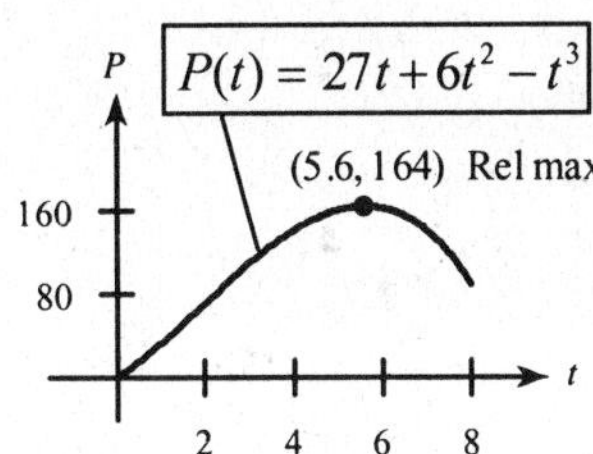

53. $\overline{C}(x)=5000x+125{,}000x^{-1}$

$\overline{C}'(x)=5000-125{,}000x^{-2}$

a. $\overline{C}'(x)=0$ if $x^2-25=0$

Critical value: $x=5$

b. $\overline{C}'(x)<0$ if $x<5$.

c. $\overline{C}'(x)>0$ if $x>5$.

55. a. Increasing at $x = 150$, decreasing at $x = 350$, changing from increasing to decreasing at $x=250$.

b. $R'(x)>0$ if $x<250$

c. $x=250$ units

57. $R(t)=\frac{50t}{t^2+36}$

$R'(t)=\frac{(t^2+36)(50)-50t(2t)}{(t^2+36)^2}=\frac{1800-50t^2}{(t^2+36)^2}$

a. $R'(t)=0$ if $50(36-t^2)=0$ or $t=6$.

(−6 is not in the domain.)

b.

t		6	
$R'(t)$	+	0	−

Revenue will increase for 6 weeks.

59. $P(t)=\frac{13t}{t^2+100}+0.18$

$P'(t)=\frac{(t^2+100)(13)-13t(2t)}{(t^2+100)^2}=\frac{13(100-t^2)}{(t^2+100)^2}$

Domain = $\{t:\ t\ge0\}$. Critical value is $t = 10$.

t		10	
$P'(t)$	+	0	−

a. Ten months after the campaign starts the recognition is a maximum.

b. Begins on Jan. 1.

61. a. $f(x)=-9.3458x^2+34.0863x+235.8625$

b. $f'(x)=-18.6916x+34.0863$

$f'(x)=0$ if $x=\frac{34.0863}{18.6916}=1.8236$

$f'(x)$ changes from + to − at $x\approx1.8236$.

Maximum deficit from model is in 1992.
Maximum deficit from data is in 1992.

63. a. cubic

b. $y=-0.000316x^3+0.0174x^2$
$-0.0688x+4.313$

c. The graph indicates a maximum at $x=34.61$ or in 1984.

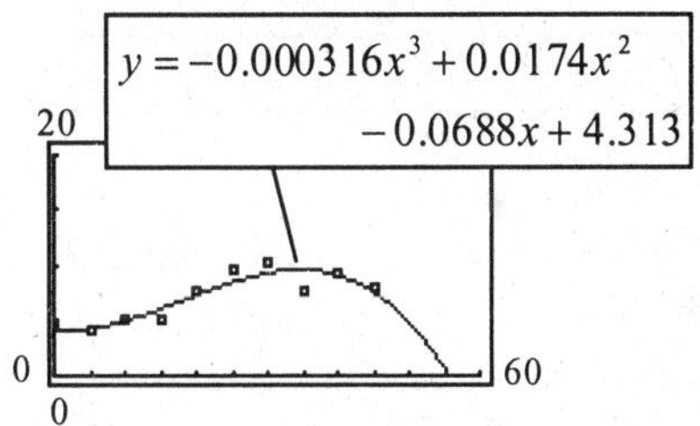

65. a. Use a cubic model.

b. $y=6.1636x^3-90.5303x^2$
$+373.5975x+202.867$

c. $(2.97, 675)$

d. In the early 90's the maximum grant was approximately \$675 million.

e. Absolute max of \$1050 million in 2000.

Exercise 10.2

1. $f(x)=x^3-3x^2+1$

$f'(x)=3x^2-6x$

$f''(x)=6x-6=6(x-1)$

a. $f''(-2)<0$ concave downward

b. $f''(3)>0$ concave upward

3. $f(x)=2x^3+4x-8$

$f'(x)=6x^2+4$

$f''(x)=12x$

a. $f''(-1)<0$ concave downward

b. $f''(4)>0$ concave upward

5. (a, c) and (d, e)

7. (c, d) and (e, f)

9. c, d and e

11. $f(x) = x^3 - 6x^2 + 5x + 6$

$f'(x) = 3x^2 - 12x + 5$

$f''(x) = 6x - 12$

$f''(x) = 0$ if $x = 2$. We have $f(2) = 0$.

x		2	
$f''(x)$	$-$	0	$+$

The point (2, 0) is a point of inflection.
$f(x)$ is concave up if $x > 2$. $f(x)$ is concave down if $x < 2$.

13. $y = f(x) = \frac{1}{4}x^4 + \frac{1}{2}x^3 - 3x^2 + 3$

$y' = x^3 + \frac{3}{2}x^2 - 6x$

$y'' = 3x^2 + 3x - 6 = 3(x+2)(x-1)$

$f(-2) = -9 \qquad f(1) = \frac{3}{4}$

x		-2		1	
y''	$+$	0	$-$	0	$+$

The points $(-2, -9)$ and $\left(1, \frac{3}{4}\right)$ are points of inflection.
$f(x)$ is concave up if $x < -2$ and if $x > 1$.
$f(x)$ is concave down if $-2 < x < 1$.

15. $y = x^2 - 4x + 2$

$y' = 2x - 4$

$y'' = 2$

Critical value: $x = 2$

$y'' > 0$ at $x = 2$.

There is no point of inflection.
There is a relative minimum at $(2, -2)$.

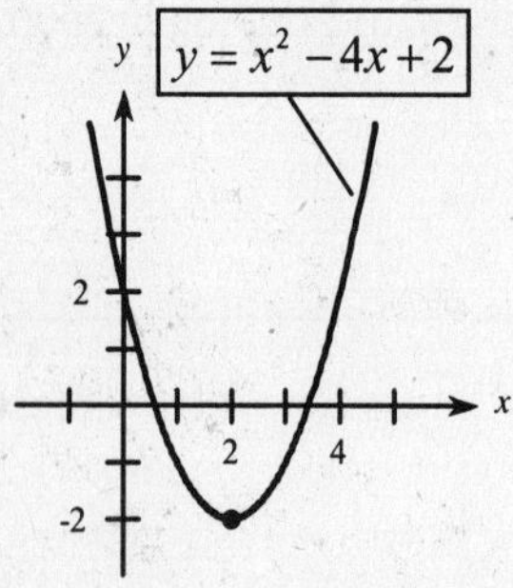

17. $y = \frac{1}{3}x^3 - 2x^2 + 3x + 2$

$y' = x^2 - 4x + 3 = (x-3)(x-1)$

$y'' = 2x - 4$

Point of inflection at $x = 2$. (concavity changes)
Relative minimum at $x = 3$ ($y'' > 0$).
Relative maximum at $x = 1$ ($y'' < 0$).

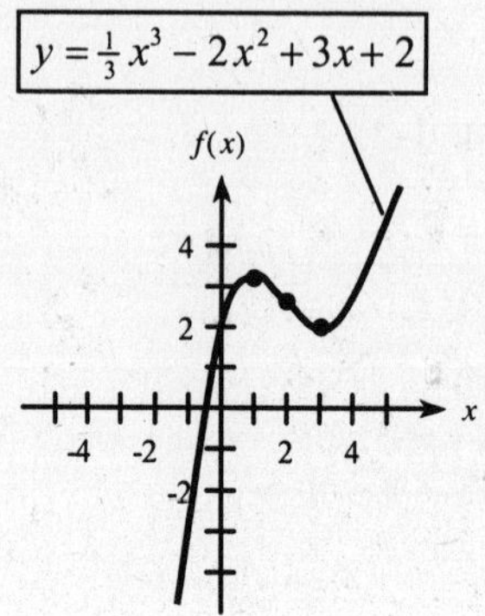

19. $y = x^4 - 16x^2$

$y' = 4x^3 - 32x = 4x(x^2 - 8)$

$y'' = 12x^2 - 32 = 4(3x^2 - 8)$

Critical values are $x = 0, \pm 2\sqrt{2}$.

Possible inflection points at $x = \pm\frac{2\sqrt{6}}{3}$

x		$-\frac{2\sqrt{6}}{3}$		$\frac{2\sqrt{6}}{3}$	
y''	$+$	0	$-$	0	$+$

$\left(\frac{-2\sqrt{6}}{3}, -\frac{320}{9}\right)$ and $\left(\frac{2\sqrt{6}}{3}, -\frac{320}{9}\right)$ are inflection points.

x		$-2\sqrt{2}$		0		$2\sqrt{2}$	
y'	$-$	0	$+$	0	$-$	0	$+$

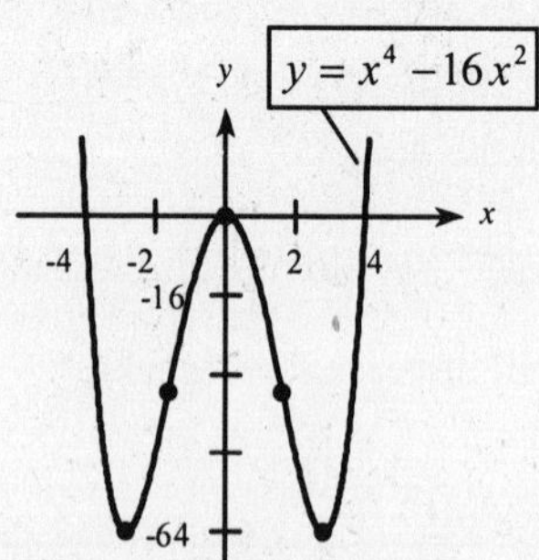

Relative maximum at (0, 0). Relative minima at $(-2\sqrt{2}, -64)$ and $(2\sqrt{2}, -64)$.

21.

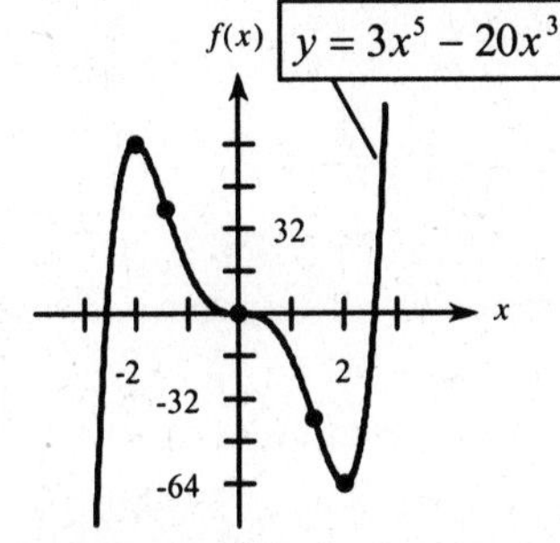

Critical values are $x = -2, 0, 2$.
Possible points of inflection at $x = -\sqrt{2}$, $x = 0$, and $x = \sqrt{2}$.

x		$-\sqrt{2}$		0		$\sqrt{2}$	
$f''(x)$	$-$	0	$+$	0	$-$	0	$+$

$(-\sqrt{2}, 39.6)$, $(0, 0)$ and $(\sqrt{2}, -39.6)$ are points of inflection.
$f''(0) = 0$ means second derivative test fails.
$f'(x) < 0$ for values near 0, both to the left and right of 0 means HPI at (0, 0).
$f''(-2) < 0$ means relative maximum at (–2, 64).
$f''(2) > 0$ means relative minimum at (2, –64).

23.

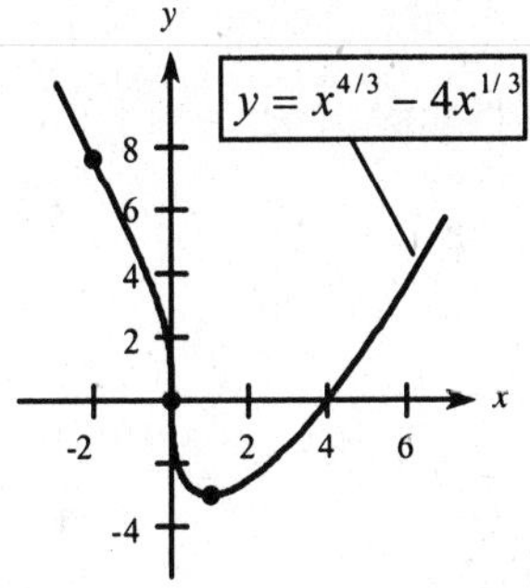

Critical values are $x = 0, 1$.
Possible points of inflection are at $x = -2, 0$.

x		-2		0	
$f''(x)$	$+$	0	$-$	*	$+$

* means $f''(0)$ is undefined.
$(0,0)$ and $(-2, 7.6)$ are points of inflection.

x		0		1	
$f'(x)$	$-$	*	$-$	0	$+$

* means $f'(0)$ is undefined
$(1, -3)$ is a relative minimum.

25. **a.** $f''(x) > 0$ if $x < 1$
$f''(x) < 0$ if $x > 1$
$f''(x) = 0$ if $x = 1$

b. $f'(x)$ [NOT $f(x)$] has rel max at $x = 1$.

c. $f(x) = -\frac{1}{3}x^3 + x^2 + 8x - 12$
$f'(x) = -x^2 + 2x + 8$
$f''(x) = -2x + 2$

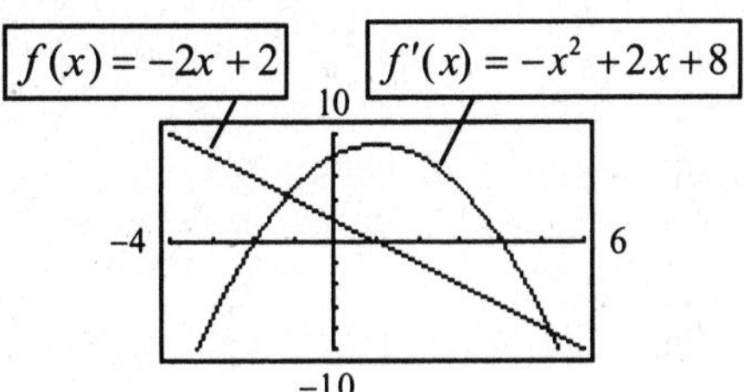

27. $f'(x) = 4x - x^2$

a. If $x < 2$, $f'(x)$ is increasing.
Thus, $f''(x) > 0$ and f is concave up.
If $x > 2$, $f'(x)$ is decreasing.
Thus, $f''(x) < 0$ and f is concave down.

b. Point of inflection at $x = 2$.

c. $f''(x) = 4 - 2x$

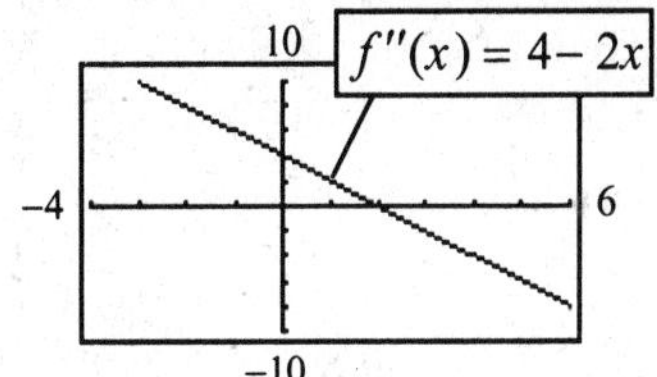

d. All graphs have the same basic form.

possible graph of $f(x)$

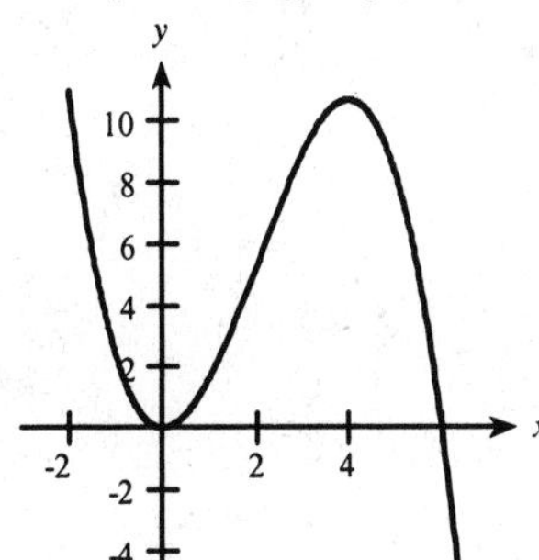

29. **a.** G
b. C
c. F
d. H
e. I

31. a. Concave up if $x < 0$; concave down if $x > 0$; point of inflection at $x = 0$.

b. Concave up if $-1 < x < 1$; concave down if $x < -1$ and if $x > 1$; points of inflection at $x = \pm 1$.

c. Concave up if $x > 0$; concave down if $x < 0$; point of inflection at $x = 0$.

33. a. $P'(t)$ or $\dfrac{dP}{dt}$

b. $P''(t) = 0$ at B.

c. C

35. a. Concavity changes at C.

b. $S''(t) > 0$ means concave up or to the right of C.

c. Yes. $S'(t)$ is the rate of change of sales. $S''(t)$ changes from – to + at C.

37. $P(t) = 27t + 12t^2 - t^3$

$P'(t) = 27 + 24t - 3t^2 = 3(9-t)(1+t)$

$P''(t) = 24 - 6t = 6(4-t)$

a.

t		9	
$P'(t)$	+	0	–

Production is maximized at $t = 9$. But the domain is $0 \le t \le 8$. Thus, production is maximized when $t = 8$.

b. $P''(t)$ changes signs at $t = 4$. Point of diminishing returns begins at $t = 4$.

39. $S(t) = 3(t+3)^{-1} - 18(t+3)^{-2} + 1$

a. $S'(t) = \dfrac{-3}{(t+3)^2} + \dfrac{36}{(t+3)^3}$

$= \dfrac{-3(t+3)+36}{(t+3)^3} = \dfrac{27-3t}{(t+3)^3}$

Critical value is $t = 9$.

t		9	
$S'(t)$	+	0	–

Sales volume is maximized after 9 days.

b. $S''(t) = \dfrac{6}{(t+3)^3} - \dfrac{108}{(t+3)^4}$

$= \dfrac{6(t+3)-108}{(t+3)^4} = \dfrac{6t-90}{(t+3)^4}$

Critical value is $t = 15$.

t		15	
$S''(t)$	–	0	+

Point of diminishing returns is $t = 15$ days.

41. $y = 0.0013x^4 - 0.0583x^3 + 0.881x^2 - 5.235x + 13.772$

$y' = 0.0052x^3 - 0.1749x^2 + 1.762x - 5.235$

$y'' = 0.0156x^2 - 0.3498x + 1.762$

The graph of y'' changes from – to + at $x = 14.78$ or in 1995.

43. a. $y = 0.0000591x^3 - 0.00675x^2 + 0.0523x + 14.147$

b. $y' = 0.0001773x^2 - 0.0135x + 0.0523$

y' changes from – to + at $x \approx 72$.

Critical point is $(72, 5)$.

c. Percentage of foreign born is a minimum of approximately 5% in 1972.

45. a. $y = -0.4052x^3 + 6.0161x^2 - 17.9605x + 52$

b. $y' = -1.2156x^2 + 12.0322x - 17.9605$

The graphing utility shows that a minimum occurs at $x \approx 1.83$ or in 1913,

c. A maximum occurs at $x \approx 8.07$ or in 1920.

Exercise 10.3

1. $f(x) = x^3 - 2x^2 - 4x + 2$, $[-1, 3]$

$f'(x) = 3x^2 - 4x - 4 = (3x+2)(x-2)$

$f'(x) = 0$ if $x = 2$ or $x = -\dfrac{2}{3}$

x	-1	$-\frac{2}{3}$	2	3
$f(x)$	3	$\frac{94}{27}$	-6	-1

Maximum is $\dfrac{94}{27}$ at $x = -\dfrac{2}{3}$.

Minimum is -6 at $x = 2$.

3. $f(x) = x^3 + x^2 - x + 1$, $[-2, 0]$

$f'(x) = 3x^2 + 2x - 1 = (3x-1)(x+1)$

$f'(x) = 0$ if $x = -1$ or $x = \dfrac{1}{3}$

x	-2	-1	0	($\frac{1}{3}$ not in domain)
$f(x)$	-1	2	1	

Maximum is 2 at $x = -1$.
Minimum is -1 at $x = -2$.

5. **a.** $R(x) = 36x - 0.01x^2$ $R'(x) = 36 - 0.02x$

$R'(x) = 0$ if $36 - 0.02x = 0$ or $x = 1800$ units

$R'(1000) > 0$; $R'(1800) = 0$; $R'(2000) < 0$

Total revenue is maximized at 1800 units.

$R(1800) = \$32,400$ is the maximum revenue.

b. $R(1500) = \$31,500$

7. $R(x) = 2000x - 20x^2 - x^3$

$R'(x) = 2000 - 40x - 3x^2$

$= (100 + 3x)(20 - x)$

$R'(x) = 0$ if $x = 20$ units (Domain ≥ 0)

$R'(10) > 0$; $R'(20) = 0$; $R'(30) < 0$

Total revenue is maximized at 20 units.

$R(20) = 40,000 - 8,000 - 8,000 = \$24,000$

9. Let x = number of people above 30.

R = (number of people)(price per person)

$= (30 + x)(10 - 0.20x)$

$= 300 + 4x - 0.20x^2$

$R'(x) = 4 - 0.40x$

$R'(x) = 0$ if $x = 10$

x	0	10	20
$R(x)$	300	320	300

Therefore, $30 + 10 = 40$ people will maximize revenue.

11. Let x = new customers

$R(x) = (1000 + x)(20 - 0.01x)$

$= 20,000 + 10x - 0.01x^2$

Since the price must be positive, we have $0 \leq x \leq 2000$.

$R'(x) = 10 - 0.02x$

$R'(x) = 0$ if $x = 500$.

x	0	500	2000
$R(x)$	$20,000$	$22,500$	0

We maximize revenue with 500 new customers and a price of \$15. The maximum revenue is $R(500) = 1500(15) = \$22,500$.

13. $R(x) = 2000x + 20x^2 - x^3$

$\overline{R}(x) = \dfrac{R(x)}{x} = 2000 + 20x - x^2$

a. $\overline{R}'(x) = 20 - 2x$

$\overline{R}'(x) = 0$ if $x = 10$.

$\overline{R}'(0) > 0$; $\overline{R}'(10) = 0$; $\overline{R}'(20) < 0$

$\overline{R}(10) = 2000 + 200 - 100 = \$2100$ is the maximum average revenue.

b. $\overline{R} = \dfrac{R(x)}{(x)}$ $\overline{R}'(x) = \dfrac{x \cdot R'(x) - R(x) \cdot 1}{x^2}$

$\overline{R}'(x) = 0$ if $xR'(x) - R(x) = 0$ or

$R'(x) = \dfrac{R(x)}{x}$.

Note: $\dfrac{R(x)}{x} = \overline{R}(x)$ and $R'(x) = \overline{MR}$.

$\overline{R}(x)$ is a maximum where $\overline{R}(x) = \overline{MR}$.

$2000 + 20x - x^2 = 2000 + 40x - 3x^2$

$2x^2 - 20x = 0$

$2x(x - 10) = 0$

$x = 10$

15. $\overline{C}(x) = \dfrac{25}{x} + 13 + x$

$\overline{C}'(x) = -\dfrac{25}{x^2} + 1$

$\overline{C}'(x) = 0$ if $x = 5$.

$\overline{C}'(4) < 0$; $\overline{C}'(5) = 0$; $\overline{C}'(6) > 0$

$\overline{C}(5) = \dfrac{25}{5} + 13 + 5 = \$23$

Minimum average cost is \$23 at 5 units of production.

17. $\bar{C}(x)=\frac{100}{x}+x$

$\bar{C}'(x)=-\frac{100}{x^2}+1$

$\bar{C}'(x)=0$ if $x=10$.

$\bar{C}'(9)<0;\ \bar{C}'(10)=0;\ \bar{C}'(11)>0$

$\bar{C}(10)=\frac{100}{10}+10=\$20$

Minimum average cost is \$20 at 10 units of production.

19. $\bar{C}(x)=\frac{(x+4)^3}{x}$

$\bar{C}'(x)=\frac{x(3)(x+4)^2-(x+4)^3\cdot 1}{x^2}$

$=\frac{(x+4)^2(2x-4)}{x^2}$

$\bar{C}'(x)=0$ if $2x-4=0$ or $x=2$.

Total number of units is $2(100)=200$.

$\bar{C}'(1)<0;\ \bar{C}'(2)=0;\ \bar{C}'(3)>0$

$\bar{C}(2)=\frac{6^3}{2}=\$108$ per 100 units.

Minimum average cost per unit is $\frac{108}{100}=\$1.08$.

21. $\bar{C}(x)=\frac{C(x)}{x}$ $\qquad \bar{C}'(x)=\frac{x\cdot C'(x)-C(x)\cdot 1}{x^2}$

$\bar{C}'(x)=0$ if $x\cdot C'(x)-C(x)=0$ or $C'(x)=\frac{C(x)}{x}$

But, $\frac{C(x)}{x}=\bar{C}(x)$ and $C'(x)=\overline{MC}$. Thus, the minimum average cost occurs where $\bar{C}(x)=\overline{MC}$. Note that this is true for all cost functions. Using the $C(x)$ of problem 15:

$\overline{MC}=13+2x=\frac{25}{x}+13+x=\bar{C}(x)$

$x^2=25$ or $x=5$

The same answer as for problem 15.

23.

a. A line from $(0,0)$ to $(x,C(x))$ has slope $\frac{C(x)}{x}=\bar{C}(x)$. This is minimized when the line has the least rise. This occurs when the line is tangent to $C(x)$.

b. For 23, the level is approximately 600 units. For 24, the level is approximately 200 units.

25. $P(x)=5600x+85x^2-x^3-200{,}000$

$P'(x)=5600+170x-3x^2=(80-x)(70+3x)$

$P'(x)=0$ if $x=80$ units, (domain: $x\ge 0$)

$P'(70)>0;\ P'(80)=0;\ P'(90)<0$

$P(80)=\$280{,}000$ is the maximum profit.

27. $P(x)=4600x-(45{,}000+100x+x^3)$

$=4500x-x^3-45{,}000$

$P'(x)=4500-3x^2$

$P'(x)=0$ if $3x^2=4500$ or if $x^2=1500$ or $x=10\sqrt{15}\approx 39$ units.

$P'(30)>0;\ P'(39)\approx 0;\ P'(40)<0$

$P(39)=\$71{,}181$ is the maximum profit.

29. $P(x)=250x-0.01x^2-(300+200x)$

$=-0.01x^2+50x-300$

$P'(x)=-0.02x+50 \qquad P'(x)=0$ if $x=2500$.

Since $P'(x)>0$ for all x in the domain ($0\le x\le 1000$), produce the maximum of 1000 units and obtain the maximum profit. That maximum profit is $P(1000)=\$39{,}700$.

31. **a.** B

b. B

c. B ($P'=0$ if $R'-C'=0$)

d. P is a max when $P'=0$ or $\overline{MR}=\overline{MC}$

33. Let $x=$ number of vacant units.

$P(x)=(720+20x)(50-x)-12(50-x)$

$=35{,}400+292x-20x^2$

$P'(x)=292-40x$

$P'(x)=0$ if $x=7.3$ units

x	0	7	8	50
$P(x)$	35,400	36,464	36,456	0

Using $x=7$ units, maximum profit is obtained if rent is $720+20(7)=\$860$.

35. $C(x)=45{,}000+100x+x^2$

$R(x)=1600x$

$P(x)=R(x)-C(x)=-x^2+1500x-45{,}000$

$P'(x)=-2x+1500$

$P'(x)=0$ if $x=750$, but this is not in the domain.

x	0	600
$P(x)$	−45,000	495,000

Maximum profit is at 600 units due to limited production. $P(600)=\$495{,}000$

37. $R(x) = p \cdot x = 600x - \frac{1}{2}x^2$

$C(x) = \overline{C}(x) \cdot x = 300x + 2x^2$

$P(x) = \left(600x - \frac{1}{2}x^2\right) - (300x + 2x^2)$

$= 300x - \frac{5}{2}x^2$

$P'(x) = 300 - 5x$

$P''(x) = -5 < 0$ for all x.

$P'(x) = 0$ if $x = 60$.

a. Maximum profit is at $x = 60$.

b. Selling price is $p = 600 - \frac{1}{2} \cdot 60 = \$570$.

c. Maximum profit is $P(60) = \$9000$.

39. $R(x) = p \cdot x = 1960x - \frac{1}{3}x^3$

$C(x) = \overline{C}(x) \cdot x = 1000x + 2x^2 + x^3$

$P(x) = 960x - 2x^2 - \frac{4}{3}x^3, 0 \le x \le 10$

1000 units = 10 hundreds

$P'(x) = 960 - 4x - 4x^2$

$= 4(16 + x)(15 - x)$

$P'(x) = 0$ if $x = 15$, but this is not in the domain.

a. Maximum profit at 1000 units ($x = 10$) due to limited production.

b. $P(10) = \$8066.67$

41. $R(x) = p \cdot x = 120x - 0.015x^2$

$C(x) = \overline{C}(x) \cdot x$

$= 10,000 + 60x - 0.03x^2 + 0.00001x^3$

$P(x) = -10,000 + 60x + 0.015x^2 - 0.00001x^3$

$P'(x) = 60 + 0.03x - 0.00003x^2$

$P'(x) = 0$ if

$$x = \frac{-0.03 \pm \sqrt{0.0009 + 0.0072}}{-0.00006}$$

$$= \frac{-0.03 \pm 0.09}{-0.00006}$$

$= 2000$ (positive value only)

$P'(1090) > 0; P'(2000) = 0; P'(2010) < 0$

Maximum profit is at $x = 2000$ units.

Price is $p = 120 - 0.015(2000) = \$90$ per unit.

Maximum profit is $P(2000) = \$90,000$.

43. a. $y = 0.0002524x^3 - 0.02785x^2 + 1.6300x + 2.1550$

b. $y' = 0.0007572x^2 - 0.05570x + 1.6300$

$y'' = 0.0015144x - 0.05570$

y'' changes signs (equals zero) at

$x = \frac{0.05570}{0.0015144} = 36.78$.

Point of inflection is $(36.78, 36.99)$.

c.

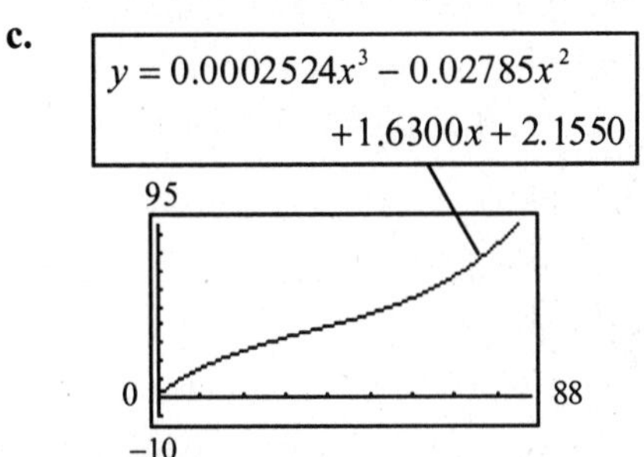

d. The rate of change (y') was decreasing until 1987 (from b). After 1987, the rate of change has been increasing ($y'' > 0$). Since 1987 the number of beneficiaries has been increasing at an increasing rate.

45. a. Absolute max was reached in May.

b. Absolute min was reached in September. This was triggered by the 9-11 terrorist attack.

47. a. From the graph, the absolute maximum for $f(t)$ is 16.5.

b. From the graph, the absolute minimum for $f(t)$ is 1.9.

c. The graph suggests that Social Security taxes might rise in the early twenty-first century in order that fewer people can pay in enough to cover those already retired.

Exercise 10.4

1. $S_1 = 30 + 20x_1 - 0.4x_1^2$

$S_2 = 20 + 36x_2 - 1.3x_2^2$

a. $S_1' = 20 - 0.8x_1$

$S_1' = 0$ if $x_1 = \$25$ million

$S_2' = 36 - 2.6x_2$

$S_2' = 0$ if $x_2 \approx \$13.846$ million

b. Total money needed is $25 + 13.846 = \$38.846$ million.

3. $P(x) = 800x - x^2$

$P'(x) = 800 - 2x$

$P'(x) = 0$ if $x = 400$

$P'(390) > 0;\ P'(400) = 0;\ P'(410) < 0$

Thus, 400 trees per acre will maximize the profit.

5. $y = 70t + \frac{1}{2}t^2 - t^3$

$y' = 70 + t - 3t^2 = (5 - t)(14 + 3t)$

a. $y' = 0$ if $t = 5$.

$y'(4) > 0$

$y'(5) = 0$

$y'(6) < 0$

Production is maximized after 5 hours.

b. $y(5) = 350 + 12.5 - 125 = 237.5$

7. $E(p) = 10{,}000p - 100p^2$

$E'(p) = 10{,}000 - 200p$

$E'(p) = 0$ if $p = 50$

$E'(40) > 0,\ E'(50) = 0,\ E'(60) < 0$

Expenditure is greatest at $p = \$50$.

9. $R = \frac{c}{2}m^2 - \frac{m^3}{3}$

$R' = cm - m^2$

$R' = 0$ if $m = 0$ or $m = c$.

$R'(0) = 0;\ R'\left(\frac{c}{2}\right) > 0;\ R'(c) = 0;\ R'(2c) < 0$

Reaction is a max when $m = c$.

11. $S(t) = \frac{200t}{(t+1)^2}$

$S'(t) = \frac{(t+1)^2(200) - 200t \cdot 2 \cdot (t+1)}{(t+1)^4} = \frac{200(1-t)}{(t+1)^3}$

$S'(t) = 0$ if $t = 1$

$S'(0) > 0,\ S'(1) = 0,\ S'(2) < 0$

Sales are maximized after one week.

13. $p(t) = \frac{6.4t}{t^2 + 64} + 0.05$

$p'(t) = \frac{(t^2 + 64)(6.4) - (6.4t)(2t)}{(t^2 + 64)^2}$

$= \frac{6.4(64 - t^2)}{(t^2 + 64)^2}$

$p'(t) = 0$ if $t = 8$.

$p'(7) > 0;\ p'(8) = 0;\ p'(9) < 0$

Awareness is maximized after 8 days.
Maximum percentage that identify 2 defendants is $p(8) = 45\%$.

15. Quantity being minimized: $F = 4x + 3y$

Another equation: $A = xy = 1200$ or $y = \frac{1200}{x}$

Substituting: $F(x) = 4x + \frac{3600}{x}$

$F'(x) = 4 - \frac{3600}{x^2}$

$F'(x) = 0$ if $x = 30$.

$F'(29) < 0;\ F'(30) = 0;\ F'(31) > 0$

Fence needed is $F(30) = 120 + 120 = 240$ feet.

17. Quantity being minimized: $C = 20x + 5(2y)$

Another equation: $A = xy = 45{,}000$ or $y = \frac{45{,}000}{x}$

Substituting: $C(x) = 20x + \frac{450{,}000}{x}$

$C'(x) = 20 - \frac{450{,}000}{x^2}$

$C'(x) = 0$ if $x^2 = 22{,}500$ or $x = 150$

$C'(140) < 0;\ C'(150) = 0;\ C'(160) > 0$

Then $y = \frac{45{,}000}{150} = 300$.

Dimensions are 150 ft for fence parallel to the river and 300 ft for each of the other sides.

19. Quantity being maximized: $A = xy$

Another equation:

$800 = 10(2x) + 10(2y) + 20(2y)$

$800 = 20x + 60y$ or $x = 40 - 3y$

Substituting: $A(y) = y(40 - 3y) = 40y - 3y^2$

$A'(y) = 40 - 6y = 0$ if $y = \frac{20}{3}$

$A'(6) > 0;\ A'\left(\frac{20}{3}\right) = 0;\ A'(7) < 0$

Then $x = 40 - 3\left(\frac{20}{3}\right) = 20$.

Dimensions are $\frac{20}{3}$ ft for the side parallel to the dividers and 20 ft for the other side.

21. $256 = x(2x)h$

$h = \frac{128}{x^2}$

$C = 10(x)(2x) + 5(x)(2x) + 5(2)(h)2x + 5(2x)(h)$

$= 30x^2 + 30xh$

$C(x) = 30x^2 + 30x\left(\frac{128}{x^2}\right) = 30x^2 + \frac{3840}{x}$

$C'(x) = 60x - \frac{3840}{x^2}$

$C'(x) = 0$ if $60x^3 = 3840$ or

$x^3 = 64$ or $x = 4$

$C'(3) < 0;\ C'(4) = 0;\ C'(5) > 0$

Minimum cost is at $x = 4$. Then $h = \frac{128}{16} = 8$.

Dimensions are 8" by 4" by 8" (high).

23. $C(x) = \frac{1,500,000}{x}(600) + 1,500,000(15) + \frac{x}{2} \cdot 2$

$C'(x) = \frac{-900 \times 10^6}{x^2} + 1$

$C'(x) = 0$ if $x^2 = 900 \times 10^6$ or $x = 30,000$.

$C'(1) < 0$, $C'(30,000) = 0$, $C'(40,000) > 0$

30,000 items should be produced in each run.

25. $C(x) = \frac{150,000}{x}(360) + 150,000(7) + \frac{x}{2} \cdot \frac{3}{4}$

$C'(x) = \frac{-54,000,000}{x^2} + \frac{3}{8}$

$C'(x) = 0$ if $x^2 = \frac{8}{3}(54,000,000)$ or $x = 12,000$

$C'(11,000) < 0;\ C'(12,000) = 0;\ C'(13,000) > 0$

12,000 items should be produced in each run.

27. $V(x) = x(12 - 2x)^2$

$V'(x) = x(2)(12 - 2x)(-2) + (12 - 2x)^2(1)$

$= (12 - 2x)(12 - 6x)$

$V'(x) = 0$ if $x = 2$

Domain: $0 < x < 6$

$V'(1) > 0;\ V'(2) = 0;\ V'(3) < 0$

Volume is a maximum at $x = 2$.

29. Let x = time in weeks to pick oranges.

Quantity being maximized:

$R(x) = (8 - 0.5x)(5 + 0.5x), 0 \le x \le 5$

$R'(x) = (8 - 0.5x)(0.5) + (5 + 0.5x)(-0.5)$

$= 1.5 - 0.50x$

$R'(x) = 0$ if $1.5 - 0.5x = 0$ or $x = \frac{1.5}{0.5} = 3$

x	0	3	5
$R(x)$	40	42.25	41.25

Thus, 3 weeks from now oranges should be picked.

31. Let x = number of plates and y = number of impressions.

Quantity being minimized:

$C = 2x + 12.50\left(\frac{y}{1000}\right) = 2x + 0.0125y$

Another equation: $xy = 100,000$ or $y = \frac{100,000}{x}$

Substituting: $C(x) = 2x + \frac{1250}{x}$

$C'(x) = 2 - \frac{1250}{x^2}$

$C'(x) = 0$ if $x = 25$.

$C'(24) < 0;\ C'(25) = 0;\ C'(26) > 0$

Thus, 25 plates will minimize the cost.

33. The graphing calculator shows a maximum at $t = 5.3$ or in 2005.

Exercise 10.5

1. a. $x=2$
   b. $\lim_{x\to\infty} f(x)=1$
   c. $y=1$
   d. $\lim_{x\to-\infty} f(x)=1$

   The denominator of f is 0 when $x=2$, but the numerator is not 0 when $x=2$, so $x=2$ is a vertical asymptote.

   $$\lim_{x\to\infty}\frac{x-4}{x-2}=\lim_{x\to\infty}\frac{1-\frac{4}{x}}{1-\frac{2}{x}}=\frac{1-0}{1-0}=1$$

   $$\lim_{x\to-\infty}\frac{x-4}{x-2}=\lim_{x\to-\infty}\frac{1-\frac{4}{x}}{1-\frac{2}{x}}=\frac{1-0}{1-0}=1$$

3. a. $x=2$
   b. $\lim_{x\to\infty} f(x)=1$
   c. $y=1$
   d. $\lim_{x\to-\infty} f(x)=1$

   When $x=2$, the denominator of f is 0, but the numerator is not, so $x=2$ is a vertical asymptote.

   $$\lim_{x\to\infty}\frac{x^2}{x^2-4x+4}$$
   $$=\lim_{x\to\infty}\frac{1}{1-\frac{4}{x}+\frac{4}{x^2}}=\frac{1}{1-0+0}=1$$

   Likewise, $\lim_{x\to-\infty}\frac{x^2}{x^2-4x+4}=1$.

5. $y=\frac{2x}{x-3}$

   $$\lim_{x\to\infty}\frac{2x}{x-3}=\lim_{x\to\infty}\frac{2}{1-\frac{3}{x}}=2$$

   Horizontal asymptote: $y=2$
   Vertical asymptote: $x=3$

7. $y=\frac{x+1}{(x+2)(x-2)}$

   $$\lim_{x\to\infty}\frac{x+1}{x^2-4}=\lim_{x\to\infty}\frac{\frac{1}{x}-\frac{1}{x^2}}{1-\frac{4}{x^2}}=\frac{0}{1}=0$$

   Horizontal asymptote: $y=0$
   Vertical asymptotes: $x=\pm 2$

9. $y=\frac{3x^3-6}{x^2+4}$

   Vertical asymptote: None since $x^2+4\neq 0$
   Horizontal asymptote: None since degree of numerator > degree of denominator.

11. $f(x)=\frac{2x+2}{x-3}$

    $$f'(x)=\frac{(x-3)2-(2x+2)}{(x-3)^2}=\frac{-8}{(x-3)^2}$$

    VA; $x=3$

    $$\lim_{x\to\infty} f(x)=\lim_{x\to\infty}\frac{2+\frac{2}{x}}{1-\frac{3}{x}}=2$$

    HA : $y=2$
    No critical values. Thus, there can be no maximum or minimum points.

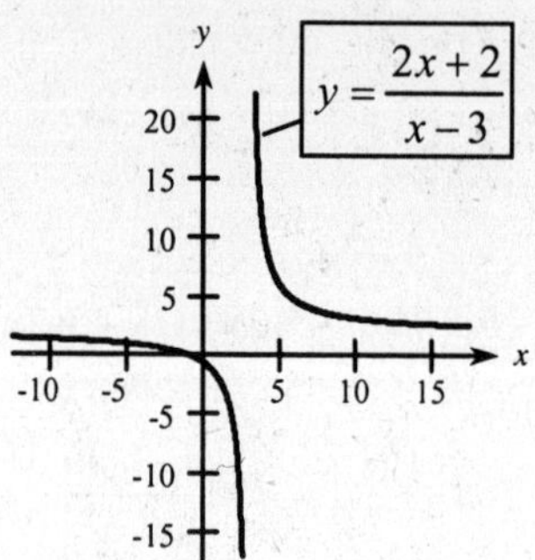

13. $y=\frac{x^2+4}{x}$

    $$\frac{dy}{dx}=\frac{x(2x)-(x^2+4)}{x^2}$$
    $$=\frac{(x+2)(x-2)}{x^2}$$

    Critical values: $x=-2, 2$

x		-2		0		2	
y'	+	0	−	*	−	0	+

    * means y' is undefined at $x=0$. In fact, 0 is not in the domain.
    Relative maximum point: $(-2,-4)$
    Relative minimum point: $(2,4)$
    $y>0$ if $x>0$ and $y<0$ if $x<0$.
    VA: $x=0$
    HA: None (Deg x^2+4 > Deg x)

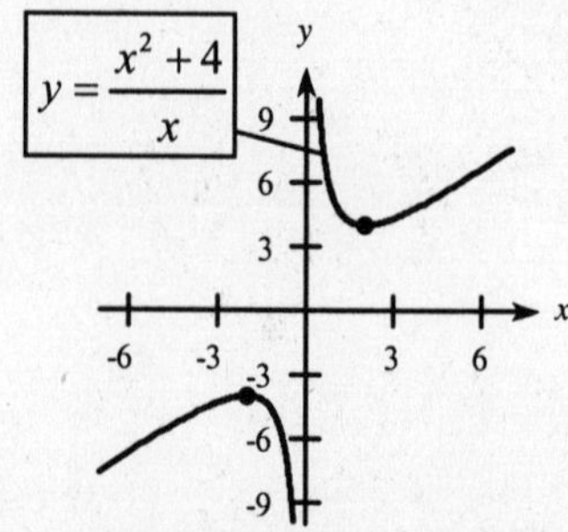

15. $y = \dfrac{27x^2}{(x+1)^3}$

$$y' = \frac{(x+1)^3 \cdot 54x - 27x^2 \cdot 3(x+1)^2}{(x+1)^6}$$

$$= \frac{27x(x+1)^2[2(x+1)-3x]}{(x+1)^6} = \frac{27x(2-x)}{(x+1)^4}$$

VA: $x = -1$

$$\lim_{x\to\infty} y = \lim_{x\to\infty} \frac{27x^2}{x^3+3x^2+3x+1}$$

$$= \lim_{x\to\infty} \frac{\frac{27}{x}}{1+\frac{3}{x}+\frac{3}{x^2}+\frac{1}{x^3}} = \frac{0}{1} = 0$$

HA: $y = 0$

Critical values: $x = 0, 2$

x		-1		0		2	
y'	$-$	$*$	$-$	0	$+$	0	$-$

*means y' is undefined at $x = -1$. In fact, -1 is not in the domain.

Relative minimum point: (0, 0)

Relative maximum point: (2, 4)

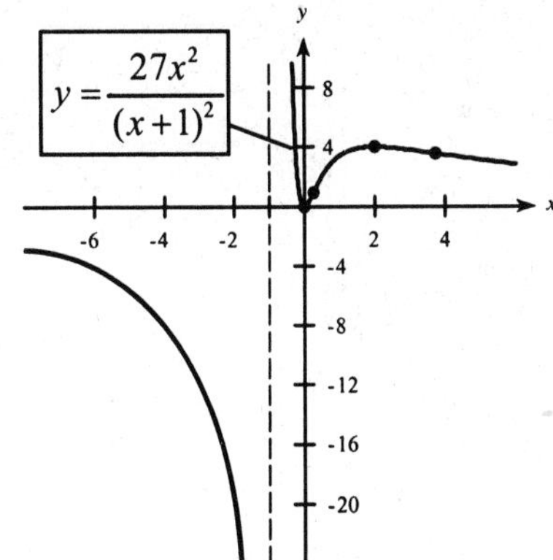

17. $f(x) = \dfrac{16x}{x^2+1}$

$$f'(x) = \frac{(x^2+1)16 - 16x(2x)}{(x^2+1)^2} = \frac{16-16x^2}{(x^2+1)^2}$$

Critical values: $x = -1, 1$

x		-1		1	
$f'(x)$	$-$	0	$+$	0	$-$

Relative maximum point: (1, 8)

Relative minimum point: (−1, −8)

VA: None since $x^2+1 \neq 0$. HA: $y = 0$ since degree $(x^2+1) >$ degree $(16x)$.

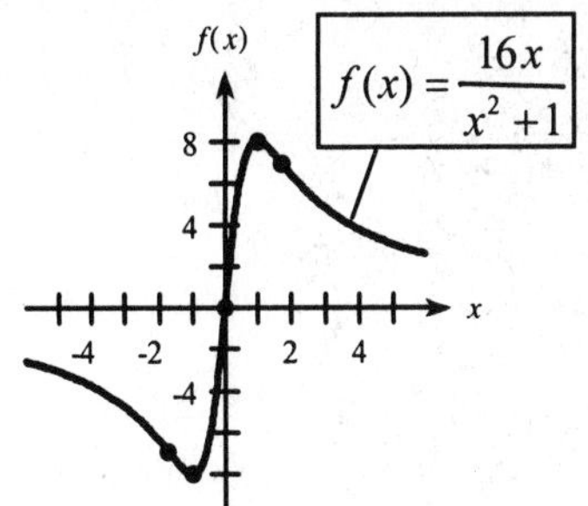

19. Critical value is $x = -1$.

($x = 1$ is not in the domain of the function.)

x		-1		1	
y'	$-$	0	$+$	$*$	$-$

* means y' is undefined at $x = 1$. In fact, 1 is not in the domain.

Relative minimum point: $\left(-1, -\dfrac{1}{4}\right)$

$y > 0$ if $x > 0$ and $y < 0$ if $x < 0$.

(0, 0) is an x-intercept.

$$\lim_{x\to\infty} \frac{x}{(x-1)^2} = 0$$

HA: $y = 0$, VA: $x = 1$

Point of inflection at $x = -2$ (y'' changes signs)

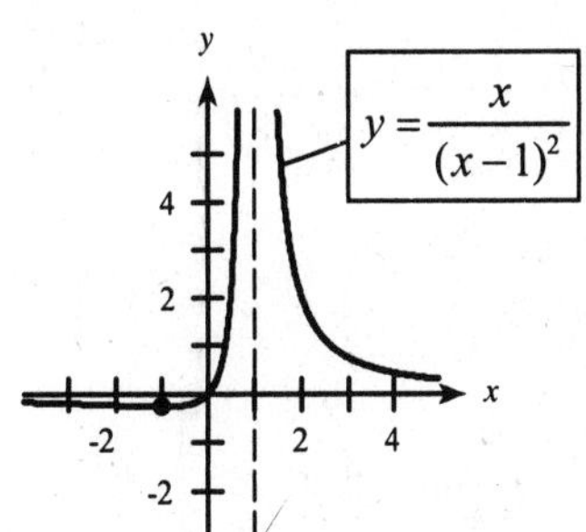

21. VA: $x = 3$, Critical values: $x = 2, 4$

x		2		3		4	
$f'(x)$	+	0	−	*	−	0	+

* means $f'(3)$ is undefined. In fact, 3 is not in the domain.
Relative maximum at $x = 2$.
Relative minimum at $x = 4$.
$y'' < 0$ when $x < 3$, and
$y'' > 0$ when $x > 3$ but $x = 3$ is not in the domain of the function, so there are no inflection points.

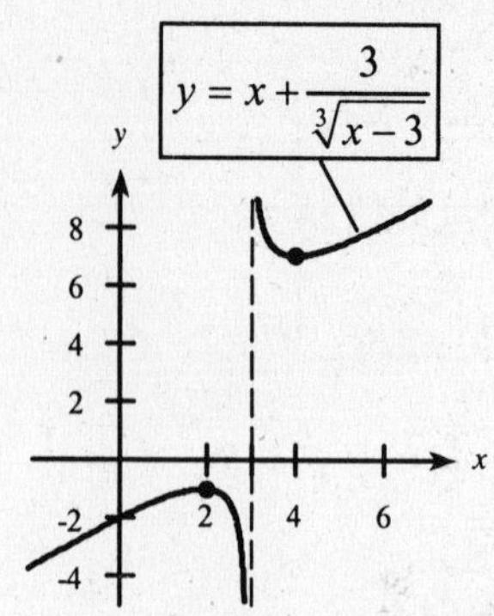

23. VA: $x = 0$

$$\lim_{x\to\infty} f(x) = \lim_{x\to\infty} \frac{9(x-2)^{2/3}}{(x^3)^{2/3}} = \lim_{x\to\infty} 9\left(\frac{x-2}{x^3}\right)^{2/3} = \lim_{x\to\infty} 9\left(\frac{\frac{1}{x^2}-\frac{2}{x^3}}{1}\right)^{2/3} = 9\cdot\left(\frac{0}{1}\right)^{2/3} = 0$$

HA: $y = 0$, Critical values: $x = 3$ and $x = 2$

x		0		2		3	
$f'(x)$	+	*	−	*	+	0	−

* means $f'(0)$ and $f'(2)$ are undefined. In fact, 0 is not in the domain.
Relative minimum: (2, 0)
Relative maximum: (3, 1)
Possible inflection points: $x = 2$ and

$$x = \frac{42 \pm \sqrt{42^2 - 4(7)(54)}}{2(7)} = \frac{42 \pm 6\sqrt{7}}{14} = 3 \pm \frac{3\sqrt{7}}{7} \approx 1.87 \text{ and } 4.13$$

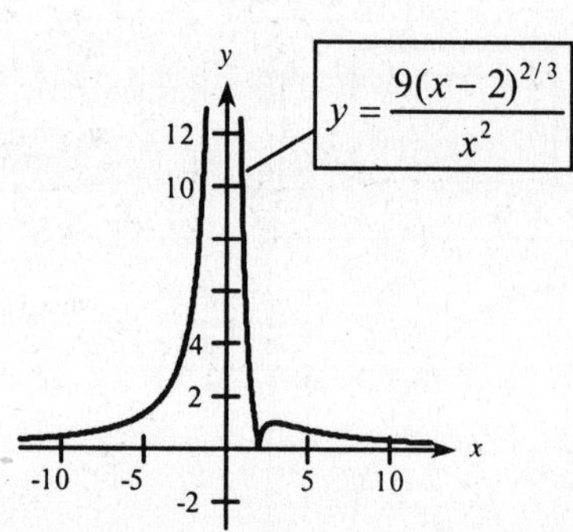

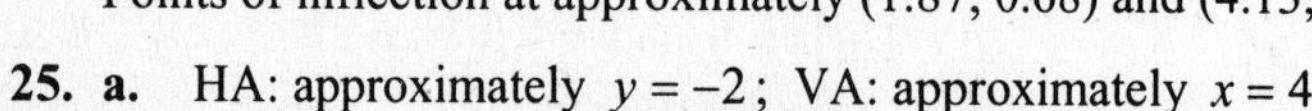

x		0		$3-\frac{3\sqrt{7}}{7}$		2		$3+\frac{3\sqrt{7}}{7}$	
$f''(x)$	+	*	+	0	−	*	−	0	+

* means $f''(0)$ and $f''(2)$ are undefined.
Points of inflection at approximately (1.87, 0.68) and (4.13, 0.87)

25. **a.** HA: approximately $y = -2$; VA: approximately $x = 4$

b. $f(x) = \frac{9x}{17-4x} = \frac{9}{\frac{17}{x}-4}$, HA: $y = -\frac{9}{4}$; VA: $x = \frac{17}{4}$

27. **a.** HA: approximately $y = 2$; VA: approximately $x = \pm 2.5$

b. $f(x) = \frac{20x^2+98}{9x^2-49} = \frac{20+\frac{98}{x^2}}{9-\frac{49}{x^2}}$, HA: $y = \frac{20}{9}$; VA: $x = \pm\frac{7}{3}$

29. a. Standard Viewing Window

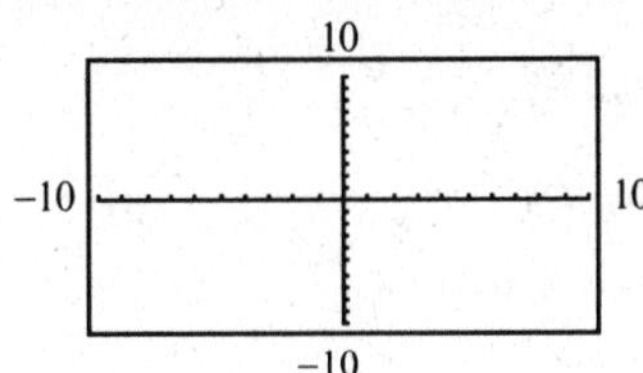

b. $f(x) = \dfrac{x+25}{x^2+1400}$

HA: $y = 0$

degree $(x^2+1400) >$ degree $(x+25)$

$$f'(x) = \frac{(x^2+1400)\cdot 1-(x+25)2x}{(x^2+1400)^2}$$

$$= \frac{(70+x)(20-x)}{(x^2+1400)^2}$$

x		-70		20	
$f'(x)$	$-$	0	$+$	0	$-$

Relative minimum at $x = -70$. Relative maximum at $x = 20$.

c.

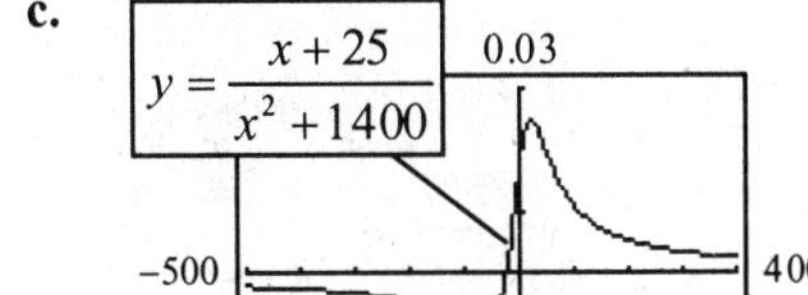

31. a. Standard Viewing Window

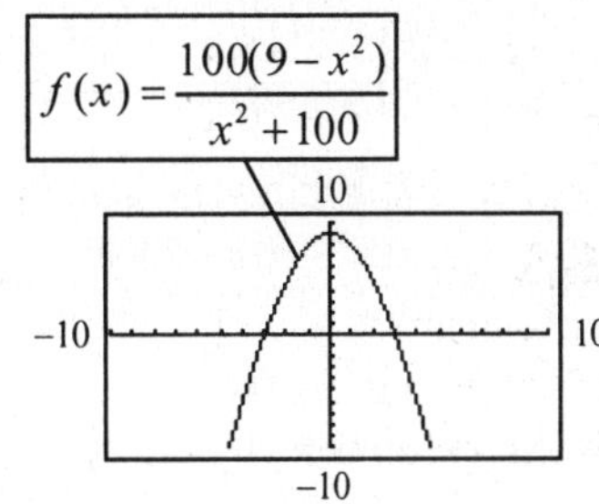

b. $f(x) = \dfrac{100(9-x^2)}{x^2+100}$

$$\lim_{x\to\infty} f(x) = \lim_{x\to\infty} \frac{100\left(\frac{9}{x^2}-1\right)}{1+\frac{100}{x^2}} = -100$$

HA: $y = -100$

$$f'(x) = 100\left[\frac{(x^2+100)(-2x)-(9-x^2)2x}{(x^2+100)^2}\right]$$

$$= \frac{100(-218x)}{(x^2+100)^2}$$

x		0	
$f'(x)$	$+$	0	$-$

Relative maximum: (0, 9)

For viewing window try: x: –75 to 75

y: –120 to 20.

c.

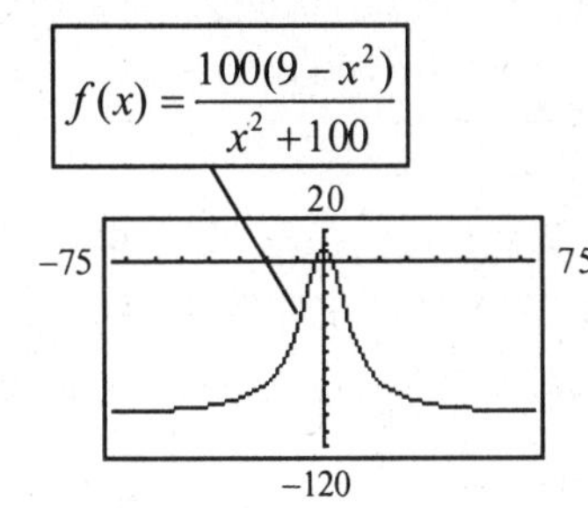

33. a. Standard viewing window: See graphing section below problem.

b. $f(x)=\dfrac{1000(x-4)}{(x-50)(x+40)}$ $\lim\limits_{x\to\infty} f(x)=\lim\limits_{x\to\infty}\dfrac{\frac{1000}{x}-\frac{4000}{x^2}}{1-\frac{10}{x}-\frac{2000}{x^2}}=\dfrac{0}{1}=0$

Horizontal asymptote: $y=0$, Vertical asymptotes: $x=-40$, $x=50$

$$f'(x)=1000\cdot\frac{(x^2-10x-2000)\cdot 1-(x-4)(2x-10)}{[(x-50)(x+40)]^2}=\frac{1000(-x^2+8x-2040)}{[(x-50)(x+40)]^2}$$

Numerator is never zero. No relative maximum or relative minimum.
For viewing window, try x: –200 to 200, y: –200 to 200

a.

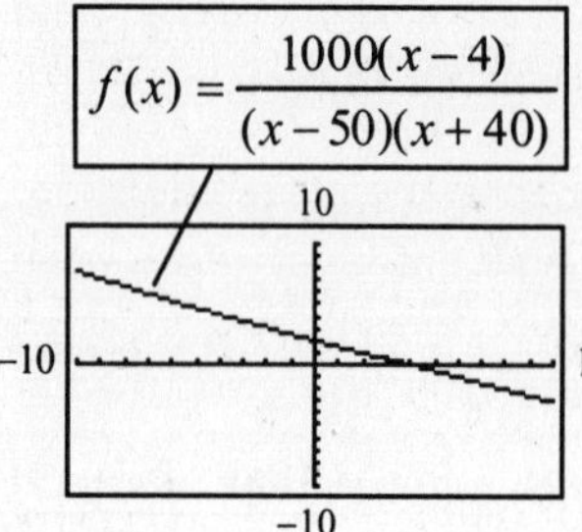

c.

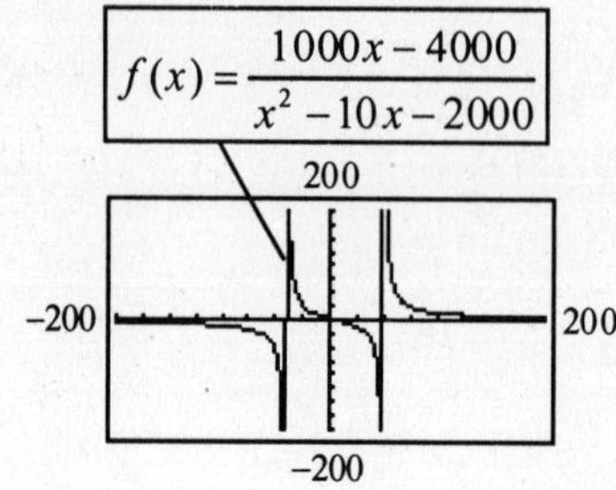

35. $p(C)=\dfrac{100C}{7300+C}$

$$p'(C)=\frac{(7300+C)100-100C(1)}{(7300+C)^2}=\frac{730000}{(7300+C)^2}$$

a. Domain = $\{C : C\geq 0\}$. Thus $p(C)$ exists for all C in the domain.

b. $p'(C)$ is always positive. Thus, $p(C)$ is increasing if $C\geq 0$.

c. $\lim\limits_{C\to\infty}\dfrac{100C}{7300+C}=100$

Horizontal asymptote: $p=100$

d. No

37. $R(t)=\dfrac{50t}{t^2+36}$

$$R'(t)=\frac{(t^2+36)50-50t(2t)}{(t^2+36)^2}=\frac{50(36-t^2)}{(t^2+36)^2}$$

Critical value: $t=6$

$R'(5)>0,\ R'(6)=0,\ R'(7)<0$

Horizontal asymptote: $y=0$ (for $t>0$)

a.

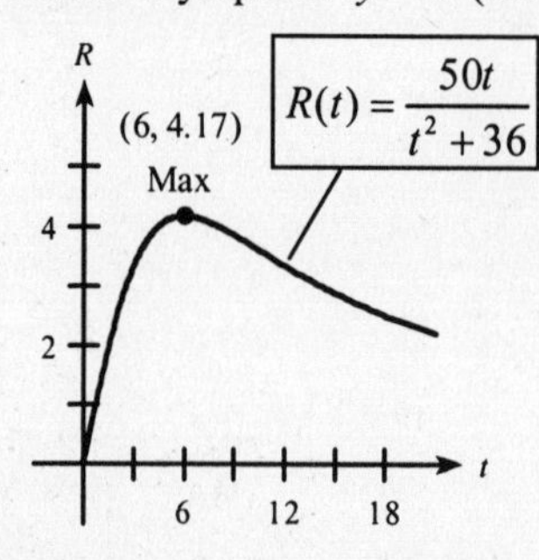

b. Revenue is maximized at $t=6$.

c. Revenue is decreasing for $t>6$. The video will be released in $6+4+12=22$ weeks from date of original release.

39. $f(x)=\dfrac{289.173-58.5731x}{x+1}$

a. Yes; $x=-1$

b. No; Domain: $x\geq 5$

c. Yes; $y=-58.5731$

d. At high wind speeds any additional wind will have little effect on the wind-chill. Interpretation is meaningful.

41. a. $P=C$

b. C

c. 0

d. 0

43. a. $\lim\limits_{t\to\infty} f(t)=0$ (degree denominator > degree numerator)

b. The percentage of workers in farm occupations approaches 0.

c. No. Denominator >0 $(b^2-4ac<0)$.

45. a. No. Barometric pressure is always greater than zero. It can drop off the scale as shown.

b. A blizzard struck the East Coast, causing \$3-6 billion in damages and approximately 270 deaths.

Review Exercises

1. $y = -x^2$

$y' = -2x$

Critical value is at $x = 0$. Critical point is (0, 0).

$y'(-1) > 0,\ y'(0) = 0,\ y'(1) < 0$

(0, 0) is a maximum point.

2. $p = q^2 - 4q - 5$

$p' = 2q - 4$

Since $p' = 0$ for $q = 2$, the point (2, –9) is a critical point. To the left of (2, –9) we have $p' < 0$ and to the right of (2, –9) we have $p' > 0$. The point (2, –9) is a minimum point.

3. $f(x) = 1 - 3x + 3x^2 - x^3$

$f'(x) = -3 + 6x - 3x^2$

$= -3(x^2 - 2x + 1)$

$= -3(x-1)^2$

$f'(x) = 0$ if $x = 1$

$f(1) = 1 - 3 + 3 - 1 = 0$

Critical point is (1, 0).

x		1	
$f'(x)$	–	0	–

(1, 0) is a horizontal point of inflection.

4. $f(x) = \dfrac{3x}{x^2+1}$

$f'(x) = \dfrac{(x^2+1)(3) - (3x)(2x)}{(x^2+1)^2}$

$= \dfrac{-3x^2 + 3}{(x^2+1)^2}$

$= \dfrac{-3(x+1)(x-1)}{(x^2+1)^2}$

$f'(x) = 0$ if $x = -1$ or $x = 1$. Critical points are $\left(-1, -\dfrac{3}{2}\right)$ and $\left(1, \dfrac{3}{2}\right)$.

$f'(-2) < 0$; $f'(-1) = 0$; $f'(0) > 0$;

$f'(1) = 0$; $f'(2) < 0$ means that we have a relative minimum at $\left(-1, -\dfrac{3}{2}\right)$ and a relative maximum at $\left(1, \dfrac{3}{2}\right)$.

5. $f(x) = x^3 + x^2 - x - 1$

$f'(x) = 3x^2 + 2x - 1 = (3x-1)(x+1)$

a. Critical points are (–1, 0) and $\left(\dfrac{1}{3}, -\dfrac{32}{27}\right)$.

x		–1		$\frac{1}{3}$	
$f'(x)$	+	0	–	0	+

b. Relative maximum: (–1, 0)

Relative minimum: $\left(\dfrac{1}{3}, -\dfrac{32}{27}\right)$

c. No horizontal points of inflection.

d.

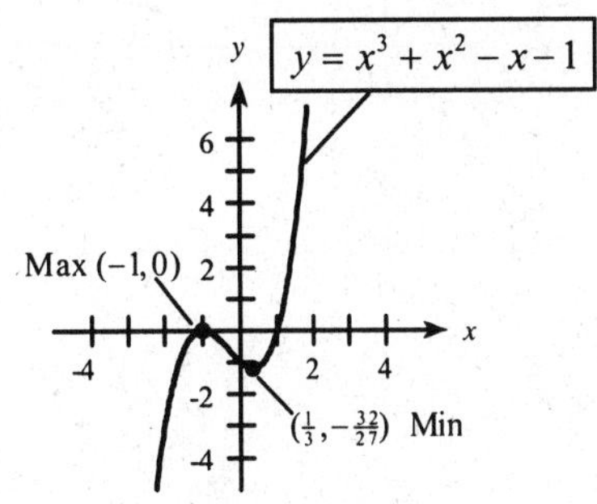

6. $f(x) = 4x^3 - x^4$

a. $f'(x) = 12x^2 - 4x^3 = 4x^2(3-x)$

Critical values: 0, 3

b. $f'(x) > 0$ when $x < 3$, $x \neq 0$.

$f'(x) < 0$ when $x > 3$.

Relative maximum at (3, 27).

c. Horizontal point of inflection: (0, 0)

d.

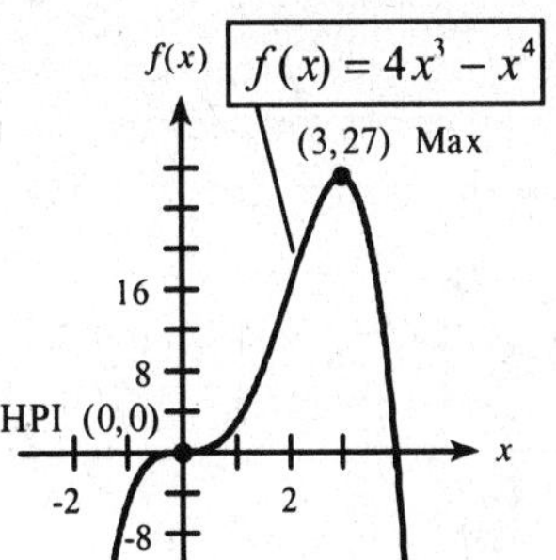

7. $f(x) = x^3 - \frac{15}{2}x^2 - 18x + \frac{3}{2}$

$f'(x) = 3x^2 - 15x - 18 = 3(x-6)(x+1)$

$f'(x) = 0$ at $x = -1, 6$.

a. Critical points: (–1, 11) and (6, –160.5).

x		-1		6	
$f'(x)$	+	0	–	0	+

b. Relative maximum: (–1, 11)
Relative minimum: (6, –160.5)

c. No horizontal points of inflection.

d.

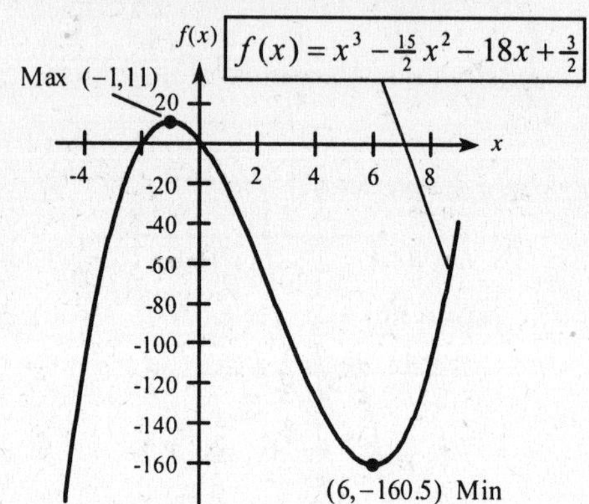

8. $y = f(x) = 5x^7 - 7x^5 - 1$

a. $y' = f'(x) = 35x^6 - 35x^4$

$= 35x^4(x+1)(x-1)$

Critical values: 0, –1, 1

b. $f'(x) > 0$ when $x < -1$ and when $x > 1$.

$f'(x) < 0$ when $-1 < x < 1$, $x \neq 0$.

Relative maximum: (–1, 1)
Relative minimum: (1, –3)

c. Horizontal point of inflection: (0, –1)

d.

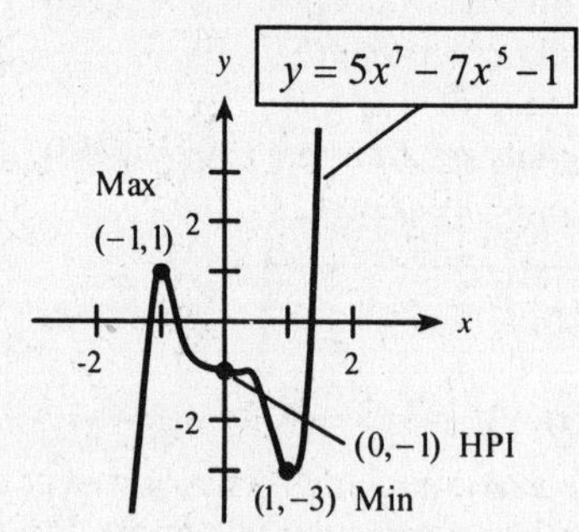

9. $y = x^{2/3} - 1$

$y'(x) = \frac{2}{3x^{1/3}}$

a. Critical value is at $x = 0$.

$y'(-1) < 0$, $y'(0)$ is undefined, $y'(1) > 0$

b. (0, –1) is a minimum point.

c. No horizontal points of inflection.

d.

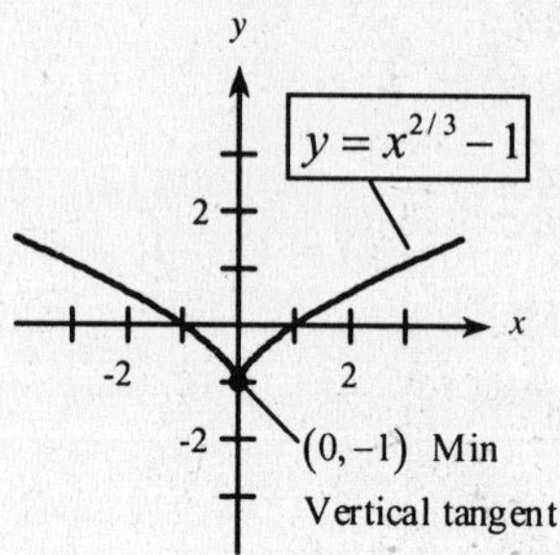

10. $y = f(x) = x^{2/3}(x-4)^2$

a. $f'(x) = x^{2/3} \cdot 2(x-4) + (x-4)^2 \cdot \frac{2}{3}x^{-1/3}$

$= \frac{2}{3}(x-4)x^{-1/3}(4x-4)$

Critical values: 0, 1, 4

b. $f'(x) > 0$ when $0 < x < 1$ and when $x > 4$.

$f'(x) < 0$ when $x < 0$ and when $1 < x < 4$.

Relative minimum: (0, 0); Relative minimum: (4, 0); Relative maximum: (1, 9)

c. No horizontal points of inflection.

d.

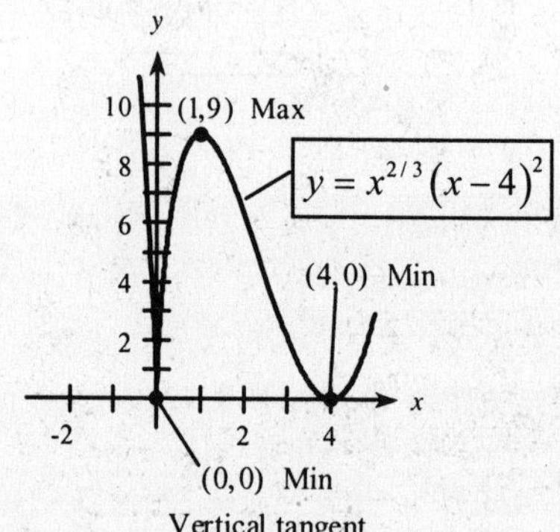

11. $y = x^4 - 3x^3 + 2x - 1$

$y' = 4x^3 - 9x^2 + 2$

$y'' = 12x^2 - 18x$

At $x = 2$ we have $y'' = 12$. Thus, the curve is concave up.

12. $y = f(x) = x^4 - 2x^3 - 12x^2 + 6$

$f'(x) = 4x^3 - 6x^2 - 24x$

$f''(x) = 12x^2 - 12x - 24$

$= 12(x^2 - x - 2)$

$= 12(x-2)(x+1)$

$f''(x) > 0$ when $x > 2$ and when $x < -1$.

$f''(x) < 0$ when $-1 < x < 2$.

$f(x)$ is concave up when $x > 2$ and when $x < -1$;

$f(x)$ is concave down when $-1 < x < 2$.

Points of inflection: (–1, –3) and (2, –42).

13. $y = x^3 - 3x^2 - 9x + 10$

$y' = 3x^2 - 6x - 9 = 3(x-3)(x+1)$

Critical values are $x = -1$ and $x = 3$.

$y'' = 6x - 6$. So, possible point of inflection is at $x = 1$.

x	–1	1	3
y''	–	0	+

There is a point of inflection at (1, –1).

$f''(-1) < 0$ means there is a relative maximum at (–1, 15).

$f''(3) > 0$ means there is a relative minimum at (3, –17).

14. $y = x^3 - 12x$

$y' = 3x^2 - 12$

$y' = 0$ when $x = 2$ and when $x = -2$.

Critical points: (2, –16) and (–2, 16)

$y'' = f''(x) = 6x$

$f''(2) > 0$ means that (2, –16) is a relative minimum. $f''(-2) < 0$ means that (–2, 16) is a relative maximum. $y'' = 0$ when $x = 0$. $y'' < 0$ when $x < 0$ and $y'' > 0$ when $x > 0$. So, (0, 0) is a point of inflection.

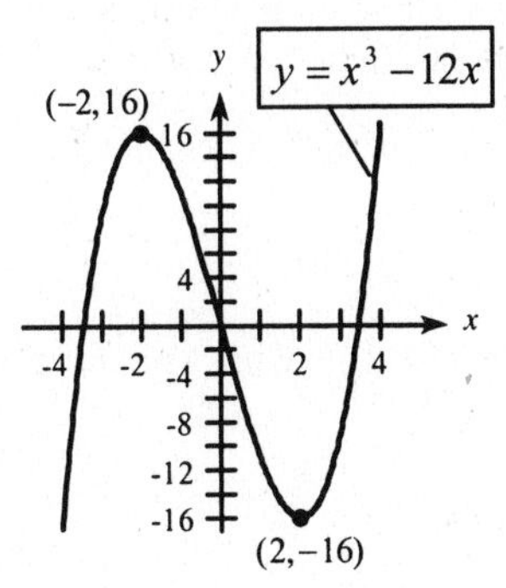

15. $y = 2 + 5x^3 - 3x^5$

$y' = 15x^2 - 15x^4 = 15x^2(1 - x^2)$

Critical values: $x = 0, 1, -1$

x		–1		0		1	
y'	–	0	+	0	+	0	–

Relative minimum: (–1, 0)

Relative maximum: (1, 4)

$y'' = 30x - 60x^3 = 30x(1 - 2x^2)$

Possible points of inflection at $x = 0, \frac{1}{\sqrt{2}}, -\frac{1}{\sqrt{2}}$.

x		$-\frac{1}{\sqrt{2}}$		0		$\frac{1}{\sqrt{2}}$	
y''	+	0	–	0	+	0	–

There are points of inflection at $x = 0, \pm\frac{1}{\sqrt{2}}$.

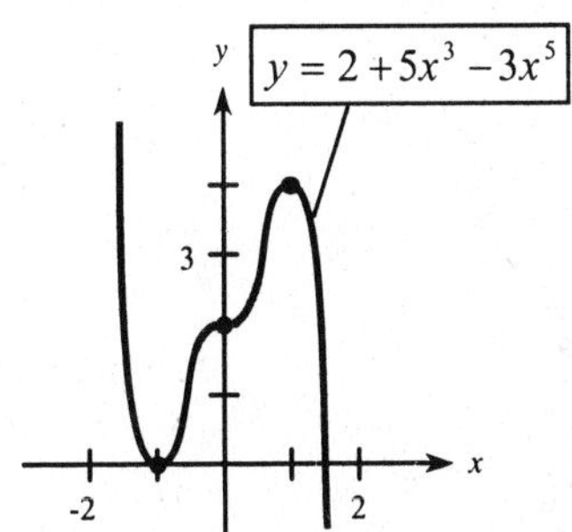

16. $R(x) = 280x - x^2$

a. $0 \le x \le 200$

$R'(x) = 280 - 2x$

$R'(x) = 0$ when $x = 140$

$R(0) = 0$, $R(140) = 19{,}600$,

$R(200) = 16{,}000$

Absolute maximum: (140, 19,600)

Absolute minimum: (0, 0)

b. $0 \le x \le 100$

Absolute maximum: (100, 18,000)

Absolute minimum: (0, 0)

17. $y = 6400x - 18x^2 - \frac{x^3}{3}$

$y' = 6400 - 36x - x^2 = (100 + x)(64 - x)$

Absolute maximum and minimum occur at critical values or at endpoints of the domain.

a. No critical values.

$y(50) \approx 233{,}333$ absolute maximum

$y(0) = 0$ absolute minimum

b. Critical value is at $x = 64$.

$y(64) \approx 248{,}491$ absolute maximum

$y(0) = 0$ absolute minimum

$y(100) \approx 126{,}667$

18. a. Vertical asymptote: $x = 1$

b. Horizontal asymptote: $y = 0$

c. $\lim_{x\to\infty} f(x) = 0$

d. $\lim_{x\to-\infty} f(x) = 0$

19. a. $x = -1$

b. $y = \frac{1}{2}$

c. $\frac{1}{2}$

d. $\frac{1}{2}$

20. $y = \frac{3x+2}{2x-4}$

Vertical asymptote: $x = 2$

Horizontal asymptote:

$$y = \lim_{x\to\infty} \frac{3x+2}{2x-4} = \lim_{x\to\infty} \frac{3+\frac{2}{x}}{2-\frac{4}{x}} = \frac{3}{2}$$

21. $y = \frac{x^2}{1-x^2}$

$$\lim_{x\to\infty} \frac{x^2}{1-x^2} = \lim_{x\to\infty} \frac{1}{\frac{1}{x^2}-1} = -1$$

Horizontal asymptote: $y = -1$

Vertical asymptotes: $x = \pm 1$

22. $y = \frac{3x}{x+2}$

a. Vertical asymptote: $x = -2$

Horizontal asymptote: $y = \lim_{x\to\infty} \frac{3x}{x+2} = 3$

b. $y' = \frac{(x+2)(3)-(3x)}{(x+2)^2} = \frac{6}{(x+2)^2}$

y' is never zero, no maximum nor minimum.

c.

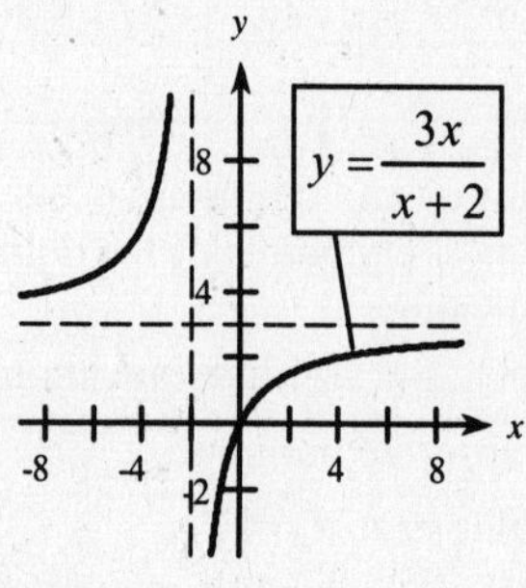

23. $y = \frac{8x-16}{x^2}$

$$\frac{dy}{dx} = \frac{x^2 \cdot 8 - (8x-16)2x}{x^4} = \frac{8(4-x)}{x^3}$$

Critical value at $x = 4$.

$y'(3) > 0,\ y'(4) = 0,\ y'(5) < 0$

a. Horizontal asymptote: $y = 0$
Vertical asymptote: $x = 0$

b. Relative maximum: (4, 1)

c.

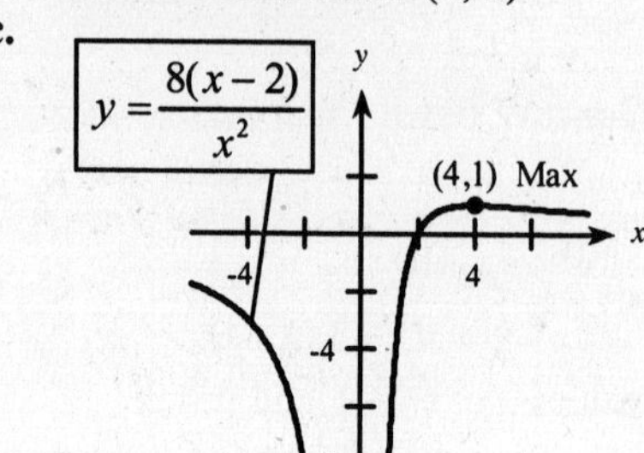

24. $y = \frac{x^2}{x-1}$

a. Vertical asymptote: $x = 1$
Horizontal asymptote: none

b. $y' = \frac{(x-1)(2x)-x^2}{(x-1)^2} = \frac{x^2-2x}{(x-1)^2}$

$y' = 0$ when $x = 0$ and when $x = 2$.

Critical points: (0, 0), (2, 4)

$y' = f'(-1) > 0$, $f'(0.5) < 0$, $f'(1.5) < 0$, $f'(3) > 0$

Relative maximum: (0, 0)

Relative minimum: (2, 4)

c.

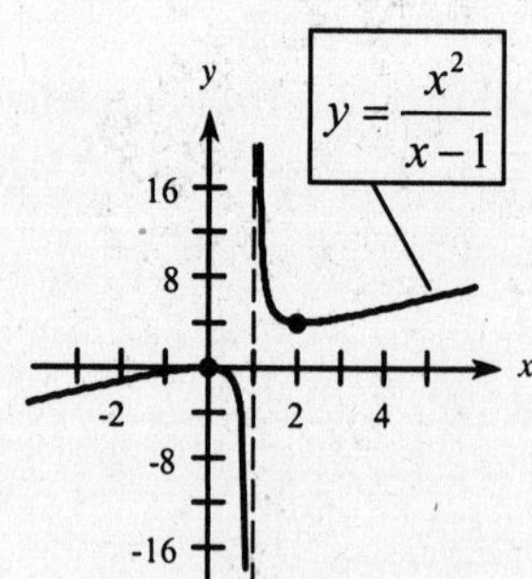

25. a. $f'(x) > 0$ if $x < \frac{2}{3}$ and if $x > 2$.
$f'(x) < 0$ if $\frac{2}{3} < x < 2$.
$f'(x) = 0$ if $x = \frac{2}{3}$ and if $x = 2$.

b. $f''(x) > 0$ if $x > \frac{4}{3}$.
$f''(x) < 0$ if $x < \frac{4}{3}$.
$f''(x) = 0$ if $x = \frac{4}{3}$.

c.

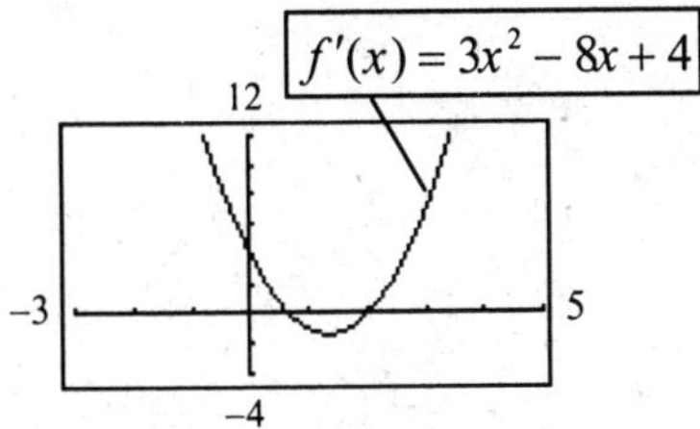

d.

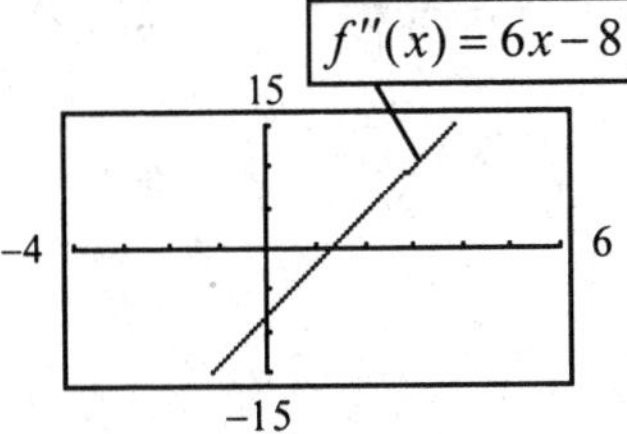

26. $f(x) = 0.0025x^4 + 0.02x^3 - 0.48x^2 + 0.08x + 4$

a. From the graph, the estimated values are
$f'(x) > 0$ when $-13 < x < 0$ and $x > 7$.
$f'(x) < 0$ when $x < -13$ and $0 < x < 7$.
$f'(x) = 0$ when $x = -13$, $x = 0$ and $x = 7$.

b. Estimates from the graph:
$f''(x) > 0$ when $x < -8$ and $x > 4$.
$f''(x) < 0$ when $-8 < x < 4$.
$f''(x) = 0$ when $x = -8$ and $x = 4$.

c.

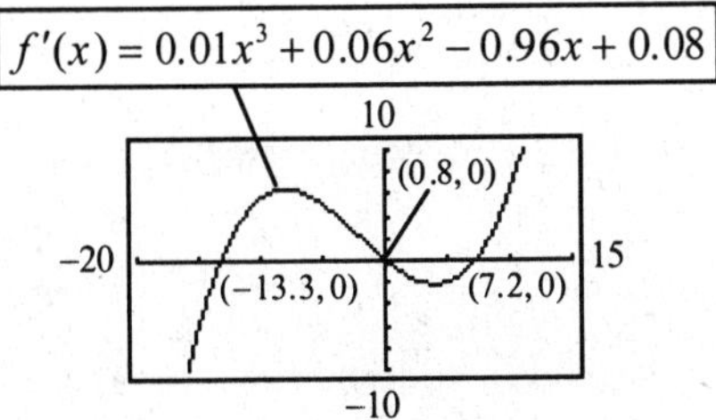

d.

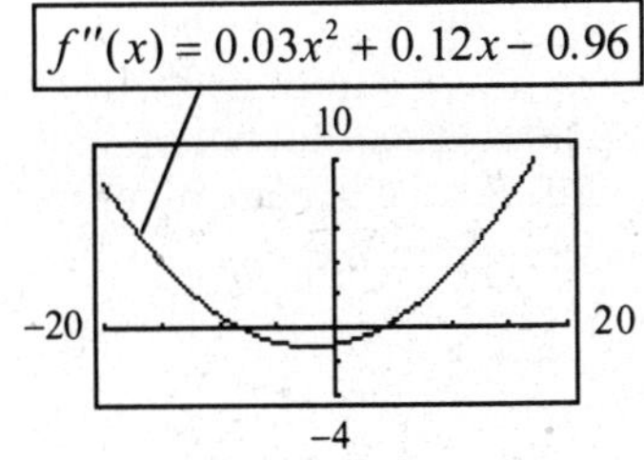

27. a. $f(x)$ is increasing when $f'(x) > 0$.
$f(x)$ is decreasing when $f'(x) < 0$.
Increasing if $x < -5$ and if $x > 1$.
Decreasing if $-5 < x < 1$.
Relative maximum at $x = -5$ (+ to –)
Relative minimum at $x = 1$ (– to +)

b. If $f'(x)$ is increasing, then $f''(x) > 0$.
If $f'(x)$ is decreasing, then $f''(x) < 0$.
$f''(x) > 0$ if $x > -2$, $f''(x) < 0$ if $x < -2$
$f''(x) = 0$ if $x = -2$.

c. $f(x) = \dfrac{x^3}{3} + 2x^2 - 5x \quad f'(x) = x^2 + 4x - 5$

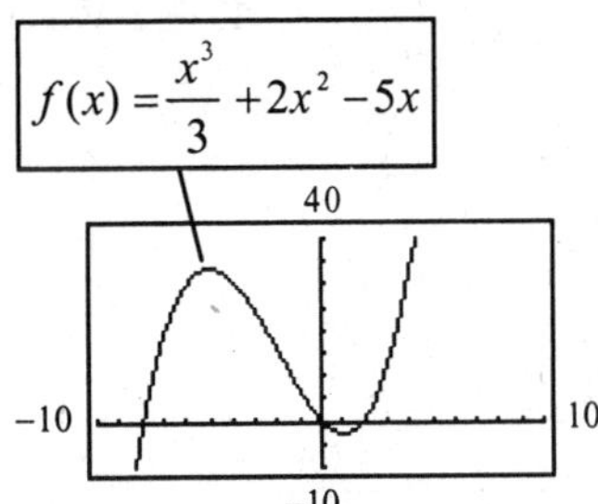

d.

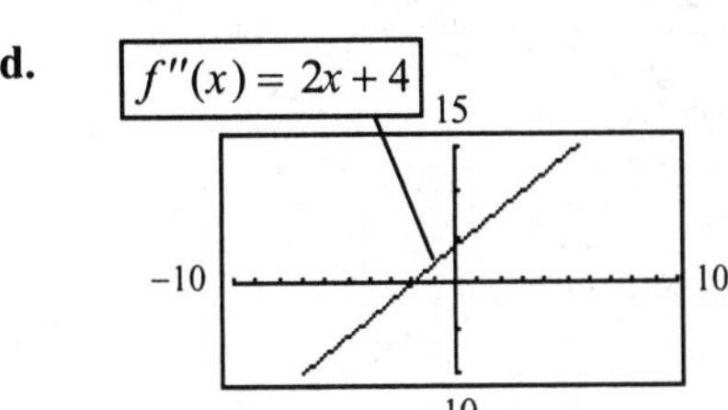

28. $f'(x) = 6x^2 - x^3$

a. From the graph,
$f(x)$ is increasing when $x < 6$, $x \neq 0$.
$f(x)$ is decreasing when $x > 6$.
$f(x)$ has a relative maximum at $x = 6$.
$f(x)$ has a horizontal point of inflection at $x = 0$.

b. From the graph,
$f''(x) > 0$ when $0 < x < 4$.
$f''(x) < 0$ when $x < 0$ and when $x > 4$.
$f''(x) = 0$ when $x = 0$ and $x = 4$.

c. $f(x) = 2x^3 - \frac{x^4}{4}$, $f'(x) = 6x^2 - x^3$

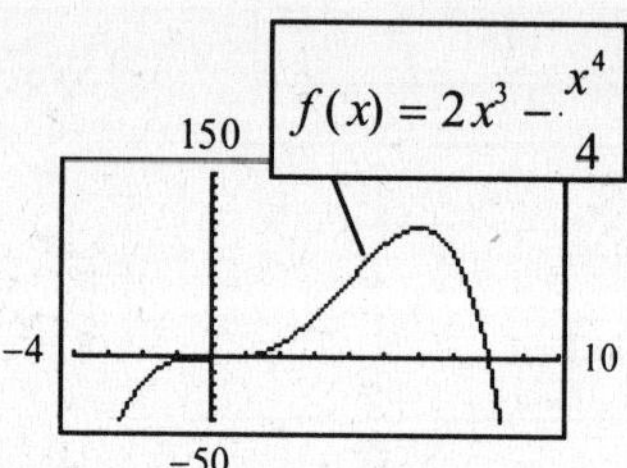

d. $f''(x) = 12x - 3x^2 = 3x(4 - x)$
Graph of f''

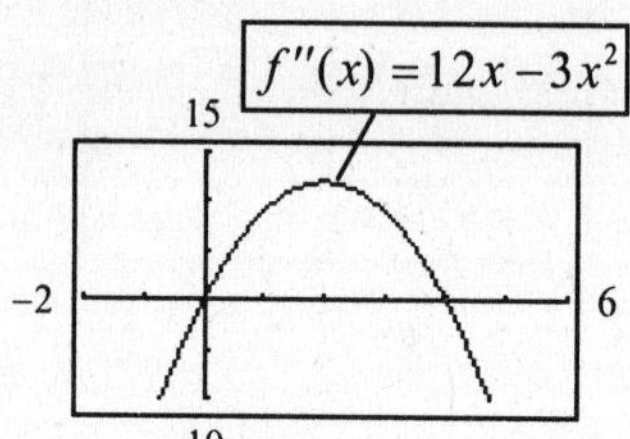

29. a. $f(x)$ is concave up if $f''(x) > 0$.
$f(x)$ is concave down if $f''(x) < 0$.
Concave up if $x < 4$.
Concave down if $x > 4$.
There is a point of inflection at $x = 4$.

b. $f(x) = 2x^2 - \frac{x^3}{6}$, $f'(x) = 4x - \frac{1}{2}x^2$,
$f''(x) = 4 - x$

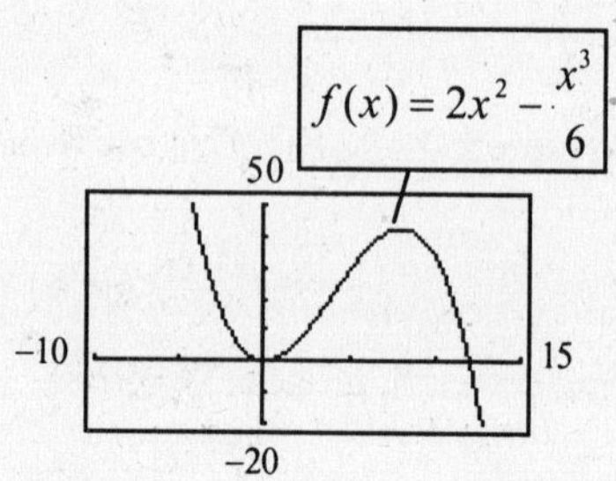

30. $f''(x) = 6 - x - x^2$

a. From the given graph,
$f(x)$ is concave up when $-3 < x < 2$.
$f(x)$ is concave down when $x < -3$, and when $x > 2$. $f(x)$ has a point of inflection at $x = -3$ and at $x = 2$.

b. $f(x) = 3x^2 - \frac{x^3}{6} - \frac{x^4}{12}$
$f'(x) = 6x - \frac{1}{2}x^2 - \frac{1}{3}x^3$
$f''(x) = 6 - x - x^2$
Graph of *f*:

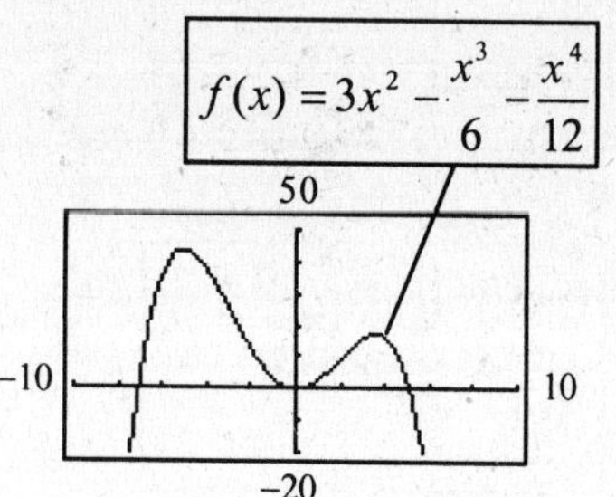

31. $\overline{C}(x) = 3x + 15 + \frac{75}{x}$

$\overline{C}'(x) = 3 - \frac{75}{x^2}$

$\overline{C}'(x) = 0$ if $x^2 = 25$ or $x = 5$.

x	5
$C'(x)$	− 0 +

$\overline{C}(5) = 15 + 15 + 15 = 45$
Minimum average cost is $45 at 5 units.

32. $R(x) = 32x - 0.01x^2$

a. $R'(x) = 32 - 0.02x$
$R'(x) = 0$ when $x = 1600$.

x	1600
$R'(x)$	+ 0 −

$R(1600) = \$25{,}600$
The maximum revenue is $25,600 when $x = 1600$ units are produced.

b. If production is limited to 2500, the maximum revenue still occurs when $x = 1600$. So, the maximum revenue is $25,600.

33. $P(x) = 1080x + 9.6x^2 - 0.1x^3 - 50,000$

$P'(x) = 1080 + 19.2x - 0.3x^2$

$P'(x) = 0$ if $-0.3(x^2 - 64x - 3600)$ or if $(x-100)(x+36) = 0$.

x		100	
$P'(x)$	+	0	−

Maximum profit is at $x = 100$.

$P(100) = \$54,000$ is the maximum profit.

34. $R(x) = 46x - x^2$

$C(x) = 5x^2 + 10x + 3$

$P(x) = R(x) - C(x) = -6x^2 + 36x - 3$

$P'(x) = -12x + 36$

$P'(x) = 0$ when $x = 3$

Since $P''(x) = -12 < 0$ for all *x*, the profit is maximized when $x = 3$.

35. $P(x) = 80x - \frac{1}{4}x^2 - (800 + 4x)$

$= -\frac{1}{4}x^2 + 76x - 800$

$P'(x) = -\frac{1}{2}x + 76$

$P'(x) = 0$ if $x = 152$.

x		152	
$P'(x)$	+	0	−

Since 152 is not in the domain and *P* is increasing for $0 \le x \le 150$, the maximum profit is at 150 units.

36. $C = 2x^2 + 54x + 98$

$\overline{C} = 2x + 54 + \frac{98}{x}$

$\overline{C}' = 2 - \frac{98}{x^2} = \frac{2x^2 - 98}{x^2}$

$\overline{C}' = 0$ when $x = 7$.

Since $\overline{C}'' = \frac{196}{x^3} > 0$ for all $x > 0$, the average cost is minimized when $x = 7$.

37. Profit is maximized when $\overline{MP}$ changes from + to −. This occurs at $x = 500$ units.

38. $P'(t)$ is to be maximized (not $P(t)$).

$$P(t) = \frac{95t^2}{t^2 + 2700} + 5$$

$$P'(t) = \frac{(t^2 + 2700) \cdot 190t - 95t^2 \cdot 2t}{(t^2 + 2700)^2} = \frac{513,000t}{(t^2 + 2700)^2}$$

$$P''(t) = \frac{(t^2 + 2700)^2 \cdot 513,000 - 513,000t \cdot 2(t^2 + 2700) \cdot 2t}{(t^2 + 2700)^4}$$

$$= \frac{513,000(t^2 + 2700)\left[(t^2 + 2700) - 4t^2\right]}{(t^2 + 2700)^4} = \frac{513,000(2700 - 3t^2)}{(t^2 + 2700)^3}$$

$P''(t)$ changes from + to − at $t = 30$. Point of diminishing returns is at $t = 30$ hours.

39. a. Diminishing returns or point of inflection occurs at $x = 60$.

b. $m = \frac{f(I) - 0}{I - 0} = \frac{f(I)}{I} =$ Average output

c. Maximum average output occurs when slope of average output is closest to $f'(I)$. This occurs at $x = 70$.

40. $R = (54 + 10x)(385 - 25x) = 20.790 + 2500x - 250x^2$

$R' = 2500 - 500x$

R' changes from + to − at $x = 5$. Revenue is maximized with selling price of $385 - 25(5) = \$260$.

41. $P = 20,790 + 2500x - 250x^2 - (200 \cdot 10x) = 20,790 + 500x - 250x^2$

$P' = 500 - 500x$

Profit is maximized when $P' = 0$ or $x = 1$. The selling price will be $385 - 25(1) = \$360$.

42. Equilibrium price means supply = demand.
$1200-2x=200+2x$ gives equilibrium quantity $x=250$. Equilibrium price is $p=200+2(250)=700$.
Revenue = 700x. Cost $=12,000+50x+x^2$.
$P(x)=700x-(12,000+50x+x^2)=-12,000+650x-x^2$
$P'(x)=650-2x$
$P'(x)=0$ if $x=325$.

x		325	
$P'(x)$	+	0	−

$P(325)=-12,000+211,250-105,625=\$93,625$

43. $\bar{C}=200+x,\ C=200x+x^2,\ p=800-x,\ R=800x-x^2$

a. $P(x)=800x-x^2-200x-x^2=600x-2x^2$
$P'(x)=600-4x$
$P'(x)=0$ when $x=150$.
$P''(x)=-4<0$ for all x means that $x=150$ maximizes profit.

b. The selling price is $p=800-150=\$650$.

44. $R(x)=7000x-10x^2-\dfrac{x^3}{3}$ $\quad C(x)=40,000+600x+8x^2$

$P(x)=-\dfrac{x^3}{3}-18x^2+6400x-40,000$
$P'(x)=-x^2-36x+6400=(64-x)(100+x)$
Profit is a maximum at $x=64$. $(P'(64)=0)$

x		64	
$P'(x)$	+	0	−

$P(64)=-87,381.33-73,728+409,600-40,000=\$208,490.67$

45. $R(x)=x^2\left(500-\dfrac{x}{3}\right)=500x^2-\dfrac{x^3}{3}$ $\quad R'(x)=1000x-x^2$
$R'(x)=0$ when $x=0$ and when $x=1000$.
$R''(x)=1000-2x$ and $R''(1000)<0$. So, the maximum reaction occurs at $x=1000$.

46. $N(t)=4+3t^2-t^3$ $\quad N'(t)=6t-3t^2=3t(2-t)$
Critical values: $t=0, 2$
$N''(t)=6-6t, N''(2)=-6<0$
Thus, maximum at $t=2$. Maximum production occurs at $8:00+2$ hrs =10:00 a.m.

47. $P=300+10t-t^2$, $t=0$ is 2000 $\ 0\le t\le 10$
$P'=10-2t=0$ when $t=5$.

t		5	
$P'(t)$	+	0	−

When $t=5$, $P=325$. The largest graduating class is the class of 2005 with 325 graduates.

48. $b(x) = \frac{8k}{x^2} + \frac{k}{(30-x)^2}$

$b'(x) = \frac{-16k}{x^3} - \frac{2k}{(30-x)^3}(-1)$

Set $b'(x) = 0$ and simplify.

$x^3 = 8(30-x)^3$

$x = 2(30-x)$

$3x = 60$ or $x = 20$

x	20
$b'(x)$	− 0 +

Build the observatory 20 miles from A.

49. Maximize $A = xy$ where $2x + 2y = 16$.

$A = x(8-x) = 8x - x^2 \qquad A' = 8 - 2x$

$A' = 0$ when $x = 4$.

Since $A'' < 0$ for all x, a 4×4 playpen gives a maximum area.

50. Quantity to be minimized: $A = (x+2)\left(y + \frac{7}{4}\right)$

Another equation: $xy = 56$ or $y = \frac{56}{x}$

Substituting: $A = (x+2)\left(\frac{56}{x} + \frac{7}{4}\right)$

$A = 56 + \frac{7}{4}x + \frac{112}{x} + \frac{7}{2}$

$A' = \frac{7}{4} - \frac{112}{x^2}$

Set $A' = 0$ and simplify. $7x^2 = 448$

$x^2 = 64$ or $x = 8$

x	8
A'	− 0 +

So, $x = 8$ and $y = 7$ minimizes A.

Page dimensions are (8 + 2) by $\left(7 + 1\frac{3}{4}\right)$ or 10″ by $8\frac{3}{4}''$.

51. $R(x) = x^2\left(500 - \frac{x}{3}\right) = 500x^2 - \frac{x^3}{3}$

$R'(x) = 1000x - x^2$

$R''(x) = 1000 - 2x$

$R''(x) = 0$ when $x = 500$.

x	500
$R''(x)$	+ 0 −

So, the dosage $x = 500$ maximizes sensitivity.

52. $f(x) = 145x^2 - 30x^3$

$f'(x) = 290x - 90x^2$

$f''(x) = 290 - 180x$

Critical value for $f'(x)$ is where $f''(x) = 0$.

So, the critical value is $x = \frac{29}{18}$.

$f'''(x) < 0$ for all x. Rate of change is maximized at $x = \frac{29}{18}$.

53. Total production cost: $\left(\dfrac{288{,}000}{x}\right)(1500)+(288{,}000)(30)$ Total storage cost: $\left(\dfrac{x}{2}\right)(1.5)$

$$C(x)=\left(\frac{288{,}000}{x}\right)(1500)+(288{,}000)(30)+\frac{1.5x}{2}=\frac{432{,}000{,}000}{x}+8{,}640{,}000+0.75x$$

$C'(x)=-\dfrac{432{,}000{,}000}{x^2}+0.75$ $C'(x)=0$ when $x^2=576{,}000{,}000$ or when $x=24{,}000$.

$C''(x)=\dfrac{864{,}000{,}000}{x^3}>0$ for $x>0$.

So, the minimum cost occurs for a run of $x=24{,}000$.

Chapter Test

1. $f(x)=x^3+6x^2+9x+3$

$f'(x)=3x^2+12x+9=3(x+3)(x+1)$

x		-3		-1	
$f'(x)$	+	0	−	0	+

Relative maximum at $x=-3$ $(-3, 3)$
Relative minimum at $x=-1$ $(-1,-1)$
Polynomials have no asymptotes.

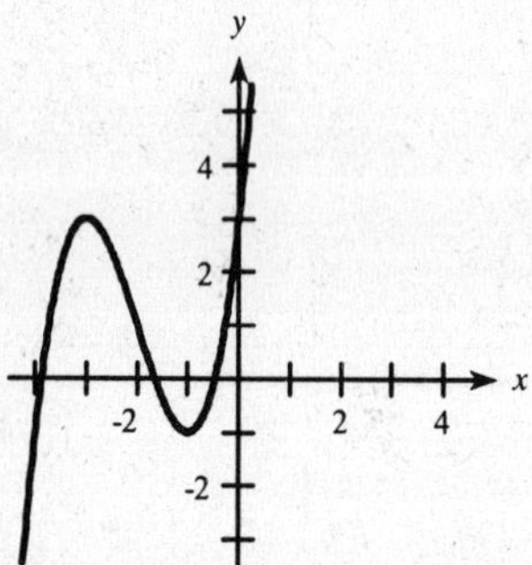

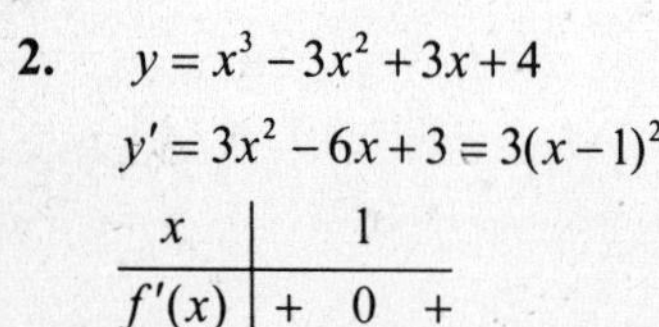

2. $y=x^3-3x^2+3x+4$

$y'=3x^2-6x+3=3(x-1)^2$

x		1	
$f'(x)$	+	0	+

Horizontal tangent and inflection point at (1, 5)
Polynomials have no asymptotes

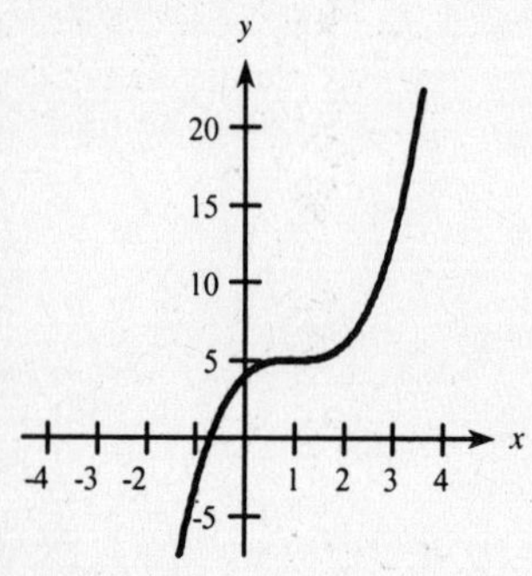

3. $y=\dfrac{x^2-3x+6}{x-2}$

$$y'=\frac{(x-2)(2x-3)-(x^2-3x+6)(1)}{(x-2)^2}=\frac{x(x-4)}{(x-2)^2}$$

x		0		2		4	
y'	+	0	−	*	−	0	+

* means y' is undefined when $x=2$.
In fact, $x=2$ is not in the domain.
Relative maximum at $x=0$ $(0,-3)$
Relative minimum at $x=4$ $(4, 5)$
Vertical asymptote: $x=2$

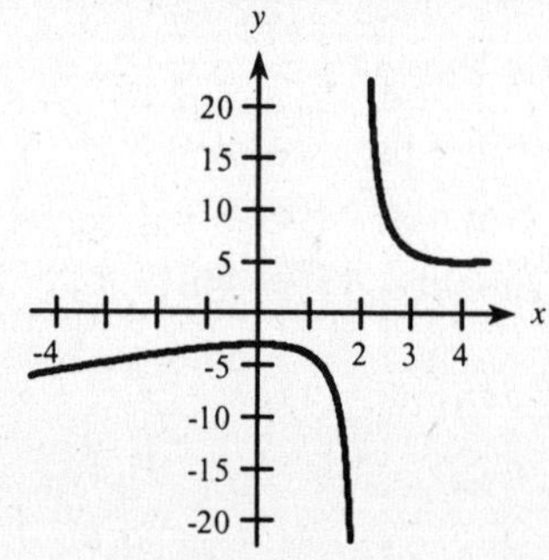

4–6. $y=3x^5-5x^3+2$

$y'=15x^4-15x^2=15x^2(x+1)(x-1)$

$y''=60x^3-30x$

$=30x(2x^2-1)$

$=30x\left[2\left(x+\dfrac{1}{\sqrt{2}}\right)\left(x-\dfrac{1}{\sqrt{2}}\right)\right]$

4. Concave up means $y''>0$.

$\left(\dfrac{1}{\sqrt{2}}\approx 0.707\right)$ $-\dfrac{1}{\sqrt{2}}<x<0,\ x>\dfrac{1}{\sqrt{2}}$

5. y'' changes signs for inflection points.

$x = 0, \pm \frac{1}{\sqrt{2}} \approx \pm 0.707$

6. Relative maximum and minimum occur at $x = \pm 1$.

$y'' > 0$ at $x = 1$ gives a relative minimum (1, 0).

$y'' < 0$ at $x = -1$ gives a relative maximum (–1, 4).

7. $f(x) = 2x^3 - 15x^2 + 3$; [–2, 8]

$f'(x) = 6x^2 - 30x = 6x(x-5)$

x	–2	0	5	8
$f(x)$	–73	3	–122	67

Absolute maximum of 67 at $x = 8$.
Absolute minimum of –122 at $x = 5$.

8. $f(x) = \frac{200x - 500}{x + 300} = \frac{200 - \frac{500}{x}}{1 + \frac{300}{x}}$

Horizontal asymptote: $y = 200$

Vertical asymptote: $x = -300$

9.

Point	f	f'	f''
A	–	+	–
B	+	–	0
C	+	0	+

$f(x) < 0$ if below x-axis

$f'(x)$ is positive if $f(x)$ is increasing.

$f''(x)$ is positive if $f(x)$ is concave up.

10. a. $\lim_{x \to -\infty} f(x) = -2$ (Horizontal asymptote)

b. $x = -1$ is the vertical asymptote

11. $f(6) = 10, f'(6) = 0, f''(6) = -3$

By the 2nd Derivative Test there is a local maximum at (6, 10).

12. $y = 0.19x^2 - 16.59x + 1038.29$

$y' = 0.38x - 16.59$

There is a critical value at $x = \frac{16.59}{0.38} \approx 43.66$.

x		43.66	
y'	–	0	+

a. The number is a minimum during 1943 (1900 + 43.7).

b. This is the inverse of (a). Thus the ratio of priests to Catholics was highest in 1943.

13. $R(x) = 164x \qquad C(x) = 0.01x^2 + 20x + 300$

$P(x) = 164x - (0.01x^2 + 20x + 300)$

$= 144x - 0.01x^2 - 300$

a. $P'(x) = 144 - 0.02x$

Critical value at $x = \frac{144}{0.02} = 7200$.

x		7200	
$P'(x)$	+	0	–

$x = 7200$ gives maximum profit

b. $P(7200) = \$518,100$

14. $C(x) = 100 + 20x + 0.01x^2$

$\overline{C}(x) = \frac{C(x)}{x} = \frac{100}{x} + 20 + 0.01x$

$\overline{C}'(x) = -\frac{100}{x^2} + 0.01$ Critical value at $x = 100$.

x		100	
$\overline{C}'(x)$	–	0	+

Minimum occurs when $\overline{C}'(x) = 0$, or $x = 100$.

15. Let x = number of additional units sold.

$R(x) = (100 + x)(300 - 2x) = 30,000 + 100x - 2x^2$

$R'(x) = 100 - 4x$

$R'(x) = 0$ if $x = 25$

x		25	
R ¢(x)	+	0	-

Price to get maximum revenue
= 300 - 2(25) = \$250.

16.

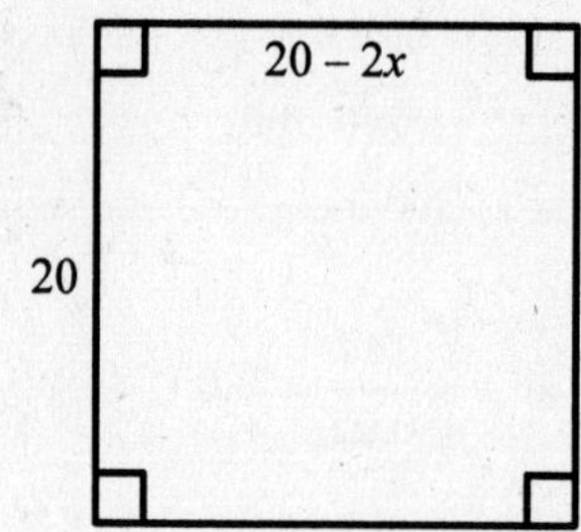

x = length of removed square.

$$V = x(20-2x)^2$$

$$V'(x) = x(2)(20-2x)(-2)+(20-2x)^2(1)$$
$$= (20-2x)[-4x+(20-2x)]$$
$$= (20-2x)(20-6x)$$

Critical value occurs if

$V'(x) = 0$ and $x = \frac{20}{6} = \frac{10}{3}$ cm.

($x = 10$ is not in domain)

x	$\frac{10}{3}$
$V'(x)$	+ 0 –

$x = \frac{10}{3}$ gives maximum volume.

17. Inventory problem

x = number in each production run

$$C = \tfrac{784000}{x}(2500) + 784000(420) + \frac{x}{2}(5)$$

Production cost Storage cost

$$C' = \frac{-784000(2500)}{x^2} + \frac{5}{2}$$

$$C' = 0 \text{ if } x^2 = \frac{2(2500)(784000)}{5}$$

Critical value occurs when x = 28,000 units.

x	28000
C'	– 0 +

$x = 28,000$ gives minimum costs.

18. a. $y = -0.0001295x^3 - 0.0165x^2 + 0.4323x + 3.198$

b. $y' = -0.0003885x^2 - 0.0330x + 0.4323$

$y' = 0$ and gives a max at $x \approx 11.5$ or in 2002.

Chapter 11: Derivatives Continued

Exercise 11.1

1. $f(x) = 4 \ln x$

$f'(x) = 4 \cdot \frac{1}{x} = \frac{4}{x}$

3. $y = \ln 8x$

$\frac{dy}{dx} = \frac{1}{8x}(8) = \frac{1}{x}$

5. $y = \ln x^4$

$\frac{dy}{dx} = \frac{1}{x^4} \cdot 4x^3 = \frac{4}{x}$

7. $f(x) = \ln(4x+9)$

$f'(x) = \frac{1}{4x+9}(4) = \frac{4}{4x+9}$

9. $y = \ln(2x^2 - x) + 3x$

$\frac{dy}{dx} = \frac{1}{2x^2 - x}(4x-1) + 3$

$= \frac{4x-1}{2x^2-x} + 3$

11. $p = \ln(q^2 + 1)$

$\frac{dp}{dq} = \frac{1}{q^2+1} \cdot 2q = \frac{2q}{q^2+1}$

13. a. $y = \ln x - \ln(x-1)$

$\frac{dy}{dx} = \frac{1}{x} - \frac{1}{x-1} = \frac{x-1-x}{x(x-1)} = \frac{-1}{x(x-1)}$

b. $y = \ln \frac{x}{x-1}$

$\frac{dy}{dx} = \frac{1}{\frac{x}{x-1}}\left[\frac{(x-1)(1) - x(1)}{(x-1)^2}\right]$

$= \frac{x-1}{x}\left[\frac{-1}{(x-1)^2}\right] = \frac{-1}{x(x-1)}$

15. a $y = \frac{1}{3}\ln(x^2 - 1)$

$\frac{dy}{dx} = \frac{1}{3}\left[\frac{1}{x^2-1} \cdot 2x\right] = \frac{2x}{3(x^2-1)}$

b. Using properties of logs.

$y = \ln \sqrt[3]{x^2 - 1} = \ln(x^2-1)^{1/3} = \frac{1}{3}\ln(x^2-1)$

Thus, $\frac{dy}{dx} = \frac{2x}{3(x^2-1)}$.

17. a. $y = \ln(4x-1) - 3\ln x$

$\frac{dy}{dx} = \frac{1}{4x-1} \cdot 4 - 3 \cdot \frac{1}{x}$

$= \frac{4x - 3(4x-1)}{x(4x-1)} = \frac{3-8x}{x(4x-1)}$

b. $y = \ln\left(\frac{4x-1}{x^3}\right) = \ln(4x-1) - 3\ln x$

$\frac{dy}{dx} = \frac{1}{4x-1} \cdot 4 - 3 \cdot \frac{1}{x}$

$= \frac{4}{4x-1} - \frac{3}{x} = \frac{3-8x}{x(4x-1)}$

19. $p = \ln\left(\frac{q^2-1}{q}\right)$

$= \ln(q^2 - 1) - \ln q$

$\frac{dp}{dq} = \frac{2q}{q^2-1} - \frac{1}{q} = \frac{2q^2 - q^2 + 1}{q(q^2-1)} = \frac{q^2+1}{q(q^2-1)}$

21. $y = \ln\left(\frac{t^2+3}{\sqrt{1-t}}\right) = \ln(t^2+3) - \frac{1}{2}\ln(1-t)$

$\frac{dy}{dt} = \frac{2t}{t^2+3} - \frac{-1}{2(1-t)} = \frac{4t - 4t^2 + t^2 + 3}{2(1-t)(t^2+3)}$

$= \frac{3 + 4t - 3t^2}{2(1-t)(t^2+3)} = \frac{-(3t^2 - 4t - 3)}{2(1-t)(t^2+3)}$

23. $y = \ln\left(x^3\sqrt{x+1}\right) = 3\ln x + \frac{1}{2}\ln(x+1)$

$\frac{dy}{dx} = 3 \cdot \frac{1}{x} + \frac{1}{2} \cdot \frac{1}{x+1} \cdot 1 = \frac{3}{x} + \frac{1}{2(x+1)}$ or

$= \frac{3(2)(x+1) + x \cdot 1}{2x(x+1)} = \frac{7x+6}{2x(x+1)}$

25. $y = x - \ln x$

$\frac{dy}{dx} = 1 - \frac{1}{x}$

27. $y = \frac{\ln x}{x}$

$\frac{dy}{dx} = \frac{x \cdot \frac{1}{x} - (\ln x)(1)}{x^2} = \frac{1 - \ln x}{x^2}$

29. $y = \ln(x^4+3)^2 = 2\ln(x^4+3)$

$\frac{dy}{dx} = 2 \cdot \frac{1}{x^4+3} \cdot 4x^3 = \frac{8x^3}{x^4+3}$

31. $y = (\ln x)^4$

$$\frac{dy}{dx} = 4(\ln x)^3 \cdot \frac{1}{x} = \frac{4(\ln x)^3}{x}$$

33. $y = \left[\ln(x^4+3)\right]^2$

$$\frac{dy}{dx} = 2\left[\ln(x^4+3)\right] \cdot \frac{1}{x^4+3} \cdot 4x^3$$

$$= \frac{8x^3 \ln(x^4+3)}{x^4+3}$$

35. $y = \log_4 x = \frac{\ln x}{\ln 4} = \frac{1}{\ln 4}(\ln x)$

$$\frac{dy}{dx} = \frac{1}{\ln 4} \cdot \frac{1}{x} = \frac{1}{x \ln 4}$$

37. $y = \log_6(x^4 - 4x^3 + 1) = \frac{1}{\ln 6}\ln(x^4 - 4x^3 + 1)$

$$\frac{dy}{dx} = \frac{1}{\ln 6} \cdot \frac{1}{x^4 - 4x^3 + 1} \cdot (4x^3 - 12x^2)$$

$$= \frac{4x^3 - 12x^2}{(x^4 - 4x^3 + 1)\ln 6}$$

39. $y = x \ln x$

$y' = x \cdot \frac{1}{x} + (\ln x)(1) = 1 + \ln x$

Set $y' = 0$.

$1 + \ln x = 0$ gives $\ln x = -1$ or $x = e^{-1}$

$y = e^{-1} \ln e^{-1} = e^{-1}(-1) = -e^{-1}$

Rel min at $(e^{-1}, -e^{-1})$.

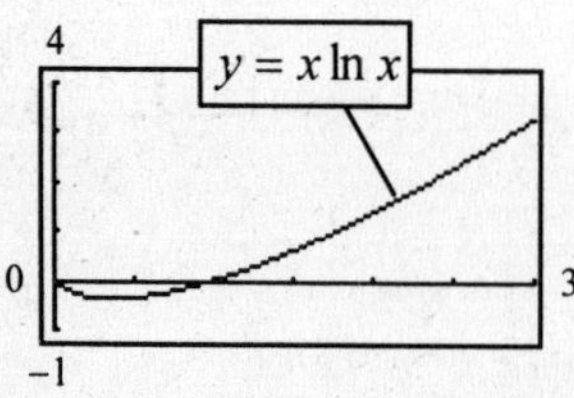

41. $y = x^2 - 8\ln x$

$y' = 2x - 8 \cdot \frac{1}{x}$

$2x^2 - 8 = 0$ or $x^2 - 4 = 0$ gives $x = \pm 2$.

Note: $\ln(-2)$ is not defined.

$x = 2$ gives $y = 2^2 - 8\ln 2 = 4 - 8\ln 2$

Rel min at $(2,\ 4 - 8\ln 2)$.

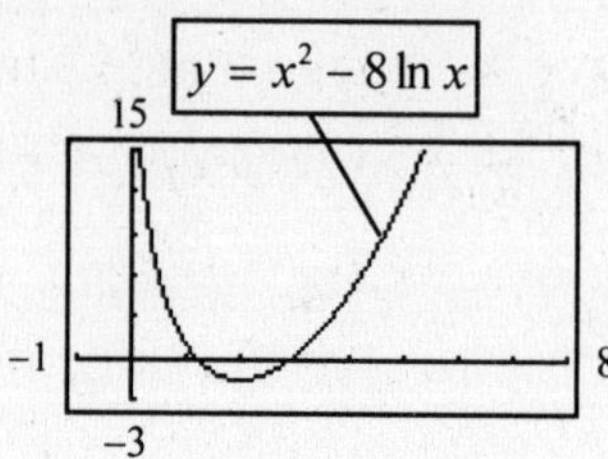

43. $C(x) = 1500 + 200\ln(2x+1)$

a. $C'(x) = 200 \cdot \frac{1}{2x+1} \cdot 2 = \frac{400}{2x+1}$

b. $C(200) = \frac{400}{401} \approx \$1.00$

It will cost approximately \$1.00 to make the next unit.

c. $\overline{MC} > 0$. Yes.

45. $R(x) = \frac{2500x}{\ln(10x+10)}$

a. $$R'(x) = \frac{2500\ln(10x+10) - 2500x\left(\frac{1}{10x+10} \cdot 10\right)}{(\ln(10x+10))^2} = \frac{2500\ln(10x+10) - \frac{2500x}{x+1}}{(\ln(10x+10))^2}$$

$$= \frac{2500(x+1)\ln(10x+10) - 2500x}{(x+1)\left[\ln(10x+10)\right]^2} = \frac{2500\left[(x+1)\ln(10x+10) - x\right]}{(x+1)\left[\ln(10x+10)\right]^2}$$

b. $$R'(100) = \frac{2500(101)\ln 1010 - 250{,}000}{101(\ln 1010)^2} = 309.67$$

Selling one additional unit yields \$309.67.

47. $p(x) = \dfrac{4000}{\ln(x+10)}$ $\quad \dfrac{dp}{dx} = \dfrac{(\ln(x+10))(0) - \frac{4000}{x+10}}{(\ln(x+10))^2} = -\dfrac{4000}{(x+10)(\ln(x+10))^2}$

a. When $x = 40$, $\dfrac{dp}{dx} = -\dfrac{4000}{(40+10)(\ln(40+10))^2} = -5.23$

b. When $x = 90$, $\dfrac{dp}{dx} = -\dfrac{4000}{(90+10)(\ln(90+10))^2} = -1.89$

c. $p''(x) = \dfrac{0 + 4000\left[(x+10)2\ln(x+10)\cdot\frac{1}{x+10} + (\ln(x+10))^2(1)\right]}{\left[(x+10)(\ln(x+10))^2\right]^2} = \dfrac{4000\left[2\ln(x+10) + (\ln(x+10))^2\right]}{\left[(x+10)(\ln(x+10))^2\right]^2}$

When $x = 40$, $p''(x) > 0$. Thus, $p'(x)$ is increasing at 40 units.

49. $R(r) = A\ln r - Br$

$R'(r) = \dfrac{A}{r} - B$

$R'(r) = 0$ for $r = \dfrac{A}{B}$.

$R''(r) = -\dfrac{A}{r^2} < 0$ for all r.

Thus, $R(r)$ is a maximum for $r = \dfrac{A}{B}$.

51. $R = \dfrac{1}{\ln 10}(\ln I - \ln I_0)$

$\dfrac{dR}{dI} = \dfrac{1}{\ln 10}\left(\dfrac{1}{I} - 0\right) = \dfrac{1}{I\ln 10}$

53. a. $y = 105.175 + 2847.788\ln x$

b. $y'(x) = \dfrac{2847.788}{x}$

$y'(30) = \$94.93$ (increase per year)

Exercise 11.2

1. $y = 5e^x - x$

$y' = 5e^x - 1$

3. $f(x) = e^x - x^e$

$f'(x) = e^x - ex^{e-1}$

5. $y = e^{x^3}$

$\dfrac{dy}{dx} = e^{x^3}\cdot 3x^2 = 3x^2e^{x^3}$

7. $y = 6e^{3x^2}$

$\dfrac{dy}{dx} = 6e^{3x^2}\cdot 6x = 36xe^{3x^2}$

9. $y = 2e^{(x^2+1)^3}$

$\dfrac{dy}{dx} = 2e^{(x^2+1)^3}\cdot 3(x^2+1)^2(2x)$

$= 12x(x^2+1)^2e^{(x^2+1)^3}$

11. $y = e^{\ln x^3} = x^3$

$y' = 3x^2$

13. $y = e^{-1/x} = e^{-(x^{-1})}$

$y' = e^{-(x^{-1})}\cdot(1x^{-2}) = \dfrac{e^{-1/x}}{x^2}$

15. $y = e^{-1/x^2} + e^{-x^2} = e^{-(x^{-2})} + e^{-x^2}$

$y' = e^{-(x^{-2})}(2x^{-3}) + e^{-x^2}(-2x) = \dfrac{2e^{-1/x^2}}{x^3} - 2xe^{-x^2}$

17. $s = t^2e^t$

$s' = t^2(e^t) + e^t(2t) = te^t(t+2)$

19. $y = e^{x^4} - (e^x)^4 = e^{x^4} - e^{4x}$

$y' = e^{x^4}\cdot 4x^3 - e^{4x}(4) = 4x^3e^{x^4} - 4e^{4x}$

21. $y = \ln(e^{4x} + 2)$

$y' = \dfrac{1}{e^{4x}+2}\cdot e^{4x}(4) = \dfrac{4e^{4x}}{e^{4x}+2}$

23. $y = e^{-3x}\ln(2x)$

$y' = e^{-3x}\cdot\dfrac{1}{2x}\cdot 2 + \ln(2x)\cdot e^{-3x}(-3)$

$= \dfrac{e^{-3x}}{x} - 3e^{-3x}\ln(2x)$

25. $y = \frac{1+e^{5x}}{e^{3x}} = \frac{1}{e^{3x}} + e^{5x-3x} = e^{-3x} + e^{2x}$

$y' = e^{-3x}(-3) + e^{2x}(2) = 2e^{2x} - 3e^{-3x}$

27. $y = (e^{3x}+4)^{10}$

$y' = 10(e^{3x}+4)^9(e^{3x}\cdot 3) = 30e^{3x}(e^{3x}+4)^9$

29. $y = 6^x$

$y' = 6^x \cdot \ln 6$

31. $y = 4^{x^2}$

$y' = 4^{x^2}(2x\ln 4)$

33. **a.** $y = xe^{-x}$

$y' = xe^{-x}(-1) + e^{-x}(1) = e^{-x} - xe^{-x}$

$y'(1) = e^{-1} - 1e^{-1} = 0$

b. If $x = 1$, then $y = 1e^{-1} = e^{-1}$.

$y - e^{-1} = 0(x-1)$

$y = e^{-1}$

35. **a.** $y = \frac{1}{\sqrt{2\pi}}e^{-z^2/2}$

$y' = \frac{1}{\sqrt{2\pi}}e^{-z^2/2}(-z)$

$y' = 0$ at $z = 0$.

Maximum occurs at $z = 0$.

b.

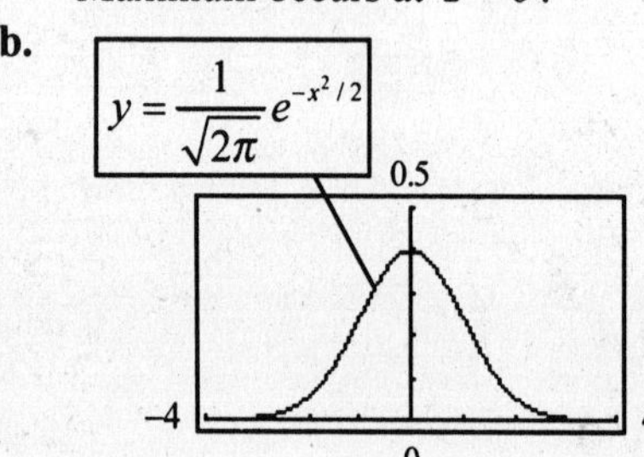

37. $y = \frac{e^x}{x}$

$y' = \frac{xe^x - e^x(1)}{x^2} = \frac{e^x(x-1)}{x^2}$

$y' = 0$ if $e^x(x-1) = 0$.

Since $e^x \neq 0$ for any x, the critical value is $x = 1$.
Relative minimum at $(1, e)$.

39. $y = x - e^x$

$y' = 1 - e^x$

$1 - e^x = 0$ gives $x = 0$ as the critical value.

Relative maximum at $(0, -1)$.

41. $S = Pe^{0.1n}$

a. $\frac{dS}{dn} = Pe^{0.1n}(0.1)$

b. When $n = 1$, $\frac{dS}{dn} = 0.1Pe^{0.1}$

c. Yes, since $e^{0.1} > 1$.

43. $S(t) = 100{,}000e^{-0.5t}$

a. $S'(t) = 100{,}000e^{-0.5t}(-0.5) = -50{,}000e^{-0.5t}$

b. The function is decay exponential. The derivative is always negative.

45. $C(x) = 10{,}000 + 20xe^{x/600}$

$C'(x) = 20x \cdot e^{x/600} \cdot \frac{1}{600} + e^{x/600}(20)$

$= e^{x/600}\left(\frac{x}{30} + 20\right)$

$C'(600) = e(20+20) = 40e \approx \$108.73/\text{unit}$

47. $y = 100(1 - e^{-0.462t}) = 100 - 100e^{-0.462t}$

a. $y' = -100e^{-0.462t}(-0.462) = 46.2e^{-0.462t}$

b. $y'(1) = 46.2e^{-0.462(1)} = 29.107$

49. $y = \frac{10{,}000}{1 + 9999e^{-0.99t}}$

$y' = \frac{0 - 10{,}000(9999e^{-0.99t})(-0.99)}{(1+9999e^{-0.99t})^2}$

$= \frac{98{,}990{,}100e^{-0.99t}}{(1+9999e^{-0.99t})^2}$

51. $x(t) = 0.05 + 0.18e^{-0.38t}$

$x'(t) = (0.18)e^{-0.38t}(-0.38) = -0.068e^{-0.38t}$

53. $N = N_0(1+r)^t$

The function is of the form $y = a^u$.

$\frac{dy}{dx} = a^u \cdot \frac{du}{dx} \cdot \ln a$

$\frac{dN}{dt} = N_0(1+r)^t \ln(1+r)$

55. $\frac{I}{I_0} = 10^R$

Multiplying both sides by I_0 to be in the form $y = a^u$, we obtain $I = I_0 10^R$, so

$I' = I_0 10^R \ln 10$.

If $I_0 = 1$, $\frac{dI}{dR} = 10^R \ln 10$.

57. $H(t) = 28.8e^{0.0994t}$

$H'(t) = 28.8e^{0.0994t} \cdot (0.0994)$

$= 2.28627e^{0.0994t}$

$H'(42) = \$186.15$ billion

59. $d(t) = 1.63e^{0.0836t}$

a. $d'(t) = 0.13627e^{0.0836t}$

b. $d'(40) \approx \$3.86$ billion per year

$d'(100) \approx \$582.23$ billion per year

c. 1999-2000: $\text{Avg.} = \dfrac{5674.2 - 5656.3}{1}$

$= \$17.9$ billion per year

2000-2001: $\text{Avg.} = \dfrac{5807.5 - 5674.2}{1}$

$= \$133.3$ billion per year

Change the model since average rates are smaller than predicted rates.

d. The terrorist attacks of 9-11-01 changed many programs.

61. $P(x) = 3.974e^{-0.0527x}$

a. $P'(x) = -0.20943e^{-0.0527x}$

$P'(40) \approx -0.025$

b. The purchasing power of \$1 will decrease \$0.025 during 2000.

c. 1999-2000: Avg. = −0.019 dollars per year

63. a. $y = 246.47e^{0.0689x}$

b. $y'(x) = 16.9818e^{0.0689x}$

$y'(60) \approx \$1060.12$

Federal taxes will increase \$1060.12 during 2010.

65. a. $P(x) = \dfrac{2,008,060}{1 + 5.4912e^{-0.12776x}}$

b. $P(20) = 1,407,645$ means that the model predicts this number of prisoners in 2000.

$P'(x) = \dfrac{1,408,765.963}{\left(1 + 5.4912e^{-0.12776x}\right)^2}$

$P'(20) \approx 53,773$ means that the number of prisoners will increase by this amount in 2001. (Your calculator may give an overflow error.)

Exercise 11.3

1. $x^2 - 4y - 17 = 0$

$2x - 4y' = 0$

At $(1, -4)$, $y' = \dfrac{1}{2}$.

3. $xy^2 = 8$

$x \cdot 2yy' + y^2 = 0$

At (2, 2), $8y' + 4 = 0$.

$8y' = -4$

$y' = -\dfrac{1}{2}$

5. $x^2 + 3xy - 4 = 0$

$2x + 3[xy' + y(1)] = 0$

At (1, 1), $2(1) + 3[(1)y' + (1)(1)] = 0$

$2 + 3y' + 3 = 0$

$3y' = -5$

$y' = -\dfrac{5}{3}$

7. $x^2 + 2y^2 - 4 = 0$

$2x + 4yy' - 0 = 0$

$y' = -\dfrac{2x}{4y} = -\dfrac{x}{2y}$

9. $x^2 + 4x + y^2 - 3y + 1 = 0$

$2x + 4 + 2yy' - 3y' + 0 = 0$

$y' = -\dfrac{2x + 4}{2y - 3}$

11. $x^2 + y^2 = 4$

$2x + 2yy' = 0$

$y' = -\dfrac{x}{y}$

13.
$$xy^2 - y^2 = 1$$
$$x \cdot 2yy' + y^2(1) - 2yy' = 0$$
$$y'(2xy - 2y) = -y^2$$
$$y' = \frac{-y^2}{2xy - 2y}$$
$$= \frac{-y}{2(x-1)}$$

15.
$$p^2q = 4p - 2$$
$$p^2(1) + q \cdot 2p\frac{dp}{dq} = 4\frac{dp}{dq}$$
$$(2qp - 4)\frac{dp}{dq} = -p^2$$
$$\frac{dp}{dq} = \frac{p^2}{4 - 2qp}$$

17.
$$3x^5 - 5y^3 = 5x^2 + 3y^5$$
$$15x^4 - 15y^2\frac{dy}{dx} = 10x + 15y^4\frac{dy}{dx}$$
$$\frac{dy}{dx}(-15y^2 - 15y^4) = 10x - 15x^4$$
$$y' = -\frac{5(2x - 3x^4)}{15(y^2 + y^4)} = -\frac{2x - 3x^4}{3(y^2 + y^4)}$$

19. $x^4 + 2x^3y^2 = x - y^3$
$$4x^3 + 2\left[x^3 \cdot 2y\frac{dy}{dx} + y^2 \cdot 3x^2\right] = 1 - 3y^2\frac{dy}{dx}$$
$$3y^2\frac{dy}{dx} + 2x^3 \cdot 2y\frac{dy}{dx} = 1 - 4x^3 - 6x^2y^2$$
$$\frac{dy}{dx} = \frac{1 - 4x^3 - 6x^2y^2}{3y^2 + 4x^3y}$$

21.
$$x^4 + 3x^3y^2 - 2y^5 = (2x + 3y)^2$$
$$4x^3 + 3x^3 \cdot 2y\frac{dy}{dx} + y^2 \cdot 9x^2 - 10y^4\frac{dy}{dx} = 2(2x + 3y)\left(2 + 3\frac{dy}{dx}\right)$$
$$(6x^3y - 10y^4)\frac{dy}{dx} = -4x^3 - 9x^2y^2 + 8x + 12x\frac{dy}{dx} + 12y + 18y\frac{dy}{dx}$$
$$(6x^3y - 10y^4 - 12x - 18y)\frac{dy}{dx} = -4x^3 - 9x^2y^2 + 8x + 12y$$
$$\frac{dy}{dx} = \frac{-4x^3 - 9x^2y^2 + 8x + 12y}{6x^3y - 10y^4 - 12x - 18y} = \frac{4x^3 + 9x^2y^2 - 8x - 12y}{10y^4 + 12x + 18y - 6x^3y}$$

23. $x^2 + 4x + y^2 + 2y - 4 = 0$
$$2x + 4 + 2yy' + 2y' = 0$$
$$2(y + 1)y' = -2(x + 2)$$
$$y' = -\frac{x + 2}{y + 1}$$
At $(1, -1)$, y' is undefined.

25. $x^2 + 2xy + 3 = 0$
$$2x + 2xy' + y(2) = 0$$
$$2xy' = -2x - 2y$$
$$y' = \frac{-2x - 2y}{2x} = -\frac{x + y}{x}$$
At $(-1, 2)$, $y' = 1$.

27. $x^2 - 2y^2 + 4 = 0$
$$2x - 4yy' = 0$$
$$y' = \frac{x}{2y}$$
At (2, 2) we have $y' = \frac{2}{4} = \frac{1}{2}$.
The equation of the tangent line is
$y - 2 = \frac{1}{2}(x - 2)$ or $y = \frac{1}{2}x + 1$.

29. $4x^2 + 3y^2 - 4y - 3 = 0$
$$8x + 6yy' - 4y' = 0$$
At $(-1, 1)$, we have $8(-1) + 6(1)y' - 4y' = 0$ or $2y' = 8$ or $y' = 4$. So, $m = 4$.
The equation of the tangent line
$$y - 1 = 4(x - (-1))$$
$$y - 1 = 4x + 4$$
$$y = 4x + 5.$$

31. $\ln x = y^2$

$\frac{1}{x} = 2y\frac{dy}{dx}$

$\frac{dy}{dx} = \frac{1}{2xy}$

33. $y^2 \ln x = 4$

$y^2\left(\frac{1}{x}\right) + \ln x(2yy') = 0$

$y' = \frac{-y^2}{2xy\ln x} = \frac{-y}{2x\ln x}$

35. $x^2 + \ln y = 4$

$2x + \frac{1}{y}\cdot\frac{dy}{dx} = 0$

$\frac{dy}{dx} = -2xy$

At $(2, 1)$, $\frac{dy}{dx} = -2(2)(1) = -4.$

37. $xe^y = 6$

$xe^y\cdot\frac{dy}{dx} + e^y(1) = 0$

$\frac{dy}{dx} = \frac{-e^y}{xe^y} = -\frac{1}{x}$

39. $e^{xy} = 4$

$e^{xy}\left(x\cdot\frac{dy}{dx} + y\cdot 1\right) = 0$

$\frac{dy}{dx} = \frac{-ye^{xy}}{xe^{xy}} = -\frac{y}{x}$

41. $ye^x - y = 3$

$y\cdot e^x + e^x\cdot\frac{dy}{dx} - \frac{dy}{dx} = 0$

$\frac{dy}{dx} = \frac{ye^x}{1-e^x}$

43. $y(x) = xe^{-x}$

$y'(x) = x\cdot e^{-x}(-1) + e^{-x}(1) = e^{-x}(1-x)$

$y'(1) = e^{-1}(0) = 0$

45. From Problem 37, $\frac{dy}{dx} = -\frac{1}{x}$

At $x = 3, \frac{dy}{dx} = -\frac{1}{3}.$

$y - 0 = -\frac{1}{3}(x-3)$ or $y = -\frac{1}{3}x + 1$

47. $x^2 + 4y^2 - 4x - 4 = 0$

$2x + 8yy' - 4 = 0$

$y' = \frac{4-2x}{8y} = \frac{2-x}{4y}$

a. There is a horizontal tangent if $y' = 0$ or $x = 2$. Then,

$4 + 4y^2 - 8 - 4 = 0$

$4y^2 = 8$

$y^2 = 2$ or $y = \pm\sqrt{2}.$

Horizontal tangents at $(2, \sqrt{2})$ and at $(2, -\sqrt{2})$.

b. There is a vertical tangent when $y = 0$. Then, $x^2 - 4x - 4 = 0$ gives $x = 2 \pm 2\sqrt{2}.$ Vertical tangents at $(2+2\sqrt{2}, 0)$ and at $(2-2\sqrt{2}, 0)$.

49. $y' = -\frac{x}{y} = \frac{-x}{y}$

a. $y'' = \frac{y(-1) - (-x)y'}{y^2} = \frac{-y + xy'}{y^2}$

b. $y'' = \frac{-y + x\left(-\frac{x}{y}\right)}{y^2}\cdot\frac{y}{y}$

$= \frac{-y^2 - x^2}{y^3} = \frac{-\left(x^2 + y^2\right)}{y^3}$

c. Yes. $x^2 + y^2 = 4$

51. $\sqrt{x} + \sqrt{y} = 1$

$\frac{1}{2}x^{-1/2} + \frac{1}{2}y^{-1/2}y' = 0$

$y' = -\frac{x^{-1/2}}{y^{-1/2}} = -\frac{y^{1/2}}{x^{1/2}}$

$y'' = \frac{-x^{1/2}\cdot\frac{1}{2}y^{-1/2}y' - (-y^{1/2})\cdot\frac{1}{2}x^{-1/2}}{x}$

$= \frac{\frac{y^{1/2}}{x^{1/2}} - \frac{x^{1/2}}{y^{1/2}}y'}{2x} = \frac{\frac{y^{1/2}}{x^{1/2}} - \frac{x^{1/2}}{y^{1/2}}\cdot\frac{-y^{1/2}}{x^{1/2}}}{2x}$

$= \frac{\frac{y^{1/2}}{x^{1/2}} + 1}{2x}\cdot\frac{x^{1/2}}{x^{1/2}} = \frac{y^{1/2} + x^{1/2}}{2x^{3/2}}$

Since $y^{1/2} + x^{1/2} = 1$ we have

$y'' = \frac{1}{2x^{3/2}} = \frac{1}{2x\sqrt{x}}.$

53. $x^2+y^2-9=0$

$2x+2yy'=0$

$y'=\frac{-2x}{2y}=-\frac{x}{y}$

$-\frac{x}{y}=0$ gives $x=0$.

Max at $x=0$, $y=3$, and min at $x=0$, $y=-3$.

55. $xy-20x+10y=0$

$xy'+y(1)-20+10y'=0$

$y'=\frac{20-y}{x+10}$

At $x=10$, we have $y=10$.

Thus $y'=\frac{20-10}{10+10}=\frac{1}{2}=0.5.$

57. $(x+1)^{3/4}(y+2)^{1/3}=384$

$(x+1)^{3/4}\cdot\frac{1}{3}(y+2)^{-2/3}\cdot y'+(y+2)^{1/3}\cdot\frac{3}{4}(x+1)^{-1/4}\cdot 1=0$

Multiplying both sides by $\left(12(x+1)^{1/4}(y+2)^{2/3}\right)$ and simplifying and solving for y' we have $y'=\frac{-9(y+2)}{4(x+1)}$.

At (255, 214) we have $y'=\frac{-9(216)}{4(256)}=-\frac{243}{128}\approx-1.898.$

59. $p(q+1)^2=200{,}000$

$p\cdot 2(q+1)\frac{dq}{dp}+(q+1)^2\cdot 1=0$

$\frac{dq}{dp}=-\frac{q+1}{2p}$

At $p=80$, we have $q=49$.

Thus, $\frac{dq}{dp}=\frac{-50}{160}=\frac{-5}{16}$.

If the price is increased by \$1.00, the demand will decrease by $\frac{5}{16}$ unit.

61. $-0.000436t=\ln y-\ln(100)$

$-0.000436=\frac{1}{y}\cdot\frac{dy}{dt}$

So, $\frac{dy}{dt}=-0.000436y$

63. $\text{THI}=t-0.55(1-h)(t-58)$

$0=1-0.55\left[(1-h)(1)+(t-58)\left(-\frac{dh}{dt}\right)\right]$

$0.55(t-58)\frac{dh}{dt}=-1+0.55(1-h)$

$\frac{dh}{dt}=\frac{-0.55h-0.45}{0.55(t-58)}$

At $t=70$, $\frac{dh}{dt}=\frac{-0.55h-0.45}{0.55(12)}=\frac{-h}{12}-\frac{3}{44}$

Exercise 11.4

1. $y=x^3-3x$

$\frac{dy}{dt}=3x^2\frac{dx}{dt}-3\frac{dx}{dt}$

If $x=2$ and $\frac{dx}{dt}=4$,

$\frac{dy}{dt}=3(2)^2(4)-3(4)=36$

3. $xy=4$ gives $y=\frac{4}{x}$

$\frac{dy}{dt}=-\frac{4}{x^2}\cdot\frac{dx}{dt}$

If $x=8$ and $\frac{dx}{dt}=-2$,

$\frac{dy}{dt}=-\frac{4}{8^2}(-2)=\frac{1}{8}$

5. $x^2 + y^2 = 169$

$\frac{d}{dt}(x^2 + y^2) = \frac{d}{dt}(169)$

$2x \cdot \frac{dx}{dt} + 2y \cdot \frac{dy}{dt} = 0$

$\frac{dx}{dt} = -\frac{y}{x} . \frac{dy}{dt}$

$\frac{dx}{dt} = -\frac{12}{5} \cdot 2 = -\frac{24}{5}$

7. $y^2 = 2xy + 24$

$\frac{d}{dt}(y^2) = \frac{d}{dt}(2xy + 24)$

$2y \cdot \frac{dy}{dt} = 2x \cdot \frac{dy}{dt} + 2y \cdot \frac{dx}{dt} + 0$

$\frac{dx}{dt} = \frac{y - x}{y} \cdot \frac{dy}{dt}$

$\frac{dx}{dt} = \frac{12 - 5}{12} \cdot 2 = \frac{7}{6}$

9. $x^2 + y^2 = z^2$

$\frac{d}{dt}(x^2 + y^2) = \frac{d}{dt}(z^2)$

$2x \cdot \frac{dx}{dt} + 2y \cdot \frac{dy}{dt} = 2z \cdot \frac{dz}{dt}$

$\frac{dy}{dt} = \frac{z \cdot \frac{dz}{dt} - x \cdot \frac{dx}{dt}}{y}$

$x^2 + y^2 = z^2$ yields $3^2 + 4^2 = z^2$.

So, $z^2 = 25$ or $z = \pm 5$.

When $z = 5$, $\frac{dy}{dt} = \frac{5 \cdot 2 - 3 \cdot 10}{4} = -5.$

When $z = -5$, $\frac{dy}{dt} = \frac{-5 \cdot 2 - 3 \cdot 10}{4} = -10.$

11. $y = -4x^2$

$\frac{dy}{dt} = -8x \cdot \frac{dx}{dt} = -8(5)(2) = -80$ units/sec.

13. $A = \pi r^2$; $\frac{dA}{dt} = A'(t) = 2\pi r \cdot \frac{dr}{dt}$

$A'(t) = 2\pi \cdot 3 \cdot 2 = 12\pi$ sq ft/min

15. $V = x^3$; $V'(t) = 3x^2 \cdot \frac{dx}{dt}$

$64 = 3 \cdot 36 \cdot \frac{dx}{dt}$

$\frac{dx}{dt} = \frac{64}{108} = \frac{16}{27}$ in./sec

17. $P = 180x - \frac{1}{1000}x^2 - 2000$

$P'(t) = 180 \cdot \frac{dx}{dt} - \frac{1}{500} x \cdot \frac{dx}{dt}$

If $x = 100$ and $\frac{dx}{dt} = 10$, then

$P'(t) = 180(10) - \frac{1}{500}(100)(10) = \$1798/\text{day}$

19. $p = \frac{1000 - 10x}{400 - x}$

$\frac{dp}{dt} = \frac{(400 - x)(-10) - (1000 - 10x)(-1)}{(400 - x)^2} \cdot \frac{dx}{dt}$

If $\frac{dx}{dt} = -20$ and $x = 20$,

then $\frac{dp}{dt} = \frac{380(-10) - 800(-1)}{(380)^2}(-20)$

$= \frac{(-3000)(-20)}{380 \cdot 380} = 0.42$ dollars/day.

21. $x = 30y + 20y^2$

$\frac{dx}{dt} = 30\frac{dy}{dt} + 40y\frac{dy}{dt}$

If $y = 10$ and $\frac{dy}{dt} = 1$, then

$\frac{dx}{dt} = 30(1) + 40(10)(1) = 430$ units/month

Note: y is the number of thousands.

23. $V = \frac{4}{3}\pi r^3$

$\frac{dV}{dt} = 3 \cdot \frac{4}{3}\pi r^2 \cdot \frac{dr}{dt}$

Substituting $\frac{dr}{dt} = -1$ and $r = 3$, we have

$\frac{dV}{dt} = 4\pi \cdot 3^2(-1) = -36\pi$ mm^3/month.

V is decreasing at the rate of 36π mm^3/month.

25. $W = kL^3$

$\frac{dW}{dt} = 3kL^2\frac{dL}{dt}$

Percentage rate of change:

$\frac{\frac{dW}{dt}}{W} = \frac{3kL^2\frac{dL}{dt}}{W}$

$\frac{\frac{dW}{dt}}{W} = \frac{3kL^2\frac{dL}{dt}}{kL^3} = \frac{3\frac{dL}{dt}}{L} = 3\left(\frac{\frac{dL}{dt}}{L}\right)$

27. $C = 0.11W^{1.54}$

$$\frac{dC}{dt} = 1.54\left(0.11W^{0.54}\right)\cdot\frac{dW}{dt}$$

$$= 0.1694W^{0.54}\cdot\frac{dW}{dt}$$

$$\frac{\frac{dC}{dt}}{C} = \frac{0.1694W^{0.54}\cdot\frac{dW}{dt}}{C}$$

$$= \frac{0.1694W^{0.54}\frac{dW}{dt}}{0.11W^{1.54}} = 1.54\left(\frac{\frac{dW}{dt}}{W}\right)$$

29. $V = \frac{4}{3}\pi r^3$

$$\frac{dV}{dt} = \frac{4}{3}\pi\cdot 3r^2\cdot\frac{dr}{dt}$$

Substituting $\frac{dV}{dt} = 4,\ r = 2$, we have

$4 = 4\pi\cdot 2^2\cdot\frac{dr}{dt}$ or $\frac{dr}{dt} = \frac{1}{4\pi}$ micrometers/day.

31. $V = \frac{4}{3}\pi r^3$

$$\frac{dV}{dt} = \frac{4}{3}\pi\cdot 3r^2\cdot\frac{dr}{dt}$$

Substituting $\frac{dV}{dt} = 5 \text{ in}^3/\text{min}$, $r = 5$ gives

$5 = 4\pi\cdot 5^2\cdot\frac{dr}{dt}$ or $\frac{dr}{dt} = \frac{5}{100\pi} = \frac{1}{20\pi}$ in./min.

33. Let y be the height of the ladder on the wall and let x be the distance from the wall to the bottom of the ladder.

$\frac{dx}{dt} = 1$ ft/sec

$30^2 = x^2 + y^2$

$$2x\frac{dx}{dt} + 2y\frac{dy}{dt} = 0$$

When $x = 18,\ y = \sqrt{576} = 24.$

$$2(18)(1) + 2(24)\frac{dy}{dt} = 0$$

$48\frac{dy}{dt} = -36$ or $\frac{dy}{dt} = -\frac{36}{48} = -0.75$ ft/sec

The ladder is sliding down at the rate of $\frac{3}{4}$ ft/sec.

35. Let x be the horizontal distance from the plane to the observer and let y be the direct distance.

$\frac{dx}{dt} = -300$ mph

$y^2 = x^2 + 1^2$

$$2y\frac{dy}{dt} = 2x\frac{dx}{dt}$$

When $y = 5,\ x = \sqrt{24}$.

$2(5)\frac{dy}{dt} = 2\sqrt{24}(-300)$ or

$\frac{dy}{dt} = -\sqrt{24}(60) \approx -293.94$ mph .

The plane is approaching the observer at the rate of 293.94 mph.

37. Let x be the distance of car A from the intersection, let y be the distance of car B from the intersection, and let z be the distance between them.

$$\frac{dx}{dt} = -40,\ \frac{dy}{dt} = -55$$

$z^2 = x^2 + y^2$

$$2z\frac{dz}{dt} = 2x\frac{dx}{dt} + 2y\frac{dy}{dt}$$

When $x = 15$ and $y = 8$, $z = 17$.

$2(17)\frac{dz}{dt} = 2(15)(-40) + 2(8)(-55)$ or

$\frac{dz}{dt} \approx -61.18$ mph.

The distance between the cars is decreasing at the rate of 61.18 mph.

39. $V = 10(25)(r) = 250r$, $\frac{dV}{dt} = 10$

Find $\frac{dr}{dt}$ when $r = 4$.

$\frac{dV}{dt} = 250\frac{dr}{dt}$; $10 = 250\frac{dr}{dt}$ or $\frac{dr}{dt} = \frac{1}{25}$ ft/hr.

Exercise 11.5

1. $p+4q=80$

$1+4\frac{dq}{dp}=0$ gives $\frac{dq}{dp}=-\frac{1}{4}$

a. At (10, 40) we have $\eta=-\frac{40}{10}\left(-\frac{1}{4}\right)=1$.

b. Since the demand is unitary, there will be no change in the revenue with a price increase.

3. $p^2+2p+q=49$

$2p+2+\frac{dq}{dp}=0$ gives $\frac{dq}{dp}=-2p-2$

a. At (1, 6) we have $\eta=-\frac{6}{1}(-14)=84$.

b. Since the demand is elastic, an increase in price will decrease the total revenue.

5. $pq+p=5000$

a. $p\cdot\frac{dq}{dp}+q\cdot 1+1=0$ gives $\frac{dq}{dp}=\frac{-(q+1)}{p}$.

At (99, 50) the elasticity is

$\eta=-\frac{50}{99}\left(-\frac{100}{50}\right)=\frac{100}{99}$.

b. $\eta>1$, so demand is elastic.

c. A price increase will decrease the total revenue.

7. $pq+p+100q=50{,}000$

a. $p\cdot\frac{dq}{dp}+q\cdot 1+1+100\frac{dq}{dp}=0$ gives

$\frac{dq}{dp}=\frac{-(q+1)}{p+100}$. At $p=401$ we have $q=99$.

Thus, elasticity is $\eta=-\frac{401}{99}\left(-\frac{100}{501}\right)\approx 0.81$

at $(99,401)$.

b. $\eta<1$, so demand is inelastic.

c. An increase in price will result in an increase in total revenue.

9. $p=\frac{1}{2}[\ln(5000-q)-\ln(q+1)]$

a. $1=\frac{1}{2}\left[\frac{1}{5000-q}(-1)\cdot\frac{dq}{dp}-\frac{1}{q+1}\cdot\frac{dq}{dp}\right]$

$2=-\left[\frac{q+1+5000-q}{(5000-q)(q+1)}\right]\cdot\frac{dq}{dp}$

$\frac{dq}{dp}=\frac{-2(5000-q)(q+1)}{5001}$

At $p=3.71$, we were given that $q=2$.

Elasticity is $\eta=-\frac{3.71}{2}\left(\frac{-29{,}988}{5001}\right)\approx 11.1$.

b. Demand is elastic.

11. $p=120\sqrt[3]{125-q}$

a. $1=40(125-q)^{-2/3}(-1)\cdot\frac{dq}{dp}$ gives

$\frac{dq}{dp}=\frac{-(125-q)^{2/3}}{40}$

Elasticity is

$\eta=-\frac{p}{q}\cdot\frac{dq}{dp}$

$=-\frac{120(125-q)^{1/3}}{q}\cdot\frac{-(125-q)^{2/3}}{40}$

$=\frac{3(125-q)}{q}$

b. $\eta=1$ gives $1=\frac{375-3q}{q}$

$4q=375$ or $q=93.75$ for unitary elasticity.

Inelastic means $\frac{375-3q}{q}<1$ or $q>93.75$.

Elastic means $\frac{375-3q}{q}>1$ or $0<q<93.75$.

c. Revenue increases when $\eta>1$ or $0<q<93.75$.
Revenue decreases when $\eta<1$ or $q>93.75$.
Revenue is maximized at $q=93.75$.

d. $R=pq=120q\sqrt[3]{125-q}$

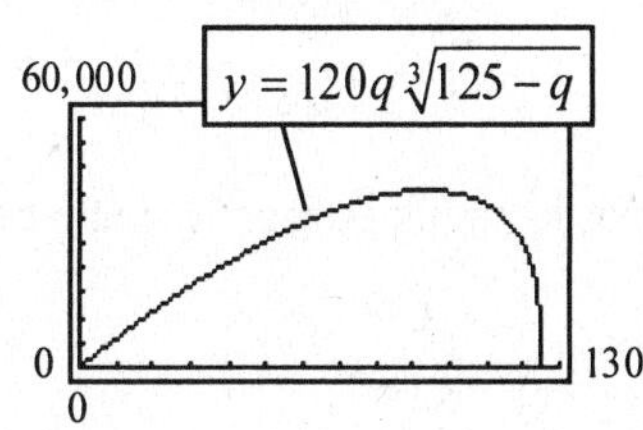

Yes.

13. After taxation the supply function is $p = 6 + 2q + t$.
Then, $6 + 2q + t = 30 - q$ yields $t = 24 - 3q$.
Total tax revenue: $T = t \cdot q = 24q - 3q^2$.
$T'(q) = 24 - 6q = 0$ if $q = 4$.
Since $T''(q) < 0$, the tax revenue is maximized at $q = 4$ and $t = 24 - 12 = \$12$.

15. After taxation the supply function is $p = 100 + 0.5q + t$.
Then, $100 + 0.5q + t = 800 - 2q$ yields $t = 700 - 2.5q$.
Total tax revenue: $T = t \cdot q = 700q - 2.5q^2$.
$T'(q) = 700 - 5q = 0$ if $q = 140$.
Since $T''(q) < 0$, the tax revenue is maximized at $q = 140$ and $t = 700 - 2.5(140) = \$350$.

17. After taxation the supply function is $p = 20 + 3q + t$. Then, $20 + 3q + t = 200 - 2q^2$ yields $t = 180 - 3q - 2q^2$.
Total tax revenue: $T = t \cdot q = 180q - 3q^2 - 2q^3$.
$T'(q) = 180 - 6q - 6q^2$.
$T'(q) = 0$ if $6(6+q)(5-q) = 0$ and $T''(q) < 0$.
Total tax revenue is maximized at $q = 5$ and $t = 180 - 15 - 50 = \$115$ / item.

19. After taxation the supply function is $p = 0.02q^2 + 0.55q + 7.4 + t$.
Then, $0.02q^2 + 0.55q + 7.4 + t = 840 - 2q$ yields $t = 832.6 - 2.55q - 0.02q^2$.
$T(q) = 832.6q - 2.55q^2 - 0.02q^3$ and $T'(q) = 832.6 - 5.1q - 0.06q^2$.
$T'(q) = 0$ if $q = \dfrac{5.1 \pm \sqrt{26.01 + 199.824}}{-0.12} = \dfrac{5.1 \pm 15.03}{-0.12} \approx 83.$
Tax per item is $t = 832.6 - 2.55(83) - 0.02(83)^2 = \$483.17$ / item.
Total tax revenue (to nearest dollar) = \$40,100.

21. After taxation the supply function is $p = 300 + 5q + 0.5q^2 + t$.
Then, $300 + 5q + 0.5q^2 + t = 2100 - 10q - 0.5q^2$ gives $t = 1800 - 15q - q^2$.
Total tax revenue $T = t \cdot q = 1800q - 15q^2 - q^3$.
$T'(q) = 1800 - 30q - 3q^2 = 3(30+q)(20-q) = 0$ if $q = 20$. Note $T''(q) < 0$.
Tax per unit that will maximize total tax revenue is $t = 1800 - 300 - 400 = \$1100$.

Review Exercises

1. $y = e^{3x^2-x}$

$\frac{dy}{dx} = e^{3x^2-x}(6x-1)$

2. $y = \ln e^{x^2} = x^2 \ln e = x^2$

$y' = 2x$

3. $p = \ln q - \ln(q^2-1)$

$\frac{dp}{dq} = \frac{1}{q} - \frac{2q}{q^2-1} = \frac{q^2-1-2q^2}{q(q^2-1)} = \frac{-1-q^2}{q(q^2-1)}$

4. $y = xe^{x^2}$

$\frac{dy}{dx} = x\left(e^{x^2} \cdot 2x\right) + e^{x^2}(1) = e^{x^2}(2x^2+1)$

5. $f(x) = 5e^{2x} - 40e^{-0.1x} + 11$

$f'(x) = 5e^{2x} \cdot 2 - 40\left(e^{-0.1x}\right)(-0.1)$

$= 10e^{2x} + 4e^{-0.1x}$

6. $g(x) = \left(2e^{3x+1} - 5\right)^3$

$g'(x) = 3\left(2e^{3x+1} - 5\right)^2\left(2e^{3x+1} \cdot 3\right)$

$= 18e^{3x+1}\left(2e^{3x+1} - 5\right)^2$

7. $y = \ln\left(3x^4 + 7x^2 - 12\right)$

$\frac{dy}{dx} = \frac{1}{3x^4+7x^2-12}\left(12x^3+14x\right)$

$= \frac{12x^3+14x}{3x^4+7x^2-12}$

8. $s = \frac{3}{4}\ln\left(x^{12} - 2x^4 + 5\right)$

$\frac{ds}{dx} = \frac{3}{4} \cdot \frac{1}{x^{12}-2x^4+5}\left(12x^{11} - 8x^3\right)$

$= \frac{3\left(3x^{11} - 2x^3\right)}{x^{12} - 2x^4 + 5}$

9. $y = 3^{3x-4}$

$\frac{dy}{dx} = 3^{3x-4}(3)\ln 3 = 3^{3x-3}(\ln 3)$

10. $y = 1 + \log_8\left(x^{10}\right) = 1 + \frac{10}{\ln 8}\ln x$

$\frac{dy}{dx} = \frac{10}{\ln 8} \cdot \frac{1}{x} = \frac{10}{x \ln 8}$

11. $y = \frac{\ln x}{x}$

$\frac{dy}{dx} = \frac{x \cdot \frac{1}{x} - \ln x \cdot 1}{x^2} = \frac{1 - \ln x}{x^2}$

12. $y = \frac{1+e^{-x}}{1-e^{-x}}$

$\frac{dy}{dx} = \frac{(1-e^{-x})(-e^{-x}) - (1+e^{-x})(e^{-x})}{(1-e^{-x})^2}$

$= \frac{-2e^{-x}}{(1-e^{-x})^2}$

13. $y = 4e^{x^3}$

$y' = 4e^{x^3}(3x^2)$

At $x = 1$, $y' = 12e$

Tangent line: $y - 4e = 12e(x-1)$ or $y = 12ex - 8e$.

14. $y = x \ln x$ (If $x = 1$, $y = 0$)

$\frac{dy}{dx} = x \cdot \frac{1}{x} + \ln x(1) = 1 + \ln x$

At $x = 1$, $\frac{dy}{dx} = 1$.

Thus, $y - 0 = 1(x-1)$ or $y = x - 1$.

15. $y \ln x = 5y$

$y \cdot \frac{1}{x} + \ln x \cdot \frac{dy}{dx} = 5\frac{dy}{dx}$ gives

$\frac{dy}{dx} = \frac{y}{x} \div (5 - \ln x) = \frac{y}{x(5-\ln x)}$.

16. $e^{xy} = y$

$e^{xy}\left(x\frac{dy}{dx} + y\right) = \frac{dy}{dx}$

$e^{xy} y = \frac{dy}{dx}(1 - xe^{xy})$

$\frac{dy}{dx} = \frac{ye^{xy}}{1 - xe^{xy}}$

17. $y^2 = 4x - 1$

$2y\frac{dy}{dx} = 4$ gives $\frac{dy}{dx} = \frac{2}{y}$

18. $x^2+3y^2+2x-3y+2=0$

$$2x+6y\frac{dy}{dx}+2-3\frac{dy}{dx}=0$$

$$\frac{dy}{dx}(6y-3)=-(2x+2)$$

$$\frac{dy}{dx}=-\frac{2x+2}{6y-3}=\frac{2(x+1)}{3(1-2y)}$$

19. $3x^2+2x^3y^2-y^5=7$

$$6x+2x^3\cdot 2y\cdot\frac{dy}{dx}+y^2(6x^2)-5y^4\cdot\frac{dy}{dx}=0$$

gives $\dfrac{dy}{dx}=\dfrac{6x+6x^2y^2}{5y^4-4x^3y}=\dfrac{6x(1+xy^2)}{y(5y^3-4x^3)}$

20. $x^2+y^2=1$

$$2x+2y\frac{dy}{dx}=0$$

$$\frac{dy}{dx}=\frac{-x}{y}$$

$$y''=\frac{y(-1)-(-x)y'}{y^2}$$

$$=\frac{-y-\frac{x^2}{y}}{y^2}=-\frac{y^2+x^2}{y^3}=-\frac{1}{y^3}$$

21. $x^2+4x-3y^2+6=0$

$2x+4-6y\dfrac{dy}{dx}=0$ gives $\dfrac{dy}{dx}=\dfrac{x+2}{3y}$. At $(3,3)$,

we have $\dfrac{dy}{dx}=\dfrac{5}{9}$.

22. $x^2+4x-3y^2+6=0$

$$2x+4-6y\frac{dy}{dx}=0$$

$$\frac{dy}{dx}=\frac{-2x-4}{-6y}=\frac{x+2}{3y}$$

$\dfrac{dy}{dx}=0$ when $x=-2$.

When $x=-2$, $y=\pm\sqrt{\dfrac{2}{3}}$.

So, horizontal tangents at $\left(-2,\pm\sqrt{\dfrac{2}{3}}\right)$.

23. $3x^2-2y^3=10y$

$$6x\frac{dx}{dt}-6y^2\frac{dy}{dt}=10\frac{dy}{dt}$$

$$\frac{dy}{dt}=\frac{6x\cdot\frac{dx}{dt}}{10+6y^2}$$

At $(10,5)$, we have $\dfrac{dy}{dt}=\dfrac{6(10)2}{10+6(25)}=\dfrac{120}{160}=\dfrac{3}{4}$.

24. $A=\dfrac{1}{2}xy$

$$\frac{dA}{dt}=\frac{1}{2}x\frac{dy}{dt}+y\left(\frac{1}{2}\frac{dx}{dt}\right)$$

Use $\dfrac{dx}{dt}=2$, $\dfrac{dy}{dt}=5$, $x=4$, and $y=1$.

$$\frac{dA}{dt}=\frac{1}{2}\cdot 4\cdot 5+1\cdot\frac{1}{2}\cdot 2=11\text{ units}^2/\text{min}$$

25. $y=-3.91435+2.62196\ln t$

a. $y'(t)=\dfrac{2.62196}{t}$

b. $y(50)\approx 6.343$ is the predicted number of hectares of deforestation in 2000. $y'(50)\approx 0.05244$ is the predicted increase in the number of hectares of deforestation in 2001.

26. $S(n)=1000e^{0.12n}$

$S'(n)=120e^{0.12n}$

$S'(1)=120(1.1275)\approx \$135.30/\text{yr}$

27. $S=1000e^{0.12n}$

a. $\dfrac{dS}{dn}=1000e^{0.12n}(0.12)$

$=1000e^{0.12(2)}(0.12)$

$\approx \$152.55/\text{yr}$

b. At $n=1$,

$\dfrac{dS}{dn}=e^{0.12(1)}(0.12)\approx \$135.30/\text{yr}$

At the end of 2 years the value is growing $\dfrac{152.55}{135.50}\approx 1.13$ times as fast.

28. $A(t) = A_0 e^{-0.00002876t}$

a. $A'(t) = A_0 e^{-0.00002876t}(-0.00002876)$

$A'(0) = -0.00002876 A_0$

b. $A'(1) = -0.00002876 A_0$

Note that $e^{-0.00002876} \approx e^0 = 1$.

c. The rate of decay when $t = 24101$ is approximately -1.44×10^{-5} and the rate when $t = 1$ is approximately -2.88×10^{-5}.

29. $\overline{C} = 600e^{x/600}$

$C = 600xe^{x/600}$

$\overline{MC} = C' = 600xe^{x/600} \cdot \dfrac{1}{600} + 600e^{x/600}$

$= e^{x/600}(x + 600)$

When $x = 600$, $\overline{MC} = 1200e$.

30. $P(t) = 20,000e^{-0.0495t}$

$P'(t) = -990e^{-0.0495t}$

$P'(10) = -990e^{-0.495}$

$= -990(0.60957)$

$= -\$603.48/\text{yr}$

31. $V = \dfrac{4}{3}\pi r^3$

$\dfrac{dV}{dt} = 4\pi r^2 \dfrac{dr}{dt}$

Using $r = 2.5$ and $\dfrac{dV}{dt} = -1$ gives

$-1 = 4\pi(2.5)^2 \dfrac{dr}{dt}$ $\dfrac{dr}{dt} = -\dfrac{1}{25\pi}$ mm/min

32. Let y be the height of the sign above the workers' hands and let z be the length of the guide line.

$\dfrac{dy}{dt} = 2\text{ft}/\text{min}$

$z^2 = y^2 + 7^2$

$2z\dfrac{dz}{dt} = \dfrac{dy}{dt}(2y) + 0$

If $z = 25$, $y = 24$.

$2(25)\dfrac{dz}{dt} = 2(24)(2)$ gives $\dfrac{dz}{dt} = \dfrac{48}{25}$ ft/min.

33. $S = kA^{1/3}$

$\dfrac{dS}{dt} = \dfrac{1}{3} \cdot k \cdot \dfrac{1}{A^{2/3}} \cdot \dfrac{dA}{dt}$

$\dfrac{\frac{dS}{dt}}{S} = \dfrac{1}{S} \cdot \dfrac{k}{3} \cdot \dfrac{1}{A^{2/3}} \cdot \dfrac{dA}{dt}$

$= \dfrac{1}{kA^{1/3}} \cdot \dfrac{k}{3} \cdot \dfrac{1}{A^{2/3}} \cdot \dfrac{dA}{dt} = \dfrac{1}{3}\left(\dfrac{\frac{dA}{dt}}{A}\right)$

34. Yes.

35. After taxation, the supply function is $p = 400 + 2q + t$.

$400 + 2q + t = 2800 - 8q - \dfrac{1}{3}q^2$ gives

$t = 2400 - 10q - \dfrac{1}{3}q^2$.

$T(q) = tq = 2400q - 10q^2 - \dfrac{1}{3}q^3$

$T'(q) = 2400 - 20q - q^2 = (60 + q)(40 - q)$

$T(q)$ is maximized at $q = 40$ and

$t = 2400 - 400 - \dfrac{1600}{3} = \$1466.67$ /unit.

36. After taxation, the supply function is $p = 40 + 20q + t$.

$40 + 20q + t = \dfrac{5000}{q+1}$ gives $t = \dfrac{5000}{q+1} - 40 - 20q$

$T = tq = \dfrac{5000q}{q+1} - 40q - 20q^2$ and

$T'(q) = \dfrac{(q+1)5000 - 5000q(1)}{(q+1)^2} - 40 - 40q$

Set $T'(q) = 0$. Then, $5000 - 40(q+1)^3 = 0$ or $q + 1 = 5$ or $q = 4$.

So, $t = \dfrac{5000}{5} - 40 - 80 = \$880/\text{unit}$.

37. a. $pq = 27$

$p \cdot \dfrac{dq}{dp} + q \cdot 1 = 0$ or $\dfrac{dq}{dp} = -\dfrac{q}{p}$

$\eta = -\dfrac{p}{q} \cdot \dfrac{dq}{dp} = -\dfrac{p}{q} \cdot \left(-\dfrac{q}{p}\right) = 1$

So at $(9,3)$ we have $\eta = 1$.

b. Since $\eta = 1$, there is no change in total revenue with a price increase.

38. $p^2(2q+1) = 10{,}000$

$p^2 \cdot 2 \cdot \dfrac{dq}{dp} + (2q+1)2p = 0$

$\dfrac{dq}{dp} = \dfrac{-(2q+1)}{p}$

$\eta = -\dfrac{p}{q} \cdot \dfrac{dq}{dp} = -\dfrac{20}{12}\left(-\dfrac{25}{20}\right) = \dfrac{25}{12}$

39. $p = 100e^{-0.1q}$

$1 = 100e^{-0.1q}(-0.1)\dfrac{dq}{dp}$ or $\dfrac{dq}{dp} = \dfrac{-e^{0.1q}}{10}$

At $p = 36.79$, $q = 10$ we have

$\eta = -\dfrac{36.79}{10}\left(-\dfrac{e}{10}\right) \approx 1.$

40. $p = 100 - 0.5q$

a. $1 = -0.5\dfrac{dq}{dp}$ or $\dfrac{dq}{dp} = -2$

$\eta = -\dfrac{p}{q}(-2) = \dfrac{2(100-0.5q)}{q}$

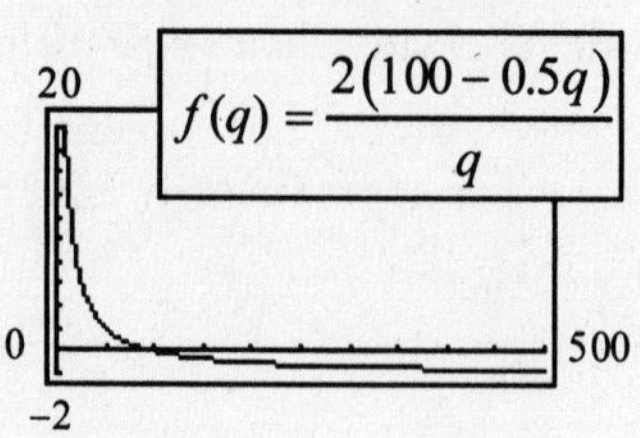

b. $\dfrac{2(100-0.5q)}{q} - 1 = 0$ if $q = 100.$

c. $R(q) = 100q - 0.5q^2$

Max revenue occurs at vertex.

$q = \dfrac{-b}{2a} = \dfrac{-100}{-1} = 100$

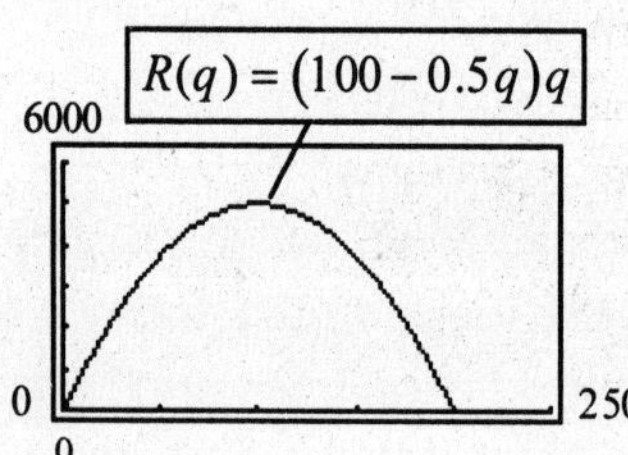

d. Maximum revenue occurs when $\eta = 1$.

Chapter Test

1. $y = 5e^{x^3} + x^2$

$y' = 5 \cdot e^{x^3}(3x^2) + 2x = 15x^2e^{x^3} + 2x$

2. $y = 4\ln(x^3+1)$

$y' = 4 \cdot \dfrac{1}{x^3+1} \cdot 3x^2 = \dfrac{12x^2}{x^3+1}$

3. $y = \ln(x^4+1)^3 = 3\ln(x^4+1)$

$y' = 3 \cdot \dfrac{1}{x^4+1} \cdot 4x^3 = \dfrac{12x^3}{x^4+1}$

4. $f(x) = 10(3^{2x})$

$f'(x) = 10(3^{2x} \cdot 2\ln 3) = 20(3^{2x})\ln 3$

5. $S = te^{t^4}$

$S'(t) = t\left(4t^3e^{t^4}\right) + e^{t^4}(1) = e^{t^4}(4t^4+1)$

6. $y = \dfrac{e^{x^3+1}}{x}$

$y' = \dfrac{x(3x^2e^{x^3+1}) - e^{x^3+1}(1)}{x^2} = \dfrac{e^{x^3+1}(3x^3-1)}{x^2}$

7. $y = \dfrac{\ln x}{x}$

$y' = \dfrac{x \cdot \frac{1}{x} - \ln x(1)}{x^2} = \dfrac{1 - \ln x}{x^2}$

8. $g(x) = 2\log_5(4x+7)$

$g'(x) = 2 \cdot \dfrac{1}{(4x+7)} \cdot \dfrac{4}{\ln 5} = \dfrac{8}{(4x+7)\ln 5}$

9. $3x^4 + 2y^2 + 10 = 0$

$12x^3 + 4yy' = 0$

$y' = \dfrac{-3x^3}{y}$

10. $x^2 + y^2 = 100, \frac{dx}{dt} = 2, x = 6, y = 8$

$$2x \cdot \frac{dx}{dt} + 2y\frac{dy}{dt} = 0$$

$$2(6) \cdot 2 + 2(8)\frac{dy}{dt} = 0$$

$$\frac{dy}{dt} = -\frac{3}{2}$$

11. $xe^y = 10y$

$$x(e^y y') + e^y(1) = 10y'$$

$$y' = \frac{e^y}{10 - xe^y}$$

12. $R(x) = 300x - 0.001x^2$

$C(x) = 4000 + 30x$

$P(x) = -0.001x^2 + 270x - 4000$

$$\frac{dP}{dt} = -0.002x\frac{dx}{dt} + 270\frac{dx}{dt}$$

If $x = 50, \frac{dx}{dt} = 5$ then

$$\frac{dP}{dt} = (-0.002)(50)(5) + 270(5) = \$1349.50 \text{ /week}$$

13. $p^2 + 3p + q = 1500$

$p = 30$ gives $q = 510$.

$q = 1500 - p^2 - 3p$

$\frac{dq}{dp} = -2p - 3$ If $p = 30, \frac{dq}{dp} = -63$.

$$\eta = -\frac{p}{q} \cdot \frac{dq}{dp} = \frac{-30}{510}(-63) \approx 3.71$$

Revenue decreases.

14. $(p+1)q^2 = 10{,}000$

Find $\frac{dq}{dp}$ if $p = 99$ $(q = 10)$.

$$(p+1)\left(2q\frac{dq}{dt}\right) + q^2(1) = 0$$

$$\frac{dq}{dt} = \frac{-q}{2(p+1)}$$

$$\frac{dq}{dt} = \frac{-10}{2(99+1)} = -\frac{1}{20} \text{ units/dollar}$$

15. $S = 80{,}000e^{-0.4t}$

$S'(t) = 80{,}000(e^{-0.4t} \cdot (-0.4)) = -32{,}000e^{-0.4t}$

$S'(10) = -32{,}000e^{-4} = -586$ sales/day

16. $y = 130.4e^{0.007163(x-1500)}$

$y'(x) = 0.9341e^{0.007163(x-1500)}$

a. $y'(1930) \approx 20.325$ million people per year

b. $y'(2001) \approx 33.799$ million people per year

17. $D: p = 1100 - 5q \quad S: p = 20 + 0.4q$

New supply: $p = 20 + 0.4q + t$ (t = tax/unit)

Break-even: $1100 - 5q = 20 + 0.4q + t$

$$t = 1080 - 5.4q$$

Total tax revenue: $T = tq = 1080q - 5.4q^2$

$T'(q) = 1080 - 10.8q$

Tax revenue is maximized when $T' = 0$ or $q = 100$.

Then $t = 1080 - 5.4(100) = \$540$ / unit.

18. a. $S(t) = -197.03 + 56.65\ln t$

b. $S'(t) = \frac{56.65}{t}$

c. $S(100) = 63.85$ The model predicted that in 2000 the percent of paved streets was 63.85%.

$S'(100) = 0.5665$ The model predicted that in 2001 the percent of paved streets would increase by 0.5665%.

19. $P(t) = 3.974(0.94864)^t$

$P'(t) = 3.974(0.94864)^t \cdot \ln(0.94864)$

$P'(45) = -0.0195$

Chapter 12: Indefinite Integrals

Exercise 12.1

1. $f'(x) = 4x^3$

$f(x) = \frac{4x^{3+1}}{4} + C = x^4 + C$

3. $f'(x) = x^6$

$f(x) = \frac{x^{6+1}}{7} + C = \frac{x^7}{7} + C$

5. $\int x^7 dx = \frac{x^8}{8} + C$

7. $8\int x^5 dx = 8\left(\frac{x^6}{6}\right) + C = \frac{4}{3}x^6 + C$

9. $\int (3^3 + x^{13})dx = 27x + \frac{x^{14}}{14} + C$

11. $\int (3 - x^{3/2})dx = 3x - \frac{x^{5/2}}{\frac{5}{2}} = 3x - \frac{2}{5}x^{5/2} + C$

13. $\int (x^4 - 9x^2 + 3)dx = \frac{x^5}{5} - \frac{9x^3}{3} + 3x + C$

$= \frac{1}{5}x^5 - 3x^3 + 3x + C$

15. $\int (2 + 2\sqrt{x})dx = \int (2 + 2x^{1/2})dx$

$= 2x + 2 \cdot \frac{x^{3/2}}{\frac{3}{2}} + C$

$= 2x + \frac{4}{3}x^{3/2} + C$

$= 2x + \frac{4}{3}x\sqrt{x} + C$

17. $\int 6\sqrt[4]{x}\, dx = \int 6x^{1/4} dx = 6 \cdot \frac{x^{5/4}}{\frac{5}{4}} + C$

$= \frac{24}{5}x^{5/4} + C$

$= \frac{24}{5}x\sqrt[4]{x} + C$

19. $\int \frac{5}{x^4} dx = \int 5x^{-4} dx = 5 \cdot \frac{x^{-3}}{-3} + C$

$= -\frac{5}{3}x^{-3} + C$

$= -\frac{5}{3x^3} + C$

21. $\int \frac{dx}{2\sqrt[3]{x^2}} = \frac{1}{2}\int x^{-2/3} dx = \frac{1}{2} \cdot \frac{x^{1/3}}{\frac{1}{3}} + C$

$= \frac{3}{2}x^{1/3} + C = \frac{3}{2}\sqrt[3]{x} + C$

23. $\int \left(x^3 - 4 + \frac{5}{x^6}\right)dx = \frac{x^4}{4} - 4x + 5 \cdot \frac{x^{-5}}{-5} + C$

$= \frac{1}{4}x^4 - 4x - \frac{1}{x^5} + C$

25. $\int \left(x^9 - \frac{1}{x^3} + \frac{2}{\sqrt[3]{x}}\right)dx = \frac{x^{10}}{10} - \frac{x^{-2}}{-2} + 2 \cdot \frac{x^{2/3}}{\frac{2}{3}} + C$

$= \frac{1}{10}x^{10} + \frac{1}{2x^2} + 3x^{2/3} + C$

27. $\int (x+5)^2 x\, dx = \int (x^3 + 10x^2 + 25x)dx$

$= \frac{x^4}{4} + \frac{10x^3}{3} + \frac{25x^2}{2} + C$

29. $\int (4x^2 - 1)^2 x^3 dx = \int (16x^4 - 8x^2 + 1)(x^3)dx$

$= \int (16x^7 - 8x^5 + x^3)dx$

$= 16 \cdot \frac{x^8}{8} - 8 \cdot \frac{x^6}{6} + \frac{x^4}{4} + C$

$= 2x^8 - \frac{4}{3}x^6 + \frac{1}{4}x^4 + C$

31. $\int \frac{x+1}{x^3} dx = \int \left(\frac{1}{x^2} + \frac{1}{x^3}\right)dx$

$= \int (x^{-2} + x^{-3})dx$

$= \frac{x^{-1}}{-1} + \frac{x^{-2}}{-2} + C$

$= -\frac{1}{x} - \frac{1}{2x^2} + C$

33. $\int (2x+3)dx = 2 \cdot \frac{x^2}{2} + 3x + C = x^2 + 3x + C$

$f(x) = x^2 + 3x + C$
($C = -8, -4, 0, 4,$ and 8)

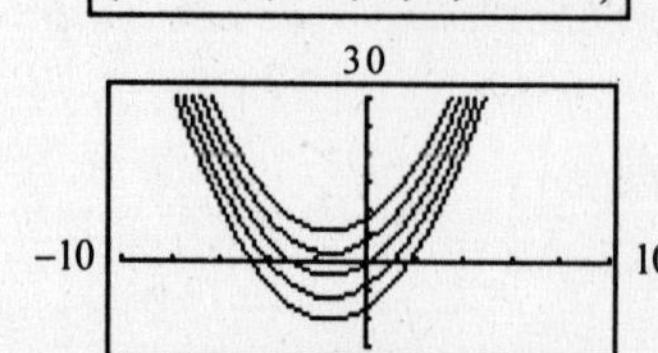

35. $F(x) = 5x - \frac{x^2}{4} + C$

$F'(x) = 5 - \frac{1}{4} \cdot 2x = 5 - \frac{1}{2}x$

$\int \left(5 - \frac{1}{2}x\right) dx$

37. $F(x) = x^3 - 3x^2 + C$

$F'(x) = 3x^2 - 6x$

$\int (3x^2 - 6x) dx$

39. $\overline{MR} = -0.4x + 30$

$R(x) = \int (-0.4x + 30) dx$

$= -0.4 \cdot \frac{x^2}{2} + 30x + C$

$C = 0$ since $R(0) = 0$.

$R(x) = -0.2x^2 + 30x$

41. $\overline{MR} = -0.3x + 450$

$R(x) = \int (-0.3x + 450) dx$

$= -0.3 \cdot \frac{x^2}{2} + 450x + C$

$C = 0$ since $R(0) = 0$.

$R(50) = \frac{-0.3(50)^2}{2} + 450(50) = \$22{,}125$

43. $\int (t^3 + 4t^2 + 6) dt = \frac{t^4}{4} + 4 \cdot \frac{t^3}{3} + 6t + C$

When $t = 0$, $C = 0$. So, $P(t) = \frac{1}{4}t^4 + \frac{4}{3}t^3 + 6t$.

45. $\frac{dx}{dt} = \frac{1}{600}t^{3/4}$

a. $x = \frac{1}{600} \cdot \frac{t^{7/4}}{\frac{7}{4}} + C$

$t = 0$ means $x = 0$. Thus $C = 0$ and

$x(t) = \frac{1}{1050}t^{7/4}$.

b. $x(52) = \frac{1}{1050}(52)^{7/4}$

$\approx \frac{1}{1050}(1006.95)$

≈ 0.96 tons

47. $\overline{C}'(x) = \frac{1}{4} - \frac{100}{x^2} = \frac{1}{4} - 100x^{-2}$

a. $\overline{C}(x) = \frac{1}{4}x - \frac{100x^{-1}}{-1} + K = \frac{1}{4}x + \frac{100}{x} + K$

Since $\overline{C}(20) = 40$, we have

$40 = \frac{1}{4}(20) + \frac{100}{20} + K$. So, $K = 30$. Thus,

$\overline{C}(x) = \frac{1}{4}x + \frac{100}{x} + 30.$

b. $\overline{C}(100) = \frac{1}{4}(100) + \frac{100}{100} + 30 = \$56$

49. $\frac{dB}{dt} = -2.1136t + 8.2593$

a. $\frac{dB}{dt} = 0$ if $-2.1136t + 8.2593 = 0$

$t \approx 3.91$

Balance begins to decrease 3.9 years after 2000.

b. $B = \int (-2.1136t + 8.2593) dt$

$= \frac{-2.1136}{2}t^2 + 8.2593t + C$

$B(0) = 74.07$ gives $C = 74.07$

$B = -1.0568t^2 + 8.2593t + 74.07$

c. Graphing $B(t)$ shows that $B = 0$ about 13 years after 2000.

51. $\frac{dA}{dt} = 160.869t^{0.5307}$

a. Yes. $\frac{dA}{dt} > 0$ also means A is increasing.

b. $A = \int 160.869t^{0.5307} dt$

$= 160.869\left(\frac{t^{1.5307}}{1.5307}\right) + C$

$1234.5 \text{ (billion)} = 105.095 \cdot \left(5^{1.5307}\right) + C$

gives $C = 0$.

$A = 105.095t^{1.5307}$

53. $\frac{dH}{dt} = -0.0042t^2 + 2.100t - 8.349$

a. $H(t) = -0.0014t^3 + 1.050t^2 - 8.349t + C$

$H(0) = 26.7$ gives $C = 26.7$. Thus,

$H(t) = -0.0014t^3 + 1.050t^2 - 8.349t + 26.7$.

b. $H(45) = \$1649.67$ billion

Exercise 12.2

1. Let $u = x^2 + 3$. Then, $du = 2xdx$

$$\int (x^2+3)^3(2x\,dx) = \int u^3 du = \frac{u^4}{4} + C = \frac{1}{4}(x^2+3)^4 + C$$

3. Let $u = 15x^2 + 10$. Then, $du = 30xdx$

$$\int (15x^2+10)^4(30x\,dx) = \int u^4 du = \frac{u^5}{5} + C = \frac{1}{5}(15x^2+10)^5 + C$$

5. Let $u = 3x - x^3$. Then, $du = (3-3x^2)dx$.

$$\int (3x-x^3)^2(3-3x^2)dx = \int u^2 du = \frac{u^3}{3} + C = \frac{1}{3}(3x-x^3)^3 + C$$

7. Let $u = x^2 + 5$. Then, $du = 2x\,dx$.

$$\int (x^2+5)^3 x\,dx = \int (x^2+5)^3 \cdot \frac{1}{2}(2x\,dx) = \frac{1}{2}\int u^3 du = \frac{1}{2}\cdot\frac{u^4}{4} + C = \frac{1}{8}(x^2+5)^4 + C$$

9. $$\int 7(4x-1)^6 dx = 7\cdot\frac{1}{4}\int (4x-1)^6(4dx) = \frac{7}{4}\cdot\frac{(4x-1)^7}{7} + C = \frac{1}{4}(4x-1)^7 + C$$

11. $$\int (x^2+1)^{-3} x\,dx = \frac{1}{2}\int (x^2+1)^{-3}(2x\,dx) = \frac{1}{2}\cdot\frac{(x^2+1)^{-2}}{-2} + C = -\frac{1}{4}(x^2+1)^{-2} + C$$

13. $$\int (x-1)(x^2-2x+5)^4 dx = \frac{1}{2}\int (2x-2)(x^2-2x+5)^4 dx = \frac{(x^2-2x+5)^5}{10} + C$$

15. $$\int 2(x^3-1)(x^4-4x+3)^{-5} dx = \frac{1}{2}\int (4x^3-4)(x^4-4x+3)^{-5} dx = \frac{1}{2}\cdot\frac{(x^4-4x+3)^{-4}}{-4} + C = -\frac{1}{8}(x^4-4x+3)^{-4} + C$$

17. $$\int 7x^3\sqrt{x^4+6}\,dx = 7\cdot\frac{1}{4}\int 4x^3(x^4+6)^{1/2} dx = \frac{7}{4}\cdot\frac{(x^4+6)^{3/2}}{\frac{3}{2}} + C = \frac{7}{6}(x^4+6)^{3/2} + C$$

19. $$\int (x^3+1)^2(3x)dx = \int (x^6+2x^3+1)(3x)dx = \int (3x^7+6x^4+3x)dx = \frac{3}{8}x^8 + \frac{6}{5}x^5 + \frac{3}{2}x^2 + C$$

21. $$\int (3x^2-1)^2(8x^2)dx = \int (9x^4-6x^2+1)(8x^2)dx = \int (72x^6-48x^4+8x^2)dx = \frac{72}{7}x^7 - \frac{48}{5}x^5 + \frac{8}{3}x^3 + C$$

23. $$\int \sqrt{x^3-3x}(x^2-1)dx = \frac{1}{3}\int (x^3-3x)^{1/2}(3x^2-3)dx = \frac{1}{3}\cdot\frac{(x^3-3x)^{3/2}}{\frac{3}{2}} + C = \frac{2}{9}(x^3-3x)^{3/2} + C$$

25. $$\int \frac{x^2dx}{(x^3-1)^2} = \frac{1}{3}\int (x^3-1)^{-2}(3x^2)dx = \frac{1}{3}\cdot\frac{(x^3-1)^{-1}}{-1} + C = \frac{-1}{3(x^3-1)} + C$$

27. $\displaystyle\int \frac{3x^4\,dx}{(2x^5-5)^4} = 3\cdot\frac{1}{10}\int (2x^5-5)^{-4}(10x^4)dx$

$\displaystyle = \frac{3}{10}\cdot\frac{(2x^5-5)^{-3}}{-3}+C$

$\displaystyle = -\frac{1}{10}(2x^5-5)^{-3}+C$

$\displaystyle = \frac{-1}{10(2x^5-5)^3}+C$

29. $\displaystyle\int \frac{x^3-1}{(x^4-4x)^3}dx = \int (x^4-4x)^{-3}(x^3-1)dx$

$\displaystyle = \frac{1}{4}\int (x^4-4x)^{-3}(4x^3-4)dx$

$\displaystyle = \frac{1}{4}\cdot\frac{(x^4-4x)^{-2}}{-2}+C$

$\displaystyle = \frac{-1}{8(x^4-4x)^2}+C$

31. $\displaystyle\int \frac{x^2-4x}{\sqrt{x^3-6x^2+2}}dx$

$\displaystyle = \frac{1}{3}\int (x^3-6x^2+2)^{-1/2}\cdot 3(x^2-4x)dx$

$\displaystyle = \frac{1}{3}\cdot\frac{(x^3-6x^2+2)^{1/2}}{\frac{1}{2}}+C$

$\displaystyle = \frac{2}{3}(x^3-6x^2+2)^{1/2}+C$

33. a. $\displaystyle\int x(x^2-1)^3dx = \frac{1}{2}\int 2x(x^2-1)^3dx$

$\displaystyle = \frac{1}{2}\cdot\frac{(x^2-1)^4}{4}+C = \frac{1}{8}(x^2-1)^4+C$

b.

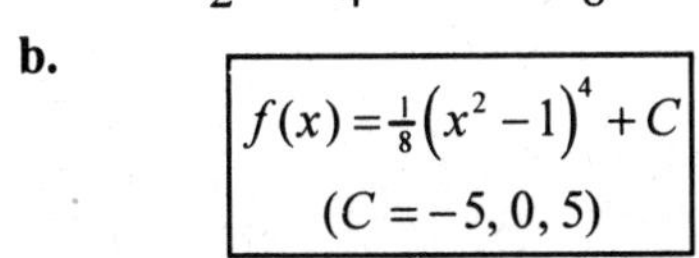

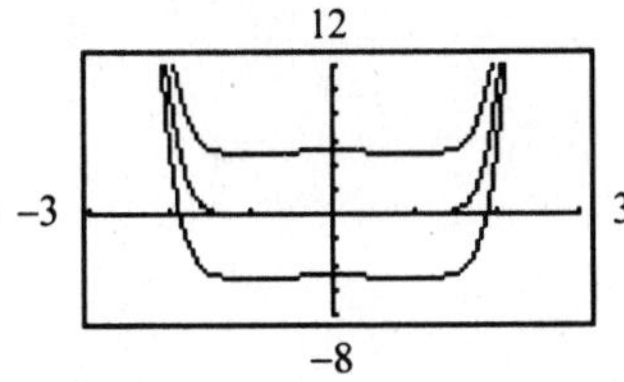

35. a. $\displaystyle\int \frac{3\,dx}{(2x-1)^{3/5}} = 3\cdot\frac{1}{2}\int (2x-1)^{-3/5}\cdot 2\,dx$

$\displaystyle = \frac{3}{2}\cdot\frac{(2x-1)^{2/5}}{\frac{2}{5}}+C$

$\displaystyle = \frac{15}{4}(2x-1)^{2/5}+C$

b. When $x=0$, $y=2$, so we have

$\displaystyle 2 = \frac{15}{4}(-1)^{2/5}+C$

$\displaystyle 2 = \frac{15}{4}+C$

$\displaystyle -\frac{7}{4} = C$

$\displaystyle F(x) = \frac{15}{4}(2x-1)^{2/5}-\frac{7}{4}$

$F(x) = \frac{15}{4}(2x-1)^{2/5} - \frac{7}{4}$

10

−6 8

−2

c. $f(x)$ is not defined at $x = \frac{1}{2}$.

d. The tangent line to $F(x)$ at $x = \frac{1}{2}$ is vertical.

37. $F(x) = (x^2-1)^{4/3}+C$

$\displaystyle F'(x) = \frac{4}{3}(x^2-1)^{1/3}(2x) = \frac{8}{3}x(x^2-1)^{1/3}$

The indefinite integral that gives the family is $\displaystyle\int \frac{8x(x^2-1)^{1/3}}{3}dx.$

39. Only (b) can be integrated using methods we have studied:

b. $\displaystyle\int 7x^2\left(x^3+4\right)^{-2}dx = \frac{7}{3}\frac{(x^3+4)^{-1}}{-1}+C$

$\displaystyle = \frac{-7}{3\left(x^3+4\right)}+C$

d. One form is $\int x\left(x^3+4\right)^{-p}dx$. ($p>0$)

41. $\overline{MR} = \dfrac{-30}{(2x+1)^2} + 30$

$$\int(-30(2x+1)^{-2} + 30)dx = \frac{-30}{2} \cdot \frac{(2x+1)^{-1}}{-1} + 30x + C$$
$$= \frac{15}{2x+1} + 30x + C$$

$R(0) = 0$ gives $C = -15$. Thus,

$$R(x) = \frac{15}{2x+1} + 30x - 15.$$

43. $P(x) = \int 90(x+1)^2\,dx$

$$= \frac{90(x+1)^3}{3} + C$$
$$= 30(x+1)^3 + C$$

$P(0) = 0$ or $0 = 30(1)^3 + C$ or $C = -30$ and

$P(x) = 30(x+1)^3 - 30.$

$P(4) = 30(5)^3 - 30 = 3720$

45. a. $\int 5(x+1)^{-1/2}\,dx = 5 \cdot \dfrac{(x+1)^{1/2}}{\frac{1}{2}} + C$

$$= 10(x+1)^{1/2} + C$$

$10 = 10(0+1)^{1/2} + C$ gives $C = 0$.

Thus, $s(x) = 10\sqrt{x+1}$.

b. $s(24) = 10\sqrt{24+1} = 50$

47. $A(t) = \int[-100(t+10)^{-2} + 2000(t+10)^{-3}]\,dt$

$$= \frac{-100(t+10)^{-1}}{-1} + \frac{2000(t+10)^{-2}}{-2} + C$$
$$= \frac{100}{t+10} - \frac{1000}{(t+10)^2} + C$$

$A(0) = 0$ or $0 = 10 - 10 + C$ or $C = 0$.

a. $A(t) = \dfrac{100}{t+10} - \dfrac{1000}{(t+10)^2}$

b. $A(10) = \dfrac{100}{20} - \dfrac{1000}{400} = 2.5$ (million)

49. $N(x) = \int -300(x+9)^{-1/2}\,dx$

$$= -300 \cdot \frac{(x+9)^{1/2}}{\frac{1}{2}} + C$$
$$= -600(x+9)^{1/2} + C$$

$8000 = -600(0+9)^{1/2} + C$ or $C = 9800$

$N(x) = -600(x+9)^{1/2} + 9800$ and

$N(7) = -600(16)^{1/2} + 9800 = 7400.$

51. $\dfrac{dp}{dt} = -0.006633(t+60)^2 + 1.1006(t+60) - 44.93$

a. $p(t) = -0.002211(t+60)^3 + 0.5503(t+60)^2 - 44.93t + C$

$p(30) = 33.6$ gives $C = -1464.111$.

Thus, $p(t) = -0.002211(t+60)^3 + 0.5503(t+60)^2 - 44.93t - 1464.111$

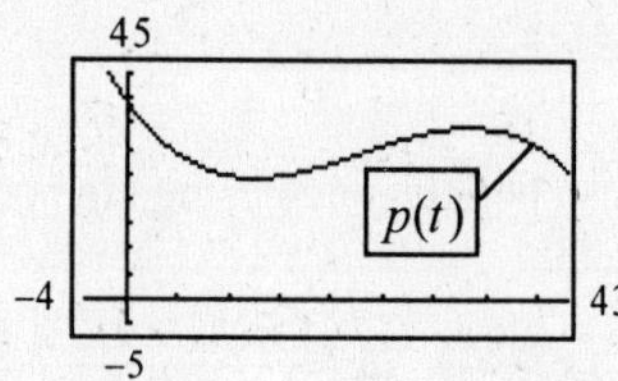

b.

45

p(t)

−4 43

−5

c. There is a good fit except for the mid 90's.

53. $\frac{dU}{dt} = 0.111(0.1t+2)^2 - 1.72(0.1t+2) + 6.08$

a. $U(t) = 0.37(0.1t+2)^3 - 8.6(0.1t+2)^2 + 6.08t + C$

$U(65) = 0$ gives $C = 16.92375$

Thus, $U(t) = 0.37(0.1t+2)^3 - 8.6(0.1t+2)^2 + 6.08t + 16.92375$.

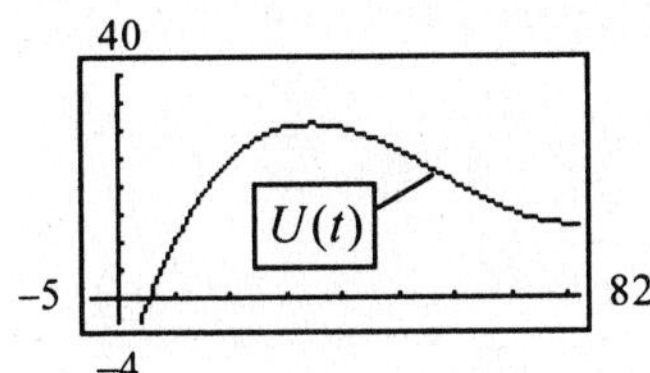

b.

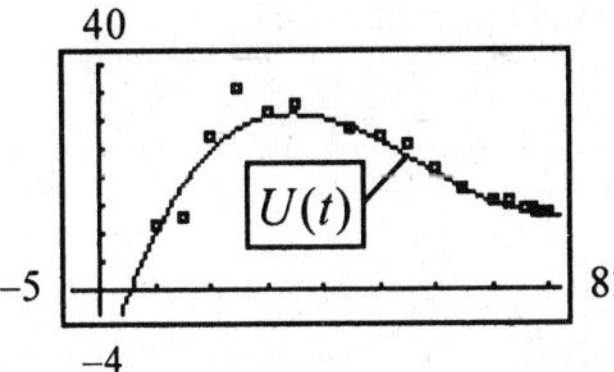

c. There is a good fit, except for the maximum point.

Exercise 12.3

1. Let $u = 3x$. Then $du = 3dx$.

$\int e^{3x}(3dx) = \int e^u du = e^u + C = e^{3x} + C$

3. Let $u = -x$. Then $du = (-1)dx$.

$$\int e^{-x}dx = (-1)\int e^{-x}(-1dx)$$
$$= -1\int e^u du$$
$$= (-1)e^u + C$$
$$= -e^{-x} + C$$

5. $$\int 1000e^{0.1x}dx = \frac{1000}{0.1}\int e^{0.1x}(0.1)dx$$
$$= 10,000e^{0.1x} + C$$

7. $$\int 840e^{-0.7x}dx = \frac{840}{-0.7}\int e^{-0.7x}(-0.7)dx$$
$$= -1200e^{-0.7x} + C$$

9. Let $u = 3x^4$. Then, $du = 12x^3dx$.

$\int x^3e^{3x^4}dx = \frac{1}{12}\int e^{3x^4}(12x^3dx) = \frac{1}{12}e^{3x^4} + C$

11. $\int \frac{3}{e^{2x}}dx = \frac{3}{-2}\int e^{-2x}(-2)dx = -\frac{3}{2}e^{-2x} + C$

13. $$\int \frac{x^5}{e^{2-3x^6}}dx = \frac{1}{18}\int e^{3x^6-2}(18x^5dx)$$
$$= \frac{1}{18}e^{3x^6-2} + C$$

15. $$\int\left(e^{4x} - \frac{3}{e^{x/2}}\right)dx$$
$$= \frac{1}{4}\int e^{4x}(4)dx - 3(-2)\int e^{-x/2}\left(-\frac{1}{2}\right)dx$$
$$= \frac{1}{4}e^{4x} + 6e^{-x/2} + C$$

17. Let $u = x^3 + 4$. Then, $du = 3x^2dx$.

$\int \frac{3x^2}{x^3+4}dx = \int \frac{du}{u} = \ln|u| + C = \ln|x^3+4| + C$

19. $\int \frac{dz}{4z+1} = \frac{1}{4}\int(4z+1)^{-1}(4dz) = \frac{1}{4}\ln|4z+1| + C$

21. Let $u = x^4 + 1$. Then, $du = 4x^3 dx$.

$$\int \frac{x^3}{x^4+1}dx = \frac{1}{4}\int \frac{4x^3}{x^4+1}dx = \frac{1}{4}\int \frac{du}{u} = \frac{1}{4}\ln|u| + C = \frac{1}{4}\ln|x^4+1| + C$$

23. Let $u = x^2 - 4$. Then, $du = 2x\,dx$.

$$\int \frac{4x}{x^2-4}dx = 2\int \frac{2x}{x^2-4}dx = 2\int \frac{du}{u} = 2\ln|u| + C = 2\ln|x^2-4| + C$$

25. Let $u = x^3 - 2x$. Then, $du = (3x^2 - 2)dx$.

$$\int \frac{3x^2-2}{x^3-2x}dx = \int \frac{du}{u} = \ln|u| + C = \ln|x^3 - 2x| + C$$

27. $$\int \frac{z^2+1}{z^3+3z+17}dz = \frac{1}{3}\int \frac{3z^2+3}{z^3+3z+17}dz = \frac{1}{3}\ln|z^3+3z+17| + C$$

29.

$$\begin{array}{r} x^2 \\ x-1\overline{)x^3 - x^2 + 1} \\ \underline{x^3 - x^2} \\ \text{Rem } 1 \end{array}$$

$$\int \frac{x^3-x^2+1}{x-1}dx = \int\left(x^2 + \frac{1}{x-1}\right)dx = \frac{1}{3}x^3 + \ln|x-1| + C$$

31. $$\int \frac{x^2+x+3}{x^2+3}dx = \int\left(\frac{x^2+3}{x^2+3} + \frac{x}{x^2+3}\right)dx = \int\left(1 + \frac{x}{x^2+3}\right)dx = x + \frac{1}{2}\ln|x^2+3| + C$$

33. $h(x) = f(x) = -e^{-(1/2)x}$

$$\int f(x)dx = \int(-e^{-(1/2)x})dx = 2\int e^{-(1/2)x}\left(-\frac{1}{2}\right)dx = 2e^{-(1/2)x} + C = g(x) + C$$

35. $\int \frac{1}{3-x}dx = -\ln|3-x| + C = F(x)$

$0 = -\ln|3-0| + C$ gives $C = \ln 3$.

$F(x) = \ln 3 - \ln|3-x|$

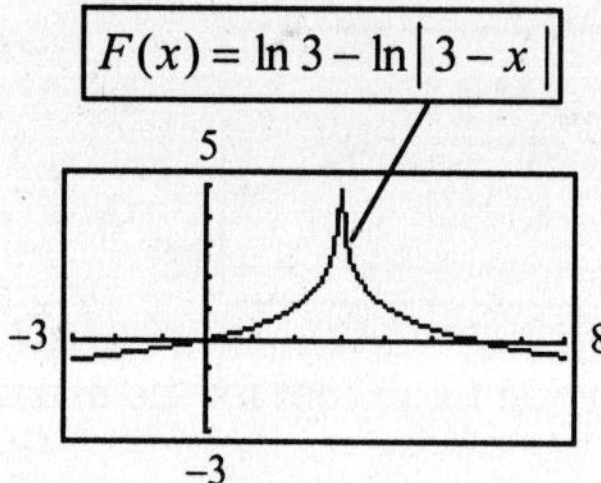

37. $F'(x) = f(x)$;

$f(x) = 1 + \frac{1}{x}$;

$F(x) = \int\left(1 + \frac{1}{x}\right)dx$

39. $F'(x) = f(x)$;

$f(x) = 5x(-e^{-x}) + e^{-x}(5)$;

$F(x) = \int(5e^{-x} - 5xe^{-x})dx$

41. Only (c) and (d) can be integrated using methods we have studied:

c. $\frac{1}{3}\int \frac{3(x^2+2x)}{x^3+3x^2+7}dx = \frac{1}{3}\ln\left(x^3+3x^2+7\right) + C$

d. $\frac{5}{8}\int e^{2x^4}\left(8x^3\right)dx = \frac{5}{8}e^{2x^4} + C$

43. $R(x) = 6\int e^{0.01x}dx = \frac{6}{0.01}\int e^{0.01x}(0.01dx) = 600e^{0.01x} + K$

$R(0) = 0$ gives $K = -600$. Thus,

$R(x) = 600e^{0.01x} - 600$.

$R(100) = 600e^{0.01(100)} - 600 = 600e - 600 = \$1030.97$

45. $n = \int n_0(-K)e^{-Kt}dt = n_0\int e^{-Kt}(-K\,dt) = n_0e^{-Kt}$

(Given $C = 0$.)

47. $v(t) = \int \frac{40}{t+1}dt = 40\ln(t+1) + C$

$v(0) = 0$ gives $C = 0$. Recall $\ln 1 = 0$.

$v(3) = 40\ln 4 \approx 55.45$ or 55 words.

49. a. $S(n) = P\int e^{0.1n}(0.1dn) = Pe^{0.1n} \quad (C = 0)$

b. $S = 2P$

$2P = Pe^{0.1n}$

$0.1n = \ln 2$

$n = \frac{\ln 2}{0.1} \approx 6.9$ years

51. a. $p = 46.645\int e^{-0.491t}(-1dt)$

$= \frac{46.645}{0.491}\int e^{-0.491t}(-0.491dt)$

$= 95e^{-0.491t} + C$

$p = 95$ when $t = 0$ gives $C = 0$.

b. $p(0.1) = 95e^{-0.0491} \approx 90.45$

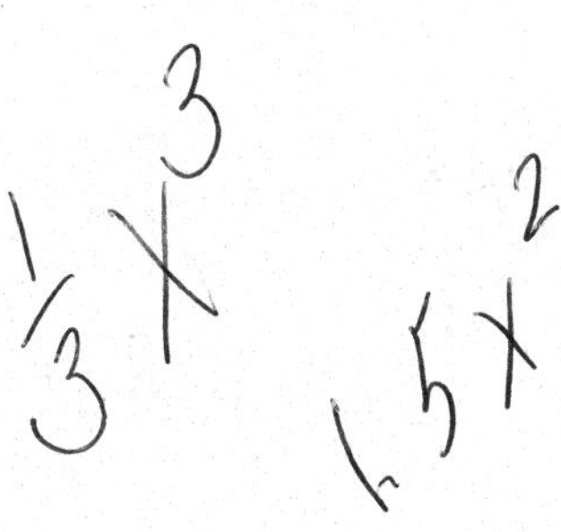

53. $C'(t) = \frac{14.180}{t+20}$

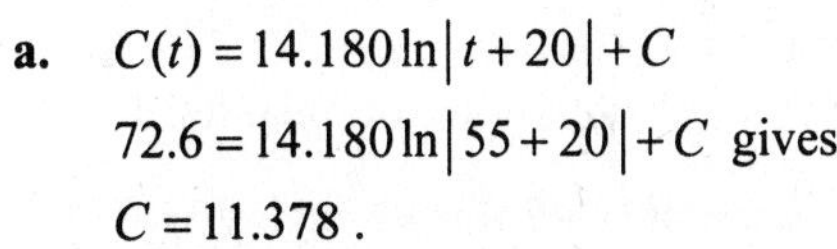

a. $C(t) = 14.180\ln|t+20| + C$

$72.6 = 14.180\ln|55+20| + C$ gives

$C = 11.378$.

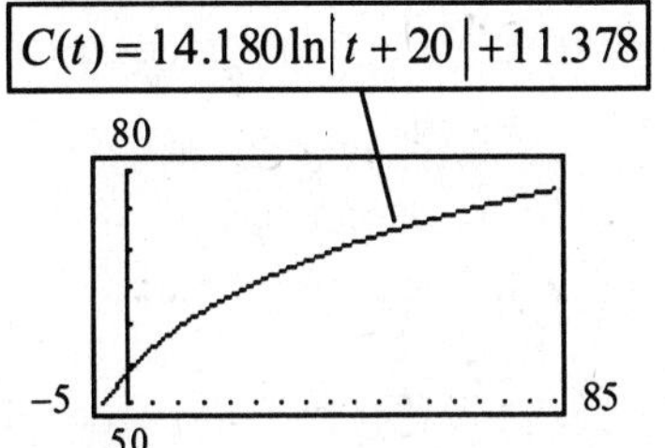

b.

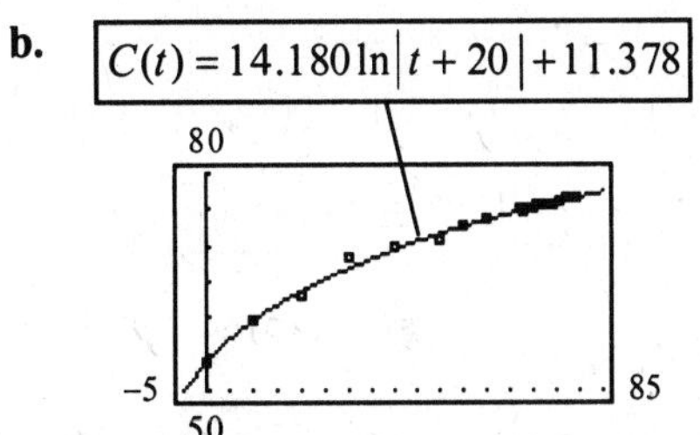

c. The model is a really good fit.

55. $T'(t) = 16.984e^{0.06891t}$

a. Yes. This is a growth exponential (positive and increasing) which corresponds with the data.

b. $T(t) = \frac{16.984}{0.06891}e^{0.06891t} + C$

$1375.84 = 246.466e^{0.06891(25)} + C$ gives

$C = -4.349$

$T(t) = 246.466e^{0.06891t} - 4.349$

c. $T(55) \approx 10{,}904 \quad T'(55) \approx 752$

The model predicts that in 2005, federal tax per capita will be \$10,904 and will increase \$752 for next year.

Exercise 12.4

1. $C(x) = \int(2x+100)dx = x^2 + 100x + K$

From the fact that $C(0) = 200$ we have

$200 = 0 + 0 + K$ or $K = 200$.

$C(x) = x^2 + 100x + 200$

3. $C(x) = \int(4x+2)dx = 2x^2 + 2x + K$

$C(10) = 300 = 200 + 20 + K$ or $K = 80$.

$C(x) = 2x^2 + 2x + 80$

5. $C(x) = \int(4x+40)dx = 2x^2 + 40x + K$

$C(25) = 3000 = 1250 + 1000 + K$ gives $K = 750$.

$C(30) = 1800 + 1200 + 750 = \$3750$

7. a. Optimal level is when $\overline{MR} = \overline{MC}$.
Thus, $44 - 5x = 3x + 20$
$$24 = 8x$$
So, $x = 3$ units is the optimal level.

b. $C(x) = \int (3x + 20)dx = \frac{3}{2}x^2 + 20x + K$
Now, $C(80) = 11,400 = 9600 + 1600 + K$ gives $K = 200$.
$$R(x) = \int (44 - 5x)dx = 44x - \frac{5}{2}x^2$$
$$P(x) = R(x) - C(x)$$
$$= 44x - \frac{5}{2}x^2 - \left(\frac{3}{2}x^2 + 20x + 200\right)$$
$$= 24x - 4x^2 - 200$$

c. $P(3) = 72 - 36 - 200 = -\$164$ (a loss)

9. $C(x) = \int 30(x+4)^{1/2}dx = \frac{30(x+4)^{3/2}}{\frac{3}{2}} + K$
$$= 20(x+4)^{3/2} + K$$
$C(0) = 1000$ gives $1000 = 20(8) + K$ or $K = 840$.
$$R(x) = \int 900\,dx = 900x$$
$$P(x) = 900x - 20(x+4)^{3/2} - 840$$

a. $P(5) = 4500 - 540 - 840 = \$3120$

b. $30\sqrt{x+4} = 900$
$$\sqrt{x+4} = 30$$
$$x + 4 = (30)^2 = 900$$
Thus, $x = 896$ units will yield maximum profit.

11. $\overline{C}(x) = \int \left(-6x^{-2} + \frac{1}{6}\right)dx$
$$= \frac{-6x^{-1}}{-1} + \frac{1}{6}x + K$$
$$= \frac{6}{x} + \frac{x}{6} + K$$
$\overline{C}(6) = 10 = \frac{6}{6} + \frac{6}{6} + K$ gives $K = 8$.

a. $\overline{C}(x) = \frac{6}{x} + \frac{x}{6} + 8$

b. $\overline{C}(12) = \frac{6}{12} + \frac{12}{6} + 8 = \$10.50$

13. a. $C(x) = \int 1.05(x+180)^{0.05}dx$
$$= 1.05 \cdot \frac{(x+180)^{1.05}}{1.05} + K$$
$$= (x+180)^{1.05} + K$$
$C(0) = 180^{1.05} + K = 200$ or $K \approx -33.365$.
$$C(x) = (x+180)^{1.05} - 33.365$$
$$R(x) = \int \left(\frac{1}{\sqrt{0.5x+4}} + 2.8\right)dx$$
$$= \int ((0.5x+4)^{-1/2} + 2.8)dx$$
$$= 2 \cdot \frac{(0.5x+4)^{1/2}}{\frac{1}{2}} + 2.8x + C$$
$$= 4(0.5x+4)^{1/2} + 2.8x + C$$
$R(0) = 0 = 4(0+4)^{1/2} + 0 + C$ or $C = -8$
$$R(x) = 4(0.5x+4)^{1/2} + 2.8x - 8$$

b.

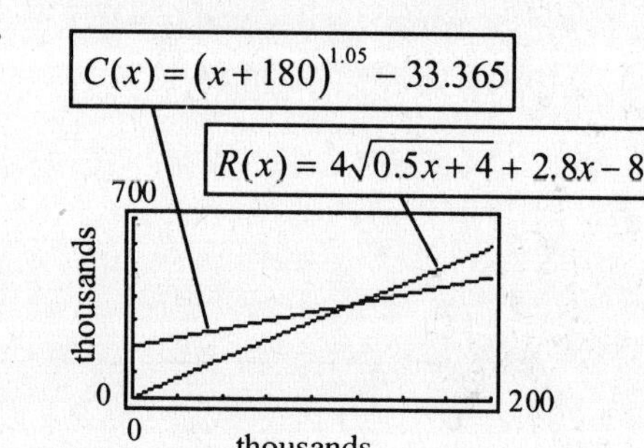

c. Using the graph, the maximum profit is at $x = 200$ thousand units.
$$P(x) = 4\sqrt{0.5x+4} + 2.8x - 8$$
$$-(x+180)^{1.05} + 33.365$$
$$P(200) \approx \$114.743 \text{ thousand}$$

15. $\frac{dC}{dy} = 0.80 \quad C(y) = \int 0.80dy = 0.80y + K$
$C(0) = 7 = 0.80(0) + K$ gives $K = 7$.
So, $C(y) = 0.80y + 7$.

17. $C(y) = \int [0.3 + 0.2(y^{-1/2})]dy$
$$= 0.3y + 0.2 \cdot \frac{y^{1/2}}{\frac{1}{2}} + K$$
$C(0) = 8$ gives $K = 8$. Thus,
$C(y) = 0.3y + 0.4\sqrt{y} + 8$.

19. $C(y) = \int [(y+1)^{-1/2} + 0.4]dy$

$= 2(y+1)^{1/2} + 0.4y + K$

$C(0) = 6$ gives $K = 4$.

Thus, $C(y) = 2(y+1)^{1/2} + 0.4y + 4$.

21. $C(y) = \int (0.7 - e^{-2y})dy$

$= 0.7y + \frac{1}{2}e^{-2y} + K$

$C(0) = 5.65$ gives $5.65 = 0 + 0.5 + K$ or $K = 5.15$.

Thus, $C(y) = 0.7y + \frac{1}{2}e^{-2y} + 5.15$.

23. $\frac{dS}{dy} = 1 - \frac{dC}{dy}$

So, $0.15 = 1 - \frac{dC}{dy}$ or $C'(y) = 0.85$.

Then, $C(y) = \int 0.85\,dy = 0.85y + K$

$C(0) = 5.15$ means $5.15 = 0 + K$ or $K = 5.15$.

Thus, $C(y) = 0.85y + 5.15$.

25. $0.2 - \frac{1}{\sqrt{3y+7}} = 1 - C'(y)$ or

$C'(y) = 0.8 + (3y+7)^{-1/2}$

$C(y) = \int [0.8 + (3y+7)^{-1/2}]dy$

$= 0.8y + \frac{2}{3}(3y+7)^{1/2} + K$

$C(0) = 6 = 0 + \frac{2}{3}(0+7)^{1/2} + K$ gives $K \approx 4.24$.

Thus, $C(y) = 0.8y + \frac{2}{3}(3y+7)^{1/2} + 4.24$.

Exercise 12.5

1. $4y - 2xy' = 0 \qquad y = x^2$ gives $y' = 2x$

$4(x^2) - 2x(2x) = 0$

$4x^2 - 4x^2 = 0$ ✓

3. $y = 3x^2 + 1 \qquad y' = 6x$

$2y\,dx - x\,dy = 2\,dx$ or $(2y - 2)\,dx = x\,dy$

Then, $\frac{dy}{dx} = \frac{2y-2}{x}$

$6x = \frac{2(3x^2+1)-2}{x} = \frac{6x^2+2-2}{x} = 6x$ ✓

5. $dy = xe^{x^2+1}dx$

$y = \int xe^{x^2+1}dx = \frac{1}{2}\int e^{x^2+1}(2x)dx = \frac{1}{2}e^{x^2+1} + C$

7. $2y\,dy = 4x\,dx$

$\int 2y\,dy = \int 4x\,dx$

$2\cdot\frac{y^2}{2} = 4\cdot\frac{x^2}{2} + C$

$y^2 = 2x^2 + C$

9. $3y^2dy = (2x-1)dx$

$\int 3y^2dy = \int (2x-1)dx$

$3\cdot\frac{y^3}{3} = 2\cdot\frac{x^2}{2} - x + C$

$y^3 = x^2 - x + C$

11. $y' = e^{x-3}$

$y = \int e^{x-3}dx$

$y = e^{x-3} + C$

$y(0) = e^{0-3} + C = 2$ gives $C = 2 - e^{-3}$.

So, $y = e^{x-3} + 2 - e^{-3}$

13. $dy = \left(\frac{1}{x} - x\right)dx$

$y = \int\left(\frac{1}{x} - x\right)dx = \int (x^{-1} - x)dx$

$y = \ln|x| - \frac{x^2}{2} + C$

$y(1) = \ln 1 - \frac{(1)^2}{2} + C = 0$ or $0 - \frac{1}{2} + C = 0$ or

$C = \frac{1}{2}$. So, $y = \ln|x| - \frac{x^2}{2} + \frac{1}{2}$.

15. $\frac{dy}{dx}=\frac{x^2}{y}$

$y\,dy = x^2 dx$

$\int y\,dy = \int x^2 dx$

$\frac{y^2}{2}=\frac{x^3}{3}+C$

17. $dx = x^3 y\,dy$

$x^{-3}dx = y\,dy$

$\int x^{-3}dx = \int y\,dy$

$\frac{x^{-2}}{-2}=\frac{y^2}{2}+\bar{C}$

or $\frac{1}{2x^2}+\frac{y^2}{2}=C \quad (C=-\bar{C})$

19. $dx = (x^2y^2 + x^2)dy$

$dx = x^2(y^2+1)dy$

$x^{-2}dx = (y^2+1)dy$

$\int x^{-2}dx = \int (y^2+1)dy$

$\frac{x^{-1}}{-1}=\frac{y^3}{3}+y+\bar{C}$

or $C=\frac{1}{x}+y+\frac{1}{3}y^3$

21. $y^2dx = x\,dy$

$x^{-1}dx = y^{-2}dy$

$\int x^{-1}dx = \int y^{-2}dy$

$\ln|x| = \frac{y^{-1}}{-1}+C$

$C=\frac{1}{y}+\ln|x|$

23. $\frac{dy}{dx}=\frac{x}{y}$

$x\,dx = y\,dy$

$\int x\,dx = \int y\,dy$

$\frac{x^2}{2}=\frac{y^2}{2}+\bar{C}$

$x^2 - y^2 = C$

25. $(x+1)\frac{dy}{dx}=y$

$y^{-1}dy = (x+1)^{-1}dx$

$\int y^{-1}dy = \int (x+1)^{-1}dx$

$\ln|y| = \ln|x+1| + C$

$\ln|y| = \ln|x+1| + \ln\bar{C}$

$\ln|y| = \ln|\bar{C}(x+1)|$

$y=\bar{C}(x+1)$

$(\ln\bar{C}=C)$

27. $e^{2x}y\,dy = (y+1)dx$

$e^{-2x}dx = \frac{y}{y+1}dy$

$-\frac{1}{2}\int e^{-2x}(-2)dx = \int\frac{y}{y+1}dy$

$-\frac{1}{2}e^{-2x} = \int\left(1-\frac{1}{y+1}\right)dy$

$-\frac{1}{2}e^{-2x} = y-\ln|y+1|+C$

(C can be on either side.)

29. $y^3dy = x^2dx$ gives $\frac{1}{4}y^4 = \frac{1}{3}x^3 + C.$

At $x=1$, $y=1$, we have

$\frac{1}{4}=\frac{1}{3}+C$ or $C=-\frac{1}{12}.$

Using this value for C and clearing fractions, we have $3y^4 - 4x^3 + 1 = 0.$

31. $\frac{2\,dx}{x^2}=\frac{3\,dy}{y^2}$ gives $\frac{2x^{-1}}{-1}=\frac{3y^{-1}}{-1}+C$ or

$-\frac{2}{x}+\frac{3}{y}=C.$ At $x=2$, $y=-1$, we have

$-1+(-3)=C$ or $C=-4$.

Thus, $\frac{3}{y}-\frac{2}{x}+4=0$ or $y=\frac{3x}{2-4x}.$

33. $e^{2y}dy = \frac{x^3+1}{x^2}dx$

$\frac{1}{2}(e^{2y}2\,dy) = (x + x^{-2})dx$

$\frac{1}{2}e^{2y} = \frac{x^2}{2} + \frac{x^{-1}}{-1} + C$

At $(1,0)$ we have $\frac{1}{2}e^0 = \frac{1}{2} - \frac{1}{.1} + C$ which gives $C = 1$. $(e^0 = 1)$

Thus, $\frac{1}{2}e^{2y} = \frac{1}{2}x^2 - \frac{1}{x} + 1$ or $xe^{2y} = x^3 - 2 + 2x$.

35. $\frac{2y}{y^2+1}dy = \frac{dx}{x}$ gives

$\ln(y^2+1) = \ln|x| + \ln C = \ln(C|x|)$ or

$y^2 + 1 = C|x|$

At $(1,2)$, we have $4 + 1 = C\cdot 1$ or $C = 5$.

Thus, $y^2 + 1 = 5|x|$.

37. $\frac{dy}{y} = k\frac{dx}{x}$ gives $\ln|y| = k\ln|x| + \ln|C| = \ln|Cx^k|$ or $y = |Cx^k|$.

39. **a.** $\frac{dx}{dt} = rx;\ r = 0.06$

$\frac{dx}{x} = r\,dt$

$\ln x = rt + C$

$x = e^{rt+C} = e^{rt}\cdot e^C = Ke^{rt}$

At $t = 0$, $x = 10{,}000$ and therefore, $x = 10{,}000e^{rt}$.

b. $x = 10{,}000e^{0.06} = \$10{,}618.37$

$x = 10{,}000e^{0.30} = \$13{,}498.59$

c. $20{,}000 = 10{,}000e^{0.06t}$

$2 = e^{0.06t}$

$\ln 2 = 0.06t$

$t = \frac{\ln 2}{0.06} \approx 11.55$ years

41. $\frac{dy}{y} = k\,dt$ gives $\ln|y| = kt + C$ or $y = Ae^{kt}$

At $t = 0$, we have $y = 10{,}000$ and, therefore, $A = 10{,}000$. Thus, $y = 10{,}000e^{kt}$. At $t = 2$, we have $y = 30{,}000$ and, therefore, $3 = e^{2k}$ or $2k = \ln 3$ or $k \approx 0.55$. Then, if $y = 1{,}000{,}000$, we have $100 = e^{0.55t}$ or $\ln 100 = 0.55t$ or $t \approx 8.4$ hr.

43. $\frac{dy}{dt} = -kt$ gives, after several steps, $y = Ae^{-kt}$.

At $t = 0$, $y = 100$ and, thus, $A = 100$. So, $y = 100e^{-kt}$. At $t = 10$, $0.9997(100) = 99.97$ pounds remain. Solving $99.97 = 100e^{-10k}$ for k gives $k = 0.00003$. With $y = 50$ (half the original 100), we have $50 = 100e^{-0.00003t}$. Solving, we have $-0.00003t = \ln(0.5)$ or $t \approx 23{,}100$ years.

45. $\frac{dx}{dt} = 5(0.06) - 5\cdot\frac{x}{100} = \frac{3}{10} - \frac{x}{20} = \frac{6-x}{20}$

Thus, $\frac{dx}{6-x} = \frac{dt}{20}$ gives $-\ln|6-x| = \frac{t}{20} + C$ or

$\ln|6-x| = -\frac{t}{20} - C$. So, $6 - x = ke^{-t/20}$ or

$x = 6 - ke^{-t/20}$. At $t = 0$, we have $x = 0$ and, thus, $k = 6$. Then, $x = 6 - 6e^{-t/20}$.

47. $\frac{dx}{dt} = 5(0.1) - 5\cdot\frac{x}{200} = \frac{20-x}{40}$

Thus, $\frac{dx}{20-x} = \frac{dt}{40}$ gives $-\ln|20-x| = \frac{t}{40} + C$ or

$\ln|20-x| = -\frac{t}{40} - C$ So, $20 - x = ke^{-t/40}$.

At $t = 0$, we have $x = 10$ and, thus, $k = 10$.

Then, $x = 20 - 10e^{-t/40}$. $\left(\frac{t}{40} = 0.025t\right)$

49. $\frac{dy}{dp} = -\frac{2}{5}\left(\frac{y}{p+8}\right)$

$\frac{dy}{y} = -\frac{2}{5}\frac{dp}{p+8}$

$\ln y = -\frac{2}{5}\ln|p+8| + \ln C = \ln[(p+8)^{-2/5}C]$

$y = C(p+8)^{-2/5}$

At $p = 24$, $y = 8$, we have $8 = C(24+8)^{-2/5}$ or $C = 2^3(32)^{2/5} = 2^5 = 32$. Thus, $y = 32(p+8)^{-2/5}$

51. $\frac{dV}{V} = 0.2e^{-0.1t}dt$ gives $\ln|V| = -2e^{-0.1t} + C.$
At $t = 0$, $V = 1.86$ and $\ln 1.86 = -2(1) + C$ or
$C = 2 + \ln 1.86.$ So, $\ln|V| = -2e^{-0.1t} + 2 + \ln 1.86$
or $\ln\left(\frac{|V|}{1.86}\right) = 2 - 2e^{-0.1t}$. Thus, $V = 1.86e^{2-2e^{-0.1t}}$.

53. $\frac{dV}{dt} = kV^{2/3}$ or $V^{-2/3}dV = k\,dt$ gives
$\frac{V^{1/3}}{\frac{1}{3}} = kt + C$ or $3V^{1/3} = kt + C.$ At $t = 0$,
$V = 0$ and $3(0)^{1/3} = k(0) + C$ gives $C = 0$. So,
$3V^{1/3} = kt$ or $V^{1/3} = \frac{kt}{3}$ or $V = \frac{k^3t^3}{27}$.

55. $\frac{du}{u - T} = k\,dt$ gives $\ln|u - T| = kt + C$ and after several steps $u - T = Ae^{kt}$.
At $t = 0$, $u - T = 0 - 20 = A \cdot 1$ gives $A = -20.$
Thus, $u = 20 - 20e^{kt}$ $(T = 20)$
At $t = 1$, $u = 8$, we have

$$8 = 20 - 20e^{k(1)}$$
$$-12 = -20e^{k}$$
$$e^{k} = 0.6$$

$\ln 0.6 = k \ln e$ or $k \approx -0.51$
Thus, $u = 20 - 20e^{-0.51t}$. If $u = 18$, then
$18 = 20 - 20e^{-0.51t}$ gives $t \approx 4.5$ hours.

57. a. $\frac{dE}{dt} = 0.043E$ or $\frac{dE}{E} = 0.043\,dt$ gives
$\ln E = 0.043t + C$ or $E = ke^{0.043t}$.
Then, $0.1 = ke^{0}$ gives $k = 0.1$.
Thus, $E(t) = 0.1e^{0.043t}$

b.

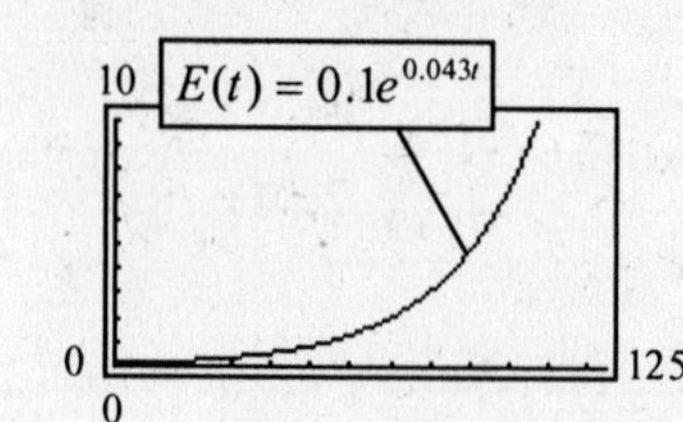

The two graphs have a good comparison.

59. a. $\frac{dP}{dt} = -0.05P$ or $\frac{dP}{P} = -0.05dt$ gives
$\ln P = -0.05t + C$ or $P = ke^{-0.05t}$. Then
$50{,}000 = ke^{0}$ gives $k = 50{,}000$. Thus,
$P(t) = 50{,}000e^{-0.05t}$.

b. $P(30) \approx \$11{,}156$

Review Exercises

1. $\int x^6 dx = \frac{1}{7}x^7 + C$

2. $\int x^{1/2} dx = \frac{2}{3}x^{3/2} + C$

3. $\int (x^3 - 3x^2 + 4x + 5)dx$
$= \frac{1}{4}x^4 - x^3 + 2x^2 + 5x + C$

4. $\int (x^2 - 1)^2 dx = \int (x^4 - 2x^2 + 1)dx$
$= \frac{1}{5}x^5 - \frac{2}{3}x^3 + x + C$

5. $\int (x^2 - 1)^2 x\,dx = \frac{1}{2}\int (x^2 - 1)^2 (2x\,dx)$
$= \frac{1}{2} \cdot \frac{(x^2 - 1)^3}{3} + C$
$= \frac{1}{6}(x^2 - 1)^3 + C$

6. $\int (x^3 - 3x^2)^5 (x^2 - 2x)dx$
$= \frac{1}{3}\int (x^3 - 3x^2)^5 (3x^2 - 6x)dx$
$= \frac{1}{18}(x^3 - 3x^2)^6 + C$

7. $\int (x^3 + 4)^2 x\,dx = \int (x^6 + 8x^3 + 16)x\,dx$
$= \int (x^7 + 8x^4 + 16x)dx$
$= \frac{1}{8}x^8 + \frac{8}{5}x^5 + 8x^2 + C$

8. $\int (x^3 + 4)^6 x^2 dx = \frac{1}{3}\int (x^3 + 4)^6 \cdot 3x^2 dx$
$= \frac{1}{3} \cdot \frac{1}{7}(x^3 + 4)^7 + C$
$= \frac{1}{21}(x^3 + 4)^7 + C$

9. $\int \frac{x^2}{x^3 + 1} dx = \frac{1}{3}\int \frac{3x^2}{x^3 + 1} dx$
$= \frac{1}{3}\ln|x^3 + 1| + C$

10. $\int \frac{x^2}{(x^3 + 1)^2} dx = \frac{1}{3}\int \frac{3x^2}{(x^3 + 1)^2} dx$
$= \frac{-1}{3(x^3 + 1)} + C$

11. $\int (x^3 - 4)^{-1/3} x^2 dx = \frac{1}{3}\int (x^3 - 4)^{-1/3} (3x^2 dx)$
$= \frac{1}{3} \cdot \frac{(x^3 - 4)^{2/3}}{\frac{2}{3}} + C$
$= \frac{1}{2}(x^3 - 4)^{2/3} + C$

12. $\int \frac{x^2}{x^3 - 4} dx = \frac{1}{3}\int \frac{3x^2 dx}{x^3 - 4} = \frac{1}{3}\ln|x^3 - 4| + C$

13. $\int \left(\frac{x^3 + 1}{x^2}\right) dx = \int (x + x^{-2})dx$
$= \frac{x^2}{2} + \frac{x^{-1}}{-1} + C$
$= \frac{1}{2}x^2 - \frac{1}{x} + C$

14. $\int \frac{x^3 - 3x + 1}{x - 1} dx = \int \left(x^2 + x - 2 - \frac{1}{x - 1}\right) dx$
$= \frac{1}{3}x^3 + \frac{1}{2}x^2 - 2x - \ln|x - 1| + C$

15. $\int y^2 e^{y^3} dy = \frac{1}{3}\int e^{y^3} (3y^2 dy) = \frac{1}{3}e^{y^3} + C$

16. $\int (x - 1)^2 dx = \frac{1}{3}(x - 1)^3 + C$ or $\frac{1}{3}x^3 - x^2 + x + \bar{C}$
$\left(\bar{C} = C - \frac{1}{3}\right)$

17. $\int \frac{3x^2}{2x^3 - 7} dx = \frac{1}{2}\int 2(3x^2)(2x^3 - 7)^{-1} dx$
$= \frac{1}{2}\ln|2x^3 - 7| + C$

18. $\int \frac{5\,dx}{e^{4x}} = -\frac{5}{4}\int e^{-4x}(-4\,dx)$
$= -\frac{5}{4}e^{-4x} + C$
$= -\frac{5}{4e^{4x}} + C$

19. $\int (x^3 - e^{3x})dx = \int x^3 dx - \frac{1}{3}\int e^{3x}(3\,dx)$
$= \frac{1}{4}x^4 - \frac{1}{3}e^{3x} + C$

20. $\int xe^{1+x^2}\,dx = \frac{1}{2}\int e^{1+x^2}(2x\,dx)$

$= \frac{1}{2}e^{1+x^2} + C$

21. $\int \frac{6x^7}{(5x^8+7)^3}\,dx = 6\cdot\frac{1}{40}\int (5x^8+7)^{-3}(40x^7\,dx)$

$= \frac{6}{40}\cdot\frac{(5x^8+7)^{-2}}{-2}$

$= \frac{-3}{40(5x^8+7)^2} + C$

22. $\int \frac{7x^3}{(1-x^4)^{1/2}}\,dx = -\frac{7}{4}\int \frac{-4x^3}{(1-x^4)^{1/2}}\,dx$

$= -\frac{7}{4}[2(1-x^4)^{1/2}] + C$

$= -\frac{7}{2}\sqrt{1-x^4} + C$

23. $\int \left(\frac{e^{2x}}{2} + \frac{2}{e^{2x}}\right)dx$

$= \frac{1}{2}\cdot\frac{1}{2}\int e^{2x}(2\,dx) + 2\left(-\frac{1}{2}\right)\int e^{-2x}(-2\,dx)$

$= \frac{1}{4}e^{2x} - e^{-2x} + C$

24. $\int \left(x - \frac{1}{(x+1)^2}\right)dx = \frac{1}{2}x^2 + \frac{1}{x+1} + C$

25. a. $\int (x^2-1)^4(x\,dx) = \frac{1}{2}\int (x^2-1)^4(2x\,dx)$

$= \frac{1}{10}(x^2-1)^5 + C$

b. $\int (x^2-1)^{10}(x\,dx) = \frac{1}{2}\int (x^2-1)^{10}(2x\,dx)$

$= \frac{1}{22}(x^2-1)^{11} + C$

c. $\int (x^2-1)^7(3x\,dx) = \frac{3}{2}\int (x^2-1)^7(2x\,dx)$

$= \frac{3}{16}(x^2-1)^8 + C$

d. $\int (x^2-1)^{-2/3}(x\,dx) = \frac{1}{2}\int (x^2-1)^{-2/3}(2x\,dx)$

$= \frac{3}{2}(x^2-1)^{1/3} + C$

26. a. $\int \frac{2x}{x^2-1}\,dx = \ln|x^2-1| + C$

b. $\int \frac{2x}{(x^2-1)^2}\,dx = -\frac{1}{x^2-1} + C$

c. $\int \frac{3x\,dx}{(x^2-1)^{1/2}} = \frac{3}{2}\int \frac{2x\,dx}{(x^2-1)^{1/2}}$

$= 3(x^2-1)^{1/2} + C$

$= 3\sqrt{x^2-1} + C$

d. $\int \frac{3x\,dx}{x^2-1} = \frac{3}{2}\int \frac{2x\,dx}{x^2-1} = \frac{3}{2}\ln|x^2-1| + C$

27. $\frac{dy}{dt} = 4.6e^{-0.05t}$

$dy = 4.6e^{-0.05t}\,dt$

$\int dy = -\frac{1}{0.05}\int 4.6e^{-0.05t}(-0.05\,dt)$

$y = -20(4.6)e^{-0.05t} + C$

$y = C - 92e^{-0.05t}$

28. $dy = (64 + 76x - 36x^2)dx$

$\int dy = \int (64 + 76x - 36x^2)dx$

$y = 64x + 38x^2 - 12x^3 + C$

29. $\frac{dy}{dx} = \frac{4x}{y-3}$ or $(y-3)dy = 4x\,dx$ gives

$\frac{y^2}{2} - 3y = 2x^2 + C.$

Also: $\frac{(y-3)^2}{2} = 2x^2 + C_1$ or $(y-3)^2 = 4x^2 + C$

30. $t\,dy = \frac{dt}{y+1}$

$\int (y+1)dy = \int \frac{dt}{t}$

$\frac{(y+1)^2}{2} = \ln|t| + C_1$

$(y+1)^2 = 2\ln|t| + C$

31. $\frac{dy}{dx} = \frac{x}{e^y}$ or $e^y\,dy = x\,dx$ gives $e^y = \frac{x^2}{2} + C.$

32. $\frac{dy}{dt} = \frac{4y}{t}$

Write in separated form and integrate.

$\int \frac{dy}{y} = 4\int \frac{dt}{t}$

$\ln|y| = 4\ln|t| + C$

Assume $y > 0$, $\ln y = \ln t^4 + \ln C_1$ where $C = \ln C_1$.

$\ln y = \ln C_1 t^4$

$y = C_1 t^4$

33. $y' = \frac{x^2}{y+1}$ or $x^2 dx = (y+1)dy$ gives

$\frac{x^3}{3} = \frac{(y+1)^2}{2} + C$. $y(0) = 4$ or $\frac{0}{3} = \frac{(4+1)^2}{2} + C$

gives $C = -\frac{25}{2}$. Then, $3(y+1)^2 = 2x^3 + 75$.

34. $(1+2y)dy = 2x\,dx$ gives $y + y^2 = x^2 + C$.

At $(2,0)$ we have $0+0 = 4+C$ or $C = -4$.

So, $x^2 = y + y^2 + 4$.

35. $\overline{MR} = 6x + 12$

$\int (6x+12)dx = 6 \cdot \frac{x^2}{2} + 12x = 3x^2 + 12x$

When $x = 4$, revenue $= 3(4)^2 + 12(4) = 96$.

36. $p'(t) = 27 + 24t - 3t^2$

$p(t) = 27t + 12t^2 - t^3 + C$

At $t = 0$, $p = 0$ and this yields $C = 0$.

Then $p(8) = 216 + 768 - 512 = 472$.

37. $P'(t) = 400\left[\frac{5}{(t+5)^2} - \frac{50}{(t+5)^3}\right]$

$P(t) = 400\int [5(t+5)^{-2} - 50(t+5)^{-3}]dt$

$= 400\left[\frac{5(t+5)^{-1}}{-1} - \frac{50(t+5)^{-2}}{-2}\right] + C$

$= 400\left[\frac{-5}{t+5} + \frac{25}{(t+5)^2}\right] + C$

Since $P(0) = 400$, $400 = 400(-1+1) + C$ gives $C = 400$.

$P(t) = 400\left[1 - \frac{5}{t+5} + \frac{25}{(t+5)^2}\right]$

38. $p'(t) = \frac{100,000}{(t+100)^2}$

$p(t) = 100,000 \cdot \frac{(t+100)^{-1}}{-1} + C$

$p(1) = 1000$ gives $1000 = \frac{-100,000}{101} + C$ or

$C \approx 1990.099$. Thus,

$p(t) = 1990.099 - \frac{100,000}{t+100}$

39. **a.** $\frac{dy}{dt} = 2.4e^{-0.04t}$ or $dy = 2.4e^{-0.04t}dt$ gives

$y = -60e^{-0.04t} + C$

$0 = -60(1) + C$ or $C = 60$

$y = 60 - 60e^{-0.04t}$

b. $y = 60 - 60e^{-0.04(12)} \approx 23\%$

40. $R'(x) = \frac{800}{x+1}$

$R(x) = 800\ln(x+1)$ (Note $C = 0$.)

41. $\overline{MC} = 6x + 4$

a. $C(x) = \int (6x+4)dx = 3x^2 + 4x + K$

$3(100)^2 + 4(100) + K = 31,400$ gives

$K = 1000$ fixed costs.

b. $C(x) = 3x^2 + 4x + 1000$

42. $\overline{MR} = \overline{MC}$

$46 = 30 + \frac{1}{5}x$

$16 = \frac{1}{5}x$

$x = 80$

So, 80 units are needed to maximize profit.

$C(x) = 30x + \frac{1}{10}x^2 + K$. $C(0) = 200$ gives

$K = 200$. Since $R(x) = 46x$ we have

$P(x) = 16x - \frac{1}{10}x^2 - 200$.

$P(80) = 1280 - 640 - 200 = \$440$

43. $\frac{dC}{dy} = (2y+16)^{-1/2} + 0.6$ or

$dC = \frac{1}{2}(2y+16)^{-1/2}(2dy) + 0.6dy$ gives

$C = \frac{1}{2} \cdot \frac{(2y+16)^{1/2}}{\frac{1}{2}} + 0.6y + K = \sqrt{2y+16} + 0.6y + K.$

$8.5 = \sqrt{0+16} + 0 + K$ gives $K = 4.5$. So,
$C = \sqrt{2y+16} + 0.6y + 4.5.$

44. $\frac{dS}{dy} = 0.2 - 0.1e^{-2y}$

$\frac{dC}{dy} = 1 - \frac{dS}{dy}$

$\frac{dC}{dy} = 0.80 + 0.1e^{-2y}$

$C(y) = \int (0.80 + 0.1e^{-2y})dy$

$C(y) = 0.8y - \frac{0.1}{2}e^{-2y} + K$

Using $C(0) - 7.8$ to find K gives $K - 7.85$.
So, $C(y) = 0.8y - 0.05e^{-2y} + 7.85.$

45. $\frac{dW}{dL} = \frac{3W}{L}$ or $\frac{dW}{W} = 3\frac{dL}{L}$ gives

$\ln W = 3\ln L + \ln C = \ln(L^3 \cdot C).$

Thus, $W = CL^3$

46. $\frac{dx}{dt} = kx$ As before $x = Ce^{kt}$.
When $t = 0$, $x = 10$ and from this information we have $C = 10$. So, $x = 10e^{kt}$. Use the half-life, $t = 4.6$ million years and $x = 5$, to determine k. The result is $k \approx -0.15$. So, $x = 10e^{-0.15t}$.
With 20% left, $x = 2$. Now determine the corresponding t.

$2 = 10e^{-0.15t}$

$\ln 0.2 = -0.15t$

$t \approx 10.73$ million years

47. $\frac{dx}{dt} = 4 \cdot 3 - 4 \cdot \frac{x}{120} = \frac{360-x}{30}$

Thus, $\frac{dx}{360-x} = \frac{dt}{30}$ gives

$-\ln(360-x) = \frac{1}{30}t + C_1$ or

$\ln(360-x) = -\frac{t}{30} + C.$ After algebraic manipulation, we have $360 - x = Ae^{-t/30}$. Then, $360 - 0 = A \cdot 1$ or $A = 360$ and $x = 360 - 360e^{-t/30}$.

48. $\frac{dx}{dt} = 3 \cdot 2 - 3 \cdot \frac{x}{300}$

$\frac{dx}{dt} = 6 - \frac{x}{100}$

$\frac{dx}{dt} = \frac{600-x}{100}$

$\int \frac{dx}{600-x} = \frac{1}{100}\int dt$

$-\ln(600-x) = \frac{1}{100}t + C$

$x = 100$ when $t = 0$.
Using $x = 100$ when $t = 0$ gives $C = -\ln 500$.
So, $-\ln(600-x) = 0.01t - \ln 500$

$\ln\left(\frac{600-x}{500}\right) = \ln e^{-0.01t}$

$600 - x = 500e^{-0.01t}$

$x = 600 - 500e^{-0.01t}$

When $x = 500$ we have $500 = 600 - 500e^{-0.01t}$

$0.2 = e^{-0.01t}$

$\ln 0.2 = -0.01t$

$t \approx 161$ minutes

Chapter Test

1. $\int (6x^2 + 8x - 7)dx = \frac{6x^3}{3} + \frac{8x^2}{2} - 7x + C$
$= 2x^3 + 4x^2 - 7x + C$

2. $\int (4 + x^{1/2} - x^{-2})dx = 4x + \frac{x^{3/2}}{\frac{3}{2}} - \frac{x^{-1}}{-1} + C$
$= 4x + \frac{2}{3}x^{3/2} + \frac{1}{x} + C$

3. $5\int(4x^3-7)^9(x^2\,dx)=\frac{5}{12}\int(4x^3-7)^9(12x^2dx)$

$=\frac{5}{12}\frac{(4x^3-7)^{10}}{10}+C=\frac{1}{24}(4x^3-7)^{10}+C$

4. $\int(3x^2-6x+1)^9(2x-2)dx$

$=\frac{1}{3}\int(3x^2-6x+1)^9(6x-6)dx$

$=\frac{1}{30}(3x^2-6x+1)^{10}+C$

5. $\int\frac{s^3}{2s^4-5}ds=\frac{1}{8}\int\frac{8s^3}{2s^4-5}ds=\frac{1}{8}\ln\left|2s^4-5\right|+C$

6. $100\int e^{-0.01x}dx=\frac{100}{-0.01}\int e^{-0.01x}(-0.01dx)$

$=-10{,}000e^{-0.01x}+C$

7. $5\int e^{2y^4-1}(y^3dy)=\frac{5}{8}\int e^{2y^4-1}(8y^3dy)$

$=\frac{5}{8}e^{2y^4-1}+C$

8. $\int\left(e^x+\frac{5}{x}-1\right)dx=e^x+5\ln|x|-x+C$

9.
$$\begin{array}{r} x-1 \\ x+1\overline{)x^2} \\ \underline{x^2+x} \\ -x \\ \underline{-x-1} \end{array}$$

Remainder: 1

$\int\frac{x^2}{x+1}dx=\int\left(x-1+\frac{1}{x+1}\right)dx$

$=\frac{x^2}{2}-x+\ln|x+1|+C$

10. $\int f(x)dx=2x^3-x+5e^x+C$. $f(x)$ is the derivative of the right side. $f(x)=6x^2-1+5e^x$

11. $y'=4x^3+3x^2$; $y(0)=4$

$y=x^4+x^3+C$

$4=0^4+0^3+C$

$y=x^4+x^3+4$

12. $\frac{dy}{dx}=e^{4x}$; $y(0)=2$

$y=\frac{1}{4}e^{4x}+C$

$2=\frac{1}{4}(1)+C$ or $C=\frac{7}{4}$

$y=\frac{1}{4}e^{4x}+\frac{7}{4}$

13. $\frac{dy}{dx}=x^3y^2$; $\frac{dy}{y^2}=x^3dx$

$\frac{y^{-1}}{-1}=\frac{x^4}{4}+\overline{C}$ or $\frac{x^4}{4}+\frac{1}{y}=K$

$\frac{1}{y}=\frac{4K-x^4}{4}$ or $y=\frac{4}{C-x^4}$

14. $p'(t)=2000t^{1.04}$; $p(0)=50{,}000$; $p(10)=?$

$p(t)=2000\frac{t^{2.04}}{2.04}+C$

$50{,}000=0+C=C$

$p(10)=\frac{2000}{2.04}10^{2.04}+50{,}000=157{,}498$

15. $\overline{MC}=4x+50$; $\overline{MR}=500$; $C(10)=1000$

$C=2x^2+50x+K$; $R=500x$

$1000=2(10)^2+50(10)+K$ gives $K=300$

$P(x)=500x-(2x^2+50x+300)$

$=450x-2x^2-300$

16. $\frac{dS}{dy}=0.22-\frac{0.25}{\sqrt{0.5y+1}}$ Since $\frac{dC}{dy}=1-\frac{dS}{dy}$, we have $\frac{dC}{dy}=0.78+\frac{0.25}{\sqrt{0.5y+1}}$.

$C=0.78y+\sqrt{0.5y+1}+K$

$6.6=0+\sqrt{0+1}+K$ gives $K=5.6$

$C(y)=0.78y+\sqrt{0.5y+1}+5.6$

17. $\frac{dx}{dt}=kx$; $\frac{dx}{x}=k\,dt$; $\ln x=kt+C$ gives $x=e^{kt+C}=e^{kt}\cdot e^C=Ae^{kt}$. At $t=0$, $x=A$. Then at $t=100$, $x=\frac{1}{2}A$.

$\frac{1}{2}A=Ae^{100k}$ gives $k=\frac{\ln\left(\frac{1}{2}\right)}{100}\approx-0.00693$

$x=Ae^{-0.00693t}$; $x=0.1A$ gives $0.1A=Ae^{-0.00693t}$

gives $t=\frac{\ln(0.1)}{-.00693}\approx 332.3$ days .

Chapter 13: Definite Integrals; Techniques of Integration

Exercise 13.1

1. $f(x)=4x-x^2$

Width of rectangles $=\dfrac{2-0}{2}=1$.

Height of each rectangle: $f(1)=4(1)-1^2=3$

$f(2)=4(2)-2^2=4$

$A\approx 1(3+4)=7$

3. $f(x)=9-x^2$

Width of rectangles $=\dfrac{3-(-1)}{4}=1$

Height of rectangles:

$f(0)=9-0=9$ $\quad f(2)=9-4=5$

$f(1)=9-1=8$ $\quad f(3)=9-9=0$

$A=1(9+8+5+0)=22$

5. $f(x)=4x-x^2$

Width of rectangles $=\dfrac{2-0}{2}=1$

Height of each rectangle:

$f(0)=4(0)-0^2=0$

$f(1)=4(1)-1^2=3$

$A\approx 1(0+3)=3$

7. $f(x)=9-x^2$

Width of rectangles $=\dfrac{3-(-1)}{4}=1$

Height of rectangles:

$f(-1)=9-1=8$ $\quad f(1)=9-1=8$

$f(0)=9-0=9$ $\quad f(2)=9-4=5$

$1(8+9+8+5)=30$

9. $S_L(10)=\dfrac{14}{3}-\dfrac{6}{10}+\dfrac{4}{300}=\dfrac{1224}{300}=4.08$

$S_R(10)=\dfrac{14(100)+180+4}{300}=\dfrac{1584}{300}=5.28$

11. $\lim\limits_{n\to\infty} S_L=\lim\limits_{n\to\infty}\left(\dfrac{14}{3}-\dfrac{6}{n}+\dfrac{4}{3n^2}\right)$

$=\dfrac{14}{3}-0+0=\dfrac{14}{3}$

$\lim\limits_{n\to\infty} S_R=\lim\limits_{n\to\infty}\left(\dfrac{14}{3}+\dfrac{6}{n}+\dfrac{4}{3n^2}\right)$

$=\dfrac{14}{3}+0+0=\dfrac{14}{3}$

13. If a point within each subinterval is selected and the area of each rectangle is determined, then the total area A would satisfy $S_L\le A\le S_R$.

Since $\lim\limits_{n\to\infty} S_L=\lim\limits_{n\to\infty} S_R=\dfrac{14}{3}$, it follows that

$\lim\limits_{n\to\infty} A=\dfrac{14}{3}$.

15. $\sum\limits_{i=1}^{4} x_i=3+(-1)+3+(-2)=3$

17. $\sum\limits_{j=2}^{5}(j^2-3)$

$=(4-3)+(9-3)+(16-3)+(25-3)$

$=42$

No formulas are used because j begins at 2 and the formulas require that j begin at 1.

19. $\sum\limits_{j=0}^{4}(j^2-4j+1)$

$=(1)+(-2)+(-3)+(-2)+(1)=-5$

21. $\sum\limits_{j=1}^{60} 3=3(60)=180$

23. $\sum\limits_{k=1}^{30}(k^2+4k)=\dfrac{30(31)(61)}{6}+\dfrac{4(30)(31)}{2}$

$=11{,}315$

25. $\sum\limits_{i=1}^{n}\left(1-\dfrac{2i}{n}+\dfrac{i^2}{n^2}\right)\left(\dfrac{3}{n}\right)$

$=\dfrac{3}{n}\left[n-\dfrac{2}{n}\cdot\dfrac{n(n+1)}{2}+\dfrac{1}{n^2}\cdot\dfrac{n(n+1)(2n+1)}{6}\right]$

$=3-\dfrac{3}{n}(n+1)+\dfrac{(n+1)(2n+1)}{2n^2}$

$=\dfrac{2n^2-3n+1}{2n^2}$

27. a. $f(x) = x$

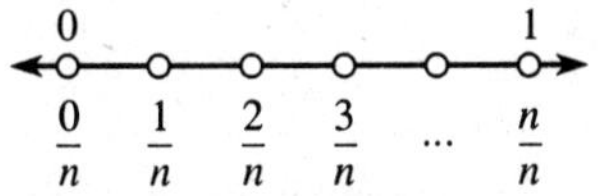

Area of 1st rectangle: $\frac{1}{n} \cdot f(0/n) = \frac{1}{n} \cdot 0$

Area of 2nd rectangle: $\frac{1}{n} \cdot f(1/n) = \frac{1}{n} \cdot \frac{1}{n}$

$\vdots$

Area of ith rectangle:

$$\frac{1}{n} \cdot f\left(\frac{i-1}{n}\right) = \frac{1}{n} \cdot \frac{i-1}{n}$$

$$S(n) = \frac{1}{n}\left[0 + \frac{1}{n} + \frac{2}{n} + \cdots + \frac{n-1}{n}\right]$$

$$= \frac{1}{n} \cdot \frac{1}{n}(0 + 1 + 2 + \cdots + (n-1)]$$

$$= \frac{n-1}{2n}$$

b. $S(1) = \frac{10-1}{20} = \frac{9}{20}$

c. $S(100) = \frac{100-1}{200} = \frac{99}{200}$

d. $S(1000) = \frac{1000-1}{2000} = \frac{999}{2000}$

e. $\lim_{n\to\infty} S(n) = \lim_{n\to\infty}\left(\frac{1}{2} - \frac{1}{2n}\right) = \frac{1}{2}$

29. $f(x) = x^2$ (See the partition in problem 27.)

a. Area of 1st rectangle: $\frac{1}{n} \cdot f(1/n) = \frac{1}{n} \cdot \frac{1}{n^2}$

Area of 2nd rectangle: $\frac{1}{n} \cdot f(2/n) = \frac{1}{n} \cdot \frac{4}{n^2}$

$\vdots$

Area of ith rectangle: $\frac{1}{n} \cdot f(i/n) = \frac{1}{n} \cdot \frac{i^2}{n^2}$

$$S(n) = \frac{1}{n}\left[\frac{1}{n^2} + \frac{4}{n^2} + \frac{9}{n^2} + \cdots + \frac{n^2}{n^2}\right]$$

$$= \frac{1}{n^3}\left[\frac{n(n+1)(2n+1)}{6}\right]$$

$$= \frac{(n+1)(2n+1)}{6n^2}$$

b. $S(10) = \frac{11 \cdot 21}{600} = \frac{77}{200}$

c. $S(100) = \frac{101 \cdot 201}{60{,}000} = \frac{6767}{20{,}000}$

d. $S(1000) = \frac{1001 \cdot 2001}{6{,}000{,}000} = \frac{667{,}667}{2{,}000{,}000}$

e. $\lim_{n\to\infty} S(n) = \lim_{n\to\infty} \frac{2n^2 + 3n + 1}{6n^2}$

$$= \lim_{n\to\infty}\left(\frac{2}{6} + \frac{1}{2n} + \frac{1}{6n^2}\right)$$

$$= \frac{2}{6} + 0 + 0 = \frac{1}{3}$$

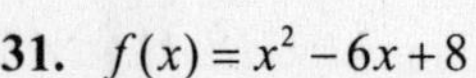

31. $f(x) = x^2 - 6x + 8$

Use right hand endpoints.

Area of 1st rectangle: $\frac{2}{n} \cdot f(2/n)$

Area of 2nd rectangle: $\frac{2}{n} \cdot f(4/n)$

$\vdots$

Area of ith rectangle: $\frac{2}{n} \cdot f(2i/n)$

$$S(n) = \frac{2}{n}\left[\left\{\left(\frac{2}{n}\right)^2 - 6\left(\frac{2}{n}\right) + 8\right\} + \left\{\left(\frac{4}{n}\right)^2 - 6\left(\frac{4}{n}\right) + 8\right\} + \cdots + \left\{\left(\frac{2n}{n}\right)^2 - 6\left(\frac{2n}{n}\right) + 8\right\}\right]$$

We are going to group like terms.

$$S(n) = \frac{2}{n}\left[\left\{\left(\frac{2}{n}\right)^2 + \left(\frac{4}{n}\right)^2 + \cdots + \left(\frac{2n}{n}\right)^2\right\} - 6\left\{\frac{2}{n} + \frac{4}{n} + \cdots + \frac{2n}{n}\right\} + \{8n\}\right]$$

$$= \frac{2}{n}\left[\left\{4\left(\frac{1}{n}\right)^2 + 4\left(\frac{2}{n}\right)^2 + \cdots + 4\left(\frac{n}{n}\right)^2\right\} - 12\left\{\frac{1}{n} + \frac{2}{n} + \cdots + \frac{n}{n}\right\} + 8n\right]$$

$$= \frac{2}{n}\left[\frac{4}{n^2}(1 + 2^2 + \cdots + n^2) - \frac{12}{n}(1 + 2 + \cdots + n) + 8n\right]$$

$$= \frac{2}{n}\left[\frac{4}{n^2} \cdot \frac{n(n+1)(2n+1)}{6} - \frac{12}{n} \cdot \frac{n(n+1)}{2} + 8n\right] = \frac{20n^2 - 24n + 4}{3n^2}$$

$$A = \lim_{n\to\infty}\left(\frac{20}{3} - \frac{24}{3n} + \frac{4}{3n^2}\right) = \frac{20}{3}$$

33. $A \approx 65 \cdot 1 + 82 \cdot 1 + 87 \cdot 1 = \$234$ billion

35. There are approximately 90 squares under the curve.

Each square represents $10 \times \frac{1 \text{ hr}}{3600} \times \frac{1 \text{ miles}}{\text{hr}} = \frac{1}{360}$ miles.

Area represents $90 \cdot \frac{1}{360} = \frac{1}{4}$ miles.

37. $A \approx 10(0 + 15 + 18 + 18 + 30 + 27 + 24 + 23) = 1550$ sq ft

Exercise 13.2

1. $\int_0^3 4x\,dx = 4 \cdot \frac{x^2}{2}\Big|_0^3 = 2\left[3^2 - 0^2\right] = 18$

3. $\int_2^4 dx = x\Big|_2^4 = 4 - 2 = 2$

5. $\int_2^4 x^3 dx = \frac{1}{4}x^4\Big|_2^4 = \frac{256}{4} - \frac{16}{4} = 60$

7. $\int_0^5 4x^{2/3} dx = \frac{12}{5}x^{5/3}\Big|_0^5$

$= \frac{12}{5}(5^{5/3} - 0)$

$= \frac{12}{5}\left[5(5)^{2/3}\right]$

$= 12\sqrt[3]{25}$

9. $\int_2^4 (4x^3 - 6x^2 - 5x)dx$

$= \left(x^4 - 2x^3 - \frac{5}{2}x^2\right)\Big|_2^4$

$= (256 - 128 - 40) - (16 - 16 - 10)$

$= 98$

11. $\int_2^3 (x-4)^2\,dx = \frac{1}{3}(x-4)^3\Big|_2^3$

$= \frac{1}{3}[-1-(-8)] = \frac{7}{3}$

13. $\int_2^4 (x^2+2)^3\,x\,dx = \frac{1}{2}\int_2^4 (x^2+2)^3(2x)dx$

$= \frac{1}{2}\cdot\frac{1}{4}(x^2+2)^4\Big|_2^4$

$= \frac{1}{8}(104{,}976 - 1296)$

$= 12{,}960$

15. $\int_{-1}^2 (x^3 - 3x^2)^3(x^2 - 2x)dx$

$= \frac{1}{3}\int_{-1}^2 (x^3 - 3x^2)^3(3x^2 - 6x)dx$

$= \frac{1}{3}\cdot\frac{1}{4}(x^3 - 3x^2)^4\Big|_{-1}^2 = \frac{1}{12}[256 - 256] = 0$

17. $\int_2^3 (x^2+3)^{1/2}x = dx$

$= \frac{1}{2}\int_2^3 (x^2+3)^{1/2}(2x\,dx)$

$= \frac{1}{2}\cdot\frac{2}{3}(x^2+3)^{3/2}\Big|_2^3$

$= \frac{1}{3}\left[12^{3/2} - 7^{3/2}\right]$

$= \frac{1}{3}\left[12(12)^{1/2} - 7(7)^{1/2}\right]$

$= \frac{1}{3}\left[12\cdot 2\sqrt{3} - 7\sqrt{7}\right]$

$= 8\sqrt{3} - \frac{7}{3}\sqrt{7}$

19. $\int_1^3 \frac{3}{y^2}dy = \int_1^3 3y^{-2}dy$

$= 3\cdot\frac{y^{-1}}{-1}\Big|_1^3 = -3\left[3^{-1} - 1^{-1}\right]$

$= -3\left[\frac{1}{3} - 1\right] = 2$

21. $\int_0^1 e^{3x}dx = \frac{1}{3}\int_0^1 e^{3x}(3\,dx)$

$= \frac{1}{3}e^{3x}\Big|_0^1 = \frac{1}{3}\left[e^3 - e^0\right] = \frac{1}{3}(e^3 - 1)$

23. $\int_1^e \frac{4}{z}dz = 4\int_1^e z^{-1}dz$

$= 4\ln|z|\Big|_1^e$

$= 4[\ln e - \ln 1] = 4$

25. $\int_4^4 \sqrt{x^2 - 2}dx = 0$

Note that we can find the value of the definite integral and yet we cannot integrate the function.

27. a. $\int_3^6 \frac{x}{3x^2+4}dx = \frac{1}{6}\int_3^6 \frac{6x}{3x^2+4}dx$

$= \frac{1}{6}\ln(3x^2+4)\Big|_3^6 = \frac{1}{6}[\ln 112 - \ln 31]$

$= \frac{1}{6}\ln\left(\frac{112}{31}\right)$

b. Graphing utility gives 0.2140853.

29. a. $\int_1^2 \frac{x^2+3}{x}dx = \int_1^2 (x + 3x^{-1})dx$

$= \left(\frac{x^2}{2} + 3\ln|x|\right)\Big|_1^2 = 2 + 3\ln 2 - \left(\frac{1}{2} + 3\ln 1\right)$

$= \frac{3}{2} + 3\ln 2 = 3.5794415$

b. Graphing utility gives 3.5794415.

31. a. A, C
b. B

33. a. $\int_0^4 \left(2x - \frac{1}{2}x^2\right)dx$

b. $A = \left[2\cdot\frac{x^2}{2} - \frac{1}{2}\cdot\frac{x^3}{3}\right]\Big|_0^4$

$= \left(x^2 - \frac{1}{6}x^3\right)\Big|_0^4 = 16 - \frac{1}{6}(64) - 0$

$= \frac{96-64}{6} = \frac{16}{3}$

35. a. $A = \int_{-1}^0 (x^3+1)dx$

b. $A = \left(\frac{x^4}{4} + x\right)\Big|_{-1}^0 = 0 - \left(\frac{1}{4} - 1\right) = \frac{3}{4}$

37. $\int_1^2 \left(-x^2+3x-2\right)dx$

$=\left(-\frac{1}{3}x^3+\frac{3}{2}x^2-2x\right)\Big|_1^2$

$=\left(-\frac{8}{3}+6-4\right)-\left(-\frac{1}{3}+\frac{3}{2}-2\right)=\frac{1}{6}$

39. $\int_1^3 e^{x^2}x\,dx=\frac{1}{2}\int_1^3 e^{x^2}\left(2x\,dx\right)$

$=\frac{1}{2}e^{x^2}\Big|_1^3=\frac{1}{2}\left(e^9-e\right)$

41. The integrals differ only in sign.

43. $\int_1^2 \left(2x-x^2\right)dx+\int_2^4 \left(2x-x^2\right)dx=(-1)\int_1^2\left(x^2-2x\right)dx+(-1)\int_2^4\left(x^2-2x\right)dx=(-1)\int_1^4\left(x^2-2x\right)dx$

Thus, we have $\int_1^2 \left(2x-x^2\right)dx+\int_2^4 \left(2x-x^2\right)dx=\frac{2}{3}+\left(-\frac{20}{3}\right)=-6$ So, $\int_1^4\left(x^2-2x\right)dx=(-1)(-6)=6$.

45. a. $D=50{,}000\cdot 5+30{,}000\cdot 5=\$400{,}000$ (RHP)

or $60{,}000\cdot 5+50{,}000\cdot 5=\$550{,}000$ (LHP)

or $\frac{1}{2}(10)\left(60{,}000+30{,}000\right)=\$450{,}000$ (trapezoid)

b. $D=\int_0^{10} 3000\left(20-t\right)dt=3000\left(20t-\frac{t^2}{2}\right)\Big|_0^{10}=3000\left(200-50\right)-0=\$450{,}000$

47. a. $\int_0^7\left(300t-3t^2\right)dt=\left(150t^2-t^3\right)\Big|_0^7=7350-343=\$7007$

b. $\int_7^{14}\left(300t-3t^2\right)dt=\left(150t^2-t^3\right)\Big|_7^{14}=\left(29{,}400-2744\right)-7007=\$19{,}649$

49. $\int_0^2 10{,}000e^{0.02t}dt=\frac{1}{0.02}\int_0^2 10{,}000e^{0.02t}\left(0.02dt\right)=\frac{10{,}000}{0.02}\cdot e^{0.02t}\Big|_0^2$

$=500{,}000\left[e^{0.04}-e^0\right]=500{,}000\left[e^{0.04}-1\right]=20{,}405.39$

51. $\int_0^{0.3}\left(0.3-3.33r^2\right)2\pi r\,dr=\int_0^{0.3}\left(0.6\pi r-6.66\pi r^3\right)dr=\left(0.6\pi\cdot\frac{r^2}{2}-6.66\pi\cdot\frac{r^4}{4}\right)\Big|_0^{0.3}$

$=\left(0.3\pi r^2-1.665\pi r^4\right)\Big|_0^{0.3}=\left[0.3\pi(0.3)^2-1.665\pi(0.3)^4\right]-\left(0-0\right)$

$=\pi\left(0.027-0.0134865\right)=0.04\text{ cm}^3$

53. $x=\int_0^5 200\left(1+\frac{400}{(t+40)^2}\right)dt=200\left(t+\frac{400(t+40)^{-1}}{-1}\right)\Big|_0^5=200\left[\left(5-\frac{400}{45}\right)-\left(0-10\right)\right]=1222.22$

55. a. $y=0.005963x^{0.701215}$

b. $d=\int_0^{9.2}.005963x^{0.70125}dx=0.005963\times\left(\frac{x^{1.701215}}{1.701215}\right)\Big|_0^{9.2}=0.153$ miles

Exercise 13.3

1. a. $\int_0^2 (4-x^2)\,dx$

b. $\int_0^2 (4-x^2)\,dx = \left(4x - \frac{x^3}{3}\right)\Big|_0^2 = 4(2) - \frac{2^3}{3} = 8 - \frac{8}{3} = \frac{24-8}{3} = \frac{16}{3}$

3. a. $A = \int_1^8 \left(x^{1/3} - 2 + x\right)dx$

b. $A = \left|\left(\frac{3}{4}x^{4/3} - 2x + \frac{x^2}{2}\right)\right|_1^8 = 28 - \left(-\frac{3}{4}\right) = 28\frac{3}{4}$

5. a. $\int_1^2 \left[(4-x^2) - \left(\frac{1}{4}x^3 - 2\right)\right]dx$

b. $\int_1^2 \left(6 - x^2 - \frac{1}{4}x^3\right)dx = \left(6x - \frac{1}{3}x^3 - \frac{1}{16}x^4\right)\Big|_1^2 = \left(12 - \frac{8}{3} - 1\right) - \left(6 - \frac{1}{3} - \frac{1}{16}\right) = \frac{131}{48}$

7. a.

$$x + 2 = x^2$$
$$x^2 - x - 2 = 0$$
$$(x-2)(x+1) = 0$$
$$x = 2 \text{ or } x = -1$$

If $x = 2$, $y = 4$. If $x = -1$, $y = 1$.

b., c. $\int_{-1}^2 \left[(x+2) - x^2\right]dx = \left(\frac{x^2}{2} + 2x - \frac{x^3}{3}\right)\Big|_{-1}^2 = \left(2 + 4 - \frac{8}{3}\right) - \left(\frac{1}{2} - 2 + \frac{1}{3}\right) = \frac{18-8}{3} - \frac{3-12+2}{6} = \frac{20}{6} + \frac{7}{6} = \frac{9}{2}$

9. a.

$$x^2 - 4x = x - x^2$$
$$2x^2 - 5x = 0$$
$$x(2x-5) = 0$$
$$x = 0 \text{ or } x = \frac{5}{2}$$

If $x = 0$, $y = 0$. If $x = \frac{5}{2}$, $y = -\frac{15}{4}$

b., c. $\int_0^{5/2} \left[(x - x^2) - (x^2 - 4x)\right]dx = \int_0^{5/2} (5x - 2x^2)\,dx = \left(\frac{5}{2}x^2 - \frac{2}{3}x^3\right)\Big|_0^{5/2} = \left(\frac{125}{8} - \frac{250}{24}\right) - 0 = \frac{125}{24}$

11. a.

$$x^3 - 2x = 2x$$
$$x^3 - 4x = 0$$
$$x(x^2 - 4) = 0$$
$$x(x+2)(x-2) = 0$$
$$x = 0, x = -2, x = 2$$

If $x = 0$, then $y = 0$. If $x = -2$, then $y = -4$. If $x = 2$, then $y = 4$.

b., c. $\int_{-2}^0 \left(x^3 - 2x - 2x\right)dx + \int_0^2 \left[2x - (x^3 - 2x)\right]dx = \int_{-2}^0 (x^3 - 4x)\,dx + \int_0^2 (-x^3 + 4x)\,dx$

$$= \left[\frac{x^4}{4} - 4\cdot\frac{x^2}{2}\right]\Big|_{-2}^0 + \left[-\frac{x^4}{4} + 4\cdot\frac{x^2}{2}\right]\Big|_0^2 = 0 - \left(\frac{16}{4} - 2(-2)^2\right) + \left(-\frac{16}{4} + 2(2)^2\right) - 0 = 8$$

13. $f(1)=3,\ \ g(1)=-1,\ \ f(x)\ge g(x)$

$$\int_0^2\left[\left(x^2+2\right)-\left(-x^2\right)\right]dx$$

$$=\left(\frac{2}{3}x^3+2x\right)\Big|_0^2=\frac{16}{3}+4=\frac{28}{3}$$

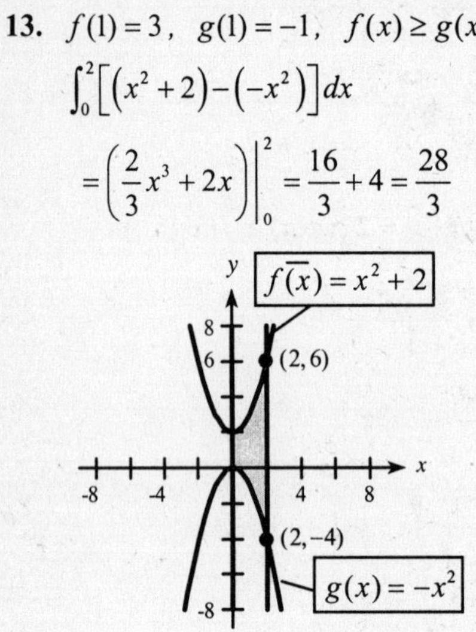

15. $x^3-1=x-1$

$$x^3-x=x(x-1)(x+1)=0$$

$$x=0,\ x=1$$

$$f\left(\frac{1}{2}\right)=\frac{1}{8}-1=-\frac{7}{8}$$

$$g\left(\frac{1}{2}\right)=\frac{1}{2}-1=-\frac{1}{2}$$

$$g(x)\ge f(x)$$

$$\int_0^1\left[(x-1)-\left(x^3-1\right)\right]dx$$

$$=\int_0^1\left(x-x^3\right)dx$$

$$=\left(\frac{x^2}{2}-\frac{x^4}{4}\right)\Big|_0^1=\frac{1}{2}-\frac{1}{4}=\frac{1}{4}$$

17. $x^2-2x=\frac{1}{2}x^2$

$$\frac{1}{2}x^2-2x=\frac{1}{2}x(x-4)$$

$$=0$$

$$x=0,\ x=4$$

$$f(1)=\frac{1}{2}\quad g(1)=1-2=-1,\ f(x)\ge g(x)$$

$$\int_0^4\left[\frac{1}{2}x^2-\left(x^2-2x\right)\right]dx$$

$$=\int_0^4\left(2x-\frac{1}{2}x^2\right)dx=\left(x^2-\frac{1}{6}x^3\right)\Big|_0^4$$

$$=\left(16-\frac{64}{6}\right)=\frac{16}{3}$$

19. $x^2=\sqrt{x}$

$$x^2-\sqrt{x}=x^{1/2}\left(x^{3/2}-1\right)$$

$$=0$$

$$x=0,\ x=1$$

$$h\left(\frac{1}{4}\right)=\frac{1}{16},\ k\left(\frac{1}{4}\right)=\frac{1}{2},\ k(x)\ge h(x)$$

$$\int_0^1\left(\sqrt{x}-x^2\right)dx=\left(\frac{2}{3}x^{3/2}-\frac{1}{3}x^3\right)\Big|_0^1$$

$$=\left(\frac{2}{3}-\frac{1}{3}\right)=\frac{1}{3}$$

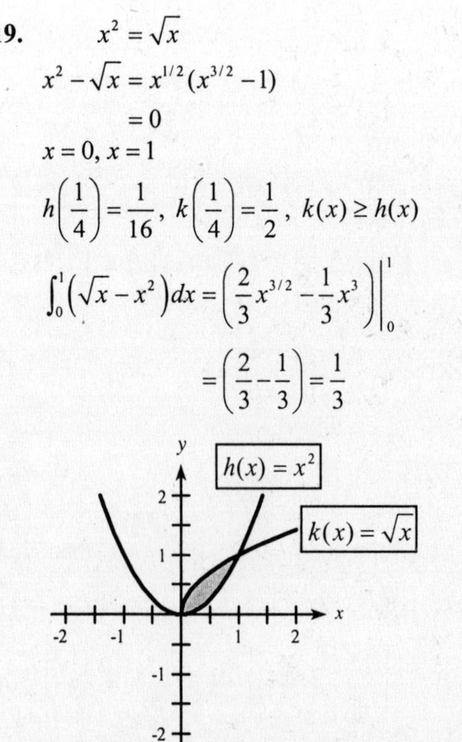

21.

$$x^3 = x^2 + 2x$$

$$x^3 - x^2 - 2x = x(x-2)(x+1) = 0$$

$$x = -1, x = 0, x = 2$$

$$f(1) = 1, \quad g(1) = 3,$$

$$g(x) \geq f(x) \text{ over } [0,2]$$

$$\int_{-1}^{0}\left[x^3 - \left(x^2 + 2x\right)\right]dx + \int_{0}^{2}\left[\left(x^2 + 2x\right) - x^3\right]dx$$

$$= \left(\frac{x^4}{4} - \frac{x^3}{3} - x^2\right)\Bigg|_{-1}^{0} + \left(\frac{x^3}{3} + x^2 - \frac{x^4}{4}\right)\Bigg|_{0}^{2}$$

$$= \left[0 - \left(\frac{1}{4} + \frac{1}{3} - 1\right)\right] + \left[\left(\frac{8}{3} + 4 - 4\right) - 0\right]$$

$$= \frac{5}{12} + \frac{8}{3} = \frac{37}{12}$$

$g(x) = x^2 + 2x$

$f(x) = x^3$

23.

$$\frac{3}{x} = 4 - x$$

$$x^2 - 4x + 3 = (x-3)(x-1) = 0$$

$$x = 1, \quad x = 3$$

$$f(2) = \frac{3}{2}, \quad g(2) = 2, \quad g(x) \geq f(x)$$

$$\int_{1}^{3}\left[(4-x) - \frac{3}{x}\right]dx$$

$$= \left(4x - \frac{x^2}{2} - 3\ln x\right)\Bigg|_{1}^{3}$$

$$= \left(12 - \frac{9}{2} - 3\ln 3\right) - \left(4 - \frac{1}{2} - 3 \cdot 0\right)$$

$$= 4 - 3\ln 3$$

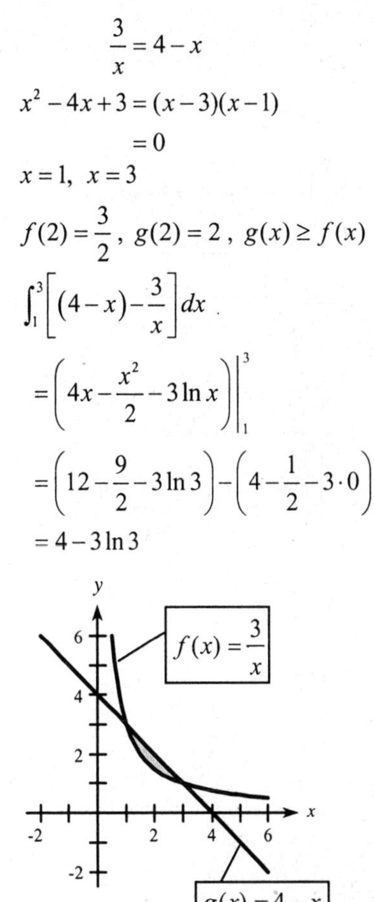

25. $\sqrt{x+3} = 2$

$x + 3 = 4$ or $x = 1$, $x = -3$

$f(0) = \sqrt{3}$, $g(0) = 2$, $g(x) \geq f(x)$

$$\int_{-3}^{1}(2 - \sqrt{x+3})dx = \left(2x - \frac{2}{3}(x+3)^{3/2}\right)\Bigg|_{-3}^{1}$$

$$= \left(2 - \frac{2}{3} \cdot 8\right) - (-6 - 0) = \frac{8}{3}$$

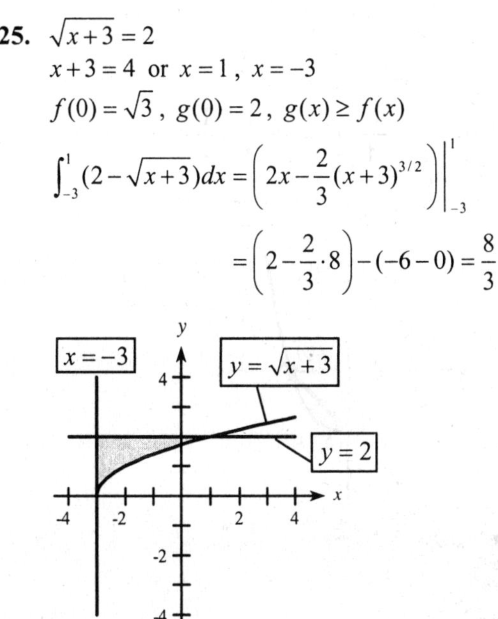

27. Avg value $= \dfrac{1}{b-a}\int_a^b f(x)dx$

$= \dfrac{1}{3-0}\int_0^3 (9-x^2)dx$

$= \dfrac{1}{3}\left(9x - \dfrac{x^3}{3}\right)\Big|_0^3 = \dfrac{1}{3}(27-9) = 6$

29. Avg value $= \dfrac{1}{b-a}\int_a^b f(x)dx$

$= \dfrac{1}{1-(-1)}\int_{-1}^1 (x^3 - x)dx$

$= \dfrac{1}{2}\left(\dfrac{x^4}{4} - \dfrac{x^2}{2}\right)\Big|_{-1}^1 = \dfrac{1}{2}\left(\dfrac{1}{4} - \dfrac{1}{4}\right) = 0$

31. Avg value $= \dfrac{1}{4-1}\int_1^4 \left(\sqrt{x} - 2\right)dx$

$= \dfrac{1}{3}\left(\dfrac{2}{3}x^{3/2} - 2x\right)\Big|_1^4$

$= \dfrac{1}{3}\left[-\dfrac{8}{3} - \left(-\dfrac{4}{3}\right)\right] = -\dfrac{4}{9}$

33. Use appropriate technology. Answer: 11.83

35. $AP = \dfrac{1}{x_1 - x_0}\int_{x_0}^{x_1} (R(x) - C(x))dx$

37. $C(x) = x^2 + 400x + 2000$

a. $\overline{C}(x) = \dfrac{C(x)}{x} = x + 400 + \dfrac{2000}{x}$

$\overline{C}(1000) = 1000 + 400 + 2 = \$1402$

b. $AV = \dfrac{1}{1000-0}\int_0^{1000} (x^2 + 400x + 2000)dx$

$= \dfrac{1}{1000}\left(\dfrac{1}{3}x^3 + 200x^2 + 2000x\right)\Big|_0^{1000}$

$= \dfrac{1000}{1000}\left(\dfrac{1{,}000{,}000}{3} + 200(1000) + 2000\right)$

$= \$535{,}333\dfrac{1}{3}$

39. a. Average $= \dfrac{1}{20-0}\int_0^{20}\left[100e^{-x^2} \cdot x + 100\right]dx$

$= \dfrac{1}{20}\left(-50e^{-x^2} + 100x\right)\Big|_0^{20}$

$= \dfrac{1}{20}[(0 + 2000) - (-50 + 0)]$

$= 102.5$

b. $\dfrac{1}{30-20}\int_{20}^{30}\left(100xe^{-x^2} + 100\right)dx$

$= \dfrac{1}{10}\left(-50e^{-x^2} + 100x\right)\Big|_{20}^{30}$

$\approx \dfrac{1}{10}(3000 - 2000) = 100$

41. Use the graphing utility and obtain 7.50%.

43. Average $= \dfrac{1}{4-0}\int_0^4 \left(30x^{18/7} - 240x^{11/7} + 480^{4/7}\right)dx$

$= \dfrac{1}{4}\left(\dfrac{7(30)}{25}\cdot x^{25/7} - \dfrac{7(240)}{18}\cdot x^{18/7} + \dfrac{7(480)}{11}\cdot x^{11/7}\right)\Big|_0^4$

$= \dfrac{1}{4}(1187.12 - 3297.55 + 2697.99) = 146.89 = 147$ mg

45. $\text{Gini coeff}_{1980} = \dfrac{\int_0^1 \left(x - 0.916x^{1.821}\right)dx}{1/2} = 2\left(\dfrac{x^2}{2} - \dfrac{0.916}{2.821}x^{2.821}\right)\Big|_0^1 = 2(0.175) = 0.350$

$\text{Gini coeff}_{1990} = \dfrac{\int_0^1 \left(x - 0.896x^{1.878}\right)dx}{\frac{1}{2}} = 2\left(\dfrac{x^2}{2} - \dfrac{0.896}{2.878}x^{2.878}\right)\Big|_0^1 = 2(0.189) = 0.377$

The difference in incomes widened.

47. Whites: $L = 0.8693x^{1.8556}$

Blacks: $L = 0.8693x^{2.0982}$

$$w: \text{Gini coeff} = \frac{\int_0^1 \left(x - 0.8693x^{1.8556}\right)dx}{\frac{1}{2}} = 2\left(\frac{x^2}{2} - \frac{0.8693x^{2.8556}}{2.8556}\right)\Bigg|_0^1 = 2(0.1956) = 0.3912$$

$$B: \text{ Gini coeff} = \frac{\int_0^1 \left(x - .8693x^{2.0982}\right)dx}{\frac{1}{2}} = 2\left(\frac{x^2}{2} - \frac{0.8693x^{3.0982}}{3.0982}\right)\Bigg|_0^1 = 2(0.2194) = 0.4388$$

The smaller coefficient gives income, more equally distributed. We have more equal distribution among whites.

Exercise 13.4

1. Total income $= \int_0^{10} 12{,}000dt$

$= 12{,}000t \Big|_0^{10} = \$120{,}000$

3. Total income $= \int_0^{12} 24{,}000e^{0.03t}dt$

$= 800{,}000e^{0.03t}\Big|_0^{12}$

$= 800{,}000(1.4333 - 1)$

$= \$346{,}664$

5. Total income $= \int_0^{10} 80e^{-0.1t}dt$

$= -800e^{-0.1t}\Big|_0^{10}$

$= -800(0.3678 - 1)$

$= \$505.70$ thousand

$=\$505{,}700$

7. Total Income $= \int_6^{12} 3000e^{0.004t}dt$

$= 750{,}000e^{0.004t}\Big|_6^{12}$

$= 750{,}000(0.0249)$

$= \$18{,}675$

9. Present Value $= \int_0^{8} 12{,}000e^{0.04t} \cdot e^{-0.08t}dt$

$= \int_0^{8} 12{,}000e^{-0.04t}dt$

$= -300{,}000e^{-0.04t}\Big|_0^{8}$

$= -300{,}000(0.72615 - 1)$

$= \$82{,}155$

11. Present Value $= \int_0^{5} 63{,}000e^{-0.07t}dt$

$= -900{,}000e^{-0.07t}\Big|_0^{5}$

$= -900{,}000(0.70469 - 1)$

$= \$265{,}781$

$\text{FV} = e^{.35}\int_0^{5} 63{,}000e^{-0.07t}dt$

$= e^{.35}(-900{,}000e^{-0.07t})\Big|_0^{5}$

$= \$377{,}161$

13. Present Value $= \int_0^{10} 97.5e^{-0.2(t+3)}e^{-0.06t}dt$

$= 97.5\int_0^{10} e^{-0.26t-0.6}dt$

$= -\frac{97.5}{0.26}e^{-0.26t-0.6}\Big|_0^{10}$

$= -\frac{97.5}{0.26}\left[e^{-3.2} - e^{-0.6}\right]$

$= 190.519$ thousand dollars

$= \$190{,}519$

By pattern in 11, $FV = e^{.6}PV$

$= \$347{,}148$

15. $\text{PV} = \int_0^7 30{,}000e^{-0.1t}$

$= -300{,}000e^{-0.1t}\Big|_0^7$

$= -300{,}000(0.4966 - 1)$

$= \$151{,}024$

Present value of gift shop is \$151,024

$\text{PV} = \int_0^7 21{,}600e^{0.08t}\left(e^{-0.1t}\right)dt$

$= \int_0^7 21{,}600e^{-0.02t}\,dt$

$= -1{,}080{,}000e^{-0.02t}\Big|_0^7$

$= -1{,}080{,}000(0.8694 - 1)$

$= \$141{,}048$

Present value of video store is \$141,048. The gift shop is the better buy.

17. $9 = 34 - x^2$ gives $x^2 = 25$ or $x = 5$.

Equilibrium point is $(5,9)$.

$CS = \int_0^5 \left(34 - x^2\right)dx - 5\cdot 9$

$= \left(34x - \frac{1}{3}x^3\right)\Big|_0^5 - 45$

$= 170 - \frac{125}{3} - 45 = \frac{250}{3} = \$83.33$

19. $p = \frac{200}{8+2} = 20$.

Equilibrium point is $(8,20)$.

$CS = \int_0^8 \frac{200}{x+2}dx - 8\cdot 20$

$= 200\ln(x+2)\Big|_0^8 - 160$

$= 200(\ln 10 - \ln 2) - 160$

$= \$161.89$

21. $x^2 + 4x + 11 = 81 - x^2$

$2x^2 + 4x - 70 = 0$

$2(x+7)(x-5) = 0$

Equilibrium point is $(5,56)$, since $p(5) = 81 - 25 = 56$.

$CS = \int_0^5 \left(81 - x^2\right)dx - 5\cdot 56$

$= \left(81x - \frac{1}{3}x^3\right)\Big|_0^5 - 280$

$= 405 - \frac{125}{3} - 280 = \$83.33$

23. $\frac{12}{x+1} = 1 + 0.2x$

$12 = 1 + 1.2x + 0.2x^2$

$2x^2 + 12x - 110 = 0$

$2(x+11)(x-5) = 0$

Equilibrium point is $(5,2)$.

$CS = \int_0^5 \frac{12}{x+1}dx - 5\cdot 2$

$= 12\ln(x+1)\Big|_0^5 - 10$

$= 12(\ln 6 - \ln 1) - 10 = \$11.50$

25. $R = px = 360x - 3x^2 - 2x^3$

$P = R - C = -2x^3 - 9x^2 + 240x - 1000$

$P'(x) = -6x^2 - 18x + 240$

$= -6(x^2 + 3x - 40)$

$= -6(x+8)(x-5)$

Maximum profit is at $x = 5$ with $p(5) = 360 - 15 - 50 = 295$

$CS = \int_0^5 \left(360 - 3x - 2x^2\right)dx - 5(295)$

$= \left(360x - \frac{3}{2}x^2 - \frac{2}{3}x^3\right)\Big|_0^5 - 1475$

$= 1800 - 37.50 - 83.33 - 1475$

$= \$204.17$

27. $422 = 4x^2 + 2x + 2$

$4x^2 + 2x - 420 = 0$

$2(2x+21)(x-10) = 0$

Equilibrium point is $(10,422)$.

$PS = 10\cdot 422 - \int_0^{10}\left(4x^2 + 2x + 2\right)dx$

$= 4220 - \left(\frac{4}{3}x^3 + x^2 + 2x\right)\Big|_0^{10}$

$= 4220 - (1333.33 + 100 + 20)$

$= \$2766.67$

29. $p(x) = 10e^{x/3}$

$p(15) = 10e^5 = 1484.13$

$PS = 15(1484.13) - \int_0^{15} 10e^{x/3}dx$

$= 22{,}261.95 - \left(30e^{x/3}\right)\Big|_0^{15}$

$= \$17{,}839.58$

31. $x^2+4x+11=81-x^2$

$2x^2+4x-70=2(x+7)(x-5)=0$

At $x=5$, we have $p=81-25=56$.

$$PS=5(56)-\int_0^5\left(x^2+4x+11\right)dx$$

$$=280-\left(\frac{1}{3}x^3+2x^2+11x\right)\Big|_0^5$$

$$=280-(41.67+50+55)=\$133.33$$

33. See problem 23 to find equilibrium point $(5,2)$.

$$PS=5\cdot 2-\int_0^5(1+0.2x)dx$$

$$=10-\left(x+0.1x^2\right)\Big|_0^5=10-(5+2.5)=\$2.50$$

35. $x^2+33x+48=144-2x^2$

$3x^2+33x-96=3(x^2+11x-32)=0$

Using the quadratic formula to solve for x and the supply function to find the equilibrium price, we have that the equilibrium point is $(2.39, 132.58)$.

$$PS=2.39(132.58)-\int_0^{2.39}\left(x^2+33x+48\right)dx$$

$$=316.87-\left(\frac{1}{3}x^3+\frac{33}{2}x^2+48x\right)\Big|_0^{2.39}$$

$$=316.87-213.52=\$103.35$$

Exercise 13.5

1. $\displaystyle\int\frac{dx}{16-x^2}=\int\frac{dx}{4^2-x^2}=\frac{1}{2(4)}\ln\left|\frac{4+x}{4-x}\right|+C$

(By formula 5)

3. $\displaystyle\int_1^4\frac{dx}{x\sqrt{9+x^2}}=\int_1^4\frac{dx}{x\sqrt{3^2+x^2}}$

$$=-\frac{1}{3}\ln\left(\frac{3+\sqrt{9+x^2}}{x}\right)\Bigg|_1^4$$

$$=-\frac{1}{3}\left[\ln 2-\ln\left(3+\sqrt{10}\right)\right]$$

$$=\frac{1}{3}\left[\ln\left(3+\sqrt{10}\right)-\ln 2\right]$$

$$=\frac{1}{3}\ln\left[\frac{3+\sqrt{10}}{2}\right]$$

(By formula 11)

5. $\displaystyle\int\ln w\,dw=w(\ln w-1)+C$

(By formula 14)

7. $\displaystyle\int_0^2\frac{q\,dq}{6q+9}=\left[\frac{q}{6}-\frac{9}{36}\ln(6q+9)\right]\Bigg|_0^2$

$$=\left(\frac{2}{6}-\frac{1}{4}\ln 21\right)-\left(0-\frac{1}{4}\ln 9\right)$$

$$=\frac{1}{3}+\frac{1}{4}(\ln 9-\ln 21)=\frac{1}{3}+\frac{1}{4}\ln\frac{3}{7}$$

(By formula 12)

9. $\displaystyle\int 3^x\,dx=3^x\log_3 e+C=\frac{3^x}{\ln 3}+C$

(By formula 3)

11. $\displaystyle\int_5^7\sqrt{x^2-25}\,dx$

$$=\frac{1}{2}\left[x\sqrt{x^2-25}-25\ln\left(x+\sqrt{x^2-25}\right)\right]\Bigg|_5^7$$

$$=\frac{1}{2}\left[7\sqrt{24}-25\ln\left(7+\sqrt{24}\right)-(0-25\ln 5)\right]$$

$$=\frac{1}{2}\left[7\sqrt{24}-25\ln\left(7+\sqrt{24}\right)+25\ln 5\right]$$

(By formula 7)

13. $\displaystyle\int w\sqrt{4w+5}\,dw=\frac{2(12w-10)(4w+5)^{3/2}}{15(16)}+C$

$$=\frac{(6w-5)(4w+5)^{3/2}}{60}+C$$

(By formula 16)

15. $\displaystyle\int x5^{x^2}\,dx=\frac{1}{2}\int 5^{x^2}(2x\,dx)$

$$=\frac{1}{2}\cdot\frac{5^{x^2}}{\ln 5}+C\text{ or }\frac{1}{2}\left(5^{x^2}\right)\log_5 e+C$$

(By formula 3)

17. $\displaystyle\int_0^3 x\sqrt{x^2+4}\,dx=\frac{1}{2}\int_0^3\left(x^2+4\right)^{1/2}(2x\,dx)$

$$=\frac{1}{2}\cdot\frac{\left(x^2+4\right)^{3/2}}{\frac{3}{2}}\Bigg|_0^3$$

$$=\frac{1}{3}\left(x^2+4\right)^{3/2}\Big|_0^3$$

$$=\frac{1}{3}\left(13^{3/2}-8\right)$$

19. $5\int\frac{dx}{x\sqrt{4-9x^2}}=-\frac{5}{2}\ln\left|\frac{2+\sqrt{4-9x^2}}{3x}\right|+C$

(By formula 9)

21. $\int\frac{dx}{\sqrt{9x^2-4}}=\frac{1}{3}\int\frac{3\,dx}{\sqrt{(3x)^2-2^2}}$

$=\frac{1}{3}\ln\left|3x+\sqrt{9x^2-4}\right|+C$

(By formula 10)

23. $\int_5^6\frac{dx}{x^2-16}=-1\int_5^6\frac{dx}{16-x^2}$

$=-1\cdot\frac{1}{8}\ln\left|\frac{4+x}{4-x}\right|\Bigg|_5^6$

$=-\frac{1}{8}[\ln 5-\ln 9]=\frac{1}{8}\ln\frac{9}{5}$

(By formula 5)

25. $\int\frac{dx}{\sqrt{(3x+1)^2+1}}=\frac{1}{3}\int\frac{3\,dx}{\sqrt{(3x+1)^2+1}}$

$=\frac{1}{3}\ln\left|3x+1+\sqrt{(3x+1)^2+1}\right|+C$

(By formula 8)

27. $\int_0^3 x\sqrt{(x^2+1)^2+9}\,dx=\frac{1}{2}\int_0^3\sqrt{(x^2+1)^2+9}\,(2x\,dx)$

$=\frac{1}{4}\left[(x^2+1)\sqrt{(x^2+1)^2+9}+9\ln\left((x^2+1)+\sqrt{(x^2+1)^2+9}\right)\right]\Bigg|_0^3$

$=\frac{1}{4}\left[10\sqrt{109}+9\ln\left(10+\sqrt{109}\right)-\sqrt{10}-9\ln\left(1+\sqrt{10}\right)\right]$

(By formula 6)

29. $\int\frac{x\,dx}{7-3x^2}=-\frac{1}{6}\int\frac{1}{7-3x^2}(-6x\,dx)=-\frac{1}{6}\ln\left|7-3x^2\right|+C$

31. $\int\frac{dx}{\sqrt{(2x)^2+7}}=\frac{1}{2}\int\frac{2\,dx}{\sqrt{(2x)^2+7}}=\frac{1}{2}\ln\left|2x+\sqrt{4x^2+7}\right|+C$ (By formula 8)

33. $\int\frac{e^{\sqrt{x-1}}}{\sqrt{x-1}}dx=2\int e^u\,du=2e^u+C$ where $u=(x-1)^{1/2}$ and $du=\frac{dx}{2\sqrt{x-1}}$.

So $\int_2^3\frac{e^{\sqrt{x-1}}}{\sqrt{x-1}}dx=2e^{\sqrt{x-1}}\Big|_2^3=2\left(e^{\sqrt{2}}-e\right)\approx 2.7899$.

35. $\int_0^1\frac{x^2\cdot x\,dx}{\left[(2x)^2+5\right]^2}=\frac{1}{8}\int_{x=0}^{x=1}\frac{(1/4)u\,du}{(u+5)^2}$

where $u=4x^2$ and $du=8x\,dx$.

$=\frac{1}{32}\left[\ln|u+5|+\frac{5}{u+5}\right]\Bigg|_{x=0}^{x=1}=\frac{1}{32}\left[\ln\left(4x^2+5\right)+\frac{5}{4x^2+5}\right]\Bigg|_0^1=\frac{1}{32}\left[\ln 9+\frac{5}{9}-\ln 5-1\right]$

$=\frac{1}{32}\left[\ln\frac{9}{5}-\frac{4}{9}\right]\approx 0.004479$

(By formula 15)

37. At $x=20$, we have $p=40+200\ln 21=648.90$

$PS=20(648.90)-\int_0^{20}\left[40+200\ln(x+1)\right]dx=12{,}978-\left[40x+200(x+1)(\ln(x+1)-1)\right]\Big|_0^{20}$

$=12{,}978-\left[800+4200(\ln 21-1)-(0+200(-1))\right]=12{,}978-800-12{,}787+4200-200=\$3391$

39. a. $C(x) = \int \sqrt{x^2+9}dx = \frac{1}{2}\left(x\sqrt{x^2+9} + 9\ln\left(x+\sqrt{x^2+9}\right)\right) + K$

(By formula 6)

With the interpretation that $C(0) = 300$, we get a K value of 295.06.

b. $C(4) = 19.888 + 295.06 = 314.95$

41. $TI = \int_0^{120} \left(10\ln(t+1) - 0.1t\right)dt = \left[10(t+1)(\ln(t+1)-1) - \frac{.1t^2}{2}\right]\Big|_0^{120}$

$= 10 \cdot 121(\ln 121 - 1) - \frac{1440}{2} - [10(-1) - 0] = \$3882.9$ thousands

(Use formula 14)

Exercise 13.6

1. $\int xe^{2x}dx = uv - \int v\,du$

$u = x \quad dv = e^{2x}dx$

$du = dx \quad v = \frac{1}{2}e^{2x}$

$\int xe^{2x}dx = \frac{1}{2}xe^{2x} - \frac{1}{2}\int e^{2x}dx$

$= \frac{1}{2}xe^{2x} - \frac{1}{4}e^{2x} + C$

3. $\int x^2 \ln x\,dx = uv - \int v\,du$

$u = \ln x \quad dv = x^2dx$

$du = \frac{1}{x}dx \quad v = \frac{1}{3}x^3$

$\int x^2 \ln x\,dx = \frac{1}{3}x^3\ln x - \frac{1}{3}\int x^3 \cdot \frac{1}{x}dx$

$= \frac{1}{3}x^3\ln x - \frac{1}{9}x^3 + C$

5. $\int_4^6 q\sqrt{q-4}dq = \left[uv - \int v\,du\right]\Big|_4^6$

$u = q \quad dv = (q-4)^{1/2}dq$

$du = dq \quad v = \frac{2}{3}(q-4)^{3/2}$

$\int_4^6 q\sqrt{q-4}dq$

$= \left[\frac{2q}{3}(q-4)^{3/2} - \frac{2}{3}\int(q-4)^{3/2}dq\right]\Big|_4^6$

$= \left[\frac{2q}{3}(q-4)^{3/2} - \frac{4}{15}(q-4)^{5/2}\right]\Big|_4^6$

$= 4\left(2^{3/2}\right) - \frac{4}{15}\left(2^{5/2}\right) - 0$

$= 4\left(2^{3/2}\right) - \frac{8}{15}\left(2^{3/2}\right)$

$= \frac{52}{15}\left(2^{3/2}\right) = \frac{104}{15}\sqrt{2}$

7. $\int \frac{\ln x}{x^2}dx = uv - \int v\,du$

$u = \ln x \quad dv = x^{-2}dx$

$du = \frac{dx}{x} \quad v = -\frac{1}{x}$

$\int \frac{\ln x}{x^2}dx = -\frac{1}{x}\ln x + \int \frac{dx}{x^2} = -\frac{1}{x}\ln x - \frac{1}{x} + C$

9. $\int_1^e \ln x\,dx = \left[uv - \int v\,du\right]\Big|_1^e$

$u = \ln x \quad dv = dx$

$du = \dfrac{dx}{x} \qquad v = x$

$\int_1^e \ln x dx = \left[x\ln x - \int dx\right]\Big|_1^e$

$= \left[x\ln x - x\right]\Big|_1^e$

$= (e\ln e - e) - (0 - 1)$

$= e - e + 1 = 1$

11. $\int x\ln(2x-3)dx = uv - \int v\,du$

$u = \ln(2x-3) \quad dv = xdx$

$du = \dfrac{2\,dx}{2x-3} \qquad v = \dfrac{x^2}{2}$

$\int x\ln(2x-3)dx$

$= \dfrac{1}{2}x^2\ln(2x-3) - \int \dfrac{x^2}{2x-3}dx$

$= \dfrac{1}{2}x^2\ln(2x-3) - \int\left(\dfrac{1}{2}x + \dfrac{3}{4} + \dfrac{9/4}{2x-3}\right)dx$

$= \dfrac{1}{2}x^2\ln(2x-3) - \dfrac{1}{4}x^2 - \dfrac{3}{4}x - \dfrac{9}{8}\ln(2x-3) + C$

13. $\dfrac{1}{2}\int q^2\left(\sqrt{q^2-3}\cdot 2q\,dq\right) = uv - \int v\,du$

$u = q^2 \qquad dv = \sqrt{q^2-3}\cdot 2q\,dq$

$du = 2qdq \qquad v = \dfrac{2}{3}(q^2-3)^{3/2}$

$\dfrac{1}{2}\int q^2\left(\sqrt{q^2-3}\cdot 2q\,dq\right)$

$= \dfrac{1}{2}\left[\dfrac{2}{3}q^2\left(q^2-3\right)^{3/2} - \dfrac{2}{3}\int\left(q^2-3\right)^{3/2}(2q\,dq)\right]$

$= \dfrac{1}{3}q^2\left(q^2-3\right)^{3/2} - \dfrac{1}{3}\cdot\dfrac{\left(q^2-3\right)^{5/2}}{5/2} + C$

$= \left(q^2-3\right)^{3/2}\left[\dfrac{1}{3}q^2 - \dfrac{2}{15}\left(q^2-3\right)\right] + C$

$= \left(q^2-3\right)^{3/2}\cdot\dfrac{3q^2+6}{15} + C$

$= \dfrac{\left(q^2-3\right)^{3/2}\left(q^2+2\right)}{5} + C$

15. $\dfrac{1}{2}\int_0^4 x^2\left(x^2+9\right)^{1/2}2x\,dx = \dfrac{1}{2}\left[uv - \int v\,du\right]\Big|_0^4$

$u = x^2 \qquad dv = \left(x^2+9\right)^{1/2}(2x\,dx)$

$du = 2xdx \qquad v = \dfrac{2}{3}\left(x^2+9\right)^{3/2}$

$\dfrac{1}{2}\int_0^4 x^2\left(x^2+9\right)^{1/2}2xdx$

$= \dfrac{1}{2}\left[\dfrac{2}{3}x^2\left(x^2+9\right)^{3/2} - \dfrac{2}{3}\int\left(x^2+9\right)^{3/2}(2x\,dx)\right]\Big|_0^4$

$= \left[\dfrac{1}{3}x^2\left(x^2+9\right)^{3/2} - \dfrac{1}{3}\cdot\dfrac{\left(x^2+9\right)^{5/2}}{5/2}\right]\Big|_0^4$

$= \left[\dfrac{16}{3}(125) - \dfrac{2}{15}(3125)\right] - \left[0 - \dfrac{2}{15}(243)\right]$

$= 666.67 - 416.67 + 32.4 = 282.4$

17. $\int x^2e^{-x}dx$

$u = x^2 \qquad dv = e^{-x}dx$

$du = 2xdx \qquad v = -e^{-x}$

$\int x^2e^{-x}dx = -e^{-x}x^2 - \int\left(-e^{-x}\right)2x\,dx$

$= -x^2e^{-x} + 2\int xe^{-x}dx$

$u = x \qquad dv = e^{-x}dx$

$du = dx \qquad v = -e^{-x}$

$\int x^2e^{-x}dx$

$= -x^2e^{-x} + 2\left(-xe^{-x} + \int e^{-x}dx\right)$

$= -x^2e^{-x} - 2xe^{-x} - 2e^{-x} + C$

$= -e^{-x}\left(x^2+2x+2\right) + C$

19. $\int_0^2 x^3e^{x^2}dx = \int_0^2 x^2e^{x^2}x\,dx$

$u = x^2 \qquad dv = e^{x^2}x\,dx$

$du = 2xdx \qquad v = \dfrac{1}{2}e^{x^2}$

$\int_0^2 x^3e^{x^2}dx = \dfrac{1}{2}x^2e^{x^2}\Big|_0^2 - \dfrac{1}{2}\int_0^2 e^{x^2}2x\,dx$

$= \left(\dfrac{1}{2}x^2e^{x^2} - \dfrac{1}{2}e^{x^2}\right)\Big|_0^2$

$= \dfrac{1}{2}(4)e^4 - \dfrac{1}{2}e^4 - \left(0 - \dfrac{1}{2}\right)$

$= 2e^4 - \dfrac{1}{2}e^4 + \dfrac{1}{2} = \dfrac{3e^4+1}{2}$

21. $\int x^3(\ln x)^2\,dx$

$u=(\ln x)^2 \qquad dv=x^3dx$

$du=\frac{2\ln x}{x}dx \qquad v=\frac{1}{4}x^4$

$\int x^3(\ln x)^2\,dx=\frac{1}{4}x^4(\ln x)^2-\frac{1}{2}\int x^3\ln x\,dx$

$u=\ln x \quad dv=x^3dx$

$du=\frac{dx}{x} \qquad v=\frac{1}{4}x^4$

$\int x^3(\ln x)^2\,dx$

$=\frac{1}{4}x^4(\ln x)^2-\frac{1}{2}\left[\frac{1}{4}x^4\ln x-\int\frac{1}{4}x^3dx\right]$

$=\frac{1}{4}x^4(\ln x)^2-\frac{1}{8}x^4\ln x+\frac{1}{32}x^4+C$

23. $\int e^{2x}\sqrt{e^x+1}dx=\int\left(e^x+1\right)^{1/2}e^xe^xdx$

$u=e^x \qquad dv=\left(e^x+1\right)^{1/2}e^xdx$

$du=e^xdx \qquad v=\frac{2}{3}\left(e^x+1\right)^{3/2}$

$\int e^{2x}\sqrt{e^x+1}dx$

$=\frac{2}{3}e^x\left(e^x+1\right)^{3/2}-\frac{2}{3}\int\left(e^x+1\right)^{3/2}e^xdx$

$=\frac{2}{3}e^x\left(e^x+1\right)^{3/2}-\frac{2}{3}\cdot\frac{2}{5}\left(e^x+1\right)^{5/2}+C$

$=\left(e^x+1\right)^{3/2}\left[\frac{2e^x}{3}-\frac{4}{15}\left(e^x+1\right)\right]+C$

$=(e^x+1)^{3/2}\left[\frac{6e^x}{15}-\frac{4}{15}\right]+C$

$=\frac{2}{15}\left(e^x+1\right)^{3/2}\left(3e^x-2\right)+C$

25. II. $\int e^{x^2}x\,dx=\frac{1}{2}\int e^u du$ where $u=x^2$ and $du=2xdx$.

$\int e^{x^2}x\,dx=\frac{1}{2}e^{x^2}+C$

27. IV. $\int\sqrt{e^x+1}\,e^xdx=\int\left(e^x+1\right)^{1/2}e^xdx=\int u^{1/2}du$ where $u=e^x+1$ and $du=e^xdx$.

$\int\sqrt{e^x+1}\,e^xdx=\frac{2}{3}\left(e^x+1\right)^{3/2}+C$

29. I. $\int_0^4\frac{t}{e^t}dt=\int_0^4 e^{-t}t\,dt$

$u=t \qquad dv=e^{-t}dt$

$du=dt \qquad v=-e^{-t}$

$\int_0^4\frac{t}{e^t}dt=-te^{-t}\Big|_0^4+\int_0^4 e^{-t}dt$

$=-te^{-t}\Big|_0^4+\left(-e^{-t}\right)\Big|_0^4$

$=-4e^{-4}+\left(-e^{-4}+e^0\right)$

$=-5e^{-4}+1$

31. At $x=30$, we have $p=30+100\ln 61=441.09$.

$PS=30(441.09)-\int_0^{30}\left[30+100\ln(2x+1)\right]dx=13{,}232.62-\left(30x+\frac{100}{2}(2x+1)\left[\ln(2x+1)-1\right]\right)\Big|_0^{30}$

$=13{,}232.62-\left\{900+50(61)\left[\ln 61-1\right]-\left[0+50\left(\ln 1-1\right)\right]\right\}$

$=13{,}232.62-900-3050\left(3.11087\right)-50=\$2794.46$

(by formula 14)

33. Present Value

$$= \int_0^5 (10{,}000 - 500t)e^{-0.1t}dt = \left[-100{,}000e^{-0.1t} - \int 500te^{-0.1t}dt\right]\Big|_0^5$$

$u = t \quad dv = e^{-0.1t}dt$

$du = dt \quad v = -10e^{-0.1t}$

(So, $\int te^{-0.1t}dt = -10te^{-0.1t} - 100e^{-0.1t}$.)

$$= \left[-100{,}000e^{-0.1t} + 5000te^{-0.1t} + 50{,}000e^{-0.1t}\right]\Big|_0^5 = \left[5000te^{-0.1t} - 50{,}000e^{-0.1t}\right]\Big|_0^5$$

$$= \left[e^{-0.1t}(5000t - 50{,}000)\right]\Big|_0^5 = 0.6065[-25{,}000] - 1[0 - 50{,}000] = \$34{,}837$$

35. Gini coeff $= \dfrac{\int_0^1 \left(x - xe^{x-1}\right)dx}{1/2}$

For xe^{x-1}, let $u = x \quad dv = e^{x-1}dx$

$du = dx \quad v = e^{x-1}$

$\int xe^{x-1}dx = xe^{x-1} - \int e^{x-1}dx = xe^{x-1} - e^{x-1}$

Gini coeff $= 2\left(\dfrac{x^2}{2} - xe^{x-1} + e^{x-1}\right)\Big|_0^1 = 2\left(\dfrac{1}{2} - 1 + 1 - \left(e^{-1}\right)\right) = 0.264$

Exercise 13.7

1. $\int_1^\infty \dfrac{dx}{x^6} = \lim\limits_{a\to\infty} \int_1^a x^{-6}dx$

$= \dfrac{1}{5}\lim\limits_{a\to\infty}\left(-x^{-5}\right)\Big|_1^a$

$= \dfrac{1}{5}\lim\limits_{a\to\infty}\left(-\dfrac{1}{a^5} - (-1)\right) = \dfrac{1}{5}$

3. $\int_1^\infty \dfrac{dt}{t^{3/2}} = \lim\limits_{a\to\infty}\int_1^a t^{-3/2}dt$

$= \lim\limits_{a\to\infty} -2t^{-1/2}\Big|_1^a$

$= \lim\limits_{a\to\infty}\left(-2\dfrac{1}{a^{1/2}} - (-2(1))\right) = 2$

5. $\int_1^\infty e^{-x}dx = (-1)\lim\limits_{b\to\infty}\int_1^b e^{-x}(-1)dx$

$= (-1)\lim\limits_{b\to\infty} e^{-x}\Big|_1^b$

$= (-1)\lim\limits_{b\to\infty}\left(\dfrac{1}{e^b} - \dfrac{1}{e}\right)$

$= (-1)\left(-\dfrac{1}{e}\right) = \dfrac{1}{e}$

7. $\int_1^\infty \dfrac{dt}{t^{1/3}} = \lim\limits_{a\to\infty}\int_1^a t^{-1/3}dt$

$= \lim\limits_{a\to\infty}\dfrac{3}{2}t^{2/3}\Big|_1^a$

$= \lim\limits_{a\to\infty}\left(\dfrac{3}{2}a^{2/3} - \dfrac{3}{2}\right)$

Thus, the integral diverges.

9. $\int_0^\infty e^{3x}dx = \lim\limits_{a\to\infty}\dfrac{1}{3}\int_0^a e^{3x}\cdot 3dx$

$= \lim\limits_{a\to\infty}\dfrac{1}{3}e^{3x}\Big|_0^a$

$= \lim\limits_{a\to\infty}\dfrac{1}{3}\left(e^{3a} - 1\right)$

Thus, the integral diverges.

11. $\int_{-\infty}^{-1} 10x^{-2}dx = \lim\limits_{a\to\infty}\int_{-a}^{-1}10x^{-2}dx$

$= \lim\limits_{a\to\infty}\dfrac{10x^{-1}}{-1}\Big|_{-a}^{-1}$

$= \lim\limits_{a\to\infty}\left(\dfrac{-10}{-1} - \dfrac{-10}{-a}\right) = 10$

13. $\int_{-\infty}^{0} x^2 e^{-x^3}\,dx = -\frac{1}{3}\lim_{a\to\infty}\int_{-a}^{0} e^{-x^3}\left(-3x^2 dx\right)$

$= -\frac{1}{3}\lim_{a\to\infty} e^{-x^3}\Big|_{-a}^{0}$

$= -\frac{1}{3}\lim_{a\to\infty}\left(1-e^{a^3}\right) = \infty$

Thus, the integral diverges.

$[-(-a)^3 = -(-a^3) = a^3]$

15. $\int_{-\infty}^{-1}\frac{6}{x}dx = \lim_{a\to\infty}\int_{-a}^{-1}\frac{6}{x}dx$

$= \lim_{a\to\infty} 6\ln|x|\Big|_{-a}^{-1}$

$= \lim_{a\to\infty}\left(6\ln 1 - 6\ln|a|\right) = -\infty$

Thus, the integral diverges.

17. $\int_{-\infty}^{\infty}\frac{2x}{x^2+1}dx = \lim_{a\to\infty}\int_{-a}^{0}\frac{2x}{x^2+1}dx + \lim_{b\to\infty}\int_{0}^{b}\frac{2x}{x^2+1}dx$

$= \lim_{-a\to\infty}\ln\left(x^2+1\right)\Big|_{-a}^{0} + \lim_{b\to\infty}\ln\left(x^2+1\right)\Big|_{0}^{b} = \lim_{-a\to\infty}\left[\ln 1 - \ln\left(a^2+1\right)\right] + \lim_{b\to\infty}\left[\ln\left(b^2+1\right) - \ln 1\right]$

Since each of these integrals diverges, the given integral diverges. (Ignore $\infty - \infty$).

19. $\int_{-\infty}^{\infty} x^3 e^{-x^4}dx = -\frac{1}{4}\left[\lim_{a\to\infty}\int_{-a}^{0} e^{-x^4}\left(-4x^3dx\right) + \lim_{b\to\infty}\int_{0}^{b} e^{-x^4}\left(-4x^3dx\right)\right]$

$= -\frac{1}{4}\left[\lim_{a\to\infty} e^{-x^4}\Big|_{-a}^{0} + \lim_{b\to\infty} e^{-x^4}\Big|_{0}^{b}\right] = -\frac{1}{4}\left[\lim_{a\to\infty}\left(e^0 - 1/e^{a^4}\right) + \lim_{b\to\infty}\left(1/e^{b^4} - e^0\right)\right] = -\frac{1}{4}(1-0+0-1) = 0$

21. $\int_{0}^{\infty}\frac{c}{e^{0.5t}}dt = 1$

$\lim_{a\to\infty}\int_{0}^{a} ce^{-0.5t}dt = c\lim_{a\to\infty} -2\int_{0}^{a} e^{-0.5t}(-0.5)dt = c\lim_{a\to\infty}(-2)e^{-0.5t}\Big|_{0}^{a} = -2c\lim_{a\to\infty}\left(e^{-0.5a}-1\right) = 2c$

So, $2c = 1$ gives $c = \frac{1}{2}$.

23. $\int_{1}^{\infty}\frac{x}{e^{x^2}}dx = \int_{1}^{\infty} e^{-x^2}x\,dx$

$= \lim_{a\to\infty} -\frac{1}{2}\int_{1}^{a} e^{-x^2}(-2)x\,dx$

$= \lim_{a\to\infty} -\frac{1}{2}\left(e^{-x^2}\right)\Big|_{1}^{a}$

$= \lim_{a\to\infty} -\frac{1}{2}\left[e^{-a^2} - e^{-1}\right] = \frac{1}{2e}$

25. $\int_{1}^{\infty}\frac{1}{\sqrt[3]{x^5}}dx = \lim_{a\to\infty}\int_{1}^{a} x^{-5/3}dx$

$= \lim_{a\to\infty} -\frac{3}{2}x^{-2/3}\Big|_{1}^{a}$

$= \lim_{a\to\infty} -\frac{3}{2}\left[a^{-2/3} - 1\right] = \frac{3}{2}$

27. $\int_{-\infty}^{\infty} f(x)dx = \int_{-\infty}^{10} f(x)dx + \int_{10}^{\infty} f(x)dx$

$= \int_{-\infty}^{10} 0\cdot dx + \int_{10}^{\infty} 200x^{-3}dx$

$= 0 + \lim_{b\to\infty}\frac{200}{-2}x^{-2}\Big|_{10}^{b}$

$= \lim_{b\to\infty}\left(-\frac{100}{b^2} - \frac{-100}{100}\right) = 1$

29. $\int_{-\infty}^{\infty} f(x)dx = \int_{-\infty}^{1} f(x)dx + \int_{1}^{\infty} f(x)dx$

$= \int_{-\infty}^{1} 0\cdot dx + \int_{1}^{\infty} cx^{-2}dx$

$= 0 + \lim_{b\to\infty} c\cdot\frac{x^{-1}}{-1}\Big|_{1}^{b}$

$= (-c)\lim_{b\to\infty}\left(\frac{1}{b} - \frac{1}{1}\right) = c$

Thus, $f(x)$ is a probability density function if $c = 1$.

31. $\int_{-\infty}^{\infty} f(x)dx = \int_{-\infty}^{0} f(x)dx + \int_{0}^{\infty} f(x)dx$

$= \int_{-\infty}^{0} 0 \cdot dx + \int_{0}^{\infty} ce^{-x/4}dx$

Therefore, $\int_{-\infty}^{\infty} f(x)dx = 0 + \lim_{b\to\infty} \frac{c}{-.25} e^{-x/4}\Big|_0^b$

$= -4c \lim_{b\to\infty}\left(\frac{1}{e^{b/4}} - e^0\right)$

Thus, $-4c(0-1) = 1$ or $c = \frac{1}{4}$.

33. Mean $= \int_{-\infty}^{\infty} x\,f(x)dx$

$= \int_{-\infty}^{10} x\,f(x)dx + \int_{10}^{\infty} x\,f(x)dx$

$= 0 + \int_{10}^{\infty} x\left(200x^{-3}\right)dx$

$= \lim_{b\to\infty} 200 \cdot \frac{x^{-1}}{-1}\Big|_{10}^{b}$

$= -200 \lim_{b\to\infty}\left(\frac{1}{b} - \frac{1}{10}\right) = 20$

35. $A = 8\int_0^{\infty} xe^{-3x}(3\,dx)$ (Alternate form used)

$u = x \quad dv = e^{-3x} \cdot 3\,dx$

$du = dx \quad v = -e^{-3x}$

$A = 8\left[-xe^{-3x}\Big|_0^{\infty} + \int_0^{\infty} e^{-3x}dx\right]$

$= 8\left[\lim_{b\to\infty}\left(-\frac{b}{e^{3b}} + 0\right) + \lim_{b\to\infty}\left(-\frac{1}{3e^{3b}} + \frac{1}{3}\right)\right]$

$= \frac{8}{3}$

If necessary ask your instructor why $\lim_{b\to\infty} \frac{b}{e^{3b}} = 0$.

37. $\int_0^{\infty} Ae^{-rt}dt = \lim_{b\to\infty}\int_0^b Ae^{-rt}dt$

$= \lim_{b\to\infty} -\frac{A}{r}e^{-rt}\Big|_0^b = -\frac{A}{r}\lim_{b\to\infty}\left(\frac{1}{e^{br}} - e^0\right)$

$= -\frac{A}{r}(0-1) = \frac{A}{r}$

39. $CV = \int_0^{\infty} 120e^{0.04t}e^{-0.09t}dt$

$= \lim_{b\to\infty}\int_0^b 120e^{-0.05t}dt$

$= \lim_{b\to\infty}\left(-2400e^{-0.05t}\right)\Big|_0^b$

$= \lim_{b\to\infty}\left(-\frac{2400}{e^{0.05b}} - \left(-2400e^0\right)\right)$

= \$2400 thousands

=\$2,400,000

41. $\int_0^{\infty} 56{,}000e^{0.02t}e^{-0.1t}dt$

$= \lim_{a\to\infty}\int_0^a 56{,}000e^{-0.08t}dt$

$= \lim_{a\to\infty} 56{,}000 \cdot \frac{1}{-0.08}\int_0^a e^{-0.08t}(-0.08)dt$

$= \lim_{a\to\infty}\left(-700{,}000e^{-0.08t}\right)\Big|_0^a$

$= -700{,}000 \lim_{a\to\infty}\left[e^{-0.08a} - 1\right]$

$= \$700{,}000$

43. a. $500\int_0^b te^{-0.03(b-t)}dt$

$u = t \quad dv = e^{-0.03(b-t)}dt$

$du = dt \quad v = \frac{1}{0.03}e^{-0.03(b-t)}$

$= \left[\frac{500te^{-0.03(b-t)}}{0.03} - \frac{500}{0.03}\int e^{-0.03(b-t)}dt\right]\Bigg|_0^b = \left[\frac{500te^{-0.03(b-t)}}{0.03} - \frac{500e^{-0.03(b-t)}}{(0.03)(0.03)}\right]\Bigg|_0^b$

$= \frac{500e^{-0.03(b-t)}}{0.03}\left(t - \frac{1}{0.03}\right)\Bigg|_0^b = \frac{500e^0}{0.03}\left(\frac{0.03b-1}{0.03}\right) - \frac{500e^{-0.03b}}{0.03}\left(0 - \frac{1}{0.03}\right) = \frac{500}{0.0009}\left(0.03b - 1 + e^{-0.03b}\right)$

b. $\lim_{b\to\infty}\int_0^b f(t)dt = \lim_{b\to\infty}\frac{500}{0.0009}\left(0.03b - 1 + \frac{1}{e^{0.03b}}\right) = \infty$

Waste is produced more rapidly than existing waste decays.

45. $P(>24)=\int_{24}^{\infty}0.08e^{-0.08t}dt=\lim\limits_{b\to\infty}\int_{24}^{b}e^{-0.08t}(-0.08)(-1)dt$

$$=(-1)\lim_{b\to\infty}e^{-0.08t}\Big|_{24}^{b}=(-1)\lim_{b\to\infty}\left(\frac{1}{e^{0.08b}}-\frac{1}{e^{1.92}}\right)=(-1)\left(0-\frac{1}{6.821}\right)=0.1466$$

Exercise 13.8

1. $[0,2]\quad n=4\quad h=\dfrac{2-0}{4}=\dfrac{1}{2}$

$x_0=0,\ x_1=\dfrac{1}{2},\ x_2=1,\ x_3=\dfrac{3}{2},\ x_4=2$

3. $[1,4]\quad n=6\quad h=\dfrac{4-1}{6}=\dfrac{1}{2}$

$x_0=1,\ x_1=\dfrac{3}{2},\ x_2=2,\ x_3=\dfrac{5}{2},\ x_4=3,\ x_5=\dfrac{7}{2},\ x_6=4$

5. $[-1,4]\quad n=5\quad h=\dfrac{4-(-1)}{5}=1$

$x_0=-1,\ x_1=0,\ x_2=1,\ x_3=2,\ x_4=3,\ x_5=4$

7. $f(x)=x^2\quad [0,3]\quad n=6\quad h=\dfrac{1}{2}$

a. $\int_0^3 x^2dx\approx\dfrac{h}{2}\left[f(0)+2f\left(\tfrac{1}{2}\right)+2f(1)+2f\left(\tfrac{3}{2}\right)+2f(2)+2f\left(\tfrac{5}{2}\right)+f(3)\right]$

$$=\frac{1}{4}\left[0+\frac{1}{2}+2+\frac{9}{2}+8+\frac{25}{2}+9\right]=\frac{1}{4}\left[\frac{73}{2}\right]=9.13$$

b. $\int_0^3 x^2dx\approx\dfrac{h}{3}\left[f(0)+4f\left(\tfrac{1}{2}\right)+2f(1)+4f\left(\tfrac{3}{2}\right)+2f(2)+4f\left(\tfrac{5}{2}\right)+f(3)\right]$

$$=\frac{1}{6}[0+1+2+9+8+25+9]=9$$

c. $\int_0^3 x^2dx=\dfrac{x^3}{3}\Big|_0^3=\dfrac{3^3}{3}-\dfrac{0^3}{3}=9$

d. Simpson's Rule is more accurate.

9. $f(x)=\dfrac{1}{x^2}\quad [1,2]\quad n=4\quad h=\dfrac{1}{4}$

a. $\int_1^2\dfrac{1}{x^2}dx\approx\dfrac{h}{2}\left[f(1)+2f\left(\tfrac{5}{4}\right)+2f\left(\tfrac{3}{2}\right)+2f\left(\tfrac{7}{4}\right)+f(2)\right]=\dfrac{1}{8}\left[1+\dfrac{32}{25}+\dfrac{8}{9}+\dfrac{32}{49}+\dfrac{1}{4}\right]\approx\dfrac{1}{8}[4.072]=0.51$

b. $\int_1^2\dfrac{1}{x^2}dx\approx\dfrac{h}{3}\left[f(1)+4f\left(\tfrac{5}{4}\right)+2f\left(\tfrac{3}{2}\right)+4f\left(\tfrac{7}{4}\right)+f(2)\right]=\dfrac{1}{12}\left[1+\dfrac{64}{25}+\dfrac{8}{9}+\dfrac{64}{49}+\dfrac{1}{4}\right]\approx\dfrac{1}{12}[6.01]=0.50$

c. $\int_1^2 x^{-2}dx=-x^{-1}\Big|_1^2=-\left(\tfrac{1}{2}-1\right)=\dfrac{1}{2}$

d. Simpson's Rule is more accurate.

11. $f(x)=x^{1/2}\quad [0,4]\quad n=8\quad h=\dfrac{1}{2}$

a. $\displaystyle\int_0^4 x^{1/2}dx \approx \frac{h}{2}\left[f(0)+2f\left(\tfrac{1}{2}\right)+2f(1)+2f\left(\tfrac{3}{2}\right)+2f(2)+2f\left(\tfrac{5}{2}\right)+2f(3)+2f\left(\tfrac{7}{2}\right)+f(4)\right]$

$\displaystyle =\frac{1}{4}[0+1.4142+2+2.4495+2.8284+3.1623+3.4641+3.7417+2]\approx\frac{1}{4}[21.0602]=5.27$

b. $\displaystyle\int_0^4 x^{1/2}dx \approx \frac{h}{3}\left[f(0)+4f\left(\tfrac{1}{2}\right)+2f(1)+4f\left(\tfrac{3}{2}\right)+2f(2)+4f\left(\tfrac{5}{2}\right)+2f(3)+4f\left(\tfrac{7}{2}\right)+f(4)\right]$

$\displaystyle =\frac{1}{6}[0+2.8284+2+4.8990+2.8284+6.3246+3.4641+7.4833+2]\approx\frac{1}{6}[31.8278]=5.30$

c. $\displaystyle\int_0^4 x^{1/2}dx=\frac{2}{3}x^{3/2}\Big|_0^4=\frac{2}{3}\left(4^{3/2}-0\right)=5.33$

d. Simpson's Rule is more accurate.

13. $f(x)=\sqrt{x^3+1}\quad [0,2]\quad n=4\quad h=\dfrac{1}{2}$

a. $\displaystyle\int_0^2\sqrt{x^3+1}\,dx\approx\frac{h}{2}\left[f(0)+2f\left(\tfrac{1}{2}\right)+2f(1)+2f\left(\tfrac{3}{2}\right)+f(2)\right]$

$\displaystyle =\frac{1}{4}[1+2.121+2.828+4.183+3]=\frac{1}{4}[13.132]=3.283$

b. $\displaystyle\int_0^2\sqrt{x^3+1}\,dx\approx\frac{h}{3}\left[f(0)+4f\left(\tfrac{1}{2}\right)+2f(1)+4f\left(\tfrac{3}{2}\right)+f(2)\right]$

$\displaystyle =\frac{1}{6}[1+4.243+2.828+8.367+3]=\frac{1}{6}[19.438]=3.240$

15. $f(x)=e^{-x^2}\quad [0,1]\quad n=4\quad h=\dfrac{1}{4}$

a. $\displaystyle\int_0^1 e^{-x^2}dx\approx\frac{h}{2}\left[f(0)+2f\left(\tfrac{1}{4}\right)+2f\left(\tfrac{1}{2}\right)+2f\left(\tfrac{3}{4}\right)+f(1)\right]$

$\displaystyle =\frac{h}{2}[1+1.879+1.558+1.140+0.368]=\frac{1}{8}[5.945]=0.743$

b. $\displaystyle\int_0^1 e^{-x^2}dx\approx\frac{h}{3}\left[f(0)+4f\left(\tfrac{1}{4}\right)+2f\left(\tfrac{1}{2}\right)+4f\left(\tfrac{3}{4}\right)+f(1)\right]$

$\displaystyle =\frac{h}{3}[1+3.758+1.558+2.279+0.368]=\frac{1}{12}[8.963]=0.747$

17. $f(x)=\ln\left(x^2-x+1\right)\quad [1,5]\quad n=4\quad h=1$

a. $\displaystyle\int_1^5\ln\left(x^2-x+1\right)dx\approx\frac{h}{2}[f(1)+2f(2)+2f(3)+2f(4)+f(5)]$

$\displaystyle =\frac{h}{2}[0+2.197+3.892+5.130+3.045]=\frac{1}{2}[14.264]=7.132$

b. $\displaystyle\int_1^5\ln\left(x^2-x+1\right)dx\approx\frac{h}{3}[f(1)+4f(2)+2f(3)+4f(4)+f(5)]$

$\displaystyle =\frac{h}{3}[0+4.394+3.892+10.260+3.045]=\frac{1}{3}[21.591]=7.197$

19. $f(x)=f(x)\quad [1,4]\quad n=5\quad h=\dfrac{3}{5}$

$$\int_1^4 f(x)dx \approx \frac{h}{2}[f(1)+2f(1.6)+2f(2.2)+2f(2.8)+2f(3.4)+f(4)]$$
$$=\frac{3}{10}[1+4.4+3.6+5.8+9.2+2.1]=\frac{3}{10}[26.1]=7.8$$

21. $f(x)=f(x)\quad [1.2,3.6]\quad n=6\quad h=0.4$

$$\int_{1.2}^{3.6} f(x)dx \approx \frac{h}{3}[f(1.2)+4f(1.6)+2f(2)+4f(2.4)+2f(2.8)+4f(3.2)+f(3.6)]$$
$$=\frac{2}{15}[6.1+19.2+6.2+8.0+5.6+22.4+9.7]=\frac{2}{15}[77.2]=10.3$$

23. $f(t)=100\dfrac{e^{0.1t}}{t+1}\quad [0,2]\quad n=4\quad h=\dfrac{1}{2}$

$$\int_0^2 f(t)dt \approx \frac{h}{3}\left[f(0)+4f\left(\tfrac{1}{2}\right)+2f(1)+4f\left(\tfrac{3}{2}\right)+f(2)\right]$$
$$=\frac{1}{6}[100+280.34+110.52+185.89+40.71]=\frac{1}{6}[717.46]=119.58$$

25. $C(x)=\left(x^2+1\right)^{3/2}+1000\quad [30,33]\quad n=3\quad h=1$

$$\frac{1}{33-30}\int_{30}^{33} C(x)dx \approx \frac{h}{6}[f(30)+2f(31)+2f(32)+f(33)]$$
$$=\frac{1}{6}[28{,}045+61{,}675+67{,}632+36{,}987]=\frac{1}{6}[194{,}339]=\$32{,}390$$

Note: This is the average total cost for that number of units.

27. $PS = p_1x_1$ – area under supply curve

$$\text{Area} \approx \frac{h}{3}[f(0)+4f(10)+2f(20)+4f(30)+2f(40)+4f(50)+f(60)]$$
$$=\frac{10}{3}[120+1040+760+1800+1080+2520+680]=\frac{10}{3}[8000]=26{,}666.67$$

$PS = 680(60)-26{,}666.67=\$14{,}133.33$

29. $N \text{ of units} \approx \dfrac{h}{2}[f(0)+2f(1)+2f(2)+2f(3)+2f(4)+f(5)]$

$$=\frac{1}{2}[250+495.2+490.8+486.6+482.6+239.5]=\frac{1}{2}[2444.70]=1222.35$$

31. $\text{Area} \approx \dfrac{h}{3}[f(0)+4f(10)+2f(20)+4f(30)+2f(40)+4f(50)+2f(60)+4f(70)+f(80)]$

$$=\frac{10}{3}[0+60+36+72+60+108+48+92+0]=\frac{10}{3}[476]=1586.67\text{ sq ft}$$

Review Exercises

1. $\sum_{k=1}^{8}(k^2+1)=\sum_{k=1}^{8}k^2+\sum_{k=1}^{8}1$

$=\dfrac{8(9)(17)}{6}+8\cdot 1$

$=204+8=212$

2. $\sum_{i=1}^{n}\dfrac{3i}{n^3}=\dfrac{3}{n^3}\sum_{i=1}^{n}i=\dfrac{3}{n^3}\cdot\dfrac{n(n+1)}{2}=\dfrac{3(n+1)}{2n^2}$

3. $f(x)=3x^2$

$\text{Area}=\dfrac{1}{6}\cdot f\left(\dfrac{1}{6}\right)+\dfrac{1}{6}\cdot f\left(\dfrac{2}{6}\right)+\cdots+\dfrac{1}{6}\cdot f\left(\dfrac{6}{6}\right)$

$=\dfrac{1}{6}\cdot\dfrac{3}{36}+\dfrac{1}{6}\cdot 3\cdot\dfrac{4}{36}+\dfrac{1}{6}\cdot 3\cdot\dfrac{9}{36}+\cdots+\dfrac{1}{6}\cdot 3\cdot\dfrac{36}{36}$

$=\dfrac{3}{216}(1+4+9+16+25+36)$

$=\dfrac{1}{72}(91)=\dfrac{91}{72}$

4. $A=\lim_{n\to\infty}3\left(\dfrac{i}{n}\right)^2\left(\dfrac{1}{n}\right)$

$=\lim_{n\to\infty}\dfrac{3}{n^3}\sum_{i=1}^{n}i^2=\lim_{n\to\infty}\dfrac{3}{n^3}\left[\dfrac{n(n+1)(2n+1)}{6}\right]$

$=\lim_{n\to\infty}\dfrac{(n+1)(2n+1)}{2n^2}=1$

5. $\text{Area}=\int_0^1 3x^2dx=x^3\Big|_0^1=1-0=1$

6. $\int_1^3\left(x^3-4x+5\right)dx=\left(\dfrac{x^4}{4}-\dfrac{4x^2}{2}+5x\right)\Big|_1^3$

$=\left(\dfrac{81}{4}-18+15\right)-\left(\dfrac{1}{4}-2+5\right)$

$=14$

7. $\int_1^4 4x^{3/2}dx=\dfrac{8}{5}x^{5/2}\Big|_1^4=\dfrac{8}{5}(32-1)=\dfrac{8}{5}(31)=\dfrac{248}{5}$

8. $\int_{-3}^{2}\left(x^3-3x^2+4x+2\right)dx$

$=\left(\dfrac{1}{4}x^4-x^3+2x^2+2x\right)\Big|_{-3}^{2}$

$=4-8+8+4-\left(\dfrac{81}{4}+27+18-6\right)$

$=8-\dfrac{237}{4}=-\dfrac{205}{4}$

9. $\int_0^5\left(x^3+4x\right)dx=\left(\dfrac{1}{4}x^4+2x^2\right)\Big|_0^5$

$=\dfrac{625}{4}+50=\dfrac{825}{4}$

10. $\int_{-2}^{3}(x+2)^2\,dx=\dfrac{1}{3}(x+2)^3\Big|_{-2}^{3}=\dfrac{125}{3}-0=\dfrac{125}{3}$

11. $\int_{-3}^{-1}(x+1)dx=\left(\dfrac{1}{2}x^2+x\right)\Big|_{-3}^{-1}$

$=\left(\dfrac{1}{2}-1\right)-\left(\dfrac{9}{2}-3\right)=-2$

12. $\int_2^3\dfrac{x^2}{2x^3-7}dx=\dfrac{1}{6}\int_2^3\dfrac{6x^2}{2x^3-7}dx$

$=\dfrac{1}{6}\ln\left(2x^3-7\right)\Big|_2^3$

$=\dfrac{1}{6}\ln 47-\dfrac{1}{6}\ln 9$

13. $\int_{-1}^{2}\left(x^2+x\right)dx=\left(\dfrac{1}{3}x^3+\dfrac{1}{2}x^2\right)\Big|_{-1}^{2}$

$=\left(\dfrac{8}{3}+2\right)-\left(-\dfrac{1}{3}+\dfrac{1}{2}\right)$

$=4\dfrac{1}{2}=\dfrac{9}{2}$

14. $\int_1^4\left(x^{-1}+x^{1/2}\right)dx=\left(\ln x+\dfrac{2}{3}x^{3/2}\right)\Big|_1^4=\ln 4+\dfrac{14}{3}$

15. $\int_0^4(2x+1)^{1/2}\,dx=\dfrac{1}{2}\cdot\dfrac{(2x+1)^{3/2}}{3/2}\Big|_0^4$

$=\dfrac{1}{3}(27-1)=\dfrac{26}{3}$

16. $\int_0^1\dfrac{x}{x^2+1}dx=\dfrac{1}{2}\int_0^1\dfrac{2x}{x^2+1}dx$

$=\dfrac{1}{2}\ln\left(x^2+1\right)\Big|_0^1=\dfrac{1}{2}\ln 2-0=\dfrac{1}{2}\ln 2$

17. $\int_0^1 e^{-2x}dx=-\dfrac{1}{2}e^{-2x}\Big|_0^1$

$=-\dfrac{1}{2}\left(\dfrac{1}{e^2}-e^0\right)=\dfrac{1}{2}\left(1-e^{-2}\right)$

18. $\int_0^1 xe^{x^2}\,dx = \frac{1}{2}\int_0^1 2xe^{x^2}\,dx = \frac{1}{2}e^{x^2}\Big|_0^1 = \frac{1}{2}e - \frac{1}{2}$

19. $x = 1:\ f(x) = x^2 - 3x + 2 \quad g(x) = x^2 + 4$

$f(1) = 0 \qquad g(1) = 5$

$$A = \int_0^5 \left[\left(x^2+4\right) - \left(x^2 - 3x + 2\right)\right]dx$$
$$= \int_0^5 (2 + 3x)\,dx$$
$$= \left(2x + \frac{3}{2}x^2\right)\Big|_0^5 = \left(10 + \frac{75}{2}\right) - 0 = \frac{95}{2}$$

20.
$$x^2 = 4x + 5$$
$$x^2 - 4x - 5 = 0$$
$$(x-5)(x+1) = 0$$
$$x = 5, -1$$
$$\int_{-1}^5 \left(4x + 5 - x^2\right)dx = \left(2x^2 + 5x - \frac{1}{3}x^3\right)\Big|_{-1}^5$$
$$= 75 - \frac{125}{3} - \left(-3 + \frac{1}{3}\right)$$
$$= \frac{108}{3} = 36$$

21. $A = \int_{-1}^0 \left(x^3 - x\right)dx$
$$= \left(\frac{1}{4}x^4 - \frac{1}{2}x^2\right)\Big|_{-1}^0 = 0 - \left(\frac{1}{4} - \frac{1}{2}\right) = \frac{1}{4}$$

22.
$$x^3 - 1 = x - 1$$
$$x^3 - x = 0$$
$$x(x+1)(x-1) = 0$$
$$x = 0, -1, 1$$
$$\int_{-1}^0 \left[x^3 - 1 - (x-1)\right]dx + \int_0^1 \left[x - 1 - \left(x^3 - 1\right)\right]dx$$
$$= \int_{-1}^0 \left(x^3 - x\right)dx + \int_0^1 \left(x - x^3\right)dx$$
$$= \left(\frac{1}{4}x^4 - \frac{1}{2}x^2\right)\Big|_{-1}^0 + \left(\frac{1}{2}x^2 - \frac{1}{4}x^4\right)\Big|_0^1$$
$$= -\frac{1}{4} + \frac{1}{2} + \frac{1}{2} - \frac{1}{4} = \frac{1}{2}$$

23. $\int \sqrt{x^2 - 4dx}$
$$= \frac{1}{2}\left(x\sqrt{x^2 - 4} - 4\ln\left|x + \sqrt{x^2 - 4}\right|\right) + C$$
(By formula 7)

24. $\int_0^1 3^x\,dx = 3^x \log_3 e\Big|_0^1 = 3\log_3 e - 1\log_3 e = 2\log_3 e$

25. $\frac{1}{2}\int \ln x^2\,(2x\,dx) = \frac{1}{2}x^2\left(\ln x^2 - 1\right) + C$
(By formula 14)

26. $\int \frac{dx}{x(3x+2)} = \frac{1}{2}\ln\left|\frac{x}{3x+2}\right| + C$

27. $\int x^5 \ln x\,dx$

$u = \ln x \qquad dv = x^5\,dx$

$du = \frac{1}{x}dx \qquad v = \frac{1}{6}x^6$

$$\int x^5 \ln x dx = \frac{1}{6}x^6 \ln|x| - \frac{1}{6}\int x^5 dx$$
$$= \frac{1}{6}x^6 \ln|x| - \frac{1}{6}\cdot\frac{x^6}{6} + C$$
$$= \frac{1}{6}x^6 \ln|x| - \frac{1}{36}x^6 + C$$

28. $\int xe^{-2x}dx = uv - \int v\,du$

$u = x \qquad dv = e^{-2x}dx$

$du = dx \qquad v = -\frac{1}{2}e^{-2x}$

$$\int xe^{-2x}dx = -\frac{1}{2}xe^{-2x} + \frac{1}{2}\int e^{-2x}dx$$
$$= -\frac{1}{2}xe^{-2x} - \frac{1}{4}e^{-2x} + C$$

29. $\int x(x+5)^{-1/2}dx$

$u = x \qquad dv = (x+5)^{-1/2}\,dx$

$du = dx \qquad v = 2(x+5)^{1/2}$

$$\int x(x+5)^{-1/2}\,dx = 2x\sqrt{x+5} - 2\int (x+5)^{1/2}\,dx$$
$$= 2x\sqrt{x+5} - \frac{4}{3}(x+5)^{3/2} + C$$

30. $\int_1^e \ln x\,dx = uv - \int v\,du$

$u = \ln x \qquad dv = dx$

$du = \frac{1}{x}dx \qquad v = x$

$$\int_1^e \ln x\,dx = x\ln|x|\Big|_1^e - \int_1^e dx$$
$$= (x\ln x - x)\Big|_1^e = e - e - (0 - 1) = 1$$

31. $\int_1^\infty \frac{1}{x}dx = \lim_{b\to\infty}\int_1^b \frac{1}{x}dx = \lim_{b\to\infty}\ln x\Big|_1^b = \lim_{b\to\infty}(\ln b - \ln 1) = \infty$

Thus, the integral diverges.

32. $\int_{-\infty}^{-1}\frac{200}{x^3}dx = \lim_{a\to\infty}\int_{-a}^{-1}\frac{200}{x^3}dx = \lim_{a\to\infty} -100x^{-2}\Big|_{-a}^{-1} = -100$

33. $\int_0^\infty 5e^{-3x}dx = \lim_{b\to\infty}\int_0^b 5e^{-3x}dx = -\frac{5}{3}\lim_{b\to\infty}e^{-3x}\Big|_0^b = -\frac{5}{3}\lim_{b\to\infty}\left(\frac{1}{e^{3b}} - e^0\right) = \frac{5}{3}$

34. $\int_{-\infty}^0 \frac{x}{(x^2+1)^2}dx = \lim_{a\to\infty}\int_{-a}^0 x(x^2+1)^{-2}dx = \lim_{a\to\infty}\frac{1}{2}\int_{-a}^0 2x(x^2+1)^{-2}dx = \lim_{a\to\infty} -\frac{1}{2}(x^2+1)^{-1}\Big|_{-a}^0 = -\frac{1}{2}$

35. a. $\int_1^3 2x^{-3}dx = -\frac{1}{x^2}\Big|_1^3 = -\left(\frac{1}{9}-1\right) = \frac{8}{9}$ or 0.889

b. $\int_1^3 \frac{2}{x^3}dx \approx \frac{h}{2}[f(1)+2f(1.5)+2f(2)+2f(2.5)+f(3)]$

$= \frac{1}{4}[2+1.185+0.500+0.256+0.074] = \frac{1}{4}[4.015] = 1.004$

c. $\int_1^3 \frac{2}{x^3}dx \approx \frac{h}{3}[f(1)+4f(1.5)+2f(2)+4f(2.5)+f(3)]$

$= \frac{1}{6}[2+2.370+0.500+0.512+0.074] = \frac{1}{6}[5.456] = 0.909$

36. $\int_0^1 \frac{4}{x^2+1}dx \approx \frac{h}{2}[f(0)+2f(0.2)+2f(0.4)+2f(0.6)+2f(0.8)+f(1)]$

$= \frac{1}{10}[4+7.692+6.897+5.882+4.878+2] = \frac{1}{10}[31.349] = 3.135$

37. $\int_1^{2.2} f(x)dx \approx \frac{h}{3}[f(1)+4f(1.3)+2f(1.6)+4f(1.9)+f(2.2)] = \frac{1}{10}[0+11.2+10.2+16.8+0.6] = \frac{1}{10}[38.8] = 3.9$

38. a. $n = 5$

b. $n = 6$

39. $M = \int_0^9 14{,}000(t+16)^{-1/2}dt = 14{,}000\cdot 2(t+16)^{1/2}\Big|_0^9 = 28{,}000(5-4) = \$28{,}000$

40. $\int_0^2 1.4e^{-1.4x}dx = -e^{-1.4x}\Big|_0^2 = -0.0608+1 = 0.9392$

So the probability that it lasts 2 years is $1-0.9392 \approx 0.061$.

41. Average $= \frac{1}{5-0}\int_0^5 1000e^{0.1t}dt = \frac{1}{5}\cdot\frac{1000}{0.1}e^{0.1t}\Big|_0^5 = 2000(1.6487-1) = \$1297.44$

42. $AI = \frac{1}{4}\int_0^4 50e^{0.2t}dt = \frac{250}{4}\int_0^4 0.2e^{0.2t}dt = \frac{250}{4}e^{0.2t}\Big|_0^4 \approx \frac{250}{4}[2.2255-1] \approx \$76.60$

43. a. $\sqrt{64-4x} = x-1$

$$64-4x = x^2-2x+1$$
$$0=(x+9)(x-7)$$

So, equilibrium is at $(7,6)$.

b. $CS = \int_0^7 (64-4x)^{1/2}dx - 7\cdot 6 = -\frac{1}{4}\cdot\frac{(64-4x)^{3/2}}{3/2}\Big|_0^7 - 42 = -\frac{1}{6}\left(36^{3/2}-64^{3/2}\right)-42$

$$= -\frac{1}{6}(-296)-42 = 49.33-42 = \$7.33$$

44. $x-1=\sqrt{64-4x}$

$$x^2-2x+1 = 64-4x$$
$$x^2+2x-63=0$$
$$(x+9)(x-7)=0$$
$$x = 7 \text{ (only positive values)}$$

$$PS = (7)(6) - \int_0^7 (x-1)dx = 42 - \left[\left(\frac{1}{2}x^2 - x\right)\Big|_0^7\right] = 42-17.5 = 24.50$$

45. $TI = \int_0^{10} 125e^{0.05t}dt = 2500e^{0.05t}\Big|_0^{10} = 2500(1.64872-1) = \$1621.803$ thousands $= \$1,621,803$

46. a. $PV = \int_0^5 150e^{-0.2t}e^{-0.08t}dt = \int_0^5 150e^{-0.28t}dt = \frac{150}{-0.28}\int_0^5 -0.28e^{-0.28t}dt$

$$= -535.71\left[e^{-0.28t}\Big|_0^5\right] = -535.71(0.2466-1) \approx 403.604 \text{ thousands or } \$403,604.$$

b. $FV = e^{0.40}(PV) = 1.4918(PV) = \$602,106$

47. Average Cost $= \frac{1}{150-0}\int_0^{150}\left(x^2+40,000\right)^{1/2}dx$

$$= \frac{1}{150}\cdot\frac{1}{2}\left[x\sqrt{x^2+40,000}+40,000\ln\left(x+\sqrt{x^2+40,000}\right)\right]\Big|_0^{150}$$
$$= \frac{1}{300}\left[150\sqrt{62,500}+40,000\ln\left(150+\sqrt{62,500}\right)-\left(0+40,000\ln\sqrt{40,000}\right)\right]$$
$$= \frac{1}{300}[150(250)+40,000(5.9915)-40,000(5.2983)] = \frac{1}{300}(65,228) = \$217.43$$

48. $PV = 9000\int_0^5 te^{-0.08t}dt$ Use integration by parts.

$$PV = 9000\left[\frac{-t}{0.08}e^{-0.08t}\Big|_0^5 + \frac{1}{0.08}\int_0^5 e^{-0.08t}dt\right] = 9000\left[\frac{-t}{0.08}e^{-0.08t}\Big|_0^5 + \frac{1}{-(0.08)^2}\int_0^5 -0.08e^{-0.08t}dt\right]$$
$$= 9000[-41.895]-1,406,250\left[e^{-0.08t}\Big|_0^5\right] = 9000[-41.895]-1,406,250[-0.32968] = \$86,557.44$$

49. $C(x)=\int[3+x\ln(x+1)]dx$

$u=\ln(x+1) \quad dv = xdx$

$du=\dfrac{dx}{x+1} \qquad v=\dfrac{1}{2}x^2$

$$\int[3+x\ln(x+1)]dx = 3x+\frac{1}{2}x^2\ln(x+1)-\frac{1}{2}\int\frac{x^2}{x+1}dx = 3x+\frac{1}{2}x^2\ln(x+1)-\frac{1}{2}\int\left(x-1+\frac{1}{x+1}\right)dx$$

$$= 3x+\frac{1}{2}x^2\ln(x+1)-\frac{1}{2}\left[\frac{1}{2}x^2-x+\ln(x+1)\right]+K$$

$$= 3x+\frac{1}{2}x^2\ln(x+1)-\frac{1}{4}x^2+\frac{1}{2}x-\frac{1}{2}\ln(x+1)+K$$

Using the concept that fixed cost is *K*, we have $C(x)=-\frac{1}{4}x^2+\frac{7}{2}x+\frac{x^2-1}{2}\ln(x+1)+2000$. Also, $C(0)=K$.

50. $\int_1^{\infty}1.4e^{-1.4x}dx=\lim_{b\to\infty}\left[-\int_1^b-1.4e^{-1.4x}dx\right]=-\lim_{b\to\infty}e^{-1.4x}\Big|_1^b=-\lim_{b\to\infty}\left(e^{-1.4b}-e^{-1.4}\right)=e^{-1.4}\approx 0.247$

51. $CV=\int_0^{\infty}120e^{.03t}\cdot e^{-0.06t}dt=\int_0^{\infty}120e^{-0.03t}dt=\lim_{a\to\infty}\int_0^a 120e^{-0.03t}dt$

$$=\frac{120}{-0.03}\lim_{a\to\infty}e^{-0.03t}\Big|_0^a=-4000\lim_{a\to\infty}\left(\frac{1}{e^{0.03a}}-1\right)=\$4000\text{ thousands}$$

52. $\int_0^2 100e^{-0.01t^2}dt\approx\frac{h}{3}[f(0)+4f(0.5)+2f(1)+4f(1.5)+f(2)]$

$$=\frac{1}{6}[100+399.00+198.01+390.10+96.08]=\frac{1}{6}[1183.19]=\$197.198\text{ (thousands)}$$

53. $\int_0^{10}\overline{MR}\,dx\approx\frac{h}{2}[f(0)+2f(2)+2f(4)+2f(6)+2f(8)+f(10)]$

$$=1[0+960+1440+1440+960+0]=\$4,800\text{ (hundreds) or }\$480,000$$

Chapter Test

1.

x_i	0	.5	1	1.5
$f(x_1)$	2	1.936	1.732	1.323

$A\approx(0.5)\sum f(x_i)$

$=(0.5)(6.991)$

$=3.496$

2. $f(x)=5-2x \quad [0,1]$

(number line from 0 to 1 with points Δx, $2\Delta x$, ..., $i\Delta x$)

$\Delta x=\dfrac{1-0}{n}=\dfrac{1}{n}$

Rect 1: $\Delta x(5-2\Delta x)$

Rect 2: $\Delta x(5-4\Delta x)$

...

Rect i: $\Delta x(5-2i\Delta x)$

a. $S=\sum_{i=1}^{n}\left(5-2i\cdot\frac{1}{n}\right)\frac{1}{n}$

$=\frac{1}{n}\left(5n-\frac{2}{n}\cdot\frac{n(n+1)}{2}\right)$

$=5-\frac{n+1}{n}$ or $4-\frac{1}{n}$

b. $\lim_{n\to\infty}S_n=\lim_{n\to\infty}\left(4-\frac{1}{n}\right)=4$

3. $\int_0^6\left(12+4x-x^2\right)dx=\left(12x+2x^2-\frac{x^3}{3}\right)\Big|_0^6=72$

4. a $\int_0^4 (9-4x)dx = \left(9x-2x^2\right)\Big|_0^4 = 4-0=4$

b. $\int_0^3 \left(8x^2+9\right)^{-1/2}(x\,dx) = \frac{1}{16}\left(\frac{\left(8x^2+9\right)^{1/2}}{1/2}\right)\Bigg|_0^3$

$= \frac{1}{8}(9-3) = \frac{3}{4}$

c. $\int_1^4 \frac{5}{4x-1}dx = \frac{5}{4}\ln(4x-1)\Big|_1^4 = \frac{5}{4}\ln 5$

d. $\int_1^\infty \frac{7}{x^2}dx = \lim_{a\to\infty}\left(-\frac{7}{x}\right)\Big|_1^a = \lim_{a\to\infty}\left(-\frac{7}{a}+7\right) = 7$

5. a. $\int 3xe^x dx = 3xe^x - \int 3e^x dx = 3xe^x - 3e^x + C$

$u = 3x \quad dv = e^x dx$

$du = 3dx \quad v = e^x$

b. $\int x\ln(2x)dx = \frac{x^2\ln 2x}{2} - \int \frac{x}{2}dx$

$= \frac{x^2\ln 2x}{2} - \frac{x^2}{4} + C$

6. $\int_1^3 f(x)dx + \int_3^4 f(x)dx = \int_1^4 f(x)dx$, so

$2\int_1^3 f(x)dx = 2\left[\int_1^4 f(x)dx - \int_3^4 f(x)dx\right]$

$= 2[3-7] = -8$

7. a. $\int \ln(2x)dx = \frac{1}{2}\int \ln(2x)(2dx)$

$= \frac{1}{2}\cdot 2x(\ln(2x)-1)$

$= x(\ln(2x)-1)+C$

(Formula 14)

b. $\int x\sqrt{3x-7}(dx)$

$= \left[\frac{2(3\cdot 3x+2\cdot 7)(3x-7)^{3/2}}{15\cdot 9}\right]+C$

(Formula 16)

8. $\int_1^4 \sqrt{x^3+10}\,dx \approx 16.089$

9. $80 = 120-0.2x$ gives an equilibrium quantity $x = 200$.

$CS = \int_0^{200}(120-0.2x)dx - 80\cdot 200$

$= \left(120x-0.1x^2\right)\Big|_0^{200} - 16000 = \$4000$

$PS = 80\cdot 200 - \int_0^{200}\left(40+0.001x^2\right)dx$

$= 16000 - \left(40x+\frac{.001}{3}x^3\right)\Bigg|_0^{200} = \frac{\$16000}{3}$

10. a. $TI = \int_0^{12} 85e^{-0.01t}dt$

$= \frac{85}{-0.01}e^{-0.01t}\Big|_0^{12} = -8500(0.8869-1)$

$= \$961.18$ thousands

b. $PV = \int_0^{12} 85e^{-0.01t}e^{-0.07t}dt$

$= 85\int_0^{12} e^{-0.08t}dt = \frac{85}{-0.08}e^{-0.08t}\Big|_0^{12}$

$= \$655.68$ thousands

c. $CV = \int_0^\infty 85e^{-0.01t}e^{-0.07t}dt$

$= \lim_{a\to\infty}\int_0^a 85e^{-0.08t}dt = \lim_{a\to\infty}\left[\frac{85}{-0.08}e^{-0.08t}\Big|_0^a\right]$

$= \frac{85}{-0.08}\lim_{a\to\infty}\left(\frac{1}{e^{0.08a}}-1\right) = \$1062.5$ thousand

11. First, find the points of intersection.

$x^2 - x = 2x+4$

$0 = x^2-3x-4 = (x-4)(x+1)$

Points are -1 and 4.

$A = \int_{-1}^4\left[(2x+4)-\left(x^2-x\right)\right]dx$

$= \int_{-1}^4\left(3x+4-x^2\right)dx = \left(\frac{3}{2}x^2+4x-\frac{x^3}{3}\right)\Bigg|_{-1}^4$

$= \frac{56}{3} - \left(-\frac{13}{6}\right) = \frac{125}{6}$

12.

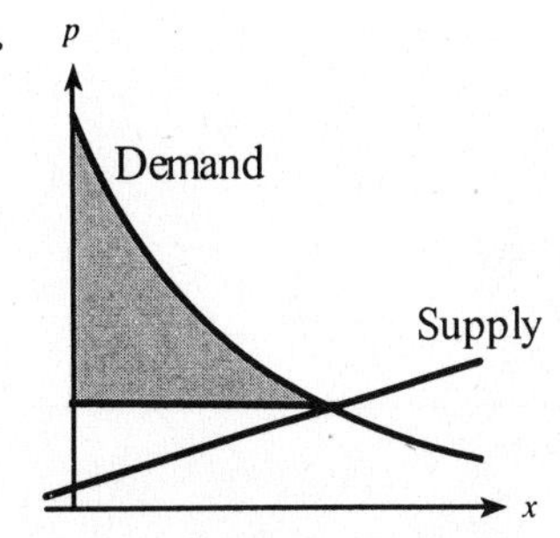

13. Gini coeff $= \dfrac{\int_0^1 \left(x - 0.998x^{2.6}\right)dx}{1/2} = 2\left(\dfrac{x^2}{2} - \dfrac{0.998}{3.6}x^{3.6}\right)\Bigg|_0^1 = 2(0.223) = 0.446$

Gini coeff $= \dfrac{\int_0^1 \left[x - \left(0.57x^2 + 0.43x\right)\right]dx}{1/2} = 2\left(\dfrac{x^2}{2} - \dfrac{0.57}{3}x^3 - \dfrac{0.43}{2}x^2\right)\Bigg|_0^1 = 2(0.095) = 0.19$

The change decreases the difference in income.

14. This problem is more easily solved by using the graphing calculator.

a. Avg. $= \dfrac{1}{20-5}\int_5^{20}\left(3.722x^3 - 145.3x^2 + 1836.8x - 5735.2\right)dx$

$= \dfrac{1}{15}\left(0.9305x^4 - 48.43x^3 + 918.4x^2 - 5735.2x\right)\Big|_5^{20}$

$= \dfrac{1}{15}\left[14{,}096 - (-11{,}188.19)\right] = 1{,}685.61\ \dfrac{\text{thousand barrels}}{\text{day}}$

b. Avg. $= \dfrac{1}{20-10}\int_{10}^{20} f(x)dx = \dfrac{1}{10}\left[14096 - (-4637)\right] = 1{,}873.3\ \dfrac{\text{thousand barrels}}{\text{day}}$

15. $\int_1^3 x\ln x\,dx \approx \dfrac{h}{2}[f(1) + 2f(1.5) + 2f(2) + 2f(2.5) + f(3)]$

$= \dfrac{1}{4}[0 + 1.216 + 2.773 + 4.581 + 3.296] = \dfrac{1}{4}[11.866] = 2.967$

16. Area $\approx \dfrac{h}{3}[f(0) + 4f(40) + 2f(80) + 4f(120) + 2f(160) + 4f(200) + f(240)]$

$= \dfrac{40}{3}[0 + 88 + 70 + 192 + 64 + 96 + 0] = \dfrac{40}{3}[510] = 6800$ sq ft

Chapter 14: Functions of Two or More Variables

Exercise 14.1

1. $z = x^2 + y^2$

Domain = $\{(x, y): x \text{ and } y \text{ are real}\}$

3. $z = \dfrac{4x-3}{y}$

Domain = $\{(x, y): x \text{ and } y \text{ are real and } y \neq 0\}$

5. $z = \dfrac{4x^3y - x}{2x - y}$

Domain =

$\{(x, y): x \text{ and } y \text{ are real and } 2x - y \neq 0\}$

7. $q = \sqrt{p_1} + 3p_2$

Domain =

$\{(p_1, p_2): p_1 \text{ and } p_2 \text{ are real and } p_1 \geq 0\}$

9. $z = x^3 + 4xy + y^2$

$x = 1,\ y = -1$

$z = 1 - 4 + 1 = -2$

11. $z = \dfrac{x-y}{x+y}$

$x = 4,\ y = -1$

$z = \dfrac{4+1}{4-1} = \dfrac{5}{3}$

13. $C(x_1, x_2) = 600 + 4x_1 + 6x_2$

$x_1 = 400,\ x_2 = 50$

$C(400, 50) = 600 + 1600 + 300 = 2500$

15. $q_1(p_1, p_2) = \dfrac{p_1 + 4p_2}{p_1 - p_2}$

$q_1(40, 35) = \dfrac{40 + 140}{40 - 35} = 36$

17. $z(x, y) = xe^{x+y}$

$z(3, -3) = 3e^0 = 3$

19. $f(x, y) = \dfrac{\ln(xy)}{x^2 + y^2}$

$f(-3, -4) = \dfrac{\ln 12}{9 + 16} = \dfrac{\ln 12}{25}$

21. $w = \dfrac{x^2 + 4yz}{xyz}$

At $(1,3,1)$ we have $w = \dfrac{1+12}{3} = \dfrac{13}{3}$.

23. $S = f(P, t) = Pe^{0.06t}$

$f(2000, 20) = 2000e^{1.2} = \$6640.23$

When \$2000 is invested for 20 years the compound amount is \$6640.23.

25. $Q = f(K, M, h) = \sqrt{\dfrac{2KM}{h}}$

$f(200, 625, 1) = \sqrt{\dfrac{2(200)(625)}{1}} = 500$

For the given numbers, the most economical order size is 500 units.

27. $S = 1.98T - 1.09(1 - H)(T - 58) - 56.9$

$A = 2.70 + 0.885T - 78.7H + 1.20TH$

Max: 97.8°F $H = 44\%$

Min: 74.7°F $H = 80\%$

Max: $S = 1.98(97.8) - 1.09(1 - 0.44)(97.8 - 58) - 56.9 = 112.5°\text{F}$

$A = 2.70 + 0.885(97.8) - 78.7(0.44) + 1.2(97.8)(0.44) = 106.3°\text{F}$

Min: $S = 1.98(74.7) - 1.09(1 - 0.8)(74.7 - 58) - 56.9 = 87.4°\text{F}$

$A = 2.70 + 0.885(74.7) - 78.7(0.8) + 1.2(74.7)(0.8) = 77.6°\text{F}$

29. **a.** $f(90, 20, 8) = \$752.80$ (Table 1, Row 5, Col 4)

When \$90,000 is borrowed for 20 years at 8%, the monthly payment needed to pay off the loan is \$752.80.

b. $f(160, 15, 9) = \$1622.82$ (Table 2, Row 9, Col 3) When \$160,000 is borrowed for 15 years at 9%, the monthly payment needed to pay off the loan is \$1622.82.

31. $U = xy^2$

If $x = 9$, $y = 6$, then $U = 9 \cdot 6^2 = 324$.

a. $324 = x \cdot 9^2$ or $x = 4$ units

b. $324 = 81 \cdot y^2$ or $y = 2$ units

c.

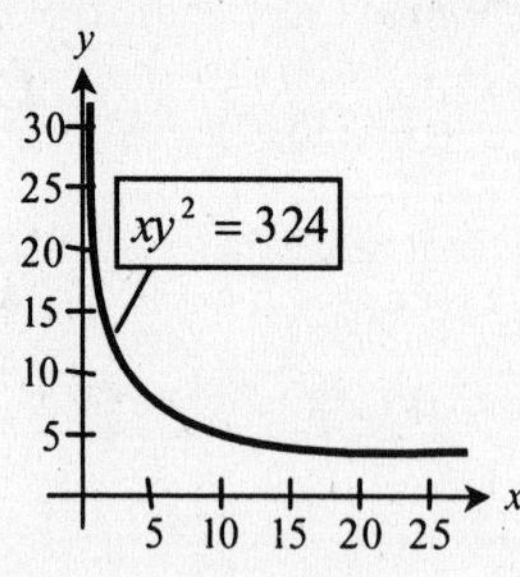

33. $Q = 30K^{1/4}L^{3/4}$

a. $Q = 30(10{,}000)^{1/4}(625)^{3/4}$
$= 30(10)(125) = 37{,}500$

b. $Q = 30(2K)^{1/4}(2L)^{3/4}$
$= 30(2^{1/4})K^{1/4}(2^{3/4})L^{3/4}$
$= 2\left[30K^{1/4}L^{3/4}\right]$

This is 2 times the original quantity.

c.

$Q = 300L^{3/4}$

Q: 6000, 5000, 4000, 3000, 2000, 1000

L: 10 20 30 40 50

35. $z = 20xy$

a. $z = 20(12)(30) = 7200$

b. $z = 20(10)(25) = 5000$

37. $C(x, y) = 20x + 200y$

$C(14{,}000,\ 20) = 20(14{,}000) + 200(20) = \$284{,}000$

Exercise 14.2

1. $z = x^4 - 5x^2 + 6x + 3y^3 - 5y + 7$

$$\frac{\partial z}{\partial x} = 4x^3 - 10x + 6$$

$$\frac{\partial z}{\partial y} = 9y^2 - 5$$

3. $z = x^3 + 4x^2y + 6y^2$

$z_x = 3x^2 + 8yx$

$z_y = 4x^2 + 12y$

5. $f(x, y) = \left(x^3 + 2y^2\right)^3$

$$\frac{\partial f}{\partial x} = 3\left(x^3 + 2y^2\right)^2 3x^2 = 9x^2\left(x^3 + 2y^2\right)^2$$

$$\frac{\partial f}{\partial y} = 3\left(x^3 + 2y^2\right)^2 4y = 12y\left(x^3 + 2y^2\right)^2$$

7. $f(x, y) = \sqrt{2x^2 - 5y^2}$

$$f_x = \frac{1}{2}\left(2x^2 - 5y^2\right)^{-1/2} 4x = \frac{2x}{\sqrt{2x^2 - 5y^2}}$$

$$f_y = \frac{1}{2}\left(2x^2 - 5y^2\right)^{-1/2}(-10y) = \frac{-5y}{\sqrt{2x^2 - 5y^2}}$$

9. $C(x, y) = 600 - 4xy + 10x^2y$

$$\frac{\partial C}{\partial x} = -4y + 20xy$$

$$\frac{\partial C}{\partial y} = -4x + 10x^2$$

11. $Q(s,t)=\dfrac{2s-3t}{s^2+t^2}$

$$\frac{\partial Q}{\partial s}=\frac{(s^2+t^2)(2)-(2s-3t)(2s)}{(s^2+t^2)^2}$$

$$=\frac{2s^2+2t^2-4s^2+6st}{(s^2+t^2)^2}=\frac{2(t^2+3st-s^2)}{(s^2+t^2)^2}$$

$$\frac{\partial Q}{\partial t}=\frac{(s^2+t^2)(-3)-(2s-3t)(2t)}{(s^2+t^2)^2}$$

$$=\frac{-3s^2-3t^2-4st+6t^2}{(s^2+t^2)^2}=\frac{3t^2-4st-3s^2}{(s^2+t^2)^2}$$

13. $z=e^{2x}+y\ln x$

$z_x=e^{2x}(2)+y\cdot\dfrac{1}{x}=2e^{2x}+\dfrac{y}{x}$

$z_y=\ln x$

15. $f(x,y)=100e^{xy}$

$\dfrac{\partial f}{\partial x}=100e^{xy}y=100ye^{xy}$

$\dfrac{\partial f}{\partial y}=100e^{xy}x=100xe^{xy}$

17. $f(x,y)=4x^3-5xy+y^2$

$f_x(x,y)=12x^2-5y$

$f_x(1,2)=12-5(2)=2$

19. $z=5x^3-4xy$

$z_x=15x^2-4y$

$z_x(1,2)=15-4(2)=7$

21. $z=e^{xy}$

$z_y=e^{xy}(x)=xe^{xy}$

$z_y(0,1)=0\cdot e^0=0$

23. $u=y^2-x^2z+4x$

a. $\dfrac{\partial u}{\partial w}=0$

b. $\dfrac{\partial u}{\partial x}=4-2xz$

c. $\dfrac{\partial u}{\partial y}=2y$

d. $\dfrac{\partial u}{\partial z}=-x^2$

25. $C=4x_1^2+5x_1x_2+6x_2^2+x_3$

a. $\dfrac{\partial C}{\partial x_1}=8x_1+5x_2$

b. $\dfrac{\partial C}{\partial x_2}=5x_1+12x_2$

c. $\dfrac{\partial C}{\partial x_3}=1$

27. $z=x^2+4x-5y^3$

$z_x=2x+4,\quad z_y=-15y^2$

a. $z_{xx}=2$

b. $z_{xy}=0$

c. $z_{yx}=0$

d. $z_{yy}=-30y$

29. $z=x^2y-4xy^2$

$z_x=2xy-4y^2,\quad z_y=x^2-8xy$

a. $z_{xx}=2y$

b. $z_{xy}=2x-8y$

c. $z_{yx}=2x-8y$

d. $z_{yy}=-8x$

31. $z=x^2-xy+4y^3$

$z_x=2x-y$

$z_{xy}=-1$

$z_{xyx}=0$

33. $f(x,y)=x^3y+4xy^4$

$\dfrac{\partial f}{\partial x}=3x^2y+4y^4$

$\dfrac{\partial^2 f}{\partial x^2}=6xy$

$\left.\dfrac{\partial^2}{\partial x^2}f(x,y)\right|_{(1,-1)}=6(1)(-1)=-6$

35. $f(x,y)=\dfrac{2x}{x^2+y^2}$

a. $\dfrac{\partial f}{\partial x}=\dfrac{(x^2+y^2)(2)-2x(2x)}{(x^2+y^2)^2}=\dfrac{2y^2-2x^2}{(x^2+y^2)^2}$

$$\frac{\partial^2 f}{\partial x^2}=\frac{(x^2+y^2)^2(-4x)-(2y^2-2x^2)2(x^2+y^2)2x}{(x^2+y^2)^4}$$

$$=\frac{-4x(x^2+y^2)\left[(x^2+y^2)+(2y^2-2x^2)\right]}{(x^2+y^2)^4}=\frac{-4x(3y^2-x^2)}{(x^2+y^2)^3}$$

$$\left.\frac{\partial^2 f}{\partial x^2}\right|_{(-1,4)}=\frac{4(47)}{17^3}=\frac{188}{4913}$$

b. $\dfrac{\partial f}{\partial y}=2x(-1)(x^2+y^2)^{-2}(2y)=\dfrac{-4xy}{(x^2+y^2)^2}$

$$\frac{\partial^2 f}{\partial y^2}=\frac{(x^2+y^2)^2(-4x)+(4xy)2(x^2+y^2)2y}{(x^2+y^2)^4}=\frac{4x(x^2+y^2)\left[(-x^2-y^2)+4y^2\right]}{(x^2+y^2)^4}=\frac{4x(3y^2-x^2)}{(x^2+y^2)^3}$$

$$\left.\frac{\partial^2 f}{\partial y^2}\right|_{(-1,4)}=\frac{-4(47)}{17^3}=-\frac{188}{4913}$$

37. $z=x^2y+ye^{x^2}$

$z_y=x^2+e^{x^2}$

$z_{yx}=2x+e^{x^2}(2x)$

$z_{yx}\big|_{(1,2)}=2+e^1(2)=2+2e$

39. $f(x,y)=x^2+e^{xy}$

$\dfrac{\partial f}{\partial x}=2x+ye^{xy},\quad \dfrac{\partial f}{\partial y}=xe^{xy}$

a. $\dfrac{\partial^2 f}{\partial x^2}=2+y^2e^{xy}$

b. $\dfrac{\partial^2 f}{\partial y\,\partial x}=yxe^{xy}+e^{xy}\cdot 1=e^{xy}(xy+1)$

c. $\dfrac{\partial^2 f}{\partial x\,\partial y}=xye^{xy}+e^{xy}\cdot 1=e^{xy}(xy+1)$

d. $\dfrac{\partial^2 f}{\partial y^2}=x^2e^{xy}$

41. $f(x,y)=y^2-\ln xy=y^2-\ln x-\ln y$

$\dfrac{\partial f}{\partial x}=-\dfrac{1}{x},\quad \dfrac{\partial f}{\partial y}=2y-\dfrac{1}{y}$

a. $\dfrac{\partial^2 f}{\partial x^2}=\dfrac{1}{x^2}$

b. $\dfrac{\partial^2 f}{\partial y\,\partial x}=0$

c. $\dfrac{\partial^2 f}{\partial x\,\partial y}=0$

d. $\dfrac{\partial^2 f}{\partial y^2}=2+\dfrac{1}{y^2}$

43. $w=4x^3y+y^2z+z^3$

$w_x=12x^2y$

a. $w_{xx}=24xy,\quad w_{xxy}=24x$

b. $w_{xy}=12x^2,\quad w_{xyx}=24x$

c. $w_{xyz}=0$

45. $R=f(A,i)$

a. For a loan of \$100,000 at 8% interest, the monthly payment is \$1289.

b. If the interest rate changes from 8% to 9% (on this \$100,000 loan), the monthly payment increases approximately \$62.51.

47. Explanation is based on business interpretation and not the mathematics of exponents.

a. $\frac{\partial Q}{\partial M}$ is the change in quantity ordered per unit change in sales. If the sales increase by one unit, then the order quantity should also increase.

b. $\frac{\partial Q}{\partial h}$ is the change in quantity ordered per unit change in holding costs. If holding costs increase, then the order quantity should decrease.

49. $f(x,y) = 10{,}000 - 6500e^{-0.01x} - 3500e^{-0.02y}$

a. $f_x = 0 - 6500e^{-0.01x}(-0.01) - 0 = 65e^{-0.01x}$

b. $f_y = 0 - 0 - 3500e^{-0.02y}(-0.02) = 70e^{-0.02y}$

51. $U = x^2y^2$

a. $\frac{\partial U}{\partial x} = 2xy^2$

b. $\frac{\partial U}{\partial y} = 2x^2y$

53. $Q = 75K^{1/3}L^{2/3}$

$\frac{\partial Q}{\partial K} = 25K^{-2/3}L^{2/3}$, $\frac{\partial Q}{\partial L} = 50K^{1/3}L^{-1/3}$

At $K = 729$, $L = 1728$ we have

$\frac{\partial Q}{\partial K} = 25 \cdot \frac{1}{81} \cdot 144 = 44.44$ and

$\frac{\partial Q}{\partial L} = 50 \cdot 9 \cdot \frac{1}{12} = 37.5.$

If labor hours remain at 1728 and K increases by \$1000, Q will increase about 44.44 thousand units. If capital expenditures remain at \$729,000 and L increases by one hour, Q will increase about 37.5 thousand units.

55. $WC = 48.064 + 0.474t - 0.020ts - 1.85s + 0.304t\sqrt{s} - 27.74\sqrt{s}$

a. $\frac{\partial WC}{\partial s} = -0.020t - 1.85 + \frac{0.152t}{\sqrt{s}} - \frac{13.87}{\sqrt{s}}$

b. At $s = 25$, $t = 10$,

$\frac{\partial WC}{\partial s} = -0.2 - 1.85 + \frac{1.52}{5} - \frac{13.87}{5} = -4.52$

At this temperature and wind speed, the wind chill temperature will decrease 4.52°F if the wind speed increases 1 mph.

Exercise 14.3

1. $C(x,y) = 30 + 3x + 5y$

$C(4,3) = 30 + 12 + 15 = \$57$

3. $C(x,y) = 30 + 2x + 4y + \frac{xy}{50}$

a. $C_x = 2 + \frac{y}{50}$

b. $C_y = 4 + \frac{x}{50}$

5. $C(x,y) = 20x + 70y + \frac{x^2}{1000} + \frac{xy^2}{100}$

a. $C_x = 20 + \frac{x}{500} + \frac{y^2}{100}$

$C_x(10,12) = 20 + \frac{1}{50} + \frac{144}{100} = \$21.46$

The total cost will increase approximately \$21.46 if raw material costs increase to \$11 per pound.

b. $C_y = 70 + \frac{xy}{50}$

$C_y(10,12) = 70 + \frac{120}{50} = \$72.40$

The total cost will increase \$72.40 if labor costs increase to \$13 per hour.

7. $C(x,y) = 30 + x^2 + 3y + 2xy$

a. $C_x = 2x + 2y$

$C_x(8,10) = 16 + 20 = \$36$

b. $C_y = 3 + 2x$

$C_y = (8,10) = 3 + 16 = \$19$

9. $C(x,y)=x\sqrt{y^2+1}$

a. $C_x=\sqrt{y^2+1}$

b. $C_y=x\cdot\frac{1}{2}\left(y^2+1\right)^{-1/2}(2y)=\frac{xy}{\sqrt{y^2+1}}$

11. $C(x,y)=1200\ln(xy+1)+10{,}000$

a. $C_x=1200\cdot\frac{1}{xy+1}\cdot y=\frac{1200y}{xy+1}$

b. $C_y=1200\cdot\frac{1}{xy+1}\cdot x=\frac{1200x}{xy+1}$

13. $z=\sqrt{4xy}=2x^{1/2}y^{1/2}$

a. $z_x=2y^{1/2}\cdot\frac{1}{2}x^{-1/2}=\sqrt{\frac{y}{x}}$

b. $z_y=2x^{1/2}\cdot\frac{1}{2}y^{-1/2}=\sqrt{\frac{x}{y}}$

15. $z=x^{1/2}\ln(y+1)$

a. $z_x=\frac{1}{2}x^{-1/2}\ln(y+1)=\frac{\ln(y+1)}{2\sqrt{x}}$

b. $z_y=x^{1/2}\cdot\frac{1}{y+1}\cdot 1=\frac{\sqrt{x}}{y+1}$

For 17 and 19, $z=\frac{11xy-0.0002x^2-5y}{0.03x+3y}$.

17. At $x=300$ and $y=500$ we have $z=\frac{11(150{,}000)-0.0002(90{,}000)-2500}{9+1500}=1092$

19. $z_x=\frac{(0.03x+3y)(11y-0.0004x)-(11xy-0.0002x^2-5y)(0.03)}{(0.03x+3y)^2}$

$$z_x(300,500)=\frac{1509(5499.88)-(1{,}650{,}000-18-2500)(0.03)}{(1509)^2}=3.62$$

21. $z=400x^{3/5}y^{2/5}$

a. $\frac{\partial z}{\partial x}=\frac{240y^{2/5}}{x^{2/5}}$

b.

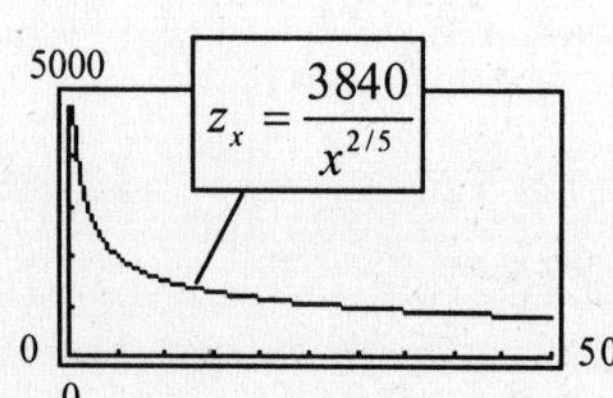

a. $\frac{\partial z}{\partial y}=\frac{160x^{3/5}}{y^{3/5}}$

b.

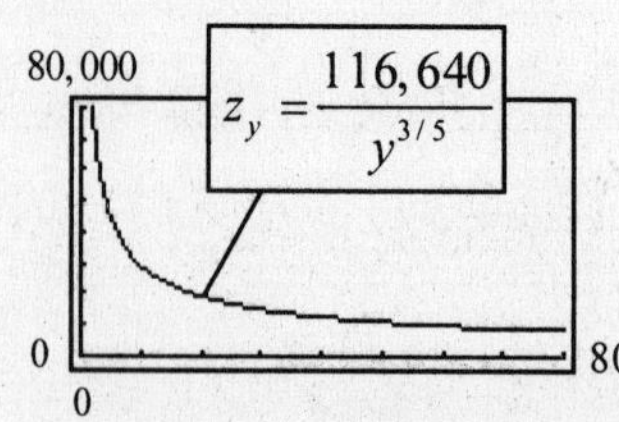

c. An increase in capital investment or work hours results in an increase in productivity since both partials are positive. As capital investment or work hours continue to increase, the increase in productivity is diminishing.

23. $q_1=300-80-32=188$

$q_2=400-50-80=270$

25. Setting $q_1=q_2$, we have

$$300-8p_1-4p_2=400-5p_1-10p_2$$
$$-8p_1+5p_1=100-10p_2+4p_2$$
$$p_1=-\frac{100}{3}+2p_2 \quad (1)$$

Any values p_1 and p_2 satisfying (1) and that make q_1 and q_2 nonnegative will be a solution. Two solutions are $(6.67,\ 20)$ and $(20,\ 26.67)$.

27. $q_A = 400 - 3p_A - 2p_B$, $q_B = 250 - 5p_A - 6p_B$

a. $\frac{\partial q_A}{\partial p_A} = -3$

b. $\frac{\partial q_A}{\partial p_B} = -2$

c. $\frac{\partial q_B}{\partial p_B} = -6$

d. $\frac{\partial q_B}{\partial p_A} = -5$

e. Complementary since (b) and (d) are negative.

29. $q_A = 5000 - 50p_A - \frac{600}{p_B + 1}$

$q_B = 10,000 - \frac{400}{p_A + 4} + \frac{400}{p_B + 4}$

a. $\frac{\partial q_A}{\partial p_A} = -50$

b. $\frac{\partial q_A}{\partial p_B} = -600(-1) \cdot \frac{1}{(p_B + 1)^2} = \frac{600}{(p_B + 1)^2}$

c. $\frac{\partial q_B}{\partial p_B} = \frac{-400}{(p_B + 4)^2}$

d. $\frac{\partial q_B}{\partial p_A} = \frac{400}{(p_A + 4)^2}$

e. Competitive since (b) and (d) are positive.

Exercise 14.4

1. $z = 9 - x^2 - y^2$

$z_x = -2x,\ z_y = -2y$

Critical point is at $x = 0$, $y = 0$.

$z_{xx} = -2,\ z_{yy} = -2,\ z_{xy} = 0$

$D = (-2)(-1) - (0)^2 > 0$ with $z_{xx} < 0$

A relative maximum at $(0, 0, 9)$.

3. $z = x^2 + y^2 + 4$

$z_x = 2x,\ z_y = 2y$

Critical point at $x = 0$, $y = 0$.

$z_{xx} = 2,\ z_{yy} = 2,\ z_{xy} = 0$

$D = (2)(2) - 0^2 > 0$ with $z_{xx} > 0$

A relative minimum at $(0, 0, 4)$.

5. $z = x^2 + y^2 - 2x + 4y + 5$

$z_x = 2x - 2,\ z_y = 2y + 4$

$z_x = 0$ at $x = 1$ and $z_y = 0$ at $y = -2$. Critical point is at $(1, -2)$.

$z_{xx} = 2,\ z_{yy} = 2,\ z_{xy} = 0$

$D = (2)(2) - 0^2 > 0$ with $z_{xx} > 0$

A relative minimum at $(1, -2, 0)$.

7. $z = x^2 + 6xy + y^2 + 16x$

$z_x = 2x + 6y + 16,\ z_y = 6x + 2y$

$z_y = 0$ gives $y = -3x$. $z_x = 0$ and $y = -3x$ gives $2x + 6(-3x) + 16 = 0$.

Thus, $x = 1$ and $y = -3(1) = -3$. So, $(1, -3)$ is a critical point.

$z_{xx} = 2,\ z_{yy} = 2,\ z_{xy} = 6$

$D = (2)(2) - 6^2 < 0$ and thus, there is neither a relative maximum nor a relative minimum at $(1, -3, 8)$. It is a saddle point.

9. $z = \frac{1}{9}(x^2 - y^2)$

$z_x = \frac{2}{9}x,\ z_y = -\frac{2}{9}y$

Critical point is at $(0, 0)$.

$z_{xx} = \frac{2}{9},\ z_{yy} = -\frac{2}{9},\ z_{xy} = 0$

$D = \left(\frac{2}{9}\right)\left(-\frac{2}{9}\right) - 0^2 < 0$ and thus, there is no relative maximum nor minimum at $(0, 0, 0)$. It is a saddle point.

11. $z = 24 - x^2 + xy - y^2 + 36y$

$z_x = -2x + y,\ z_y = x - 2y + 36$

$z_x = 0$ yields $y = 2x$. $z_y = 0$ and $y = 2x$ yields $x - 2(2x) + 36 = 0$.

Solving, we have $x = 12$ and $y = 24$ are the critical values.

$z_{xx} = -2,\ z_{yy} = -2,\ z_{xy} = 1$

$D = (-2)(-2) - 1^2 > 0$ with $z_{xx} < 0$

There is a relative maximum at $(12, 24, 456)$.

13. $z = x^2 + xy + y^2 - 4y + 10x$

$z_x = 2x + y + 10,\ z_y = x + 2y - 4$

$z_x = 0$ or $2x + y = -10$ and $z_y = 0$ or $x + 2y = 4$.

Solving these two equations, we have a critical point at $(-8, 6)$.

$z_{xx} = 2,\ z_{yy} = 2,\ z_{xy} = 1$

$D = (2)(2) - 1^2 > 0$ with $z_{xx} > 0$

There is a relative minimum at $(-8, 6, -52)$.

15. $z = x^3 + y^3 - 6xy$

$z_x = 3x^2 - 6y,\ z_y = 3y^2 - 6x$

$z_x = 0$ or $y = \frac{1}{2}x^2$. $z_y = 0$ and $y = \frac{1}{2}x^2$ gives $\frac{3}{4}x^4 - 6x = 0$. $3x^4 - 24x = 0$ yields $3x(x^3 - 8) = 0$. At $x = 0$ we have $y = 0$ and at $x = 2$ we have $y = 2$.

Thus, the critical points are $(0,0)$ and $(2,2)$.

$z_{xx} = 6x,\ z_{yy} = 6y,\ z_{xy} = -6$

At $(0,0)$: $D = (0)(0) - (-6)^2 < 0$

There is a saddle point at $(0,0,0)$.

At $(2,2)$: $D = (12)(12) - (-6)^2 > 0$ with $z_{xx} > 0$.

There is a relative minimum at $(2, 2, -8)$.

17. $\sum x = 18,\ \sum y = 97,\ \sum xy = 465,\ \sum x^2 = 86$

$$b = \frac{18(97) - 4(465)}{324 - 4(86)} = 5.7$$

$$a = \frac{97 - 5.7(18)}{4} = -1.4$$

$\hat{y} = a + bx = 5.7x - 1.4$

19. $P(x, y) = 100x + 64y - 0.01x^2 - 0.25y^2$

$P_x = 100 - 0.02x,\ P_y = 64 - 0.5y$

$P_x = 0$ if $x = 5000$ and $P_y = 0$ if $y = 128$.

Critical point is $(5000, 128)$.

$P_{xx} = -0.02,\ P_{yy} = -0.5,\ P_{xy} = 0$

$D = (-0.02)(-0.5) - 0^2 > 0$

$D > 0$ with $P_{xx} < 0$ means that there is a maximum profit when 5000 pounds of Kisses and 128 pounds of Kreams are sold.

21. $W = 20xy - x^2y - 2xy^2$

$W_x = 20y - 2xy - 2y^2,\ W_y = 20x - x^2 - 4xy$

If $W_x = 0$, then $2y(10 - x - y) = 0$. Thus, $y = 0$ or $y = 10 - x$.

If $W_y = 0$, then $x(20 - x - 4y) = 0$. Thus, $x = 0$ or $x = 20 - 4y$.

Critical points are $(0,0)$, $(0,10)$, and $(20,0)$ using $x = 0$ and $y = 0$.

Using $x = 20 - 4(10 - x)$, we have $x = \frac{20}{3}$. For $x = \frac{20}{3}$, we have $y = 10 - \frac{20}{3} = \frac{10}{3}$. So, $\left(\frac{20}{3}, \frac{10}{3}\right)$ is also a critical point.

$W_{xx} = -2y,\ W_{yy} = -4x,\ W_{xy} = 20 - 2x - 4y$

At $(0,0)$, $(0,10)$, and $(20,0)$, $D < 0$ and thus, there is no maximum or minimum there. At $\left(\frac{20}{3}, \frac{10}{3}\right)$, $D = \left(-\frac{20}{3}\right)\left(-\frac{80}{3}\right) - \left(-\frac{20}{3}\right)^2 > 0$ and since $W_{xx} < 0$, there is a maximum weight gain with $x = \frac{20}{3}$ and $y = \frac{10}{3}$.

23. $P = 3.78x^2 + 1.5y^2 - 0.09x^3 - 0.01y^3$

$P_x = 7.56x - 0.27x^2$, $P_y = 3y - 0.03y^2$

$P_x = 0$ if $0.27x(28 - x) = 0$ or $x = 0, 28$.

$P_y = 0$ if $3y(1 - 0.01y) = 0$ or $y = 0, 100$.

Critical points are $(0,0)$, $(0,100)$, $(28,0)$, and $(28,100)$.

$P_{xx} = 7.56 - 0.54x$, $P_{yy} = 3 - 0.06y$, $P_{xy} = 0$

At $(0,0)$ $D = (7.56)(3) - 0^2 > 0$. But $P_{xx} > 0$.

At $(0,100)$ $D = (7.56)(-3) - 0^2 < 0$.

At $(28,0)$ $D = (-7.56)(3) - 0^2 < 0$.

At $(28,100)$ $D = (-7.56)(-3) - 0^2 > 0$.

Since $P_{xx} < 0$, there is a maximum at $x = 28$ and $y = 100$.

25. $P = 10x - x^2 + 40y - 2y^2 - xy$

$P_x = 10 - 2x - y$, $P_y = 40 - 4y - x$

$P_x = 0$ if $y = 10 - 2x$. $P_y = 0$ if $x = 40 - 4y$.

Combining, we have $x = 40 - 4(10 - 2x)$ or $x = 0$. Thus, the critical point is $(0,10)$.

$P_{xx} = -2$, $P_{yy} = -4$, $P_{xy} = -1$

$D = (-2)(-4) - (-1)^2 > 0$.

Since $P_{xx} < 0$, there is a maximum at $(0,10)$.

27. $A = xy + 2y \cdot \dfrac{500,000}{xy} + 2x \cdot \dfrac{500,000}{xy}$

$= xy + \dfrac{1,000,000}{x} + \dfrac{1,000,000}{y}$

$A_x = y - \dfrac{10^6}{x^2}$, $A_y = x - \dfrac{10^6}{y^2}$

Combining $A_x = 0$ and $A_y = 0$, we have

$x = \dfrac{10^6}{\frac{10^{12}}{x^4}} = \dfrac{x^4}{10^6}$, $x \neq 0$.

Thus, $x^3 = 10^6$ or $x = 100$.

The critical point is $(100,100)$.

$A_{xx} = \dfrac{2 \cdot 10^6}{x^3}$, $A_{yy} = \dfrac{2 \cdot 10^6}{y^3}$, $A_{xy} = 1$

$D = (2)(2) - 1^2 > 0$ with $A_{xx} > 0$.

There is a minimum for length and width equal 100 and height equal 50.

29. $P = R - C = 12x - 6y - 2x^2 + 2xy - y^2 - 11$

$P_x = 12 - 4x + 2y$, $P_y = -6 + 2x - 2y$

$P_x = 0$ if $y = 2x - 6$ and $P_y = 0$ if $y = x - 3$.

Solving yields $(3,0)$ as the critical point.

$P_{xx} = -4$, $P_{yy} = -2$, $P_{xy} = 2$

$D = (-4)(-2) - 2^2 > 0$

Since $P_{xx} < 0$, maximum profit occurs with 3 *A*'s and 0 *B*'s.

For 31-33 use the linear regression feature of the graphing calculator.

31. a. $y = 0.113x + 4.860$

y is the median age.

b. $y = 0.113(30) + 4.860 = 8.25$

c. Higher prices and better quality are two reasons for the median age to increase.

33. a. $y = 2.097x + 59.103$ (y is in millions)

b. $y = 2.097(95) + 59.103 = 258.318$ million

c. A quadratic function is a better fit. ($y = 0.010x^2 + 0.961x + 81.8201$)

Exercise 14.5

1. The objective function is
$F(x,y,\lambda)=x^2+y^2+\lambda(x+y-6)$.
$F_x=2x+\lambda,\ F_y=2y+\lambda,\ F_\lambda=x+y-6$
Setting each of these equal to zero and combining, we have $x-y=0$ and $x+y=6$. This yields $x=3$ and $y=3$. The minimum is $z=3^2+3^2=18$.

3. The objective function is
$F(x,y,\lambda)=3x^2+5y^2-2xy+\lambda(x+y-5)$.
$F_x=6x-2y+\lambda,\ F_y=10y-2x+\lambda$
$F_\lambda=x+y-5$
Setting $F_x=0$ and $F_y=0$, we have
$6x-2y=10y-2x$ or $x=\frac{3}{2}y$.
Using $F_\lambda=0$, we have $\frac{3}{2}y+y-5=0$ or $y=2$.
Now $x=\frac{3}{2}(2)=3$.
Thus, the minimum occurs at $(3,2)$.
The minimum is $z=27+20-12=35$.

5. The objective function is
$F(x,y,\lambda)=x^2y+\lambda(x+y-6)$.
$F_x=2xy+\lambda,\ F_y=x^2+\lambda,\ F_\lambda=x+y-6$
Using $F_x=0$ and $F_y=0$, we have $x^2=2xy$ or $x(x-2y)=0$. Thus, $x=0$ or $x=2y$.
Using $F_\lambda=0$, we have $x=0$ and $y=6$.
Also, $2y+y-6=0$ yields $y=2$ and so $x=4$.
At $(0,6)$, $z=0$ and at $(4,2)$, $z=16\cdot 2=32$.
The maximum is at $(4,2)$.

7. The objective function is
$F(x,y,\lambda)=2xy-2x^2-4y^2+\lambda(x+2y-8)$.
$F_x=2y-4x+\lambda,\ F_y=2x-8y+2\lambda$
$F_\lambda=x+2y-8$
$F_x=0$ means $\lambda=4x-2y$. $F_y=0$ means $\lambda=-x+4y$. Combining we have $5x-6y=0$.
$F_\lambda=0$ gives $x+2y=8$.
These two equations yield $x=3$, $y=\frac{5}{2}$.
The maximum is $F=15-18-25=-28$.

9. The objective function is
$F(x,y,\lambda)=x^2+y^2+\lambda(2x+y+1)$.
$F_x=2x+2\lambda,\ F_y=2y+\lambda,\ F_\lambda=2x+y+1$
$F_x=0$ and $F_y=0$ yields $x-2y=0$. $F_\lambda=0$ gives $2x+y=-1$.
These two equations yield $x=-\frac{2}{5},\ y=-\frac{1}{5}$.
The minimum is $z=\frac{4}{25}+\frac{1}{25}=\frac{1}{5}$.

11. The objective function is
$F(x,y,z,\lambda)=x^2+y^2+z^2+\lambda(x+y+z-3)$.
$F_x=2x+\lambda,\ F_y=2y+\lambda,$
$F_z=2z+\lambda,\ F_\lambda=x+y+z-3$
$F_x=F_y=F_z=0$ or $x=y=z=-\frac{\lambda}{2}$.
$F_\lambda=0$ yields $-\frac{3\lambda}{2}=3$ or $\lambda=-2$.
Then, $x=y=z=1$.
The minimum is $1+1+1=3$.

13. The objective function is
$F(x,y,z,\lambda)=xz+y+\lambda(x^2+y^2+z^2-1)$.
$F_x=z+2\lambda x,\ F_y=1+2\lambda y,$
$F_z=x+2\lambda z,\ F_\lambda=x^2+y^2+z^2-1$
$F_x=F_z=0$ gives $x-2\lambda x=z-2\lambda z$ or $x(1-2\lambda)-z(1-2\lambda)=0$. So, $x=z$ or $1-2\lambda=0$. If $x=z$, then $1+2\lambda=0$. The solutions are $\lambda=\pm\frac{1}{2}$. If $\lambda=\frac{1}{2}$, then $F_y=0$ gives $y=-1$. Also, $z+x=0$ and $F_\lambda=0$ means that $x^2+1+z^2-1=0$ or $x=z=0$. The critical point is $(0,-1,0)$. If $\lambda=-\frac{1}{2}$, then $F_y=0$ gives $y=1$. From $F_x=F_z$ we get $z=x$. From $F_\lambda=0$ we get $x=z=0$. The critical point is $(0,1,0)$. At $(0,-1,0)$, we have $w=0-1=-1$. At $(0,1,0)$, we have $w=0+1=1$. The maximum is $w=1$ at $(0,1,0)$.

15. The objective function is
$F(x,y,\lambda)=x^2y+\lambda(x+2y-6)$.
$F_x=2xy+\lambda,\ F_y=x^2+2\lambda,\ F_\lambda=x+2y-6$
$F_x=F_y=0$ or $-\frac{x^2}{2}=-2xy$ or $x^2-4xy=0$ or $x(x-4y)=0$ thus $x=0$ or $x=4y$.
$F_\lambda=0$: If $x=0$, then $y=3$. If $x=4y$, then $6y-6=0$ or $y=1$. If $y=1$, then $x=4$. At $(0,3)$, we have $U=0\cdot 3=0$ and at $(4,1)$ we have $U=16\cdot 1=16$. The maximum occurs at $(4,1)$.

17. The objective function is
$F(x,y,\lambda)=x^2y+\lambda(2x+3y-120)$.
$F_x=2xy+2\lambda,\ F_y=x^2+3\lambda,\ F_\lambda=2x+3y-120$
$F_x=0$ gives $\lambda=-xy$. $F_y=0$ and $F_x=0$ give $x^2-3xy=0$ so $x=0$ or $x=3y$.
$F_\lambda=0$: If $x=0$, then $y=40$.
If $x=3y$, then $y=\frac{40}{3}$ and $x=40$.
At $(0,40)$, we have $U=0\cdot 40=0$.
At $(40,\frac{40}{3})$, we have $U=1600\cdot\frac{40}{3}=\frac{64,000}{3}$.
The maximum is at $(40,\frac{40}{3})$.

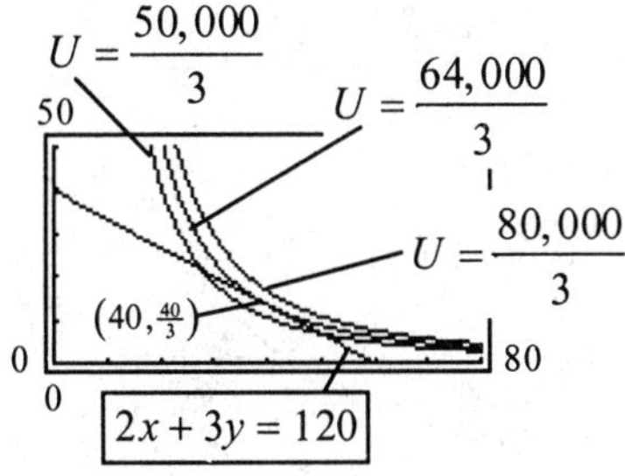

19. a. The objective function is
$F(x,y,\lambda)=400x^{0.6}y^{0.4}+\lambda(150x+100y-100,000)$
$F_x=\frac{240y^{0.4}}{x^{0.4}}+150\lambda,\ F_y=\frac{160x^{0.6}}{y^{0.6}}+100\lambda$
$F_\lambda=150x+100y-100,000$
$F_x=0$ and $F_y=0$ give $\frac{240y^{0.4}}{150x^{0.4}}=\frac{160x^{0.6}}{100y^{0.6}}$
or $x=y$.
$F_\lambda=0$: $250x-100,000=0$ or $x=400$.
$250y-100,000=0$ or $y=400$.

b. $-\lambda=\frac{240(400)^{0.4}}{150(400)^{0.4}}=\frac{8}{5}=1.6$
An additional 1.6 units are produced for each additional dollar spent on production.

c.

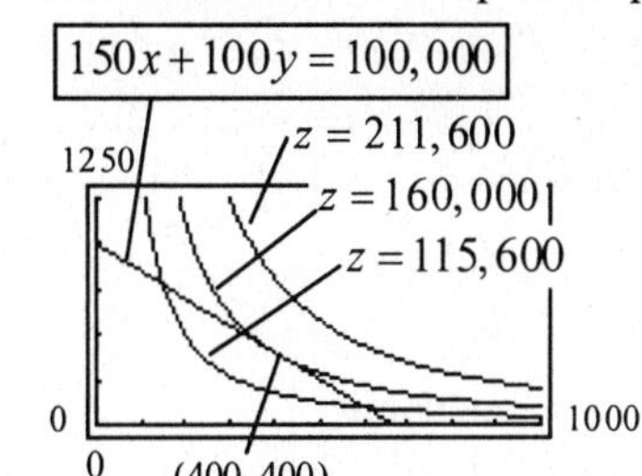

21. The objective function is
$F(x,y,\lambda)=x^2+1200+3y^2+800+\lambda(x+y-1200)$.
$F_x=2x+\lambda,\ F_y=6y+\lambda,\ F_\lambda=x+y-1200$
$F_x=F_y=0$ yields $x-3y=0$.
$F_\lambda=0$ yields $x+y=1200$.
Solving these 2 equations yields $(900,300)$.
To minimize cost let $x=900$ and $y=300$.

23. The objective function is
$F(x,y,\lambda)=20x+y^2+4xy+\lambda(x+y-30,000)$.
$F_x=20+4y+\lambda,\ F_y=2y+4x+\lambda,$
$F_\lambda=x+y-30,000$
$F_x=F_y=0$ yields $2x-y=10$.
$F_\lambda=0$ yields $x+y=30,000$.
Solving these 2 equations gives maximum revenue if $x=\$10,003.33$ and $y=\$19,996.67$.

25. The objective function is
$F(x,h,\lambda)=x^2+4xh+\lambda(x^2h-500,000)$.
$F_x=2x+4h+2xh\lambda,\ F_h=4x+x^2\lambda$
$F_\lambda=x^2h-500,000$
$F_h=0$ means $x=0$ or $x\lambda=-4$. x cannot be 0.
$F_x=0$ and $x\lambda=-4$ gives $2x+4h-8h=0$ or $x=2h$. $F_\lambda=0$ gives $4h^3=500,000$ or $h=50$. With $x=2h=100$, the dimensions are 100 by 100 by 50.

Review Exercises

1. $z = \dfrac{3}{2x - y}$

Domain = $\{(x, y): x, y \in \text{Reals and } y \neq 2x\}$.

2. $z = \dfrac{3x + 2\sqrt{y}}{x^2 + y^2}$

Domain: $\{(x, y): x$ and y are real with $y \geq 0$ and $(x, y) \neq (0, 0)\}$

3. $w(x, y, z) = x^2 - 3yz$

$w(2,3,1) = 4 - 3(3)(1) = -5$

4. $Q(K, L) = 70K^{2/3}L^{1/3}$

$Q(64{,}000,\ 512) = 70(1600)(8) = 896{,}000$

5. $z = 5x^3 + 6xy + y^2$

$z_x = 15x^2 + 6y$

6. $z = 12x^5 - 14x^3y^3 + 6y^4 - 1$

$\dfrac{\partial z}{\partial y} = -42x^3y^2 + 24y^3$

7. $z = 4x^2y^3 + \dfrac{x}{y}$

$z_x = 8xy^3 + \dfrac{1}{y}, \quad z_y = 12x^2y^2 - \dfrac{x}{y^2}$

8. $z = \left(x^2 + 2y^2\right)^{1/2}$

$z_x = \dfrac{1}{2}\left(x^2 + 2y^2\right)^{-1/2}(2x) = \dfrac{x}{\sqrt{x^2 + 2y^2}}$

$z_y = \dfrac{1}{2}\left(x^2 + 2y^2\right)^{-1/2}(4y) = \dfrac{2y}{\sqrt{x^2 + 2y^2}}$

9. $z = (xy + 1)^{-2}$

$z_x = -2(xy + 1)^{-3}(y) = -\dfrac{2y}{(xy + 1)^3}$

$z_y = -2(xy + 1)^{-3}(x) = -\dfrac{2x}{(xy + 1)^3}$

10. $z = e^{x^2y^3}$

$z_x = e^{x^2y^3} \cdot 2xy^3 = 2xy^3e^{x^2y^3}$

$z_y = e^{x^2y^3} \cdot 3x^2y^2 = 3x^2y^2e^{x^2y^3}$

11. $z = e^{xy} + y\ln x$

$z_x = e^{xy}(y) + y \cdot \dfrac{1}{x} = ye^{xy} + \dfrac{y}{x}$

$z_y = e^{xy}(x) + \ln x = xe^{xy} + \ln x$

12. $z = e^{\ln xy} = xy$

$z_x = y, \ z_y = x$

13. $f(x, y) = 4x^3 - 5xy^2 + y^3$

$f_x = 12x^2 - 5y^2$

$f_x(1, 2) = 12 - 20 = -8$

14. $z = 5x^4 - 3xy^2 + y^2$

$z_x = 20x^3 - 3y^2, \ z_x(1,\ 2) = 20 - 12 = 8$

15. $z = x^2y - 3xy$

$z_x = 2xy - 3y, \ z_y = x^2 - 3x$

a. $z_{xx} = 2y$

b. $z_{yy} = 0$

c. $z_{xy} = 2x - 3$

d. $z_{yx} = 2x - 3$.

16. $z = 3x^3y^4 - \dfrac{x^2}{y^2}$

$z_x = 9x^2y^4 - \dfrac{2x}{y^2}, \ z_y = 12x^3y^3 + \dfrac{2x^2}{y^3}$

a. $z_{xx} = 18xy^4 - \dfrac{2}{y^2}$

b. $z_{yy} = 36x^3y^2 - \dfrac{6x^2}{y^4}$

c. $z_{xy} = 36x^2y^3 + \dfrac{4x}{y^3}$

d. $z_{yx} = 36x^2y^3 + \dfrac{4x}{y^3}$

17. $z = x^2e^{y^2}$

$z_x = 2xe^{y^2}, \ z_y = x^2e^{y^2} \cdot 2y$

a. $z_{xx} = 2e^{y^2}$

b. $z_{yy} = x^2e^{y^2} \cdot 2 + 2yx^2e^{y^2} \cdot 2y$

$= 2x^2e^{y^2} + 4x^2y^2e^{y^2}$

c. $z_{xy} = 2xe^{y^2} \cdot 2y = 4xye^{y^2}$

d. $z_{yx} = 4xye^{y^2}$

18. $z = \ln(xy+1)$

$z_x = \dfrac{y}{xy+1},\ z_y = \dfrac{x}{xy+1}$

a. $z_{xx} = \dfrac{-y^2}{(xy+1)^2}$

b. $z_{yy} = \dfrac{-x^2}{(xy+1)^2}$

c. $z_{xy} = \dfrac{(xy+1) - y\cdot x}{(xy+1)^2} = \dfrac{1}{(xy+1)^2}$

d. $z_{yx} = \dfrac{(xy+1) - x\cdot y}{(xy+1)^2} = \dfrac{1}{(xy+1)^2}$

19. $z = 16 - x^2 - xy - y^2 + 24y$

$z_x = -2x - y,\ z_y = -x - 2y + 24$

$z_x = 0$ gives $y = -2x$. Using $z_y = 0$ and $y = -2x$ yields $x = -8$ and thus, $y = 16$.

Critical point at $(-8,16)$.

$z_{xx} = -2,\ z_{yy} = -2,\ z_{xy} = -1$

$D = (-2)(-2) - (-1)^2 > 0$

Since $D > 0$ and $z_{xx} < 0$, there is a relative maximum of $z = 208$ at $(-8,16)$.

20. $z = x^3 + y^3 - 3xy$

$z_x = 3x^2 - 3y,\ z_y = 3y^2 - 3x$

$z_x = 0$ gives $y = x^2$. Using $z_y = 0$ and $y = x^2$ yields $x^4 - x = 0$ or $x(x^3 - 1) = 0$.

When $x = 0$, $y = 0$ and when $x = 1$, $y = 1$.

The critical points are $(0,0)$ and $(1,1)$.

$z_{xx} = 6x,\ z_{yy} = 6y,\ z_{xy} = -3$

At (0, 0), $D = (6x)(6y) - (-3)^2 = -9 < 0$. The point $(0,0,0)$ is a saddle point.

At (1, 1), $D = (6x)(6y) - (-3)^2 = 36 - 9 = 27 > 0$ with $z_{xx} > 0$. A relative minimum at $(1,1,-1)$.

21. The objective function is

$F(x,y,\lambda) = 4x^2 + y^2 + \lambda(x+y-10)$

$F_x = 8x + \lambda,\ F_y = 2y + \lambda,\ F_\lambda = x + y - 10$

$F_x = F_y = 0$ gives $y = 4x$. $y = 4x$ and $F_\lambda = 0$ gives $5x - 10 = 0$ or $x = 2$. So, $y = 8$.

The minimum is $z = 4(4) + 64 = 80$ at $(2,8)$.

22. The objective function is

$F(x,y,\lambda) = x^4y^2 + \lambda(x+y-9)$

$F_x = 4x^3y^2 + \lambda,\ F_y = 2x^4y + \lambda,$

$F_\lambda = x + y - 9$

$F_x = F_y = 0$ gives $4x^3y^2 = 2x^4y$ or $2x^3y(2y - x) = 0$ so $x = 0$, $y = 0$, or $x = 2y$.

$F_\lambda = 0$: If $x = 0$, then $y = 9$. If $y = 0$, then $x = 9$. If $x = 2y$, then $y = 3$ and $x = 6$.

At $(0,9)$ $z = 0\cdot 81 = 0$.

At $(9,0)$, $z = 6561\cdot 0 = 0$.

At $(6,3)$, $z = 6^4\cdot 3^2 = 11{,}664$.

The maximum value is 11,664 at $(6,3)$.

23. At $x = 6$ and $y = 15$, $U = 6^2\cdot 15 = 540$.

To retain the same value of U at $y = 60$, we have $540 = x^2\cdot 60$ or $x^2 = 9$ or $x = 3$.

24. $xy = 1600$

When $x = 80$, $y = 20$.

25. a. $w = 2\sqrt{\dfrac{Th}{d}},\quad w = 2\sqrt{\dfrac{293(16.8)}{5}} = 62.8$ km

b. At a surface temperature of 287 K, an altitude of 17.1 km, and a vertical temperature gradient of 4.9 K/km, the width of the region on the ground on either side where people hear the sonic boom is about 63.3 km.

c. $\dfrac{\partial f}{\partial h} = \dfrac{T}{d\sqrt{\frac{Th}{d}}} = \dfrac{\sqrt{T}}{\sqrt{dh}} = \sqrt{\dfrac{T}{dh}}$

At $(293,16.8,5)$, $\dfrac{\partial f}{\partial h} = \sqrt{\dfrac{293}{5(16.8)}} = 1.87.$

For the given T and d, a change of 1 km in the altitude will result in an increase of 1.87 km in the width of the sonic boom path on either side.

d. $\dfrac{\partial f}{\partial d} = 2\sqrt{Th}\cdot\left(-\dfrac{1}{2}d^{-3/2}\right) = -\dfrac{\sqrt{Th}}{\left(\sqrt{d}\right)^3}$

At $(293,16.8,5)$,

$\dfrac{\partial f}{\partial d} = -\dfrac{\sqrt{293(16.8)}}{\left(\sqrt{5}\right)^3} = -6.28.$

For the given T and h, a change of 1 K/km in d will result in a decrease of 6.28 km in the width of the sonic boom path on either side.

26. $C(x,y)=x^2\sqrt{y^2+13}$

a. $C_x=2x\sqrt{y^2+13}$

$C_x(20,6)=40\sqrt{36+13}=280$

b. $C_y=\dfrac{x^2y}{\sqrt{y^2+13}}$

$C_y(20,6)=\dfrac{400\cdot 6}{\sqrt{36+13}}=\dfrac{2400}{7}$

27. $Q=80K^{1/4}L^{3/4},\ K=625,\ L=4096$

$\dfrac{\partial Q}{\partial K}=\dfrac{20L^{3/4}}{K^{3/4}},\ \dfrac{\partial Q}{\partial L}=\dfrac{60K^{1/4}}{L^{1/4}}$

For the given values $\dfrac{\partial Q}{\partial K}=81.92$ and

$\dfrac{\partial Q}{\partial L}=37.5$ (hundreds).

When work-hours are fixed at 4096, an increase of \$1000 in expenditures results in an increase of 8192 units produced. When expenditures are fixed at \$625,000 an increase of one work-hour results in an increase of 3750 units produced.

28. $q_A=400-2p_A-3p_B$

$q_B=300-5p_A-6p_B$

a. $\dfrac{\partial q_A}{\partial p_A}=-2$

b. $\dfrac{\partial q_B}{\partial p_B}=-6$

c. The two products are complementary since $\dfrac{\partial q_A}{\partial p_B}$ and $\dfrac{\partial q_B}{\partial p_A}$ are both negative.

29. $q_A=800-40p_A-2(p_B+1)^{-1}$

$q_B=1000-10(p_A+4)^{-1}-30p_B$

$\dfrac{\partial q_A}{\partial p_A}=-40$

$\dfrac{\partial q_A}{\partial p_B}=\dfrac{2}{(p_B+1)^2}$

$\dfrac{\partial q_B}{\partial p_A}=\dfrac{10}{(p_A+4)^2}$

$\dfrac{\partial q_B}{\partial p_B}=-30$

Competitive since $\dfrac{\partial q_A}{\partial p_B}$ and $\dfrac{\partial q_B}{\partial p_A}$ are both positive.

30. $P(x,y)=40x+80y-x^2-y^2$

$P_x=40-2x,\ P_y=80-2y$

$P_x=0$ when $x=20$ and $P_y=0$ when $y=40$.

Critical point at $(20,40)$.

$P_{xx}=-2,\ P_{yy}=-2,\ P_{xy}=0$

$D=(-2)(-2)-(0)^2>0$ with $P_{xx}<0$ means profit is maximized when $x=20$ and $y=40$.

31. The objective function is

$F(x,y,\lambda)=x^2y+\lambda(4x+5y-60)$

$F_x=2xy+4\lambda,\ F_y=x^2+5\lambda$

$F_\lambda=4x+5y-60$

$F_x=0$ and $F_y=0$ give $\dfrac{-xy}{2}=\dfrac{-x^2}{5}$ or

$5xy-2x^2=0$ or $x(5y-2x)=0$. Since x cannot be 0, we have $5y-2x=0$ or $y=\dfrac{2}{5}x$.

So, $4x+5\cdot\frac{2}{5}x-60=0$ or $x=10$. Then $y=\frac{2}{5}\cdot 10=4$.

32. a. The objective function is

$F(x,y,\lambda)=300x^{2/3}y^{1/3}$
$+\lambda(50x+50y-75{,}000)$

$F_x=\dfrac{200y^{1/3}}{x^{1/3}},\ F_y=\dfrac{100x^{2/3}}{y^{2/3}}$

$\dfrac{\partial F}{\partial\lambda}=50x+50y-75{,}000$

$F_x=F_y=0$ gives $\dfrac{4y^{1/3}}{x^{1/3}}=\dfrac{2x^{2/3}}{y^{2/3}}$ or $x=2y$.

$F_\lambda=0$ and $x=2y$ give $150y-75{,}000=0$ or $y=500$, so $x=1000$.

Production is maximized at $x=1000$, $y=500$.

b. $-\lambda=\dfrac{4y^{1/3}}{x^{1/3}}=\dfrac{4\sqrt[3]{500}}{\sqrt[3]{1000}}\approx 3.17$ means each additional dollar spent on production results in an additional 3.17 units produced.

c.

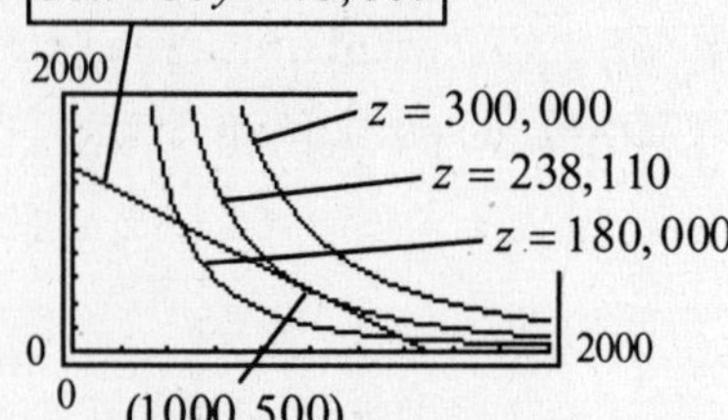

33. a. Use a graphing utility to obtain $\hat{y} = 0.127x + 8.926$

b. If $x = 7000$, $\hat{y} = \$897.926$.

34. Use a graphing utility to obtain $q = -55.09p + 34,726.46$.

Chapter Test

1. $f(x, y) = \dfrac{2x+3y}{\sqrt{x^2 - y}}$

a. Domain = $\{(x, y): x \text{ and } y \text{ are real and } y < x^2\}$

b. $f(-4, 12) = \dfrac{-8+36}{2} = 14$

2. $z = 5x - 9y^2 + 2(xy+1)^5$

$z_x = 5 + 10(xy+1)^4(y) = 5 + 10y(xy+1)^4$

$z_y = -18y + 10(xy+1)^4(x)$

$= -18y + 10x(xy+1)^4$

$z_{xx} = 40y(xy+1)^3(y) = 40y^2(xy+1)^3$

$z_{yy} = -18 + 40x(xy+1)^3(x)$

$= -18 + 40x^2(xy+1)^3$

$z_{xy} = z_{yx} = 10y \cdot 4(xy+1)^3(x) + (xy+1)^4(10)$

$= 10(xy+1)^3(5xy+1)$

3. $z = 6x^2 + x^2y + y^2 - 4y + 9$

$z_x = 12x + 2xy = 2x(6+y)$

$z_y = x^2 + 2y - 4$

$z_x = 0$ if $x = 0$ or $y = -6$

If $x = 0$ and $z_y = 0$, then $y = 2$. If $y = -6$ and $z_y = 0$, then $x^2 = 16$ or $x = \pm 4$.

Critical points are $(0,2)$, $(4,-6)$, and $(-4,-6)$.

$z_{xx} = 12 + 2y,\ z_{yy} = 2,\ z_{xy} = 2x$

$D = z_{xx} \cdot z_{yy} - (z_{xy})^2$

At $(0,2)$: $D = 16 \cdot 2 - 0 > 0$

At $(4,-6)$: $D = 0 \cdot 2 - 64 < 0$

At $(-4,-6)$: $D = 0 \cdot 2 - 64 < 0$

$D > 0$ and $z_{xx} > 0$ yield a relative minimum at $(0,2)$. $D < 0$ yields saddle points at $(4,-6)$ and $(-4,-6)$.

4. $Q = 10K^{0.45}L^{0.55},\ K = 10,\ L = 1590$

a. $Q = 10(10)^{0.45}(1590)^{0.55} = \$1625$ thousands

b. $Q_K = \dfrac{4.5L^{0.55}}{K^{0.55}},\ Q_K = \dfrac{4.5(1590)^{0.55}}{10^{0.55}} = 73.11$

If labor hours remain constant and capital investment increases to \$11,000, the monthly production should increase \$73.11 thousands.

c. $Q_L = \dfrac{5.5K^{0.45}}{L^{0.45}},\ Q_L = \dfrac{5.5(10)^{0.45}}{(1590)^{0.45}} = 0.56$

If capital investment remains constant and labor hours increase to 1591, the monthly production should increase \$0.56 thousands.

5. a. $f(94.5, 25, 7) = \$667.91$

When \$94,500 is borrowed for 25 years at 7%, compounded monthly, the monthly payment is \$667.91.

b. $\dfrac{\partial f}{\partial r}(94.5, 25, 7) = \$49.76$

If the rate increases from 7% to 8%, the monthly payment will increase \$49.76. (The \$94,500 and 25 years remain fixed.)

c. $\dfrac{\partial f}{\partial n}(94.5, 25, 7) < 0$ means that the monthly payment decreases if the time to pay off the loan increases. (The \$94,500 and 7% rate remain fixed.)

6. $f(x, y) = 2e^{x^2y^2}$

$\dfrac{\partial f}{\partial y} = 2e^{x^2y^2}(2x^2y) = 4x^2ye^{x^2y^2}$

$\dfrac{\partial^2 f}{\partial x \partial y} = 4x^2y\left(e^{x^2y^2} \cdot 2xy^2\right) + e^{x^2y^2}(8xy)$

$= 8xye^{x^2y^2}\left(1 + x^2y^2\right)$

7. Find $\dfrac{\partial q_1}{\partial p_2}$ and $\dfrac{\partial q_2}{\partial p_1}$ and compare their signs.

Both positive means the products are competitive. Both negative means the products are complementary.

$\dfrac{\partial q_1}{\partial p_2} = -5,\ \dfrac{\partial q_2}{\partial p_1} = -4$

These products are complementary.

8. $P = 915x - 30x^2 - 45xy + 975y - 30y^2 - 3500$
$P_x = 915 - 60x - 45y = 15(61 - 4x - 3y)$
$P_y = -45x + 975 - 60y = 15(65 - 3x - 4y)$
Set P_x and P_y to 0 and solve.
$3x + 4y = 65$
$4x + 3y = 61$
The solution is $x = 7$ and $y = 11$.
$P_{xx} = -60,\ P_{yy} = -60,\ P_{xy} = -45$
$D = (-60)(-60) - (-45)^2 > 0$ and $P_{xx} < 0$ means that P is a maximum at $(7, 11)$.

9. $U = x^3 y$ with budget constraint $30x + 20y = 8000$.
The objective function is
$F(x, y, \lambda) = x^3 y + \lambda(30x + 20y - 8000)$
$\frac{\partial F}{\partial x} = 3x^2 y + 30\lambda,\quad \frac{\partial F}{\partial y} = x^3 + 20\lambda$
$\frac{\partial F}{\partial \lambda} = 30x + 20y - 8000$
Set each partial to zero and solve for λ.
$\lambda = -\frac{x^2 y}{10}$ and $\lambda = -\frac{x^3}{20}$ yield $2x^2 y = x^3$ or $x^2(x - 2y) = 0$, $x \neq 0$. Substituting $x = 2y$ into $\frac{\partial F}{\partial \lambda}$ gives $80y - 8000 = 0$ or $y = 100$.
Then $x = 200$. The function is maximized at 800,000,000 when $x = 200$ and $y = 100$.

10. a. $\sum x = 66, \sum y = 129{,}680, \sum x^2 = 506,$ $\sum xy = 795{,}260$
$b = \frac{66(129{,}680) - 12(795{,}260)}{(66)^2 - 12(506)} = 573.57$
$a = \frac{129{,}680 - 573.57(66)}{12} = 7652.05$
$\hat{y} = 7652.05 + 573.57x$

b. If $x = 14$,
$\hat{y} = 7652.05 + 573.57(14) = \$15{,}682$

c. The model would not make sense since it is a model for children and not adults.

Supplementary Exercises: Chapter 0

Section 0.1

1. If $U = \{1, 2, 3, 4, 5\}$, $A = \{1, 3, 4\}$, and $B = \{2, 4, 5\}$, then $(A' \cap B)'$ is equal to which of the following?
   a. $\{4\}$ **b.** $\{1, 3\}$ **c.** $\{1, 3, 4\}$ **d.** $\{1, 2, 3, 5\}$ **e.** $\varnothing$

2. Records at Midtown University show the following about the enrollments of 200 students:
   70 take Math, 60 take History, 70 take English
   25 take Math and History
   15 take History and English
   30 take Math and English
   10 take all three
   How many take none of these three courses?
   a. 90 **b.** 60 **c.** 40 **d.** 20 **e.** 0

3.

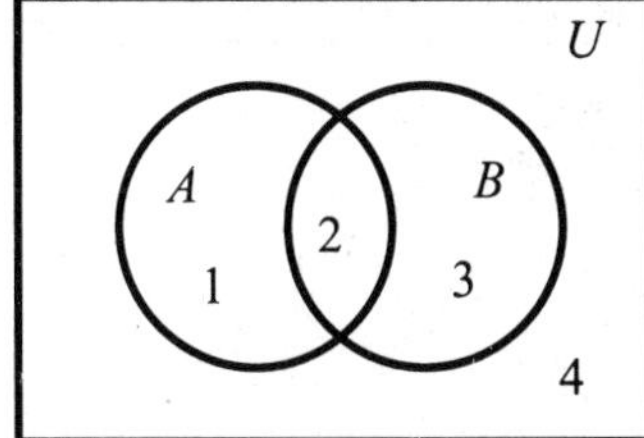

Regions 1 and 3 are represented by which of the following?
a. $(A \cup B)'$ **b.** $A' \cup B'$ **c.** $A' \cap B'$ **d.** $(A \cap B') \cup (A' \cap B)$ **e.** $(A \cup B') \cap (A' \cup B)$

4. Question to be answered: Is there enough information to tell how many students are in this dorm wing?
   Suppose that each student in a college dorm wing is taking at least one of the three courses, A, B, and C. Enrollments from the dorm are given in the table below.

Course	A only	B only	C only	A and B only	B and C only	A and C only
Number of students	6	7	8	5	4	3

(1) Can you answer the question if there exactly two students taking all three courses? (Yes/No)

(2) Can you answer the question if exactly 14 students are taking two or more of the three courses? (Yes/No)

Additional question: How many students are in this dorm wing?

Section 0.1: Solutions

1. $A' \cap B = \{2, 5\}$
   $\{2, 5\}' = \{1, 3, 4\}$
   Answer: c

2. Step 1:

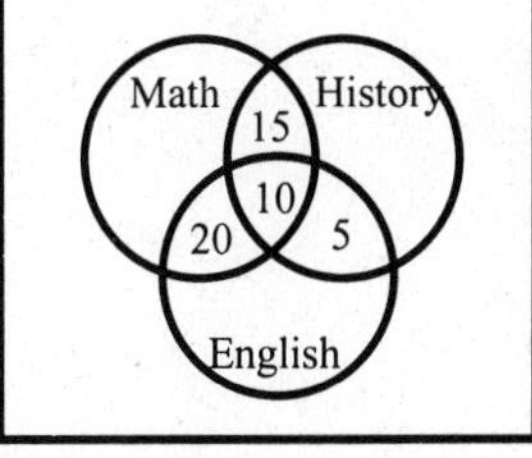

Step 2:

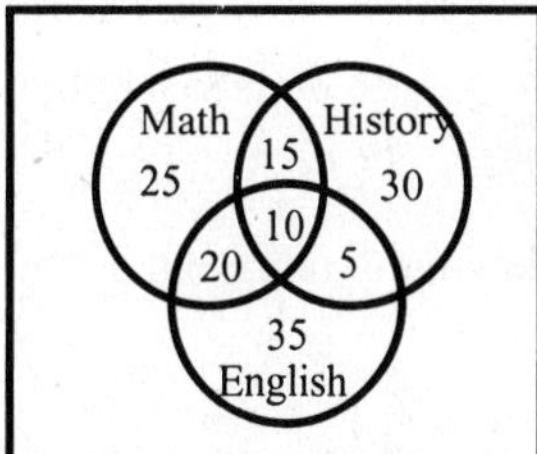

Step 3: Adding all the numbers in step 2, we have $n(M \cup H \cup E) = 140$. Thus, 60 take none of the three courses. Answer: b

3. $A \cap B' = \text{Region 1}$

   $A' \cap B = \text{Region 3}$

   Answer: d

4. 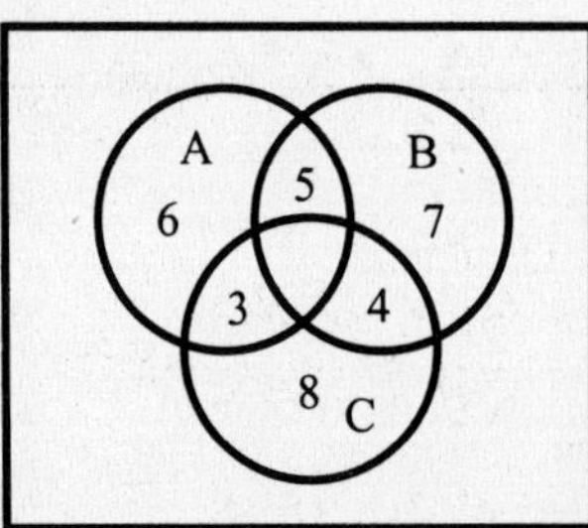

   Yes, we can answer the question with (1).

   Yes, we can answer the question with (2). There are 35 students in this dorm wing.

Section 0.2

1. The value of $\dfrac{-2^4+3\cdot 2-4}{5-3\cdot 2^2}$ is which of the following?

   a. $\dfrac{18}{7}$ **b.** $-\dfrac{18}{7}$ **c.** 2 **d.** −2 **e.** $-\dfrac{8}{7}$

2. Choose one of the following to replace □ in the statement $15^3+15^3 \square 30^3$. Do not use a calculator.

   a. > **b.** = **c.** < **d.** Cannot be determined.

3. $(-1)^2-(-1)^3=?$

   a. −2 **b.** −1 **c.** 0 **d.** 1 **e.** 2

4. $-1^3-(-1)^2=?$

   a. −2 **b.** −1 **c.** 0 **d.** 1 **e.** 2

Section 0.2: Solutions

1. $\dfrac{-2^4+3\cdot 2-4}{5-3\cdot 2^2}=\dfrac{-16+6-4}{5-12}=\dfrac{-14}{-7}=2$

   Answer: c

2. $15^3+15^3=2\cdot 15^3$

   $30^3=2^3\cdot 15^3$

   Answer: c

3. $(-1)^2-(-1)^3=1-(-1)=2$

   Answer: e

4. $-1^3-(-1)^2=-1-1=-2$

   Answer: a

Section 0.3 (no calculators)

1. Choose one of the following to replace h in the statement 8^7 **h** $2\cdot 8^6+4\cdot 8^6+8^6$.

   a. > **b.** = **c.** < **d.** Cannot be determined

2. Choose one of the following to replace h in the statement 15^2+12^2 **h** $(15+12)^2$.

   a. > **b.** = **c.** < **d.** Cannot be determined

3. Simplify: $\dfrac{x^{b(a-b)}\cdot x^{a(a+b)}}{x^{b^2-a^2}}\cdot\dfrac{1}{\left(x^{ab}\right)^2}$

4. Find the value of $(2^{-2}+3^{-1})^{-1}$.

5. Find the value of $(2^{-2}\cdot 3^{-1})^{-1}$.

Section 0.3 Solutions

1. $8^7 = 8\cdot 8^6$
$2\cdot 8^6+4\cdot 8^6+8^6 = 8^6(2+4+1) = 8^6\cdot 7$
$8^7 > 2\cdot 8^6+4\cdot 8^6+8^6$
Answer: a

2. $15^2+12^2 = 225+144 = 369$
$(15+12)^2 = 27^2 = 729$
$15^2+12^2 < (15+12)^2$
Answer: c
(Note: This problem is preparing you for the question, "Is $(x+y)^2 = x^2+y^2$?")

3. $\dfrac{x^{b(a-b)}\cdot x^{a(a+b)}}{x^{b^2-a^2}}\cdot\dfrac{1}{(x^{ab})^2}$
$=\dfrac{x^{ab-b^2}\cdot x^{a^2+ab}}{x^{b^2-a^2}}\cdot\dfrac{1}{x^{2ab}}=\dfrac{x^{a^2+2ab-b^2}}{x^{b^2-a^2}}\cdot x^{-2ab}$
$=\dfrac{x^{a^2-b^2}}{x^{b^2-a^2}}=x^{a^2-b^2-(b^2-a^2)}=x^{2a^2-2b^2}$

4. $(2^{-2}+3^{-1})^{-1}=\left(\dfrac{1}{2^2}+\dfrac{1}{3}\right)^{-1}$
$=\left(\dfrac{3+4}{12}\right)^{-1}=\left(\dfrac{7}{12}\right)^{-1}=\dfrac{12}{7}$

5. $(2^{-2}\cdot 3^{-1})^{-1} = 2^2\cdot 3 = 4\cdot 3 = 12$
(Note that different methods were required for problems 4 and 5.)

Section 0.4 (no calculators)

1. Find the value of $\left(8^{2/3}-27^{1/3}\right)^{1/2}$.

2. Find the value of $\left[\dfrac{9^{3/2}-2\cdot 8^{-2/3}}{2+\left(\frac{1}{4}\right)^{-2}}\right]^{-1}$.

3. $\dfrac{4x^{-3}}{(4x)^3}=?$

a. $\dfrac{1}{16}$ b. $\dfrac{1}{16x^6}$ c. $\dfrac{1}{16x^9}$ d. $\dfrac{1}{4x^6}$ e. $\dfrac{1}{256x^6}$

4. $\left(2x^{-2}y\right)^{-2}\left(3xy^2\right)=?$

a. $12x^5y^3$ b. $\dfrac{3x^5}{4}$ c. $\dfrac{3x^5}{y^4}$ d. $6x^5$ e. $\dfrac{6x^5}{y^4}$

Section 0.4 Solutions

1. $\left(8^{2/3}-27^{1/3}\right)^{1/2}=\left[\left(\sqrt[3]{8}\right)^2-\sqrt[3]{27}\right]^{1/2}=(4-3)^{1/2}=\sqrt{1}=1$

2. $\left[\dfrac{9^{3/2}-2\cdot 8^{-2/3}}{2+\left(\frac{1}{4}\right)^{-2}}\right]^{-1}=\left[\dfrac{\left(\sqrt{9}\right)^3-2\cdot\frac{1}{\left(\sqrt[3]{8}\right)^2}}{2+(4)^2}\right]^{-1}=\left[\dfrac{27-2\cdot\frac{1}{4}}{2+16}\right]^{-1}=\left[\dfrac{\frac{53}{2}}{18}\right]^{-1}=\left(\dfrac{53}{36}\right)^{-1}=\dfrac{36}{53}$

3. $\dfrac{4x^{-3}}{(4x)^3} = \dfrac{4x^{-3}}{4^3 \cdot x^3} = \dfrac{1}{4^2} \cdot \dfrac{1}{x^{3-(-3)}} = \dfrac{1}{16x^6}$ Answer: b

4. $\left(2x^{-2}y\right)^{-2} \cdot \left(3xy^2\right) = 2^{-2} \cdot x^4 y^{-2} \cdot 3xy^2 = \dfrac{3x^5}{4}$ Answer: b

Section 0.5

1. $(3x+5)^2 - (2x-3)(x+2) = ?$

   a. $7x^2 + 29x + 31$ **b.** $7x^2 + x + 31$ **c.** $7x^2 - x + 31$ **d.** $7x^2 - x + 19$ **e.** $7x^2 + 31x + 19$

2. $2x - 3\left[2x - (x+2)^2\right] = ?$

   a. $-x^2 - 8x + 4$ **b.** $x^2 - 8x + 4$ **c.** $-3x^2 + 4x - 12$ **d.** $3x^2 + 8x + 12$ **e.** none of these

3. If $x = 2 - 3\sqrt{2}$, then $x^2 - 4x + 1 = ?$

   a. 15 **b.** $15 + 12\sqrt{2}$ **c.** $-15 + 12\sqrt{2}$ **d.** $-21 + 12\sqrt{2}$ **e.** 33

4. $(2x-3)\left[4x - (x+2)^2\right] = ?$ (Compare with problem 2.)

5. When is $(x+y)^2 = x^2 + y^2$? >? <?

Section 0.5: Solutions

1. $(3x+5)^2 - (2x-3)(x+2) = 9x^2 + 30x + 25 - \left(2x^2 + x - 6\right) = 7x^2 + 29x + 31$ Answer: a

2. $2x - 3\left[2x - (x+2)^2\right] = 2x - 3\left[2x - (x^2 + 4x + 4)\right] = 2x - 3\left[-x^2 - 2x - 4\right]$

   $= 2x + 3x^2 + 6x + 12 = 3x^2 + 8x + 12$ Answer: d

3. $\left(2 - 3\sqrt{2}\right)^2 - 4\left(2 - 3\sqrt{2}\right) + 1 = \left(4 - 12\sqrt{2} + 18\right) - 8 + 12\sqrt{2} + 1 = 15$ Answer: a

4. $(2x-3)\left[4x - (x+2)^2\right] = (2x-3)\left[4x - (x^2 + 4x + 4)\right]$

   $= (2x-3)\left(-x^2 - 4\right) = -1\left(x^2 + 4\right)(2x-3) = -\left(2x^3 - 3x^2 + 8x - 12\right)$

5. "=" occurs when $x \cdot y = 0$.
   ">" occurs when x and y have the same sign.
   "<" occurs when x and y have opposite signs.
   (Note: The purpose of the question is for the student to know that algebraically $(x+y)^2 = x^2 + 2xy + y^2$.)

Section 0.6

1. Using factoring find the value of $103^2 - 97^2$.

2. Using factoring choose one of the following to replace h in the statement $\dfrac{3^{25} - 3^{23}}{8}$ **h** 3^{23}.

   a. > **b.** = **c.** < **d.** Cannot be determined

3. When is $\sqrt{x+y} = \sqrt{x} + \sqrt{y}$?

4. Factor: $3x^4\left(x^2 - 1\right)^3 + 4x^2\left(x^2 - 1\right)^4$

5. Factor: $8x^3\left(x^2 + 1\right)^2 + 3x^4\left(x^2 + 1\right)^4$

Section 0.6: Solutions

1. $103^2 - 97^2 = (103+97)(103-97) = 200 \cdot 6 = 1200$

2. $\dfrac{3^{25} - 3^{23}}{8} = \dfrac{3^{23}(3^2 - 1)}{8} = \dfrac{3^{23} \cdot 8}{8} = 3^{23}$

3. $\sqrt{x+y} = \sqrt{x} + \sqrt{y}$ when $x \cdot y = 0$. In general, the conclusion for algebraic purposes is $\sqrt{x+y} \neq \sqrt{x} + \sqrt{y}$.

4. $3x^4(x^2-1)^3 + 4x^2(x^2-1)^4 = (x^2-1)^3\left[3x^4 + 4x^2(x^2-1)\right]$
$$= x^2(x^2-1)^3\left[3x^2 + 4x^2(x^2-1)\right] = x^2(x^2-1)^3(7x^2-4)$$

5. $8x^3(x^2+1)^2 + 3x^4(x^2+1)^4 = (x^2+1)^2\left[8x^3 + 3x^4(x^2+1)^2\right]$
$$= x^3(x^2+1)^2\left[8 + 3x(x^2+1)^2\right] = x^3(x^2+1)^2\left[8 + 3x^5 + 6x^3 + 3x\right]$$

(Note: The factoring in problems 4 and 5 is necessary in the second course of the two course sequence.)

Section 0.7

1. $\dfrac{2x+1}{4x-2} - \dfrac{5}{2x} - \dfrac{x-1}{2x^2 - x} = ?$

   a. $\dfrac{2x^2 - 5x + 6}{2x(x-1)}$ b. $\dfrac{2x^2 - 6x + 7}{2x(x-1)}$ c. $\dfrac{2x^2 - 6x - 7}{2x(2x-1)}$ d. $\dfrac{2x^2 - 11x - 7}{2x(2x-1)}$ e. $\dfrac{2x^2 - 11x + 7}{2x(2x-1)}$

2. $\dfrac{\sqrt{x^2+4} - \frac{13}{\sqrt{x^2+4}}}{(x+3)} = ?$

   a. $x - 3$ b. $\dfrac{x-3}{\sqrt{x^2+4}}$ c. $\dfrac{x - 3\sqrt{x^2+4}}{x^2+4}$ d. Both b and c e. None of the above

3. Find the value of $(3 \cdot 2^{-2} + 3^{-2})^{-1}$.

4. Why doesn't $\dfrac{x^2 + 2x - 5}{x^2 - 5} = 2x$?

Section 0.7: Solutions

1. $\dfrac{2x+1}{2(2x-1)} - \dfrac{5}{2x} - \dfrac{x-1}{x(2x-1)} = \dfrac{2x+1}{2(2x-1)} \cdot \dfrac{x}{x} - \dfrac{5}{2x} \cdot \dfrac{(2x-1)}{(2x-1)} - \dfrac{x-1}{x(2x-1)} \cdot \dfrac{2}{2}$
$$= \frac{2x^2 + x - 10x + 5 - 2x + 2}{2x(2x-1)} = \frac{2x^2 - 11x + 7}{2x(2x-1)}$$

   Answer: e

2. $\dfrac{\sqrt{x^2+4} - \frac{13}{\sqrt{x^2+4}}}{(x+4)} \cdot \dfrac{\sqrt{x^2+4}}{\sqrt{x^2+4}} = \dfrac{x^2 + 4 - 13}{(x+3)\sqrt{x^2+4}} = \dfrac{x^2 - 9}{(x+3)\sqrt{x^2+4}} = \dfrac{x-3}{\sqrt{x^2+4}}$ Answer: b

3. $(3 \cdot 2^{-2} + 3^{-2})^{-1} = \left(\dfrac{3}{2^2} + \dfrac{1}{3^2}\right)^{-1} = \left(\dfrac{31}{36}\right)^{-1} = \dfrac{36}{31}$

4. You can divide common factors only. You cannot divide terms.

Section 6.2

1. \$8000 is invested for 8 years at 8% compounded quarterly. To the nearest dollar, what is the amount of interest earned?
a. \$1280
b. \$5120
c. \$7076
d. \$15,076
e. \$20,480

2. Given are the four sequences
R: 4, 6, 8, 18
S: 81, –54, 36, –24
T: 12, 6, 0, –6
U: $(1+i)$, $(1+i)^2$, $(1+i)^3$, $(1+i)^4$
What is the number of geometric sequences less the number of arithmetic sequences?
a. –2
b. –1
c. 0
d. 1
e. 2

3. Write the next three terms of each sequence.
a. $\frac{4}{\sqrt{2}}, 4, 4\sqrt{2}$, ___, ___, ___
b. 121, 81, 49, ___, ___, ___

4. At what rate of interest will \$2000 amount to \$14,800 in 21 years if the interest is compounded annually?

Section 6.2 Solutions

1. $P = 8000, \quad \frac{r}{m} = \frac{0.08}{4} = 0.02, \quad mt = 4 \cdot 8 = 32$

$$S = 8000(1+0.02)^{32}$$
$$= 8000(1.884541)$$
$$= \$15,076.32$$

Interest $= S - P = \$7076.32$

2. *R*: 18 makes it neither AP or GP.
S: Geometric sequence with $r = -\frac{2}{3}$.
T: Aritmetic sequence with $d = -6$.
U: Geometric sequence with $r = 1 + i$.
So, $2 - 1 = 1$. Answer: **d**

3. **a.** Geometric sequence with $r = \sqrt{2}$.
Answer: 8, $8\sqrt{2}$, 16
b. Sequence of squares of odd integers.
Answer: 25, 9, 1

4. $P = 2000, \quad S = 14{,}800, \quad n = 21, \quad i = ?$

$$14{,}800 = 2000(1+i)^{21}$$
$$(1+i)^{21} = 7.400$$
$$(1+i) = (7.400)^{1/21}$$
$$1+i = 1.09999$$
$$i = 0.09999 = 0.10$$
$$i = 10\%$$

Section 6.3 & 6.4

1. \$40 is deposited at the end of each 6 month period for 10 years. The account earns interest at 6% compounded semiannually. Which formula gives the amount in the account at the end of 10 years?

a. $A = 40 \cdot a_{\overline{20}|0.03}$ **b.** $S = 80 \cdot s_{\overline{10}|0.06}$ **c.** $S = 40 \cdot s_{\overline{20}|0.03}$ **d.** $S = 40 \cdot s_{\overline{10}|0.06}$ **e.** $S = 40(1+0.03)^{20}$

2. J.R. buys a \$12,000 car by making a \$2000 down payment and equal payments each three months for the next 4 years. The interest rate is 12% compounded quarterly. Which formula gives the size of the quarterly payment?

a. $R = 10{,}000 \cdot \dfrac{1}{s_{\overline{16}|0.03}}$ **b.** $R = 10{,}000 \cdot \dfrac{1}{a_{\overline{16}|0.03}}$ **c.** $R = 10{,}000 \cdot \dfrac{1}{s_{\overline{4}|0.12}}$

d. $R = 10{,}000 \cdot \dfrac{1}{a_{\overline{4}|0.12}}$ **e.** $R = 12{,}000 \cdot \dfrac{1}{a_{\overline{16}|0.03}}$

3. \$500 is deposited at the end of each 6 months for 6 years with interest being paid at the rate of 8% compounded semiannually. The first investment is 6 months from now. At the end of 6 years (and last deposit) the account begins paying 6% compounded annually. One year after the last deposit I begin the first of 10 annual withdrawals. After the tenth withdrawal the account has a zero balance. How much is withdrawn each year?

Section 6.3 & 6.4 Solutions

1. Each 6 months means we have an annuity. Amount at the end means we need to amount of annuity.

$i = \dfrac{0.06}{2} = 0.03,\ n = 2(10) = 20$

Thus, $S = 40 \cdot s_{\overline{20}|0.03}$

Answer: **c**

2. Equal payments means annuity. The \$10,000 is now. This is the present value of an annuity with

$i = \dfrac{0.12}{4} = 0.03$ and $n = 4 \cdot 4 = 16$.

So, $10{,}000 = R \cdot a_{\overline{16}|0.03}$ or $R = 10{,}000 \cdot \dfrac{1}{a_{\overline{16}|0.03}}$.

Answer: **b**

3.

Now		1 yr	⋯	6	7	8	⋯	16
500	500	500	⋯	500	R	R	⋯	R

$S = 500 \cdot s_{\overline{12}|0.04} = \$7096$ in the account at the end of 6 years.

Then we have $R = ?$ $n = 10$, $i = 6\%$, $A = \$7096$

So, $R = 7096 \cdot \dfrac{1}{a_{\overline{10}|0.06}} = \dfrac{7096}{7.36008}$ or $7096(0.135868) = \$964.12$.

Section 6.5 *NO SUPPLEMENTARY EXERCISES*

Supplementary Exercises: Chapter 7

Section 7.1 — *NO SUPPLEMENTARY EXERCISES*

Section 7.2

1. Does neither A nor B mean not A and not B?

2. Does either A or B mean

a. $A \cap B$
b. $A \cup B$
c. $A \cup B - (A \cap B)$

Section 7.2: Solutions

1. Yes

2. Inclusive or gives (b).

Section 7.3

1. A bowl contains 5 red marbles and 7 blue marbles. Two marbles are drawn together. What is the probability that both are blue?

a. $\frac{7}{22}$

b. $\frac{7}{24}$

c. $\frac{5}{33}$

d. $\frac{49}{132}$

e. $\frac{49}{144}$

2. In problem 1, what is the probability that at least one of the marbles is blue?

3. A bowl contains 4 red marbles and 6 white marbles. Two marbles are drawn without replacement. What is the probability that the second marble is red, given that the first marble is white?

a. $\frac{4}{15}$

b. $\frac{4}{9}$

c. $\frac{2}{3}$

d. $\frac{20}{27}$

e. 1

4. In problem 3, what is the probability that the first marble is white given that the second marble is red?

Section 7.3: Solutions

1. Together means without replacement.

$$\Pr(\text{BB}) = \Pr(\text{B}) \cdot \Pr(\text{B}|\text{B}) = \frac{7}{12} \cdot \frac{6}{11} = \frac{7}{22}$$

Answer: **a**

2.
$$\begin{aligned}\Pr(\text{At least one B}) &= 1 - \Pr(\text{RR}) \\ &= 1 - \frac{5}{12} \cdot \frac{4}{11} \\ &= \frac{28}{33}\end{aligned}$$

3. The sample space for the second marble is 5 white marbles and 4 red marbles.

$$\Pr(\text{R}) = \frac{4}{9}$$

(The question was about the second marble and not about both marbles.)
Answer: **b**

4. $\Pr(\text{W}) = \frac{6}{10} = \frac{3}{5}$

Section 7.4

1. A bowl contains 5 red marbles, 4 blue marbles, and 3 white marbles. Two marbles are drawn without replacement.

a. Construct a probability tree. Be sure to label the branches and the probabilities.

b. From the tree, find the probability that at least one blue marble is drawn.

2. In a group of people, $\frac{1}{3}$ are female and $\frac{2}{3}$ are male. Of the males, $\frac{1}{6}$ are blonde, and of the females, $\frac{1}{3}$ are blonde. Picking a person at random from this group, what is the probability that the person is a female, given that the person is a blonde?

a. $\frac{2}{3}$ **b.** $\frac{1}{2}$ **c.** $\frac{1}{3}$ **d.** $\frac{1}{6}$

3. Explain how to determine if the problem uses Bayes' formula.

Section 7.4: Solutions

1. a.

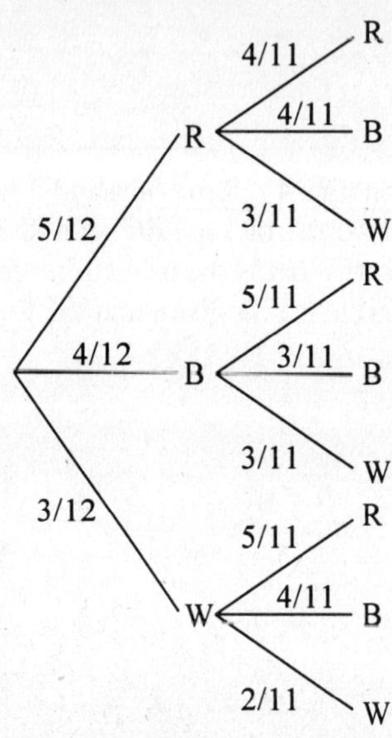

b. $\left(\frac{5}{12}\right)\left(\frac{4}{11}\right)=\frac{20}{132}$ $\left(\frac{4}{12}\right)\left(\frac{5}{11}\right)=\frac{20}{132}$

$\left(\frac{4}{12}\right)\left(\frac{3}{11}\right)=\frac{12}{132}$ $\left(\frac{4}{12}\right)\left(\frac{3}{11}\right)=\frac{12}{132}$

$\left(\frac{3}{12}\right)\left(\frac{4}{11}\right)=\frac{12}{132}$

The probability is $\frac{76}{132}=\frac{19}{33}$.

2. $\Pr(F)=\frac{1}{3}$ $\Pr(M)=\frac{2}{3}$ $\Pr(B|F)=\frac{1}{3}$ $\Pr(B|M)=\frac{1}{6}$ $\Pr(F|B)=?$

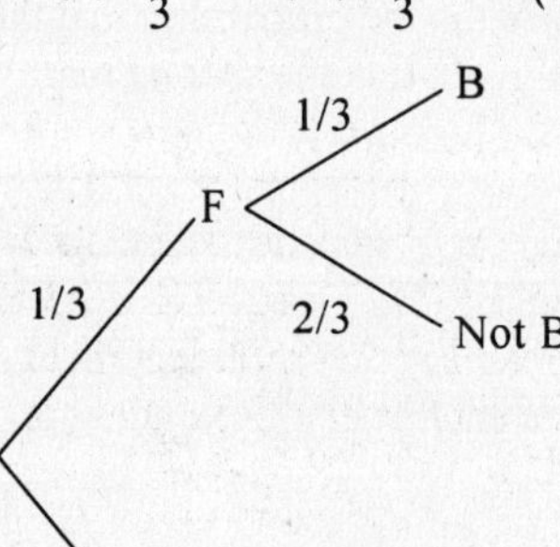

$\Pr(FB)=\frac{1}{9}$ $\Pr(MB)=\frac{1}{9}$ $\Pr(F|B)=\frac{\frac{1}{3}\cdot\frac{1}{3}}{\frac{1}{3}\cdot\frac{1}{3}+\frac{2}{3}\cdot\frac{1}{6}}=\frac{1}{2}$

Answer: **b**

Alternate method:

	B	*B'*	
F			300
M			600
			900

	B	*B'*	
F	100		300
M	100		600
	200		900

$\Pr(F|B)=\frac{n(FB)}{n(B)}=\frac{100}{200}=\frac{1}{2}$

3. Using problem 2 as a guide, if you are given $\Pr(A|B)$ and asked to find $\Pr(B|A)$, then use Bayes' formula.

Section 7.5

1. There are 6 teams in a conference. If each team plays each other team exactly once, how many games must be scheduled?
   a. 6!
   b. 36
   c. 30
   d. 15
   e. 12

2. A multiple choice test has 5 questions. Each question has 3 choices. In how many ways can this test be answered?
   a. $5 \cdot 3 = 15$
   b. $3^3 = 27$
   c. $5^3 = 125$
   d. $3^5 = 243$
   e. none of these

3. In constructing some positive numbers we are allowed to use only the digits from A, where $A = \{2,3,5,6,9\}$.
   a. How many odd digit numbers can be formed from A if repetition of digits is allowed? If repetition is not allowed?
   b. How many 3 digit numbers less than 400 can be formed from A if repetition of digits is not allowed?
   c. How may positive numbers can be formed from A if repetition of digits is not allowed?

Section 7.5: Solutions

1. No order is implied. Thus, ${}_6C_2 = \binom{6}{2} = \dfrac{6!}{2!4!} = 15.$ Answer: **d**

2. Use FCP:
   Ways to answer: $\underline{3} \cdot \underline{3} \cdot \underline{3} \cdot \underline{3} \cdot \underline{3} = 3^5$
   Question number: 1 2 3 4 5
   Answer: **d**

3. a. Restriction is on unit's digit:
   ___ ___ $\underline{\ 3\ }$; $\underline{5} \cdot \underline{5} \cdot \underline{3} = 75$
   ___ ___ $\underline{\ 3\ }$; $\underline{3} \cdot \underline{4} \cdot \underline{3} = 36$
   b. Restriction is on hundred's digit:
   $\underline{\ 2\ }$ ___ ___; $\underline{2} \cdot \underline{4} \cdot \underline{3} = 24$
   c. Success means one digit or two digits or $\cdots$ five digits. "or" means add.
   Solution: $\underline{5} + \underline{5} \cdot \underline{4} + \underline{5} \cdot \underline{4} \cdot \underline{3} + \underline{5} \cdot \underline{4} \cdot \underline{3} \cdot \underline{2} + \underline{5} \cdot \underline{4} \cdot \underline{3} \cdot \underline{2} \cdot \underline{1} = 325$

Section 7.6

1. A true-false test has 4 questions. If a student guesses at the answer, what is the probability that the student answers all 4 questions correctly?
   a. $\frac{1}{256}$ b. $\frac{1}{24}$
   c. $\frac{1}{16}$ d. $\frac{1}{8}$

2. In a five-part matching question there are five possible answers and each answer is used once. What is the probability that a student makes a correct match on each part if the student is guessing?
   a. $\frac{1}{3125}$ b. $\frac{1}{125}$ c. $\frac{1}{120}$
   d. $\frac{1}{32}$ e. $\frac{1}{25}$

3. In Supplementary problem 2 in Section 7.3, the student had to find the probability that both marbles were red or both were white by two different methods. Use the methods of this section to find the probability by a third method.

Section 7.6: Solutions

1. Success occurs 1 way.
Total ways to answer: $\underline{2} \cdot \underline{2} \cdot \underline{2} \cdot \underline{2} = 16$

$\text{Pr(All correct)} = \frac{1}{16}$

Answer: **c**

2. Success occurs 1 way.
Total ways to answer: $\underline{5} \cdot \underline{4} \cdot \underline{3} \cdot \underline{2} \cdot \underline{1} = 120$

$\text{Pr(All correct)} = \frac{1}{120}$

Answer: **c**

3. Success:
Red = $8 \cdot 8 = 64$ ways
White = $4 \cdot 4 = 16$ ways
Total number of successes = 80
Total ways to draw = $12 \cdot 12 = 144$

$\text{Pr(Win)} = \frac{80}{144} = \frac{5}{9}$

Section 7.7 — *NO SUPPLEMENTARY EXERCISES*

Supplementary Exercises: Chapter 8

Section 8.1

1. A fair coin is tossed 5 times. What is the probability that 3 heads will occur?
   a. Solve by the method of this section.
   b. Solve by the methods in the previous chapter. That is, $\Pr(3H)=\dfrac{\text{number of successes}}{\text{total number of outcomes}}$.

2. A weighted coin is tossed 5 times. The probability of a head on any toss is $\dfrac{2}{3}$.
   a. Find the probability of a head on toss 1, 3, and 5 and a tail on toss 2 and 4.
   b. Find the probability of tossing exactly 3 heads.
   c. Explain the difference between (a) and (b).

Section 8.1: Solutions

1. a. $\Pr(3H)=\binom{5}{3}\left(\frac{1}{2}\right)^3\left(\frac{1}{2}\right)^{5-3}=\frac{5!}{2!3!}\cdot\frac{1}{8}\cdot\frac{1}{4}=\frac{10}{32}=\frac{5}{16}$

   b. number of successes $=\binom{5}{3}=\frac{5!}{2!3!}=10$

   total number of outcomes $=\underline{2}\cdot\underline{2}\cdot\underline{2}\cdot\underline{2}\cdot\underline{2}=32$

   $\Pr(3H)=\frac{10}{32}=\frac{5}{16}$

2. a. The events are independent.

   $\Pr(\text{success})=\frac{2}{3}\cdot\frac{1}{3}\cdot\frac{2}{3}\cdot\frac{1}{3}\cdot\frac{2}{3}=\frac{8}{243}$

   b. $\Pr(3H)=\binom{5}{3}\left(\frac{2}{3}\right)^3\left(\frac{1}{3}\right)^{5-3}=10\cdot\frac{8}{27}\cdot\frac{1}{9}=\frac{80}{243}$

   c. Part (a) has order. In part (b), there is no order.

Section 8.2

1. The graph below shows the frequency of certain scores. Find the arithmetic mean, median and mode for this set of scores.

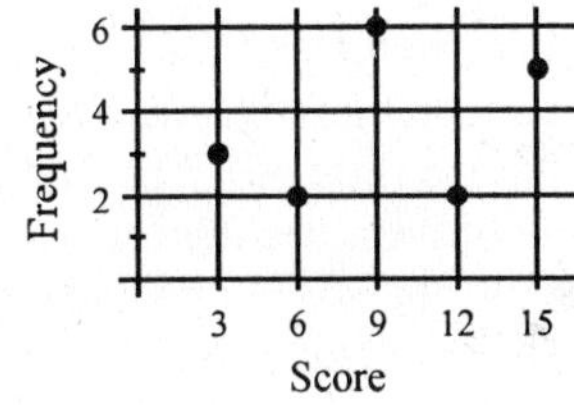

2. Using 5 numbers between 5 and 10, inclusively, construct a set of numbers in which
   a. the median is greater than the mean.
   b. the median and the mode are greater than the mean.

3. Give a weakness of the mean for a set of numbers.

4. Give a strength of the median for a set of numbers.

Section 8.2: Solutions

1. $\text{Mean} = \dfrac{3\cdot 3+2\cdot 6+6\cdot 9+2\cdot 12+5\cdot 15}{3+2+6+2+5}$

   $= 9.67$

   The median is the average of the 9th and 10th scores when the scores are arranged in numerical order. For this example the median is 9. Note that scores in the 6th position through the 11th position are 9.
   The mode is 9 since it occurs six times.

2. There is no unique answer. To build a solution consider the set of numbers 6, 7, 8, 9, 10.
   - **a.** Change the 8 to a 9. Now change the 10 to a 9. The mean is still 8 and the median is 9.
   - **b.** The set in (a) of 6, 7, 9, 9, 9 satisfies the conditions.

3. A weakness of the mean is that it is strongly influenced by unusually large (small) values in a distribution.

4. A strength of the median is that it is not influenced by a few unusually large (small) values in a distribution.

Section 8.3 — *NO SUPPLEMENTARY EXERCISES*

Section 8.4 — *NO SUPPLEMENTARY EXERCISES*

Section 8.5

1. Explain the meaning of a z-score of 1.5. If the z-score is related to the IQ of a person, what does this indicate about the IQ of the person? (Assume a normal distribution.)

2. If $\Pr(0 \le z \le y) = 0.4332$, find y.

3. In problem 2, if $\mu = 20$ and $\sigma = 4$, then $x = ?$

4. A certain curve is symmetric around 0 and the total area under the curve is 1. The area between -1 and 1 is 0.56. The curve is sketched below. Note that this is not the normal curve.

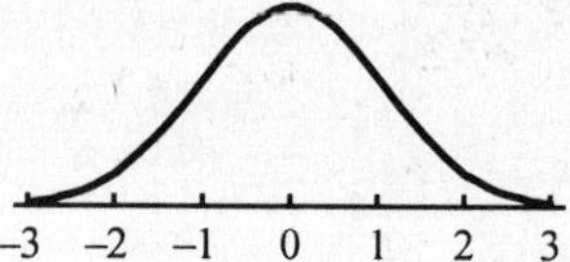

   - **a.** If possible, find the area between 0 and 1.
   - **b.** If possible, find the area to the right of 1.
   - **c.** If possible, find the area to the left of 2.

Section 8.5: Solutions

1. $z = 1.5$ states that the z-score is 1.5 standard deviations above the mean. The IQ of the person is in the top 7% of the population.

2. 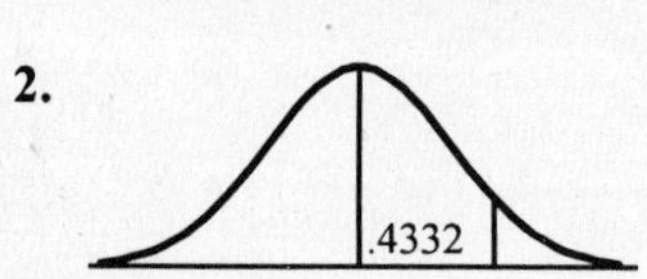

   $y = ?$

   $A = 0.4332$

   From the table, we have $y = 1.5$.

3. $z = \dfrac{x-\mu}{\sigma}$: $1.5 = \dfrac{x-20}{4}$

   $6 = x - 20$

   $26 = x$

4.
   - **a.** As a result of symmetry, $\frac{1}{2}(0.56) = 0.28$ is the area between 0 and 1.
   - **b.** The area to the right of 0 is 0.5. The area to the right of 1 is $0.5 - 0.28 = 0.22$.
   - **c.** To find the area to the left of 2 we need to know the area to the right of 2. To find the area to the right of 2 we need the area between 1 and 2. There is not enough information given to determine this area. Thus, we cannot find the area to the left of 2. The area needed $= 1 -$ area to the right of 2.

Supplementary Exercises: Chapter 9

Section 9.1

1. Find $\lim\limits_{x\to 3}\dfrac{x^2-x-6}{x^2-9}$.

a. 0 **b.** undefined **c.** $\frac{0}{0}$ **d.** $\frac{5}{6}$ **e.** $\frac{2}{3}$

2.

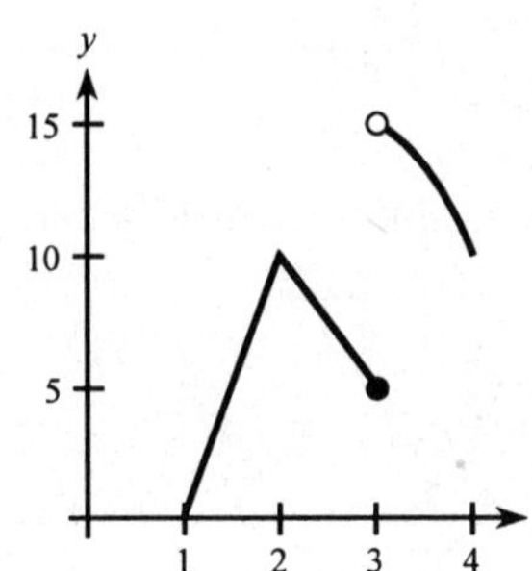

For the graph of $f(x)$ above, find

a. $\lim\limits_{x\to 3^-} f(x)$ **b.** $\lim\limits_{x\to 3^+} f(x)$ **c.** $\lim\limits_{x\to 3} f(x)$ **d.** $f(3)$

3. If $f(x)=\begin{cases} x^2+1, & x\ge 2 \\ x+4, & x<2 \end{cases}$, find

a. $\lim\limits_{x\to 2^-} f(x)$ **b.** $\lim\limits_{x\to 2^+} f(x)$ **c.** $f(2)$

4. $\lim\limits_{x\to 2}\dfrac{2^x-4}{x-2}=?$

Section 9.1: Solutions

1. $\lim\limits_{x\to 3}\dfrac{(x-3)(x+2)}{(x-3)(x+3)}=\lim\limits_{x\to 3}\dfrac{x+2}{x+3}=\dfrac{5}{6}$

Answer: d

2. **a.** 5
b. 15
c. does not exist
d. 5

3. **a.** $\lim\limits_{x\to 2^-}(x+4)=6$

b. $\lim\limits_{x\to 2^+}(x^2+1)=5$

c. $f(2)=2^2+1=5$

4. The function is of an indeterminate form and cannot be factored. A table of values gives a solution.

x	1.999	1.9999	2	2.0001	2.001
$f(x)$	2.7716	2.7725		2.7727	2.7735

$\lim\limits_{x\to 2} f(x)=2.77$ (to two decimals)

Section 9.2

In problems 1–3, none of the functions is continuous at $x = 2$. Choose from the following choices the reason that makes each function discontinuous at $x = 2$.

a. $f(2)$ is not defined.

b. $\lim_{x \to 2} f(x) = 6$

c. $f(2) = 5$

d. There are different rules for calculating $f(x)$.

e. $\lim_{x \to 2} f(x)$ does not exist.

1. $f(x) = \begin{cases} \dfrac{x^2 + 2x - 8}{x - 2} & \text{if } x \neq 2 \\ 5 & \text{if } x = 2 \end{cases}$

2. $f(x) = \begin{cases} x^2 + 1, & x \geq 2 \\ x + 4, & x < 2 \end{cases}$

3. $f(x) = \dfrac{x^2 + x - 6}{x - 2}$

4. If $f(x) = \begin{cases} \dfrac{x^2 - 4x + 3}{x - 3}, & x > 3 \\ x^2 - 7, & x < 3 \end{cases}$, is $f(x)$ everywhere continuous?

   If the answer is *no*, then give one additional definition so that it will be continuous.

Section 9.2: Solutions

1. c. $\lim_{x \to 2} f(x) = 6$ and $f(2)$ is defined, but $\lim_{x \to 2} f(x) \neq f(2)$.

2. e. $\lim_{x \to 2^-} f(x) = 6$ and $\lim_{x \to 2^+} f(x) = 5$, so, $\lim_{x \to 2} f(x)$ does not exist.

3. a.

4. $f(x)$ is not defined at $x = 3$ but $\lim_{x \to 3} f(x) = 2$. $f(x)$ is continuous if we define $f(3) = 2$.

Section 9.3

1. If $f(x) = 2x^2 - x + 3$, which of the following choices expresses $\dfrac{f(x+h) - f(x)}{h}$ in simplest form?

   a. $4x - 1$

   b. $4x$

   c. $4x + 2$

   d. $4x + 2h$

   e. $4x + 2h - 1$

2. If $R(x)$ is the revenue function and $R'(10) = -4$, what is happening to the revenue at $x = 10$?

3. For the graph of $F(x)$ shown below, give the interval(s) for which the rate of change of $F(x)$ is (a) positive, (b) zero, (c) negative.

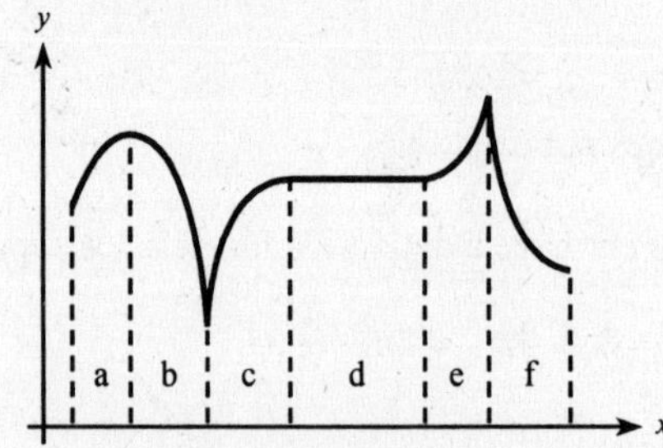

4. If $P(x)$ is the profit function and $P'(100) = -20$, does this mean the company is losing money at $x = 100$ units?

Section 9.3: Solutions

1. $\dfrac{f(x+h)-f(x)}{h}=\dfrac{\left[2(x+h)^2-(x+h)+3\right]-\left(2x^2-x+3\right)}{h}=\dfrac{4xh+2h^2-h}{h}=4x+2h-1$

   Answer: **e**

   Note that you were not asked to find $\lim\limits_{h\to 0}\dfrac{f(x+h)-f(x)}{h}$.

2. The revenue is decreasing (by \$4 per unit) at $x=10$.

3. Rate of change also means slope.
   a. a, c, e
   b. d
   c. b, f

4. Not necessarily; it means that the profit is decreasing at $x=100$.

Section 9.4

1. If $y=x^2-3x+12$, then which of the following is the equation of the tangent at $x=4$?
   a. $y=5$
   b. $y=16$
   c. $y=5x-20$
   d. $y=5x-4$
   e. $y=2x^2-11x+28$

2. If $f(x)=\dfrac{x^3-2x+4}{x}$, find $f'(x)$.

3. If $f(x)=\dfrac{x^3}{3}+\dfrac{1}{3x^3}+\dfrac{1}{3}$, find $f'(x)$.

Section 9.4: Solutions

1. $\dfrac{dy}{dx}=2x-3$

   At $x=4$, $y=16$ and $\dfrac{dy}{dx}=5$. So, $y-y_1=m(x-x_1)$ gives $y-16=5(x-4)$ or $y=5x-4$.

   Answer: **d**

2. We do not have a derivative formula for a quotient. By simplifying we have $f(x)=x^2-2+4x^{-1}$. Thus,

   $f'(x)=2x-4x^{-2}=2x-\dfrac{4}{x^2}$.

3. $f(x)=\dfrac{1}{3}x^3+\dfrac{1}{3}x^{-3}+\dfrac{1}{3}$

   $f'(x)=x^2-x^{-4}=x^2-\dfrac{1}{x^4}$

Section 9.5

1. If $f(x)=\dfrac{8}{x^{10}}$, is it correct to use the derivative of a quotient formula to find $f'(x)$?

2. In problem 1 find a more efficient way to find $f'(x)$.

3. If $f(x)=\frac{4x^3}{x^2+5}$, find the error below in finding $f'(x)$.

$$f'(x)=\frac{(x^2+5)(12x^2)-4x^3(2x)}{(x^2+5)^2}$$
$$=\frac{12x^2-8x^4}{x^2+5}$$
$$=\frac{4x^2(3-2x^2)}{x^2+5}$$

Section 9.5: Solutions

1. $f'(x)=\frac{x^{10}\cdot 0-8\cdot 10x^9}{x^{20}}=\frac{-80x^9}{x^{20}}=-80x^{-11}$

   Yes, it is correct to use the quotient formula.

2. $f(x)=\frac{8}{x^{10}}=8x^{-10}$

   Use the formula for $f(x)=cx^n$.

   $f'(x)=-10\cdot 8\cdot x^{-11}=-80x^{-11}$ is much more efficient.

3. (x^2+5) cannot be divided out from just the first term in the numerator.

   Recall the rule for factoring and fractions: $\frac{ac+bc}{xc}=\frac{c(a+b)}{xc}=\frac{a+b}{x}$.

   Correct answer: $f'(x)=\frac{4x^4+60x^2}{\left(x^2+5\right)^2}$

Section 9.6 — *NO SUPPLEMENTARY EXERCISES*

Section 9.7

1. If $f(x)=\frac{x^3}{(x^4+5x)^2}$, find $f'(x)$ using the quotient formula and chain rule. Be sure to simplify your answer.
2. If $f(x)=x^3(x^4+5x)^{-2}$, find $f'(x)$ using the product formula and chain rule. Express the answer in simplest form with positive exponents.
3. Which method in problems 1 and 2 appears to have the better chance of success? Why?

Section 9.7: Solutions

1. $f(x)=\frac{x^3}{\left(x^4+5x\right)^2}$

$$f'(x)=\frac{\left(x^4+5x\right)^2\left(3x^2\right)-x^3\cdot 2\left(x^4+5x\right)\left(4x^3+5\right)}{\left(x^4+5x\right)^4}=\frac{\left(x^4+5x\right)\left(x^2\right)\left[3\left(x^4+5x\right)-2x\left(4x^3+5\right)\right]}{\left(x^4+5x\right)^4}$$
$$=\frac{x^2\left[-5x^4+5x\right]}{\left(x^4+5x\right)^3}=\frac{5x^3\left(1-x^3\right)}{\left(x^4+5x\right)^3}$$

2. $f(x) = x^3\left(x^4+5x\right)^{-2}$

$$f'(x) = x^3(-2)\left(x^4+5x\right)^{-3}\left(4x^3+5\right)+\left(x^4+5x\right)^{-2}\left(3x^2\right) = x^2\left(x^4+5x\right)^{-3}\left[-2x\left(4x^3+5\right)+3\left(x^4+5x\right)\right]$$

$$= \frac{x^2\left[-5x^4+5x\right]}{\left(x^4+5x\right)^3} = \frac{5x^3\left(1-x^3\right)}{\left(x^4+5x\right)^3}$$

3. Most teachers feel that the method of problem 1 has the better chance of being correct. Factoring with negative exponents, as in problem 2, often results in mistakes being made.

Section 9.8

1. If $f(x) = \frac{1}{6}\left(x^2+1\right)^3$, which of the following represents $f''(x)$?

a. $x\left(x^2+1\right)^2$ **b.** $4x\left(x^2+1\right)$ **c.** $4x^2\left(x^2+1\right)$ **d.** $\left(x^2+1\right)\left(4x^2+1\right)$ **e.** $\left(x^2+1\right)\left(5x^2+1\right)$

2. Suppose the revenue for x units is given by $R(x) = 20x - x^2$.

a. How fast is the revenue changing at $x = 5$?
b. What is the revenue at $x = 5$?
c. How fast is the marginal revenue changing at $x = 5$?

3. Suppose the distance in feet that a ball travels upward in t seconds is given by the equation $s(t) = 130t - 16t^2$.

a. How far does the ball travel in 2 seconds?
b. How fast is the velocity of the ball changing at $t = 2$?
c. What is the velocity of the ball at $t = 2$?

Section 9.8: Solutions

1. $f(x) = \frac{1}{6}\left(x^2+1\right)^3$

$$f'(x) = 3\cdot\frac{1}{6}\left(x^2+1\right)^2\cdot 2x = x\left(x^2+1\right)^2$$

$$f''(x) = x\cdot 2\left(x^2+1\right)(2x)+\left(x^2+1\right)^2\cdot 1$$

$$= \left(x^2+1\right)\left[4x^2+\left(x^2+1\right)\right]$$

$$= \left(x^2+1\right)\left(5x^2+1\right)$$

Answer: **e**

2. **a.** We need $R'(x)$.

$R'(x) = 20 - 2x$

$R'(5) = 10$

b. We need $R(x)$.

$R(5) = 100 - 25 = 75$

c. We need $R''(x)$.

$R''(x) = -2$

$R''(5) = -2$

3. **a.** We need $s(t)$.

$s(2) = 260 - 64 = 196$ ft

b. We need $s''(t)$.

$s'(t) = 130 - 32t$

$s''(t) = -32$

$s''(2) = -32$ ft/ sec^2

c. We need $s'(t)$.

$s'(t) = 130 - 32t$

$s'(2) = 66$ ft/sec

The purpose of problems 2 and 3 is to determine when to use the function itself, when to use the derivative of the function, and when to use the second derivative of the function.

Section 9.9

1. If $P(x)$ is the profit function and $P'(10)=5$, then what is the meaning of $P'(10)=5$? Is the company making a profit at $x=10$?

2. A firm determines that the revenue and cost functions for x units are $R(x)=10x^2+5$ and $C(x)=4x^2+30$, respectively. Which of the following is the marginal profit at $x=5$?
a. 125 **b.** 100 **c.** 60 **d.** 40 **e.** −25

3. What do the other answers in problem 2 measure at $x=5$? (profit, cost, revenue, $\overline{MR}$, $\overline{MC}$)

Section 9.9: Solutions

1. On the sale of the 11th unit, the firm will make a profit of approximately five dollars. The company is not necessarily making a profit at $x=10$. For example, $P(10)=-55$ and $P(11)=-50$ satisfy $P'(10)=5$.

2. $P(x)=(10x^2+5)-(4x^2+30)=6x^2-25$
$P'(x)=12x$
$P'(5)=12\cdot 5=60$
Answer: **c**

3. $R(5)=10\cdot 25+5=255$
$R'(5)=20\cdot 5=100$ (Answer: **b**)
$P(5)=225-130=125$ (Answer: **a**)
$C(5)=4\cdot 25+30=130$
$C'(5)=8\cdot 5=40$ (Answer: **d**)
$P(0)=-25$ (Answer: **e**)

Supplementary Exercises: Chapter 10

Section 10.1

1. If $f(x)=3x^5-20x^3$ and $f'(x)=15x^2(x+2)(x-2)$, then $f(x)$ has a relative minimum at which of the following x-values.

2. For the graph of $f(x)$ below, which of the following represents the set of all critical values?

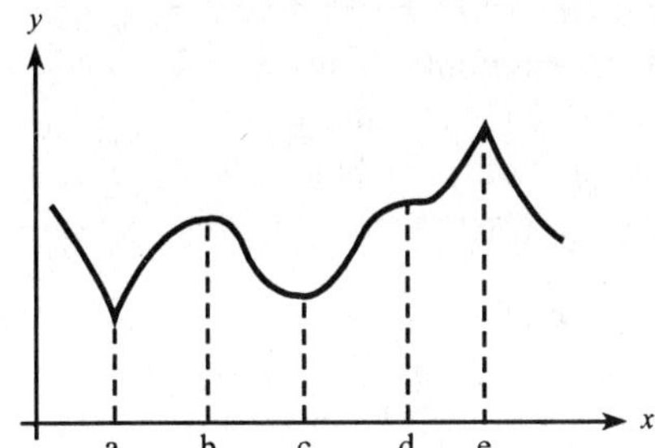

   a. {b, c}
   b. {a, e}
   c. {b, c, d}
   d. {a, b, c, d, e}
   e. {a, b, c, e}

3. Given the table below for $F'(x)$, which graph best represents the graph of $F(x)$?

x		2		4		6	
$F'(x)$	+	0	+	0	–	∞	+

a.
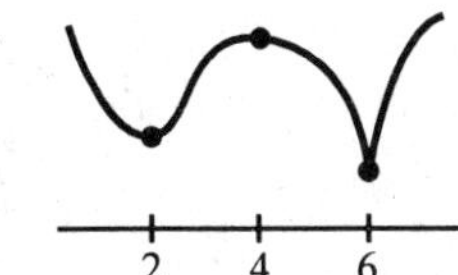

b.
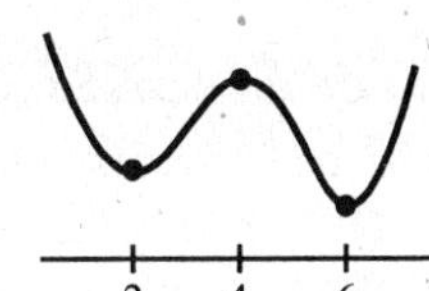

c.
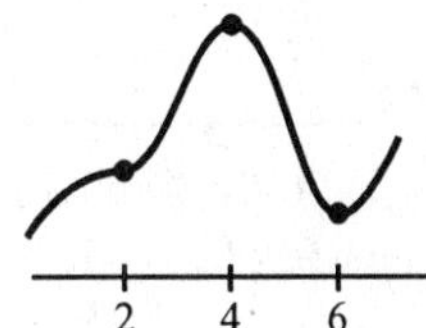

d.
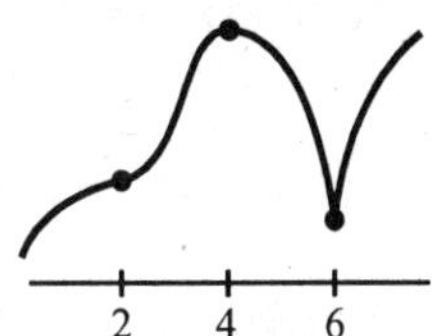

e.
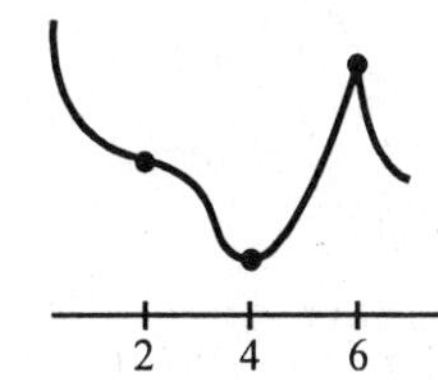

4. If $f(x)=3x^4-6x^2+2$ and $f'(x)=12x(x+1)(x-1)$, then which of the following is the relative maximum?
   a. -1
   b. 0
   c. 1
   d. 2
   e. 9

Section 10.1: Solutions

1. Critical values are $x=0,\ 2,\ -2$

x		-2		0		2	
$f'(x)$	+	0	–	0	–	0	+

Relative minimum at $x=2$.
Answer: **c**

2. Critical values mean $f'(x)=0$ or ∞.
At d there is a horizontal point of inflection.
Answer: **d**

3. $f'(x<2)>0$ means c and d are the only choices.
$f'(6)=\infty$ means d and e are the only choices.
Answer: **d**

4.

x		-1		0		1	
$f'(x)$	–	0	+	0	–	0	+

Relative maximum at $x=0$. Relative maximum is $f(0)=2$. Answer: **d**

Section 10.2

1. For a given function $f(x)$ we have $f(2)=4$, $f'(2)=0$, $f''(2)=0$, $f'(1)>0$, $f'(3)<0$, $f''(1)<0$, and $f''(3)<0$. At $x=2$, which of the following is true for $f(x)$?

a. it is increasing
b. it has a point of inflection
c. it has a relative minimum
d. it has a relative maximum
e. it cannot be determined

2. For the graph of $y=f(x)$ below, there is a point of inflection at

a. A **b.** B **c.** C **d.** D **e.** E

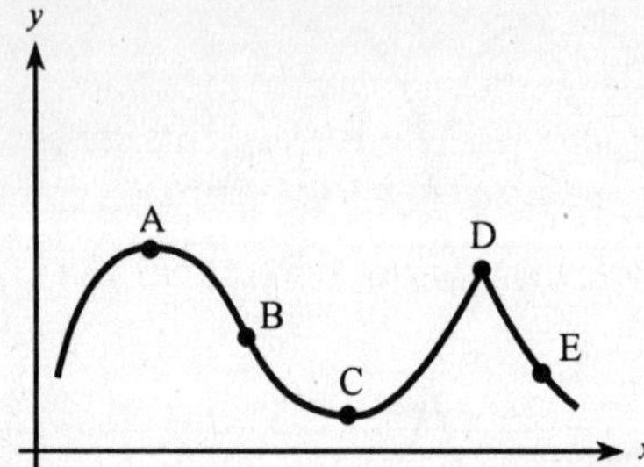

3. $f(x)=3x^5-20x^3$

$f'(x)=15x^2(x+2)(x-2)$

$f''(x)=60x(x^2-2)$

I. $f(x)$ has a relative max at $x=0$.

II. $f(x)$ has a relative min at $x=2$.

III. $f(x)$ has points of inflection.

Which one of these three statements is false?

a. I only
b. II only
c. III only
d. I and III only
e. II and III only

4. If $f(x)=3x^4-4x^3$, $f'(x)=12x^2(x-1)$, and $f''(x)=12x(3x-2)$, for what value(s) of x does $f(x)$ have a point or points of inflection?

a. 0
b. $\frac{2}{3}$
c. 0 and $\frac{2}{3}$
d. 0 and 1
f. no points of inflection

5. For the graph of $f(x)$ below, which of the following is true?

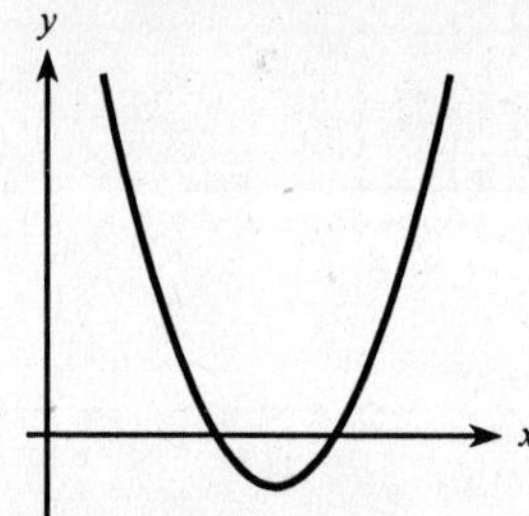

a. $f''(x)>0$
b. $f''(x)=0$
c. $f''(x)<0$
d. $f''(x)$ cannot be determined

Section 10.2: Solutions

1. $f'(1)>0$, $f'(2)=0$, $f'(3)<0$ means that $f(x)$ has a relative maximum at $x=2$.

Answer: **d**

2. **b**

3.

x		-2		0		2	
$f'(x)$	+	0	−	0	−	0	+

I is false and II is true.

x		$-\sqrt{2}$		0		$\sqrt{2}$	
$f'(x)$	−	0	+	0	−	0	+

III is true. Answer: **a**

4.

x		0		$\frac{2}{3}$	
$f''(x)$	$+$	0	$-$	0	$+$

0 and $\frac{2}{3}$ are points of inflection.

Answer: **c**

5. $f(x)$ is concave up so $f''(x) > 0$.

Answer: **a**

Section 10.3

1. JJ manages an apartment with 50 units. If the rent for each unit is \$860, all of the units will be filled. One unit will become vacant for each \$10 increase in the monthly rent. Let $x =$ the number of vacant units. Which of the following equations is used to find the maximum revenue?

a. $R = 860(50 - x) + 10x$

b. $R = 860(50 - 10x)$

c. $R = (860 - 10x)(50 - x)$

d. $R = (860 + 10x)(50 - x)$

e. $R = (860 + 10x)(50 + x)$

2. If the total cost function is $C(x) = 250 + 16x + 0.1x^2$, which of the following is the number of units that will minimize average cost per unit?

a. 250

b. 80

c. 50

d. 16

e. −80

3. Briefly explain why the average cost per unit decreases and then begins to increase.

4. Explain why the profit is a maximum when $\overline{MR} = \overline{MC}$.

Section 10.3: Solutions

1. $R =$ price per unit $\times$ number of units

Trial patterns: 860×50

870×49

880×48

Final result: $R = (860 + 10x)(50 - x)$

Answer: **d**

2. $\overline{C}(x) = \dfrac{250}{x} + 16 + 0.1x$

$\overline{C}'(x) = -\dfrac{250}{x^2} + 0.1$

$\overline{C}'(x) = 0$ if $x = 50$

$\overline{C}'(1) < 0,\ \overline{C}'(50) = 0,\ \overline{C}'(100) > 0$

The result of the above information is that the average cost will be minimized when the number of units is 50. Answer: **c**

3. As the number of units increases, production is more efficient. Fixed costs are spread over more units. At a certain point, more employees, more machinery, two shifts, or operating the machinery at a rate that causes breakdowns all contribute to increasing the cost per unit.

4. $P(x) = R(x) - C(x)$

$P'(x) = R'(x) - C'(x) = \overline{MR} - \overline{MC}$

Profit is a maximum when $P'(x)$ changes from positive to zero to negative. $P'(x) = 0$ means $\overline{MR} - \overline{MC} = 0$ or $\overline{MR} = \overline{MC}$.

Section 10.4

1. A rectangular dog pen is to be constructed as shown.

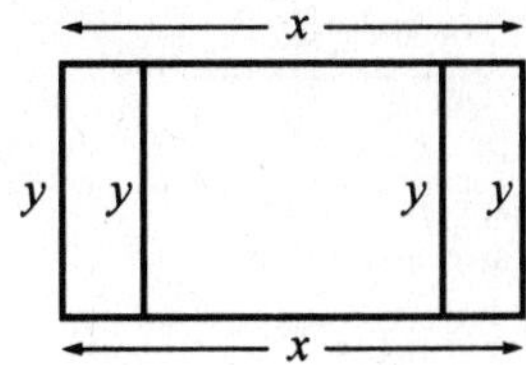

There is 600 feet of fencing available. If the area is to be maximized, which one of the following choices represents the equation for the quantity being maximized?

a. $2x + 4y = 600$

b. $A = 2x + 4y$

c. $xy = 600$

d. $A = 2x + 2y$

e. $A = xy$

2. In problem 1, what is the equation in x only for the quantity being maximized?

a. $2x+\dfrac{2400}{x}=600$ **b.** $\dfrac{600-2x}{4}$ **c.** $\dfrac{300x-x^2}{2}$ **d.** $150-x$ **e.** $2x+\dfrac{300-x}{2}=600$

3. The product of two positive numbers is 10. Find the two numbers if the sum of their squares is a minimum.

Section 10.4: Solutions

1. Quantity being maximized is the area.
$A=xy$
Answer: **e**

2. $2x+4y=600$ is the equation for the amount of fencing. So, $y=\dfrac{300-x}{2}$. Thus,
$A=\dfrac{x(300-x)}{2}$.
Answer: **c**

3. x = one number
y = other number
Quantity being minimized: $S=x^2+y^2$
Another equation $xy=10$ or $y=\dfrac{10}{x}$
Substituting: $S(x)=x^2+\dfrac{100}{x^2}$,
$S'(x)=2x-\dfrac{200}{x^3}$
$S'(x)=0$ if $x^4=100$. Thus, $x^2=10$ and $x=\sqrt{10}$. The two numbers are $\sqrt{10}$ and $\sqrt{10}$.

Section 10.5

For problems 1-3, suppose $y=\dfrac{Ax^p+A_1x^{p-1}+\cdots+A_p}{Bx^q+B_1x^{q-1}+\cdots+B_q}$

1. If $y=0$ is a horizontal asymptote, what is the relation between p and q?

2. If there is no horizontal asymptote, what is the relation between p and q?

3. If there is a horizontal asymptote different from $y=0$, what is the equation of the asymptote?

4. If $y=\dfrac{(5x-1)(2x-3)(x-1)(x-4)}{(3x-1)(x-2)(x-1)(x-6)}$, then:
   a. What is the equation of the horizontal asymptote?
   b. Without graphing, what happens to the graph at $x=1$?

Section 10.5: Solutions

1. For $y=0$ to be a horizontal asymptote, $p<q$.

2. If there is no horizontal asymptote, $p>q$.

3. If there is a horizontal asymptote other than $y=0$, then $p=q$ and $y=\dfrac{A}{B}$.

Supplementary Exercises: Chapter 11

Section 11.1

1. Find $\frac{dy}{dx}$ if $y = [\ln(3x+1)]^2 + \ln(3x+1)^2$.

2. Find $\frac{dy}{dx}$ if $y = \ln[\ln(3x+1)]$.

3. Find $\frac{dy}{dx}$ if $y = \ln(e^2+2) + \ln(x^2+2)$.

4. Find $f'(x)$ if $f(x) = \ln(x^3) - \ln(x) - \ln(x^3 - x)$.

Section 11.1: Solutions

1. $y = [\ln(3x+1)]^2 + \ln(3x+1)^2$

$$\frac{dy}{dx} = 2[\ln(3x+1)] \cdot \frac{1}{3x+1} \cdot 3 + 2 \cdot \frac{1}{3x+1} \cdot 3$$

$$= \frac{6[\ln(3x+1)]}{3x+1} + \frac{6}{3x+1}$$

2. $y = \ln[\ln(3x+1)]$

$$\frac{dy}{dx} = \frac{1}{\ln(3x+1)} \cdot \frac{1}{3x+1} \cdot 3$$

$$= \frac{3}{(3x+1) \cdot \ln(3x+1)}$$

3. $y = \ln(e^2+2) + \ln(x^2+2)$

$$\frac{dy}{dx} = 0 + \frac{1}{x^2+2} \cdot 2x = \frac{2x}{x^2+2}$$

$\ln(e^2+2)$ is a constant.

4. $f(x) = \ln(x^3) - \ln(x) - \ln(x^3 - x)$

$$f'(x) = \frac{3}{x} - \frac{1}{x} - \frac{3x^2-1}{x^3-x} = \frac{2}{x} - \frac{3x^2-1}{x^3-x}$$

$\ln(A+B) \neq \ln A + \ln B$

Section 11.2

1. Find $f'(x)$ if $f(x) = (e^x)^2 + e^{x^2}$.

2. If $f(x) = \frac{1}{4}e^{x^2}$, which of the following is $f''(x)$?

   a. $x^2 e^{x^2}$ **b.** xe^{x^2} **c.** $x^2 e^{x^2} + \frac{1}{2}e^{x^2}$ **d.** $x^2 + \frac{1}{2}e^{x^2}$ **e.** e^{x^2}

3. Write the equation of the line tangent to the graph of $y = e^x \ln x$ at $x = 1$.

4. Find $f'(x)$ if $f(x) = x^2 + e^2 + e^{x^2} + x^{e^2}$

Section 11.2: Solutions

1. $f(x) = (e^x)^2 + e^{x^2}$

$f'(x) = e^{2x} \cdot 2 + e^{x^2} \cdot 2x = 2e^{2x} + 2xe^{x^2}$

2. $f(x) = \frac{1}{4}e^{x^2}$

$f'(x) = \frac{1}{4}e^{x^2} \cdot 2x = \frac{x}{2}e^{x^2}$

$f''(x) = \frac{x}{2} \cdot e^{x^2} \cdot 2x + e^{x^2} \cdot \frac{1}{2} = x^2 e^{x^2} + \frac{1}{2}e^{x^2}$

Answer: **c**

3. $y = e^x \ln x$

At $x = 1$, $y = e \cdot \ln 1 = 0$.

$y - 0 = e(x-1)$

$y = ex - e$

$\frac{dy}{dx} = e^x \cdot \frac{1}{x} + (\ln x)(e^x)$

At $x = 1$, $\frac{dy}{dx} = e + 0 \cdot e = e$.

4. $f(x) = x^2 + e^2 + e^{x^2} + x^{e^2}$

$f'(x) = 2x + 0 + e^{x^2} \cdot 2x + e^2 \cdot x^{e^2 - 1}$

$= 2x + 2xe^{x^2} + e^2 x^{e^2 - 1}$

e^2 is a constant

x^{e^2} is in the form x^n.

Section 11.3

1. For $x^2 + y^2 = 9$, the text gave $y = +\sqrt{9 - x^2}$ and $y = -\sqrt{9 - x^2}$ as two functions that satisfied the equation. By looking at the two graphs, find two other (discontinuous) functions that satisfy the equation.

2. Instead of using implicit differentiation to find $\frac{dy}{dx}$ when $\ln(x + y) = x^2$, solve for y and find the derivative.

3. Use implicit differentiation to evaluate the derivative at $(-9, 6)$ for $\frac{2x}{x + y} = y$.

4. Use implicit differentiation to evaluate the derivative at $(-9, 6)$ for $xy + y^2 = 2x$.

5. Show algebraically that the two derivatives in 3 and 4 are always equal. Hint: Solve the original equation for x.

Section 11.3: Solutions

1. $y = \begin{cases} +\sqrt{9 - x^2} & \text{if } 0 \le x \le 3 \\ -\sqrt{9 - x^2} & \text{if } -3 \le x < 0 \end{cases}$ and

$\begin{cases} -\sqrt{9 - x^2} & \text{if } 0 \le x \le 3 \\ +\sqrt{9 - x^2} & \text{if } -3 \le x < 0 \end{cases}$ are two functions that satisfy the equation.

2. $\ln(x + y) = x^2$ means $x + y = e^{x^2}$. So, $y = e^{x^2} - x$

and $\frac{dy}{dx} = e^{x^2} \cdot 2x - 1 = 2xe^{x^2} - 1$.

3. $\frac{2x}{x + y} = y$

$\frac{(x + y)2 - 2x(1 + y')}{(x + y)^2} = y'$

$2x + 2y - 2x - 2xy' = (x + y)^2 y'$

$y' = \frac{2y}{2x + (x + y)^2}$

At $(-9, 6)$, $y' = \frac{12}{-18 + 9} = -\frac{4}{3}$.

4. $xy + y^2 = 2x$

$xy' + y \cdot 1 + 2yy' = 2$

$(x+2y)y' = 2 - y$

$$y' = \frac{2-y}{x+2y}$$

At $(-9,6)$, $y' = \frac{2-6}{-9+12} = -\frac{4}{3}$.

5. $\frac{2x}{x+y} = y$

$2x = xy + y^2$ (Clear of fractions.)

$2x - xy = y^2$ (Isolate x's.)

$$x = \frac{y^2}{2-y}$$

Substitute for x in 3:

$$y' = \frac{2y}{\frac{2y^2}{2-y} + \left(\frac{y^2}{2-y} + y\right)^2} = \frac{2y}{\frac{2y^2}{2-y} + \left(\frac{y^2+2y-y^2}{2-y}\right)^2}$$

$$= \frac{2y}{\frac{2y^2}{2-y} + \frac{4y^2}{(2-y)^2}} \cdot \frac{(2-y)^2}{(2-y)^2}$$

$$= \frac{2y(2-y)^2}{2y^2(2-y) + 4y^2}$$

$$= \frac{2y(2-y^2)}{2y^2(4-y)} = \frac{(2-y)^2}{y(4-y)}$$

Section 11.4 ——— ***NO SUPPLEMENTARY EXERCISES***

Section 11.5 ——— ***NO SUPPLEMENTARY EXERCISES***

Supplementary Exercises: Chapter 12

Section 12.1

This set of problems is designed to test the student's reasoning and analytical abilities.

1. **a.** Find $\frac{d}{dx}\left(\int(2x+3)dx\right)$.

b. From the result in (a), generalize and find $\frac{d}{dx}\left(\int f'(x)\,dx\right)$.

2. Find $\int x^2\left(3x^2-2x+1\right)dx$.

3. Find $\int\frac{4x^8-5x^2+1}{x^2}dx$.

4. If $f(x)=\left(x^3+1\right)^5$, then $f'(x)=5\left(x^3+1\right)^4\left(3x^2\right)$.

a. Find $\int 15x^2\left(x^3+1\right)^4 dx$.

b. Find $\int 5x^2\left(x^3+1\right)^4 dx$.

Section 12.1: Solutions

1. **a.** $\int(2x+3)dx=\frac{2x^2}{2}+3x+C=x^2+3x+C$

$\frac{d}{dx}\left(x^2+3x+C\right)=2x+3$

b. $\frac{d}{dx}\left(\int f'(x)dx\right)=\frac{d}{dx}\left(f(x)+C\right)=f'(x)$

If we integrate and then differentiate, the result is the original integrand.

2. There is no product formula for our use.

$$\int x^2\left(3x^2-2x+1\right)dx=\int\left(3x^4-2x^3+x^2\right)dx$$
$$=\frac{3x^5}{5}-\frac{2x^4}{4}+\frac{x^3}{3}+C$$
$$=\frac{3}{5}x^5-\frac{1}{2}x^4+\frac{1}{3}x^3+C$$

3. There is no quotient formula for our use.

$$\int\frac{4x^8-5x^2+1}{x^2}dx=\int\left(4x^6-5+x^{-2}\right)dx$$
$$=\frac{4x^7}{7}-5x+\frac{x^{-1}}{-1}+C$$
$$=\frac{4}{7}x^7-5x-\frac{1}{x}+C$$

4. **a.** $15x^2\left(x^3+1\right)^4$ is the derivative of $\left(x^3+1\right)^5$.

Thus, $\int 15x^2\left(x^3+1\right)^4 dx=\left(x^3+1\right)^5+C$.

b.
$$\int 5x^2\left(x^3+1\right)^4 dx=\int\frac{3}{3}\cdot 5x^2\left(x^3+1\right)^4 dx$$
$$=\frac{1}{3}\int 15x^2\left(x^3+1\right)^4 dx$$
$$=\frac{1}{3}\left(x^3+1\right)^5+C$$

Section 12.2

1. Evaluate

a. $\int\left(x^3+2\right)^5 x^2dx$ **b.** $\int 5x^2\left(x^3+2\right)^5 dx$ **c.** $\int\frac{x^2}{\left(x^3+2\right)^5}dx$

2. Which of the following can be evaluated using the Power Rule? This means that the others must be simplified and then use the methods in Section 12.1.

a. $\int\frac{4x^3}{\left(x^4+2\right)^2}dx$ **b.** $\int\frac{\left(x^4+2\right)^3}{4x^3}dx$ **c.** $\int\left(x^2+8x-1\right)^2\left(2x+1\right)dx$

d. $\int\left(x^2+8x-1\right)^2\left(x+4\right)dx$ **e.** $\int\frac{3x+12}{\left(x^2+8x-1\right)^{1/3}}dx$

3. How can any solution be checked?

Section 12.2: Solutions

1. **a.** Let $u = x^3 + 2$. Then, $du = 3x^2 dx$.

$$\int (x^3+2)^5 (x^2\,dx) = \int (x^3+2)^5 \cdot \frac{1}{3}(3x^2 dx)$$
$$= \frac{1}{3}\int u^5 du = \frac{1}{3}\cdot\frac{u^6}{6} + C$$
$$= \frac{1}{18}(x^3+2)^6 + C$$

b. $5\int (x^3+2)^5 (x^2\,dx) = 5\int (x^3+1)^5 \cdot \frac{1}{3}(3x^2 dx)$

$$= \frac{5}{3}\int u^5 du = \frac{5}{3}\cdot\frac{u^6}{6} + C$$
$$= \frac{5}{18}(x^3+2)^6 + C$$

c. $\int (x^3+2)^{-5} (x^2 dx) = \int (x^3+2)^{-5} \cdot \frac{1}{3}(3x^2 dx)$

$$= \frac{1}{3}\int u^{-5} du = \frac{1}{3}\cdot\frac{u^{-4}}{-4} + C$$
$$= \frac{-1}{12(x^3+2)^4} + C$$

2. **a.** Use the Power Rule. $\frac{-1}{x^4+2} + C$

b. Cube, simplify, use Section 12.1.

$$\frac{1}{40}x^{10} + \frac{1}{4}x^6 + \frac{3}{2}x^2 - x^{-2} + C$$

c. Simplify and use Section 12.1.

$$\frac{1}{3}x^6 + \frac{33}{5}x^5 + 35x^4 + 10x^3 - 7x^2 + x + C$$

d. Use the Power Rule.

$$\frac{1}{6}(x^2+8x-1)^3 + C$$

e. Use the Power Rule.

$$\frac{9}{4}(x^2+8x-1)^{2/3} + C$$

3. To check the solution, take the derivative. The result should be the original integrand.

Section 12.3

A list of formulas and a list of integrals are given. Recognition of the type of problem determines which formula is to be used. Determine which formula is associated with each integral and then evaluate the integral.

Formulas:

I. $\int \frac{u'}{u} dx$ or $\int \frac{du}{u}$ II. $\int e^u u' \,dx$ III. $\int u^n du$ IV. other

Integrals:

1. $\int \frac{4x^3}{x^4+2} dx$

2. $\int \frac{4x^3}{(x^4+2)^2} dx$

3. $\int \frac{x^4+2}{x^3} dx$

4. $\int x e^{x^2} dx$

5. $\int \frac{e^x}{(e^x+1)^2} dx$

6. $\int \frac{e^x+1}{e^x} dx$

7. $\int \frac{e^x}{e^x+1} dx$

8. $\int \frac{\ln x}{x} dx$

Section 12.3: Solutions

1. I; $\ln(x^4+2) + C$

2. III; $\frac{-1}{x^4+2} + C$

3. IV; $\frac{x^2}{2} - \frac{1}{x^2} + C$

4. II; $\frac{1}{2}e^{x^2} + C$

5. III; $\frac{-1}{e^x+1}+C$

6. IV; $x-e^{-x}+C$

7. I; $\ln\left(e^x+1\right)+C$

8. III; $\frac{(\ln x)^2}{2}+C$

Section 12.4

1. If the marginal cost for a product is $\overline{MC}=4x+2$, and the production of 10 units results in a total cost of $300, which of the following is the cost of production of 20 units?
   a. 382
   b. 600
   c. 840
   d. 920
   e. 1140

2. The marginal cost for a product is $\overline{MC}=6x+60$. The marginal revenue is $\overline{MR}=180-2x$. The total cost of producing 10 items is $1000. Which of the following represents the number of units necessary to maximize profit?
   a. 8
   b. 15
   c. 30
   d. 40
   e. none of these

3. In problem 2, what are the fixed costs?

4. In problem 2, what is the maximum profit?

5. If the average cost of a product changes at the rate $\overline{C}'(x)=2-\frac{120}{x^2}$, and if the average cost of 20 units is $60, find
   a. the average cost of 10 units.
   b. the total cost of 30 units.

Section 12.4: Solutions

1. $\overline{MC}=C'(x)=4x+2$ gives
   $C(x)=2x^2+2x+K.$
   $C(10)=300=2(100)+2(10)+K$ gives $K=80$.
   Then, $C(20)=2(400)+2(20)+80=\$920$.
   Answer: **d**

2. Maximum profit occurs when $\overline{MR}=\overline{MC}.$
   $180-2x+6x+60$ gives $x=15$.
   Answer: **b**

3. $C'(x)=6x+60$ gives $C(x)=3x^2+60x+K$.
   $C(10)=1000=3(100)+60(10)+K$ gives $K=100$.
   $C(x)=3x^2+60x+100$ means $FC=\$100$.

4. $R'(x)=180-2x$ gives $R(x)=180x-x^2+K$ and $R(0)=0$ means that $K=0$.
   $P(x)=\left(180x-x^2\right)-\left(3x^2+60x+100\right)$
   $=-4x^2+120x-100$
   $P(15)=-4(225)+120(15)-100=\$800$

5. $\overline{C}'(x)=2-\frac{120}{x^2}$
   a. $\overline{C}(x)=2x+\frac{120}{x}+K$
   $60=2(20)+\frac{120}{20}+K$ gives $K=14$.
   $\overline{C}(10)=2(10)+\frac{120}{10}+14=\$46$
   b. $C(x)=\overline{C}(x)\cdot x=2x^2+120+14x$
   $C(30)=2(900)+120+14(30)=\$2340$

Section 12.5

NO SUPPLEMENTARY EXERCISES

Supplementary Exercises: Chapter 13

Section 13.1

When the area under $f(x) = x^2$ from $x = 0$ to $x = 5$ is approximated, the formulas for the sum of n rectangles using left hand endpoints and right hand endpoints are as follows:

Left hand endpoints: $S_L = \frac{125}{3} - \frac{125}{2n} + \frac{125}{6n^2}$

Right-hand endpoints: $S_R = \frac{125}{3} + \frac{125}{2n} + \frac{125}{6n^2}$

Use these formulas in problems 1–3.

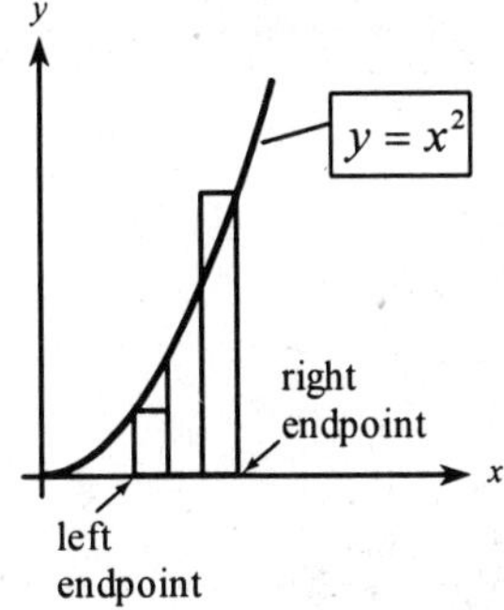

1. a. $S_L(5) = ?$
   b. $S_L(25) = ?$
   c. $S_R(5) = ?$
   d. $S_R(25) = ?$

2. a. What does the answer in 1b represent?
   b. What does the answer in 1d represent?

3. $\lim_{n\to\infty} S_L = \lim_{n\to\infty} S_R = \frac{125}{3}$ represents what?

4. $\sum_{i=1}^{100}(4i+20)$ equals which of the following?
   a. 2400
   b. 20,200
   c. 20,220
   d. 22,200
   e. none of these

5. Does $\sum_{i=1}^{100}(4i+20) = \sum_{i=1}^{90}(4i+20) + \sum_{i=91}^{100}(4i+20)$?

Section 13.1: Solutions

1. a. $S_L(5) = \frac{125}{3} - \frac{25}{2} + \frac{5}{6} = 30$
   b. $S_L(25) = \frac{125}{3} - \frac{5}{2} + \frac{1}{30} = \frac{196}{5}$
   c. $S_R(5) = \frac{125}{3} + \frac{25}{2} + \frac{5}{6} = 55$
   d. $S_R(25) = \frac{125}{3} + \frac{5}{2} + \frac{1}{30} = \frac{221}{5}$

2. a. $\frac{196}{5}$ is the area of 25 inscribed rectangles.
   b. $\frac{221}{5}$ is the area of 25 circumscribed rectangles.

3. $\frac{125}{3}$ is the area under $y = x^2$ from $x = 0$ to $x = 5$.

4. $\sum_{i=1}^{100}(4i+20) = 4\sum_{i=1}^{100} i + \sum_{i=1}^{100} 20$

   $= 4 \cdot \frac{100(100+1)}{2} + 100 \cdot 20$

   $= 22,200$

   Answer: d

5. Yes; $\sum_{i=1}^{100} f(i)$ is the sum of the first 100 terms.

   $\sum_{i=1}^{90} f(i)$ is the sum of the first 90 terms.

   $\sum_{i=91}^{100} f(i)$ is the sum of the 91st through 100th terms.

Section 13.2

1. Which of the following represents the solution of $\int_{-2}^{2}(x+2)dx$?

 a. 0
 b. 2
 c. 4
 d. 6
 e. 8

2. Explain the difference between an indefinite integral and a definite integral.

3.

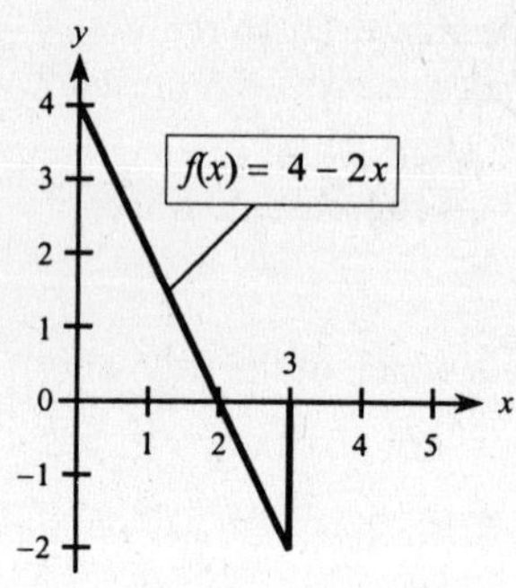

$$\int_0^2 (4-2x)dx = (4x-x^2)\Big|_0^2 = (8-4)-0 = 4$$

$$\begin{aligned}\int_2^3 (4-2x)dx &= (4x-x^2)\Big|_2^3 \\ &= (12-9)-(8-4) \\ &= -1\end{aligned}$$

$\int_0^3 (4-2x)dx = 3$ represents what?

4.

$$1\!-\!\int_a^b f(x)dx = F(x)\Big|_a^b = F(b)-F(a)$$

(labels: 2 under the lower limit a, 3 under $f(x)$, 4 under $F(x)$)

In words, state what each number represents.

Section 13.2: Solutions

1. $$\begin{aligned}\int_{-2}^{2}(x+2)dx &= \left(\frac{x^2}{2}+2x\right)\Bigg|_{-2}^{2} \\ &= (2+4)-(2-4) \\ &= 8\end{aligned}$$

 Answer: e

2. An indefinite integral is a function (an antiderivative). A definite integral is a number (the difference of the values of an antiderivative evaluated at two points).

3. 3 represents the difference between the area above the x-axis and the area below the x-axis.

4. 1 is the integral sign.
 2 is the lower limit.
 3 is the integrand.
 4 is the antiderivative.

Section 13.3

1. For the graph below, which of the following will give the area of the shaded region?

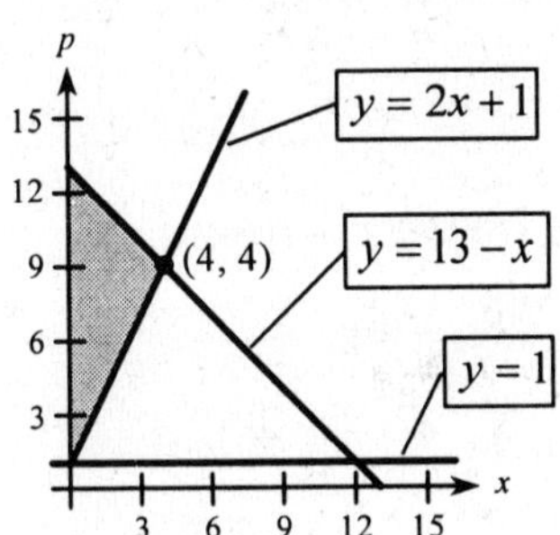

   a. $\int_0^4 (3x-12)dx$

   b. $\int_0^4 (12-3x)dx$

   c. $\int_0^4 (2x+1)dx$

   d. $\int_0^4 (13-x)dx$

   e. $\int_0^4 2x\,dx + \int_4^{12} (13-x)dx$

2. For the graph below, which of the following will give the area of the shaded region?

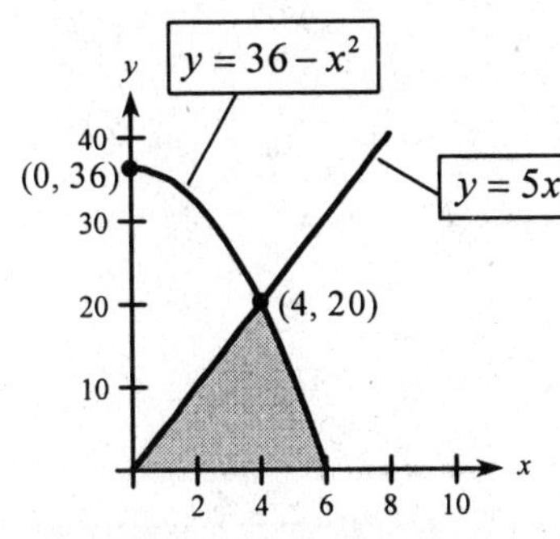

   a. $\int_0^4 (36-x^2-5x)dx$

   b. $\int_0^{20} (36-x^2-5x)dx$

   c. $\int_0^6 [5x-(36-x^2)]dx$

   d. $\int_0^4 5x\,dx + \int_4^6 (36-x^2)dx$

   e. $\int_0^6 (5x+36-x^2)dx$

3. For the graph below, each of the following integrals represents the area of region(s). Identify the region(s) associated with each integral.

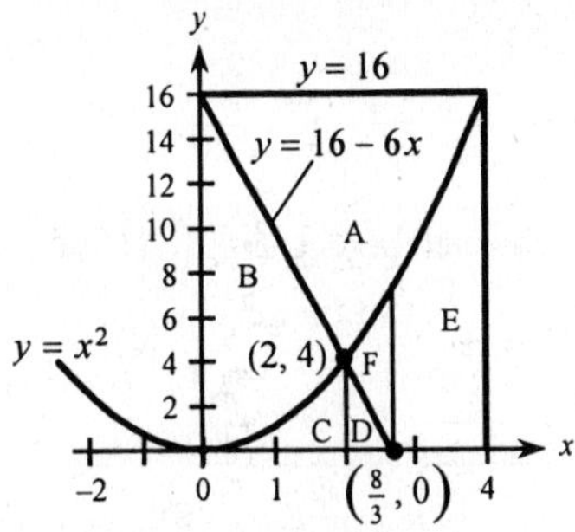

   a. $\int_0^4 (16-x^2)dx$

   b. $\int_0^2 (16-6x-x^2)dx$

   c. $\int_0^2 x^2 dx$

   d. $\int_2^{8/3} (16-6x)dx$

   e. Give the integral(s) with limits that will represent area A.

Section 13.3: Solutions

$\int_a^b (f(x) - g(x))dx$:

"a" is where area of region begins.
"b" is where area of region ends.
$f(x)$ is the curve on "top."
$g(x)$ is the curve on "bottom."

1. B

2. D

3. **a.** A, B
   b. B
   c. C
   d. D
   e. $\int_0^2 [16 - (16 - 6x)]dx + \int_2^4 (16 - x^2)dx$

Section 13.4

1. For the graph below, the producer's surplus is represented by which of the following integrals?

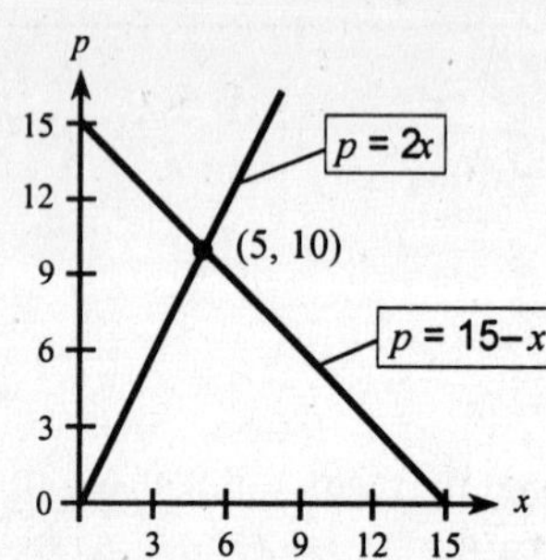

a. $\int_0^5 (15 - x)dx - 5 \cdot 10$

b. $5 \cdot 10 - \int_0^5 (15 - x)dx$

c. $\int_0^5 2x\,dx - 5 \cdot 10$

d. $5 \cdot 10 - \int_0^5 2x\,dx$

e. $\int_0^5 [(15 - x) - 2x]dx$

2. Which of the choices in problem 1 represents the consumer's surplus?

3. What does (e) represent in problem 1?

Section 13.4: Solutions

1. d

2. a

3. The area between the two curves from $x = 0$ to $x = 5$.

Section 13.5 — *NO SUPPLEMENTARY EXERCISES*

Section 13.6

1. Match each of the following integrals with the correct choice. Then integrate if possible.
   I. Integration by parts
   II. $\int e^u du$
   III. $\int \frac{du}{u}$
   IV. $\int u^n du$
   V. Cannot be integrated.

   a. $\int 4x^2 e^{x^3} dx$
   b. $\int xe^x dx$
   c. $\int e^{x^2} dx$
   d. $\int x\sqrt{x^2 - 1}\, dx$
   e. $\int x^2 \sqrt{x - 1}\, dx$

2. How do we know that the wrong choice for u and dv was made in integrating $\int \frac{3x^2}{\sqrt{x-1}}dx$?

$u=(x-1)^{-1/2}$

$dv=3x^2dx$

$du=-\frac{1}{2}(x-1)^{-3/2}dx \quad v=x^3$

$=\frac{-1}{2(x-1)^{3/2}}dx$

$\int \frac{3x^2}{\sqrt{x-1}}dx=\frac{x^3}{\sqrt{x-1}}+\int \frac{x^3}{2\left(\sqrt{x-1}\right)^3}dx$

3. In choosing u and dv to solve by integration by parts there are some guidelines. Name at least two guidelines.

Section 13.6: Solutions

1. **a.** II; $\frac{4}{3}e^{x^3}+C$

b. I; xe^x-e^x+C

c. V

d. IV; $\frac{1}{3}(x^2-1)^{3/2}+C$

e. I; $\frac{2}{3}x^2(x-1)^{3/2}-\frac{8}{15}x(x-1)^{5/2}+\frac{16}{105}(x-1)^{7/2}+C$

2. The new integrand is "messier" than the original integrand.

3. **a.** dv must be easily integrated.
b. u should decrease the powers of x.
c. If integrand has ln function, then let u = ln function and du becomes algebraic.
d. Let dv be as much of the original integrand as possible. For example, in $\int x^5\sqrt{x^3-1}\,dx$, let $dv=x^2\sqrt{x^3-1}\,dx$.

Section 13.7 — *NO SUPPLEMENTARY EXERCISES*

Supplementary Exercises: Chapter 14

Section 14.1 — NO SUPPLEMENTARY EXERCISES

Section 14.2

1. If $z = x^3 + 4y^2 - 8x^2y$ and $f(x, y) = x^3 + 4y^2 - 8x^2y + 5x - 2y$, match each derivative in the left column with the correct formula(s) in the right column.

a. $3x^2 - 16xy$	**m.** z_x
b. $-16x$	**n.** z_{xx}
c. 8	**p.** z_{xy}
d. $8y - 8x^2 - 2$	**q.** z_y
e. $6x - 16y$	**r.** z_{yx}
f. $3x^2 - 16xy - 2$	**s.** z_{yy}
	t. $\frac{\partial f}{\partial x}$
	u. $\frac{\partial^2 f}{\partial x^2}$
	v. $\frac{\partial^2 f}{\partial y \partial x}$
	w. $\frac{\partial f}{\partial y}$
	x. $\frac{\partial^2 f}{\partial y^2}$
	y. $\frac{\partial^2 f}{\partial x \partial y}$
	z. None of these

2. Does w_{xy} mean the same thing as $\frac{\partial^2 w}{\partial x \partial y}$?

3. Suppose that x is the number of Good Books sold and that y is the number of Better Books sold each week and the profit function is $P(x, y) = 60x + 30y - 0.2x^2 - 0.1y^2$.
   b. If the sale of Better Books is held constant, at what rate will profit increase for each additional sale of a Good Book?
   c. Find $P_x(10, 10)$ and interpret the answer.

4. If $f(x, y) = 3x^2 + 4xy^3 - y^2$, find
   a. $\frac{\partial f}{\partial x}$
   b. $\frac{\partial}{\partial y}\left(\frac{\partial f}{\partial x}\right)$
   c. $\frac{\partial f}{\partial y}$
   d. $\frac{\partial}{\partial x}\left(\frac{\partial f}{\partial y}\right)$

5. In problem 4, what do parts (b) and (d) compare to when taking derivatives of one variable.

Section 14.2: Solutions

1. **a.** m **b.** p, r, y
   c. s, x **d.** w
   e. n, u **f.** z

2. No; $w_{xy} = \frac{\partial}{\partial y}\left(\frac{\partial w}{\partial x}\right)$ and $\frac{\partial^2 w}{\partial x \partial y} = \frac{\partial}{\partial x}\left(\frac{\partial w}{\partial y}\right)$. For most of your work the answers will be the same. However, since the order of taking the derivative is different, it is possible to obtain different answers.

3. **a.** $P_x = 60 - 0.4x$
   b. $P_x(10, 10) = \$56$
   The sale of the next Good Book will increase the profit by approximately $56.

4. **a.** $\frac{\partial f}{\partial x} = 6x + 4y^3$
   b. $\frac{\partial}{\partial y}(6x + 4y^3) = 0 + 12y^2 = 12y^2$
   c. $\frac{\partial f}{\partial y} = 12xy^2 - 2y$
   d. $\frac{\partial}{\partial x}(12xy^2 - 2y) = 12y^2 - 0 = 12y^2$

5. There appears to be a similarity with the second derivative concept.

Section 14.3

1. Other than the fact that there are two products and two prices in the demand functions, what is the difference with previous demand functions of one product and one price?

2. If the cost C of manufacturing one item is found from the cost of labor (x) and the cost of the necessary raw material (y),
   a. when do we substitute for x and y, and
   b. when do we find C_x and then substitute for x and y, and
   c. when do we find C_y and then substitute for x and y?

3. Under typical conditions, if p_2 remains constant and p_1 increases, then is $\frac{\partial q_1}{\partial p_1}$ positive or negative? Why?

4. Let $q_1 = f(p_1, p_2)$ and $q_2 = g(p_1, p_2)$ where $q_1 =$ lbs of beef, $q_2 =$ lbs of chicken, $p_1 =$ price of one lb of beef, and $p_2 =$ price of one lb of chicken. Suppose that there is a price increase in beef and the price of chicken remains constant. Fill in the blanks with the correct words and/or symbols.

   The products are ________________ and we are using $\frac{\partial \_\_}{\partial \_\_}$. The sign of the partial derivative is ____.

Section 14.3: Solutions

1. q is a function of p now. Previously, p was a function of q. What is under study determines the independent variable(s).

2. **a.** We are told to find the cost, C.
   b. Find C_x if we are looking at how the increase in the cost of labor will change the cost of the item.
   c. Find C_y if we are looking at how the increase in the cost of raw material will change the cost of the item.

3. $\frac{\partial q_1}{\partial p_1} < 0$ because a price increase results in a decrease in demand.

4. competitive, $\frac{\partial q_2}{\partial p_1}$, $+$

Section 14.4 — *NO SUPPLEMENTARY EXERCISES*

Section 14.5 — *NO SUPPLEMENTARY EXERCISES*

NOTES

NOTES

NOTES